T0396657

Introduction to Structural Chemistry

Stepan S. Batsanov • Andrei S. Batsanov

Introduction to Structural Chemistry

Prof. Stepan S. Batsanov
Russian Academy of Sciences
Institute of Struct. Macrokinetics
and Materials Science
Chernogolovka
Moscow Region
Russia

Ph.D. Andrei S. Batsanov
Durham University
Chemistry Department
Durham
United Kingdom

ISBN 978-94-007-4770-8 ISBN 978-94-007-4771-5 (eBook)
DOI 10.1007/978-94-007-4771-5
Springer Dordrecht Heidelberg London New York

Library of Congress Control Number: 2012954252

Springer is part of Springer Science+Business Media (www.springer.com)

Foreword

The subject of structural chemistry is usually defined to include the geometry, i.e. spacial arrangement of atoms, and electronic structure of molecules and crystals. In our opinion, this definition is far too narrow. Firstly, the structures of liquids and amorphous solids are left out or mentioned only briefly. Secondly, all too often the discussion of crystal structure is restricted to its idealised model, assuming perfect periodicity and ignoring defects, atomic displacements and vibrations, which in fact are very important in materials science. Most significantly, geometrical structure should be considered not in isolation, but in unity with energetic parameters, chemical reactivity and various physical properties of the substance. Books on structural chemistry also suffer from partitions according to the aggregate state, the methods of investigation (X-ray crystallography, electron and neutron diffraction, spectroscopy, magnetic resonance, etc.) or the types of substances (inorganic, coordination, organometallic and organic), obscuring the essential unity of the subject. Finally, notwithstanding the present deluge of experimental results, many textbooks are still quoting "standard" parameters (bond distances, atomic radii) derived decades ago from the scantest experimental data then available.

The aim of the present book is to outline structural chemistry in this broader sense, within the wider background of chemistry and physics. The discussion of major concepts and models is combined with an extensive compilation of the reference data and tables of standard quantitative parameters, critically revised in the light of the most up-to-date experimental results. Obviously, it is impossible to cover systematically every area of structural chemistry in a single book—today, this would require dozens of volumes. Therefore, the discussions of the theory of crystal symmetry or the description of simple structural types, which are excellently presented in many books, are reduced here to a barest minimum. Instead, more attention is paid to novel, fast-growing areas of experimental structural chemistry, such as the crystal chemistry of high pressures and the structures of van der Waals molecules in gases, which have contributed unique information to the physics of the condensed state as well. Particularly, we intended to highlight numerous links between geometrical, energetic and optical properties, between different classes of compounds, and across the variety of aggregate states. The approach is essentially experimental and semi-empirical, for, notwithstanding the spectacular progress of quantum-chemical theory and computational techniques, structural chemistry remains essentially an experimental science. It is noteworthy that most fruitful concepts in this field, from Lewis' bonding electron pairs to electronegativity, started as semi-empirical concepts and got quantum-mechanical substantiation much later.

The book is organized into eleven chapters. The first of these describes ionization potentials, electron affinities and effective sizes of atoms (covalent, metallic and ionic radii) in relation to their electronic structure, and to properties of clusters (as pseudo-atoms). Chapter 2 outlines the development of the concept of chemical bonding and its various types: covalent, ionic, metallic and van der Waals. Vast data on bond energies are surveyed. The bonding in solids is linked to their electronic properties (band gaps) and lattice energies (using Born-Madelung theory). The concept of electronegativity is introduced here and its various systems and scales are explained. Chapter 3 deals with molecular structures, beginning with a brief outline of the experimental methods and databases, and then surveying the experimental data on inorganic, organic, organometallic molecules, clusters and coordination complexes. The VSEPR theory of molecular geometry, electron-counting rules for clusters and the concept of trans-effect are briefly explained. Chapter 4 describes a continuum of intermolecular interactions from weak van der Waals (vdW) forces to donor-acceptor and covalent bonds, as well as hydrogen bonding. An illuminating comparison of vdW interactions in crystals with those in gas-phase 'vdW molecules' is provided. Chapter 5 describes the (mostly inorganic) crystal structures, and ways of rationalizing their variety. The following chapter deals with the deviations of real crystals from the ideal lattice symmetry: thermal motion, thermal expansion (including negative), defects, isomorphous substitution and formation of solid solutions. It is shown that the amount of strain that a crystal lattice can absorb is remarkably similar in every case and agrees well with Lindemann's theory of melting. Chapter 7 describes the main structural trends in amorphous solids, melts and liquid solutions, as well as amorphization of powders due to diminishing grain size. Chapter 8 outlines the peculiarities of structure, melting points, ionization properties and dielectric permittivity ε (including giant ε and ferroelectricity) of nano-particles. Chapter 9 explains some general features of phase transitions, in additions to numerous concrete examples given in Chaps. 5 and 10. Chapter 10 is devoted to structural chemistry under extreme conditions—especially static and dynamic high pressures. Chapter 11 explains the basics of refraction of light in crystals and the applications of refractometry and IR spectroscopy to elucidating the structure and bonding mode in substances, particularly Szigeti's method of determining atomic charges and bond ionicity. Each chapter (except 8th and 9th) is accompanied by a set of supplementary tables, compiling relevant experimental information, including some previously unpublished measurements by the authors.

We are much indebted to the Royal Society of Chemistry for a Kapitza Fellowship (for S.S.B.) which allowed the work on this book to begin, and to Prof. Judith A. K. Howard of Durham University (U.K.), without whose help and encouragement it would not have been completed. We thank also Prof. Kenneth Wade of Durham University, Prof. Dipankar Datta of Indian Association for the Cultivation of Science, Prof. Pekka Pyykko of University of Helsinki, Prof. Láslo von Szentpály of Stuttgart University for helpful comments. Last but not least, we thank our family for constant and patient support during the work on this book.

Contents

Chapter 1
Atom

Today, there is hardly a textbook in general, inorganic or physical chemistry, which does not start with a description of electronic structure of atoms and some simple molecules. Therefore the authors presuppose the reader to be familiar with the basics of this subject.

1.1 Ionization Potentials and Electron Affinities

1.1.1 Ionization Potentials of Atoms

The energy required to remove an electron from an isolated atom to infinity is called the ionization potential I, that of eliminating the n-th electron is called the n-th ionization potential (I_n). Ionization of atoms (or molecules) can be caused by a collision with an electron or another ion or molecule, by strong electric fields, or by thermal emission of electrons. Spectroscopic methods can determine the first ionization potentials (I_1) of atoms or molecules with the accuracy of 0.01–0.001 eV, and occasionally as high as 0.0005 eV. For successive ionization potentials the errors increase to tenths or even units of eV [1–4]. The values of I_n for valence-shell electrons are listed in Table 1.1.

Ionization potentials of elements are periodic properties (Fig. 1.1), increasing from the left to right in the periods, and decreasing from top to bottom in the groups. This periodicity results from the periodical pattern in which the s, p, d and f orbitals are filled. For a given atom each successive ionization potential is markedly higher than the previous one (Fig. 1.2), because (i) the departing electron has to overcome the attraction from a higher net charge of the cation, (ii) each removed electron reduces the intraatomic electron-electron repulsion, and (iii) successive electrons may come from a lower-energy shell. As long as electrons are removed from the same (s or p) sub-shell,

$$I_n \approx nI_1 \tag{1.1}$$

S. S. Batsanov, A. S. Batsanov, *Introduction to Structural Chemistry*,
DOI 10.1007/978-94-007-4771-5_1, © Springer Science+Business Media Dordrecht 2012

Table 1.1 Ionization potentials I_n of isolated atoms (eV). (Based on reviews [1, 2] except where specified)

Atom	I_1	I_2	I_3	I_4	I_5	I_6	I_7	I_8
H	13.59							
He	24.59	54.42						
Li	5.39							
Be	9.32	18.21						
B	8.30	25.15	37.93					
C	11.26	24.38	47.89	64.49				
N	14.53	29.60	47.45	77.47	97.89			
O	13.62	35.12	54.94	77.41	113.90	138.12		
F	17.42	34.97	62.71	87.14	114.24	157.16	185.19	
Ne	21.56	40.96	63.45	97.12	126.21	157.93	207.28	239.1
Na	5.14							
Mg	7.65	15.04						
Al	5.99	18.83	28.45					
Si	8.15	16.35	33.49	45.14				
P	10.49	19.77	30.20	51.44	65.02			
S	10.36	23.34	34.79	47.22	72.59	88.05		
Cl	12.97	23.81	39.61	53.46	67.8	97.03	114.20	
Ar[a]	15.76	27.63	40.73	59.58	74.84	91.29	124.41	143.46
K	4.34							
Ca	6.11	11.87						
Sc	6.56	12.80	24.76					
Ti	6.83	13.58	27.49	43.27				
V	6.75	14.62	29.31	46.71	65.28			
Cr	6.77	16.48	30.96	49.16	69.46	90.63		
Mn	7.43	15.64	33.67	51.2	72.4	95.6	119.20	
Fe	7.90	16.19	30.65	54.8	75.0	99.1	124.98	151.06
Co	7.88	17.08	33.50	51.3	79.5	102.0	128.9	157.8
Ni	7.64	18.17	35.19	54.9	76.06	108	133	162
Cu	7.73	20.29						
Zn	9.39	17.96						
Ga	6.00	20.51	30.71					
Ge	7.90	15.93	34.22	45.71				
As	9.79	18.59	28.35	50.13	62.63			
Se	9.75	21.19	30.82	42.94	68.3	81.7		
Br	11.81	21.59	36	47.3	59.7	88.6	103.0	
Kr	14.00	24.36	36.95	52.5	64.7	78.5	111.0	125.8
Rb	4.18							
Sr	5.69	11.03						
Y	6.22	12.22	20.52					
Zr	6.63	13.1	22.99	34.34				
Nb	6.76	14.0	25.04	38.3	50.55			
Mo	7.09	16.16	27.13	46.4	54.49	68.83		
Tc	7.12[c]	15.26	29.54	46	55	80		
Ru	7.36	16.76	28.47	50	60	92		
Rh	7.46	18.08	31.06	48	65	97		
Pd	8.34	19.43	32.93	53	62	90	110	130
Ag	7.58							
Cd	8.99	16.91						
In	5.79	18.87	28.03					
Sn	7.34	14.63	30.50	40.73				
Sb	8.61	16.63	25.3	44.2	56			
Te	9.01	18.6	27.96	37.41	58.75	70.7		
I	10.45	19.13	33	42	51.5	74.4	87.6	

Table 1.1 (continued)

Atom	I_1	I_2	I_3	I_4	I_5	I_6	I_7	I_8
Xe	12.13	20.97	31.05	40.9	54.14	66.70	91.6	106.0
Cs	3.89							
Ba	5.21	10.0						
La	5.58	11.06	19.18					
Hf	6.82	14.9	23.3	33.33				
Ta	7.55	16.2	22	33	45			
W[b]	7.86	16.1	26.0	38.2	51.6	64.8		
Re	7.83	16.6	26	38	51	64	79	
Os	8.44	17	25	40	54	68	83	100
Ir	8.97	17.0	27	39	57	72	88	105
Pt	8.96	18.6	28	41	55	75	92	110
Au	9.22	20.5						
Hg	10.44	18.76						
Tl	6.11	20.43	29.83					
Pb	7.42	15.03	31.94	42.32				
Bi	7.28	16.70	25.56	45.3	56.0			
Po	8.41	19.4	27.3	38	61	73		
At	9.65	20.1	29.3	41	51	78	91	
Rn	10.75	21.4	29.4	44	55	67	97	110
Fr	4.07							
Ra	5.28	10.15						
Ac[d]	5.17	12.1	27.1					
Th	6.31	11.9	20.0	28.8				
Pa	5.89							
U	6.05	10.6	17.9	31.4				

[a] [3]; [b] [4]; [c] [5]; [d] I_1 for other actinides: Np 6.19, Pu 6.06, Am 5.99, Cm 6.02, Bk 6.23, Cf 6.30, Es 6.42, Fm 6.50, Md 6.58, No 6.65 eV.

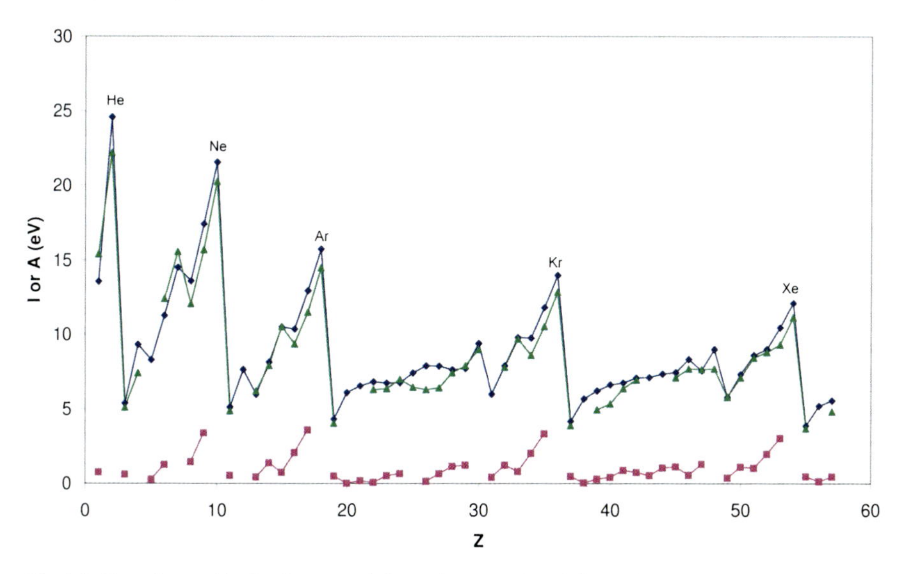

Fig. 1.1 Experimental ionization potentials I_1 of atoms (♦) and diatomic molecules (▲) and electron affinities A_1 of atoms (■)

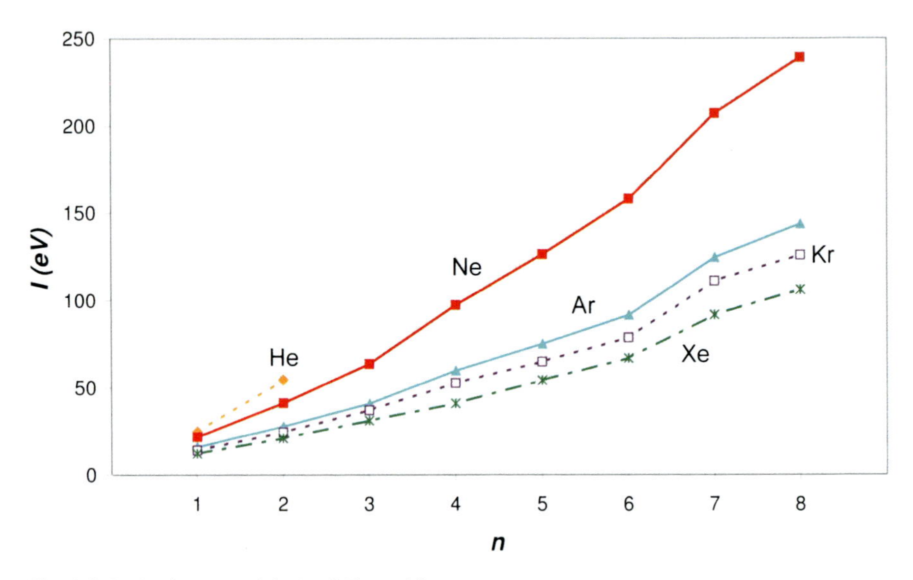

Fig. 1.2 Ionization potentials I_n of Group 18 atoms

with some deviations caused by (i) the passage from singly- to doubly-occupied *p* orbitals and (ii) the exceptionally high electron density in O, F and Ne. A transition from *p* to *s*-subshell gives a bigger jump, e.g. the I_2 of Group 3 elements exceed their I_1 approximately threefold. Generally, *d* and *f* electrons do not show such simple behavior. The theory behind these trends is explained in ref. [6].

Thus, on the whole the increase of I_n with *n* is somewhat faster than linear, and for valence electrons can be described by Eq. 1.2 with good precision (except for Group 2).

$$\frac{\sum_{1}^{n} I_n}{\sum_{1}^{N} I_n} = \left(\frac{n}{N}\right)^2 \tag{1.2}$$

Here the $i = n/N$ signifies the relative degree of ionization and the left-side term is the relative energy of such ionization, both normalized with respect to complete loss of the valence shell. This dependence (Fig. 1.3) is a continuous and smooth function, even though the ionization of an atom is a discrete process. It has been demonstrated [7] that energy functions E for atoms-in-molecules are differentiable with respect to N, even though N has only integer values for isolated atoms. This continuity is useful for analyzing the nature of the chemical bond (see below).

For ions having the same number of electrons but different Z, Glockler (1934) suggested an empirical equation,

$$I_1 = a + bZ + cZ^2 \tag{1.3}$$

where *a, b* and *c* are constants for each isoelectronic series [8]. In fact, Eq. 1.3 can be derived theoretically as well [9], but its accuracy is limited.

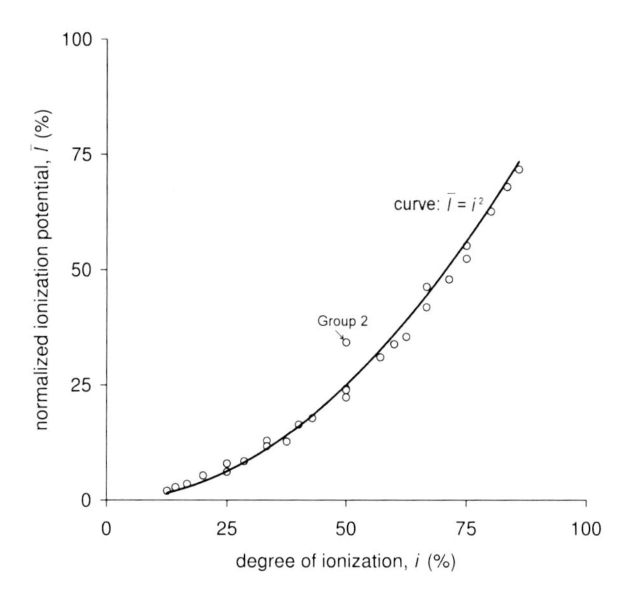

Fig. 1.3 Normalized ionization potential $\sum_1^n I_n / \sum_1^N I_n$ as a function (Eq. 1.2) of the degree of ionization $i = n/N$. Note the outlying position of Group 2 elements

The experimentally measured ionization potentials of isolated atoms refer to their ground state (I_g). The potentials for atoms in the valence state (I_v), widely used in quantum chemistry, describe the (imaginary) atoms obtained by disrupting all chemical bonds in a molecule but otherwise leaving their electron structure (accommodated to these bonds) unaltered. These potentials differ by the energy E_P of promoting an electron from the ground state to a given valence state,

$$I_v = I_g + E_P. \tag{1.4}$$

The calculations of I_v belong to quantum chemistry and lie outside the scope of this book, but it is worth mentioning that for similar valence states, the dependence of I_v on the atomic charge is similar to Eq. 1.3.

So far we discussed isolated atoms. The ionization potentials of homo-diatomic molecules A_2 (Table S1.1) are similar to those of atoms (A) and show qualitatively the same trends (Fig. 1.1). The small difference $\Delta = I_g(A_2) - I_g(A)$ depends of the details of the molecule's electronic structure. Usually $\Delta < 0$, as the electron departs from a non-bonding orbital which has higher energy than in the isolated atom because the atom in a molecule has more outer-shell electrons and the electron-electron repulsion is stronger. On the other hand, the H_2 molecule has no electrons apart from the bonding pair, while in the C=C (C_2) and N≡N (N_2) molecules all electrons of the outermost (2p) sub-shell are bonding. So it is one of these (strongly bound) electrons that has to be removed, hence in all three cases $\Delta > 0$. Two rather counter-intuitive facts are noteworthy. In the dimers of Group 18 elements (rare gases) [10], the atoms are held together only by weak van der Waals forces, their outer electron shells remain closed and little perturbed. So one may expect these atoms to behave almost as in the free state. In fact, $|\Delta|$ is larger than in many covalent molecules. It is so because the ionization of van der Waals dimers is a complicated process yielding molecular ions and cluster ions with much stronger bonding. For this reason, the small fraction of van der Waals dimers naturally present in a gas can affect its photoionization drastically; this is believed to be an important factor in the atmospheres of the outer planets of the Solar System. Another surprising feature concerns Group 1 molecules. They have no outer-shell electrons except the bonding pair, so one may expect $\Delta > 0$ (as in H_2), as two nuclei must attract these electrons stronger than one. In fact, this attraction is more than compensated by the increased electron-electron repulsion, and Δ is slightly negative. $I(A_2)$ is inversely correlated with the A–A distance d,

$$I \approx a/d^x, \tag{1.5}$$

where the parameters a and x are constant within each group of the Periodic Table, *viz*. Group 1 (Na_2 to Cs_2) has $a = 16.18$, $x = 1.05$; Group 15 (P_2 to Bi_2) $a = 22.10$, $x = 1.13$; Group 16 (O_2 to Te_2) $a = 13.18$, $x = 0.52$; Group 17 (F_2 to I_2) $a = 20.53$, $x = 0.81$ [11].

The ionization potentials in the halides of hydrogen and alkali metals indicate that the electron is removed from the halogen atom; their values also depend on the interatomic distances (Table S1.2). As the bond distance (d) and the negative charge (q) of the halogen atom increase, I(MX) decreases and ultimately approaches the electron affinity of the corresponding halogen atom (A, see below) when $q = -1$. Therefore the ionization potentials of MX molecules (M = H, Li, Na, K, Rb, Cs; X = F, Cl, Br, I) can be estimated similarly to Eq. 1.5, as

$$I - A \approx a/d^x \tag{1.6}$$

with the average $a = 11.2$ and $x = 0.87$ for all the abovementioned halogenides [11]. Hence, the ionization potentials of molecules are determined by Coulombic interactions of ions.

It is noteworthy that the NH_4 radical has the I_g (4.65 eV) [12] close to those of alkali metals (see Table 1.2) and thus can be regarded as a quasi-metal. For homo-atomic clusters of metal atoms, M_n, the ionization potentials (I_c) decreases with the increase of n and ultimately approaches the work function (Φ) of the corresponding bulk metal, according to Eq. 1.7 [13],

$$I_c = \Phi + \frac{3e^2}{8R_c} \tag{1.7}$$

where R_c is the radius of the cluster. Within the framework of the jellium (uniform positive background) model, an electron is regarded as departing from a metallic sphere of the effective radius $R_c = r_a n^{1/3}$, where r_a is the atomic radius. Eq. 1.7, or a version thereof with 1/2 substituted for 3/8 ([14, 15] and references therein), has been confirmed for a variety of metals [16–26], but it describes only the general trend. The actual curve has a number of local peaks corresponding to the closest packing modes of atoms (tetrahedron → octahedron → cubo-octahedron → dodecahedron → icosahedron) and to certain "magic" numbers of valence electrons in the cluster. This applies not only to metals, but also to such semi-metals as carbon [27], germanium and tin [28], selenium [29–31] and arsenic [32], as well as to organo-metallic clusters

Table 1.2 Ionization potentials of metal-benzene π-complexes

M	Sc	Ti	V
$M(C_6H_6)_2$	5.05	5.68	5.75
$M_2(C_6H_6)_3$	4.30	4.53	4.70
$M_3(C_6H_6)_4$	3.83	4.26	4.14

[33]. The trend of I_c toward Φ reflects the structural similarity between the *interior* part of a large cluster and the crystalline metal. One known exception to this rule is titanium: neither the I_c of the clusters up to $n = 130$, nor even an extrapolation of I_c to $n = \infty$ approaches Φ, indicating an irreducible structural difference between clusters and bulk titanium [34].

The first ionization potentials of fullerene molecules C_n decrease from 7.57 to 6.92 eV as n increases from 60 to 106 [35]. Fullerenes, especially those with $n > 60$, show a behavior similar to that of a simple charged sphere. An extrapolation of I of these fullerenes to $n \to \infty$ gives $\Phi = 5.13$ eV, i.e. the work function of the graphite monolayer (graphene). This effect is manifest also in the fullerides $M_x(C_{60})$, where M = Sc, Ti, V, Cr and $x = 1, 2, 3$; the ionization potentials are generally lower than in free fullerenes, and further decrease from 6.4 eV for $n = 1$ to 5.7 eV for $n = 3$. A similar decrease of the ionization potentials has been observed on complexation of the same metals with benzene, I decreasing as n and m in the $M_n(C_6H_6)_m$ complex increase (Table 1.2) [33],

Ionization potentials of Si_n clusters show a similar, but weaker, dependence on n [36]:

$n =$	1	2	3	4	5	6	7
I (eV) =	8.13	7.92	8.12	8.2	7.96	7.8	7.8

1.1.2 *Electron Affinity*

Another atomic characteristic very important for understanding the nature of the chemical bond is the electron affinity (A), i.e. the amount of energy *released* when one electron is added to a neutral atom. Thus, A is equivalent to the ionization potential of a negatively charged atom, or a "zero ionization potential".

Whereas the ionization potentials are always positive, electron affinity can be either positive or negative. $A < 0$ or $A \approx 0$ for atoms with closed outer-shell configurations (s^2 or s^2p^6), i.e. Groups 2, 12 and 18, and also for Mn and N where a more complicated quantum-chemical explanation is required. For most elements $A > 0$, i.e. an electron is attracted by an electrically neutral atom as if the latter carried a net positive charge, because the atom's own electrons neutralize its nuclear charge incompletely, resulting in a surplus effective nuclear charge [37]. Even for these atoms, the electron affinities usually are of the order of 1 eV, compared to tens of eV for valence-electron ionization potentials. The affinity of any atom toward the *second* electron is negative ($A_2 < 0$), i.e., no atom with multiple negative charge can exist as a separate particle. Note here a principal difference with cations. The latter can reduce their charge only by picking an electron from their environment. If none is available

Table 1.3 Electron affinities A of isolated atoms (eV). (Main sources: [1, 38])

Atom	A	Atom	A	Atom	A	Atom	A
Ag	1.304	Cu	1.236	Na	0.548	Sb	1.047
Al	0.433	F	3.401	Nb	0.894	Sc	0.189
As	0.805[a]	Fe	0.151	Ni	1.157	Se	2.021
Au	2.309	Fr	0.46	O	1.461	Si	1.390
B	0.280	Ga	0.41	Os	1.078	Sn	1.112
Ba	0.145	Ge	1.233	P	0.746	Sr	0.052
Bi	0.942	H	0.754	Pa	0.222	Ta	0.323
Br	3.364	I	3.059	Pb	0.364	Tc	0.55
C	1.262	In	0.384[b]	Pd	0.562	Te	1.971
Ca	0.043	Ir	1.564	Pt	2.125	Ti	0.079
Ce	0.700	K	0.501	Rb	0.486	Tl	0.377
Cl	3.613	La	0.47	Re	0.15	V	0.526
Co	0.663	Li	0.618	Rh	1.143	W	0.816[c]
Cr	0.676	Mo	0.747	Ru	1.046	Y	0.307
Cs	0.472	N	−0.07 (?)	S	2.077	Zr	0.427

[a][40]; [b][41]; [c][42].

(in deep vacuum), the cation can survive indefinitely, no matter how high its energy. An unstable anion, on the contrary, can easily stabilize by ejecting an electron.

Electron affinities are intrinsically much more difficult to measure than ionization potentials. In fact, all determinations before 1970 were indirect and unreliable. Today, the principal experimental technique uses the photoelectric effect. A beam of anions is crossed with a light (laser) beam and the frequency is recorded at which the anion dissociates and scattered electrons occur [38]. The most accurate experimental values of A are listed in Table 1.3. At present, there is no practical way of measuring negative electron affinities, which are only available from theoretical calculations. Thus, a universal method to calculate A_2 of atoms and molecules has been suggested by von Szentpály [39],

$$A_2 = A_1 - \frac{7}{6}\eta^0 \tag{1.8}$$

where η^0 is the chemical hardness of the corresponding neutral species,

$$\eta^0 = \frac{1}{2}(I_1 - A_1), \tag{1.9}$$

preferably for the valence state, although the difference is significant only for Groups 15 and 16.

Note that hydrogen is similar to alkali metals in its electron affinity, and to halogens in its ionization potential (see Table 1.1), in accordance with the traditionally ambiguous placing of H in the Periodic Table in either Group 1 or Group 17.

Note that even the highest A_1 (3.6 eV for Cl) is less than the lowest I_1 (3.9 eV for Cs), hence a spontaneous conversion of any pair of neutral atoms into a cation-anion pair seems thermodynamically unfavourable. However, this is true only for atoms/ions separated to *infinite* distance. For ions in contact, and especially for

Table 1.4 Charge transfer i at the contact of metal and halogen atoms (D in kJ/mol, Q in eV, i in e)

M, D(M–M)	F, D(F–F) = 155 D(MF), Q, i			Cl, D(Cl–Cl) = 240 D(MCl), Q, i			Br, D(Br–Br) = 190 D(MBr), Q, i			I, D(I–I) = 149 D(MI), Q, i		
Li, 105	577	4.63	0.956	469	3.07	0.862	419	2.81	0.840	345	2.26	0.793
Na, 74.8	477	3.75	0.915	412	2.64	0.845	363	2.39	0.823	304	1.99	0.785
K, 53.2	489	3.99	0.977	433	2.97	0.910	379	2.67	0.885	322	2.10	0.835
Rb, 48.6	494	4.06	0.992	428	2.94	0.917	381	2.71	0.898	319	2.28	0.859
Cs, 43.9	517	4.33	1.029	446	3.15	0.949	389	2.82	0.923	338	2.50	0.875
Cu, 201	427	2.58	0.733	375	1.60	0.678	331	1.40	0.655	290	1.19	0.628
Ag, 163	341	1.89	0.694	311	1.14	0.651	278	1.05	0.635	250	0.95	0.616
Au, 226	325	1.39	0.616	302	0.72	0.581	286	0.81	0.575	263	0.78	0.559

those packed in a solid, the energy gain from Coulomb attraction can more than compensate for the energy spent on electron transfer, even for charges higher than ±1. Furthermore, polyatomic molecules or clusters, as distinct from isolated atoms, *can* have positive A_2 (see below).

Equation 1.2 applies to all ionization potentials, including those of negatively charged atoms, i.e. to the electron affinities. Since an increase of the (net) negative charge on an atom reduces its ionization energy, the $\bar{A} = f(i)$ curve must be the anti-symmetric image of $\bar{I} = f(i)$, if for the former we replace the number of valence electrons, N, with the number of holes (vacancies) in the valence shell. Let us now consider the molecules of alkali halides, where $N = 1$ and therefore i equals the fraction of an electron transferred from metal to halogen. In the first approximation, the heat effect of such a transfer is

$$i^2 I = (1 - i^2)A + Q \tag{1.10}$$

where I refers to the (neutral) metal atom, A to that of halogen and Q is the balance between the energy of the covalent interaction between the initial neutral atoms and the Coulombic attraction between the resulting cation and anion, $Q = D(\text{M–X}) - 1/2 \times [D(\text{M–M}) + D(\text{X–X})]$ where D is the standard bond energy. Thus the equilibrium degree of ionization is

$$i = \sqrt{\frac{A + Q}{A + I}} \tag{1.11}$$

Earlier we have calculated [43] the charge transfer (i) on contact of neutral atoms without formation of a covalent bond between them, i.e. using the relation $i^2 I = (1 - i^2)A$. More accurate estimates by Eqs. 1.10 and 1.11, which take into account the bonding, using the values of I and A from Tables 1.3 and 1.4, are listed in Table 1.4; they are in qualitative agreement with the bond polarity changes in the LiX → CsX and MCl → MI series.

Fluorides are a special case. If we determine the correlation between the electron affinity and the atomic size for heavier halogens and extrapolate it to fluorine, we obtain for the latter $A = 4.10$ eV, which is 0.7 eV more than the observed value. The difference is due to the destabilizing effect of a high concentration of negative charge in a small volume [44–46]. As evident from Table 1.3, the same is true for other elements of the late 2nd Period: the electron affinities of O, N, C and B, respectively, are less than those of S, P, Si and Al, while for Ne the measured ionization potential is lower than the extrapolated value by 1.76 eV [47]. As the combination of the fluorine atom with any other tends to increase this volume and hence to relieve the electron-electron repulsion, the destabilizing force disappears. Therefore, a fluorine atom in a *compound* can be better characterized by the extrapolated value of A (4.10 eV).

Different isotopes of the same element show negligible differences of the electron affinity. Thus, accurate measurements give $A = 1.4611221$ eV for ^{16}O, 1.4611157 eV for ^{17}O, 1.4611129 eV for ^{18}O [48]. A replacement of deuterium for ^{1}H in XH, where X=Li, O, S, Mn, Fe, Co, as well as in CH_2, BH_3, SiH_3, $C{=}CH_2$, $N{=}CH_2$ and $C{\equiv}CH$ affects the electron affinity, on average, only by 0.006 eV [38], which is close to the experimental error.

The electron affinities of homo-nuclear diatomic molecules are listed in Table S1.3. Like isolated atoms, most molecules have positive A. The exceptions are H_2 and N_2, adding an electron to which requires an expense of energy. The electron affinities of hetero-atomic molecules can reveal the location of the negative charge (the added electron) and sometimes the bond polarity. Thus, for the NaF, NaCl, NaBr and NaI, $A = 0.520$, 0.727, 0.788 and 0.865 eV, respectively. A comparison with the A of the Na_2 and the corresponding X_2 molecules indicates that in NaX the electron is added to the sodium atom, and the magnitude of A depends on the polarizability α and the bond distance d [49],

$$A = 1.189 - \alpha/d \tag{1.12}$$

A comparison of electron affinities between Group 2 metals and their hydrides highlights the importance of the stable configurations of the outer shell: the metals (s^2 configuration) all have $A \approx 0$, while BeH, MgH, CaH, and ZnH (where the s^2 pair is broken up) have $A = 0.70$, 1.05, 0.93 and 0.95 eV, respectively. The electron affinity of the carbon atom depends strongly on the type of hybridization, increasing in the succession $sp^3 < sp^2 < sp$, viz. $A(CH_3) = 0.08$, $A(CCH_2) = 0.67$, $A(CCH) =$ 2.97eV.

The electron affinities of halides and oxides of polyvalent metals (Table S1.4) are of special interest because they characterize the oxidizing force of molecules, e.g. PtF_6 can oxidize oxygen and xenon. The second electron affinities, A_2, in polyatomic molecules are increased because the additional electron is delocalized between all the ligands. Thus, the observed A_2 are equal to: MCl_6^- 0.46 eV for M = Re and Os, 0.82 eV for Ir, 1.58 eV for Pt; MBr_6^- 0.76 eV for M = Re, 0.96 eV for Ir, 1.52 eV for Pt [50]; ZrF_6^- 2.9 eV [51], CrF_6^-, MoF_6^- < 0.58 [52]; $Re_2Cl_8^-$ 1.00 eV [53]. It is also known that the following dianions are stable: MX_3^{2-} (M = Li, Na, K; X = F, Cl) [54, 1.55], PtX_4^{2-} and PdX_4^{2-} (X = Cl, Br) [56, 57], BeF_4^{2-} and MgF_4^{2-} [58].

A variety of multiply-charged anions have been proven to be stable in the gas phase, viz. C_n^{2-} ($7 \le n \le 29$); BeC_n^{2-} ($n = 4$, 6, 8, 10); SiC_n^{2-} ($n = 6$, 8, 10); $SiOC_n^{2-}$ ($n = 4$, 6, 8); OC_n^{2-} ($5 \le n \le 14$); SC_n^{2-} ($6 < n < 18$); $O_2C_7^{2-}$, $Cr_2O_7^{2-}$, $Mo_2O_7^{2-}$ $W_2O_7^{2-}$ [59], PO_4^{3-}, $[CuPc(SO_3)_4H]^{3-}$ and even $[MPc(SO_3)_4]^{4-}$

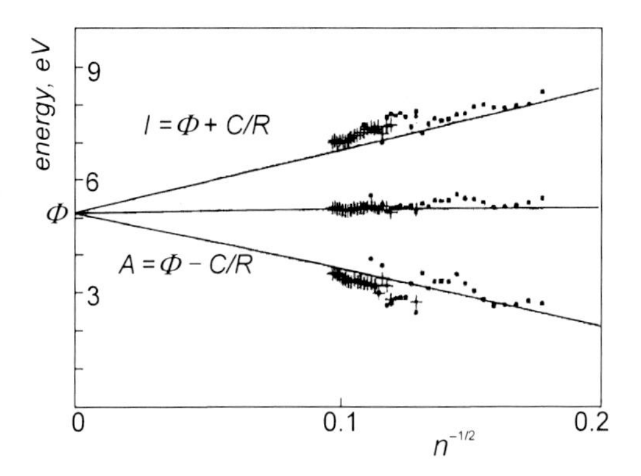

Fig. 1.4 Ionization potentials and electron affinities of fullerenes, C_n, as functions of n. (Adapted with permission from [35], Copyright 2000 American Chemical Society)

(M = Ni, Cu; Pc = phthalocyanine), see [60] and references therein. SO_4^{2-} (having $A_2 = -1.6$ eV) and $C_2O_4^{2-}$ are not stable as isolated species due to strong intramolecular Coulomb repulsion, yet both are stable if solvated by three molecules of water [61, 62]. Positive values of A_2 were also observed for some fullerenes, viz. (in eV): 0.325 for C_{76}, 0.44 for C_{78} (C_{2v}), 0.53 for C_{78} (D_3), 0.615 for C_{84} (D_{2d}), 0.82 for C_{84} (D_2) eV [63] (the symmetry notations in parentheses indicate the isomers). Nevertheless, in all these cases the negative charge per atom is much less than one electron, in agreement with Pauling's electroneutrality rule.

The electron affinities of clusters M_n (Table S1.5) exceed those of isolated atoms and diatomic molecules and increase with the cluster radius R_c, i.e. as n^{-3}. At relatively low n, the increase is steep and often periodically; then the dependence becomes smooth and A converges to the work function Φ which signifies the transformation of the molecule into bulk metal (compare with Eq. 1.7).

$$A_c = \Phi - \frac{5e^2}{8R_c} \tag{1.13}$$

A study of Co_n and Ni_n clusters shows that an increase of n causes the s- and d-bands in the electronic spectrum first to broaden and then to merge. The dependences of the electron affinities and magnetic moments on n have breaks at the values of n, 5–10 for Ni and 15–20 for Co [64, 65], which are close to 8 and 12, respectively—the common coordination numbers in metals.

Fullerene molecules C_n are quasi-spherical hollow cages allowing almost ideal delocalization of both positive and negative charge. Thus, both their I_c and A_c change linearly with $n^{-1/2}$ (which is proportional to the radius R_c of the cage). These dependences are symmetrical, so that the average $1/2\ (I_c + A_c)$, which is the electronegativity according to Mulliken (see below), remains fairly constant at 5.2(2) eV for all fullerenes (Fig. 1.4). For $n \to \infty$, both I_c and A_c converge to this value, which equals the work function Φ of bulk graphite [35, 66]. These observations agree with Eqs. 1.7 and 1.13 and seem to confirm the earlier ideas of Miedema et al. [67] who recommended to use Φ as the measure of the electronegativity of metals in solids (alloys).

The electron affinity of a solid, A_s, is defined as the difference between the vacuum energy, E_{vac}, and the minimum energy of the conduction band, E_c. This can be expressed in terms of the work function Φ, the Fermi level E_F, the upper limit of the valence band E_v, and the band gap E_g:

$$A_s = E_{vac} - E_c = \Phi + (E_F - E_v) - E_g \quad (1.14)$$

In metals $E_g \approx 0$, but in other solids there are different relationships between E_v and E_c, and therefore values of A_s can differ from Φ. Thus, in diamond the conduction band minimum is *ca.* 2.2 eV and the electron affinity was found to be *ca.* 0.8 eV for the (100) surface [68] and 0.38 eV for the (111) surface [69], while Φ of carbon equals 5.0 eV. Solid argon is believed to have $A_s = -0.4$ eV [70].

Although Miedema's approach has widely accepted, it has weak point. In general, the average ionization potentials I and electron affinities A of small clusters depend on the cluster size in a nonlinear manner, hence their experimental values cannot be simply extrapolated to the bulk work function for $n \to \infty$ [71]. Furthermore, how meaningful it is to extrapolate the properties of even the biggest clusters to a bulk solid, i.e. across at least *twenty* orders of magnitude of n? Most importantly, the work function (however determined) is a property of a bulk solid while electronegativity is an atomic property. Therefore, it is against the chemical intuition to assume that $A_c = \Phi$. Indeed, many metals, e.g. Cu, Zn, Al, Sn, Cr, Fe, etc., have Φ higher than the electron affinities of halogens and oxygen, and therefore should have reacted with brine to release chlorine gas! Chemical reactions occur between atoms rather than between bulk bodies, hence it is the atomic properties which one should compare. Hence, to derive the electron affinity of an atom in a solid metal from Φ, it is necessary to subtract the atomization energy of the negatively charged metal, which is unknown. However, recently Wilson [72–74] measured the electron affinities of almost all elements in solid state (A_s) using the technique of secondary ion mass spectrometry (SIMS), whereby negative ions (together with positive ions and neutral particles) are sputtered from a solid surface, focused into an ion beam and mass-analyzed. The results (Table 1.5) show that A_s can be determined accurately, e.g., by sputtering *different* elements from the *same* well-characterized matrix.

1.1.3 Effective Nuclear Charge

An electron in an atom is attracted to the nucleus but also repelled by other electrons, hence it does not 'feel' the full strength of the nuclear attraction. On the other hand, some atoms have positive electron affinity, i.e. attract an electron as if they possessed a net positive charge. Hence the electron shells do not completely compensate the nuclear charge Z, so that an approaching electron experiences the attraction by a certain effective nuclear charge, $Z^* > 0$. In 1930 Slater [37] suggested the following approximate solution to this problem. An electron at the radius r from the nucleus is shielded (screened) from its attraction by the electron density within the sphere of

Table 1.5 Electron affinities (eV) of solid elements according to Wilson [72–74]

Li	Be	B	C	N	O	F	Ne		
0.36	0.12	0.54	1.32	≈0	1.78	2.38	≈0		
Na	Mg	Al	Si	P	S	Cl	Ar		
0.28	≈0	0.41	1.10	1.10	2.07	2.5	≈0		
K	Ca	Sc	Ti	V	Cr	Mn	Fe	Co	Ni
0.24	0.18	0.21	0.19	0.30	0.32	≈0	0.34	0.57	0.7
Cu	Zn	Ga	Ge	As	Se	Br	Kr		
0.6	≈0	0.26	1.02	0.85	2.24	2.30	≈0		
Rb	Sr	Y	Zr	Nb	Mo	Tc	Ru	Rh	Pd
0.2	0.17	0.24	0.42	0.89	0.33		0.52	0.65	0.48
Ag	Cd	In	Sn	Sb	Te	I	Xe		
0.8	≈0	0.25	0.97	0.88	2.3	2.44	≈0		
Cs	Ba	La	Hf	Ta	W	Re	Os	Ir	Pt
	0.19	0.34	0.11	0.24	0.48	<0.12	0.66	1.31	1.70
Au	Hg	Tl	Pb	Bi		Th	U		
1.94	≈0	0.15	0.34	0.5		0.30	0.26		

this radius. This includes not only the electrons at lower energy levels, but also (to some extent) those on the same energy level as the electron in question, because every electron orbital has its radial distribution and can penetrate through the inner electron shells. Slater ascribed to each electron a constant shielding contribution (screening constant) which depends only on its energy level with respect to the electron in question. Essentially, the closer an electron is to the nucleus, the more efficiently it neutralizes the nuclear charge. Then,

$$Z^* = Z - S \tag{1.15}$$

where S is the sum of shielding contributions of all (relevant) electrons. For many years Slater's rules were applied in quantum chemistry to compute hydrogen-like wavefunction and in structural chemistry to calculate electronegativities, radii, magnetic susceptibilities and other atomic properties. These rules proved most efficient for elements without d-electrons, but for transition elements the results were rather imperfect, especially for the properties depending heavily on Z^*.

Numerous attempts to improve the technique of calculating Z^*, bear witness to the importance of this problem for theoretical chemistry [75–79]. Most of these attempts concentrated on deriving more adequate screening constants for d- and f-electrons, Slater's over-estimation of those being obvious on comparison with experimental

Table 1.6 Rules for calculating effective nuclear charges

Electron, Type	Electron screening constant, σ		
	shell	s, p	d, f
$n\ (s, p)$	$n-2$	1.00	0.95
	$n-1$	0.85[a]	0.70
	n	0.40	0.30
	$n+1$	0.10	0
$n\ (d, f)$	$n-2$	1.00	1.00
	$n-1$	0.95	0.80
	n	0.50	0.40
	$n+1$	0.20	0.10

[a] if d-orbitals of the same level are entirely or partially vacant; for fully occupied d, f-shells, $\sigma = 0.95$.

data. The results of this vast work can be summarized in the simple rules, listed in Table 1.6 [80], which agree well with recent state-of-the-art quantum-mechanical calculations [81–85]. Table 1.7 gives the effective nuclear charges for all elements in the ground state, calculated by this method.

Waldron et al. [86] proposed to use the 'screening percentage', $1 - Z^*/Z$, as a more evident indicator of the fraction of the nuclear charge that is shielded from a

Table 1.7 Effective nuclear charges of atoms in the ground state (for H, $Z^* = 1$)

Li 1.3	Be 1.9	B 2.5	C 3.1	N 3.7	O 4.3	F 4.9	Ne 5.5		
Na 2.2	Mg 2.8	Al 3.4	Si 4.0	P 4.6	S 5.2	Cl 5.8	Ar 6.4		
K 2.2	Ca 2.8	Sc 3.1	Ti 3.4	V 3.7	Cr 3.7	Mn 4.3	Fe 4.6	Co 4.9	Ni 5.2
Cu 4.4	Zn 5.0	Ga 5.6	Ge 6.2	As 6.8	Se 7.4	Br 8.0	Kr 8.6		
Rb 2.7	Sr 3.3	Y 3.6	Zr 3.9	Nb 3.9	Mo 4.2	Tc 4.8	Ru 4.8	Rh 5.1	Pd 4.3
Ag 4.9	Cd 5.5	In 6.1	Sn 6.7	Sb 7.3	Te 7.9	I 8.5	Xe 9.1		
Cs 2.7	Ba 3.3	La 3.6	Hf 4.6	Ta 4.9	W 5.2	Re 5.5	Os 5.8	Ir 6.1	Pt 6.1
Au 5.6	Hg 6.2	Tl 6.8	Pb 7.4	Bi 8.0	Po 8.6	At 9.2	Rn 9.8		

valence electron. These percentages show clear periodic trends and correlations with ionization potentials, atomic radii and electronegativities.

The electron energy (ionization potential) in Bohr's theory depends not only on Z^* but also on the *effective* principal quantum number n^* which characterizes the electron shell of an atom,

$$E = R(Z^*/n^*)^2 \tag{1.16}$$

where R is the Rydberg constant. According to Slater, n^* coincides with the period number n for Periods 1–3, whereas for Periods 4, 5 and 6 it equals 3.7, 4.0 and 4.2, respectively. Thus $n^* < n$ if the atom has d-electrons; the divergence is due to the diffuse (penetrating) character of d-orbitals. The Z^* listed in Table 1.7 correspond to atoms in the ground state, while structural chemical calculations require Z^* for the valence state, and the ground and valence states are identical only for hydrogen, Group 1 metals and halogens. The valence-state Z^* can be calculated by the same rules, but taking into account the changes of electron configuration, e.g., $s^n p^m$ hybridization or promotion of electrons from $(n–1)d$ to ns shell, etc. Besides, these Z^* refer to *isolated* atoms and their values are affected by chemical bonding in an ambivalent way. On one hand, the overlap of valence orbitals increases the total electron density around each nucleus and screens the latter more efficiently, thus reducing Z^*. On the other hand, the joint action of both partners increases Z^* of the atom in the overlap region. On the balance, bonding electrons are influenced by higher Z^*, and nonbonding electrons by lower Z^* than in an isolated atom. Obviously, the integral value of the effective nuclear charge of a given atom remains the same in any environment. A quantum chemical interpretation of the mutual influence of chemically bonded atoms on Z^* is available [87, 88].

Ionization potentials and electron affinities are widely used in structural chemistry for semi-empirical calculations. However, usually only the I_1 and A_1 are taken into account, which is justified for the latter (for reasons explained above) but not always for the former, if we deal with polyvalent elements. It is often assumed that higher-order ionization potentials can be easily calculated from I_1 by the Glockler equation. Unfortunately, the coefficients of Eq. 1.3 vary from element to element. Moreover, using only the first ionization potentials, one cannot adjust the calculations to the actual valence of an atom. Indeed, while in a MX_n molecule all *bonding* electrons of the metal are equivalent, in a MX crystal (provided the coordination number ≥ 4) all the valence-shell electrons of both elements can be involved in the bonding and through hybridization become completely averaged in their properties. In this case, the average ionization potential $\bar{I}$ of all the valence-shell electrons proved to be much more informative than I_1 [89]. The applications of $\bar{I}$ are discussed below.

1.2 Absolute Dimensions of Atoms

There is no consensus about how to define atomic sizes. As a quantum object, the atom has no clear-cut boundary and no definite 'size': electron density of an isolated atom drops to zero only at infinite distance. However, nearly all this electron density

Table 1.8 Orbital radii (Å) of the outer (valence) orbitals of elements

Li	Be	B	C	N	O	F			
1.586	1.040	0.776	0.620	0.521	0.450	0.396			
Na	Mg	Al	Si	P	S	Cl			
1.713	1.279	1.312	1.068	0.919	0.810	0.725			
K	Ca	Sc	Ti	V	Cr	Mn	Fe	Co	Ni
2.162	1.690	1.570	1.477	1.401	1.453	1.278	1.227	1.181	1.139
Cu	Zn	Ga	Ge	As	Se	Br			
1.191	1.065	1.254	1.090	1.001	0.918	0.851			
Rb	Sr	Y	Zr	Nb	Mo	Tc	Ru	Rh	Pd
2.287	2.836	1.693	1.593	1.589	1.520	1.391	1.410	1.364	1.318
Ag	Cd	In	Sn	Sb	Te	I			
1.286	1.184	1.382	1.240	1.193	1.111	1.044			
Cs	Ba	La	Hf	Ta	W	Re	Os	Ir	Pt
2.518	2.060	1.915	1.476	1.413	1.360	1.310	1.266	1.227	1.221
Au	Hg	Tl	Pb	Bi	Po	At			
1.187	1.126	1.319	1.215	1.295	1.212	1.146			

can be found within several angstroms from the nucleus, so an effective size can be physically meaningful. Furthermore, many measurable atomic properties, e.g. ionization potentials, electron affinities, polarizabilities, diamagnetic susceptibilities and atomic capacitances, are size-related. Hence some effective atomic sizes can be derived from them, but these sizes do not—and should not—be the same for different properties. Several quantum-based sets of atomic radii, based mainly on wave-function averages rather than on experimental observables, have been reported [90]. What can be computed by quantum mechanics most accurately is the orbital radius, i.e. the distance from the atomic nucleus at which an orbiting electron can be found with the maximum probability[1] [91–99]. The orbital radius of the hydrogen atom, so-called Bohr radius $a_0 = 0.529177$ Å, serves also as the atomic unit (a. u.) of distance. The orbital radii of other elements are listed in Table 1.8; these refer to the outermost occupied orbital [91].

However, in structural chemistry it is more important to have the radii describing, however imperfectly, the outer boundary of atoms. Electronic polarizability, α, of an isolated atom is one of the size-related properties (see above), hence Nagle [100] defined its radius as $r_a = \alpha^{1/3}$. Bohórquez and Boyd [90] derived atomic radii r_a as

$$r_a = a_0\sqrt{I_H/I} \tag{1.17}$$

[1] Note, however, that the electron density in the usual sense, i.e. the number of electrons per unit of volume, has no maximum in this area—it decays exponentially with the distance from the nucleus.

where a_o is the Bohr radius and I_H and I are the ionization potentials of hydrogen and the element in question, respectively. To obtain the boundary radius, these radii can be multiplied by the factor of 3.024, obtained by dividing the free H atom radius of 1.60 Å (Chap. 2) by a_o.

Since isolated atoms are generated by dissociation of a chemical bond, their radii can be roughly estimated from the critical value of the interatomic distance on the verge of dissociation [101], using the universal equation of state (EOS) [102],

$$E(d) = E_o E^*(d^*) \tag{1.18}$$

where $E(d)$ is the binding energy as a function of the bond length, E_o is the equilibrium binding energy, and the parameters d^* and l (see below) are the scaled lengths. Since $P = -\partial E/\partial V$, then from the universal form of the energy,

$$P = -(E_o/4\pi B_o r_{WS})E^{*\prime} d \tag{1.19}$$

where B_o is the bulk modulus, V_o is the mole volume, d is the interatomic distance, $r_{WS} = (3V_o/4\pi)^{1/3}$ is the radius of the Wigner—Seitz sphere containing the average volume per atom, and $E^{*\prime}$ is the derivative of the scaled energy with respect to d^*, i.e. the scaled force. Eqs 1.18 and 1.19 describe quantitatively all known experimental data and first-principles calculations for all classes of metals and covalent condensed molecules. This EOS was used to determine the negative pressure, P_R, required to disperse solid metal into free atoms [103]. Taking into account that $E = P\Delta V$ and equating E to the atomization energy E_a and P to P_R,

$$\Delta V_R = E_a/P_R. \tag{1.20}$$

Hence to disrupt a chemical bond, it must be stretched by a factor of

$$q_R = \left(\frac{V_o + \Delta V_R}{V_o}\right)^{1/3}. \tag{1.21}$$

The values of q_R calculated from experimental E_a, B_o and P_R [103] are listed in Table 1.9, where $V_R = V_o + \Delta V_R$.

The universal EOS can be used to estimate the maximum possible interatomic distances in metals at the boiling temperature T_b, i.e. when the cohesion energy becomes equal to the thermal energy, $E_{Tb} = RT_b$. Defining d^* for the condition $E(d) = E_{Tb}$ and knowing l and d_o, the bond-stretching factor can be found (see Table 1.9) [101] as

$$q_T = (d^* l + d_o)/d_o. \tag{1.22}$$

The average $q_R = 1.50(8)$ and $q_T = 1.48(9)$ are equal and can be used to define a system of radii for isolated atoms, as $r_M \times q$, where r_M is half the M–M distance in the metal. Table S1.6 contains the radii obtained by the above mentioned techniques, as $r = \sqrt[3]{\alpha}$ and also by the following simple consideration. In the first approximation, the *lower limit* of the radius of a free atom (r_m) can be estimated assuming than the

Table 1.9 Volume ratios (V_R/V_o), thermal energies (E_{Tb}, kJ/mol) and bond stretching factors under negative pressure (q_R) and on boiling (q_T)

M	V_R/V_o	q_R	E_{Tb}	q_T	M	V_R/V_o	q_R	E_{Tb}	q_T
Li	5.378	1.75	13.43	1.75	Si	4.148	1.61	29.42	1.65
Na	4.329	1.63	9.61	1.64	Ge	3.975	1.58	25.82	1.62
K	4.022	1.59	8.58	1.56	Sn			23.90	1.49
Rb	3.847	1.57	7.99	1.52	Pb	2.839	1.42	16.81	1.42
Cs	3.914	1.58	7.85	1.54	V	3.601	1.53	30.60	1.53
Cu	3.389	1.50	23.57	1.46	Nb	3.511	1.52	41.71	1.54
Ag	2.968	1.44	20.25	1.40	Ta	3.458	1.51	47.65	1.52
Au	2.651	1.38	26.02	1.35	Cr	3.056	1.45	24.48	1.46
Be	4.383	1.64	22.80	1.60	Mo	3.025	1.45	40.84	1.44
Mg	3.041	1.45	11.33	1.42	W	3.147	1.46	48.46	1.46
Ca	3.756	1.55	14.61	1.50	Mn	3.110	1.46	19.40	1.41
Sr	3.564	1.53	13.76	1.48	Tc			42.82	1.42
Ba	3.872	1.57	18..04	1.50	Re	2.825	1.41	48.80	1.40
Zn	2.607	1.38	9.81	1.33	Fe	3. 415	1.51	25.06	1.50
Cd	2.396	1.34	8.65	1.29	Co	3.346	1.50	26.61	1.47
Hg			5.24	1.26	Ni	3.306	1.49	26.49	1.49
Sc			25.85	1.52	Ru	2.882	1.42	36.77	1.42
Y	3.962	1.58	30.00	1.56	Rh	3.087	1.46	32.99	1.42
La			31.07	1.73	Pd	2.826	1.41	26.90	1.37
Al	3.707	1.55	23.21	1.50	Os	2.986	1.44	43.94	1.41
Ga			20.59	1.55	Ir	2.761	1.40	39.09	1.39
In	3.451	1.51	19.50	1.46	Pt	2.790	1.41	34.07	1.39
Tl	3.155	1.47	14.52	1.40	Th	4.122	1.60	42.08	1.57
Ti	3.686	1.54	29.6	1.52	U			36.62	1.50
Zr	3.974	1.58	38.93	1.55					
Hf	3.542	1.52	40.54	1.52					

highest occupied orbital extends no further than the lowest unoccupied one, although in fact some interpenetration of these shells is always present. Since the orbital radius depends on the principal quantum number n,

$$r_o = a_o \frac{n^2}{Z^*} \tag{1.23}$$

Table 1.10 Radii of isolated atoms (in Å)

Li	Be	B	C	N	O	F			
2.7	1.9	1.9	1.7	1.6	1.6	1.5			
Na	Mg	Al	Si	P	S	Cl			
2.9	2.3	2.2	1.95	1.8	1.7	1.6			
K	Ca	Sc	Ti	V	Cr	Mn	Fe	Co	Ni
3.3	2.8	2.45	2.3	2.2	2.1	2.05	2.0	1.95	1.9
Cu	Zn	Ga	Ge	As	Se	Br			
2.0	1.9	2.1	2.0	1.8	1.8	1.7			
Rb	Sr	Y	Zr	Nb	Mo	Tc	Ru	Rh	Pd
3.4	2.9	2.65	2.45	2.3	2.2	2.1	2.05	2.0	2.0
Ag	Cd	In	Sn	Sb	Te	I			
2.1	2.0	2.3	2.15	2.05	2.0	1.9			
Cs	Ba	La	Hf	Ta	W	Re	Os	Ir	Pt
3.7	3.1	2.9	2.4	2.25	2.1	2.05	2.0	1.95	1.9
Au	Hg	Tl	Pb	Bi	Th	U			
2.0	1.9	2.25	2.2	2.2	2.8	2.5			

where n refers to the highest occupied orbital, hence (neglecting the small change of Z^*),

$$r_m = r_o \left[\frac{(n+1)}{n} \right]^2. \tag{1.24}$$

Replacing the covalent radii for r_o (see Sect. 1.4.1) and multiplying the right part of Eq. 1.24 by $k = 0.9 + 0.05n$ [37] in order to convert the radius of the maximum electron density r_o to the radius of its minimum, r_m, which can be regarded as the atomic boundary. The results are listed in Table S1.6. The difference between these approaches for calculating radii of free atoms to day allows to present the average values to within 0.1 Å (Table 1.10).

1.3 Radii of Atoms in Molecules and Crystals

1.3.1 Historical Outline

The concept of the atomic radius was introduced in 1920 by Bragg [104], who estimated the radii for some 40 elements from the few structures then known. Later, Slater [105] used a more extensive experimental basis (1200 crystal structures of various chemical types) to compile a table of atomic radii for 95 elements, which

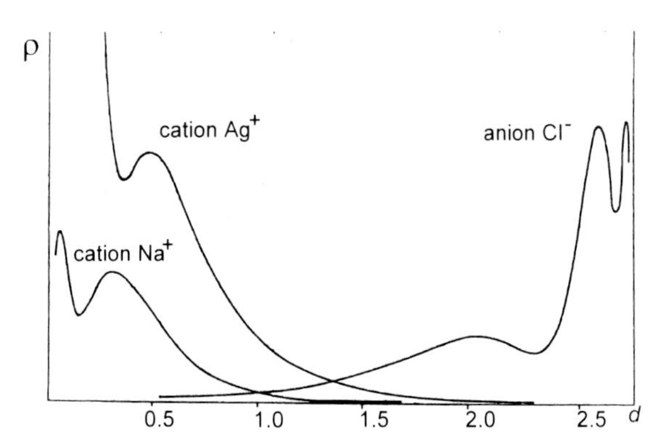

Fig. 1.5 Electron density radial distributions ρ in ions, plotted against interatomic distances d (in Å) in AgCl and NaCl crystals

described the observed bond distances in continuous solids of elements and inorganic compounds with the mean precision of ± 0.12 Å. Slater has clarified the physical meaning of the atomic radius, showing its similarity (correlation) with the distance between the nucleus and the maximum of electron density on the bonding orbital, known as the 'orbital radius' (see above, Table 1.8).

It should be noted, that the term 'radius' implies that atoms are considered as hard spheres, which touch each other when atoms are bonded, but can not penetrate or deform each other. This image seems to contradict the concepts of quantum mechanics, as the electron clouds of atoms have no clear-cut boundaries and must overlap to form a chemical bond. However, one must keep in mind that (*i*) the repulsion which prevents closer approach of atoms, is due mainly to Pauli exclusion which forbids electrons with similar spins to occupy the same space, and (*ii*) the electron cloud of an atom (other than H or He) comprises two distinct parts: the dense core and the much more diffuse valence electron clouds, with a steep jump of electron density between the former and the latter (Fig. 1.5). For these reasons, the repulsive force increases very steeply with the decrease of d, and the overlap of valence shells of chemically bonded atoms is limited in practice to area where the electron density does not exceed *ca.* 10 % of the maximum [37]. Furthermore, since core electrons are unaffected by chemical bonding, the atomic radius can be regarded in the first approximation as the constant of a given element, invariant against the composition and structure of the solid.

As structural data grew more abundant and accurate, it became clear that bond distances are substantially affected by the structure and the type of bonding. Thus the universal system of atomic radii has been replaced by a number of specific systems, each of them intended to describe a particular chemical or structural class. Today one can find in the literature a bewildering variety of such systems, not infrequently leading to inconsistent or misplaced usage. Actually, these radii serve two purposes: to make a rough prediction of a bond distance in an unknown structure (e.g. while solving a crystal structure from diffraction data) or to provide a standard length of an 'ideal' bond, a comparison of which with the experimental value can give some insight into the character of the bonding. For the former purpose, it is sufficient to have a system of radii describing the distances in a group of kindred structures with moderate precision. The latter not only requires more precise radii, but those with *external* foundations, whether theoretical (first principles) or empirical (structures approaching an ideal type). A 'good agreement' between a bond distance and the radii derived from that very distance, of course gives no new knowledge.

In the first place, atomic radii have been diversified into metallic and covalent radii, the latter applied to all structures other than metallic. These structures being very diverse in character, the radii obtained by different authors vary widely depending on the data basis used and the simplifying assumptions made. In fact, the metallic and covalent bonds are basically similar, both implying complete sharing of valence electrons; indeed, the metallic bonding is often described as non-directional covalent. As will be shown below, there is a basic unity between all systems of metallic and covalent radii; most of the differences between them can be attributed to the differences in the coordination number (N_c), bond polarity and valence (oxidation state). The atomic radius always increases with N_c; the increase is related to the parallel decrease of the ionization potential.

The decrease of Z^* in a solid is caused by multi-particle (extra valence) interaction of atoms within the coordination sphere, which can be described as external screening [106],

$$\Delta Z^* = s(N_c - v)\frac{r_1}{2r_N} \tag{1.25}$$

where r_1 is the normal radius for $N_c = 1$, r is the distance to the nearest atom in the coordination sphere, and s is the constant of external screening, which equals 0.10 for s- and p-electrons [37]. Then, taking into account Eq. 1.23,

$$r_N/r_1 = Z^*/\left(Z^* - \Delta Z^*\right) = 1 + c/Z^* \tag{1.26}$$

where $c = 0.1\ (N_c - v)/2$ (from Eq. 1.25). Eq. 1.26, which takes into account the size and electronic structure of the given atom, will be used below for calculating atomic radii for different coordination numbers.

The empirical concept of atomic radius has found an explanation in quantum chemistry. Thus, *ab initio* calculations have shown that the radial function of electron density of an atom has a minimum, which gives a physically meaningful boundary surface between the inner (core) and outer (valence) regions [107–112]. The distance from the nucleus at which the chemical potential is equal to the electrostatic potential of a free atom, has been calculated for a number of elements and proved to be proportional to the covalent radius [113]. The general theory of atoms in molecules (AIM), developed by Bader, has demonstrated that atoms in molecules or solids can be divided by physically meaningful boundary surfaces, rather than regarded as extending to infinity [114–117].

1.3.2 Metallic Radii

The crystals of metals usually adopt body-centered cubic structures with $N_c = 8$ or various motifs of close-packing of spheres with $N_c = 12$ (see Chap. 5). The first

Table 1.11 Atomic [105] (*upper lines*) and metallic [119–121] (*lower lines*) radii (in Å)

Li	Be	B	C	N	O	F			
1.45	1.05	0.85	0.70	0.65	0.60	0.50			
1.55	1.12	0.98							
Na	Mg	Al	Si	P	S	Cl			
1.80	1.50	1.25	1.10	1.00	1.00	1.00			
1.90	1.60	1.43	1.37	1.28	1.27	1.26			
K	Ca	Sc	Ti	V	Cr	Mn	Fe	Co	Ni
2.20	1.80	1.60	1.40	1.35	1.40	1.40	1.40	1.35	1.35
2.35	1.97	1.62	1.47	1.34	1.28	1.27	1.26	1.25	1.24
Cu	Zn	Ga	Ge	As	Se	Br			
1.35	1.35	1.30	1.25	1.15	1.15	1.15			
1.28	1.38	1.40	1.44	1.48	1.40	1.41			
Rb	Sr	Y	Zr	Nb	Mo	Tc	Ru	Rh	Pd
2.35	2.00	1.80	1.55	1.45	1.45	1.35	1.30	1.35	1.40
2.48	2.15	1.80	1.60	1.46	1.39	1.36	1.34	1.34	1.37
Ag	Cd	In	Sn	Sb	Te	I			
1.60	1.55	1.55	1.45	1.45	1.40	1.40			
1.44	1.51	1.58	1.62	1.66	1.60	1.62			
Cs	Ba	La	Hf	Ta	W	Re	Os	Ir	Pt
2.60	2.15	1.95	1.55	1.45	1.35	1.35	1.30	1.35	1.35
2.67	2.21	1.87	1.58	1.46	1.39	1.37	1.35	1.35	1.38
Au	Hg	Tl	Pb	Bi	Po				
1.35	1.50	1.90	1.80	1.60	1.90				
1.44	1.51	1.60	1.70	1.78					

system of metallic radii has been calculated by Goldschmidt simply as halves of the shortest interatomic distances in the latter [118]. For other N_c he introduced correction factors derived from the changes of interatomic distances actually observed on polymorphous transformations. A complete system of metallic radii for $N_c = 12$ together with the atomic radii by Slater is given in Table 1.11.

Some metals (e.g. Mn, U, Np, Pu) have less symmetrical structures, where bond lengths vary widely and atoms have different coordination numbers. Therefore Zachariasen [122] suggested to derive averaged metallic radii from the (experimental) crystal volumes per atom (V_a),

$$r_{12} = 0.5612\ V_a^{1/3} \tag{1.27}$$

thereby obtaining the metallic radii for d-metals and, most importantly, lanthanides and actinides (Table S1.6); later Eq. 1.27 was used by Trömel [123, 124].

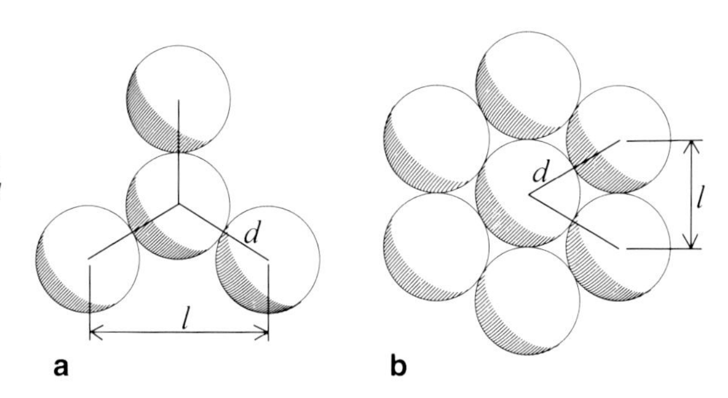

Fig. 1.6 Representation of homoatomic structures as packing of spheres: **a** planar-trigonal coordination, $l = 2.362\ d$, **b** a cross-section of a close-packed motif, $l = d$

It is noteworthy that whilst for most elements the electron configuration of the neutral atom is an unequivocal function of the atomic number, in actinides the 5*f*, 6*d* and 7*s* levels are so close in energy that various polymorph modifications (and other phases) of the same element can have different electron configurations. This has strong effects on the radius, which correlates with the total number (N_e) of the outer non-*f* electrons [122]. Among lanthanides, the same applies to Ce, which has $N_e = 4$ in the α-form ($r = 1.704$ Å) and $N_e = 3$ in the β- and γ-forms ($r = 1.827$ and 1.824 Å, respectively). Most of the lanthanides have $N_e = 3$ and the radii ranging from 1.747–1.828 Å, while Eu has $N_e = 2$ and $r = 2.041$ Å.

Assuming the structures of metals as arrays of rigid spherical atoms contacting one another at single points, the fraction of crystal space occupied by these spheres can be calculated as

$$\rho = \frac{4}{3}\pi r^3 / V_a \tag{1.28}$$

where V_a is the unit cell volume per atom. Note that ρ, known as the packing coefficient, increases with increase of N_c while *r* also increases. This apparent paradox is explained by the 'diagonal' (second-shortest) distances *l* becoming relatively shorter (see Fig. 1.6). Thus, $l/d = 2.362$ for graphite, 1.633 for diamond, 1.366 for β-Sn (white tin), 1.155 for *bcc,* 1.000 for *fcc* structures [125]. Structures with the same N_c can have substantially different ρ: compare the structures of white tin (ρ = 0.52) and α-Po (ρ = 0.56), both having $N_c = 6$. A random array of equal spheres has the average ρ = 0.64 [126] compared to 0.74 for the closest regular packing. However, spheres of *different* radii can achieve even denser packing, smaller spheres filling the gaps between larger ones [127].

The packing coefficient ρ should not be confused with the space-filling coefficient *k* in molecular crystals, which is calculated on the assumption that atoms contact each other at van der Waals radii, much larger than bond radii (see Sect. 4.3). In fact, in crystals comprising separate molecules, chains or layers (e.g. graphite), each atom participates in two different types of contacts, described by metallic/covalent and van der Waals radii, respectively. In this case the term 'radius' should not be taken literally. Thus, the crystal structures of P, As, Sb and Bi ($N_c = 3$) have ρ = 0.23 but $k \approx 0.7$.

1.3.3 Covalent Radii

The three widely used types of covalent radii are the normal (r_{nor}), the tetrahedral (r_{te}), and the octahedral (r_{oc}) ones; r_{nor} is defined as half the single-bond distance in a homo-atomic molecule with $N_c = v$, and r_{te} as half the bond distance in a diamond-like structure, hence $r_{nor} = r_{te}$ for tetravalent elements. The systems of tetrahedral and octahedral radii have been first introduced by Huggins and Pauling [128–131, 139–142], who observed that $r_{te} \leq r_{nor}$ for nonmetals while $r_{te} > r_{nor}$ for metals. The difference can be qualitatively explained by the fact that metals have fewer than 4 outer electrons, hence an increase of N_c from 1 to 4 is bound to reduce the number of electrons per bond, while nonmetals have enough outer electrons to provide for all these bonds.

At first, r_{te} and r_{oc} have been calculated only for the elements, for which the corresponding coordination is more typical, i.e. r_{te} for the Group 11–17 elements and Be, r_{oc} for the rest. Later, Van Vechten and Phillips [132] have determined r_{te} and r_{oc} for the same elements and observed that always $r_{te} < r_{oc}$, and that either radius remains practically constant within the following series of elements, (i) Si, P, S, Cl; (ii) Cu, Zn, Ga, Ge, As, Se, Br; (iii) Ag, Cd, In, Sn, Te, I, probably because an increase of Z^* along a period is compensated by an increase of inter-electron repulsion. Note that the systems of r_{te} derived by Pauling and Phillips, differ substantially, as do the r_{nor} published by different authors [133–139]. This is so because that additivity of the radii is perfect only for purely covalent (non-polar) bond, while polar bonds are shorter. This tendency, first rationalized by Schomaker and Stevenson [140], can be described by equations

$$d = r_A + r_B - 0.09\ \Delta\chi \tag{1.29}$$

$$d = r_A + r_B - 0.085\ \Delta\chi^{1.4} \tag{1.30}$$

$$d = r_A + r_B - a\Delta\chi - 0.08\ \log n \tag{1.31}$$

where d is the length of an A–B bond, n is the bond order, $\Delta\chi$ is the difference between electronegativities (EN, in Pauling's scale) of the atoms A and B, r_A and r_B are the covalent radii of these atoms, and the factor a varies from 0.02 to 0.08 for different elements.

Now it is clear why sums of covalent radii cannot represent adequately the variety of polar bonds. Optimizing the radii can only 'tune' the system to a certain range of polarities. The r values will vary depending on the choice of the reference structures and the optimization procedure, especially when metal atoms (with low ENs) are involved. Thus, Slater's atomic radii of the most electronegative elements (F, O, N) are underestimated. However, a simple additive model can work where all atoms happen to be of similar EN, as in many organic molecules.

Batsanov [143–145] comprehensively revised the system of covalent radii was using the extensive structural data now available. As far as possible, the normal radii were derived directly from interatomic distances in homo-atomic molecules, elemental solids or compounds containing homo-atomic moieties. The only hetero-atomic bond distances used were the M–CH_3 and M–H in metal alkyls and hydrides,

Table 1.12 Normal (*upper lines*) and crystalline[a] (*lower lines*) covalent radii (Å)

Li	Be	B	C	N	O	F	H		
1.33	1.02	0.85	0.77	0.73[b]	0.72[b]	0.71[b]	0.37[b]		
1.58	1.07	0.87	0.77	0.74	0.75[c]	0.75[c]	0.46[c]		
Na	Mg	Al	Si	P	S	Cl			
1.65	1.39	1.26	1.16	1.11	1.03	0.99			
1.84	1.49	1.28	1.16	1.12	1.07[c]	1.03[c]			
K	Ca	Sc	Ti	V	Cr	Mn	Fe	Co	Ni
1.96	1.71	1.48	1.36	1.34	1.22	1.19	1.16	1.11	1.10
2.18	1.83	1.55	1.40	1.39	1.27	1.24	1.21	1.16	1.14
Cu	Zn	Ga	Ge	As	Se	Br			
1.12	1.18	1.24	1.21	1.21	1.16	1.14			
1.16	1.20	1.25	1.21	1.22	1.19[c]	1.18[c]			
Rb	Sr	Y	Zr	Nb	Mo	Tc	Ru	Rh	Pd
2.10	1.85	1.63	1.54	1.47	1.38	1.28	1.25	1.25	1.20
2.29	1.96	1.70	1.58	1.53	1.41	1.31	1.30	1.30	1.26
Ag	Cd	In	Sn	Sb	Te	I			
1.28	1.36	1.42	1.40	1.40	1.36	1.33			
1.32	1.38	1.43	1.46	1.41	1.39[c]	1.37[c]			
Cs	Ba	La	Hf	Ta	W	Re	Os	Ir	Pt
2.32	1.96	1.80	1.52	1.46	1.37	1.31	1.29	1.22	1.23
2.53	2.08	1.88	1.55	1.50	1.40	1.33	1.33	1.26	1.27
Au	Hg	Tl	Pb	Bi	Po	At	Ac	Th	U
1.24	1.33	1.44	1.44	1.51	1.45	1.47	1.86	1.75	1.70
1.27	1.35	1.45	1.44	1.52	1.48[c]	1.51[c]	1.94	1.79	1.75

[a]r_{te} are given for Group 11 to 17 elements and Be, r_{oc} for other elements; [b]derived as ½ d(A–A), i.e. as--suming purely covalent single bonds; [c]r_{oc}, for these elements $r_{te} = r_{nor} + 0.02$ Å

these being the least polar of metal-ligand single bonds. These values also agree well with those from homoatomic bond distances, except for Na where the 'hetero-atomic' radius exceeds the 'homoatomic' one by 0.1 Å, probably due to *sp* hybridization of the bonding electrons, as the orbital radius for *p* electrons is by 0.25 Å larger than for *s*. The discrepancy between the covalent radii derived from M–C and M–H distances is small, especially if corrected for polarity using Eq. 1.29. Recently, self-consistent systems of single-bond (normal) covalent radii was derived by Pyykkö and Atsumi [146] from both experimental (E–E, E–H and E–CH_3 distances) and theoretical data for all elements with $Z = 1$–118, and by Cordero et al. [147] from statistical analysis of crystallographic data for elements with $Z = 1$–96, interpolating the radii for the few elements for which experimental data is still lacking (e.g. rare gases) on the basis of robust periodic dependence of these radii. The dependence of r_{nor} on the oxidation state of an atom has been determined [148] taking into account the hybridization of orbitals and the influence of unpaired electrons (Table S1.7) in good agreement with the experiment.

Octahedral and tetrahedral covalent radii can be either calculated by additivity from the structures with corresponding N_c, or derived from r_{nor} using Pauling's equation [119] converted from bond distances to radii,

$$r_1 - r_n = c \ln n \quad (1.32)$$

and assuming the bond order, n, to equal the ratio of the valence v to the coordination number N_c (see Sect. 5.5). The drawback is that a uniform correction is applied to all atoms, irrespective of their sizes and electronic structures, although both r_{te} and r_{oc} are known to depend on the electronic structures of the elements [130]. Thus the c factor in Eq. 1.32 was found to vary widely, depending on the type of compounds [149]. If the type of structure changes, the *ratio* of interatomic distances gives more informative results [2]. For these reasons, Eq. 1.26 is preferable for calculating r_{te} and r_{oc}, as it takes into account both factors affecting the covalent radii when N_c changes; the results of these calculations are presented in Table 1.12 [150]. Lengths of double and triple bonds can be described by special systems of covalent radii, see Table 1.13.

1.4 Radii of Ions in Molecules and Crystals

It is known since the dawn of X-ray crystallography in the early twentieth century that crystal structures of highly polar compounds, e.g. metal halides or oxides, can be conveniently described as a close packing of rigid spherical *ions* with closed electron shells. The radius of an ion is, in the first approximation, a constant for a given element and formal charge, independent of the environment. The bond distances are thus the sums of element-specific increments. By a rule of thumb, this ionic model works well (i.e. additivity of bond distances holds) if the electronegativities of the two elements differ by ≥ 2 on Pauling's scale (see Sect. 2.4). In fact, even in such compounds there is a significant amount of covalent bonding present. Purely ionic bond, unlike purely covalent one, is only an ideal archetype – isolated ions exist only in plasma. Historically, the ionic model was often applied even to compounds of moderate polarity, in competition with the polar-covalent scheme, whereby bond lengths were expressed as sums of covalent radii, minus a correction for polarity.

1.4.1 Methods of Estimating Ionic Radii

The covalent radius can be simply calculated as half of the shortest homonuclear distance in the structure. Finding ionic radii is not so straightforward, because in ionic structures, ions of similar charge are never the nearest neighbours. Thus, at

Table 1.13 The double-bond (*upper lines*) and triple-bond (*lower lines*) covalent radii (Å). (From [2, 151, 152])

Be	B	C	N	O	F			
0.90	0.78	0.67	0.62	0.60	0.54			
0.85	0.73	0.60	0.55					
Mg	Al	Si	P	S	Cl			
1.32	1.13	1.07	1.02	0.94	0.89			
1.27	1.11	1.02	0.94	0.87				
Ca	Sc	Ti	V	Cr	Mn	Fe	Co	Ni
1.47	1.16	1.23	1.12	1.11	1.05	1.09	1.03	1.01
1.33		0.97	1.06	1.03		1.02	0.96	
Zn	Ga	Ge	As	Se	Br			
	1.17	1.13	1.14	1.08	1.04			
	1.03	1.06	1.06					
Sr	Y	Zr	Nb	Mo	Tc	Ru	Rh	Pd
1.57	1.30	1.27	1.25	1.21	1.20	1.14	1.10	1.17
1.39	1.24	1.21	1.16	1.13	1.10	1.30	1.30	1.12
Cd	In	Sn	Sb	Te	I			
	1.36	1.30	1.33	1.28	1.23			
			1.17					
Ba	La	Hf	Ta	W	Re	Os	Ir	Pt
1.61	1.39	1.28	1.26	1.20	1.19	1.16	1.15	1.12
1.49		1.22	1.19	1.15	1.10	1.09	1.07	
Hg	Tl	Pb	Bi	Po	At	Ac	Th	U
	1.42	1.35	1.41	1.35	1.38	1.53	1.43	1.34
			1.33			1.40	1.36	1.18

least one (reference) radius must be found by some other means. If in a series of ionic binary solids, the same anion is combined with cations of progressively smaller size, the anion-anion distances contract up to a certain point, and then become indifferent to further diminution of the cation. One can assume that from this point the structure is defined solely by the close packing of anions (with cations occupying the voids) whose radii therefore can be determined as halves the shortest anion-anion distances. This assumption has been used by Landé [153] to derive the first system of ionic radii, including F^- 1.31, Cl^- 1.78, Br^- 1.96, I^- 2.13 Å; O^{2-} 1.31, S^{2-} 1.84, Se^{2-} 1.92, and Te^{2-} 2.26 Å. Ladd [154] followed Landé's idea and compiled a more extensive system of ionic radii. The drawback is that the smallest cations have the highest ENs, hence the key radii are determined on the structures with the *least* ionic bond character. As an alternative, Wasastjerna [155, 156] calculated the radii of Group 1 and 2 metal cations, using radii of O^{2-} (1.32 Å) and F^- (1.33 Å) derived from their polarizabilities according to Klausius–Mossotti theory. This approach was further developed by Kordes [157, 158] who obtained 1.35 Å and 1.38 Å for the radii of O^{2-} and F^-, respectively, and by Vieillard [159] who calculated the radii of ions from their electron polarizabilities, obtaining 1.49 Å for O^{2-}. A purely empirical system of ionic radii has been derived by Goldschmidt [160] from observed interatomic distances in crystals, using Wasastjerna's radius of O^{2-} as the key. For a long time, this system was regarded as canonic, although attempts to improve it continued. Thus, Zachariasen [161, 162] recalculated ionic radii using wider experimental data and the radii of F^- (1.33 Å), Cl^- (1.81 Å) and O^{2-} (1.40 Å) as key ones; he also suggested a system of ionic radii for actinides [163]. Shannon and Prewitt [164–168] used least-squares optimization of the radii against an enormous set of structural data for binary and ternary ionic compounds, assuming $r(O^{2-}) = 1.40$ Å and $r(F^-) = 1.33$ Å for $N_c = 6$ and using theoretical estimates for multi-charged ions [169, 170]; this system is in general use today (Table S1.7). It gives the radii of cations for various coordination numbers which are found in solids. According to Shannon and Prewitt, anion radii depend on N_c much less than those of cations, or not at all, in the same way as the covalent radii of nonmetals are less affected than those of metals. Cation radii (r^+) are smaller and anion radii (r^-) larger than the covalent radii of the corresponding neutral atoms, except for Au^+ and Tl^+. At the same time, according to the ionic model of Pauling and Zachariasen, both anion and cation radii increase with N_c as a function of the Born exponent, n, which equals 5 for He, Li^+, H^-, 7 for Ne, Na^+, F^-, 9 for Ar, K^+, Cu^+, Cl^-, 10 for Kr, Rb^+, Ag^+, Br^-, and 12 for Xe, Cs^+, Au^+. For a crystal of mixed-ion type, mean n is to be used. A comparison of bond distances in structures with different coordination numbers shows that as N_c increases the difference of interatomic separations in polymorphic modifications widens in the successions MCl → MI; MS → MTe; MP → MSb, whereas it would remain constant if *only* the cation radii were affected by the change of N_c. A similar conclusion follows from the Bond Valence Equation:

$$v_{ij} = \exp\left[(R_{ij} - d_{ij})/b\right] \tag{1.33}$$

where v_{ij} is the bond valence of atoms i and j, $b = 0.37$, and R_{ij} is the bond-valence parameter. This parameter changes with j in the above mentioned series [171], although according to Shannon and Prewitt it should not.

The first theoretical estimate of ionic radii has been made by Pauling [172] who partitioned interatomic distances in the crystals of alkali halides comprising *iso-electronic* ions (i.e. $Na^+ F^-$, K^+Cl^-, Rb^+Br^- and Cs^+I^-) in an inverse relation to their effective nuclear charges, in accordance with Eq. 1.23, and obtained the following radii: K^+ 1.33, Cl^- 1.81, Rb^+ 1.48, Br^- 1.95, Cs^+ 1.69 and I^- 2.13 Å; the radii of F^- (1.36 Å) and O^{2-} (1.40 Å) were derived by additivity. These values agree with the empirical radii, especially the latest system of Brown [173]. Batsanov [174] applied Pauling's idea to molecular and crystalline halides MX_n (with iso-electronic M^{n+} and X^- ions) and calculated the halide radii for different N_c. This approach was applied also to the molecular and crystalline iso-electronic oxides and chalcogenides (Table 1.14).

Table 1.14 Radii of anions (Å) for different N_c, calculated by Pauling's method

N_c	1	2	3	4	6	8
F	1.09	1.18	1.24		1.36	1.39
Cl	1.48	1.60	1.66		1.81	1.88
Br	1.59	1.65		1.85	1.92	1.99
I	1.79		2.01	2.12	2.13	2.20
O	1.19	1.17	1.30	1.35	1.44	1.56
S	1.52		1.76		1.86	1.98
Se			1.75		1.92	2.03
Te			1.84		2.13	2.25

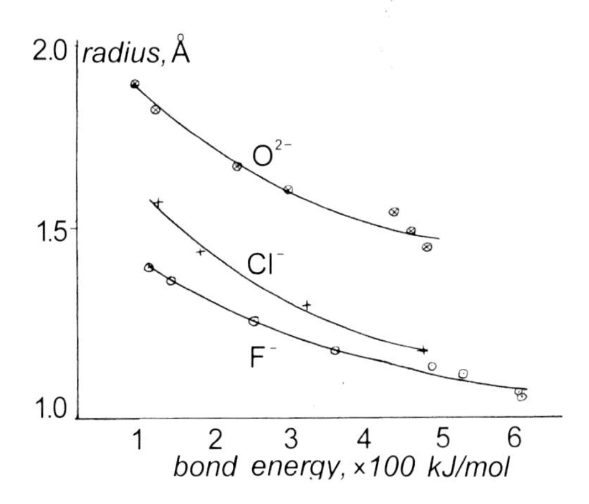

Fig. 1.7 Dependence of ionic radii on the bond energies, in molecules and crystals of NaF, MgF_2, AlF_3, SiF_4; KCl, $CaCl_2$, $ScCl_3$; Na_2O, MgO, Al_2O_3 and SiO_2

However, much more general is the dependence of the ionic radius on the bond energy (Fig. 1.7). For example, N_c(Cl) remains 1 in the molecules KCl, $CaCl_2$, and $ScCl_3$, but the r(Cl) changes from 1.542 to 1.489 to 1.451 Å, as the bond energy increases (433, 448, and 470 kJ/mol, respectively). Similarly, the ionic radii in the vdW complexes M^+ Rg and X^-Rg are larger than in solids, even though N_c increases from 1 in the complex to 6 or 8 in the solid [174].

The $r(X^-) = f(E)$ dependence being the same for molecular and crystalline halides, the radii of non-isoelectronic ions can be determined, provided the bond energies are known. The ionic radii for MX molecules and crystals are listed in Table 1.15, from which it follows that cationic and anionic radii change with N_c to a similar degree. The radii of F^- (1.21 Å), Cl^- (1.63 Å), Br^- (1.71 Å), and I^- (1.90 Å) have been calculated from the Morse potential energy curves (calculated from the spectroscopic data) for the dissociation of molecular ions, $X_2^- \rightarrow X^- + X$, assuming the additivity of covalent and ionic radii [175]. These values are by 0.07 Å larger than the radii obtained by Pauling's method from bond lengths in the NaF, KCl, RbBr and CsI molecules. Other sources of information are the structures of ionic vdW complexes. Thus, the Ar–Ag distances in Ar · AgCl and Ar · AgBr give $r(Ag^+) = 0.65$ Å [176].

An independent method of calculating ionic radii [177] uses Sanderson's principle of equalization of atomic electronegativities (see Sect. 2.4.5), it shows the metal

Table 1.15 Ionic radii (Å) in molecules and crystals, determined by the $r(X^-) = f(E)$ dependence

State	M	F		Cl		Br		I	
		r^+	r^-	r^+	r^-	r^+	r^-	r^+	r^-
Molecule	Li	0.48	1.08	0.55	1.47	0.57	1.60	0.58	1.82
	Na	0.79	1.14	0.85	1.51	0.88	1.62	0.90	1.84
	K	1.06	1.11	1.13	1.54	1.21	1.61	1.20	1.85
	Rb	1.16	1.11	1.29	1.50	1.30	1.64	1.32	1.85
	Cs	1.24	1.10	1.42	1.49	1.45	1.61	1.48	1.84
Crystal	Li	0.67	1.34	0.76	1.80	0.85	1.90	0.87	2.16
	Na	0.94	1.36	0.99	1.82	1.05	1.93	1.06	2.17
	K	1.30	1.36	1.31	1.81	1.36	1.93	1.37	2.16
	Rb	1.46	1.35	1.46	1.82	1.52	1.92	1.50	2.16
	Cs	1.64	1.37	1.62	1.85	1.68	1.93	1.71	2.13
	M	O		S		Se		Te	
		r^{2+}	r^{2-}	r^{2+}	r^{2-}	r^{2+}	r^{2-}	r^{2+}	r^{2-}
Crystal	Mg	0.67	1.44	0.68	1.92	0.74	1.99	0.73	2.23
	Ca	0.96	1.44	0.98	1.86	1.04	1.92	1.01	2.17
	Sr	1.14	1.44	1.15	1.86	1.20	1.92	1.16	2.17
	Ba	1.33	1.44	1.33	1.86	1.39	1.90	1.37	2.13

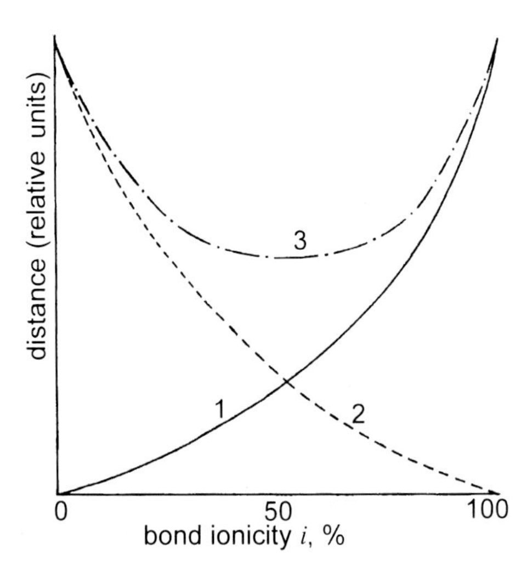

Fig. 1.8 Change of atomic radii with the increase of bond ionicity: 1 – non-metal radius, 2 – metal radius, 3 – bond distance

radius sharply decreasing as the bond polarity i increases (Fig. 1.8), so that even at $i = 70\,\%$ the radius is close to the cationic one. This explains why ionic radii correctly predict interatomic distances even in such inorganic compounds where the character of bonding is far from fully ionic. It is evident from Fig. 1.8 that an increase of polarity decreases the bond length in predominantly covalent (e.g. organic) compounds (see Eq. 1.26) but increases it in predominantly ionic compounds [178].

All these approaches are based on the concept of hard ions. However, in reality one must take into account polarization effects. The theory of ionic polarization, developed by Fajans [179–181], explains how a decrease of the cation radius must

Table 1.16 Averaged bonded radii (Å) in crystalline M_2SiO_4, from [203]

M	r_b (M)	r_b (O_M)	r_b (Si)	r_b (O_{Si})
Mn	1.06	1.11	0.66	0.98
Fe	1.08	1.08	0.68	0.96
Co	1.05	1.08	0.68	0.95

shorten interatomic distances. Because

$$d^k(M - X) < r^k(\mathrm{M}) + r^k(\mathrm{X})$$

for $k > 1$, the polarization effect can be ascribed to the change of k. The best agreement with experiment for alkali halide crystals was obtained for $k = 5/3$, i.e. assuming a substantial deformation (overlap) of the ionic spheres, hence this approach is known as 'soft-sphere radii' [182]. The difference between the sum of ionic radii and the actual interatomic distance in molecules and crystals of alkali halides was shown [183] to correlate with the dissociation energies and ionic charges.

1.4.2 *Experimental (bonded) Ionic Radii*

Experimental determination of electron density distribution in crystals by means of X-ray diffraction allow to determine the ionic radii (known as 'effective distribution radii', or 'bonded radii') as the distances between the corresponding nuclei and the minimum point of the electron density along the internuclear vector. Thus, Witte and Wölfel [184] found that in the NaCl crystal this minimum lies 1.17 Å from the Na and 1.64 Å from the Cl nucleus. Studies of other alkali halides gave the following values (Å): Li^+ 0.94, Na^+ 1.17, K^+ 1.17, Rb^+ 1.63, Cs^+ 1.86, F^- 1.16, Cl^- 1.64, Br^- 1.80, I^- 2.05 [185]. Maslen [186] showed that the ionic radii in crystals can be estimated assuming either the minimum overlap of atomic spheres (which gives traditional ionic radii) or the minimum electron density on the bond line (which gives bonded radii), confirming these conclusions by comparing the Li^+, Na^+, K^+, F^-, Cl^- and Br^- radii calculated by him with Pauling's values and the experimental bond radii. Bonded radii (r_b) of cations in oxides are: Be 0.56 Å [187]; Mg 0.92, Ca 1.27, Mn 1.15, Co 1.09, Ni 1.08, Al 0.91 Å [188]; Si 0.68 Å in coesite [189] and 0.82 Å in stishovite [190], Mg 1.28 Å in MgS [191], Fe 1.02 and S 1.18 in Fe_3S_4 [192]. In hydrides of lithium and magnesium the bonded radii are: Li 0.92, H 1.24 Å [193], Mg 0.95 and H 1.00 Å [194]. Bonded ionic radii in silicates are close to those in oxides [195–202]; the results for M_2SiO_4 are summarized in Table 1.16. Bonded radii of cations are larger than the conventional crystal-chemical radii, while the opposite is true for oxide anions (see Table S1.7), this reveals the limitations of the traditional ionic model.

The difficulty of experimental determination of atomic radii comes from several reasons, mainly the blurring effect of thermal vibrations of atoms and the extreme complexity of theoretical interpretation of the experimental data [204]. Johnson [205, 206] noted that electron density of an atom in a metallic structure shows a minimum at $d = 0.64\ r_M$ (where r_M is the metallic radius), which gives the r_b of this cation in the structure of the corresponding compound. Using this expression, Johnson calculated r_b of all elements in different valence states; however, generally the coefficient of 0.64 cannot be constant since the same metal in different compounds actually has different bond radii. Besides, the metallic (r_M) and ionic (r_i) radii were shown to correlate linearly,

$$r_M = a + br_i,$$

where constants a and b depend on the place of the elements in the Periodic Table [123, 124].

At contacts of a cation (c) and an anion (a), the outer electrons are affected by the nuclear charges resulting in the energies Z_c^*/r_c and Z_a^*/r_a, respectively. Because the former ratio is always larger than the latter, the electron cloud is shifted toward the cation. As a result, Z_c^*/r_c decreases (as r_c increases) and Z_a^*/r_a increases (as r_a decreases); this process will occur until the mutual influence of both ions is balanced. The equalization can be achieved in three ways [207]:

1. by changing the effective charges of ions until $Z_c^* = Z_a^*$; at a constant bond length this gives $r_c = r_a$, i.e. the bonded radii are equal to half the corresponding interatomic distances [208];
2. by changing the bond length from the experimental value d to the sum of the covalent radii; this allows to calculate the coefficients $C = a_0 n^2$ in Eq. 1.23 for the ionic and covalent states and at $Z_c^* = Z_a^*$ makes possible to define the bond radii by the equation $C_c r_c = C_a\ (d - r_c)$;
3. by changing the interaction energies of ions Z^*/r, calculated from the linearly interpolated Z^* and C values and the corresponding radii, until $Z_c^*/r_c = Z_a^*/r_a$.

All three methods give similar results. The average values of radii presented in Table S1.9 are in accordance with the independent estimations of r_b [161, 162, 202, 203]; while Table S1.10 shows r_b of halogens for various N_c. Experimental bonded radii of halogens depend on both the bond polarity and N_c. Thus, r_b(Cl) decreases from 1.57 Å in KCl ($N_c = 6$) to 1.40 Å in $CaCl_2$ ($N_c = 3$), and 1.27 Å in $ScCl_3$ ($N_c = 2$), and from 1.16 Å in CuCl ($N_c = 4$) to 1.10 Å in $ZnCl_2$ ($N_c = 2$).

1.4.3 Energy-Derived Ionic Radii

There are several methods for determining ionic radii from physical characteristics of atoms and crystals. Thus, Fumi and Tosi [209] derived ionic radii (similar to the bonded ones) for alkali halides, using the Born model of crystal lattice energy with experimental interatomic distances, compressibilities and polarizabilities. Rosseinsky [210] calculated ionic radii from ionization potentials and electron affinities of atoms, his results were close to Pauling's. Important conclusions can also be drawn from the behaviour of solids under pressure. Considering metal as an assembly of cations immersed into electron gas, its compressibility at extremely high pressures

Table 1.17 Ultimate radii (Å) of anions

Anion	[205]	[213]	[214]
F^-	1.90	1.60	1.99
Cl^-	2.35	2.17	2.33
Br^-	2.53	2.35	2.42
I^-	2.76	2.64	2.60
O^{2-}	2.46	1.68	1.96
S^{2-}	2.68	2.21	2.26
Se^{2-}	2.80	2.38	2.33
Te^{2-}	3.01	2.65	2.47

can be rationalized in terms of direct cation-cation contacts in a *bcc* or *fcc* lattice [211], this gives the following radii: Li^+ 0.86 Å, Na^+ 1.18 Å, K^+ 1.55 Å, Rb^+ 1.65 Å, Cs^+ 1.74 Å, Ba^{2+} 1.37 Å (for $N_c = 8$); Cu^+ 1.00 Å, Ag^+ 1.18 Å, Au^+ 1.20 Å (for $N_c = 12$). It was found that a M–X distance in an ultimately-compressed M_nX_m crystals can be described as the sum of the *ionic* radius of M and *covalent* radius of X [212]. The mean ionic radii calculated from the compressibilities of M_nX_m compounds (Table S1.11) are in excellent agreement with the crystal-chemical radii of cations according to Shannon and Prewitt.

Table 1.18 Ideal ionic radii (Å) for $N_c = 6$

+1	+2	+3	+4	+5	+6	+7	−1	−2
Li	Be	B	C	N	O	F	F	O
0.61	0.36	0.22	0.13	0.10	0.07	0.06	1.80	1.82
Na	Mg	Al	Si	P	S	Cl	Cl	S
0.82	0.58	0.43	0.32	0.30	0.23	0.22	2.25	2.24
K	Ca	Sc	Ti	V	Cr	Mn		
1.10	0.88	0.59	0.48	0.43	0.44	0.37		
Cu	Zn	Ga	Ge	As	Se	Br	Br	Se
0.62	0.59	0.50	0.42	0.37	0.34	0.31	2.38	2.36
Rb	Sr	Y	Zr	Nb	Mo	Tc		
1.22	0.94	0.72	0.58	0.51	0.40	0.45		
Ag	Cd	In	Sn	Sb	Te	I	I	Te
0.92	0.76	0.64	0.55	0.48	0.45	0.42	2.62	2.56
Cs	Ba	La	Hf	Ta	W	Re		
1.34	1.08	0.82	0.57	0.51	0.48	0.42		
Au	Hg	Tl	Pb	Bi	Po	At		
0.88	0.82	0.70	0.62	0.61	0.54	0.50		

1.4.4 Ultimate Ionic Radii

By extrapolating the dependence of ionic radius on inter-ion separations, Johnson [205, 206] determined the anion radii at zero influence of cations, i.e. the radii for the purely ionic state (Table 1.17). Alternatively, these characteristics can be derived from the structures of solid solutions. For example, in the KBr–KI solid solution the ability of a bromine atom to draw electrons from potassium increases with the concentration of iodine, hence the K–Br bond becomes more ionic and the Br^- radius increases. Similarly, in the KBr–KCl system, substituting Br for Cl increases the ionicity and the length of the K–Cl bond. In the infinitely diluted MX–MY solid solution, the ionic radii of X and Y should converge. The same reasoning applies to solid solutions with mixed cations, e.g. RbX–KX, where $r(Rb^+)$ and $r(K^+)$ converge. According to Goldschmidt [118], isomorphous substitution of ions is possible if their radii differ by ≤ 15 %, hence the standard radii of anions should be multiplied by 1.15 and those of cations by 0.85 to obtain their ultimate ionic radii [213]; later these factors were adjusted to 1.2 and 0.8, as wider limits isomorphism were found. The resulting ultimate radii of anions are listed in Table 1.17 [213], together with the ultimate radii of anions calculated from the anion–anion distances in the structures of RbX and BaX containing the largest cations [214].

The above mentioned data allows to compile a system of ideal ionic radii (Table 1.18) which can be used to interpret bond lengths in real crystalline compounds. For the purposes of interpreting the character of chemical bonds, a system of ionic radii for molecules was created [179] by applying Pauling's corrections for the variation of r^+ with N_c [94] (see Table S1.12).

1.4.5 Concluding Remarks

Systems of ionic radii were derived almost a century ago, to predict (roughly) interatomic distances in crystals. Today this is no longer necessary, since the structures of practically all binary compounds are already known and easily accessible through databases, so the ionic (as well as covalent) radii are used mainly as standard reference points for interpreting the observed interatomic distances. At the same time, for rough estimations of interatomic distances one can use the standard ionic radii and the corrections in bond lengths for different coordination numbers, derived from the structures of polymorphous inorganic compounds (see Sect. 5.2):

$N_c \rightarrow N'_c$:	$1 \rightarrow 2$	$1 \rightarrow 3$	$1 \rightarrow 4$	$1 \rightarrow 6$	$1 \rightarrow 8$
$d(N_c)/d(N'_c)$:	1.10	1.15	1.17	1.22	1.26

Thus, for practical purposes one can use the radii (listed in Tables S1.7, S1.8 and S1.12) of cations with charges +4 or less and of anions with charges −1 and −2, the ions with higher charges do not exist in reality.

Appendix

Supplementary Tables

Table S1.1 Experimental ionization potentials I of diatomic molecules (A_2) in comparison with isolated atoms, in eV

A_2	$I(A_2)$	$I(A)$	Ref.	A_2	$I(A_2)$	$I(A)$	Ref.	A_2	$I(A_2)$	$I(A)$	Ref.
Li_2	5.11	5.39	1.1	Rh_2	7.1	7.46	1.12	P_2	10.53	10.49	1.1
Na_2	4.89	5.14	1.1	Pd_2	7.7	8.34	1.7	As_2	9.69	9.79	1.23
K_2	4.06	4.34	1.1	Pt_2	8.68	8.96	1.13	Sb_2	8.43	8.61	1.24
Rb_2	3.9	4.18	1.2	Cu_2	7.90	7.73	1.11	Bi_2	7.34	7.28	1.24
Cs_2	3.7	3.89	1.3	Ag_2	7.66	7.58	1.14	O_2	12.07	13.62	1.1
Be_2	7.42	9.32	1.4	Au_2	9.2	9.22	1.14	S_2	9.36	10.36	1.1
Y_2	4.96	6.22	1.5	Zn_2	9.0	9.39	1.11	Se_2	8.6	9.75	1.25
La_2	4.84	5.58	1.6	Cd_2	7.7	8.99	1.15	Te_2	8.8	9.01	1.26
Ti_2	6.3	6.83	1.7	Hg_2	9.56	10.44	1.7	H_2	15.42	13.59	1.1
Zr_2	5.35	6.63	1.8	Al_2	6.21	5.99	1.16	F_2	15.70	17.42	1.1
V_2	6.36	6.75	1.9	In_2	5.8	5.79	1.17	Cl_2	11.48	12.97	1.1
Nb_2	6.37	6.76	1.9	Tl_2	5.24	6.11	1.18	Br_2	10.52	11.81	1.1
Cr_2	7.00	6.77	1.10	C_2	12.4	11.26	1.19	I_2	9.31	10.45	1.1
Mo_2	6.95	7.09	1.10	Si_2	7.92	8.15	1.20	He_2	22.20	24.59	a
Mn_2	≤ 6.47	7.43	1.11	Ge_2	7.8	7.90	1.21	Ne_2	20.27	21.56	a
Fe_2	6.30	7.90	1.11	Sn_2	7.1	7.34	1.7	Ar_2	14.51	15.76	1.27
Co_2	≤ 6.42	7.88	1.11	Pb_2	6.1	7.42	1.22	Kr_2	12.87	14.00	1.27
Ni_2	7.43	7.64	1.11	N_2	15.58	14.53	1.1	Xe_2	11.13	12.13	1.27

a Calculated from thermodynamic cycles using the data from Tables 1.1, S2.5 and S4.1.

Table S1.2 Ionization potentials I, electron affinities A (in eV) and parameters a and x in the $I - A = a/d^x$ equation for MX molecules

MX	I	$I - A$	d	a	MX	I	$I - A$	d	a
				x					x
HF	15.8	12.4	0.917		HBr	11.6	8.26	1.414	
LiF	10.9	7.5	1.564		LiBr	9.2	5.84	2.170	
NaF	9.5	6.1	1.926	11.3	NaBr	8.4	2.50	5.036	11.1
KF	9.1	5.7	2.171	0.90	KBr	8.0	4.64	2.825	0.84
RbF	8.9	5.5	2.269		RbBr	7.8	4.44	2.944	
CsF	8.7	5.3	2.345		CsBr	7.7	4.34	3.068	
HCl	12.74	9.13	1.275		HI	10.4	7.32	1.609	
LiCl	9.8	6.2	2.021		LiI	8.6	5.54	2.392	
NaCl	8.9	5.3	2.360	11.3	NaI	7.6	4.54	2.739	11.1
KCl	8.3	4.7	2.668	0.88	KI	7.3	4.24	3.050	0.85
RbCl	8.2	4.6	2.785		RbI	7.2	4.14	3.170	
CsCl	8.1	4.5	2.907		CsI	7.1	4.04	3.315	

Table S1.3 Electron affinities A (eV) of diatomic molecules. (According to [1.1])

Li_2 0.437							B_2[c] > 1.3	C_2 3.269		O_2 0.450	F_2 3.08
Na_2 0.430							Al_2[d] 1.46	Si_2 2.201	P_2 0.589	S_2 1.670	Cl_2 2.38
K_2 0.497	Cr_2 0.505		Fe_2 0.902	Co_2 1.110	Ni_2 0.926	Cu_2 0.836		Ge_2[c] 2.074	As_2[d] 0.739	Se_2 1.94	Br_2 2.55
Rb_2 0.498					Pd_2 1.685	Ag_2 1.023	In_2[e] 1.27	Sn_2 1.962	Sb_2 1.282	Te_2 1.92	I_2 2.524
Cs_2 0.469	W_2 1.460[a]	Re_2 1.571			Pt_2 1.898	Au_2 1.938	Tl_2[d] 0.95	Pb_2 1.366	Bi_2 1.271		

[a][1.28], [b][1.29], [c][1.30], [d][1.31]

Table S1.4 Electron affinities (eV) of oxides and halides

MF_n	*A*	MX_n	*A*	MO_n	*A*	MO_n	*A*
NaF	0.52	LiCl	0.59	NO	0.14	NO_2	2.27
SiF	0.81	NaCl	0.73	SnO	0.60	FeO_2	2.36
GeF	1.02	NaBr	0.79	PbO	0.72	TaO_2	2.40[h]
SF	2.285	NaI	0.865	TaO	1.07[h]	CrO_2	2.43[i]
ClF	2.86	KCl	0.58	PO	1.09	GeO_2	2.50
CF	3.2	KBr	0.64	SO	1.125	ReO_2	2.5
CF_2	0.18	KI	0.73	CrO	1.22[i]	PtO_2	2.68
CF_3	*ca* 2.01	RbCl	0.54	VO	1.23	NiO_2	3.04
SiF_3	2.41	CsCl	0.455	MoO	1.285	PdO_2	3.09
SF_3	3.07	AgCl	1.59[c]	AsO	1.29	PO_2	3.42
SeF_4	1.7	AgBr	1.63[c]	TiO	1.30	CuO_2	3.46
MoF_4	2.3	AgI	1.60[c]	SeO	1.46	AlO_2	4.23
SF_4	2.35	$LiCl_2$	5.92[d]	NiO	1.47	BO_2	4.32
TiF_4	2.5	$LiBr_2$	5.42[d]	FeO	1.49	CO_3	2.69
VF_4	3.55	LiI_2	4.88[d]	MgO	1.63[j]	FeO_3	3.26
CrF_4	3.60	$NaCl_2$	5.86[d]	PdO	1.67[k]	WO_3	3.33
OsF_4	3.75	$NaBr_2$	5.36[d]	TeO	1.70	ReO_3	3.53[l]
IrF_4	4.63	NaI_2	4.84[d]	CsO	1.84	CrO_3	3.66[i]
RuF_4	4.75	KF_2	5.61[d]	ZnO	2.09[j]	NO_3	3.94
RhF_4	5.43	$CuCl_2$	4.35[e]	PtO	2.17[k]	TiO_3	4.20
PtF_4	5.50	$CuBr_2$	4.35[e]	FO	2.27	ClO_3	4.25
MnF_4	5.53	$AuCl_2$	4.60[f]	ClO	2.28	VO_3	4.36
CoF_4	6.38	$AuBr_2$	4.46[f]	BrO	2.35	KO_4	2.8[m]
FeF_4	6.6	AuI_2	4.18[f]	IO	2.38	CsO_4	2.5[m]
PF_5	0.75	$MnCl_3$	5.07[g]	BO	2.51	NaO_4	3.1[m]
SeF_5	5.1	$MnBr_3$	5.03[g]	AlO	2.60	LiO_4	3.3[m]
RuF_5	5.34	$FeCl_3$	4.22[g]	SO_2	1.11	FeO_4	3.30
CrF_5	6.10	$FeBr_3$	4.26[g]	TiO_2	1.59	VO_4	4.00
SF_6	1.07[a]	$CoCl_3$	4.7[g]	ZrO_2	1.64	MnO_4	4.80
SeF_6	2.9[a]	$CoBr_3$	4.6[g]	SeO_2	1.82	CrO_4	4.98[i]
WF_6	3.09[b]	$NiCl_3$	5.20[g]	VO_2	2.03	ClO_4	5.25
TeF_6	3.3[a]	$NiBr_3$	4.94[g]	SiO_2	2.1	ReO_4	5.58[l]
MoF_6	3.82	$FeCl_4$	6.00[g]	ClO_2	2.14	NaO_5	3.2
ReF_6	4.49[b]	$FeBr_4$	5.50[g]	HfO_2	2.14	CrO_5	4.4[i]
RuF_6	6.53[b]	$ScCl_4$	6.89	TeO_2	> 2.2		
OsF_6	5.76[b]	$ScBr_4$	6.13				
IrF_6	5.85[b]						
PtF_6	6.95[b]						
AuF_6	8.01[b]						

From [1.1, 1.30], except for: [a][1.32], [b][1.33], [c][1.34], [d][1.35], [e][1.36], [f][1.37], [g][1.38], [h][1.39], [i][1.40], [j][1.41], [k][1.42], [l][1.43], [m][1.44]

Table S1.5 Electron affinities (eV) of homonuclear clusters

Cu[a]		Al[b]		Ti[c]		Ti[c]		C[d]		Ni[g]	
n	*A*, eV	*n*	*A*, eV	*n*	*A*, eV	*n*	*A*, eV	*n*	*A*, eV	*n*	*A*, eV
3	2.40	3	1.90	3	1.13	39	2.29	3	1.95	3	1.44
4	1.40	4	2.20	4	1.18	40	2.34	4	3.70	5	1.57
5	1.92	5	2.25	5	1.15	41	2.33	5	2.80	7	1.86
6	1.95	6	2.63	6	1.28	42	2.32	6	4.10	9	1.96
7	2.15	7	2.43	7	1.11	43	2.36	7	3.10	11	2.06
8	1.59	8	2.35	8	1.47	44	2.39	8	4.42	13	2.16
9	2.40	9	2.85	9	1.56	45	2.40	9	3.70	15	2.37
10	2.05	10	2.70	10	1.70	46	2.39	11	4.00	17	2.50
11	2.43	11	2.87	11	1.72	47	2.41	∞	4.7	19	2.51
12	2.12	12	2.75	12	1.71	48	2.44	Si[e]		21	2.58
13	2.33	13	3.62	13	1.87	49	2.40	3	2.29	23	2.65
14	2.01	14	2.60	14	1.87	50	2.38	4	2.13	25	2.71
15	2.47	15	2.90	15	2.00	51	2.41	5	2.59	27	2.75
16	2.55	16	2.87	16	1.96	52	2.41	7	1.85	29	2.84
17	2.92	17	2.90	17	2.01	53	2.41	∞	4.2	32	2.94
18	2.62	18	2.57	18	1.97	54	2.45	Ge[f]		35	3.04
19	2.60	19	3.12	19	1.93	55	2.51	3	2.33	38	3.08
20	2.00	20	2.86	20	2.06	56	2.48	4	1.94	41	3.11
21	2.30	21	3.30	21	1.98	57	2.49	5	2.51	44	3.15
22	2.25	22	3.22	22	2.04	58	2.47	6	2.06	47	3.19
23	2.58	23	3.45	23	2.08	59	2.49	7	1.80	50	3.20
24	2.43	24	2.80	24	1.97	60	2.51	8	2.41	52	3.20
25	2.78	25	3.34	25	2.07	61	2.55	9	2.86	54	3.24
26	2.43	26	2.88	26	2.10	62	2.57	10	2.5	56	3.25
27	2.53	27	3.32	27	2.14	63	2.57	11	2.5	58	3.28
28	2.50	28	2.90	28	2.14	64	2.60	12	2.4	60	3.30
29	2.76	29	3.32	29	2.12	65	2.55	∞	4.85	62	3.35
30	2.42	30	3.05	30	2.16	66	2.58	Sn[f]		65	3.36
31	2.65	31	3.22	31	2.19	70	2.55	3	2.24	68	3.35
32	2.61	32	2.96	32	2.24	75	2.63	4	2.04	71	3.38
33	2.87	33	3.40	33	2.21	80	2.68	5	2.65	74	3.42
34	2.38	34	3.12	34	2.22	90	2.66	6	2.28	77	3.44
35	2.75	35	3.37	35	2.24	100	2.73	7	1.95	80	3.36
36	2.77	36	2.58	36	2.24	110	2.75	8	2.48	82	3.42
37	3.07	37	3.45	37	2.27	120	2.75	9	2.8	85	3.45
38	2.82	38	2.95	38	2.28	130	2.83	10	3.08	90	3.45
39	3.07	39	3.40			∞	4.0[h]	11	2.82	95	3.48
40	2.46	40	3.26					12	3.3	100	3.52
∞	4.40[h]	∞	4.25[h]					∞	4.38[h]	∞	4.50[h]

[a][1.45], [b][1.30], [c][1.46], [d][1.47], [e][1.48], [f][1.49], [g][1.50], [h][1.51]

Table S1.6 Radii (Å) of isolated atoms. Top to bottom: from refs. [1.52], [1.53], [1.54], and from r_{Mq} (Eqs. 1.21, 1.22 and Table 1.9)

Li	Be	B	C	N	O	F			
2.66	1.82								
2.86	1.77								
2.54	1.94	2.05	1.76	1.55	1.60	1.41			
2.84	2.17	1.81	1.64	1.56	1.54	1.52			
Na	Mg	Al	Si	P	S	Cl			
3.03	2.29	2.18	1.91						
2.88	2.22	2.06	1.77	1.54	1.42	1.30			
2.61	2.13	2.42	2.06	1.82	1.84	1.63			
2.93	2.47	2.24	2.06	1.97	1.83	1.76			
K	Ca	Sc	Ti	V	Cr	Mn	Fe	Co	Ni
3.63	3.02	2.49	2.24	2.01	1.81	1.96	1.86	1.86	1.85
3.52	2.90	2.61	2.44	2.31	2.26	2.11	2.03	1.96	1.89
2.83	2.38	2.30	2.26	2.27	2.27	2.16	2.10	2.10	2.13
3.21	2.80	2.42	2.23	2.19	2.00	1.95	1.90	1.82	1.81
Cu	Zn	Ga	Ge	As	Se	Br			
1.89	1.88	2.09	1.96						
2.24	1.92[a]	2.01	1.80	1.63	1.56	1.45			
2.13	1.92	2.42	2.10	1.89	1.89	1.71			
1.84	1.93	2.04	1.98	1.98	1.90	1.87			
Rb	Sr	Y	Zr	Nb	Mo	Tc	Ru	Rh	Pd
3.82	3.24	2.83	2.51	2.18	1.96	1.94	1.90	1.93	1.92
3.63	3.06	2.83	2.61	2.50	2.34	2.25	2.12	2.05	2.00
2.88	2.48	2.37	2.29	2.27	2.21	2.19	2.18	2.16	2.05
3.32	2.93	2.58	2.44	2.33	2.19	2.02	1.98	1.98	1.90
Ag	Cd	In	Sn	Sb	Te	I			
2.05	2.06	2.46	2.29						
2.25	1.93[a]	2.17	1.97[a]	1.87	1.77	1.70			
2.14	1.97	2.45	2.18	2.02	1.97	1.82			
2.02	2.16	2.24	2.22	2.22	2.16	2.11			
Cs	Ba	La	Hf	Ta	W	Re	Os	Ir	Pt
4.15	3.33	3.23	2.40	2.17	2.00	1.94	1.93	1.89	1.94
3.90	3.36	3.14	2.53	2.36	2.23	2.13	2.04	1.97	1.87
2.99	2.59	2.50	2.26	2.14	2.10	2.11	2.03	1.97	1.97
3.63	3.07	2.66	2.38	2.29	2.14	2.05	2.02	1.91	1.92
Au	Hg	Tl	Pb	Bi	Th	U			
1.97	2.04	2.46	2.48		2.85	2.37			
2.02	1.79[a]	1.91	1.89	1.95	3.18	2.73			
1.94	1.82	2.38	2.16	2.19	2.35	2.37			
1.94	2.08	2.25	2.25	2.36	2.74	2.66			

[a]from [1.44], using the theoretical values of α

Table S1.7 Ionic radii (Å) for $N_c = 6$ according to Shannon and Prewitt

+1	+2	+3	+4	+5	+6	+7	−1	−2
Li	Be	B	C	N	O	F	F	O
0.76	0.45	0.27	0.16	0.13	0.09	0.08	1.33	1.40
Na	Mg	Al	Si	P	S	Cl	Cl	S
1.02	0.72	0.54	0.40	0.38	0.29	0.27	1.81	1.84
K	Ca	Sc	Ti	V	Cr	Mn		
1.38	1.00	0.74	0.60	0.54	0.44	0.46		
Cu	Zn	Ga	Ge	As	Se	Br	Br	Se
0.77	0.74	0.62	0.53	0.46	0.42	0.39	1.96	1.98
Rb	Sr	Y	Zr	Nb	Mo	Tc		
1.52	1.18	0.90	0.72	0.64	0.50	0.56		
Ag	Cd	In	Sn	Sb	Te	I	I	Te
1.15	0.95	0.80	0.69	0.60	0.56	0.53	2.20	2.21
Cs	Ba	La	Hf	Ta	W	Re		
1.67	1.35	1.03	0.71	0.64	0.60	0.53		
Au	Hg	Tl	Pb	Bi	Po	At		
1.10[a]	1.02	0.88	0.78	0.76	0.67	0.62		

[a]Corrected value

Table S1.8 Additions to Shannon and Prewitt's system of ionic radii (Å) for $N_c = 6$

1+/2+		2+/3+		3+		4+			
Ag^+	1.05	Cr^{2+}	0.77	As^{3+}	0.60	V^{4+}	0.55	Mn^{4+}	0.49
Au^+	1.10	Mn^{2+}	0.77	Sb^{3+}	0.80	Nb^{4+}	0.65	Re^{4+}	0.57
Tl^+	1.30	Fe^{2+}	0.72	Bi^{3+}	0.98	Ta^{4+}	0.66	Ru^{4+}	0.59
Cu^{2+}	0.70	Co^{2+}	0.69	Cr^{3+}	0.62	S^{4+}	0.42	Pd^{4+}	0.62
Zn^{2+}	0.69	Ni^{2+}	0.66	Mn^{3+}	0.60	Se^{4+}	0.53	Os^{4+}	0.57
Cd^{2+}	0.88	Pd^{2+}	0.80	Fe^{3+}	0.60	Te^{4+}	0.80	Ir^{4+}	0.55
Hg^{2+}	0.98	Sc^{3+}	0.70	Co^{3+}	0.59	Cr^{4+}	0.50	Pt^{4+}	0.57
Sn^{2+}	1.04	Y^{3+}	0.85	Ru^{3+}	0.60	Mo^{4+}	0.57	Th^{4+}	0.99
Pb^{2+}	1.12	V^{3+}	0.62	Rh^{3+}	0.59	W^{4+}	0.60	U^{4+}	0.91
				Ir^{3+}	0.64				

Table S1.9 Bond radii (Å) in binary inorganic compounds

Cation	r_b	Cation	r_b	Cation	r_b	Anion	r_b	Anion	r_b
Na^+	1.16	Mg^{2+}	1.05	Al^{3+}	0.98	F^-	1.14	O^{2-}	1.00
K^+	1.52	Ca^{2+}	1.33	Sc^{3+}	1.22	Cl^-	1.57	S^{2-}	1.40
Rb^+	1.66	Sr^{2+}	1.50	Si^{4+}	0.89	Br^-	1.73	Se^{2-}	1.58
Cs^+	1.84	Ba^{2+}	1.69	Zr^{4+}	1.24	I^-	1.94	Te^{2-}	1.80
Cu^+	1.15	Zn^{2+}	1.19	Hf^{4+}	1.22				
Ag^+	1.34	Cd^{2+}	1.29						

Table S1.10 Dependence of the halide bonded radii (Å) on coordination numbers

N_c	2	3	4	6	8	Δr_{8-2}
F^-	0.90	1.00		1.15	1.18	0.28
Cl^-	1.26	1.36		1.57	1.62	0.36
Br^-	1.32		1.58	1.72	1.78	0.46
I^-	1.42	1.68	1.83	1.92	1.98	0.56

Table S1.11 Cationic radii (Å) at high pressures (r_p) and crystal-chemical radii (r_{cc})

Cation	r_p	r_{cc}	Cation	r_p	r_{cc}	Cation	r_p	r_{cc}
Li^+	0.75	0.76	Mg^{2+}	0.70	0.72	Sc^{3+}	0.74	0.74
Na^+	0.98	1.02	Ca^{2+}	1.03	1.00	Y^{3+}	0.88	0.90
K^+	1.37	1.38	Sr^{2+}	1.15	1.18	Cr^{3+}	0.67	0.62
Rb^+	1.52	1.52	Ba^{2+}	1.38	1.35	Mn^{3+}	0.66	0.64
Cs^+	1.63	1.67	Zn^{2+}	0.76	0.74	Fe^{3+}	0.66	0.64
Cu^+	0.78	0.77	Cd^{2+}	0.95	0.95	Th^{4+}	1.07	1.05
Ag^+	1.17	1.15	Pb^{2+}	1.23	1.19	U^{4+}	0.97	1.00
Tl^+	1.44	1.50	B^{3+}	0.38	0.27	Zr^{4+}	1.06	0.84
Be^{2+}	0.47	0.45	Al^{3+}	0.63	0.54	Hf^{4+}	0.90	0.83

Table S1.12 Cationic radii for molecules

A^{n+}	r, Å	A^{n+}	r, Å	A^{n+}	r, Å	A^{n+}	r, Å	A^{n+}	r, Å
Li^+	0.48	Al^{3+}	0.40	Th^{4+}	0.80	Cr^{3+}	0.49	Fe^{2+}	0.62
Na^+	0.76	Ga^{3+}	0.49	U^{3+}	0.87	Cr^{4+}	0.44	Fe^{3+}	0.51
K^+	1.10	In^{3+}	0.66	U^{4+}	0.72	Cr^{6+}	0.35	Co^{2+}	0.59
Rb^+	1.24	Tl^+	1.00[a]	N^{5+}	0.08	Mo^{4+}	0.53	Co^{3+}	0.49
Cs^+	1.42	Tl^{3+}	0.75	P^{5+}	0.28	Mo^{6+}	0.41	Ni^{2+}	0.55
Cu^+	0.61	Sc^{3+}	0.59	As^{5+}	0.37	W^{4+}	0.56	Ni^{3+}	0.48
Cu^{2+}	0.58	Y^{3+}	0.74	Sb^{5+}	0.49	W^{6+}	0.51	Ru^{2+}	0.66
Ag^+	0.94	La^{3+}	0.88	Bi^{5+}	0.65	Mn^{2+}	0.66	Ru^{4+}	0.59
Au^+	0.93	Ac^{3+}	0.98	V^{3+}	0.59	Mn^{3+}	0.51	Rh^{2+}	0.64
Au^{3+}	0.72	C^{4+}	0.10	V^{5+}	0.43	Mn^{4+}	0.42	Rh^{4+}	0.57
Be^{2+}	0.29	Si^{4+}	0.30	Nb^{3+}	0.59	Mn^{7+}	0.36	Pd^{2+}	0.70
Mg^{2+}	0.53	Ge^{4+}	0.42	Nb^{5+}	0.52	Tc^{4+}	0.52	Pd^{4+}	0.59
Ca^{2+}	0.80	Sn^{2+}	0.70	Ta^{3+}	0.61	Tc^{7+}	0.46	Os^{2+}	0.69
Sr^{2+}	0.97	Sn^{4+}	0.56	Ta^{5+}	0.54	Re^{4+}	0.54	Os^{4+}	0.61
Ba^{2+}	1.15	Pb^{2+}	1.01	O^{6+}	0.06	Re^{7+}	0.45	Ir^{2+}	0.69
Zn^{2+}	0.59	Pb^{4+}	0.66	S^{6+}	0.22	F^{7+}	0.05	Ir^{4+}	0.60
Cd^{2+}	0.78	Ti^{4+}	0.48	Se^{6+}	0.34	Cl^{7+}	0.20	Pt^{2+}	0.68
Hg^{2+}	0.87	Zr^{4+}	0.59	Te^{6+}	0.46	Br^{7+}	0.31	Pt^{4+}	0.60
B^{3+}	0.17	Hf^{4+}	0.60	Cr^{2+}	0.64	I^{7+}	0.43		

[a]Calculated from bond lengths in molecular halides

Supplementary References

1.1 Lide DR (ed) (2007–2008) Handbook of chemistry and physics, 88th edn, CRC Press, New York
1.2 Kappes MM, Schumacher E (1985) Surf Sci 156: 1
1.3 Kappes MM, Radi P, Schar M, Schumacher E (1985) Chem Phys Lett 113: 243
1.4 Antonov IO, Barker BJ, Bondybey VE, Heaven MC (2010) J Chem Phys 133: 074309
1.5 Knickelbein MB (1995) J Chem Phys 102: 1
1.6 Liu Y, Zhang C.-H, Krasnokutski SA, Yang D-S (2011) J Chem Phys 135: 034309
1.7 Glushko VP (ed) (1981) Thermochemical constants of substances. USSR Academy of Sciences, Moscow (in Russian)
1.8 Doverstal M, Karlsson I, Lingren B, Sassenberg U (1998) J. Phys B31: 795
1.9 James AM, Kowalczyk P, Larglois E et al (1994) J Chem Phys 101: 4485
1.10 Simard B, Lebeault-Dorget M-A, Marijnissen A, Meulen JJ (1998) J Chem Phys 108: 9668
1.11 Gutsev G, Bauschlicher ChW (2003) J Phys Chem A107: 4755
1.12 Cocke DL, Gingerich KA (1974) J Chem Phys 60: 1958
1.13 Taylor S, Lemire GW, Hamrick YM et al (1988) J Chem Phys 89: 5517
1.14 Beutel V, Krämer HG, Bhale GL et al (1993) J Chem Phys 98: 2699
1.15 Rademann K, Ruppel M, Kaiser B (1992) Ber Bunsenges Phys Chem 96: 1204
1.16 Fu Z, Lemire GW, Hamrick YM et al (1988) J Chem Phys 88: 3524
1.17 DeMaria G, Drowart J, Inghram MG (1959) J Chem Phys 31: 1076
1.18 Saito Y, Yamauchi K, Mihama K, Noda T (1982) Jpn J Appl Phys 21: 396
1.19 Naqvi A, Hamdan A (1992) Canad J Appl Spectr 37:29
1.20 Kostko O, Leone SR, Duncan MA, Ahmed M (2009) J Phys Chem A114:3176
1.21 Neckel A, Sodeck G (1972) Monats Chem 103:367
1.22 Saito Y, Yamauchi K, Mihama K, Noda T (1982) Jpn J Appl Phys 21:396
1.23 Yoo RK, Ruscic B, Berkowitz J (1992) J Chem Phys 96:6696
1.24 Yoo RK, Ruscic B, Berkowitz J (1993) J Chem Phys 99:8445
1.25 Potts AW, Novak I (1983) J Electron Spectrosc Relat Phenom 28:267
1.26 Saha B, Viswanathan R, Baba MS, Mathews CK (1988) High Temp High Pres 20:47
1.27 Kiser RW (1960) J Chem Phys 33:1265
1.28 Weidele H, Kreisle D, Recknagel E et al (1995) Chem Phys Lett 237:425
1.29 Reid CJ (1993) Int J Mass Spectrom Ion Proc 127:147
1.30 Rienstra-Kiracofe JC, Tschumper GS, Schaefer HF III (2002) Chem Rev 102:231
1.31 Gausa M, Gantefö G, Lutz HO, Meiwes-Broer K-H (1990) Int J Mass Spectrom Ion Proc 102:227
1.32 Kennedy RA, Mayhew CA (2001) Phys Chem Chem Phys 3:5511
1.33 Liu J, Sprecher D, Jungen C et al (2010) J Chem Phys 132:154301
1.34 Wu X, Xie H, Qint Z et al (2011) J Phys Chem A115:6321
1.35 Wang X-B, Ding C-F, Wang L-S (1999) J Chem Phys 110:4763
1.36 Wang X-B, Wang L-S, Brown R et al (2001) J Chem Phys 114:7388
1.37 Schröder D, Brown R, Schwerdtfeger P et al (2003) Angew Chem Int Ed 42:311
1.38 Yang X, Wang X-B, Wang L-S et al (2003) J Chem Phys 119:8311
1.39 Zheng W, Li X, Eustis S, Bowen K (2008) Chem Phys Lett 460:68
1.40 Gutsev GL, Jena P, Zhai H-J, Wang L-S (2001) J Chem Phys 115:7935
1.41 Kim JH, Li X, Wang L-S et al (2001) J Phys Chem A105:5709
1.42 Ramond TM, Davico GE, Hellberg F et al (2002) J Mol Spectr 216:1
1.43 Chen W-J, Zhai H-J, Huang X, Wang L-S (2011) Chem Phys Lett 512:49
1.44 Zhai H-J, Yang X, Wang X-B et al (2002) J Am Chem Soc 124:6742
1.45 Leopold DG, Ho J, Lineberger WC (1987) J Chem Phys 86: 1715; Pettiette CL, Yang SH, Crayscraft MJ et al (1988) J Chem Phys 88:5377
1.46 Liu S-R, Zhai H-J, Castro M, Wang L-S (2003) J Chem Phys 118:2108
1.47 Yang S, Taylor KJ, Crayscraft MJ et al (1988) Chem Phys Lett 144:431

1.48 Wang L-S, Cheng H-S, Fan J, Neumark DM (1998) J Chem Phys 108:1395
1.49 Moravec VD, Klopcic SA, Jarold CC (1999) J Chem Phys 110:5079
1.50 Liu S-R, Zhai H-J, Wang L-S (2002) J Chem Phys 117:9758
1.51 Dritz ME (2003) Properties of elements. Metals, Moscow (in Russian)
1.52 Batsanov SS (2011) J Mol Struct 990:63
1.53 Nagle JK (1990) J Am Chem Soc 112:4741
1.54 Bohorquez HJ, Boyd RJ (2009) Chem Phys Lett 480:127

References

1. Lide DR (ed) (2007–2008) Handbook of chemistry and physics, 88th edn. CRC Press, New York
2. Batsanov SS (2008) Experimental foundations of structural chemistry. Moscow Univ Press, Moscow
3. Saloman EB (2010) Energy levels and observed spectral lines of ionized argon, Ar-II through Ar-XVIII. J Phys Chem Ref Data 39: 033101
4. Kramida AE, Shirai T (2009) Energy levels and spectral lines of tungsten, W-III through W-LXXIV. Atom Data Nucl Data Table 95: 305–474
5. Mattolat C, Gottwald T, Raeder S et al (2010) Determination of the first ionization potential of technetium. Phys Rev A81:052–513
6. Pyper NC, Grant IP (1978) The relation between successive atomic ionization potentials. Proc Roy Soc London A359:525–543
7. Fuentealba P, Parr RG (1991) Higher-order derivatives in density-functional theory, especially the hardness derivative $\partial\eta/\partial N$. J Chem Phys 94:5559–5564
8. Glockler G (1934) Estimated electron affinities of the light elements. Phys Rev 46:111–114
9. Crossley RJS, Coulson CA (1963) Glockler's equation for ionization potentials and electron affinities. Proc Phys Soc 81: 211–218
10. Smirnov BM (1992) Cluster ions and van der Waals molecules. Gordon and Breach Science Publishers, Philadelphia
11. Batsanov SS (2007) Ionization, atomization, and bond energies as functions of distances in inorganic molecules and crystals. Russ J Inorg Chem 52:1223–1229
12. Signorell R, Palm H, Merkt F (1997) Structure of the ammonium radical from a rotationally resolved photoelectron spectrum. J Chem Phys 106:6523–6533
13. Wood DM (1981) Classical size dependence of the work function of small metallic spheres. Phys Rev Lett 46:749
14. Perdew JP (1988) Energetics of charged metallic particles: from atom to bulk solid. Phys Rev B37:6175–6180
15. Seidl M, Perdew JP, Brajczewska M, Fiolhais C (1998) Ionization energy and electron affinity of a metal cluster in the stabilized jellium model. J Chem Phys 108:8182–8189
16. Joyes P, Tarento RJ (1989) On the electronic structure of Hg_n and Hg^{+}_{n} aggregates. J Physique 50:2673–2681
17. Yang S, Knickelbein MB (1990) Photoionization studies of transition metal clusters: ionization potentials for Fe_n and Co_n. J Chem Phys 93:1533–1539
18. Göhlich H, Lange T, Bergmann T et al (1991) Ionization energies of sodium clusters containing up to 22000 atoms. Chem Phys Lett 187:67–72
19. Pellarin M, Vialle JL, Lerme J et al (1991) Production of metal cluster beams by laser vaporization. J Physique IV 1:C7-725- C7-728
20. Pellarin M, Baguenard B, Broyer M et al (1993) Shell structure in photoionization spectra of large aluminum clusters. J Chem Phys 98:944–950
21. Yamada Y, Castlemann AW (1992) The magic numbers of metal and metal alloy clusters. J Chem Phys 97:4543–4548

22. Kietzmann H, Morenzin J, Bechthold PS et al (1998) Photoelectron spectra of Nb_n^- clusters. J Chem Phys 109:2275–2278
23. Sakurai M, Watanabe K, Sumiyama K, Suzuki K (1999) Magic numbers in transition metal (Fe, Ti, Zr, Nb, and Ta) clusters. J Chem Phys 111:235–238
24. Wrigge G, Hoffmann MA, Haberland BI (2003) Ultraviolet photoelectron spectroscopy of Nb_4^- to Nb^-_{200}. Eur Phys J D24:23–26
25. Morokhov ID, Petinov VI, Trusov LI, Petrunin VF (1981) Structure and properties of fine metallic particles. Sov Phys Uspekhi 24:295–317
26. Pargellis AN (1990) Estimating carbon cluster binding energies from measured C_n distributions. J Chem Phys 93:2099–2108
27. Saunders WA (1989) Transition from metastability to stability of Ge_n^{2+} clusters. Phys Rev B40:1400–1402
28. Yoshida S, Fuke K (1999) Photoionization studies of germanium and tin clusters in the energy region of 5.0–8.8 eV. J Chem Phys 111:3880–3890
29. Becker J, Rademann K, Hensel F (1991) Ultraviolet photoelectron studies of the molecules Se_5, Se_6, Se_7 and Se_8 with relevance to their geometrical structure. Z Phys D19:229–231
30. Tribollet B, Benamar A, Rayane D, Broyer PM (1993) Experimental studies on selenium cluster structures. Z Phys D26:352–354
31. Brechignac C, Cahuzac Ph, Kebaili N, Leygnier J (2000) Photothermodissociation of selenium clusters. J Chem Phys 112:10197–10203
32. Reid CJ, Ballantine J, Rews SW, Harris F et al (1995) Charge inversion of ground-state and metastable-state C_2^+ cations formed from electroionised C_2H_2 and C_2N_2, and a re-evaluation of the carbon dimer's ionisation energy. Chem Phys 190:113–122
33. Nakajima A, Kaya K (2000) A novel network structure of organometallic clusters in the gas phase. J Phys Chem A104:176–191
34. Liu S-R, Zhai H-J, Castro M, Wang L-S (2003) Photoelectron spectroscopy of Ti_n^- clusters. J Chem Phys 118:2108–2115
35. Boltalina OV, Ioffe IN, Sidorov LN et al (2000) Ionization energy of fullerene. J Am Chem Soc 122:974–9749
36. Kostko O, Leone SR, Duncan MA, Ahmed M (2010) Determination of ionization energies of small silicon clusters with vacuum ultraviolet radiation. J Phys Chem A114:3176–3181
37. Slater JC (1930) Atomic shielding constants. Phys Rev 36:57–64
38. Rienstra-Kiracofe JC, Tschumper GS, Schaefer HF et al (2002) Atomic and molecular electron affinities: photoelectron experiments and theoretical computations. Chem Rev 102:231–282
39. Szentpály L von (2010) Universal method to calculate the stability, electronegativity, and hardness of dianions. J Phys Chem A114:10891–10896
40. Walter CW, Gibson ND, Field RL et al (2009) Electron affinity of arsenic and the fine structure of As^- measured using infrared photodetachment threshold spectroscopy. Phys Rev A80:014–501
41. Walter CW, Gibson ND, Carman DJ et al (2010) Electron affinity of indium and the fine structure of In- measured using infrared photodetachment threshold spectroscopy. Phys Rev A82:032–507
42. Lindahl AO, Andresson P et al (2010) The electron affinity of tungsten. Eur Phys J D60:219–222
43. Batsanov SS (2010) Simple semi-empirical method for evaluating bond polarity in molecular and crystalline halides. J Mol Struct 980:225–229
44. Politzer P (1969) Anomalous properties of fluorine. J Am Chem Soc 91:6235–6237
45. Politzer P (1977) Some anomalous properties of oxygen and nitrogen. Inorg Chem 16:3350–3351
46. Politzer P, Huheey JE, Murray JS, Grodzicki M (1992) Electronegativity and the concept of charge capacity. J Mol Struct Theochem 259:99–120
47. Balighian ED, Liebman JF (2002) How anomalous are the anomalous properties of fluorine? Ionization energy and electron affinity revisited. J Fluor Chem 116:35–39

48. Blondel C, Delsart C, Valli C et al (2001) Electron affinities of ^{16}O,^{17}O,^{18}O, the fine structure of $^{16}O^-$, and the hyperfine structure of $^{17}O^-$. Phys Rev A64:052504
49. Miller TA, Leopold D, Murray KK, Lineberger WC (1986) Electron affinities of the alkali halides and the structure of their negative ions. J Chem Phys 85:368–2375
50. Wang X-B, Wang L-S (1999) Photodetachment of free hexahalogenometallate doubly charged anions in the gas phase. J Chem Phys 111:4497–4509
51. Wang X-B, Wang L-S (2000) Experimental observation of a very high second electron affinity for ZrF_6 from photodetachment of gaseous ZrF_6^{2-} doubly charged anions. J Phys Chem A104:4429–4432
52. Miyoshi E, Sakai Y, Murakami A et al (1988) On the electron affinities of hexafluorides CrF_6, MoF_6, and WF_6. J Chem Phys 89:4193–4198
53. Wang X-B, Wang L-S (2000) Probing the electronic structure and metal−metal bond of $Re_2Cl_8^{2-}$ in the gas phase. J Am Chem Soc 122:2096–2100
54. Scheller MK, Cederbaum LS (1992) Existence of doubly-negative charged ions and relation to solids. J Phys B25:2257–2266
55. Scheller MK, Compton RN, Cederbaum LS (1995) Gas-phase multiply charged anions. Science 270:1160–1166
56. Wang X-B, Wang L-S (1999) Experimental search for the smallest stable multiply charged anions in the gas phase. Phys Rev Lett 83:3402–3405
57. Wang X-B, Wang L-S (2000) Photodetachment of multiply charged anions: the electronic structure of gaseous square-planar transition metal complexes PtX_4^{2-}. J Am Chem Soc 122:2339–2345
58. Middleton R, Klein J (1999) Experimental verification of the existence of the gas-phase dianions BeF_4^{2-} and MgF_4^{2-}. Phys Rev A60:3515–3521
59. Franzreb K, Williams P (2005) Small gas-phase dianions produced by sputtering and gas flooding. J Chem Phys 123:224312
60. Feuerbacher S, Cederbaum LS (2006) A small and stable covalently bound trianion. J Chem Phys 124:044320
61. Wang X-B, Nicholas JB, Wang L-S (2000) Electronic instability of isolated SO_4^{2-} and its solvation stabilization. J Chem Phys 113:10837–10840
62. Wang X-B, Yang X, Nicholas JB, Wang L-S (2003) Photodetachment of hydrated oxalate dianions in the gas phase, $C_2O_4^{2-}$ $(H_2O)_n$. J Chem Phys 119:3631–3640
63. Wang X-B, Woo H-K, Yang J et al (2007) Photoelectron spectroscopy of singly and doubly charged higher fullerenes at low temperatures. J Phys Chem C111:17684–17689
64. Liu S-R, Zhai H-J, Wang L-S (2002) Evolution of the electronic properties of small Ni_n^- ($n = 1-100$) clusters by photoelectron spectroscopy. J Chem Phys 117:9758–9765
65. Liu S-R, Zhai H-J, Wang L-S (2002) s-d hybridization and evolution of the electronic and magnetic properties in small Co and Ni clusters. Phys Rev B65:113–401
66. Boltalina OV, Dashkova EV, Sidorov LN (1996) Gibbs energies of gas-phase electron transfer reactions involving the larger fullerene anions. Chem Phys Letters 256:253–260
67. Miedema AR, de Boer FR, de Chatel PF (1973) Empirical description of the role of electronegativity in alloy formation. J Phys F3:1558–1576
68. Van Der Weide J, Zhang Z, Baumann PK et al (1994) Negative-electron-affinity effects on the diamond (100) surface. Phys Rev B50:5803–5806
69. Cui JB, Ristein J, Ley L (1998) Electron affinity of the bare and hydrogen covered single crystal diamond. Phys Rev Lett 81:429–432
70. Savchenko EV, Grogorashchenko ON, Ogurtsov AN et al (2002) Photo- and thermally assisted emission of electrons from rare gas solids. Surf Sci 507–510:754–761
71. Seidl M, Meiwes-Broer K-H, Brack M (1991) Finite-size effects in ionization potentials and electron affinities of metal clusters. J Chem Phys 95:1295–1303
72. Wilson RG (1988) Secondary ion mass spectrometry sensitivity factors versus ionization potential and electron affinity for many elements in HgCdTe and CdTe using oxygen and cesium ion beams. J Appl Phys 63:5121–5124

73. Wilson RG, Novak SW (1991) Systematics of secondary-ion-mass spectrometry relative sensitivity factors versus electron affinity and ionization potential for a variety of matrices determined from implanted standards of more than 70 elements. J Appl Phys 69:466–474
74. Wilson RG (2004) Secondary ion mass spectrometry. Univ of Florida, Gainesville. http://pearton.mse.ufl.edu/rgw/
75. Clementi E, Raimondi DL (1963) Atomic screening constants from SCF functions. J Chem Phys 38:2686–2689
76. Mullay J (1984) Atomic and group electronegativities. J Am Chem Soc 106:5842–5847
77. Reed JL (1997) Electronegativity: chemical hardness. J Phys Chem A101:7396–7400
78. Reed JL (2002) Electronegativity: atomic charge and core ionization energies. J Phys Chem A106:3148–3152
79. Reed JL (1999) The genius of Slater's rules. J Chem Educat 76:802–804
80. Batsanov SS (1964) System of geometrical electronegativities. J Struct Chem 5:263–269
81. Koseki S, Schmidt MW, Gordon MS (1992) MCSCF/6–31G(d, p) calculations of one-electron spin-orbit coupling constants in diatomic molecules. J Phys Chem 96:10768–10772
82. Koseki S, Gordon MS, Schmidt MW, Matsunaga N (1995) Main group effective nuclear charges for spin-orbit calculations. J Phys Chem 99:12764–12772
83. Koga T, Kanayama K (1997) Noninteger principal quantum numbers increase the efficiency of Slater-type basis sets: singly charged cations and anions. J Phys B30:1623–1632
84. Koga T, Kanayama K, Thakkar AJ (1997) Noninteger principal quantum numbers increase the efficiency of Slater-type basis sets. Int J Quant Chem 62:1–11
85. Koseki S, Schmidt MW, Gordon MS (1998) Effective nuclear charges for the first- through third-row transition metal elements in spin-orbit calculations. J Phys Chem A102:10430–10435
86. Waldron KA et al (2001) Screening percentages based on Slater effective nuclear charge as a versatile tool for teaching periodic trends. J Chem Educat 78:635–639
87. Nalewajsky RF, Koninski M (1984) Atoms-in-a-molecule model of the chemical bond. J Phys Chem 88:6234–6240
88. De Proft F, Langenacker W, Geerlings P (1995) A non-empirical electronegativity equalization scheme. Theory and applications using isolated atom properties. J Mol Struct Theochem 339:45–55
89. Martynov AI, Batsanov SS (1980) New approach to calculating atomic electronegativities. Russ J Inorg Chem 25:1737–1740
90. Bohorquez HJ, Boyd RJ (2009) Is the size of an atom determined by its ionization energy? Chem Phys Lett 480:127–131
91. Waber JT, Cromer DT (1965) Orbital radii of atoms and ions. J Chem Phys 42:4116–4123
92. Simons G, Bloch AN (1973) Pauli-force model potential for solids, Phys Rev B7:2754–2761
93. Zhang SB, Cohen ML, Phillips JC (1987) Relativistic screened orbital radii. Phys Rev B36:5861–5867
94. Zhang SB, Cohen ML, Phillips JC (1988) Determination of diatomic crystal bond lengths using atomic s-orbital radii. Phys Rev B38:12085–12088
95. Zunger A, Cohen M (1979) First-principles nonlocal-pseudopotential approach in the density-functional formalism. II. Application to electronic and structural properties of solids. Phys Rev B20:4082–4108
96. Zhang SB, Cohen ML (1989) Determination of AB crystal structures from atomic properties. Phys Rev B39:1077–1080
97. (a) Ganguly P (1995) Orbital radii and environment-independent transferable atomic length scales. J Am Chem Soc 117:1777–1782; (b) Ganguly P (1995) Relation between interatomic distances in transition-metal elements, multiple bond distances, and pseudopotential orbital radii. J Am Chem Soc 117:2655–2656
98. Chanty TK, Ghosh SK (1996) New scale of atomic orbital radii and its relationship with polarizability, electronegativity, other atomic properties, and bond energies of diatomic molecules. J Phys Chem 100:17429–17433
99. Ganguly P (2009) Atomic sizes from atomic interactions. J Mol Struct 930:162–166

100. Nagle JK (1990) Atomic polarizability and electronegativity. J Am Chem Soc 112:4741–4747
101. Batsanov SS (2011) Thermodynamic determination of van der Waals radii of metals. J Mol Struct 990:63–66
102. Vinet P, Rose JH, Ferrante J, Smith JR (1989) Universal features of the equation of state of solids. J Phys Cond Matter 1:1941–1964
103. Rose JH, Smith JR, Guinea F, Ferrante J (1984) Universal features of the equation of state of metals. Phys Rev B29:2963–2969
104. Bragg WL (1920) The arrangement of atoms in crystals. Phil Mag 40:169–189
105. Slater JC (1964) Atomic radii in crystals. J Chem Phys 41:3199–3204
106. Batsanov SS, Zvyagina RA (1966) Overlap integrals and the problem of effective charges, vol 1. Nauka, Novosibirsk (in Russian)
107. Politzer P, Parr RG (1976) Separation of core and valence regions in atoms. J Chem Phys 64:4634–4637
108. Boyd RJ (1977) Electron density partitioning in atoms. J Chem Phys 66:356–358
109. Sen KD, Politzer P (1989) Characteristic features of the electrostatic potentials of singly negative monoatomic ions. J Chem Phys 90:4370–4372
110. Prasad M, Sen KD (1991) Upper bound to approximate ionic radii of atomic negative ions in terms of r^2. J Chem Phys 95: 1421–1422
111. Deb BM, Singh R, Sukumar N (1992) A universal density criterion for correlating the radii and other properties of atoms and ions. J Mol Struct Theochem 259:121–139
112. Gadre SR, Sen KD (1993) Radii of monopositive atomic ion. J Chem Phys 99:3149–3150
113. Balbas LC, Alonso JA, Vega LA (1986) Density functional theory of the chemical potential of atoms and its relation to electrostatic potentials and bonding distances. Z Phys D1:215–221
114. Bader RFW (1990) Atoms in molecules. Oxford Sci Publ, Oxford
115. Bader RFW (2006) An experimentalist's reply to "What is an atom in a molecule?" J Phys Chem A110:6365–6371
116. Bader RFW (2011) Worlds apart in chemistry: a personal tribute to J. C. Slater. J Phys Chem A115:12667–12676
117. Bultinck P, Vanholme R, Popelier PLA et al (2004) High-speed calculation of AIM charges through the electronegativity equalization method. J Phys Chem A108:10359–10366
118. Goldschmidt VM (1926) Geochemische verteilungsgesetze der elemente. Oslo Bd1, Oslo
119. Pauling L (1939) The nature of the chemical bond. Cornell Univ Press, New York Ithaca
120. Pauling L, Kamb B (1986) A revised set of values of single-bond radii derived from the observed interatomic distances in metals by correction for bond number and resonance energy. Proc Nat Acad Sci USA 83:3569–3571
121. Batsanov SS (1994) Metallic radii of nonmetals. Russ Chem Bull 43:199–201
122. Zachariasen WH (1973) Metallic radii and electron configurations of the 5f-6d metals. J Inorg Nucl Chem 35:3487–3497
123. Trömel M (2000) Metallic radii, ionic radii, and valences of solid metallic elements. Z Naturforsch 55b:243–247
124. Trömel M, Hübner S (2000) Metallradien und ionenradien. Z Krist 215:429–432
125. Batsanov SS (1994) Equalization of interatomic distances in polymorphous transformations under pressure. J Struct Chem 35:391–393
126. Berryman JG (1983) Random close packing of hard spheres and disks. Phys Rev A27:1053–1061
127. Santiso E, Müller EA (2002) Dense packing of binary and polydisperse hard spheres. Mol Phys 100:2461–2469
128. Huggins ML (1922) Atomic radii. Phys Rev 19:346–353
129. Huggins ML (1923) Atomic radii. Phys Rev 21:205–206
130. Huggins ML (1926) Atomic radii. Phys Rev 28:1086–1107
131. Pauling L, Huggins ML (1934) Covalent radii of atoms and interatomic distances in crystals. Z Krist 87:205–238
132. Van Vechten J, Phillips JC (1970) New set of tetrahedral covalent radii. Phys Rev B2:2160–2167

133. Pyykkö P (2012) Refitted tetrahedral covalent radii for solids. Phys Rev B85:024115
134. Sanderson RT (1983) Electronegativity and bond energy. J Am Chem Soc 105:2259–2561
135. Luo Y-P, Benson S (1989) A new electronegativity scale. J Phys Chem 93:7333–7335
136. Gillespie RJ, Hargittai I (1991) The VSERP model of molecular geometry. Allyn Bacon, Boston
137. O'Keeffe M, Brese NE (1991) Atom sizes and bond lengths in molecules and crystals. J Am Chem Soc 113:3226–3229
138. O'Keeffe M, Brese NE(1992) Bond-valence parameters for anion-anion bonds in solids. Acta Cryst B48:152–154
139. Bergman D, Hinze J (1996) Electronegativity and molecular properties. Angew Chem Int Ed 35:150–163
140. Schomaker V, Stevenson DP (1941) Some revisions of the covalent radii and the additivity rule for the lengths of partially ionic single covalent bonds. J Am Chem Soc 63:37–40
141. Blom R, Haaland A (1985) A modification of the Schomaker-Stevenson rule for prediction of single bond distances. J Mol Struct 128:21–27
142. Mitchell KAR (1985) Analysis of surface bond lengths reported for chemisorption on metal surfaces. Surface Sci 149:93–104
143. Batsanov SS (1991) Atomic radii of elements. Russ J Inorg Chem 36:1694–1706
144. Batsanov SS (1995) Experimental determination of covalent radii of elements. Russ Chem Bull 44:2245–2250
145. Batsanov SS (1998) Covalent metallic radii. Russ J Inorg Chem 43:437–439
146. Pyykkö P, Atsumi M (2009) Molecular double-bond covalent radii for elements Li–E112. Chem Eur J 15:186–197
147. Cordero B, Gómez V, Platero-Prats AE et al (2008) Covalent radii revisited. Dalton Trans 2832–2838
148. Batsanov SS (2002) Covalent radii of atoms as a function of their oxidation state. Russ J Inorg Chem 47:1005–1007
149. Brown ID (2009) Recent developments in the methods and applications of the bond valence model. Chem Rev 109:6858–6919
150. Batsanov SS (2010) Dependence of the bond length in molecules and crystals on coordination numbers of atoms. J Struct Chem 51:281–287
151. Pyykkö P, Riedel S, Patzschke M (2005) Triple-bond covalent radii. Chem Eur J 11:3511–3520
152. Pyykkö P, Atsumi M (2009) Molecular double-bond covalent radii for elements Li–E112. Chem Eur J 15:12770–12779
153. Landé A (1920) Üder die größe die atome. Z Physik 1:191–197
154. Ladd MFC (1968) The radii of spherical ions. Theor Chim Acta 12:333–336
155. Wasastjerna JA (1923) On the radii of ions. Comm Phys-Math Soc Sci Fenn 1(38):1–25
156. Kordes E (1939) A simple relationship between ion refraction, ion radius and the reference number of the elements. Z phys Chem B44:249–260
157. Kordes E (1939) Identification of atomic distances from refraction. Z phys Chem B44:327–343
158. Kordes E (1940) Berechnung der ionenradien mit hilfe atomphysicher groβen. Z phys Chem B48:91–107
159. Vieillard P (1987) Une nouvelle echelle des rayons ioniques de Pauling. Acta Cryst B43:513–517
160. Goldschmidt VM (1929) Crystal structure and chemical constitution. Trans Faraday Soc 25:253–283
161. Zachariasen WH (1928) Crystal radii of the heavy elements. Phys Rev 73:1104–1105
162. Zachariasen WH (1931) A set of empirical crystal radii for ions with inert gas configuration, Z Krist 80:137–153
163. Zachariasen WH, Penneman RA (1980) Application of bond length-strength analysis to 5f element fluorides. J Less Common Met 69:369–377
164. Shannon RD, Prewitt CT (1969) Effective ionic radii in oxides and fluorides. Acta Cryst B25:925–946

165. Shannon RD, Prewitt CT (1970) Revised values of effective ionic radii. Acta Cryst B26:1046–1048
166. Shannon RD (1976) Revised effective ionic radii and systematic studies of interatomic distances in halides and chalcogenides. Acta Cryst A32:751–767
167. Shannon RD (1981) Bond distances in sulfides and a preliminary table of sulfide crystal radii. Structure and Bonding 2:53–70
168. Prewitt CT (1985) Crystal chemistry: past, present, and future. Am Miner 70:443–454
169. Ahrens LH (1952) The use of ionization potentials: ionic radii of the elements. Geochim Cosmochim Acta 2:155–169
170. Ahrens LH (1954) Shielding efficiency of cations. Nature 174:644–645
171. Brese NE, O'Keefe M (1991) Bond-valence parameters for solids. Acta Cryst B47:192–197
172. Pauling L (1928) The sizes of ions and their influence on the properties of salt-like compounds. Z Krist 67:377–404
173. Brown ID (1988) What factors determine cation coordination numbers? Acta Cryst B44:545–553
174. Batsanov SS (2001) Relationship between the covalent and van der Waals radii of elements. Russ J Inorg Chem 46:1374–1375
175. Chen ECM, Wentworth WE (1985) Negative ion states of the halogens. J Phys Chem 89:4099–4105
176. Evans CJ, Gerry MCL (2000) The microwave spectra and structures of Ar–AgX. J Chem Phys 112:1321–1329
177. Batsanov SS (1956) Calculation of atomic polarizabilities in polar bonds. Zhurnal Fizicheskoi Khimii 30:2640–2648 (in Russian)
178. Batsanov SS (1966) Refractometry and chemical structure. Van Nostrand, Princeton
179. Fajans K (1953) Chemical binding forces. Shell Development Co, Emeryville
180. Fajans K (1957) Polarization. In: Clark GL (ed) Encyclopedia of chemistry. Reinhold, New York
181. Fajans K (1959) Quantikel-theorie der chemischen Bindung. Chimia 13:349–366
182. Collin RJ, Smith BC (2005) Ionic radii for Group 1 halide crystals and ion-pairs. Dalton Trans 702–705
183. Ignatiev V (2002) Relation between interatomic distances and sizes of ions in molecules and crystals. Acta Cryst B58:770–779
184. Witte H, Wölfel E (1955) Röntgenographische Bestimmung der Elektronenverteilung in Kristallen: die Elektronenverteilung im Steinsalz. Z phys Chem NF 3:296–329
185. Gourary BS, Adrian FJ (1960) Wave functions for electron-excess color centers in alkali halide crystals. Solid State Physics 10:127–247
186. Maslen VW (1967) Crystal ionic radii. Proc Phys Soc 91:259–260
187. Downs JW, Gibbs GV (1987) An exploratory examination of the electron density and electrostatic potential of phenakite. Amer Miner 72:769–777
188. Sasaki S, Fujino K, Takeuchi Y, Sadanaga R (1980) On the estimation of atomic charges by the X-ray method for some oxides and silicates. Acta Cryst A36:904–915
189. Downs JW (1995) Electron density and electrostatic potential of coesite. J Phys Chem 99:6849–6856
190. Kirfel A, Krane H-G, Blaha P et al (2001) Electron-density distribution in stishovite, SiO_2: a new high-energy synchrotron-radiation study. Acta Cryst A57:663–677
191. Takeuchi Y, Sasaki S, Bente K, Tsukimura K (1993) Electron density distribution in MgS. Acta Cryst B49:780–781
192. Gibbs GV, Cox DF, Rosso KM et al (2007) Theoretical electron density distributions for Fe- and Cu-sulfide earth materials. J Phys Chem B111:1923–1931
193. Vidal-Valat G, Vidal J-P (1992) Evidence on the breakdown of the Born-Oppenheimer approximation in the charge density of crystalline ^{7}LiH/D. Acta Cryst A48:46–60
194. Noritake T, Towata S, Aoki M et al (2003) Charge density measurement in MgH_2 by synchrotron X-ray diffraction. J Alloys Comp 356–357:84–86

195. Fujino K, Sasaki S, Takeuchi Y, Sadanaga R (1981) X-ray determination of electron distributions in forsterite, fayalite and tephroite. Acta Cryst B37:513–518
196. Sasaki S, Takeuchi Y, Fujino K, Akimoto S (1982) Electron-density distributions of three orthopyroxenes. Z Krist 158:279–297
197. Takazawa H, Ohba S, SaitoY (1988) Electron-density distribution in crystals of dipotassium tetrachloropalladate(II) and dipotassium hexachloropalladate(IV), $K_2[PdCl_4]$ and $K_2[PdCl_6]$. Acta Cryst B44:580–585
198. Liao M, Schwarz W (1994) Effective radii of the monovalent coin metals. Acta Cryst B50:9–12
199. Gibbs GV, Tamada O, Boisen MB (1997) Atomic and ionic radii: a comparison with radii derived from electron density distributions. Phys Chem Miner 24:432–439
200. Sasaki S (1997) Radial distribution of electron density in magnetite, Fe_3O_4. Acta Cryst B53:762–766
201. Belokoneva EL (1999) Electron density and traditional structural chemistry of silicates. Russ Chem Rev 68:299–316
202. Kirfel A, Lippmann T, Blaha P et al (2005) Electron density distribution and bond critical point properties for forsterite, Mg_2SiO_4. Phys Chem Miner 32:301–313
203. Gibbs GV, Downs RT, Cox DF et al (2008) Experimental bond critical point and local energy density properties determined for Mn–O, Fe–O, and Co–O bonded interactions for tephroite, Mn_2SiO_4, fayalite, Fe_2SiO_4, and olivine, Co_2SiO_4 and selected organic metal complexes. J Phys Chem A112:8811–8823
204. Gibbs GV, Boisen MB Jr, Hill FC et al (1998) SiO and GeO bonded interactions as inferred from the bond critical point properties of electron density distributions. Phys Chem Miner 25:574–584
205. Johnson O (1973) Ionic radii for spherical potential ion. Inorg Chem 12:780–785
206. Johnson O (1981) Electron density and electron redistribution in alloys: electron density in elemental metals, J Phys Chem Solids 42:65–76
207. Batsanov SS (2003) Bond radii of atoms in ionic crystals. Russ J Inorg Chem 48:533–536
208. Donald KJ, Mulder WH, von Szentpály L (2004) Influence of polarization and bond-charge on spectroscopic constants of diatomic molecules. J Phys Chem A108:595–606
209. Fumi FG, Tosi MP (1964) Ionic sizes and born repulsive parameters in the NaCl-type alkali halides. J Phys Chem Solids 25:31–43
210. Rosseinsky DR (1994) An electrostatics framework relating ionization potential (and electron affinity), electronegativity, polarizability, and ionic radius in monatomic species. J Am Chem Soc 116:1063–1066
211. Batsanov SS (2004) Determination of ionic radii from metal compressibilities. J Struct Chem 45:896–899
212. Batsanov SS (2006) Mechanism of metallization of ionic crystals by pressure. Russ J Phys Chem 80:135–138
213. Batsanov SS (1978) About ultimate ionic radii. Doklady AN USSR 238:95–97 (in Russian)
214. Batsanov SS (1983) Some crystal-chemical characteristics of simple inorganic halides. Russ J Inorg Chem 28:470–474

Chapter 2
Chemical Bond

2.1 Historical Development of the Concept

The concept of the *chemical bond* is central to modern chemistry. Its classical form, which gradually and painstakingly developed in the course of the 19th century, described molecules as a combinations of linked atoms. The idea proved extremely useful for interpreting, systematizing and predicting chemical facts, although for a long time it developed without any understanding of the underlying physics. This 'black box' situation began to change towards the close of the century. G. J. Stoney in 1881 calculated the elementary charge of electricity and in 1891 named it 'electron'. In 1894, W. Weber suggested that the atom consists of positive and negative electric charges. In 1897, W. Wiechert, J. J. Thomson, and J. S. Townsend measured the charge of the electron. In 1902–1904, William Thomson (Lord Kelvin) and J. J. Thomson developed the 'plum cake' atomic model, with electrons distributed within the homogenous sphere of positive electricity. In 1904, H. Nagaoka suggested that the positive charge is located in the center of the atom, the electrons orbiting around it. Finally, in 1911 E. Rutherford proved this planetary model experimentally.

In 1904, R. Abegg proposed that the valence of an atom corresponds to the number of electrons it lost or gained, the sum of which must be equal to 8 and the highest positive valence to the Group (column) number in the Periodic Table. In 1908, J. Stark postulated that chemical properties of an atom are defined by its outer ('valence') electrons, and W. Ramsay in his essay *Electron as the element* already mentioned the electronic nature of the bond between atoms. Finally, in 1913, N. Bohr proposed the model where the majority of the electrons in a molecule are located around the nuclei as in isolated atoms, and only their outer electrons rotate around the axes connecting atoms, forming the chemical bond. In 1916 W. Kossel explained the formation of ions by the transfer of electrons from one atom to another to complete the outer electronic shells of both to the stable 8-electron configurations; he also introduced the important idea that there is a gradual transition from purely polar compounds (e.g., HCl) to typically non-polar ones (e.g., H_2) [1]. In the same year Lewis described the formation of the covalent bond by two identical atoms sharing their electrons to acquire stable octets [2]. Langmuir developed the theory of Lewis, postulating that electrons in the atom are distributed in layers, with the 'cells' for

S. S. Batsanov, A. S. Batsanov, *Introduction to Structural Chemistry*,
DOI 10.1007/978-94-007-4771-5_2, © Springer Science+Business Media Dordrecht 2012

2 electrons in the first layer, 8 in the second, 18 in the third and 32 in the fourth [3, 4, 5]. For a long time, the octet rule was regarded as the norm of chemical bonding, and deviations from it as exceptions. However, later these exceptions became more and more numerous, until their explanation required the introduction of new ideas which will be discussed below. The independent impulse to the development of the electronic theory was given by the Periodic Law (D. I. Mendeleev, 1869) which got its physical explanation in the Bohr-Rutherford model, the quantum theory and, finally, the Pauli exclusion principle, which explained the electronic structure of the atom and thereby the cellular model of Langmuir. The development of these approaches led to the creation of quantum chemistry. Though the discussion of latter is beyond scope of this book, it should be noted that in the fundamental equation of E. Schrödinger (1926), $H\Psi = E\Psi$, where H is the Hamilton operator, E is the total energy of the system, the wavefunction Ψ (or, more exactly, its square) defines the probability of finding an electron in a certain part of space. Because of the uncertainty principle, it is not possible to describe the electron's orbit precisely, but only in terms of probability; hence we speak of the 'electronic cloud'. Schrödinger's equation cannot be solved precisely for any system containing more than one electron, therefore the application of quantum mechanics to chemistry is essentially the quest for suitable approximations.

The region defined by a wavefunction is termed an atomic orbital, which can be defined uniquely by three quantum numbers. The principal quantum number n is the number of the electron shell, the orbital quantum number l defines the sub-shell, and the shape of the orbital. Thus, atomic s-orbitals with the quantum number $l = 0$ are spherically symmetrical, whereas p-orbitals (with $l = 1$) are dumbbell-shaped, are directed along the three Cartesian axes (hence their designation as p_x, p_y and p_z) and tend to form bonds in these directions. The directionality of a (non-spherical) orbital is defined by the magnetic quantum number m_l. Averaging (hybridization) of one ns and three np orbitals leads to a tetrahedral arrangement of bonds (for instance, in diamond), other combinations of *s, p and d* electrons lead to other types of hybridization and geometrical configurations. The modern state of the calculations and 'experimental measurements' (reconstruction) of orbitals is discussed in the fine essay of Schwartz [6], who points that 'orbitals' are concepts which are useful to approximately describe the structures, properties, and processes of real molecules, crystals, *etc*. Correspondingly, although orbitals are essentially determined by the nature of the molecules, they can be defined in different ways for different purposes. The relation between the wavefunction Ψ and the corresponding orbital Φ is defined by the equation

$$\Psi(X_1, \ldots X_N) \approx \Phi(X_1, \ldots X_N) \tag{2.1}$$

Whether Φ is a satisfactory orbital approximation of Ψ, depends on the type of the molecule, on its state, and also on the nature of the problem. Hence there are no such things as *the* orbitals in a molecule, just as there are no uniquely defined charges of atoms in a molecule. There are many kinds of atomic charges (see below), and similarly, there are many different types of orbitals, appropriate for different physical phenomena. Firstly, the exact wavefunction Ψ and the exact energy E can be generated from a simple orbital product function Φ by several theoretically well-defined operators. Secondly, the popular density functional approaches of Kohn and Sham (KS) all aim at the calculation of highly *reliable* molecular energies with the help of a product wavefunction of 'KS orbitals' of different kinds. Thirdly, the most famous orbital approach for approximate energies is the first-principles, self-consistent field model of Hartree and Fock. There are also many semi-empirical varieties, such as the iterative extended Hückel, CNDO, AM1, *etc.*

In developing the theory of the chemical bond the great contributions were made by Coulson, Hückel, Hund, Slater, Mulliken (see [7]) and, especially, by Pauling who played the major role in the formation of modern structural chemistry: he has formulated such concepts as the hybridization, the polarity and strength of a bond, the degree of the double-bond character, the principle of the local electro-neutrality of atoms, the effective valence, i.e., has created that language of the given area of science on which experimenters began to speak and think. The valence-bond (VB) theory, developed by Pauling, generally followed the (implicit) idea of nineteenth-century chemists that atoms persist in a molecule as recognisable entities. Later, with the triumph of the theory of molecular orbitals, came the widespread view that in a molecule there are no atoms, only nuclei and electrons (orbitals). However, it is worth noting that the total energy of a benzene molecule, i.e. the energy required to split it into six nuclei of charge + 6, six protons and 42 electrons, all at infinite separations, amounts to 607837 kJ/mol (from MO calculations). The (experimental) atomization energy of benzene, i.e. the energy required to split the molecule into six carbon and six hydrogen atoms, is only 5463(3) kJ/mol, or less than 1 % of the former. For comparison, the sublimation enthalpy of crystalline benzene, which is the measure of intermolecular cohesion, is 44 kJ/mol. Thus, atoms in a molecule are no less "real" than molecules in a crystal. Indeed, later Bader [8] and Parr [9] brought back the concept of atoms in molecules (AIM), now on modern quantum-mechanical basis. Still, notwithstanding all the successes of quantum chemistry, structural chemistry remains a predominantly experimental science.

2.2 Types of Bonds: Covalent, Ionic, Polar, Metallic

The traditional classification of chemical bonds into ionic, covalent, donor-acceptor, metallic and van der Waals corresponds to extreme types, but a real bond is always a combination of some, or even all of these types (Fig. 2.1). Purely covalent bonding can be found only in elemental substances or in homonuclear bonds in symmetric molecules, which comprise a tiny fraction of the substances known. Purely ionic bonds do not exist at all (although alkali metal halides come close) because some degree of covalence is always present. Nevertheless, to understand real chemical bonds it is necessary to begin with the ideal types. In this Section we will consider mainly the experimental characteristics of different chemical bonds and only briefly the theoretical aspects of interatomic interactions.

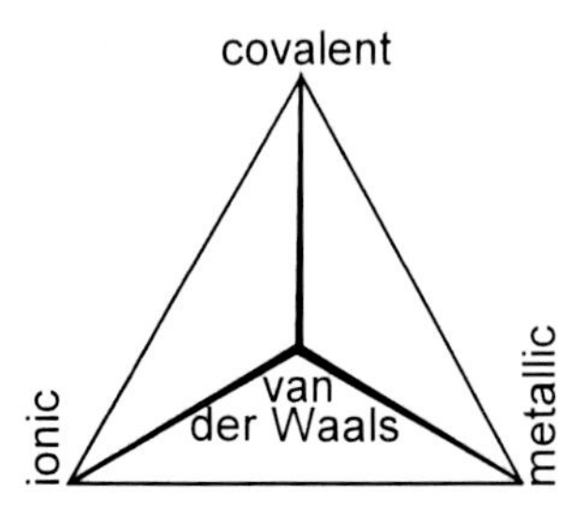

Fig. 2.1 Tetrahedron of chemical bond types. The nature of a given bond can be described by a point within the tetrahedron

2.2.1 *Ionic Bond*

The ionic bond results from the Coulomb attraction of oppositely charged ions. Its strength is characterised by the electrostatic energy; in MX ionic crystals it is the crystal lattice energy U(MX), which can be determined experimentally from the Born-Haber cycle or calculated theoretically from the known net charges of ions (Z, not to be confused with nuclear charges!) and inter-ionic distances (d), as

$$U = k_M \frac{Z^2}{d}\left(1 - \frac{1}{n}\right) \tag{2.2}$$

where k_{M} is the Madelung constant and n is the Born repulsion factor (Table 2.1), in good agreement with the experiment. The ionic theory explains many facts of structural inorganic chemistry. Thus, in many ionic structures, larger ions (anions) form a close packing motif while smaller ions (cations) occupy the voids in it. As this motif contains only tetrahedral and octahedral voids, this explains why cations usually have the coordination number, N_c, of 4 or 6. Coulomb interactions being strong, ionic crystals have high fusion (melting) temperatures and high atomization energies, but dissolve in polar liquids (e.g. water) due to high solvation (hydration) heat. The absence of electrons in the inter-ionic space results in low refractive indices and high atomic polarizations, wide band gaps, and insulator properties.

As noted above, Kossel introduced the idea that the transition from ionic to covalent substances is gradual, the covalence increasing with the mutual polarizing influence of ions. This idea was developed by Fajans and his school who defined the polarizabilities of ions and estimated the polarizing action of cations (Z/r^2), but ultimately failed to create a quantitative theory. The reason is obvious [10]: there are no completely ionic substances, only intermediate cases, more or less approaching this type. Hence the parameters of ideal ions are not available experimentally, the more so since ionic radii cannot be uniquely defined from interatomic distances (see Chap. 1). Thus the polarization concept remained only qualitative. However, the contribution in the bond energy of the polarizing effect of atoms can be described in the form that has proven itself for the van der Waals interaction (see Sect. 4.4), where the deviation of the A $\cdots$ B distance from the mean of A $\cdots$ A and B $\cdots$ B distances is a function of the difference of the atomic polarizabilities

$$p_\alpha = \left[\frac{(\alpha_{\mathrm{A}} - \alpha_{\mathrm{B}})}{\alpha_{\mathrm{A}}}\right]^{2/3} \tag{2.3}$$

Table 2.1 Hardness parameters of ions

Electron configuration of ion	He	Ne	Ar (and Cu^+)	Kr (and Ag^+)	Xe (and Au^+)
n	5	7	9	10	12
f_n	1.250	1.167	1.125	1.111	1.091

turning from distances to volumes, this function takes the form

$$p_a = \left[\frac{(\alpha_A - \alpha_B)}{\alpha_A}\right]^2 \tag{2.4}$$

Taking into account the interaction of effective charges of atoms, the total 'energetic' polarizing effect is

$$q = p_a \frac{(Zi)^2}{d} \tag{2.5}$$

Evidently, the smaller the atom the stronger its polarizing effect. If the smaller ion is the cation (as is usually the case) then it reduces the total α of the substance, if the anion then α increases. Such simple approach allows calculating polarizability of inorganic compounds with good accuracy [11].

The ionic model is widely used to predict the coordination numbers, N_c, in crystal structures. Evidently, the higher the r_c/r_a ratio, the more anions can be accommodated around a given cation. The Magnus-Goldschmidt rules, dating back to nineteen-twenties [12], predicted from simple geometrical condiderations the following succession. For $r_c/r_a \leq 0.15$ the stable configuration can only be linear ($N_c = 2$), from 0.15 to 0.22 it should be equilateral triangle ($N_c = 3$), from 0.22 to 0.41 a tetrahedron ($N_c = 4$), from 0.41 to 0.73 an octahedron ($N_c = 6$), above 0.73 a cube ($N_c = 8$). However, even for crystals with essentially ionic bonds these rules often fail. Thus, in $MgAl_2O_4$ the large Mg^{2+} ion has $N_c = 4$ and the smaller Al^{3+} has $N_c = 6$ whereas it should be the other way round [13]. Crystal structures of MX_n also often confound the simple ionic model [14]. Obviously, one would expect the cation to adopt higher N_c with smaller F^- anion than with other, bulkier, halogens (X = Cl, Br, I). In fact, $N_c(MF) \leq N_c(MX)$ and $N_c(MF_2) \approx N_c(MX_2)$, and only for $n = 3$ or 4 it is $N_c(MF_n) \geq N_c(MX_n)$. A striking case is CsF and CsI: their r_c/r_a of 1.25 and 0.76 both predict $N_c = 8$. This is correct for CsI, but CsF with the higher ratio has a NaCl-type (B1) structure with $N_c = 6$!

These failures show that the simple ionic model is a rather imperfect approximation. Firstly, the charges of ions are assumed to equal the formal oxidation states of the corresponding elements. Secondly, the ions are regarded as absolutely hard spheres, whose spatial distribution is governed only by their relative sizes and the quest for the densest possible packing and the nearest possible contacts between oppositely-charged ions. The agreement with the experiment can be improved by assigning to ions more realistic effective charges, e^* (see below, Table 2.2). For alkali halides, for instance, the charges of anions in different compounds with identical cations are related as $(e^*_{MF}/e^*_{MCl})^2 = 1.245$, $(e^*_{MBr}/e^*_{MCl})^2 = 0.923$, $(e^*_{MI}/e^*_{MCl})^2 = 0.826$.

Table 2.2 Effective charges of hydrogen and halogen atoms in molecules

HF	H_2O	H_2S	NH_3	C_2H_2	C_2H_4	CH_4	CH_3I	CH_3Br	CH_3Cl	CH_3F
0.41	0.33	0.11	0.23	0.35	0.16	0.11	0.13	0.33	0.47	0.95
		CS_2	GeH_4	SiH_4	SnH_4	$GeBr_4$	HCl	$ZnBr_2$		
		0	0.09	0.10	0.12	0.17	0.20	0.25		

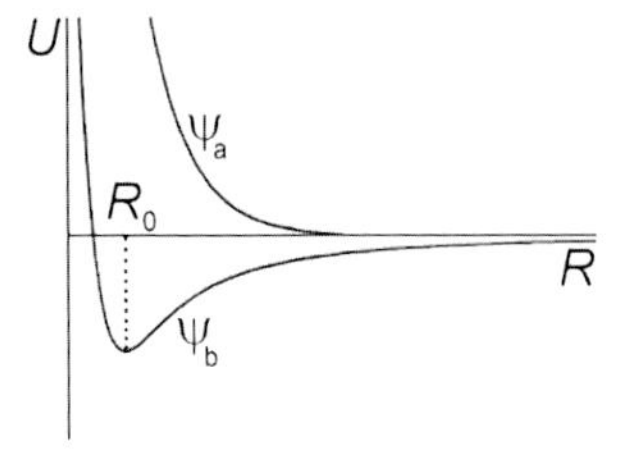

Fig. 2.2 Potential energy (U) of the bonding and antibonding orbitals of a diatomic molecule as functions of the interatomic distance R

Multiplying the ultimate radii of anions from Table 1.17 and assuming $e^*_{\text{MCl}} = 1$, by these ratios we obtain the effective, or 'energetic' ionic radii r^* (see Sect. 1.6), viz. F^- 2.30, Cl^- 2.25, Br^- 2.20 and I^- 2.17 Å [11]. It follows that the effective, rather than formal, ratio of the cation and anion radii ($r_c{}^*/r_a{}^*$) in CsF is smaller than in CsI. Hardness of ions is not infinite and varies from ion to ion. In the Born-Landé theory it is defined by the repulsion coefficient n and the rigidity factor $f_n = (n - 1)/n$ (see Table 2.1). The product $f_n \times r^*$ then gives the radii of spheres with absolutely identical properties. Their ratio,

$$\overline{R} = \frac{r^*_+}{r^*_-} \frac{f_n^+}{f_n^-} \tag{2.6}$$

which equals 0.68 for CsF, 0.72 for CsCl, 0.745 for CsBr and 0.77 for CsI, now describes the changes of N_c correctly. The agreement with reality can be further improved by taking into account the partly covalent character of the bonding in ionic compounds [14]. The effects of polarization and deformation of ions on ionic crystal structures were surveyed by Madden and Wilson [15] who concluded that the ionic model with formal charges has wider applicability than is often supposed, but covalent anomalies (layered structures, bent bonds, *etc*) can be quantitatively explained by ionic polarization.

2.2.2 *Covalent Bond*

Usually, a covalent bond between two atoms is formed by two electrons, one from each atom. These electrons tend to be partly localized in the region between the two nuclei. If the orbitals of these electrons are Ψ_1 and Ψ_2, the molecular orbital of the bonded atoms must be their linear combination, symmetric $\Psi_b = \Psi_1 + \Psi_2$ and antisymmetric $\Psi_a = \Psi_1 - \Psi_2$. As illustrated in Fig. 2.2, the former orbital has a minimum of energy at certain distance and generally has lower energy than the latter, therefore the former orbital is bonding and the latter antibonding. Since an orbital can be occupied by no more than two electrons, this picture was in fact anticipated by the Lewis' model (in 1916—a decade before the beginning of quantum mechanics!) which regarded bonds as shared electron pairs. Lewis also noted that in stable molecules, each atom usually has 8 electrons in its valence shell (except H which has two), counting both the bonding and the unshared electron pairs and taking no account of the bond polarity. This *octet rule* for a long time was regarded as a law of chemistry, apparently resulting from the fact that there are only one *s* and three *p* orbitals in an electron shell, which can accommodate a maximum of 8 electrons between them. Compounds which did not conform to this rule were regarded as special classes of compounds, hypervalent (with >8 electrons) and hypovalent (with <8 electrons). Alternatively, exceptions were explained away by including *d* orbitals into hybridization, or by invoking bond polarity and net atomic charges and assuming that the octet rule applies to the effective number of electrons around each atom (contrary to Lewis' own approach). Today it is clear that this rule, although pedagogically useful, has numerous exceptions which show no extraordinary properties [16]. Particularly, comparison of bonds in hypervalent molecules with those in octet molecules reveals no fundamental difference in their nature. Likewise, saving the octet rule by assigning net charges to atoms, contributes nothing to understanding the structures and properties of molecules. Thus, the nitrogen atom in an a protonated amine or pyridine has a formal charge of +1 but shows no corresponding contraction of the bond distances; in fact, this atom has a small negative charge!

It seems that the belief in the octet rule and in *misread* quantum-mechanical concepts helped for a long time to discourage the search for compounds of 'inert' gases, although von Antropoff [17] and Pauling [18] have predicted that these might be chemically awaken by powerful oxidizing agents with high electronegativity. This prediction was confirmed in 1962 when xenon compounds were discovered [19–23]. Since then, over 500 compounds of the 'rare' (formerly 'inert' and later 'noble') gases were synthesized [24] and dozens of them characterized by X-ray diffraction. Finally, solid xenon was converted into metallic state under ultra-high pressure [25, 26]

Another note of caution is necessary. In every modern textbook, the explanation of covalent bonding begins (and sometimes also ends) with the H_2 molecule, because this is the simplest case and historically the first one explored. Unfortunately, it is by no means a typical one—the fact often not appreciated by the students. In fact, H_2 shows many unique properties because it has no non-bonding electrons, so that the bonding pair has also to 'take responsibility' for intermolecular interactions, both attractive and repulsive. As a result, neither job is done properly. The interatomic distance and atomization energy, compressibility and other molecular-physical properties are consistent with the bond order much lower than the conventional 1 [27]. Thus, the covalent radius of hydrogen determined from its bond distances with different atoms (which can compensate the lack of electron density in the bond from their own) is 0.30 Å [28], hence the actual H–H bond length of 0.74 Å in H_2 molecule corresponds to the bond order of only 0.57 (see below, Eq. 2.8). Usually the ionization potentials of A_2 molecules are lower than those of isolated atoms, and the bond dissociation energies of ${A_2}^+$ are higher than those of neutral molecules, H_2 shows the opposite relations in both respects (see Chap. 1). Hydrides of the elements of Groups 15 and 16 violate the general VSEPR rules of molecular geometry (see Sect. 3.2.2): H–A–H angles are smaller than F–A–F, whereas usually the angles widen with the decrease of the ligand electronegativity; this effect was also explained by the electron pair in the A–H bond doubling as the non-bonding pair of the H ligand [29]. The observed van der Waals radius of hydrogen is *ca.* 0.25 Å lower than the value obtained by extrapolating the radii of halogens [27]. This indicates the weakness of the intermolecular repulsion, while the simultaneous weakness of intermolecular attraction can be seen from the large deviation of H_2 from Trouton's law: the enthalpy of evaporation has a constant ratio to the boiling point temperature. For H_2 this ratio is 44.8 J mol^{-1} K^{-1}, compared to the average of 88 J mol^{-1} K^{-1}.

Let us consider the relation between of the dissociation energy and the bond length. Morse [30] introduced the formula which describes well the experimental electronic energy (E) of a diatomic molecule as a function of the inter-nuclear distance near the equilibrium state (d_e)

$$E(d) = D_e\{1 - \exp[-a(d - d_e)]^2\} \tag{2.7}$$

where D_e is the dissociation energy. Assuming that the attractive energy is proportional to the bond order q [31], from Eq. 2.7 it is possible to derive Pauling's equation,

$$d_q = d_1 - b \ln q \tag{2.8}$$

where d_1 and d_q are the lengths of a standard single bond and a bond of the order q, respectively, and b is an (essentially empirical) constant. Pauling himself used the $0.6 \times \log q$ term, which corresponds to $b = 0.26$ if the natural logarithm is used. However, the proportionality between bond energy and bond order is a rather crude simplification: the experimental energies of the C–C and C≡C bonds relate as 1: 2.2, while those of N–N and N≡N bonds as 1: 4.5.

Parr and Borkman [32] have shown that for many diatomic molecules the bond energy at distances (d) near the equilibrium can be described as

$$E = E_o + \frac{E_1}{d} + \frac{E_2}{d^2} \tag{2.9}$$

where the second term reflects the Coulomb interactions, and the third one accounts for the overlap of atomic orbitals. According to the pseudo-potential method, E is proportional to d^{-2} [33], whereas in Phillips' theory it is proportional to $d^{-2.5}$(see Sect. 2.3). The lengths of typical carbon-carbon bonds correlate linearly with bond dissociation energies (E) in the full range of single, double, triple, and highly strained bonds, with E ranging from 16 to 230 kcal mol^{-1}. The equation

$$d = 1.748 - 0.002371E \tag{2.10}$$

(for d in Å, E in kcal/mol) has been tested on 41 typical carbon-carbon bonds, ranging in length from 1.20 to 1.71 Å. This sets a maximum bond length limit of 1.75 Å for carbon-carbon bonds [34].

By far the most-studied type of chemical bonding is known as 'aromatic'. The concept of aromaticity was introduced in the 1860's by Kekulé and Erlenmeyer to describe cyclic molecules (of which benzene is the seminal one) for which the classical theory suggested alternating single and double bonds, but which were much more stable than implied by such a formula, or than their open-chain analogues actually were. In the twentieth century, diffraction methods confirmed Kekulé's insight that benzene ring has sixfold symmetry (D_{6h}) with all six C–C bonds equivalent (1.3983 Å, cf. 1.422 Å in graphite) and intermediate in character between single and double bonds (e.g. in butadiene, 1.467 and 1.349 Å) [35]. At the dawn of quantum chemistry, Pauling [36] introduced the idea of resonance between two or more valence-bond structures as the source of energy gain. For benzene, he estimated the resonance energy (RE) as the difference between the heats of formation or of hydrogenation of benzene and of the (hypothetical) cyclohexatriene. Both comparisons gave essentially the same RE, *ca.* 150 kJ/mol (< 3 % of the atomization energy of benzene). However, other reference reactions would yield different RE; one also must take into account different energies of steric strain (repulsion between non-bonded atoms) in the real and model molecules, which are of the same order of magnitude. Thus, modern estimates of RE of benzene range from 85 to 312.6 kJ/mol [35]. In the MO theory, aromatic molecules were described in terms of circular π-electron delocalization. Hückel [37] postulated that cyclic planar π-electron systems with $(4n + 2)$ π-electrons must be stabilized (aromatic) and those with $4n$ π-electrons (e.g. cyclobutadiene) must be destabilized (anti-aromatic) with localized single and double bonds. Hückel's rule was vindicated by the discovery of the aromaticity of cyclopropenyl cation ($n = 0$), cyclopentadienyl anion and cycloheptatrienyl cation ($n = 1$) and various heterocycles, while Pauling's failure to explain anti-aromaticity [38] in terms of simple VB theory discredited the latter [39]. However, today it is clear that high-precision versions of VB and MO methods (but not the severely simplified versions of the pre-computer era!) give essentially the same results. Furthermore, it is realized that the symmetry of the benzene ring is due to σ-bond equalization: π-bonds 'left to themselves' would have yielded a localized Kekulé structure [40].

Aromatic rings provide the circuits in which circular electric currents can be induced by external magnetic field, hence magnetic susceptibility χ_m of aromatic molecules is (i) highly anisotropic and (ii) higher than the sum of atomic increments χ_a, its 'exhaltation' provides one of the quantitative measures of aromaticity [41]:

$$\Lambda = \chi_m - \sum \chi_a > 0 \tag{2.11}$$

Other aromaticity indices are based on equalization of bond lengths, bond orders or the peculiar proton shifts in NMR spectra (also due to the exhalted magnetism of aromatics) [42].

Today, it is clear that aromaticity is possible in 3-dimensional, as well as planar, systems, such as quasi-spherical cages of fullerenes [43, 44] and polyhedral boranes (e.g. $B_{12}H_{12}^{2-}$) [44, 45], in carbon nanotubes and in some metal clusters (e.g., Au_5Zn^+, Au_{20}) [46]. The $2(n + 1)^2$ rule proposed by Hirsch [47] and successfully applied to design various novel aromatic compounds, serves as the 3-dimensional counterpart of the Hückel's rule for planar systems. However, deviations from this rule are found that indicate the need for further refinement. This vast area has been comprehensively reviewed in two thematic issues of Chemical Reviews [48, 49].

2.2.3 *Polar Bond, Effective Charges of Atoms*

The term 'ionic substance' is often used in inorganic chemistry, but although the reality of ions is manifest in the ionic conductivity in molten state, and in some cases in the solid state, in fact there are not many compounds which can be regarded even as *practically* ionic, and none with purely ionic bonding. Monoatomic cations are always smaller than anions (except for F^- being smaller than K^+, Rb^+, Cs^+) and tend to polarize the latter, causing a displacement of the anion's electron density towards the cation. The ionization potentials of metals being higher than the electron affinities of nonmetals (see Chap. 1) has similar effect. Thus even in the most ionic crystals the charges must be less then the oxidation numbers. How these can be determined? Dozens of experimental and theoretical methods have been suggested for the determination of atomic charges [50]. For some AX_n or AH_n molecules, the bond polarities and hence the effective charges of ligands are known from IR or XR spectra [51–54]. These values, always <1, are listed in Table 2.2. The most poplular method of estimating the atomic charges in molecules is based on the dipole moments and will be considered in details in Chap. 11 (Table 11.2).

The effective charges of atoms in crystals, rather than molecules, can be determined by Szigeti's spectroscopic method [55], which is also described in details in Chap. 11, using the formula

$$e^* = \frac{3\nu_t}{Ze(n^2+2)}\left[\pi\left(\varepsilon - n^2\right)\overline{m}V\right]^{\frac{1}{2}} \tag{2.12}$$

where ν_t is the transverse vibration frequency of the lattice, n is the refractive index, ε is the dielectric constant, V is the molar volume, Z is the valence of an atom, and $\overline{m}$ is the reduced mass of a vibrating atom. The results are listed in Tables 2.3 and 2.4, in the conventional form e^*/v where v is the formal valence of the atom (i.e. the values are the relative bond ionicities). As we can see, the absolute values of effective charges of O in MO-type oxides are always > 1, whereas in molecules they are < 1. As discussed in Sect. 1.1.2, the $O^- + e^- \rightarrow O^{2-}$addition requires an expense of energy, but in crystals this is compensated by the Madelung energy, which makes the higher negative charges thermodynamically possible. The charges increase with the coordination number: the phase transition in HgS raises the effective charge from 0.20 to 0.28 as N_c changes from 2 to 4, in MnS a transition with $N_c = 4 \rightarrow 6$ increases e^* from 0.35 to 0.44. An unexpected feature is the greater effective charges in halides of Groups 11 and 12 elements. A study of the band structure of CuX and AgX has revealed that the metals have effective valences exceeding 1, due to participation of d-electrons. Similar increases of the metal valences were observed also in some compounds of the MX_2 type (see below). The problem of the effective valences of

Table 2.3 Effective atomic charges (e^*/v) in MX crystals by Szigeti's method (from [56], except where specified)

M (v = 1)	F	Cl	Br	I
Li	0.81	0.77	0.74	0.54
Na	0.83	0.78	0.75	0.74
K	0.92	0.81	0.77	0.75
Rb	0.97	0.84	0.80	0.77
Cs	0.96	0.85	0.82	0.78
Cu		0.98	0.96	0.91
Ag	0.89	0.71	0.67	0.61
Tl		0.88	0.84	0.83
M (v = 2)	O	S	Se	Te
Cu	0.54			
Be	0.55			0.26[a]
Mg	0.59	0.49	0.39	
Ca	0.62	0.52	0.36	
Sr	0.64	0.54	0.50	
Ba	0.74	0.65	0.52	
Zn	0.60	0.44	0.40	0.39
Cd	0.59	0.45	0.42	0.38
Hg	0.57	0.28[b]	0.27	0.26
Eu	0.67	0.55	0.53	0.50
Sn		0.33	0.28[c]	0.26
Pb	0.58	0.36	0.35	0.28
Mn	0.55	0.44[d]	0.42	0.33
Fe	0.46[e]			
M (v = 3)	N	P	As	Sb
B	0.38	0.25		
Al	0.41	0.26	0.21	0.16
Ga	0.41	0.19	0.17	0.13
In		0.22	0.18	0.14

[a][57], [b]$N_c = 4$, for $N_c = 2$ $e^* = 0.20$, [c][58], [d]$N_c = 6$, for $N_c = 4$ $e^* = 0.35$, [e]$e^* = 0.44$ for CoO and 0.41 for NiO.

the metallic elements of Groups 11–14 will be discussed later; here it is sufficient to note that in MoS_2 and $MoSe_2$ we assume v = 2 for the chalcogen and v = 4 for the metal. Spectroscopic studies of alkali halides [60], MF_2 [61, 62], ZnS and GaAs [63] have shown that their effective charges decrease on heating, signifying that the bond covalency increases.

Another spectroscopic method of measuring bond polarity (or ionicity) f_i in solids was developed by Phillips and Van Vechten (PVV) [64–67], using the equation

$$f_i = \frac{C^2}{E_g^{\,2}} \tag{2.13}$$

where E_g is the band gap and C is the Coulomb component of the bond energy. Numerical values of f_i according to PVV and charges according to Szigeti do not coincide because of different dimensionality, but can be related thus

$$f_i = \frac{(e^*)^2}{n^2} \tag{2.14}$$

Table 2.4 Effective atomic charges (e^*/v) in M_nX_m crystals, by Szigeti's method

MX_2	$e^*/2$	MX_2	$e^*/2$	MX_2	$e^*/2$	M_nX_m	e^*/v
MgF_2	0.76	$FeCl_2$	0.64	$SnSe_2$	0.25	UO_2	0.60
CaF_2	0.84	$FeBr_2$	0.58	ZrS_2	0.44	CeO_2	0.56
SrF_2	0.85	CoF_2	0.74	HfS_2	0.50	ScF_3	0.76
$SrCl_2$	0.76	$CoCl_2$	0.57	$HfSe_2$	0.45	YF_3	0.76
BaF_2	0.87	$CoBr_2$	0.52	MoS_2	0.06	LaF_3	0.74
ZnF_2	0.76	NiF_2	0.68	$MoSe_2$	0.04	AlF_3	0.60
CdF_2	0.80	$NiCl_2$	0.51	MnS_2	0.42	GaF_3	0.60
$CdCl_2$	0.74	$NiBr_2$	0.46	$MnSe_2$	0.38	InF_3	0.61
$CdBr_2$	0.69	Na_2S	0.58	$MnTe_2$	0.30	YH_3	0.50[a]
CdI_2	0.63	Cu_2O	0.29	FeS_2	0.30	Y_2O_3	0.62
HgI_2	0.38	TiO_2	0.60	RuS_2	0.36	Y_2S_3	0.40
EuF_2	0.84	TiS_2	0.39	$RuSe_2$	0.38	La_2O_3	0.62
PbF_2	0.87	$TiSe_2$	0.18	OsS_2	0.40	La_2S_3	0.40
$PbCl_2$	0.90	SiO_2	0.60	$OsSe_2$	0.38	Al_2O_3	0.59
PbI_2	0.72	GeO_2	0.54	$OsTe_2$	0.38	Cr_2O_3	0.49
MnF_2	0.81	GeS_2	0.18	PtP_2	0.28	Fe_2O_3	0.45
$MnCl_2$	0.69	$GeSe_2$	0.17	$PtAs_2$	0.24	As_2S_3	0.20
$MnBr_2$	0.66	SnO_2	0.57	$PtSb_2$	0.26	As_2Se_3	0.14
FeF_2	0.78	SnS_2	0.32	ThO_2	0.58	$RuTe_2$	0.39

[a] [59]

where n is the refractive index. Originally the PVV theory was applied only to structures of B1 and B3 types, but subsequently, owing to the works of Levin [68–70] and others [71–73], it was expanded to other structural types. The value of f_i is affected only slightly by the nature of the anion, but sharply by a change of N_c. Thus, GeO_2 has $f_i = 0.51$ in its quartz-like modification, but $f_i = 0.73$ in the rutile-like form. Phillips used this as the criterion of polymorphism; taking 0.785 as the critical value of f_i for the B3 → B1 transition. This method revealed the evolution of atomic charges under varying thermodynamic conditions, in particular a reduction of f_i on compression of crystals (see below). It is worth mentioning that PVV were anticipated by Hertz, Link and Bokii [74–77], [523] who calculated the bond ionicity

$$i = \frac{P_a}{P_e} \tag{2.15}$$

as the ratio of atomic (P_a) and electronic (P_e) polarizabilities of substances; this parameter can be related to the PVV polarity through the Mossotti-Clausius formula. Since

$$P_o = P_M - P_e = V\left(\frac{\varepsilon - 1}{\varepsilon + 2} - \frac{n^2 - 1}{n^2 + 2}\right) \tag{2.16}$$

then for low-polarity substances, where $\varepsilon \approx n^2$, we obtain

$$P_a = V\left(\frac{\varepsilon - n^2}{n^2 + 2}\right). \tag{2.17}$$

Combination of Eq. 2.17 with

$$P_e = V\left(\frac{n^2-1}{n^2+2}\right) \tag{2.18}$$

gives

$$i \approx \frac{\varepsilon - n^2}{n^2-1} \tag{2.19}$$

apparently similar to the equation:

$$i = \frac{\varepsilon - n^2}{\varepsilon - 1} \tag{2.20}$$

which follows from Eq. 2.13 and the basic formulae of the dielectric theory,

$$n^2 = 1 + \left(\frac{h\nu_{\mathrm{p}}}{E_{\mathrm{g}}}\right)^2 \quad \text{and} \quad \varepsilon = 1 + \left(\frac{h\nu_{\mathrm{p}}}{C}\right)^2 \tag{2.21}$$

X-ray spectroscopy (XRS) gives important information on the bond polarity. Experiments have shown that the binding energy of inner electrons of an atom (E_{BIE}) depends on the external electronic environment, i.e. on the effective charges of atoms: a positive net charge increases and negative one reduces E_{BIE}. Therefore, knowing the values of E_{BIE} in different crystalline compounds, one can define the magnitudes and signs of the atomic charges, and how they vary with the composition and structure changes. Thus, effective atomic charges in MX crystals were found to increase with N_{c} and $\Delta\chi$ [78], while in the succession MnS, MnO, MnO_2, MnF_2, the MnK_α-edge of the X-ray absorption band shifts to higher energies by 1, 3, and 3.6 eV, respectively [79]. In the succession $Au_2O_3 \rightarrow AuCl_3 \rightarrow AuCN \rightarrow Au \rightarrow CsAu \rightarrow M_3AuO$ the energies of the AuL_{I} and AuL_{III} absorption edges consistently decrease, passing through $e^* = 0$ for the pure metal, which indicates that in CsAu and M_3AuO the Au atoms bear negative charges, explicable by exceptionally high electron affinity of gold (2.3 eV) [80]. A study of the electron density distribution in BaAu suggested a $Ba^{2+}\ \bar{e}Au^-$electron structure.

The most reliable charge determinations by XRS [81–85] are compiled in Table 2.5. The effective charges decrease when the valence of the central atom increases or when the electronegativity of the ligands decreases. The effective charges of S, P, Si and Cl atoms in organic compounds were determined by the shifts of the K_α-line in comparison with the same atoms in the elemental solids [86, 87]. The drawback of this method is the smallness of ΔK_α in comparison with the absolute binding energies, but its advantage is that the the volume to which the charge refers is known precisely, as electronic transitions are localized within the atom; the values ΔK_α can be scaled against the effective atomic charges calculated from electronegativities [88].

Most structural methods 'see' either the position of the atomic nucleus (neutron diffraction, NMR spectroscopy), or the atomic center of mass (optical and microwave

Table 2.5 Effective atomic charges from X-ray spectroscopy

M_nX_m	$e^*(M)/v$	MX_n	$e^*(M)/v$	M_nX_m	$e^*(M)/v$
NaF	0.95	SiF_4	0.35	GeSe	0.17
NaCl	0.92	$SiCl_4$	0.25	Y_2O_3	0.54
NaBr	0.83	SiO_2	0.23	Al_2O_3	0.25
NaI	0.75	SiC	0.12	$Al(OH)_3$	0.26
Na_2O	0.90	SnF_2	0.83	AlN	0.21
CuF_2	1.0	$SnCl_2$	0.76	In_2S_3	0.24
CuO	0.51	SnI_2	0.42	In_2Se_3	0.17
Cu_2O	0.39	SnSe	0.36	As_2S_3	0.16
$CdCl_2$	0.70	$SnCl_4$	0.23	As_2Se_3	0.11
$CdBr_2$	0.60	$SnBr_4$	0.20	As_2Te_3	0.09
CdI_2	0.44	SnI_4	0.15	Sb_2S_3	0.30
CdS	0.34	SnS_2	0.33	Sb_2Se_3	0.28
CdSe	0.28	$SnSe_2$	0.24	PF_3	0.27
CdTe	0.22	GeS	0.20	PCl_3	0.14

spectroscopy) or the maximum of the electrostatic potential (electron diffraction) which practically coincide with the nucleus. On the other hand, X-rays are scattered mainly by electrons (with a negligible contribution from the nucleus) and therefore can, in principle, inform about the actual distribution of electrons in crystals. Debye had foreseen such possibility as early as 1915 [89], but its realization took a better part of the twentieth century. The present state of the problem is comprehensively described in books [90, 91] and reviews [92–94].

In principle, a map of the electron density can be calculated by a Fourier series, the amplitudes of which are related in a simple way to the intensities of the diffraction peaks ('reflections'). Unfortunately, we also need to know the *phases* of diffracted beams, which are not measurable and have to be deduced. Secondly, a good-resolution map requires very extensive (ideally — infinite) Fourier series, but the number of measured reflection is of necessity limited. The Fourier map is therefore too crude to extract from it chemically meaningful information. To make X-ray crystallography really informative, the diffraction data is fitted into certain models, which are ultimately rooted in quantum mechanics. The simplest of these is the *spherical-atom approximation*, according to which X-ray scattering of a crystal is the sum of scattering by the spherically-symmetrical, ground-state atoms (usually calculated by Hartree-Fock method). The coordinates of these atoms and the parameters describing their thermal vibrations, are then refined by least-squares technique, until the difference between the calculated and observed scattering intensities ('R-factor') is minimized. It is then tacitly assumed that the resulting 'atomic' positions are those of the atomic nuclei. In most cases the latter is true within 0.01 Å; the exceptions are triple-bonded C, N and O atoms with their strongly non-spherical electron shells, and especially H, which has no non-bonding electrons. The H atom position determined by X-ray method is usually shifted towards the chemically bonded atom by 0.1 Å or more, especially if the latter atom is electronegative.

Probably, 99.99 % of all X-ray structure determinations to-date have been done on this approximation. Of course, the real distribution of the electron density in an actual crystal/molecule is different from such model (often called pro-crystal/pro-molecule); its topology can be best understood within the framework of the AIM theory, developed by Bader [8]. Electron density is concentrated between atoms which are linked by a covalent bond, and is depleted between atoms which participate in closed-shell interactions (ionic or van der Waals). A good quantitative measure of such effects is the Laplacian of the electron density ($\nabla^2\rho$), equal to the sum of its principal curvatures (second derivatives) at a given point:

$$\nabla^2\rho = \partial^2\rho/\partial x + \partial^2\rho/\partial y + \partial^2\rho/\partial z \tag{2.22}$$

According to the Virial Theorem, the Laplacian of ρ_e is related to the densities of kinetic (G) and potential (V) energies of the electrons,

$$2G + V = \frac{h^2}{16m\pi^2}\nabla^2\rho_e \tag{2.23}$$

where m is the mass of the electron. A positive Laplacian indicates a local depletion of ρ_e and a negative one a local accumulation (this does not imply a local *peak*!). If two atomic nuclei are linked by a line along which ρ_e is enhanced, this gives a clear indication of covalent bonding ('bond path', BP). The one-dimensional minimum of ρ_e on the BCP (bond critical point) signifies the contact between the atoms, while in three dimensions the *atomic basin* of the electron density is enclosed by surfaces of zero flux of ρ_e.

In fact, the difference between ρ_e of a molecule and of its pro-molecule (deformation electron density) is not as big as a Lewis diagram may seem to imply: a bonding electron pair is localized *on* two atoms, not *between* them. Thus, in a H_2 molecule, for which precise *ab initio* calculations are available, the additional accumulation of electron density between the nuclei (compared to the pro-molecule) is only 16 % of the sole electron pair—although the H–H bond in one of the strongest bonds known! In modern X-ray experiments, R-factor usually does not exceed several per cent—another proof of the smallness of the deformation ρ_e. Charge-density studies require much more extensive and accurate sets of experimental data than ordinary, atomic-approximation, studies. Such experiments were practically impossible before 1970s, remained prohibitively long until area detectors and synchrotron radiation came into wide use in 1990's, and even today are far from routine. It is very difficult to distinguish the deformation of the electron density due to a chemical interaction, from the 'smearing' due to thermal motion of atoms; the most sure solution is to eliminate

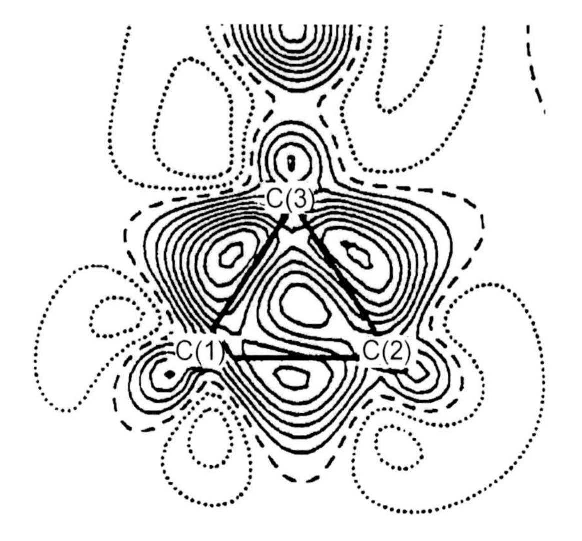

Fig. 2.3 Experimental deformation charge density in the cyclopropane ring of 7-dispiro[2.0.2.1]heptane carboxylic acid. Reproduced with permission from [95], copyright 1996 International Union of Crystallography

thermal motion physically, by collecting the data at liquid-nitrogen, or better still, at liquid-helium temperatures. Nevertheless, today the electron density data is reproducible on a sufficient level of precision, and the main difficulty has shifted to its interpretation. The results depend crucially on the model, and if the latter is inadequate or ambiguous, then the parameters will be biased or indeterminate. The parameterization retains an element of arbitrariness. Worse, often the same data can be fitted equally well (in mathematical sense) to very different sets of variables. More often than is acknowledged, researchers proceed by testing several models and choosing the one that gives the most physically meaningful outcome. And while Laplacian is very efficient in revealing subtle features of the electron density topology, it by the same token magnifies greatly the noise and bias of the original function. Partly for this reason, different tools of topological analysis often give contradictory results.

At present, charge density can be mapped with the precision of 0.05 $e/\text{Å}^3$. The experiments have consistently revealed peaks of the deformation density which can be identified with bonding and non-bonding (lone) electron pairs, as envisaged by Lewis and the VSEPR theory. The charge density at the bond critical point was found to correlate with the strength of the bond, and inversely correlates with bond length. In cyclo-propane rings, the peaks of a (bond) deformation density do not lie on the direct C–C lines, but are shifted outwards (Fig. 2.3). The ellipticity of the bond electron density, i.e. its deviation from cylindrical symmetry, reflects the π-character of the bond. The experimental charge density can serve as the basis of calculating various molecular properties, such as electrostatic potential at the molecular surface (indicating the areas favorable for electrophilic and nucleophilic attacks) or dipole moments. Some polar compounds have their dipole moments much enhanced in the crystal compared to the isolated molecule (e.g. for HCN, 4.4 D *vs* 2.5

Table 2.6 'XRD' effective atomic charges in binary compounds

MX_n	$r^{\cdot}_{M}$	$e^{\cdot}_{M}$/v	MX_n	$r^{\cdot}_{M}$	$e^{\cdot}_{M}$/v	MX	$r^{\cdot}_{M}$	$e^{\cdot}_{M}$/v
LiF		0.88	CaO	1.32	1.00	AlP		0.09
LiH	0.92	0.86	BaO	1.49	1.00	AlAs		0.07
NaCl	1.17	0.88	MnO	1.15	0.62[a]	AlSb		0.05
KCl	1.45	0.97	CoO	1.09	0.74[a]	GaP		0.08
KBr	1.57	0.80	NiO	1.08	0.46	GaAs		0.05
Cu_2O		0.61	Al_2O_3	1.01	0.55	InP		0.06
MgF_2		0.95	Cr_2O_3		0.50	InAs		0.04
MgH_2	0.95	0.95	Sb_2O_3		0.38[b]	InSb		0.02
CaF_2		0.86	SiO_2		0.25[c]	YH_3		0.5[1]
MnF_2		0.90[a]	SiO_2	0.72[d]	0.63[d]	$CaSO_4$		1.0[g]
CoF_2		0.86[a]	TiO_2	0.60	0.75	$CaSO_4$		0.4[h]
MgO	0.93	0.68	BN	0.74	0.15	Fe_3O_4	1.19[e]	0.74[e]
MgS	1.28	0.75	AlN		0.20	Fe_3O_4	1.13[f]	0.64[f]

[a][96], [b][97], [c]quartz, [d]stishovite [81], [e]Fe^{II}, [f]Fe^{III}, [g]Ca, [h]S

D, respectively). On the other hand, estimates of atomic charges proved very model-dependent. Thus, for $NH_4H_2PO_4$ a variety of refinements, fitting the experimental data equally good, yielded the ammonium cation charges varying all the way from 0 to +1 [96]. Observations of 'bond paths', of hydrogen bonds and even weaker intermolecular interactions, attracted criticism [94], since the electron density in intermolecular areas is generally low, close to the level of the experimental error, which makes topological analysis extremely unreliable. Thus, charge-density studies have confirmed many effects which structural chemists suspected for a long time. However, so far they delivered relatively few results which were really unexpected and would have remained unknown without this method.

Many works are devoted to the determination by XRD of charges of atoms in inorganic crystals. Table 2.6 lists the most reliable values of effective charges and atomic radii in binary crystalline compounds, obtained in these studies [97–113]. Similar data for complex compounds can be found in Table S2.1. It is remarkable that in MgH_2 the charge on Mg was estimated as +1.91 e, and on hydrogen as −0.26 e [114] with 1.4 e per formula unit missing. This charge may be delocalized in interatomic voids, but the material is an insulator with $E_g = 5.6$ eV [115].

Bond polarities in organic carboxylate salts, R-CO_2M (M = H, Be, B, C, N, O, Al, Si, P, Mn^{II}, Fe^{II}, Fe^{III}, Co^{II}, Ni^{II}, Cu^{II}, Zn, Gd) were determined, in good agreement with spectroscopic data and EN-based estimates, by comparing the lengths of the two C-O bonds in the carboxy group, from crystal structures [116]. These bonds must be identical (with the bond order $n = 1.5$) in the fully ionic case but different ($n = 1$ and 2) covalent case:

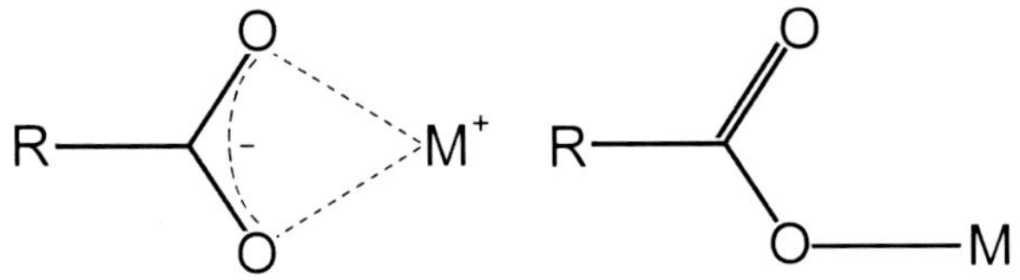

It can be concluded that in the majority of crystalline halides and oxides, a degree of ionicity is between 0.5 and 1.0. The effective sizes (radii) of atoms change with ionization in a non-linear manner (see Sect. 1.5), this translates into $\leq 10\,\%$ deviation from the perfectly ionic radii, which explains the efficiency of the ionic radii in inorganic crystal chemistry.

2.2.4 Metallic Bond

The major feature of the electron structure of metals is the availability of freely moving electrons (formerly valence electrons) shared by all atoms. This model was first formulated by Drude who applied the kinetic theory of gases to an 'electron gas' in metals, assuming that there exist charged carriers moving about between the ions with a given velocity and that they collide with one another in the same manner as do molecules in a gas. The metallic bond can be regarded as a non-directional covalent bond. Indeed, a crystal-chemical approach suggests that a transition from covalent to metallic bonding can be linked with the increase of the coordination number, so that valence electrons become increasingly delocalized and finally transfer from the valence into the conduction band. For the metallic bond to form, atoms' valence electrons must be removed from them to move freely in the interatomic voids of the crystal space. This requires the condition $E(\text{A–A}) + I(\text{A}) < E(\text{A}^+ \cdots e^-)$. When the $\text{A}^+ \cdots e^-$ interaction becomes more favourable than the A–A bond, a dielectric → metal transition occurs.

In the early theory of metals it was supposed that all valence electrons in atoms become free and the metal structure is a lattice of cations immersed in an 'electron sea'. Now it is known that only a part of the outer electrons of atoms are free, since the metallic radii are larger than those of cations (see Chap. 1). Some studies of electron density distribution in metals estimated the metallic/core radii ratio as 1:0.64 [117, 118]. The effective radii of the atomic cores in metallic structures are close to the bond radii of the same metals in crystalline compounds (see Chap. 1) which correspond to atoms with charges not exceeding ± 1. It should be noted that work functions of bulk metals are always smaller than the first ionization potentials of the corresponding atoms (see Sect. 1.1.2) and therefore there is no reason to suppose the ionization of two or more electrons from an atom.

The crystal-chemical mechanism of the metallization in ionic crystals of MX-type was studied under high pressures [119]. Assuming that the metallization of a material under pressure occurs when the chemical bond is destroyed, i.e. the compression energy becomes equal to $E(\text{M–X})$, it was concluded that the interatomic distances for an ultimately compressed MX crystal are equal to the sum of the *cationic* radius of M^+ and the normal *covalent* radius of X. Thus, metallic binary compounds differ from pure metals in having a sub-lattice of neutral nonmetal atoms X^o. From here it follows that in MX under high pressures the M atom can be the donor of electrons if the bonding is covalent, M°–X°, as $I_M < I_X$. If the substance is ionic, M^+X^-, then X^-anion must be the donor, because $A_X < I_M$. The polar character of the M–X bond prior to pressure-induced metallization can be determined experimentally. The proposed mechanism of metallization implies the availability not only of mobile electrons—for in aromatic molecules they are also mobile to a considerable extent—but also of certain structural voids which these electrons can occupy. Because metallic structures have high N_c (usually, 12), a cluster should have at least 13 metal atoms to acquire metallic properties. Measurements of the photoelectron spectra in clusters of mercury [120, 121] and magnesium [122] showed that their *s-p* band gaps are closed when the number of Hg atoms reaches 18, indicating the onset of the metallic behavior.

It has been shown [123] that the metal sub-lattice in crystal structures of the ZnS, NaCl, NiAs and CsCl types has the same (or similar) N_c of metal and M–M distances (d_{MM}), as the structure of pure metal (d°_{MM}); hence the degree of metallic bonding can be defined as

$$m = c\frac{d^o_{MM}}{d_{MM}} \tag{2.24}$$

As the next logical step, it was suggested [124] to apportion the distribution of the covalent electron density (q) in an MX structure in proportion to the strengths of M–M and M–X bonds,

$$q = \frac{N_{MM}E_{MM}}{N_{MM}E_{MM} + N_{MX}E_{MX}} \tag{2.25}$$

where $N_{MM,MX}$ and $E_{MM,MX}$ are the coordination numbers and energies of the M–M and M–X bonds, respectively. Taking into account the proportionality between the energies and overlap integrals of bonds, we obtain

$$m = uS_{MM}/S^o_{MM} \tag{2.26}$$

here u is the electron concentration (population) in the metal orbitals, S^o and S are the overlap integrals of the M–M bonds at distances $d^o{}_{MM}$ and d_{MM}. The problem of partitioning of the covalent electron density between M–M and M–X bonds is solved in [11] using a simpler model, viz.

$$q = \frac{\chi_M}{\chi_M + \chi_X} \tag{2.27}$$

where χ_M and χ_X are the electronegativities of the M and X atoms in the MX crystals. If $d^o{}_{MM}$ and d_{MM} are close, the degree of bond metallicity can be estimated as $m = cq$, otherwise it can be determined from experimental data as

$$m = cq\frac{d^o_{MM}}{d_{MM}} \tag{2.28}$$

In Table S2.2 the values of metallicity calculated by Eqs. 2.25 (m_1) and 2.28 (m_2) are listed. Good agreement of the results prove that for an approximate estimation, it is not necessary to take into account the differences between the metal bond lengths in compounds and elemental solids.

As mentioned above, the metal sub-lattices in the crystal structures of compounds are usually the same, or similar to, the structures of the corresponding pure metals. This problem was considered in depth by Vegas et al. [125] who studied the genesis of structures in metals, alloys and their derivatives. Thus, MBO_4 compounds of the $CrVO_4$ structural type have the metal lattice like the MB structure. The same situation exists in ternary oxides of the MAO_n type, where A = S or Se, and $n = 3$ or 4, and also in $MLnO_3$. This means that the metal skeleton is the basis of the structure of the compound, and atoms of oxygen are simply included into the voids between cations. Such inheriting of the structure of the parent substance by its derivatives can be explained by the minimization of work required to create the new structure on the basis of the metal lattice, although there are also more complex reasons [126].

One more experimental method of characterizing the metallic state is to compare the volumes and refractions (R) of solids. As the refractive indices of metals are very great, the Lorentz-Lorenz function (Eq. 2.18) is close to 1 and $R \approx V$. According to the Goldhammer-Herzfeld [127, 128] criterion, $V \to R$ when a dielectric converts into a metal. As the measure of bond metallicity, the ratio

$$\frac{R}{V} = \frac{n^2 - 1}{n^2 + 2} \tag{2.29}$$

can be considered [129, 130]. The pressure at which $V = R$, has been often regarded as the pressure of metallization. However, both during isomorphic compression and at phase transitions under high pressures the refractive index also changes (see Chap. 11). Therefore the Goldhammer–Herzfeld criterion is not absolutely correct, although for rough estimations of pressures of metallization it is valid. A more rigorous approach was used in [131, 132] where the changes of $R(CH_4)$ and $R(SiH_4)$ under pressure were studied, revealing a large increase in the R/V ratios at 288 and 109 GPa, respectively, which indicates phase transformations of the insulator–semiconductor type in these materials.

There is one more question to be answered. It is known that phase transition enthalpies (ΔH_{tr}) constitute only a small part of the atomization energy (E_a). Thus, the graphite-diamond transition with change of N_c from 3 to 4 has the $\Delta H_{tr} = 2$ kJ/mol; the transition from 4- to 6-coordinate Sn has 3 kJ/mol, of 6- to 8-coordinate Bi has 0.45 kJ/mol, and that of 8- to 12-coordinate Li has 54 J/mol. In each case, $\Delta H_{tr} \leq 0.01 E_a$, whereas on transition from the Sn_2 molecule ($N_c = 1$) to α-Sn ($N_c =$ 4) the E_a increases 3.2 times. Similar transformation for Group 1 or Group 11 metals from $N_c = 1$ to $N_c = 8$ or 12 results in a 3.4-fold increase, and for other metals the changes are even bigger. Note also that the E_a(MX) in crystals exceed those in molecules by a factor of $\approx$4.3 (see Sect. 2.3), but further increases of N_c in crystals make very little difference, as indicated by small changes in Madelung's constants, from $k_M = 1.748$ at $N_c = 6$ to $k_M = 1.764$ at $N_c = 8$, while some increase in interatomic distances at the B1 $\to$ B2 transition, compensates the small increase of k_M. The reason of this effect consists in the multi-particle interaction of atoms in crystals, in the Coulomb interactions of cations with anions or free electrons.

For a long time the crystal-chemical approach seemed sufficient to describe of the nature of the metallic bond. However, physically more general approach is to consider the band structures of substances, namely that the conduction band containing electrons must be only partly filled [133]. Thus, with a full band the compound $K_2Pt(CN)_4$ is an insulator and the Pt–Pt distance along the chain is 3.48 Å. However, the non-stoichiometric compound $K_2Pt(CN)_4Br_{0.3} \cdot 3H_2O$ is metallic and, since electrons have been removed from the top of the band where maximum anti-bonding interactions are found, it has a much shorter Pt–Pt distance of 2.88 Å. Thus, the partial oxidation of $K_2Pt(CN)_4$, when the band of Pt^{IV} is filled only partly, transforms this compound into a metal. At full oxidation, $K_2Pt(CN)_4Br_2$ is an insulator. The same picture is observed in $La_{2-x}Sr_xCuO_4$ where the metal conductivity is observed at $x > 0.05$.

There is one more way of formation of the metal state in molecular substances without their transition to structures with high coordination numbers. On compression of the condensed molecular H_2, O_2, N_2 and halogens, they acquire metallic properties (see Sect. 5.2 and the review [134]) which result from strengthening of electronic interactions upon shortening of intermolecular distances. These are so-called ‘molecular metals’. As the molecules approach one another, three-center A$\cdots$A–A orbitals or even chain-like structures are formed, where an increase of N_c from 1 (A_2) to 2 (–A–) leads to a linear delocalization of valence electrons.

2.2.5 *Effective Valences of Atoms*

The concept of valence (v) is one of the cornerstones of chemistry. According to IUPAC Compendium of Chemical Terminology, the valence of a chemical element is defined as the number of hydrogen atoms that one atom of this element is able to bind in a compound or to replace in other compounds. However, in solid-state physics and structural chemistry this term usually means the bonding power of atoms and then v may have a non-integer value (‘effective valence’), which is derived from physical properties. Thus, there is a widespread opinion [28, 135, 136] that metals of Group 11 (Cu, Ag, and Au) in the solid state have effective valences v^* much higher than 1, which explains the big difference between Group 1 and Group 11 metals of the same period, in physical properties, viz. melting temperatures (T_m), densities (ρ), and bulk moduli B_o (Table 2.7). However, the difference between Groups 1 and 11 goes beyond the solid state, and manifests itself in the structures and properties of gaseous molecules of these elements. Moreover, a certain parallelism is observed in the variation of characteristics of elements in both states. According to spectroscopic data [137], the bonds in molecules Cu_2, Ag_2, and Au_2 are single; i.e. $v = 1$. This agrees with the proximity of the M–M half-distance to the covalent radius of M determined as the length of an M–H bond (definitely single) or an M–CH_3 bond minus the covalent radius of hydrogen or carbon [138]. The same conclusion can

Table 2.7 Properties of Groups 1 and 11 metals in the solid state [138]

M	K	Cu	Rb	Ag	Cs	Au
T_m,°C	63.4	1085	39.3	961	28.4	1064
ρ, g/cm^3	0.86	8.93	1.53	10.5	1.90	19.3
B_0, GPa	3.0	133	2.3	101	1.8	167
v^*, Pauling	1	5.5	1	5.5	1	5.5
v^*, Brewer	1	4	1	4	1	4
v^*, Trömel	1	3	1	3	1	3

be drawn by comparing the ratios between the bond energies and bond lengths in solids and gas-phase molecules, between the atomization energies of solid metals and the dissociation energies of molecules M_2, and between force constants (f) in molecules M_2 and metals M of Groups 1 and 11 elements. Table 2.8 shows that the averaged ratios (k) of these properties for all elements are similar, averaging 1.706 ± 2.2 %, 1.154 ± 1.4 %, and 0.075 ± 5.1 %, respectively. Thus, although the absolute values of physical properties of Groups 1 and 11 elements differ widely, the relative changes (from solid to molecule) are practically identical. Table 2.9 shows the simplest estimation of the electronic energies of isolated atoms, as proportional to $\varepsilon = Z^*/r_o$ where Z^* is the effective nuclear charge (from Table 1.7) and r_o is the orbital radius (from Table 1.8), and the experimental atomization energies (in kJ/mol) of the three pairs of metals. One can see that the energies of isolated atoms are correlated with the energies of atoms in solid metals, e.g. the bond strengths of elements are determined by the properties of isolated atoms. Thus, there are no physical grounds for ascribing the exaggerated 'metallic' valences to Cu, Ag, and Au in the solid state.

Physical properties of the crystalline halides MX of the Groups 1 and 11 metals also strongly differ: the temperatures of melting (T_m) and band gaps (E_g) of alkali halides decrease in the succession MCl → MI, but in halides of Cu, Ag and Tl in the same succession they increase or change little (Table S2.2), although d(M−X) increases in all cases from MCl to MI. Experimental effective charges in alkali halides on average are smaller than in halides of the Group 11 elements (Table 2.3), although the difference of electronegativities $\Delta\chi = \chi(X) - \chi(M)$ is smaller in the case of Cu, Ag and Tl. This fact has been explained [139] by the formation of additional (dative) M → X bonds involving the (n–1)d-electrons of the metals and vacant nd-orbitals of the halogens, resulting in an increase of the atomic valences of Groups 11, 12 and 13 elements on average by 1.5, 2.4 and 3.1, respectively. It should be noted that Lawaetz [140] and Lucovsky and Martin [141] showed that to reconcile the band structure of CuX and AgX with experimental data, one can assume that v^*(Cu, Ag) = 1.5. Robertson [142–145] obtained good results in calculation of the PbI_2 band structure under condition of a 41 % Pb s orbital contribution to the upper valence band state $A^+{}_1$. Wakamura and Arai [146] also obtained $v^* = 2.8$, 2.6 and 2.6 for crystalline compounds of divalent Mn, Co and Ni, respectively. The crystal-chemical estimations of v^* for divalent Sn, Pb, Cr, Mn, Fe, Co, Ni give 2.45 ± 0.05.

Liebau and Wang [147, 148] demonstrated that the classical valence term as introduced by Frankland [149] and the term valence as used by solid-state physicists

Table 2.8 Comparison of energies (kJ/mol), distances (Å), and force constants (mdyne/Å) of M–M bonds in solid metals and molecules

M	K	Cu	Rb	Ag	Cs	Au
E_a(M)	89.0	337.4	80.9	284.6	76.5	368.4
E_b(M_2)	53.2	201	48.6	163	43.9	221
k_E	1.673	1.731	1.665	1.746	1.753	1.667
d(M)	4.616	2.556	4.837	2.889	5.235	2.884
d(M_2)	3.924	2.220	4.170	2.530	4.648	2.472
k_d	1.176	1.151	1.160	1.142	1.126	1.167
f(M)	0.007	0.108	0.006	0.093	0.005	0.154
f(M_2)	0.10	1.33	0.08	1.18	0.07	2.12
k_f	0.072	0.081	0.074	0.079	0.071	0.072

Table 2.9 Comparison of energies in Groups 1 and 11 elements for molecular and solid states

M	Z*	r_0	ε	$q_ε$	E_a	q_E
K	2.2	2.162	1.02		89.0	
Cu	4.4	1.191	3.69	3.62	337.5	3.79
Rb	2.2	2.287	0.96		80.9	
Ag	4.9	1.286	3.81	3.97	284.6	3.52
Cs	2.7	2.518	1.07		76.6	
Au	5.6	1.187	4.72	4.41	368.4	4.81

and crystallographers, are different in nature, and suggested to call them *stoichiometric valence* and *structural valence*, respectively. For the majority of crystalline structures, the difference between these values is <5 %, but for *p*-block atoms with one lone electron pair, the differences of up to 30 % have been reported.

Quantum-chemical estimations show also that the ability to form additional bonds in halides of the Group 11 metals increases from chlorides to iodides. A comparison of the observed ionization potentials and electron affinities of halogens [139] shows that it requires a smaller expense of energy to add the second electron to an I^- ion than to Cl^-. The X^{2-} ions are not found yet, but if they are ever observed in mass spectra, the lifetime of I^{2-} can be predicted to exceed that of F^{2-}.

2.3 Energies of the Chemical Interaction of Atoms

2.3.1 Bond Energies in Molecules and Radicals

Energy characteristics of atoms also define, to a large extent, the strengths of their bonds in molecules, polyatomic ions and radicals. The work required to disrupt a chemical bond, e.g. to separate chemically bonded atoms from the equilibrium distance to a practically infinite one (in the ground state) is called bond energy (E_b). In case of the A_2 and AX molecules, E_b is equal to the dissociation energy of the molecule (D_e) which can be determined by thermochemical, calorimetric, kinetic, mass-spectroscopic and molecular spectroscopic techniques. By definition, D_e characterizes atoms in molecules at the equilibrium state with zero-point energy,

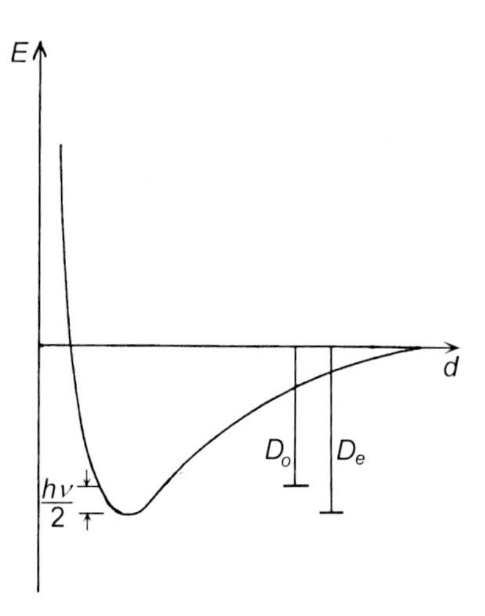

Fig. 2.4 Characteristics of the potential curve

ε (Fig. 2.4). Because $\varepsilon = \frac{1}{2} h\nu_o > 0$, even at 0 K the measurements give

$$D_o = D_e - \frac{1}{2}h\nu_o \tag{2.30}$$

where D_e is the dissociation energy calculated at the very bottom of the potential energy well. The zero-point energy is highest in H_2 (26 kJ/mol) and somewhat less in molecules with heavier atoms, therefore the difference between D_o and D_e can be ignored for structural-chemistry purposes.

Thermochemical determinations of the bond energy are based on the measurements of the heats of reaction (Q) at constant pressure

$$Q = (E_2 + PV_2) - (E_1 + PV_1) \tag{2.31}$$

where $E_1 + PV_1$ and $E_2 + PV_2$ represent the initial and final states of the system. The enthalpy being $H = E + PV$, it follows that $Q = \Delta H$ at constant P.

Heats of reactions can be measured by the calorimetric and kinetic methods, using photo- and mass-spectrometry. Bond dissociation enthalpy calculated from the thermal effect of the reaction at ambient pressure is close to the bond energy because PV is small, for example for the hydrogen molecule $PV \approx 2.5$ kJ/mol. Finally, the difference between the dissociation energy at 0 K and that at room temperature is also very small; for the hydrogen molecule the difference is $\Delta E \approx 1$ kJ/mol.

Thus measurements of bond energies by different methods usually diverge by several kJ/mol; for this reason the experimental bond energies cited in the present and the next sections, are rounded up to integer kJ/mol, except where independent measurements give better agreement. The bond energies of diatomic molecules and radicals presented here, have been compiled using as the starting-point, such reference books as the NBS Tables of Chemical Thermodynamic Properties (1976–1984), JANAF Thermochemical Tables (1980–1995), Thermochemical Data of Pure Substances (1995), Handbook of Chemistry and Physics (2007–2008), and Thermodynamic Properties of Compounds (electronic version, 2004, in Russian). These data

Table 2.10 Values of n in the Mie equation for molecules MX and M_2

M	n	M	n	M	n	M	n	M	n
Ag	3.7	Cd	4.1	In	3.8	Ni	3.1	Th	2.5
Al	3.3	Cr	4.4	K	2.9	Pb	4.7	Ti	4.2
As	3.8	Co	3.1	La	2.7	Pt	3.6	Tl	3.0
Au	4.2	Cs	3.2	Li	2.2	Rb	2.9	U	3.1
B	2.6	Cu	3.2	Mg	3.7	Sb	3.8	V	3.9
Ba	3.2	Fe	4.0	Mn	3.5	Sc	2.6	W	4.5
Be	3.2	Ga	2.5	Mo	4.1	Sn	3.8	Y	3.5
Bi	4.4	Hg	5.0	Na	2.6	Sr	3.3	Zn	3.9
Ca	3.3	Hf	3.3	Nb	3.4	Ta	3.2	Zr	3.6

have been critically compared, corrected and updated using recent original publications, for which references are given. Where several independent measurements by the same method are available, the preference is given to the more recent works and more authoritative researchers, while results of equal reliability have been averaged.

Bond energies of diatomic molecules are listed in Tables S2.2 and S2.4. Evidently, dissociation energies of hetero-nuclear diatomic molecules increase together with the bond polarity, i.e. from iodides to fluorides and from tellurides to oxides of the same metals. Therefore, $D(\text{M}-\text{X})$ in halides and chalcogenides are always larger than the additive value, i.e. the half-sum of $D(\text{M}-\text{M})$ and $D(\text{X}-\text{X})$. This fact has been first noticed by Pauling, who formulated the dependence of the bond energy on its polarity in terms of electronegativity (see Sect. 2.4). For halides of polyvalent metals, or other elements, e.g. H, B, C, the dissociation energy of a hetero-atomic bond does not necessarily exceed the additive value, because the bonds under comparison may differ not only by polarity, but also by the type of bonding orbitals and the bond order.

Also of paramount importance is the relation between the energy E_b of a bond and its length d. This question has three distinct, but ultimately connected, aspects, viz. (i) the potential curve for a chemical bond of a given order between given types of atoms, (ii) the bond length/bond order relation for a given pair of atoms, and (iii) the correlation between energies and lengths of bonds formed by different elements. However, it is probably impossible to establish a universal dependence $E_b = f(d)$, because the inter-nuclear, nucleus-electron and inter-electron interactions all change with distance in different fashions, and the combined energy curve can be highly specific. The most general form of the potential energy (E) for two interacting atoms is given by the Mie equation:

$$\mathrm{E} = -\frac{a}{d^n} + \frac{b}{d^m} \tag{2.32}$$

where a and b are constants of the substance, d is the bond length, and $m > n$. Here the first term defines the attraction and the second one the repulsion of atoms. The sum of m and n and their product ($m \times n$) can be derived from various physical properties [150], but there is very limited experimental information about the values of m and n separately. In the equilibrium state Eq. 2.32 transforms into

$$E = \frac{E_e}{n-m}\left[n\left(\frac{d_e}{d}\right)^m - m\left(\frac{d_e}{d}\right)^n\right] \tag{2.33}$$

From here, supposing $m = 2n$ and thereby transforming the Mie equation into the Morse function (see Eq. 2.7) which describes energies of molecules very well, we finally obtain

$$n = d\sqrt{f/2E_e} \tag{2.34}$$

where f is the force constant. The calculations of n from experimental data for molecules at normal thermodynamic conditions [151] are presented in Table 2.10.

As mentioned above, at $m = 2n$ the Mie equation transforms into the Morse function which describes well not only covalent bonds but also van der Waals interactions. This function was used to estimate van der Waals radii [152] and to rationalize the properties of Zn_2, Cd_2 and Hg_2 molecules [153]. Very often E is estimated using the simplified equation

$$E = \frac{a}{d^n} \tag{2.35}$$

i.e. neglecting the repulsion term. Having made this assumption, and using experimental data for transition metals, Wade et al. found that $n \approx 5$ for C–O bond, 3.3 for C–C, and 2–7 for M–O bonds [154–156]. The bond energies and lengths in many molecular and crystalline compounds have been estimated using Eq. 2.35 [157]. The results are briefly as follows. In molecules of univalent elements, Na_2 to Cs_2 and Cl_2 to I_2, $n = 1.2$ and 1.6 while according to Harrison [158] the bond covalent energy depends on the interatomic distance as d^{-2}. In the successions $P_2 \rightarrow Bi_2$ and $S_2 \rightarrow Te_2$, $n = 2.6$ and 1.8, i.e. somewhat less than the canonical factors of 3 and 2, respectively; in molecules of hydrohalides HX, alkali halides, CuX, and SnX or PbX, $n = 1.6$, 2.1, 1.9 and 2.5, respectively. Interestingly, in van der Waals molecules Zn_2, Cd_2, Hg_2, the bond energies vary as $d^{-2.4}$, although according to London's theory of van der Waals interactions, E must be the function of d^{-6}. The values of n in crystalline compounds are smaller than in the corresponding molecules, by 15–30 %. From these n values one can deduce the absence of ideal types of chemical bonds in most molecules and crystals. The bond character in solid metals is especially varied; several authors explain this variability by the fact that the effective and formal valences of atoms are different.

The chemical bond strength usually increases when the bond length decreases. Noteworthy exceptions are the N–N, O–O and F–F single bonds, which are weaker than the longer P–P, S–S and Cl–Cl bonds, respectively. This effect can be explained by the strong repulsion between the bonding and the lone electron pairs at shorter distances, which agree also with lower electron affinities of N, O and F compared with P, S and Cl (see Table 1.3). The dissociation energies of the N–N, O–O and F–F bonds, estimated by extrapolating the D *vs.* X–X curve (derived for larger atoms), exceed the experimental values by 230, 250 and 210 kJ/mol respectively. This effect should be taken into account at additive thermochemical calculations.

Another example of electron-electron repulsion affecting bond dissociation energies can be observed on monofluorides of certain elements. Thus, M–F bond energies in LiF (573 kJ/mol) and BeF (575 kJ/mol) are almost equal, notwithstanding substantially shorter bond in the latter, due to the repulsion between the nonbonding *s*-electron of Be and the electron pair of the Be–F bond, which compensates for the increased charge on the metal atom. On the contrary, in BF the bond energy is much higher (742 kJ/mol) as the two nonbonding electrons of the boron atom form a closed s^2-pair interacting weakly with the bond electrons. Introduction of another isolated electron in CF, again reduces the dissociation energy down to 548 kJ/mol.

The interaction of isolated electrons of an M atom with the bond electrons obviously would decrease as the latter shift toward the X atom, i.e. as the bond polarity increases. This explains why the increase of bond energy from M–I to M–F is higher for alkali earth metals (which have uncoupled *s*-electrons) than for alkali metals (which don't), on average 285 *vs.* 185 kJ/mol. Note that the bond energies in the MX_2 molecules (where the alkali earth metal has no non-bonding *s*-electrons) of the same series differ from the corresponding energies of alkali halides by only *ca.* 25 kJ/mol. Exceptionally low dissociation energies of M_2 molecules of alkali-earth metals are due to the formation of stable outer s^2-electron configurations that prevent the formation of covalent bonds; the interaction is rather of van der Waals character and its energy is correspondingly small (see below).

Peculiarities of the electron structures of atoms in molecules become particularly conspicuous when the dissociation energies of MX or M_2 molecules are compared with those of positively charged MX^+ or ${M_2}^+$ radicals. In agreement with Hess' law

$$E(\mathrm{M}-\mathrm{M}) + I(\mathrm{M}) = I(\mathrm{M}_2) + E(\mathrm{M}-\mathrm{M}^+) \tag{2.36}$$

Hence, the difference $I(\mathrm{M})–I(\mathrm{M}_2)$ determines the relationship between $E(\mathrm{M}_2)$ and $E({\mathrm{M}_2}^+)$. Textbooks usually give few examples thereof, and often the single one of H_2 *vs* ${H_2}^+$, where the dissociation energy decreases from 436 to 256 kJ/mol as one electron in this molecule is removed on ionization. However, this example is rather atypical. Tables S2.5 and S2.8 list all the currently known dissociation energies of positively charged diatomic radicals. For halogens, the picture is easy to interpret. An electron is removed from a non-bonding orbital which is destabilized (compared to isolated atom) by electron-electron repulsion in these electron-rich molecules, hence $I(\mathrm{A}) > I(\mathrm{A}_2)$. Ionization reduces this repulsion, hence $E(\mathrm{A}_2) < E({\mathrm{A}_2}^+)$. Both relationships are reversed for hydrogen, as it has only bonding electrons which are attracted by both nuclei and therefore are bound stronger than in the atom. The loss of one of these electrons, naturally, weakens the bonding.

One might expect Group 1 metals to be similar to hydrogen in this respect, as they have no non-bonding electrons in the outer shell. The closed shells evidently lie too low in energy (the second ionization potential exceeds the first by an order of magnitude), hence ionization in this case also means the loss of a *bonding* electron. Nevertheless, for all these metals $I(\mathrm{A}) > I(\mathrm{A}_2)$ and $E(\mathrm{A}_2) < E({\mathrm{A}_2}^+)$, i.e. apparently two nuclei attract an electron weaker than one and a single electron holds the atoms together stronger than a Lewis pair! The one plausible explanation can be derived from the 'magic formula' of Mulliken (see below, Eq. 2.46) although there are more complicated models also [159–162].

In a modified form, Eq. 2.36 can be extended to dications,

$$E(A_2) + 2I(A) = I_1(A_2) + I_2(A_2) + E({A_2}^{2+}), \tag{2.37}$$

where I_1 and I_2 are the first and the second ionization potentials, or indeed to bulk solids,

$$E_a(A) + I(A) = \Phi(A) + E_a(A^+) \tag{2.38}$$

where $E_a(A)$ and $E_a(A^+)$ are the atomization energies of the neutral and charged solids, respectively, and $\Phi(M)$ is the work function, which is the ionization potential of a bulk solid. Since $I(M) > \Phi(M)$ always, it follows that $E_a(M^+) > E_a(M)$. However, for alkali metals, $E_a(M^+)$ corresponds to the dissociation of an imaginary solid consisting of metal cations without any valence electrons. Surely, such a system must be altogether unbound!?

This paradox can be resolved as follows. The (molar) ionization potential is the energy required to ionize every atom (or molecule) in a mole of a substance, while Φ is the minimum energy required to remove the *first* electron from a neutral solid, whereas subsequent electrons would require ever greater amounts of energy. To estimate this energy, let us assume as the first approximation, that the I_2/I_1 ratio for a molecule is the same as for an atom with the equal number of valence electrons. This assumption seems not unreasonable in view of the recent observation [163] that successive ionization potentials (at least, for the outer electron shell) for atoms of all elements can be described by a single simple function. Thus, Group 1 diatomic molecules can be 'modeled' by Group 2 or Group 12 atoms. As shown in Table S2.6, for these atoms the I_2/I_1 ratio is fairly constant, averaging 1.9. Taking into account that for metals $I(A) \approx I(A_2)$, Eq. 2.37 can be reduced to

$$E({A_2}^{2+}) \approx E(A_2) - 0.9I(A). \tag{2.39}$$

From the data in Table S2.7 it is obvious that $E({A_2}^{2+}) \ll 0$, i.e. the ${A_2}^{2+}$ cation is strongly unbound, as indeed could be expected for a molecule completely stripped of valence electrons. This molecule can also serve as a model for a bulk metal deprived of its electron gas (Eq. 2.38), to estimate the ionization potentials for bulk metals. As mentioned above, Φ is the energy required to strip the first electrons from a neutral solid. When each atom in the solid is surrounded by charged atoms, $E_a(M^+)$ can be found by an expression analogous to Eq. 2.39,

$$E_a(A^+) \approx E_a(A) - 0.9\Phi(A). \tag{2.40}$$

As shown in Table S2.7, $E_a(A)$ is always smaller than 0.9 $\Phi(A)$, i.e. the structure is unbound.

Comparison of Tables S2.3 and S2.5 shows that removing an electron from a multiple bond (in N_2, P_2, As_2) reduces the dissociation energy, while removing an electron from the outer shell in molecules of O_2, chalcogens, halogens, alkali metals strengthens the bond by reducing electron-electron repulsion. Removing an electron from diatomic molecules of Groups 2, 12 and 18 elements, which have the closed s^2 or s^2p^6 outer shells, transforms the van der Waals interaction into the normal chemical bond and therefore these cations become more bound than the corresponding neutral molecules. It is interesting that removing an *np*-electron from Tl ($5s^25p$ shell), on the contrary, transforms the normal covalent bond in the Tl_2 molecule into a weak one, similar to van der Waals interaction, in the Tl_2^+ cation.

Ionization of halides of univalent metals, as well as oxides and chalcogenides of divalent metals, drastically weakens their bonds, because the electron is removed from the negatively charged atom, thus eliminating the Coulomb component of the energy. On ionization of a radical which comprises a multivalent atom and a halogen (or chalcogen), the unpaired electron is removed from the electropositive atom or, if none is present there, from the electronegative atom. In the former case the bond is strengthened, in the latter weakened.

During the last decade the information became available concerning the alteration of bond strengths in A_2 molecules on *negative* ionization. Thus, dissociation energies of Sn_2^- (265 kJ/mol) and Pb_2^- (179 kJ/mol) [164] are higher than those of the corresponding neutral molecules (187 and 87 kJ/mol, respectively). Among transition metals, M_2^- anions with M = Ni, Cu, Pt, Ag and Au have lower bond energies than the neutral molecules and only Pd_2^- has a higher one than Pd_2 [165]. The bond in the As_2^- radical is slightly stronger than in the As_2 molecule [166]. These data on the effects of positive and negative ionization on the bond energies and distances comprise excellent material for quantum chemistry, which still has to be fully utilized.

For a diatomic molecule, determining bond energy invokes no ambiguity: it is exactly equal to the dissociation energy. However, for a larger molecule, e.g. MX_n, the experiment can give only the energies of successive dissociations of the bonds $X_{n-1}M{-}X$, $X_{n-2}M{-}X$, *etc.*, until the last X atom is eliminated. A difference of the consecutive dissociation energies in a polyatomic molecule may be very significant: thus, D (CH_3–H) = 439, D(CH_2–H) = 462, D(CH–H) = 424 and D(C–H) = 338 kJ/mol. Although the mean bond energy in MX_n cannot be measured directly, it is generally used in structural chemistry, because in most MX_n molecules all M–X bonds are equivalent and should have equal strength. The average of all these dissociation energies gives the mean bond energy E,

$$E(M{-}X) = \frac{\sum D(M{-}X)}{n} \tag{2.41}$$

There is a certain parallel here with the concept of the mean ionization potential; the similarity extends also to some applications of both parameters (see below). In principle, the mean bond energy in polyatomic molecules may be directly determined, if the strong energetic influence atomizes a molecule completely: $MX_n \rightarrow M + nX$, however such experiments are not yet available. The energies of successive dissociations differ because after each atom is eliminated, the electron structure of the remaining fragment is rearranged, the difference itself being the measure of this reorganization [167]:

$$E_R = D - E \tag{2.42}$$

The reorganization energy E_R, particularly important in quantum chemistry, will not be discussed here. Nevertheless, the effects of electron reorganization accompanying bond rupture in polyatomic molecules have to be taken into account in structural chemistry also, the dissociation energy being substantially affected by the structure and composition of the molecule.

While the total dissociation energy always increases with the bond order, this energy normalized by the number of bonding electron pairs sometimes reveals the opposite trend, e.g. decreasing in the succession E(C–C) in ethane > ½ E(C=C) in ethylene > 1/2 E(C≡C) in acetylene (357, 290, 262 kJ/mol, respectively). This *relative* weakening of the multiple bond is compensated by strengthening of the adjacent σ bond: extracting the first H atom from C_2H_6, C_2H_4 or C_2H_2 requires respectively 423, 459 and 549 kJ/mol, extracting the second one requires 163, 339, 487 kJ/mol [168, 169]. Such complementarity occurs because the effective nuclear charge of a given atom is screened by the constant number of its electrons and an accumulation of electrons in one bond naturally reduces the screening in other directions. Later we shall observe other manifestations of such compensation.

Experimental values of the mean bond energies for some elements are listed in Tables 2.11 and S2.9. The compilation was based on the above mentioned thermodynamics reference books, revised and (where reference are given) updated using new original publications. The energies of similar hetero-polar bonds in di- and polyatomic molecules (compare Tables S2.3 and 2.11) are only slightly different; this fact contradicts the ionic model. Indeed, if the bonds in the BaF_2, LaF_3 and HfF_4 molecules were purely ionic, their (Coulomb) energies would be higher than in the CaF, LaF and HfF radicals by the same factors as the charges of the metal atoms, i.e. 2-, 3- and 4-fold, respectively. In fact, bond energies in mono- and poly-fluorides differ only by ± 10 %. Given that the Coulomb attraction certainly gives the major contribution to the bond energy (see below), one has to assume that as the valence (v) of the metal increases, the bond ionicity should decrease proportionally to $1/\sqrt{v}$ [196]. Below it will be shown how this simple rule agrees with the experimental data.

Thus, in polyatomic molecules with different ligands, bonds of the same type have different energies depending on the composition and structure of the molecule. Therefore the values listed in Table 2.9 are strictly applicable only to the molecules for which they have been determined, and only tentatively to other compounds with similar bonds. Table S2.9 illustrates how the environment of a given bond affects its energy. Leroy et al. [197] have also demonstrated that the energies of homo-atomic single bonds, calculated from the heats of formation of organo-element compounds (C–C 357, P–P 211 and S–S 265 kJ/mol), are close to the energies of the corresponding bonds in elemental substances (adjusted for the number of bonds in the structures), viz. diamond (358 kJ/mol), P_4 (201 kJ/mol) and S_8 (264 kJ/mol). The energies of single covalent bonds, determined by this method from the parameters of elemental substances, are included in Table 2.12.

Table 2.11 Average bond energies (kJ/mol) in MX_n-type molecules. (Bond energies of halides and hydrides of Groups 13–15 elements are calculated using the data from [170, 171] and are given without references)

M	F	Cl	Br	I	M	F	Cl	Br	I
Halide molecules MX_2					Halide molecules MX_3				
Cu	383[a]	302	258		Sb	437[b]	313[b]	264[b]	192[b]
Be	629[b]	463[b]	388[b]	299[b]	Bi	380[b]	279[b]	215[b]	170[b]
Mg	518[b]	391[b]	339[b]	262[b]	Cr	476	336	299	
Ca	558[b]	448[b]	395[b]	321[b]	Mo	494			
Sr	542[b]	438[b]	396[b]	321[b]	W	569	420	359	285
Ba	570[b]	460[b]	412[b]	353[b]	S	339[i]			
Zn	393	320	270	205	Mn	435	319		
Cd	328	274	238	192	Fe	462	345	292	222
Hg	257	227	185	145	Th	669[k]	514[k]	427[k]	
B	657	426	349	250	U	611[k]	477[k]	414[k]	
Al	563	387	321	250	Halide molecules MX_4				
Ga	431	308	258	196	C	487	321	258	199
In	398	279	232	174	Si	595[e]	399[e]	331[e]	246
Tl	360	253	203	146	Ge	471[q]	340[q]	273[q]	209[q]
Ti	690	456	436	344[c]	Sn	409	323[h]	261	210
Zr	636[d]	494[d]	423[d]	340[d]	Pb	327	249[h]	199	164
Hf	644	588	567	470	Ti	585[r]	430[r]	360[r]	295[r]
C	530	367	311	255	Zr	647[d]	488[d]	423[d]	346[d]
Si	600[e]	426[e]	365[e]	293	Hf	658[d]	496	447	355[e]
Ge	551[f]	392[f]	341[f]	269[f]	V		382[h]		
Sn	468	386[g]	323	254	Nb	574[s]	426[s]	372[s]	293[s]
Pb	388	304[g]	262	209	Cr	448	333	269	
V		453[h]	445	375	W	552	405	343	270
N	305	223	150	85.8	S	339[i]	204		
P	478	308	249	181	Th	672[k]	511[k]	448[k]	
As	431	288	235	172	U	609[k]	463[k]	398[k]	
Sb	392	264	213	148	Halide molecules MX_5				
Bi	358	231	180	116	P	461[i]	260		
Cr	507	387	340	249	As	387	253		
Mo	492			375	Nb	566[s]	406[s]	344[s]	
W	603	464	403	335	Ta	600	430	365	258
O	192	207			W	530	374	322	247
S	357[i]	271			S	316[i]			
Se	353	256	240		U	571[k]	412[k]	351[k]	

Table 2.11 (continued)

Te	377	284		
Mn	483	397[j]	342	278
Fe	463	398[j]	343	272
Co	477	382[j]	325	268
Ni	457	369[j]	316	252
Ru	433			
Pt	418			
Th	677[k]	517[k]	402[k]	
U	606[k]	460[k]	406[k]	
Halide molecules MX_3				
B	642	442[l]	366	285
Al	588[m]	422[m]	346[m]	284[m]
Ga	475[m]	355[m]	322[m]	245[m]
In	443[m]	322[m]	285[m]	225[m]
Sc	629[n]	470[n]	382[n]	337[n]
Y	643[o]	490[p]	432[p]	353[p]
La	641[o]	513[p]	456[p]	378[p]
C	477	335	273	205
Si	562	374	303	227
Ge	457	351	285	215
Sn	391	293	238	170
Pb	349	258	210	142
Ti	608	445	380	311
V	555	413[h]	369	
N	280	202	179	169
P	504[i]	329	259	177
As	438[b]	307[b]	252[b]	194[b]

Halide molecules MX_6				
S	329	182	117	42
Se	322	182	128	119
Te	343	204	145	87
W	531	364	301	231
AgF_6	AuF_6	MoF_6	TcF_6	
120[v]	176[w]	447[v]	387[v]	
ReF_6	RuF_6	RhF_6	PdF_6	
430[w]	325[v]	174[v]	118[v]	
OsF_6	IrF_6	PtF_6	UF_6	
382[w]	331[w]	259[w]	524	
Hydride and oxide molecules				
H_2O	H_2S	H_2Se	H_2Te	
459[t]	362[t]	320[t]	266[u]	
NH_3	PH_3	AsH_3	SbH_3	BiH_3
391	322	297	258	215
BH_3	AlH_3	GaH_3	InH_3	TlH_3
376	287	255	225	184
CH_3	SiH_3	GeH_3	SnH_3	PbH_3
408	301	268	235	192
CH_4	SiH_4	GeH_4	SnH_4	PbH_4
416	322	288	253	209
CO_2	SO_2	SeO_2	TeO_2	CrO_2
804	536	422	385	494
MoO_2	WO_2	SO_3	SeO_3	TeO_3
582	636	473	364	348
CrO_3	MoO_3	WO_3	RuO_3	OsO_4
479	584	630	492	530

a[172], b[173], c[174], d[175], e[176, 177], f[178], g[179], h[180], i[181], j[182], k[183], l[184], m[185], n[186], o[187], p[188], q[189], r[190], s[191], t[192], u[193], v[194], w[195]

Table 2.12 Energies of single covalent homo-atomic bonds M–M (kJ/mol)

M	*E*	M	*E*	M	*E*	M	*E*	M	E^c
Li	105[a]	Zn	64[c]	Ge	187[d]	O	144[d]	Tc	263
Na	75[a]	Cd	55[c]	Sn	151[d]	S	264[d]	Re	293
K	53[a]	Hg	33[c]	Pb	73[c]	Se	216[d]	Fe	204
Rb	49[a]	B	286[b]	Ti	175[c]	Te	212[d]	Co	210
Cs	44[a]	Al	168[c]	Zr	225[c]	Cr	185[c]	Ni	210
Cu	201[a]	Ga	135[c]	Hf	232[c]	Mo	263[c]	Ru	319
Ag	163[a]	In	103[c]	N	212[b]	W	341[c]	Rh	273
Au	226[a]	Tl	64[c]	P	211[b]	H	436[a]	Pd	182
Be	119[b]	Sc	161[c]	As	176[d]	F	155[a]	Os	387
Mg	102[c]	Y	181[c]	Sb	142[d]	Cl	240[a]	Ir	326
Ca	87[c]	La	184[c]	Bi	98[d]	Br	190[a]	Pt	277
Sr	80[c]	C	358[d]	V	232[c]	I	149[a]	Th	224
Ba	94[c]	Si	225[d]	Nb	325[c]	Mn	121	U	213
				Ta	354[c]				

[a]see Table S2.3, [b]see Table S2.9, [c][198], [d][199]

Multiple bonds in structural chemistry are traditionally described as combinations of σ- and π-bonds, the proof being the two-step ionization of the C=C bond. The π-bond energy is conventionally calculated by simply subtracting the σ-bond energy from the experimental energy of a multiple bond. The π-bond energies obtained in this way are listed in Table S2.10. As one can see, dissociation energies of multiple bonds in the same molecules, reported by different authors, often show discrepancies exceeding the experimental error by an order of magnitude, because different techniques (of both measurement and calculation) involve the products of dissociation in different valence states. Large discrepancies between the results of different authors are caused also by inherent difficulty of determining a small value as the difference of two large ones. In any case, the simple additive scheme is not applicable here, because the standard energy of the C–C σ-bond (357 kJ/mol) refers to its equilibrium length of 1.54 Å, whilst the carbon-carbon distance in the double bond is only 1.34 Å. A rough estimate of how a contraction by 0.2 Å will affect the energy of the C–C bond can be obtained from the experimental compressibility of diamond. The Universal equation of state (see Sect. 10.6),

$$P(x) = 3B_o[(1 - x)/x^2]\exp[\eta(1 - x)] \tag{2.43}$$

where P is the pressure, B_o is the bulk modulus, $x = (V/V_o)^{1/3}$ (V and V_o are the initial and the final volumes of the body), $\eta = 1.5(B_o'-1)$ and B_o' is the pressure derivative of B_o, permits to calculate the pressure P required to shorten the bond distance in diamond from 1.54 to 1.34 Å. Taking into account that diamond has $B_o = 456$ GPa and $B_o' = 3.8$, we obtain $P = 405.6$ GPa (!). The work of compression is

$$W_c \approx \frac{1}{2} P \Delta V \tag{2.44}$$

Because for diamond, $V_o = 3.417$ cm^3/mol, for $x = 0.87$ we obtain $\Delta V = 1.166$ cm^3 and $W_c = 236.6$ kJ/mol. A ratio of the atomization energy of diamond (717 kJ/mol) to its elastic energy ($B_o V_o = 1558$ kJ/mol) allows to calculate the potential part of W_c as $0.46 \times 236.6 = 108.8$ kJ/mol. Since in the diamond structure $N_c = 4$ and every C–C bond involves two atoms, shortening of a C–C bond requires 54 kJ/mol. Hence, the actual energy of the σ-bond must be reduced from 357 to 303 kJ/mol, but the π-component correspondingly increased from the conventional 262 (mean from Table S2.10) to 316 kJ/mol. Thus the σ- and π-bonds in ethylene have in fact simlar energies, rather than $E(\pi) << E(\sigma)$ of the conventional description. Thus the conventional breakdown of the bond energy into σ- and π-components is essentially formal. Nevertheless it is useful, particularly in highlighting *relative* trends. Thus, it is evident that π-bond energies decrease in the succession O > N > C > S > P > Si, both absolutely and relatively to the σ-components, because the overlap of the valence orbitals decreases in the multiple bonds. In the cases of O and N, the π-bond energy is higher than that of the σ-bond, because the formation of a π-bond reduces the repulsion between σ-bonded and lone electron pairs, which is especially strong in electron-rich O and N atoms (see above).

Theoretical calculation of the bond energy is the task of quantum chemistry. So far, satisfactory quantitative solutions have been achieved only for lighter elements. Nevertheless, in principle the problem has been treated more than half a century ago by Mulliken [200] who generalized the results of the molecular-orbital and valence-bonds methods and derived his 'magic formula',

$$E_b = \sum X_{ij} - \frac{1}{2}\sum Y_{kl} + \frac{1}{2}\sum K_{mn} - PE + E_i \tag{2.45}$$

where ΣX_{ij} is the exchange interaction of bonding electrons, ΣY_{kl} is the repulsion of non-bonding electron pairs, ΣK_{mn} is the exchange interaction of lone electron pairs, *PE* is the promotion energy and E_i is the ionic interaction. Since the exchange integrals are proportional to the products of wave functions (the overlap integrals) and the exchange energy of the bonding electrons is proportional to their ionization potentials, the first term of Eq. 2.45 can be transformed to

$$\Sigma X_{ij} = k\bar{I}_{ij}\frac{S_{ij}}{1 + S_{ij}} \tag{2.46}$$

where k is an empirical coefficient (usually of the order of 1), $\bar{I}_{ij}$ is the geometrical mean of the ionization potentials of the atoms i and j, and S_{ij} is the overlap integral. Taking into account that the overlap integral defines the fraction of the outer electron cloud which belongs to both atoms, it follows from Eq. 2.46 that the covalent bond energy is always less than half the ionization potential of the bonded atom (because $S \leq 1$). The actual E_b/I ratio for A_2 molecules with single bonds is 0.2 ± 0.1, for AB molecules it varies from 0.3 to 0.6, the higher energy shown by polar bonds.

Returning to the problem of positively charged alkali dimers (see above) we can make use of Eq. 2.46. Indeed, a transition from the two-electron bond in M_2-molecules to the $M_2{}^+$-cations yields one-half of S, but $\bar{I}$ for M^+–M increases much more. E.g., $I_1(\mathrm{Li}) = 5.39$ but $\bar{I} = 20.20$ eV, for Na respectively 5.14 and 15.59 eV, etc., whereas for Cu, $I_1 = 7.72$ but $\bar{I} = 12.52$ eV and the bond energy in Cu_2 is 201 kJ/mol compared to only 155 kJ/mol in Cu_2^+. Thus, the relation between the first and second ionization potentials governs the change of bond energies at the positive ionization in molecules M_2.

Mulliken [201–205] calculated the overlap integrals as functions of two parameters:

$$p = \frac{d}{a_o}\frac{\mu_A + \mu_B}{2} \text{ and } t = \frac{\mu_A - \mu_B}{\mu_A + \mu_B} \tag{2.47}$$

where $\mu = Z^*/n^*$ and a_o is the Bohr radius. However, calculations have been carried out only for the integer values of n^* and $p \leq 8$. Later, the integrals were calculated for $n^* = 3.7$ and 4.2 up to $p = 8$ [206], or to $p = 20$ [207], which is enough for all the real bond lengths in molecules and crystals.

Other additive descriptions of the energy of a hetero-atomic chemical bond have been suggested, besides Eq. 2.45. The earliest, as mentioned above, was Pauling's equation [18]

$$E\,(\text{M–X}) = E_{\text{cov}} + E_{\text{ion}} \tag{2.48}$$

where $E_{\text{cov}} = ½\,[E(\text{M–M}) + E(\text{X–X})]$ and E_{ion} is the extra ionic energy, equaled to Q, which is the heat of reaction, $½\,M_2 + ½X_2 = MX$. Later Pauling suggested, as an alternative, to calculate E_{cov} as $[E(\text{M–M}) \cdot E(\text{X–X})]^{½}$. Fereira [208] described the energy of a M–X bond as the sum of three components,

$$E(\text{M–X}) = E_{\text{cov}} + E_{\text{ion}} + E_{\text{tr}} \tag{2.49}$$

representing the covalent, ionic and electron-transfer contributions, respectively; components of this equation have been modified in numerous works [209–217]. However, for empirical estimations one can use Eq. 2.48.

2.3.2 *Bond Energies in Crystals*

Bond energy (E_{cr}) in a crystalline compound MX_n can be calculated from the average bond energy in the corresponding molecule (E_{mol}) and the sublimation energy (ΔH_s, see Tables 9.4–9.6),

$$E_{\text{cr}} = E_{\text{mol}} + \Delta H_s/n \tag{2.50}$$

where n is the number of atoms with the lower valence in the formula unit (i.e. $n = 2$ for SiO_2 or Li_2O). Alternatively, the bond dissociation energy in MX_n crystals can be calculated directly from heats of formation $\Delta H_f(MX_n)$,

$$E_{\text{cr}}(\text{M–X}) = [\Delta H_f(\text{M}) + n\Delta H_f(\text{X}) - \Delta H_f(\text{MX}_n)]/n \tag{2.51}$$

Average bond energies for crystalline compounds are listed in Table 2.13.

Table 2.13 Mean bond energies (kJ/mol) in crystalline compounds M_nX_m (v = valence of M). (Bond energies for ZnSe, ZnTe, HgTe, SnSe, SnTe, FeS, CoS are the averaged values from handbooks and original articles)

v_M	M	X							
		F	Cl	Br	I	O	S	Se	Te
I	Li	853	682	614	529	580	519	484	430
	Na	761	642	581	518	438	425	393	365
	K	731	656	595	527	393	418	391	365
	Rb	717	639	584	515	374	398	402	
	Cs	711	648	591	534	374	396	412	
	Cu		596	557	512	546	514	484	460
	Ag	571	545	477	455	423	438	421	408
	Au		526	494	501				
	Tl	582	508	464	425	390	364	343	322
II	Cu	514	412	329		741	668	603	578
	Be	743	530	453	364	1172	824	725	661
	Mg	710	512	452	364	998	739	662	571
	Ca	776	604	544	460	1062	925	776	662
	Sr	764	612	554	467	1002	910	776	660
	Ba	768	639	584	514	984	918	800	707
	Zn	526	395	342	276	727	602	532[a]	456[a]
	Cd	480	372	326	264	618	538	484	402
	Hg	325	267	227	190	400	391	344	295[a]
	Ge	607			334		715[b]	642	589[c]
	Sn	552	453	392	328	833	682	619[d]	568[e]
	Pb	508	397	350	292	662	570	523[d]	465[e]
	V		576	551	492	1195			
	As					481[f]	389[f]	352[f]	312[f]
	Cr	687	518	460	384		824[g]		
	Mn	642	506	447	382	917	773	680	585
	Fe	621	502	447	358	932	792[h]	707	650
	Co	633	495	433	364	937	805[h]	721	
	Ni	616	488	429	359	915	797	728	660
	Pd	501	409	358	282	741	724		626
	Pt	550	462	429	415		921		
III	Sc	754	573	474	415	1137	919	789	704
	Y	790	597	530	445	1167	951		
	La	788	622	556	472	1134	972	822	765
	U	766	537	500	433		918		
	B	650	455	386	306	1040	707		
	Al	688	462	380	317	1027	738	632	541
	Ga	560	382	340	276	794	631	542	489
	In	555	376	333	265	718	554	491	438
	Tl		287			501			
	Ti	690	505	439	328	1068	617		
	V	655	484	432	325	998			
	As	453	325	274	226	669	534	471[f]	425
	Sb	471	336	290	223	666	501	454	407
	Bi	447	318	252	209	580	464	422	377
	Cr	560	411	378	307	891	651	558[i]	
	W	672[j]	525[j]	465[j]	396[j]				

Table 2.13 (continued)

ν_M	M	X							
		F	Cl	Br	I	O	S	Se	Te
	Mn	530				758[k]			
	Re		469	429					
	Fe	547	392	339		810			
	Co	485	319						
	Ir		426	394					
	Pt		366	344					
IV	Ti	610	444	377	320	933	715	650	556
	Zr	706	516	452	378	1096	866	737	663[l]
	Hf	720	522	473	384	1131			
	Si	601	410	341	263	930	610	522	
	Ge	479	351	288	231	725	526[b]	454[d]	348
	Sn	445	336	275	230	689	505	438[d]	
	Pb	364	255			484			
	Th	757	573	497	423	1161[m]	890		
	U	689	516	454	369	1057[m]	808		
	V	559	392			863			
	Te		255	212		517			
	Mo	531	407	355		873	742[g]	659[g]	586[g]
	W	595[j]	454[j]	393[j]	323[j]	971	826[g]	743[n]	709
	Ru					727	705	664	
	Os					791	740	640	
	Pt		320	293	268		615		
MX	Sc	Y	La	B	Al	Ga	In	Th	U
N	1290	1197	1205	1299[o]	1113	901[p]	852[q]	1426[r]	1286[r]
P	1062[q]	1051	1050	984	827	697	630	1270[r]	1107[r]
As	945	1043	1039		745	667	606[s]		1046
Sb	774[t]	911[t]	913[t]		652[u]	587	542		930
C	1335[r]							1435[r]	1344

[a][218, 219, 220, 221], [b][222], [c][223], [d][224], [e][225], [f][226], [g][227], [h][228], [i][229], [j][230], [k][231], [l][232], [m][233], [n][234], [o][235], [p][236], [q][237], [r][238], [s][239], [t][240], [u][241]

2.3.3 Crystal Lattice Energies

In Sect. 2.3.1 and 2.3.2 we have considered the dissociation of bonds, molecules and solids into electroneutral atoms. Alternatively, one can imagine them dissociating into oppositely charged ions. Although such process is always less favorable thermodynamically in vacuum (see above), it can occur in a polar solvent. In any case, ionic description of the crystal proved a very fruitful model in structural chemistry, and historically the earliest one. The energy required to convert a solid ionic material into its independent gaseous ions, is known as the crystal lattice energy (U). It can be experimentally determined from the Haber-Born thermodynamic cycle,

$$U_{298} = -\sum \Delta H_{298}\,(\mathrm{M}) + \sum \Delta H_{298}\,(\mathrm{X}) - \Delta H_{298}\,(\mathrm{M_nX_m}) + \sum I\,(\mathrm{M}) - \sum A\,(\mathrm{X}) \tag{2.52}$$

where $\Delta H_{298}(M_nX_m)$ is the heat of formation of a crystalline M_nX_m compound from the elements under standard conditions, $\Sigma\Delta H_{298}(M)$ and $\Sigma\Delta H_{298}(X)$ are the sums of the heats of formation of n isolated atoms of the metal M and m atoms of the nonmetal X from the elements under standard conditions, I is the ionization potential and A is the electron affinity.

All the parameters featuring in Eq. 2.52 can, in principle, be experimentally determined, provided that the ions in question can exist as individual particles. However, no monoatomic anion with a charge exceeding -1 can exist, as they have $A < 0$. Hence, experimental determination of U is possible, in practice, only for metal halides. For compounds with polyvalent anions, like oxides, chalcogenides, nitrides, *etc.*, U can be only calculated theoretically. The history and bibliography on this topic can be found in reviews [242, 243]. When it was realized that inorganic compounds do not have purely ionic bonds, the interest in the concept and values of the crystal lattice energies declined, therefore only a brief outline of this field is given here.

The lattice energy can be expressed as the difference of two terms,

$$U = U_a - U_r \tag{2.53}$$

U_a representing the Coulomb attraction between oppositely charged ions and U_r the repulsion between similarly charged ions. The attractive term is determined easily,

$$U_a = K_\mathrm{M}\frac{z^2}{d} \tag{2.54}$$

where K_M is the Madelung constant, which depends on the structure type, stoichiometry of the compound and charges of the ions, z is the ionic charge, d is the interionic distance. There are several ways of expressing U_r as a function of d, of which the best are the approaches of Born and Landé [244–246] and Born and Mayer [247], who expressed the lattice energies in the form of Eqs 2.55 and 2.56, respectively.

$$U_\mathrm{BL} = -K_\mathrm{M}\frac{z^2}{d} + \frac{c}{d^n} \tag{2.55}$$

$$U_\mathrm{BM} = -K_\mathrm{M}\frac{z^2}{d} + \frac{C}{e^{d/\rho}}. \tag{2.56}$$

At the equilibrium interatomic distance $\partial^2 U/\partial d^2 = 0$, i.e. the attractive and the repulsive forces are equal. From here we obtain the well known Born-Landé and Born-Mayer equations,

$$U_\mathrm{BL} = -K_\mathrm{M}\frac{z^2}{d_\mathrm{o}}/\left(1 - \frac{1}{n}\right) \tag{2.57}$$

$$U_\mathrm{BM} = -K_\mathrm{M}\frac{z^2}{d_\mathrm{o}}/\left(1 - \frac{\rho}{d_\mathrm{o}}\right). \tag{2.58}$$

Table 2.14 Madelung constants K_M

MX	K_M	MX_n	K_M	MX_n	K_M	M_nO_m	K_M
HgI	1.277	$HgCl_2$	3.958	$AuCl_3$	7.471	Cu_2O	4.442
HgBr	1.290	$BeCl_2$	4.086	$SbBr_3$	7.644	VO_2	17.57
HgCl	1.311	$PdCl_2$	4.109	BiI_3	7.669	SiO_2	
TlF	1.318	$ZnCl_2$	4.268	$MoCl_3$	7.673	β-quartz	17.61
HgF	1.340	$TiCl_2$	4.347	AuF_3	7.954	α-quartz	17.68
CuCl	1.638	$CdCl_2$	4.489	SbF_3	7.985	tridymite	18.07
NaCl	1.748	$CrCl_2$	4.500	AsI_3	8.002	TiO_2	
CsCl	1.763	CrF_2	4.540	$FeCl_3$	8.299	brookite	18.29
AuI	1.988	CuF_2	4.560	$AlCl_3$	8.303	anatase	19.20
ZnO	5.994	FeF_2	4.624	YCl_3	8.312	rutile	19.26
PbO	6.028	$SrBr_2$	4.624	VF_3	8.728	SnO_2	19.22
BeO	6.368	CdI_2	4.710	FeF_3	8.926	PbO_2	19.26
ZnS	6.552	$CaCl_2$	4.731	YF_3	9.276	ZrO_2	20.16
CuO	6.591	$PbCl_2$	4.754	LaF_3	9.335	MoO_2	18.27
MgO	6.990	NiF_2	4.756	BiF_3	9.824	Al_2O_3	25.03
		MgF_2	4.762	SnI_4	12.36	V_2O_5	44.32
		MnF_2	4.766	UCl_4	13.01		
		CoF_2	4.788	$ThBr_4$	13.03		
		α-PbF_2	4.807	$ThCl_4$	13.09		
		CaF_2	5.039	PbF_4	13.24		
		$AlBr_3$	7.196	SnF_4	13.52		
		BCl_3	7.357	SiF_4	14.32		
		BI_3	7.391	ZrF_4	14.36		

Many other expressions for the crystal lattice energy have been proposed, none of which has any real advantage over the abovementioned methods, which therefore remain in general use.

Born's repulsion coefficient n depends on the type of the electron shell (see Table 2.1). For an MX compound, n is calculated as ½ $[n(M^+) + n(X^-)]$. The ρ coefficient is less variable, averaging 0.35(5). For this reason, Eq. 2.58 is more frequently used for calculations. Assuming $n = 9$ and the interatomic distance $d = 3$ Å, the repulsion energy can be estimated as *ca.* 10 % of the crystal lattice energy.

Equation 2.48 can be improved by adding the third term, which accounts for the van der Waals forces,

$$E^W = \left(\frac{T_{cat} + T_{an}}{(r_{cat} + r_{an})^6} \right) \tag{2.59}$$

where r_{cat} and r_{an} are the radii of the cation and the anion, T_{cat} and T_{an} characterizes the van der Waals attractions of the cations and anions, respectively [248]. To recognize the importance of this contribution, compare NaCl and AgCl. They have similar structures and bond distances, but the van der Waals energy of the latter is 6 times greater due to higher polarizability [249].

Madelung constants are important in many other areas of physical chemistry. Their values for the shortest distances in the most common structural types are listed in Table 2.14 [250–252]. Note that these constants vary widely, from 1.28 to 44.3.

Their rigorous theoretical computation is a demanding task: to obtain an accurate result, the contributions of tens of thousands of ions must be taken into account [253–257] (Madelung constants for organic salts are presented in [258]). This stimulated the quest for more economic methods of calculating K_M; the most successful one has been suggested by Kapustinskii [259, 260], who related K_M to the number (m) and valence (Z) of ions which comprise the formula unit of the crystal:

$$k_M = \frac{2K_M}{m z_M z_X} \qquad (2.60)$$

where k_M is the new (reduced) Madelung constant, the values of which are listed in Table S2.11. As one can see, k_M has the average value of 1.55 and varies by ±10 %, i.e. much less than K_M. The deviation of k_M from 1 is due to the crystal field, which in energy terms is characterized by the sublimation heat of the solid. Thus, for the halides of univalent metals, the ratio of the bond energies in the crystal and molecular states is equal to this number (1.55). Kapustinskii suggested that crystal lattice energies can be approximately estimated using for all structures the same $k_M = 1.745$ and the bond distances equal to the sums of ionic radii r_M and r_X (calculated for $N_c = 6$). Merging all constant factors into one, he obtained the expression

$$U = 256\frac{m z_M z_X}{r_M + r_X} \qquad (2.61)$$

for U in kcal/mol and r in Å [259] which later was modified [260], in accordance with the Born-Mayer equation, to the formula

$$U = 287\frac{m z_M z_X}{r_M + r_X}\left(1 - \frac{0.345}{r_m + r_X}\right). \qquad (2.62)$$

This method applies to complexes, as well as to binary compounds. By minimizing the discrepancy between the calculation and the experiment, one can determine the so-called 'thermochemical radii' of complex ions. These radii correspond to the (imaginary) spherical ions, iso-energetically replacing the real complex ions in the crystal structure. This problem has been discussed in great detail by Yatsimirskii [261] and later studied by Jenkins et al. [262–265]. Kapustinskii's equation has been successfully applied to the fluorides of mono- and divalent metals and to solid solutions of the LnF_3–MF_2 type [266]. A comparison of the experimental crystal lattice energies of CH_3COOM (M = alkali metals), XCH_2COOM (M = Li, Na; X = Cl, Br, I) and $ClCH(CH_3)COOM$ (M = Li, Na) with those calculated by Kapustinskii's equation [267] showed small differences in the lattice energies of all these compounds, i.e. M–O bonds define the crystal lattice energies of these compounds and all differences are due only to their bond distances.

The lattice energies for a variety of mineral and syntetic complex compounds that can be classified as double salts, were calculated by summing the lattice energies of the constituent simple salts [268]. A comparison with the lattice energies obtained from the Born-Haber or other thermodynamic cycles using the Madelung constant or

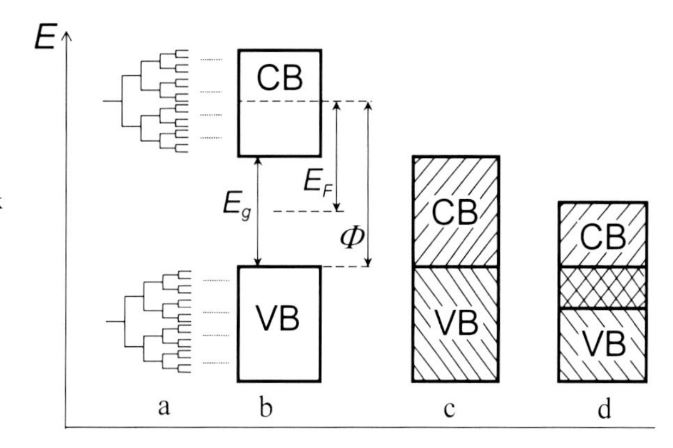

Fig. 2.5 Energy bands in crystals: **a** formation from atomic energy levels; **b** in dielectric, CB—conduction band, VB—valence band, E_g—band gap width, E_F—Fermi energy, Φ—work function; **c** in semiconductor with $E_g \geq 0$ (semi-metal); **d** in metal ($E_g \leq 0$)

more approximately through the Kapustinskii equation shows that this approximation reproduces these values generally to within 1.2 %, even for compounds that have considerable covalent character. Application of this method to the calculation of the lattice energies of silicates, using the sum of the lattice energies of the constituent oxides are, on average, within 0.2 % of the value calculated from the experimental enthalpies of formation.

Glasser and Jenkins [269] have formulated the general (but very simple!) procedures to make thermodynamic prediction for condensed phases, both ionic and organic/covalent, principally via formula unit volumes (or density). Their volume-based approach gives a new thermodynamic tool for such assessments, as it does not require detailed knowledge of crystal structures and is applicable to liquids and amorphous materials, as well as to crystalline solids. The next step was made in the work of Glasser and von Szentpály [270] who used the fundamental principle of electronegativity equalization to calculate the lattice energies for diatomic MY crystals, taking into account ionic and covalent contributions to the chemical bond. This method was applied to Groups 1 and 11 monohalides and hydrides, as well as to alkali metals. A limitation of the model occurs for the coinage metals, Cu, Ag, and Au, where *d* orbitals are strongly involved in the metallic bonding, while the homonuclear molecular bond is dominated by *s* orbitals.

2.3.4 Band Gaps in Solids

The structures of inorganic crystals usually comprise infinite chains, two- and three-dimensional networks of atoms linked by strong ionic or covalent bonds, and the energetic properties of atoms are influenced by all structural units of the crystal. Therefore the narrow electron energy level, characteristic of an isolated atom, is split into as many components as there are bonds in a crystal; each resulting level is also widened due to perturbations from adjacent atoms, the result being a broad band of continuous values of energy in the crystal (Fig. 2.5).

Table 2.15 Band gaps (eV) in MX—type compounds

M	X			
	F	Cl	Br	I
Li	12.5	9.4	7.9	6.1
Na	11.0	8.9	7.4	5.9
K	10.8	8.7	7.4	6.2
Rb	10.3	8.3	7.4	6.1
Cs	9.9	8.2	7.3	6.2
Cu		3.2	2.9	2.95
Ag	2.8	3.6	3.05	2.8
Tl		3.4	3.0	2.8

M	X			
	N	P	As	Sb
Sc	2.26[m]	1.1	0.7	
Y	1.5[n]	1.0[o]		
La		1.45[o]		0.8[n]
B	6.1[p]	2.1[q]	1.4[q]	
Al	6.23[r]	3.63[r]	3.10[r]	2.39[r]
Ga	3.51[r]	2.89[r]	1.52[r]	0.81[r]
In	1.99[r]	1.42[r]	0.42[r]	0.23[r]

M	X			
	O	S	Se	Te
Cu	1.95[a]			
Be	10.6[b]	5.5	4.2	2.8
Mg	7.3[a]	6.0	5.7[c]	4.2
Ca	6.9[d]	5.3[d]	5.0[c]	4.1
Sr	5.8[d]	4.8[d]	4.7[c]	3.7
Ba	4.0[d]	3.9[d]	3.6[d]	3.4
Zn	3.4[a]	3.7[e]	2.7[e]	2.2[e]
Cd	2.3[a]	2.4[e]	1.7[e]	1.5[e]
Hg	2.8[f]	2.0[g]	0.4[h]	0.1
Sn	4.2[i]	1.3[G]	0.9[j]	0.3
Pb	2.8[a]	0.4[g]	0.3[k]	0.2
Mn	3.8[a]	2.8	2.5	1.3
Fe	2.4[l]			
Co	2.7[a]	0.94[s]		
Ni	3.8[a]	0.5	0.3	0.2
Pd	2.4[l]			
Pt	1.3[l]			

[a][271], [b][272], [c][273], [d][274], [e][275], [f][276], [g][277], [G][278], [h][279], [i][280, 281], [j][282], [k][283], [l][284], [m][285], [n][286], [o][287], [p]for c-BN [288, 289], for h-BN $E_g \approx 5.5$ eV [290, 291], [q][292], [r]for w-phases [293], [s][294], [t][295]

Notwithstanding this qualitative difference between the energy spectra of an atom and a crystal, there are also some broad similarities. Just as an atom has certain permitted orbitals and the areas where the presence of electrons is forbidden, so a crystal has bands of permitted states: valence band and conduction band, separated by a band gap (forbidden zones), where no energy states are allowed. In an atom, the outer-shell electrons are chiefly responsible for chemical bonding—in a crystal the same role is played by the valence band. On ionization of an atom, an electron is removed from the valence shell (ideally—to infinity) in a crystal the equivalent process consists in the transfer of an electron from the valence band into the conduction band.

From the viewpoint of the conventional band theory, the band gap is absent in metals and has positive width E_g in dielectrics. The latter can be divided into dielectrics proper, with $E_g > 4$ eV, and semiconductors, with $0 < E_g < 4$ eV. Since E_g defines the energy required to transform a dielectric into a conducting (metallic) state, this parameter is widely used for various physical and chemical purposes and correlations. Tables 2.15, S2.12 and S2.13 comprise the most reliable experimental measurements of E_g.

Theoretical calculations of the band structure of crystals belong to solid state physics and are not discussed here. Quantitative *ab initio* prediction of a band gap is a problem of great complexity. However, empirical and semi-empirical estimates of E_g, using the concepts of structural chemistry, are sufficient for most purposes of physical chemistry and materials science. Indeed, since the valence band of a compound usually involves primary orbitals of the anions (nonmetal atoms), and the conduction band involves primary orbitals of the cations (metal atoms), the energy of the transition between the two (i.e., E_g) must be related to some atomic properties.

The structural-chemical approach has been pioneered by Welkner [296], who observed that E_g depends on the chemical bond energy and the effective atomic charges. The former relation is described by a linear equation [297–300],

$$E_g(\mathrm{MX}) = a\,[E(\mathrm{M-X}) - b]\,. \tag{2.63}$$

Among structurally similar compounds, E_g increases together with the difference of electronegativities ENs (see next Section) of the bonded atoms ($\Delta\chi$). The form of the correlation is not certain. Thus, for binary compounds Duffy [301, 302] suggested a linear dependence (on the optical EN),

$$E_g = a\Delta\chi \tag{2.64}$$

while Di Quarto et al. [280, 281] have recommended

$$E_g = a\Delta\chi^2 + b \tag{2.65}$$

where the constants a and b are different for the main-group (s, p) and transition (d) elements. On the other hand, the band gap decreases with the increase of the mean principal quantum number of the components, $\bar{n}$, as the interaction of the valence electrons with the nucleus becomes weaker. The dependence of E_g from both $\Delta\chi$ and $\bar{n}$ has been mapped by Mooser and Pearson [303] and later expressed in the analytical form by Makino [304],

$$E_g = a\sqrt{\frac{\Delta\chi}{\bar{n}}} - b \tag{2.66}$$

which gave satisfactory agreement with the experimental data for binary crystalline compounds. Finally, Villars [305, 306] presented a 3D map of E_g, with $\Delta\chi$ and the electron density of atoms as the coordinates. Historical reviews of this approach see in [112, 307].

The resort to graphical representation of the empirical correlations shows the difficulties of the analytical description, due to the multiplicity of factors influencing the electronic structure of crystals. The task can be simplified, making use of the additive character of E_g. Hooge [308, 309] expressed the band gaps of binary compounds as the sums of atomic increments, the increment of each element being constant and depending only on its EN,

$$E_g(\mathrm{MX}) = E_g(\mathrm{M}) + E_g(\mathrm{X}). \tag{2.67}$$

These increments are computational parameters only, but it is also possible to express the band gap of a compound through the sum of the *observed* band gaps of the component elements corrected by two additional terms, accounting for the ionicity and metallicity of the bonds, respectively. The alternative equation has been suggested [310, 311],

$$E_g(\mathrm{MX}) = E_g(\mathrm{M}) + E_g(\mathrm{X}) + a\Delta\chi_{\mathrm{MX}} - b\bar{n} \tag{2.68}$$

where a and b are constants. Band gaps of elements are, of course, different in various allotropic modifications. Therefore in Eq. 2.68 one should use the E_g values of those modifications which are structurally most similar to the compound concerned, e.g. white phosphorus for phosphides, diamond for carbides, *etc*. The development of the additive approach naturally encouraged the measurements of band gaps in elemental solids, which have been carried out for boron, iodine and the elements of the Group 4, 5 and 6. All of them fit the equation

$$E_g = k\frac{I}{n} - c \tag{2.69}$$

where I is the potential ionization, n is the principal quantum number and k and l are structure-related constants. For metals $k = 0.8$, for materials with a continuous covalent network $k = 1.2$, for molecular crystals $k = 1.6$, whilst $c = 1.7$ in all cases.

It is now evident that the conventional view of all metals having the constant $E_g = 0$ is inconsistent both with the above mentioned relations and with the variability of E_g in dielectrics and semiconductors. The difficulties can be resolved on the simple assumption that metals have band gaps of variable *negative* width, equal to the overlap between the valence and the conductivity bands (Fig. 2.5). In fact, a negative E_g has been found experimentally in InN_xSb_{1-x} [313]. Equation 2.69 also gives $E_g < 0$ for metals. Furthermore, in this interpretation the sign of E_g correlates with the thermal dependence of electric conductivity, which in semiconductors increases on heating ($E_g > 0$) and in metals decreases ($E_g < 0$). Table S2.14 lists all the currently available experimental band gap widths for elements, together with the values calculated according to Eq. 2.69.

In another variety of the additive approach, the band gap of a compound is represented by the sum of the covalent and the ionic terms, the former determined by the geometrical properties of the component elements and the latter by $\Delta\chi$ [314]. Phillips [315], by way of quantum-mechanical reasoning, has arrived to a similar additive representation of the band gap,

$$E_g^2 = E_h^2 + C^2 \tag{2.70}$$

where the covalent component E_h depends on the atomic radius and the Coulomb contribution C on $\Delta\chi$. It is noteworthy that Welkner's, Duffy's and the various additive approaches are intrinsically related, because (according to Pauling) the energy of a chemical bond comprises an ionic and a covalent contribution, the latter depending on $\Delta\chi$.

Table 2.16 Band gaps in bulk and nano phases

Substance	E_g, eV		D, nm	Substance	E_g, eV		D, nm
	bulk	nano			bulk	nano	
graphite[a]	0	0.65	0.4	Si[h]	1.1	3.5	1.3
CdS[b]	2.5	3.85	0.7	Ga_2O_3[i]	4.9	5.9	14
CdSe[c]	1.7	2.2	7	CeO_2[l]	3.2	3.45	nano
SnS[C]	1.0	1.8	7	ZrO_2[j]	5.2	6.1	7
SnSe[d]	1.3	1.7	19	SnO_2[k]	3.6	4.7	3
PbS[e]	0.41	1.0	4.5	WO_3[l]	2.6	3.25	9
Sb_2S_3[f]	2.2	3.8	20	HfO_2[m]	5.5	5.5	5
CdI_2[g]	3.1	3.6	< 250	diamond[n]	5.5	3.4	4.5

[a][320], [b][321], [c][322], [C][323], [d][324], [e][325], [f][326], [g][327], [h][328], [i][329], [l][330], [j][331], [k][332], [l][333], [m][334], [n][335, 336]

The problem of band gaps has been challenged on completely fresh basis by Nethercot [316] who exploited the similarity between electron transfer from an M to an X atom on formation of a compound, and electric conductivity in a solid. Hence the EN can be a measure of the latter, as well as the former, process. Using the ENs according to Mulliken and assuming the EN of a compound to be the geometrical mean of the elements' ENs (in accordance with Sanderson's theory, see next Section), Nethercot determined the Fermi energy as

$$E^{F}{}_{MX} = c(\chi_M \chi_X)^{1/2}. \tag{2.71}$$

Then the electron work function can be calculated as

$$\Phi = E^{F} + \frac{1}{2} E_g. \tag{2.72}$$

The work functions, calculated by Eq. 2.72, agree with the experimental results, the average discrepancy being 3.5 %. For pure metals (for $E_g = 0$) Eqs 2.71 and 2.72 give a linear dependence $\chi = 0.35\Phi$.

Nethercot's approach was based on Sanderson's theory and encouraged more extensive applications of the latter, to determine work functions of metals and compounds [317]. The results are in good agreement with the experiment, e.g. for CaF_2, SrF_2 and BaF_2 the calculated $\Phi = 11.52$, 10.95, 10.48 eV and the observed $\Phi = 11.96$, 10.96, 10.69 eV, respectively. Similar calculations have been repeated later with equal success [318], using the ENs according to Mulliken. Notwithstanding the obvious efficiency of this method [319], it is noteworthy that good results have been obtained for solids with predominantly polar bonds, where bond metallicity could be neglected. No universal rule, linking E_g directly with EN, atomic charges, bond energies, work functions, etc., is currently known. Similar alterations of anions can result in opposite changes of the bang gap with different cations. Thus, for example the band gap in AgCl is wider than in AgF, and in the zinc and cadmium sulfides wider than in the oxides of the same elements (see Table 2.16), although the bonds are stronger in the latter compounds. A satisfactory agreement with the experiment can be achieved only by taking into account bond polarity and metallicity, as well as *d*-electrons' participation in valence interactions (see above).

Obviously, the above cited values of band gaps correspond to large (ideally, infinite) samples, and can increase substantially for microscopic particles and clusters, which contain a significant fraction of surface atoms with lower coordination number and begin to resemble a molecule, with a correspondingly more covalent character of bonding (cf., k in Eq. 2.69 increasing from 1.2 to 1.6). Measurements of band gaps in clusters of varying diameter (D) confirmed this conclusion.

Polymorphs which do not differ in the coordination number, have similar band gaps, viz. for anatase, rutile and amorphous TiO_2 these are, respectively, 3.5, 3.2 and 3.8 eV for direct, or 3.2, 2.9 and 3.0 eV for indirect transitions. Crystals of ZnS, CdS and CdSe, on transition from wurtzite to cubic forms change band gaps from 3.9 to 3.7 eV, from 2.50 to 2.41 eV, and from 1.70 to 1.74 eV, respectively. At the same time, a transformation of diamond into graphite decreases E_g from 5.5 eV to zero.

2.4 Concept of Electronegativity

Effective charges of atoms are known only for a small minority of polar molecules and crystals, therefore it is important to find a dependence of these values on such characteristics of atoms which allows to estimate the polarity of bonds *a priori*. Such characteristic is the electronegativity of atom (EN) which, according to Pauling who introduced this concept in 1932, is the measure of the power of an atom in a molecule to attract electrons.

2.4.1 *Discussion About Electronegativity*

For 80 years the concept of electronegativity has been applied and modified in chemistry. This concept is used to explain such chemical properties as acidity of solvents, mechanisms of reactions, electron distributions and bond polarities. The difference of EN ($\Delta\chi$) allows to classify chemical compounds as ionic when $\Delta\chi > 1.7$, or covalent when $\Delta\chi < 1.7$. Metal elements have, as a rule, $\chi \leq 2.0$, nonmetals ≥ 2.0. These aspects are present in all textbooks of general chemistry published in recent decades (e.g. [337]). Thus it may seem amasing today that from the start the EN was a topic of arguments of uncommon intensity. Thus, Fajans ([338] and private communications) pointed that in the succession $HC{\equiv}CCl \rightarrow H_2C{=}CHCl \rightarrow H_3C{-}CH_2Cl$, the charge of the chlorine atom changes sign from $+\delta$ to 0 to $-\delta$, which contradicts the notion of a constant EN of the carbon atom. In fact, χ(C) depends on the state of hybridization, being 2.5 for sp^3, 2.9 for sp^2 and 3.2 for sp. With $\chi(Cl) = 2.9$ or 3.0; this explains the reversion of the charge. In the letter to Fajans in 1959 one of the authors (S.S.B.) attracted his attention to this fact. Hückel critisized the dimensionality of EN, the square root of energy, as physically meaningless [339], to which it was replied [340] that the parameter used to calculate the bond ionicity, was actually $\Delta\chi^2$ with the dimensionality of energy, just as for the ψ-function the square of its modulus was linked to observables. As early as 1962–1963 it was argued that the idea of EN had run its course and cannot explain new data [341], that it contains actual mistakes [342–345] or that it uses the 'atoms in molecules' approach supposedly contradicting the philosophy of quantum mechanics [346]. The analysis of this critique, exposing its irrational nature, can be found in [347, 348]. Later, more criticism was directed at the problem of dimensionality of EN [349], usually without any account of the earlier discussions. The arguments in favour of EN [350–353] can be summarized thus. The fact that EN is defined through different observed properties and so has a non-unique dimensionality, merely reflects the multi-faceted nature of the chemical bond. Indeed, this can be an asset rather than liability, as EN can serve as nodal point connecting various physical characteristics of a substance, hence its wide usage in chemistry. A certain 'fuzziness' of the concept is in fact typical for chemistry, cf. the notions of metallicity, acidity, etc. Half a century later, it is evident that EN is indispensable in structural chemistry, crystallography, molecular spectroscopy, and various fields of physical and inorganic chemistry; it was even suggested to use EN as the third coordinate of the Periodic Table [354, 355].

2.4.2 *Thermochemical Electronegativities*

Pauling derived the first quantitative scale of EN using bond energies,

$$\Delta\chi_{\mathrm{MX}} = \chi_{\mathrm{M}} - \chi_{\mathrm{X}} = c\Delta E_{\mathrm{MX}}{}^{\frac{1}{2}} \tag{2.73}$$

where

$$\Delta E_{\mathrm{MX}} = E(\mathrm{M{-}X}) - \frac{1}{2}[E(\mathrm{M{-}M}) + E(\mathrm{X{-}X})] \tag{2.74}$$

and $c = 0.102$ for E measured in kJ/mol. This formula gives only the differences of ENs, and to obtain the absolute values it was necessary to postulate the EN of one 'key' element. For this role, Pauling chosed hydrogen, initially assigning it $\chi = 0$ and later $\chi = 2.05$, to avoid negative χ for most metals.

Obviously, Eq. 2.73 makes sense only if $\Delta E_{\mathrm{MX}} > 0$, which is true for all bonds but a few, such as alkali hydrides which have exceptionally weak M–H bonds, while H–H is the strongest σ-bond known. To overcome this inconsistency, Pauling replaced the geometrical for the arithmetical mean in Eq. 2.74. As $[E(\mathrm{M{-}M}) \cdot E(\mathrm{X{-}X})]^{½} < ½\,[E(\mathrm{M{-}M}) + E(\mathrm{X{-}X})]$ for purely mathematical reasons, this change restored the condition $\Delta E_{\mathrm{MX}} > 0$, albeit at the cost of depriving the formula of the clear physical meaning. This approach gives practically the same values of EN as the previous one, if the factor $c = 0.089$ is used in Eq. 2.73. The geometrical mean for the dependence of the bond energy on electronegativities was later suggested by Matcha [212] and Reddy [215].

Pauling's work initiated numerous determinations of the ENs of elements in various valence states, which were based of more extensive and precise sets of experimental data (for historical reviews see refs. [29, 355, 356]). Most important advances of this 'thermochemical' approach were made in the works of Pauling [28], Allred [357], Reddy et al. [215], Leroy et al. [358–362], Ochterski et al. [363], Murphy et al. [199], Smith [364, 365], Matsunaga et al. [366].

Reddy and Murphy showed that Pauling's equation is valid only for a limited range of molecules where $\Delta\chi$ is small, and substitution of the arithmetical mean by the geometrical mean makes little improvement. A better correlation is found if the 'extra ionic energy' (EIE) is expressed as $k\Delta\chi$ rather than as $k\Delta\chi^2$. The EIE may be represented by a quasi-Coulombic expression based on the Born-Mayer equation, thus Eq. 2.74 transforms into

$$E_{AB} = \frac{1}{2}(E_{AA} + E_{BB}) + a\frac{q_A q_B}{d_{AB}}\left(\frac{1-\rho}{d_{AB}}\right) \tag{2.75}$$

where q is the fractional charge, d is the bond distance, a and ρ are constants. Because according to Bratsch [213, 214]

$$q = \frac{\chi_A - \chi_B}{\chi_A + \chi_B} \tag{2.76}$$

substitution of this expression into Eq. 2.75 gives an expression where EIE is proportional to $\Delta\chi^2$. Pauling's approach requires a qualification: the energy of a bond depends not only on its polarity, but also on its length. Neglecting this in Eq. 2.48 can be justified by the low polarity of the bonds in question, i.e. on the assumption that purely covalent and slightly polar bonds have the same lengths. Allred [357] assumed Eq. 2.73 to be valid if $\Delta\chi \leq 1.8$, but this criterion has not been sufficiently substantiated and any extension of the database by adding the energies for bonds of unspecified polarity can alter both the absolute EN values of elements and the order of their succession. Ionov et al. [367] suggested to remedy this shortcoming by a principal alteration of Eq. 2.73, so as to utilize both thermodynamic and geometrical data. However, the basic correctness of Pauling's insight has been confirmed by other physical methods, hence it is more sensible to account for the geometrical factor by adding a correcting term to Eq. 2.74, rather than by altering its philosophy. This has been achieved by using in this equation a variable parameter c which takes into account the principal quantum numbers, bond distance and valences of atoms, $c = f(n^*, d, v)$ [368]. This correction reflects the fact that (other things being equal) an elongation of a bond lowers its polarity, by reducing the overlap of the valence orbitals, and an increase of valence also reduces the bond polarity. However, the contribution of all these factors is by an order of magnitude smaller than the major (bond-energy) term. In Table S2.15 are compared a few systems of thermochemical ENs, and the averaged results are listed in Table 2.17.

A prominent feature of the thermochemical system is the outstandingly high EN of oxygen, nitrogen and, especially, fluorine, which are often difficult to reconcile with

Table 2.17 Thermochemical electronegativities of atoms in molecules; $\chi(\mathrm{H}) = 2.2$

Li	Be	B	C	N	O	F			
1.0	1.5	2.0	2.55	2.9	3.4	3.9			
Na	Mg	Al	Si	P	S	Cl			
0.9	1.3	1.6	1.9	2.15	2.6	3.1			
K	Ca	Sc	Ti	V	Cr	Mn	Fe	Co	Ni
0.75	1.0	1.35	1.6	1.7[a]	1.7[b]	1.7[c]	1.7[d]	1.75[e]	1.8[e]
Cu	Zn	Ga	Ge	As	Se	Br			
1.7[f]	1.6	1.7	2.0	2.1	2.5	2.9			
Rb	Sr	Y	Zr	Nb	Mo	Tc	Ru	Rh	Pd
0.7	0.95	1.2	1.6[g]	1.6[a]	2.2[h]	1.9	2.2	2.2	2.2
Ag	Cd	In	Sn	Sb	Te	I			
1.8	1.7	1.7	1.9[i]	2.0	2.1	2.6			
Cs	Ba	La	Hf	Ta	W	Re	Os	Ir	Pt
0.6	0.85	1.1	1.5[j]	1.5	2.2[k]	1.9	2.2	2.2	2.2
Au	Hg	Tl	Pb	Bi	Th[k]	U[k]			
2.2	1.9	1.3[l]	2.1[m]	2.0	1.5	1.6			

[a] $v = 3$, [b] $v = 3$, for $v = 2$ $\chi = 1.5$, for $v = 4$ $\chi = 2.0$, [c] $v = 3$, for $v = 2$ $\chi = 1.5$, [d] $v = 2$, for $v = 3$ $\chi = 2.0$, [e] $v = 2$, [f] $v = 1$, for $v = 2$, $\chi = 2.0$, [g] $v = 4$, for $v = 2$ $\chi = 1.4$, [h] $v = 4$, for $v = 2$ $\chi = 2.0$, [i] $v = 4$, for $v = 2$ $\chi = 1.6$, [j] $v = 4$, for $v = 2$ $\chi = 1.3$, [k] $v = 4$, [l] $v = 1$, for $v = 3$ $\chi = 1.8$, [m] $v = 4$, for $v = 2$ $\chi = 1.7$

the physical and chemical properties of these atoms in polyatomic molecules and crystals. Thus, fluorine is a surprisingly poor acceptor of hydrogen bonds [369, 370]. However, the apparent dissociation energies of the F–F, O–O and N–N bonds are lower than the intrinsic bond energies because of the electronic destabilization, i.e. the energetically unfavorable effect of the high electron concentration in a small volume (see Sect. 2.3.1). The underestimation of $E(\mathrm{X-X})$ in Eq. 2.48 leads to an overestimation of ΔE_{MX} and hence of χ, as Bykov and Dobrotin were the first to notice in the case of fluorine [371]. Later, Batsanov [372] calculated the electron destabilization energies for a number of compounds and re-evaluated the ENs of fluorine, oxygen and nitrogen as 3.7, 3.2 and 2.7, respectively (cf. the conventional values of 4.0, 3.5 and 3.0). The thermochemical method has been significantly improved by Finemann [373], who generalized Eq. 2.74 to cover radicals (R),

$$\Delta E_{MR} = E(\mathrm{M - R}) - \frac{1}{2}[E(\mathrm{M - M}) + E(\mathrm{R - R})] \tag{2.77}$$

Equation 2.77 was later used to calculate the ENs for radicals of various composition; the averaged values are listed in Table S2.16. It is evident that the presence of multiple bonds in radicals substantially affects the atomic ENs. Equation 2.77 was shown to give the ENs which describe quite accurately the homolytic bond dissociation enthalpies of common covalent bonds (including highly polar ones) with an average error of *ca.* 5 kJ/mol; by this method the dissociation enthalpies were calculated for more than 250 bonds, including 79 for which experimental values are not available [374]. The weakness of Pauling's approach can be seen in that the electronegativity of hydrogen (unique in this respect of all elements) is not constant but depends substantially on the atom or group (R) connected to it [374]. Thus, $\chi(\mathrm{H}) = 1.95$ for R = Me, 2.06 for Et, 2.16 for OH, 2.20 for Cl, 2.26 for F and Ph, 2.27 for ONO_2, and 2.50 for C≡CH. Unique behavior of hydrogen is not uncommon in chemistry and Pauling noted that hydrogen's electronegativity 'misbehaves'. Using an average value of $\chi(\mathrm{H}) = 2.2$, as recommended by Pauling, gives generally the correct trends in *D*(H–A), but the overall accuracy is much lower than that obtained for all other bonds. Therefore Datta and Singh chose $\chi(\mathrm{OH}) = 3.500$ as the reference value [374]. They also sugested to use geometrical means of single bond energies in organic compounds for calculating the ENs of radicals.

Note that the energies of single covalent bonds (Table 2.12) change regularly in each subgroup of the Periodic Table. Therefore within each subgroup the ENs of elements are proportional to the square roots of the energies of the corresponding homonuclear bonds. Hence, the entire system of thermochemical ENs can be derived from these energies and the known ENs of the top elements in a group, provided that the F–F, O–O and N–N bond energies are corrected for the electron destabilization by adding 220, 170 and 150 kJ/mol, respectively. Now we can, by comparing the energies of σ- and π-bonds, find out how the EN is affected by the bond order. This problem has been explored for carbon [375] and other elements capable of multiple bonding [376], using Eq. 2.78, which follows the philosophy of Eqs 2.73 and 2.74,

$$(\Delta\chi)^2 = \Delta E = \frac{1}{n} E(\mathrm{A} \sim \mathrm{A}) - E(\mathrm{A} - \mathrm{A}) \tag{2.78}$$

where $\Delta\chi$ is the difference between the ENs displayed by the *same* element A in the directions of the single (A–A) and the multiple (A~A) bond, *n* is the order of the latter bond, *E* is the corresponding energy. Using Eq. 2.78 and the data from Table S2.10, we can find that the ENs of C, Si, P and S in double bonds are *lower* that the standard values, by 0.42, 0.49, 0.43 and 0.34, respectively, but for O and N the 'double-bond' ENs are *higher* that the standard ones, by 0.43 and 0.34. One-third of the energy of the triple C≡C bond (262 kJ/mol) is lower than the energy of one single bond (357 kJ/mol) hence the 'triple-bond' EN is by 0.50 lower than the 'single-bond' EN. The opposite is true for nitrogen: *E* (N≡N)/3 = 315 kJ/mol > *E* (N–N) = 212 kJ/mol, hence the formation of a triple bond rises the EN of nitrogen by 0.52. The ENs of elements in the most common multiple bonds are:

(C=) 2.2	(Si=) 1.4	(P=) 1.8	(S=) 2.2	(N=) 3.1	(O=) 3.6
(C≡) 2.1		(P≡) 1.7		(N≡) 3.3	

In a single bonds *adjacent* to a multiple bonds, the EN of the same atom changes in a compensatory manner. Thus, the ENs of carbon, displayed in the central (single) C–C bond in the CH_3CH_2–CH_2CH_3, CH_2 = CH–CH = CH_2, and HC≡C–C≡CH molecules, are 2.6, 3.1 and 3.4, respectively.

Since the change of the bond order usually implies the change of the coordination number, it is useful to consider from this viewpoint the transformation of a molecular structure into a continuous network of covalently bonded atoms in the solid state. Thermodynamically, the depth of the structural rearrangements during gas → crystal transition is characterized by the heat of sublimation ΔH_s (see Chap. 9) from which it is natural to calculate the crystal-state ENs (χ*) [377]

$$\Delta\chi^* = a\sqrt{\Delta E + \Delta H_s} \tag{2.79}$$

Usually, ΔH_s of nonmetals (which retain the molecular structure in crystal) are small compared to that of metals, where the crystal growth implies the formation of new chemical bonds. Then, assuming the sublimation heat to be additive, almost the entire heat effect of crystallization of a compound can be related to the EN of the metallic component. The same conclusion follows from simple crystallographic reasoning. When molecules assemble into a crystal structure, both the metal and the nonmetal atoms increase their coordination numbers. For the nonmetal this means engaging previously nonbonding electron pairs into chemical bonds, which increases the mean ionization potential and hence the bond energy, according to Eq. 2.46 (see below for details). No such increase of the ionization potentials occurs for the metal, which provides the same number of bonding electrons in the molecular and in the solid state. For both the metal and the nonmetal, the covalent component of the bond (the overlap of the wave functions) is smaller in the crystal, where the bonds are somewhat longer than in the molecules. For the nonmetal the latter effect subtracts from the increase of the bond energy, whilst for the metal, it produces a net decrease. Thus, the ENs of nonmetals in crystalline compounds are close to those in molecules, while those of metals are always lower.

The system of ENs for the crystal state has been developed [378, 379] by comparing the atomization energies of the MX-type compounds with the energies of the M–M and X–X bonds in the solid state, corrected for the difference of the bond distances in the molecules and solids. The crystalline ENs of the same metal, calculated from different halides, practically coincide. Thus the obtained values are reproducible and can be recommended for general use in structural chemistry. Other ENs, tailored for thermodynamic or structural characteristics of crystals, were suggested by Vieillard and Tardy [380] and by Ionov and Sevastianov [381]. For most elements, their results are close to the thermochemical crystalline ENs, they are presented in Table 2.18.

2.4.3 *Ionization Electronegativities*

Pauling's pioneering paper [18] was soon followed by the work of Mulliken [382, 383], who approached ENs from the viewpoint of quantum mechanics. He proved that ENs can be calculated as

$$\chi = \frac{1}{2}(I_v + A_v) \tag{2.80}$$

Table 2.18 Thermochemical electronegativities of atoms in crystals

Li 0.65	Be 1.15	B 1.4	C 2.5	N 2.7	O 3.2	F 3.7			
Na 0.6	Mg 1.0	Al 1.3	Si 1.9	P 2.1	S 2.5	Cl 3.0			
K 0.5	Ca 0.75	Sc 1.1	Ti 1.55[c]	V 1.4[e]	Cr 1.25[f]	Mn 1.2[f]	Fe 1.4[f]	Co 1.45[f]	Ni 1.5[f]
Cu 1.15[a]	Zn 1.3	Ga 1.4	Ge 2.0	As 2.1	Se 2.5	Br 2.8			
Rb 0.45	Sr 0.7	Y 1.5	Zr 1.4	Nb 1.6	Mo 1.75	Tc	Ru	Rh	Pd 1.35[f]
Ag 1.3	Cd 1.35	In 1.55	Sn 1.9[d]	Sb 2.0	Te 2.1	I 2.5			
Cs 0.4	Ba 0.65	La 1.0	Hf 1.4	Ta 1.5	W 1.75	Re	Os	Ir	Pt 1.7[f]
Au 1.4	Hg 1.6	Tl 1.1[b]	Pb 2.15[d]	Bi 2.0	Th[g] 1.4	U[g] 1.3			

[a]for Cu^{II} $\chi = 1.6$, [b]for $v = 1$, [c]$v = 4$, for $v = 2$ $\chi = 1.1$, [d]$v = 4$, for $v = 2$ $\chi = 1.4$, [e]for $v = 3$, [f]for $v = 2$, [g]for $v = 4$

where I_v is the valence-state ionization energy and A_v is the electron affinity of the atom. Mulliken's ENs (χ_M) are close to Pauling's values (χ_P) multiplied by a factor of 3 ± 0.2. The most remarkable advantage of Mulliken's method is the opportunity to calculate ENs for various valence states. As *ns* electrons have higher ionization energy than *np* ones, an increase of the *s*-character of an orbital rises the EN of an atom in the succession $sp^3 < sp^2 < sp$ [384], in agreement with the results of the thermochemical method (see above). Pritchard and Skinner [385–388] calculated ENs of atoms in various valence states from spectroscopic data. They obtained good agreement with the thermochemical EN and thus were able, by combining the methods of Pauling and Mulliken, to determine the type of hybridization of the bonds in transition metal compounds. Batsanov [375, 389] calculated the ENs of sp^2 and sp hybridized carbon atoms from the experimental values of the ionization potential. The planar-trigonal *olefinic* (sp^2) carbon atom has the EN of 2.3 in the double bond and 2.6 in the single bond, whilst the linear *acetylenic* (sp) atom (–C≡) displays the EN of 2.0 in the triple bond and 2.8 in the single bond, also in accordance with thermochemical data.

The theory of EN has been substantially advanced by Iczkowski and Margrave [390], who have shown that within the same shell the ionization energy is a function of the charge q (the number of removed electrons),

$$E(q) = \alpha q + \beta q^2 + \gamma q^3 + \cdots \quad (2.81)$$

where α, β and γ are constants. Neglecting the third and successive members of the series, we obtain for a hydrogen-like atom

$$(\partial E/\partial q)_{q=1} = \frac{1}{2}(I_v + A_v) \quad (2.82)$$

Thus, the assumption that EN is the derivative of the energy by the charge follows Mulliken's formula. Hinze and Jaffe [391–393] regarded EN as the ability of an atom to attract electrons into a given orbital, and therefore introduced the term 'orbital EN' (simultaneously the same term was introduced by Pritchard and Skinner [388]). Having calculated the orbital ENs exhibited by several elements in single and multiple bonds, they obtained for the tetrahedral C (sp^3) $\chi = 2.48$, for the trigonal C (sp^2) $\chi = 2.75$ in the single bond and 1.68 in the double bond, for the linear acetylenic carbon $\chi = 3.29$ in the single and $\chi = 1.69$ in the triple bond.

Mulliken's method, like that of Pauling, tends to overestimate the ENs of fluorine, oxygen and nitrogen, and essentially for the same reason: neglecting the inter-electron repulsion (see above). The quantum-mechanical approach was further developed using the electron density functional theory [394, 395], according to which EN is the negative chemical potential μ,

$$\chi = -\mu = -\left(\frac{\partial E}{\partial N}\right) \quad (2.83)$$

where E is the ground-state energy as a function of the number of electrons (N), for a given potential μ affecting the system. The electron chemical potential has the same tendency towards equalization as the macroscopic (thermodynamic) potential: electrons move from the areas of high potential (μ_h) to those of low potential (μ_l), whereupon μ_l increases and μ_h decreases until they become equal. In the DFT formalism, Mulliken's equation can be derived on the assumption that the energy of the system is a quadric function of the number of electrons [396, 397]. An outline of this approach can be found in the *Structure and Bonding* edition [398], comprising contributions from all the major researchers in this field and in excellent reviews by Allen [399] and Cherkasov et al. [355]. Details of the theoretical calculations are outside the scope of the present book, which is devoted to experimental aspects of structural chemistry. The reader can consult a review by Bergmann and Hinze [400] on the quantum-mechanical calculations of the ENs of elements from the ionization energies. A purely empirical formula linking EN with the ionization potential I and electron affinity A has been suggested by Sacher and Currie [401]. Further development of the Iczkowski-Margrave model is given in [402].

Pearson [403, 404] used the ground-state ionization energy and electron affinity of an atom (I_o and A_o) for calculating 'the absolute electronegativity' by Eq. 2.80. Since I_o and A_o are known for all elements and for all steps of oxidation, Pearson's approach became popular, although it does not conform rigorously to Mulliken's original definition of EN. Pearson's method is now widely used to calculate atomic and molecular electronegativities; selected ENs from this system (normalized by $\chi(H) = 2.2$) are presented in Table S2.16. In general, Pearson's ENs follow the expected trends in the Periodic Table, increasing from left to right in periods and decreasing from top to bottom in groups. However, there are some strikingly unrealistic values: Cl is assigned higher EN than O and N, Br is on par with O and more electronegative than N, H is as electronegative as N and more so than C or S. The errors disappear if valence-state ionization energies and electron affinities are used. Unfortunately, there is serious ambiguity in specifying the valence state; for instance, for three-coordinate N atom one has to choose from seven possible valence states [406–408].

The ionization energies of ground-state atoms are considerably larger than their electron affinities, hence EN is defined mostly by I_0. Therefore Allen et al. [406–408] introduced the atomic electronegativity scale based upon the spectroscopic (averaged) ionization energies of the valence electrons in a ground-state free atom:

$$\chi = \frac{m\varepsilon_p + n\varepsilon_s}{m + n} \tag{2.84}$$

where m and n are the numbers of p and s valence electrons, ε_p and ε_s are the ionization energies of the p- and s-electrons, determined from atomic spectra. These characterisrics became known as 'spectroscopic electronegativities', SEN. Selected values of SEN, normalized by $\chi(H) = 2.2$, are listed in Table S2.17. SENs are closer to the thermochemical ENs than Pearson's values. According to Allen, SENs characterize the atom's ability to absorb (or to retain, in the case of rare gases) electrons, they do not depend on the valence and coordination number and are specific parameters of elements, which can be regarded as the third dimension of the Periodic Table. SENs correlate with Lewis acidity, defined as $S_a = v/N_c$, where v is the valence and N_c is the average coordination number of an element in its compounds with oxygen [409]. Politzer et al. [405] calculated absolute electronegativities on different levels of MO theory; these magnitudes of ENs are also given in Table S2.17.

All the abovementioned systems of ENs have been normalized to Pauling's thermochemical scale. However, the thermochemical and ionization (except Allen's system) ENs have different dimensionalities, viz. square root of energy and energy, respectively. This reflects the fundamental difference, that Pauling's method uses *mean* bond energies, thus treating all electrons of the central atom as equivalent, while Mulliken's method uses the *first* ionization potentials, thus singling out one electron. To compare the thermochemical and the ionization methods correctly, the energy of valence electrons in the latter should be characterized by the average, rather than the first, ionization potential ($\bar{I}$) of all outer electrons. This gives the simple formula

$$\chi = k\sqrt{\bar{I}} \tag{2.85}$$

where $k = 0.39$ [410]. Significantly, Eq. 2.85 permits to determine ENs for different oxidation states by averaging the corresponding number of successive ionization potentials. This equation gives ENs in accordance with Pauling's scale for sp-elements (a-subgroups), but for transition elements the calculated ENs are somewhat lower than the thermochemical values, since d-electrons from the previous shell can participate in the bonding. To account for this, in the case of d-elements the values of χ

Table 2.19 Ionization electronegativities of elements (for H, $\chi = 2.2$)

Li	Be	B	C	N	O	F			
0.90	1.45	1.90	2.37	2.85	3.31	3.78			
Na	Mg	Al	Si	P	S	Cl			
0.88	1.31	1.64	1.98	2.32	2.65	2.98			
K	Ca	Sc	Ti	V	Cr	Mn	Fe	Co	Ni
0.81	1.17	1.50	1.25[b]	1.60[c]	1.33[b]	1.32[b]	1.35[b]	1.38[b]	1.40[b]
			1.86	1.92[d]	1.63[c]	1.70[c]	1.66[c]	1.72[c]	1.76[c]
				2.22	1.97[d]	2.02[d]			
					2.58	2.93			
Cu	Zn	Ga	Ge	As	Se	Br			
1.48	1.64	1.84	2.09	1.70[c]	2.61	2.88			
1.66[b]				2.35					
Rb	Sr	Y	Zr	Nb	Mo	Tc	Ru	Rh	Pd
0.80	1.13	1.40	1.22[b]	1.52[c]	1.92[d]	1.93[d]	1.35[b]	1.39[b]	1. 45[b]
			1.71	2.02	2.36		1.97[d]	1.99[d]	2.08[d]
Ag	Cd	In	Sn	Sb	Te	I			
1.57	1.65	1.80	1.29[b]	1.60[c]	2.46	2.70			
			2.01	2.24					
Cs	Ba	La	Hf	Ta	W	Re	Os	Ir	Pt
0.77	1.07	1.35	1.28[b]	1.52[c]	1.83[d]	1.83[d]	1.39[b]	1.40[b]	1.45[b]
			1.73	1.94	2.28	2.48	1.85[d]	1.87[d]	1.92[d]
Au	Hg	Tl	Pb	Bi	Po	Th	U		
1.78	1.79	0.96[a]	1.31[b]	1.58[c]	2.50	1.60[d]	1.58[d]		
		1.89	2.07	2.26					

[a] $v = 1$, [b] $v = 2$, [c] $v = 3$, [d] $v = 4$

calculated by Eq. 2.85 must be increased by the term

$$\Delta\chi = 0.1\frac{n}{v} \qquad (2.86)$$

where n is the principal quantum number and v is the group or the intermediate valence. The resulting ENs are listed in Table 2.19.

A comparison of the ionization and thermochemical ENs of elements reveals the largest discrepancies for Cu, Ag, Au and smaller ones for Zn, Cd and Hg, due to d-electrons participating in the bonding. For Cu, a comparison of the thermochemical EN with the χ calculated for the s- and d-electrons, revealed a 23 % participation of the 3d-electrons in the Cu–X bonds [387].

It has been proposed [411, 412] to transform Pauling's ENs into Mulliken's, by equalizing their dimensionalities accordingly. However, in these works only the first ionization potentials were used, thus reproducing the shortcoming of Mulliken's original approach. Ionization potentials have also been used in these works and in [413] to determine ENs for groups of atoms (radicals), the mean values of which are listed in Table S2.18.

To calculate EN for crystals, it is sensible to use the work function (Φ), i.e. the energy of removing an electron from a solid, which can be regarded as the ionization potential of the solid (see above). This has been first attempted by Stevenson and Trasatti, who suggested the simple dependence $\chi^* = k\Phi$, where $k = 0.355$ [414] or 0.318 [415, 416]. Eq. 2.87 gives the best agreement with the thermochemical scale of crystalline EN for metals.

$$\chi^* = k\Phi + k\left(\frac{v}{n^*} - 1\right) \tag{2.87}$$

where $k = 0.32$ and other symbols as above. The values of χ^* calculated by this technique using modern values of Φ [417] are listed in Table 2.20; for elements of Groups 1 to 4 and 11 to 14, using the group valences, for other metals the lowest oxidation numbers. It is noteworthy that the heats of formation of inter-metallic compounds can be calculated according to Miedema's theory assuming $\chi^* = \Phi$ [418–420]. However, EN is an atomic property and cannot be adequately derived from bulk properties (see Sect. 1.1.2)

ENs of ions also can be calculated by Mulliken's method in the same way as for neutral atoms, by substituting the ionization potential and electron affinity of the corresponding ion into Eq. 2.80. Thus, to calculate the EN for a cation with the + 1 charge, one should use the *second atomic* ionization potential as the first cationic I, and the first atomic ionization potential for A. For an anion charged –1 the first atomic A should be used for I, and the second atomic A for the electron affinity. The ionic ENs thus calculated [376, 421, 422], are listed in Table S2.19. Bratsch [423] has made a rough estimate that the EN of a neutral atom doubles when it acquires the +1 charge and becomes zero when –1. The latter statement has been since confirmed, whilst the real increase of the EN for cations proved several times, or even an order of magnitude, higher.

2.4.4 Geometrical Electronegativities

Electronegativity being a qualitative property which describes the power of an atom in a molecule to attract the bonding electrons, it can be defined by the ratio of the effective nuclear charge to the covalent radius, Z^*/r^n. Many authors have proposed different values of n in order to reconcile the geometrical and thermochemical systems of EN, see reviews [56, 355, 356, 423]. A brief history of these attempts is presented in Table S2.20. From the Z^* and r, EN can be calculated by the formulae

Table 2.20 Work functions (eV, *upper lines*) and crystal electronegativities (*lower lines*)

Li	Be	B	C						
2.38	3.92	4.5	5.0						
0.60	1.25	1.6	1.92						
Na	Mg	Al	Si						
2.35	3.64	4.25	4.8[a]						
0.54	1.06	1.36	1.64						
K	Ca	Sc	Ti	V	Cr	Mn	Fe	Co	Ni
2.22	2.75	3.3	4.0[b]	4.40	4.58	4.52	4.31	4.41	4.50
0.48	0.73	1.00	1.31	1.35	1.32	1.30	1.23	1.26	1.29
Cu	Zn	Ga	Ge	As	Se				
4.40	4.24	4.19	4.85	5.11	4.72				
1.17	1.21	1.28	1.58	1.75	1.71				
Rb	Sr	Y	Zr	Nb	Mo	Tc	Ru	Rh	Pd
2.16	2.35	3.3	4.0	3.99	4.29	4.4	4.6	4.75	4.8
0.45	0.59	0.98	1.28	1.20	1.37	1.41	1.31	1.36	1.38
Ag	Cd	In	Sn	Sb	Te				
4.30	4.10	3.8	4.38	4.08	4.73				
1.14	1.15	1.14	1.40	1.38	1.67				
Cs	Ba	La	Hf	Ta	W	Re	Os	Ir	Pt
1.81	2.49	3.3	3.20[c]	4.12	4.51	4.99	4.7	4.7	5.32
0.34	0.63	0.96	1.01	1.23	1.40	1.55	1.34	1.34	1.53
Au	Hg	Tl	Pb	Bi	Th	U			
4.53	4.52	3.70	4.0	4.4	3.3	2.2			
1.21	1.28	1.09	1.26	1.47	1.04	0.69			

[a]p-Si, for n-Si $\Phi = 4.8$ eV, [b]α-Ti, for β-Ti $\Phi = 3.65$ eV, [c]α-Hf, for β-Hf $\Phi = 3.53$ eV

of Cottrell and Sutton (Eq. 2.88) [424], Pritchard and Skinner (Eq. 2.89) [425] and Allred and Rochow (Eq. 2.90) [426].

$$\chi_1 = a\left(\frac{Z^*}{r}\right)^{1/2} + b \tag{2.88}$$

$$\chi_2 = c\frac{Z^*}{r} + d \tag{2.89}$$

$$\chi_3 = e\left(\frac{Z^* - f}{r^2}\right) + g. \tag{2.90}$$

where *a, b, c, d, e, f,* and *g* are constants. Most of these constants are composition-dependent and therefore Eqs 2.88–2.90 are of limited utility in structural chemistry.

In view of this, it is expedient to modify these equations, by making the constants universal and including additional terms depending explicitly on the nature of the elements concerned. Thus, Eq. 2.88 was reduced [427] to

$$\chi_1 = \gamma \left(\frac{Z^*}{r} \right)^{\frac{1}{2}} \tag{2.91}$$

where γ is the function of the Group number and the effective principal quantum number. Better agreement between the calculated and thermochemical data can be attained by reducing Eq. 2.90 to

$$\chi_3 = e \frac{Z^*}{(r + \beta)^2} + g \tag{2.92}$$

ENs have been calculated for all elements in different valence states according to Eqs 2.91, 2.89 and 2.92, showing good agreement with the thermochemical characteristics.

Electronegativity can be also calculated by the method proposed by Sanderson [429–431], who has established a correlation between EN and the ‘relative electron stability’, $S = \rho_a / \rho_{rg}$ of the atom; $\rho_a = N_e/V$, where N_e is the number of electrons in the given atom, V is the its volume, and ρ_{rg} is the same for the iso-electronic atom of the rare-gas type. Sanderson found that the electronic stability, or “compactness”, is a good measure of electronegativity:

$$\chi^{1/2} = aS + b. \tag{2.93}$$

Individual values of S have been revised from time to time, and refined data of χ by Sanderson’s method, are given in [431, 432]. This method gives only a qualitative agreement with thermochemical values, because the electron densities in the core and valence-shell regions are very different, hence the integral approach cannot give adequate results (see [433]). It makes more sense to calculate χ in terms of the electron density of the outer (valence) shell of the atom, ρ_e. To do this, the number of valence electrons v should be divided by the volume of the outer shell, $V_e = V_a - V_c$ where V_a is the atomic volume, V_c is the core volume, so $\rho_e = v/V_e$. Assuming that the outer electrons in the atom are identical, one can treat them as a Fermi gas. Then the energy of these electrons is $E_e \sim \rho_e^{2/3}$. Since χ is proportional to $E^{1/2}$, we obtain

$$\chi_4 = C\rho_e^{1/2} \tag{2.94}$$

[434, 435]. Note, however, that treating the valence electrons as a Fermi gas means that these electrons are similar to those in metals. In any group of elements in the Periodic Table, the metallicity of bonding increases on going down the column to peak in Period 6. Therefore, ρ_e in Eq. 2.94 should be normalized against these elements. For example, the ratio of the effective principal quantum number of a given period to n^* is 4.2. The work equation will then appear as

$$\chi_4{}^* = 2.65 \left(\frac{n^* \rho_e}{4.2} \right)^{1/3}. \tag{2.95}$$

Table 2.21 Geometrical electronegativities of atoms in the group valences, in molecules (*upper lines*) and crystals (*lower lines*)

Li	Be	B	C	N	O	F			
1.01	1.54	2.05	2.61	3.08	3.44	3.90			
0.38	0.98	1.71	1.97	2.86	3.15	3.48			
Na	Mg	Al	Si	P	S	Cl			
0.99	1.28	1.57	1.89	2.20	2.58	2.91			
0.37	0.70	1.32	1.47	2.09	2.45	2.69			
K	Ca	Sc	Ti	V	Cr	Mn	Fe[a]	Co[a]	Ni[a]
0.83	1.07	1.36	1.60	1.89	2.10	2.33	1.80	1.86	1.92
0.32	0.58	0.92	1.27	1.06[a]	1.40[b]	1.44[b]	1.17	1.21	1.25
Cu	Zn	Ga	Ge	As	Se	Br			
1.62	1.72	1.89	2.07	2.28	2.53	2.82			
0.71	1.14	1.60	1.64	2.16	2.42	2.63			
Rb	Sr	Y	Zr	Nb	Mo	Tc	Ru[b]	Rh[b]	Pd[b]
0.82	1.01	1.26	1.50	1.80	1.98	2.20	1.86	1.90	1.92
0.32	0.56	0.86	1.16	1.03[a]	1.35[b]	1.38[b]	1.46	1.49	1.51
Ag	Cd	In	Sn	Sb	Te	I			
1.49	1.56	1.65	1.86	1.97	2.15	2.41			
0.55	1.07	1.41	1.47	1.90	2.08	2.28			
Cs	Ba	La	Hf	Ta	W	Re	Os[b]	Ir[b]	Pt[b]
0.75	0.96	1.19	1.54	1.81	1.99	2.26	1.89	1.95	1.96
0.27	0.54	0.81	1.20	1.02	1.37[b]	1.44[b]	1.50	1.54	1.54
Au[c]	Hg	Tl[d]	Pb	Bi	Th[d]	U[d]			
1.55	1.66	1.65	1.81	1.93	1.40	1.44			
1.02	1.12	1.41	1.43	1.84	0.98	1.00			

[a]for $v = 3$, [b]for $v = 4$, [c]for $v = 3$, $\chi = 2.12$ (molecule) and 1.78 (crystal), [d]for $v = 1$, $\chi = 1.47$ (molecule) and 0.56 (crystal)

The satisfactory agreement among χ calculated by the four equations, allows to set up a scale of the averaged geometrical ENs for atoms in molecules in different valences (see [428]) which correspond to σ bonding (upper rows in Table 2.21). To calculate ENs of atoms with π bonds, the covalent radii of atoms for double and triple bonds must be used (Sect. 1.4). From these we obtain $\chi_{C=} = 2.2$, $\chi_{N=} = 4.2$, $\chi_{O=} = 5.2$ and $\chi_{S=} = 2.2$. Hence, formation of π bonds lowers the EN of C or S but increases that of N or O. The dependence of χ(C) on the bond order was established in [436–438].

Strictly speaking, it is incorrect to use the classical (molecular) ENs to interpret the structures and properties of crystalline inorganic compounds. Therefore, systems of ENs were derived specifically for atoms in crystals [428, 439]. Geometrical ENs in this case should be defined in terms of crystal covalent radii (Sect. 1.4.3). Furthermore, it is necessary to take into account the dependence of χ on the bond order, $q = v/N_c$, which changes as N_c increases on transition from molecules to crystals. Since the bond order figures in the expression for energy, while Pauling's EN is proportional to $\sqrt{E}$, the data calculated by Eqs 2.92, 2.94 and 2.95 should be multiplied by $\sqrt{q}$ to obtain the atomic ENs for crystals. For elements of Groups 14 through 17, which have enough electrons to form four or six bonds in the coordination sphere, this correction is not needed. The lower lines of Table 2.21 list averaged crystalline atomic ENs for the group-number oxidation states, except for metals of Groups 5–10, where ENs refer to their usual oxidation states. For Au and Tl, the ENs also are given for the oxidation states +3 and +1, respectively [428].

Crystalline ENs were also determined by Phillips [440–444]. Assuming that the outer electrons of an atom can be treated as a Fermi gas, he obtained

$$\chi = 3.6\left(\frac{Z}{r}\right) f + 0.5 \tag{2.96}$$

where f is the screening factor according to Thomas–Fermi. Constants 3.6 and 0.5 were chosen for consistency with Pauling's ENs for C and N, while for elements of Groups 11–14 the obtained values were close to the crystalline ENs considered above. Li and Xue [445] calculated crystalline ENs using ionic radii (r_{ion}) for different coordination numbers, as

$$\chi^* = \frac{an^*\sqrt{\bar{I}}}{r_{ion}} + b \tag{2.97}$$

where n^* is the effective quantum number, $\bar{I}$ is the ionization potential of the given ion normalized by $I(\text{H}) = 13.6$ eV, a and b are the constants. ENs calculated by Eq. 2.97 are presented in Table S2.21. Later it was proposed [446] to calculate crystalline ENs using covalent radii of elements in crystals, as

$$\chi^* = \frac{cn_e}{r_{\text{cov}}} \tag{2.98}$$

where c is the constant, n_e is the number of the valent electrons and r_{cov} is the crystalline covalent radius of the atom. The authors assumed that in any covalent bond the contributions of the two atoms are inversely proportional to their respective coordination numbers, N_{cA} and N_{cB}. Using the idea of EN equalization on bonding (see Sect. 2.4.6), the bond EN can be defined as the mean of the electron-holding energy of the bonded atoms,

$$\chi^*_{AB} = \left(\frac{\chi_A}{N_{cA}}\frac{\chi_B}{N_{cB}}\right)^{1/2}. \tag{2.99}$$

These ENs were used to rationalize the properties of new superhard materials. They show good agreement with Pettifor's 'chemical scale' of EN [447–449], which

Table 2.22 Recommended values of electronegativities for atoms in molecules

Li	Be	B	C	N	O	F			
0.95	1.5	2.0	2.5	3.0	3.4	3.9			
Na	Mg	Al	Si	P	S	Cl			
0.90	1.3	1.6	1.9	2.2	2.6	3.0			
K	Ca	Sc	Ti	V	Cr	Mn	Fe^{II}	Co^{II}	Ni^{II}
0.80	1.05	1.35	1.75	2.0[d]	2.3[f]	2.6[i]	1.5	1.6	1.6
Cu	Zn	Ga	Ge	As	Se	Br			
1.6[a]	1.7	1.8	2.0	2.25	2.5	2.85			
Rb	Sr	Y	Zr	Nb	Mo	Tc	Ru^{IV}	Rh^{IV}	Pd^{IV}
0.75	1.0	1.3	1.7	1.9[e]	2.2[g]	2.4[j]	2.0	2.0	2.1
Ag	Cd	In	Sn	Sb	Te	I			
1.65	1.6	1.7	1.9[c]	2.1	2.2	2.6			
Cs	Ba	La	Hf	Ta	W	Re	Os^{IV}	Ir^{IV}	Pt^{IV}
0.70	0.95	1.3	1.7	1.9[e]	2.2[h]	2.2[j]	2.0	2.0	2.05
Au	Hg	Tl^{I}	Pb	Bi	Th^{IV}	U^{IV}			
1.85[b]	1.8	1.2	1.9[c]	2.1	1..5	1.6			

[a]$v = 1$, for $v = 2$ $\chi = 1.9$, [b]$v = 1$, for $v = 3$ $\chi = 2.2$, [c]$v = 4$, for $v = 2$ $\chi = 1.5$, [d]$v = 5$, for $v = 3$ $\chi = 1.6$, [e]$v = 5$, for $v = 3$ $\chi = 1.6$, [f]$v = 6$, for $v = 3$ $\chi = 1.7$, [g]$v = 6$, for $v = 4$ $\chi = 1.9$, [h]$v = 6$, for $v = 4$ $\chi = 1.85$, [i]$v = 7$, for $v = 3$ $\chi = 1.7$, [j]$v = 7$, for $v = 4$ $\chi = 1.9$, [k]$v = 7$, for $v = 4$ $\chi = 1.8$

adequately explains the structural properties of crystalline substances. These electronegativity data help to understand the fundamental difference in bonding between inorganic molecules and crystals. In the former, bonds vary widely in polarity; in the latter, bonds are less different and more polar, hence the ionic radii describe the interatomic distances well.

2.4.5 Recommended System of Electronegativities of Atoms and Radicals

As we have seen, values of EN obtained by different methods are consistent, this allows us to recommend the generalized systems of EN for molecules (Table 2.22) and crystals (Table 2.23), taking into account all the available data.

2.4.6 Equalization of Electronegativities and Atomic Charges

The principle of electronegativity equalization (ENE) proposed by Sanderson [211, 429], states that ‘when two or more atoms with different electronegativity

Table 2.23 Recommended values of electronegativities for atoms in crystals

Li 0.55	Be 1.1	B 1.5	C 2.0[a]	N 2.9	O 3.2	F 3.5			
Na 0.50	Mg 0.9	Al 1.25	Si 1.5[a]	P 2.1	S 2.5	Cl 2.7			
K 0.40	Ca 0.7	Sc 1.0	Ti 1.3	V^{III} 1.3	Cr^{II} 1.0	Mn^{II} 1.0	Fe^{II} 1.05	Co^{II} 1.1	Ni^{II} 1.1
Cu 1.0	Zn 1.15	Ga 1.3	Ge 1.6[a]	As 2.0	Se 2.4	Br 2.6			
Rb 0.40	Sr 0.6	Y 0.95	Zr 1.2	Nb^{III} 1.2	Mo^{IV} 1.3	Tc^{IV} 1.4	Ru^{IV} 1.4	Rh^{IV} 1.4	Pd^{IV} 1.45
Ag 0.95	Cd 1.1	In 1.25	Sn 1.3	Sb 1.65	Te 1.9	I 2.3			
Cs 0.35	Ba 0.6	La 0.9	Hf 1.2	Ta^{III} 1.1	W^{IV} 1.3	Re^{IV} 1.5	Os^{IV} 1.4	Ir^{IV} 1.45	Pt^{IV} 1.5
Au 1.15	Hg 1.3	Tl 0.8	Pb 1.2	Bi 1.65	Th^{IV} 1.3	U^{IV} 1.4			

combine, they become adjusted to the same intermediate EN within the compound'. This approach became very popular and was applied in numerous empirical [213, 214, 433–435, 450–453] and quantum-chemical [454–472] studies. It allows fast calculation of atomic charges for large series of molecules and crystals, which agree well with *ab initio* calculations and experimental results. According to Parr, the EN of an atom can be treated as the chemical potential (see Eq. 2.83)

$$\chi = -\mu = -\left(\frac{\partial E}{\partial N}\right),$$

so the equalization principle corresponds to equalization of chemical potentials of atoms in a compound. The problem is that in isolated atoms the number of electrons, N, must be integer; hence E is not a continuous function of N. However, if Eq. 2.83 is applied to an individual atom in a molecule, fractional N are acceptable. The mathematics of treating $E(N)$ as being continuous function have been discussed [473].

A method of calculating the molecular electron compactness by Sanderson as

$$EC_{MX} = \sqrt{EC_M EC_X} \tag{2.100}$$

allows us to calculate the atomic charges in molecules by comparing the molecular and atomic EC. Sanderson has postulated (assuming the bond ionicity $q = 0.75$ in NaCl) that one positive or negative charge on atom A will change its EC by the increment $\Delta q = \pm a\sqrt{EC_A}$. The coefficient was estimated as 2.08, later corrected to 1.56 [472]. Thereby it is possible to calculate *EC* for any cations and anions, and from them to calculate bond ionicities

$$q_A = \frac{EC_{AB} - EC_A}{EC_{A^+} - EC_A}. \tag{2.101}$$

Sanderson has applied this principle indiscriminately, assuming *EC* to equalize for all atoms even in such species as K_2SO_4, where K and S play quite different chemical roles and have different valences. Later it was suggested [474, 475] to equalize *EC* in separate bonded pairs of atoms, rather than throughout the entire molecule. It was also observed that total equalization in organic molecules would give different *EC* for isomers of the same composition, and a novel, rather efficient, method of calculating *EC* for isomers was proposed instead [451–453].

One must keep in mind that different scales of EN have different dimensionality, viz. energy (or potential) in Mulliken's scale, square root of energy in Pauling's scale, relative electron density in Sanderson's, whereas Parr et al. defined the absolute EN as the electronic chemical potential. There is no unique method to calculate EN, for every scale has its own calculation scheme, as it is done by Bratsch for Pauling's scale [213, 214]. ENs of atoms M and X in a M–X bonds can be equalized using the simple rule

$$\chi_M \times f = \frac{\chi_X}{f} \tag{2.102}$$

where f is the equalization factor, $f = \sqrt{\chi_X/\chi_M}$, and χ is the Mulliken electronegativity of atoms (see Eq. 2.80). EN equalization will influence the interatomic distance, decreasing the M size in the M–X separation:

$$r_{q+} = \frac{r_o}{f} \tag{2.103}$$

where r_o is the orbital radius of the electroneutral atom and r_{q+} is the radius of the same atom with a charge of $q+$ [476]. As the first approximation, the atomic radii of the metal atoms in molecules with fractional charges can be calculated by linear interpolation between the radii of neutral atoms and corresponding cations, which gives the bond ionicity (see Table 2.24) [477],

$$i = \frac{r_o - r_{q+}}{r_o - r_{cat}}. \tag{2.104}$$

Bond ionicities in solids can also be calculated in this manner, taking into account the real valence states of atoms. Table 2.25 contains the χ(X) for the tetragonal (te, sp^3) and octahedral (oc, sp^5) hybridization of bonds in structures of the ZnS and NaCl types, together with the standard Mulliken's values of χ(M), and the calculated r_{q+} and i_{cr}, in crystalline compounds MX [477].

Table 2.26 gives the comparison of the bond ionicities in molecules and crystals (i_{mol} and i_{cr}, from Tables 2.24 and 2.25) with the bond polarities calculated as $p = \mu/d$ from the dipole moments (μ) and bond lengths (d), and with the effective charges (e^*)

Table 2.24 Bond ionicities (as fractions of e) in MX molecules

M^I	H	F	Cl	Br	I	M^{II}	O	S	Se	Te
Li	0.40	0.57	0.49	0.46	0.44	Be	0.35	0.24	0.21	0.18
Na	0.44	0.62	0.54	0.51	0.48	Mg	0.43	0.32	0.29	0.25
K	0.58	0.76	0.68	0.65	0.62	Ca	0.61	0.49	0.46	0.42
Rb	0.63	0.83	0.74	0.71	0.68	Sr	0.69	0.57	0.54	0.50
Cs	0.71	0.91	0.82	0.78	0.75	Ba	0.80	0.67	0.64	0.59
Cu	0.29	0.54	0.42	0.38	0.35	Zn	0.40	0.26	0.22	0.17
Ag	0.37	0.68	0.53	0.48	0.44	Cd	0.54	0.37	0.33	0.27
Au	0.22	0.67	0.46	0.39	0.32	Hg	0.50	0.28	0.23	0.15

Table 2.25 Electronegativities (in Mulliken's scale), orbital atomic radii (in Å) and the bond ionicity in MX crystals

M	χ(M)	F_{oc} (15.82)[a]		Cl_{oc} (11.22)		Br_{oc} (10.52)		I_{oc} (9.51)	
		r_{q^+}	i_{cr}	r_{q^+}	i_{cr}	r_{q^+}	i_{cr}	r_{q^+}	i_{cr}
Li	3.005	0.691	0.64	0.821	0.55	0.848	0.53	0.891	0.50
Na	2.844	0.726	0.69	0.862	0.59	0.891	0.57	0.938	0.54
K	2.421	0.846	0.84	1.004	0.74	1.037	0.72	1.091	0.68
Rb	2.332	0.878	0.91	1.043	0.80	1.077	0.78	1.132	0.74
Cs	2.183	0.935	0.99	1.111	0.88	1.147	0.86	1.206	0.82
M	χ	F_{te} (17.63)		Cl_{te} (12.15)		Br_{te} (11.46)		I_{te} (10.26)	
Cu	4.477	0.600	0.68	0.723	0.54	0.744	0.52	0.787	0.47
Ag	4.439	0.645	0.85	0.777	0.68	0.800	0.65	0.846	0.59
Au	5.767	0.679	0.92	0.818	0.67	0.842	0.62	0.890	0.54
M	χ	O_{oc} (12.56)		S_{oc} (9.04)		Se_{oc} (8.64)		Te_{oc} (7.83)	
Mg	4.11	0.732	0.53	0.862	0.404	0.882	0.385	0.927	0.341
Ca	3.29	0.865	0.72	1.019	0.583	1.043	0.562	1.095	0.517
Sr	3.07	0.908	0.81	1.070	0.665	1.094	0.645	1.150	0.596
Ba	2.79	0.971	0.91	1.144	0.769	1.171	0.746	1.230	0.697
M	χ	O_{te} (14.02)		S_{te} (9.84)		Se_{te} (9.48)		Te_{te} (8.52)	
Be	4.65	0.599	0.49	0.715	0.36	0.728	0.35	0.768	0.30
Zn	4.99	0.635	0.57	0.758	0.41	0.773	0.39	0.815	0.33
Cd	4.62	0.680	0.74	0.811	0.55	0.826	0.53	0.872	0.46
Hg	5.55	0.708	0.77	0.846	0.52	0.862	0.49	0.909	0.40

[a]electronegativities of non-metals are given in parentheses

of atoms, determined by Szigeti's method (see Chap. 11). In each case, $i_c > i_{mol}$, in accordance with chemical experience, and the calculated i agrees qualitatively with the empirical values of p and e^*. At the same time, p varies non-monotonically, e.g. for fluorides as LiF < NaF > KF > RbF > CsF, and for iodides as LiI < NaI < KI < RbI > CsI, because of two competing effects: (i) EN of the metal atom decreases with the increase of its size, but (ii) bond ionicity is reduced by the polarizing influence of anions on cations, which increases with the cation size. For this reason, bond ionicity in CsX is always lower than in RbX.

Table 2.26 Calculated and empirical values of the bond ionicity in molecules and crystals MX

MX-type compounds									
M	Property[a]	X = F		X = Cl		X = Br		X = I	
Li	p, i_{mol}	0.84	0.57	0.73	0.49	0.70	0.46	0.65	0.44
	e^*, i_{cr}	0.81	0.64	0.77	0.55	0.74	0.53	0.54	0.50
Na	p, i_{mol}	0.88	0.62	0.79	0.54	0.79	0.51	0.71	0.48
	e^*, i_{cr}	0.83	0.69	0.78	0.59	0.75	0.57	0.74	0.54
K	p, i_{mol}	0.82	0.76	0.80	0.68	0.78	0.65	0.74	0.62
	e^*, i_{cr}	0.92	0.84	0.81	0.74	0.77	0.72	0.75	0.68
Rb	p, i_{mol}	0.78	0.83	0.78	0.74	0.77	0.71	0.75	0.68
	e^*, i_{cr}	0.97	0.91	0.84	0.80	0.80	0.78	0.77	0.74
Cs	p, i_{mol}	0.70	0.91	0.74	0.81	0.73	0.78	0.73	0.75
	e^*, i_{cr}	0.96	0.99	0.85	0.88	0.82	0.86	0.78	0.82
Cu	p, i_{mol}	0.69	0.54	0.53	0.42				
	e^*, i_{cr}		0.68	0.66	0.54	0.64	0.52	0.60	0.47
Ag	p, i_{mol}	0.65	0.68	0.55	0.53				
	e^*, i_{cr}	0.89	0.86	0.71	0.68	0.67	0.65	0.61	0.59
MO-type compounds[b]									
	M	Be	Mg	Ca	Sr	Ba	Zn	Cd	Hg
	i_{mol}	0.35	0.43	0.61	0.69	0.80	0.40	0.54	0.50
	i_{cr}	0.49	0.53	0.72	0.81	0.91	0.57	0.74	0.77
	$e^*/2$	0.55	0.59	0.62	0.64	0.74	0.60	0.59	0.57

[a] p and e^* in the left sub-columns, i_{mol} and i_{cr} in the right ones, [b] p are not given, because for oxides the measurements of μ are few and unreliable

Table 2.27 Electronegativities, empirical atomic radii (Å) and the bond ionicity in molecules MX

M	χ	H (7.176)		F (12.20)		Cl (9.35)		Br (8.63)		I (8.00)	
		r_{q^+}	i_{mol}	r_{q^+}	i_{mol}	r_{q^+}	i_{mol}	r_{q^+}	i_{mol}	r_{q^+}	i_{mol}
Li	3.005	1.721	0.43	1.320	0.62	1.508	0.53	1.570	0.50	1.630	0.47
Na	2.844	1.907	0.50	1.463	0.69	1.671	0.60	1.739	0.57	1.807	0.54
K	2.421	2.108	0.60	1.617	0.80	1.847	0.70	1.923	0.68	1.997	0.64
Rb	2.332	2.178	0.64	1.670	0.83	1.908	0.74	1.986	0.71	2.062	0.68
Cs	2.183	2.289	0.68	1.755	0.88	2.005	0.79	2.087	0.76	2.168	0.73
Cu	4.477	1.493	0.31	1.145	0.58	1.308	0.46	1.361	0.41	1.414	0.37
Ag	4.439	1.612	0.40	1.236	0.73	1.412	0.58	1.470	0.52	1.527	0.47
Au	5.767	1.766	0.20	1.354	0.59	1.547	0.41	1.610	0.35	1.673	0.29

Since the atomic size is not uniquely defined (see Chap. 1), it is important to assess how much this uncertainty affects the calculations. Table 2.27 illustrates the calculation of the polarities using Pearson's ENs, the empirical radii of neutral isolated atoms [478] and their molecular cations (see Chap. 1). Comparison with the results in Table 2.24 reveals the average variation of 5.6 %, which is acceptable for the purposes of structural chemistry.

In conclusion of this section we should note that the concept of EN has been created by Pauling, first of all, to estimate the bond ionicity (i), i.e. the displacement of valence electrons towards one of the atoms. Experimental values of i(H–X), defined as the ratio of the dipole moment to the bond length, have been approximated by

Table 2.28 Dependence of bond ionicity (%) on differences of electronegativities

$\Delta\chi$	molecule	crystal	$\Delta\chi$	molecule	crystal	$\Delta\chi$	molecule	crystal
0.1	1	4	1.1	23	39	2.1	54	66
0.2	2	8	1.2	26	42	2.2	58	69
0.3	3	12	1.3	29	45	2.3	61	71
0.4	5	16	1.4	32	48	2.4	64	73
0.5	7	20	1.5	35	51	2.6	70	77
0.6	9	23	1.6	38	54	2.8	75	81
0.7	11	26	1.7	41	57	3.0	80	85
0.8	14	29	1.8	44	59	3.2	84	88
0.9	17	32	1.9	47	61	3.4	88	91
1.0	20	36	2.0	51	64	3.6	91	94

Pauling [479] as

$$i = 1 - e^{-A} \tag{2.105}$$

where $A = c\Delta\chi^2$ (in the beginning was accepted $c = 0.25$, and then 0.18). This formula agrees with observations and is often used in the structural and quantum chemistry to estimate bond ionicity in molecules. The change of the bond ionicity upon transition from molecules to a solid (which from the structural viewpoint is principally a change of N_c) can be considered either using 'crystalline EN' or changing an exponent in Eq. 2.105 by $1/N_c$ [18]. The values of i in molecules and crystals as functions of $\Delta\chi$, detrmined by all available experimental methods, are summarized in Table 2.28.

As noted in Chap. 1, a change of the atomic valence has only a slight ($\leq$10 %) effect on the bond energy. Apparently, an increase of the net charge on the metal atom in a polar molecule or in a solid, caused by the increasing valence, should correspondingly enhance the Coulomb component, which is the major part of the bond energy. However this does not occur. There are two alternative explanations of this: either the electric charges on two atoms act only within an orbital and charges of other bonds (accordingly, the total charge on the metal atom) have no effect on the strength of the given bond or, as the valence changes, atomic charges vary so that their product (hence, Coulomb's energy) is invariant. In [313] both alternatives were considered and the latter proved consistent with available experimental data, implying that the effective atomic charge varies in inverse proportion to the oxidation number of the atom. As a first approximation, we assume that as the valence v(M) in MX_n increases in the succession $1 \rightarrow 2 \rightarrow 3 \ldots \rightarrow 8$, the product of the effective charges of atoms M and X remains invariant. Expressing the charge of an atom in a single bond as e^*, we obtain:

$$(e^*)^2 = \frac{(ve^*)_M}{m}\frac{(e^*)_X}{m} \tag{2.106}$$

and, hence, $v = m^2$, where m is the number by which the effective atomic charge (bond ionicity) must be divided for the Coulomb energy to remain constant. Therefore, as

$v(M)$ increases and the ligands remain the same, the effective charge of an atom decreases in proportion to $\sqrt{v}$, i.e. as $\sqrt{1} \rightarrow \sqrt{2} \rightarrow \sqrt{3} \ldots \rightarrow \sqrt{8}$. From here, if we know $\chi(M)$ and e^* of the M–X bond for one v (usually for a low-valence state, as these are better studied), we can find the bond ionicities for other valences and, using the dependence $i = f(\Delta\chi)$ and data from Table 2.28, to define χ. Such values, obtained for molecular and crystalline halides, are close to empirical ENs [313].

The values of i calculated from Eq. 2.105 on $v(M)$ in different molecules, give the average (for 700 molecular halides and chalcogenides) of $e^* = \pm 0.5e$, in agreement with Pauling's famous Electroneutrality Rule [480, 481] which states that net charges of atoms in stable molecules and crystals should not exceed ± ½, even though later he softened this limitation to ±1 [475]. This principle later has been proved theoretically, confirmed experimentally and now plays a key role in the description of electronic structure of molecules and crystals.

2.5 Effective Charges of Atoms and Chemical Behavior

In this section we shall consider only the acid-basic properties and redox reactions, as the processes most closely connected with the electronic structure of substances. According to the Brönsted-Lewis theory, the acidity of oxygen-containing molecules depends on the effective charge on the oxygen atom. Sanderson [482] has shown that values of *EC* of oxides are inversely proportional to the *pH* of their aqueous solutions. Reed [483] has shown that *pK* of hydrates and amino-complexes of transition metals also depends on their atomic charges. The EN concept allows also to explain the acid-base properties of organic substances: higher acidity of aromatic compounds in comparison with aliphatic molecules is caused by higher positive charge on the H. For similar reasons, phenols are more acidic than aliphatic alcohols. Apart from the effects of the proximity of multiple bonds, the acidic properties of organic compounds with C–OH bonds depend on other atoms enhancing the EN of the carbon atom and, therefore, the effective positive charge on H. For this reason, Cl_3CCOOH is a stronger acid than H_3CCOOH.

Let us now consider redox reactions from the chemical bonding viewpoint, which is important for physical and structural chemistry. For this purpose we again must return to the concept of atomic charge. This term is used to describe two basically different things: the 'intrinsic charge of atom, q_i' (ICA) and the 'coordination charge of atom, Ω' (CCA). The first type is a deficit (positive charge) or an excess (negative charge) of electrons inside the closed shells of the bonded atoms in comparison with those in the isolated state. This q_i defines the Coulomb energy, is responsible for the IR absorption, causes the atomic polarization bands, and affects the binding energy of the internal electrons in the atom. However, what matters for redox reactions is the electron density in the interatomic space, i.e. the CCA [436]. Suchet, to highlight the same distinction, introduced the terms 'physical' and 'chemical' charges [484, 485]. The CCA of the M and X atoms in a MX crystal are

$$\Omega_M = +Z - cNN_c \text{ and } \Omega_X = -Z + NN_c, \qquad (2.107)$$

where Z is the formal charge (valence), c and N are the covalency and the order (multiplicity) of the bonds, N_c is the coordination number. These Ω_M can be compared with the charges determined by X-ray spectroscopy, in which an electron is promoted from an internal shell and into the region of chemical bonding [54]. Table S2.21 contains the Ω_M of several transition metals in complex compounds, experimental and calculated by the EN method. According to these calculations, in crystalline compounds with low bond polarity, Ω_M can even become negative if $N_c > Z$. This prediction has been confirmed by physical methods, e.g. XRD studies of the electron density in PbS, PbSe, and PbTe have shown that within the Pb atom region, limited by $r = 1.66$ Å, there is a negative net charge of -0.4, -0.9, and $-1.1e$, respectively [486]. Gold compounds provide another proof. Thus, CsAu crystallizes in a CsCl-type structure and has $E_g = 2.6$ eV [487], indicating the ionic (rather than inter-metallic) character of the solid, with Au acting as an anion. Note also the structural similarity between $K_3\mathit{Br}O$ and $K_3\mathit{Au}O$ [488]. XRS studies [489] revealed that the AuL_I and AuL_{III} absorption edge energies monotonically decrease in the succession Au_2O_3, $AuCl_3$, AuCN, Au, CsAu and M_3AuO, hence in the last two compounds the charge of Au must be negative. Additional argument in favour of Au^- anion is the dissociation of $M_7Au_5O_2$ compounds into Au^+ and Au^- ions [490]. Besides, ESCA measurements [80, 491] have shown that the electronic structure of BaAu, $BaAu_2$ and $BaAu_{0.5}Pt_{0.5}$ compounds can be formulated as $Ba^{2+}[e^-] \cdot [Au^-]$, $Ba^{2+}(Au^-)_2$ and $[Ba^{2+} \cdot 0.5e^-] \cdot [Au^-{}_{0.5} \cdot Pt^{2-}{}_{0.5}]$, respectively. Such behavior of Au is caused by its having the highest electron affinity of all metals ($A = 2.31$ eV). Platimum takes the second place with $A = 2.12$ eV and, accordingly, Cs_2Pt has the electronic structure $Cs_2{}^+ Pt^{2-}$, i.e. can be considered as an analogue of alkali metal chalcogenides, M_2X [492]. These recent results confirmed negative charges on Pt and Au, predicted from 1959 onwards on the basis of electronegativities [493–495]. This prediction has an important chemical corollary: oxidation of certain compounds of Au and Pt will rise the metal valence (without replacing 'anions') and yield a salt with mixed ligands, e.g.

$$PtI_2 + Cl_2 \rightarrow PtI_2Cl_2$$

For other metals the result will be substitution of the halogeno anions, which does not happen here because the halogen atoms are *not* anions. Similarly, all possible mixed tetra-halides and di-chalcogenides of Pt, tri-halides of Au, and di-halides of Cu were synthesized by Batsanov et al., see reviews in [356, 496]. Similar results were obtained for other high-EN metals, such as Hg, Tl, Sn, Mn, and by other researchers for Fe, Sb, Cr, Re, W, U, with a variety of ligands, such as halogens, chalcogens, SCN, N_3, NO_3, CO_3, SO_4 and methyl. Mixed halides of Pt^{IV} formed different isomers, depending on the order in which of halogens were added, e.g. PtX_2 has the motif of squares with shared vertices, but additional halogens complete these to octahedra in PtX_2Y_2; such compounds were named square-coordinate isomers [493]. TlSeBr also showed different properties depending on the route of the synthesis: Se + TlBr $\rightarrow$ Se=Tl–Br or 2Tl + Se_2Br_2 $\rightarrow$ 2 Tl–Se–Br, where Tl had different valences. The

Table 2.29 Change of atomic charges ($-\partial e^*/\partial P$, 10^{-4} GPa^{-1}) in crystals MX under $P = 10$ GPa

M^I	F	Cl	Br	I
Li	5.2	1.5	0	1.9
Na	7.0	2.4	1.5	4.0
K	11.7	5.9	5.4	4.1
Rb	10.6	3.9	2.4	1.7
Cs	10.4	6.7	8.3	7.6
Cu		−8.4	−4.7	6.9
Ag	−3.7	−5.6	−3.4	−6.6
Tl	−12.6	−12.5	−11.6	−10.9
M^{II}	O	S	Se	Te
Be	0	−1.6	1.6	3.7
Mg	0.9	0	0	5.6
Ca	3.7	2.2	3.3	7.5
Sr	7.0	5.9	7.2	9.6
Ba	16.5	16.5	18.0	24.6
Zn	0	−2.2	0	4.4
Cd	0	−2.8	0	5.9
Hg	−5.9	−2.8	0.6	10.0
Sn	−12.2	−2.8	0	0
Pb	−7.8	−3.7	0	4.7
Mn	−0.4	−4.0		1.6
M^{III}	N	P	As	Sb
B	0.6	9.7	4.7	
Al	0.6	5.3	0.6	0
Ga	2.2	9.4	2.5	2.8
In	6.9	10.0	1.9	1.9
La		15.3	14.7	14.4
Th	2.5	7.5		
U	0	3.4		

given structural formulae were confirmed by IR-spectroscopy and these compounds were named the valence isomers [497]. Dehnicke [498] discovered the reactions of 'chemical annihilation of charges'. For example, chloro ligands bear negative charges in $SbCl_5$ but positive in ClF, thus a reaction between these compounds yields $SbFCl_4$ and Cl_2. Similar reactions was carried out with hydrides, viz. $MBH_4 + HX = MBH_{4-n}X_n + H_2$ [499].

2.6 Change of Chemical Bond Character under Pressure

Distribution of the electron density in molecules and crystals depends on thermodynamic parameters. Spectroscopic methods [500–507] show that in crystals under high pressures, e^* usually decreases, although in AgI, TlI, HgTe, AlSb, GaN, InAs, PbF_2 it increases with pressure. High-pressure XRD studies of SiO_2 also indicated an increase of bond ionicity [508]. Studies of Se and GaSe under pressure showed that under compression, bonding electrons are displaced from covalent to intermolecular regions [502, 504] with shortening of the intra- and intermolecular distances.

However, difficulties of measurements of ε, n and Ω under high pressures, increasing anharmonicy of vibrations and deformation of IR absorption bands limits the choice of investigated substances and reduces the precision. Therefore, it is desirable to have independent methods of determining the effective charges of atoms in compressed crystals. It has been proposed [376, 509] to derive atomic charges in crystalline compounds under high pressures from the physical properties of the components of the system. Suppose that a reaction $M + X \rightarrow MX$ has the thermal effect Q at ambient thermodynamic conditions. For such system under pressure, the compression work (W_c) of the initial reagents and the final product can be calculated as

$$W_c = 9V_oB_o/\eta^2\{[\eta(1-x)-1]\exp[\eta(1-x)]+1\} \quad (2.108)$$

deduced by integrating the 'universal equation of state' (EOS) of Vinet-Ferrante [510]

$$P(x) = 3B_o\left[\frac{(1-x)}{x^2}\right]\exp[\eta(1-x)] \quad (2.109)$$

where V_o and V are the starting and final molar volumes, respectively; $x = (V/V_o)^{1/3}$; B_o is the bulk modulus; $\eta = 1.5(B_o'-1)$ and B_o' is the pressure derivative of B_o. Obviously, if W_c (mixture) – W_c (compound) > 0, then ΔW_c should be subtracted from the standard heat effect to yield the Q corresponding to high pressures, and *vice versa*. Usually $\Delta W_c > 0$, hence under pressure Q and $\Delta\chi$ decrease. The results of such approach qualitatively agree with the experiment, except for alkali hydrides, where calculations predict e^* to fall to 0 already at several tens of GPa while in fact no change of electronic structure occurred up to 100 GPa and even beyond [511, 512]. This contradiction can be resolved by taking into account that the compression work only partly goes into changing the chemical bonds [513]. Studies of A_2 molecules and chalcogens in condensed state, revealed that the compression initially (or mainly) results in the contraction of intermolecular distances and only after the bond equalization, i.e., the transformation of the molecular structure into a monatomic one, the covalent bonds begin to shorten. Therefore W_c calculated using Eq. 2.109 must be multiplied by the ratio of vdW energy (ΔH_s) to the A–A bond energy (E_b) in order to obtain the 'efficiency factor', Φ, of high pressure. For metals and semi-metals $\Phi = E_a/B_oV_o$ where E_a is the atomization energy and B_oV_o is the compression energy reduced to $P = 0$, and the product $\Phi \times W_c$ characterizes the compression energy (E_c) spent on altering the chemical bonding.

Thus, a comparison of E_c of mixtures and compounds allows to define a change of Q (and hence of the ENs of atoms) on variation of P, and from Table 2.28 to find the effective charges of atoms. The decrease of Q and the bond polarity under pressure is observed in crystals of AB-type, viz. Group 1– Group 17, Group 2– Group 16, Group 13–Group 15 compounds. Table 2.29 shows that $\partial e^*/\partial P$ for these crystals decrease by 10^{-4} to 10^{-3} GPa^{-1}. Szigeti's method predicts the same signs and similar absolute values, $\partial e^*/\partial P = 1$ to 3.3×10^{-4} GPa^{-1}. Remarkably, the increase of $Q(P)$ in CuX, AgX, and TlX under high pressures indicates an increase of polarity; unfortunately, corresponding data by Szigeti's method are not available.

Chalcogenides of bivalent metals can be divided into two classes: compounds of the Group 2 metals, crystallizing in the B1 structures, become less ionic under compression, while compounds of the Group 12 metals, crystallizing in the B3 structure, become more ionic. On compression of crystalline compounds of the Group 13 and Group 15 elements, the effective charges decrease in agreement with the results of Szigeti's method. Such behavior of substances under pressure can be explained assuming the additive character of compressibility of compounds. If the anion is softer than the cation (e.g. in halides of Cu, Ag and Tl) it will be compressed more strongly. The electronegativity of the anion, being inversely related to the atomic size, will increase and so will $\Delta\chi$, as $\chi_X > \chi_M$. In the case of softer cation (e.g., in alkali halides) χ_M on compression will increase more strongly than χ_X, and ionicity will decrease. However, under stronger compression, as calculated from the experimental EOS, ΔW_c does not change to $P \approx 100$ GPa and on further compression even decreases, as has been shown experimentally by the shock-wave technique [514]. Hence, the situation when $\Delta W_c = Q$, i.e. when the compound must dissociate to neutral atoms (elements), cannot be reached at any pressure. However, metallization of ionic crystals under pressure has been proven experimentally. As noted above, the volumes of MX crystals under P when $\Delta W_c = E_a$, i.e. when chemical bonds are destroyed and valence electrons delocalized, correspond to distances $d(\text{M–X}) = r(\text{M}^+) + r(\text{X}^o)$ [515]. It means that if on compression of MX bond polarity decreases, as in alkali halides, ZnO and GaAs [517], then the donor of electrons must be M^o, since $I(\text{M}) < I(\text{X})$. If the bond polarity increases, as in SiC [516], SiO_2 [508], ZnS [63], or remains nearly constant, as in AlN and GaN [507], then X^- must be the donor, since $A(\text{X}) < I(\text{M})$.

The behavior of metals under pressure is remarkable. At ambient conditions, the metal atoms are ionized (by releasing itinerant electrons), but only partly so. Under compression the rest of the outer-shell electrons are 'squeezed out', and the degree of ionization of atoms increases. Ultimately, the atomic cores become cations and the crystal structure of a metal will correspond to a close-packing of cations. Stabilization of such a system requires very high pressures, to counterbalance the repulsion of cations. The parameters of such ultimate states have been calculated [517], see Table S2.22. The internuclear distances are expected to equal the sums of cationic radii.

There have been other attempts to estimate the change of EN and bond polarity under compression. The principal difficulty is that intra- and intermolecular distances change differently. In [518] the increase of covalent radii was calculated as the inverse of the reduction of vdW-radii of elements, and in [519] as being proportional to the ratio of energies of the chemical and vdW-bonds. For the metallic state, the following radii were obtained (Å): F 1.00, Cl 1.25, Br 1.41 and I 1.61, whereas experimental values are Br 1.41 Å [523] and I 1.62 Å [521]; for fluorine and chlorine data are unavailable. Certainly, this approach is not rigorous: it makes no allowance for polymorphic transformations, at which the material changes its properties by a jump. Nevertheless, the obtained values of $\partial e^*/\partial P$, summarized in Table S2.23, are close to experiemental results. The problem of changing bond lengths under pressure has been explored theoretically in terms the bond-valence model [522], yielding for ionic crystals a quantitative dependence

$$\frac{\Delta d_o}{\Delta P} = 10^{-4}\frac{\Delta d_o^4}{B}, \quad (2.110)$$

where d_o is the initial bond length, $B = 1/b - 2/d_o$ and $b = 0.37$. This relationship allows to compute the effects of pressure on bond lengths and force constants.

2.7 Conclusions

The formation of chemical bonds in molecules or crystals release the energy equal to a few tenths, and more often not exceeding 0.1, of the ionization potentials of the individual (isolated) atoms involved. The ionization potentials of atoms in molecules decrease by similar amounts compared to the free state. Bond energies themselves are determined by the ionization potentials of the isolated atoms, according to Mulliken's theory. Thus the major part of the energy of any chemical system in any aggregate state depends on the nature of the component atoms, and the remaining energy is mostly defined by the immediate atomic environment, or the short-range order in a crystal structure. Inversely the ionization potential, i.e. the electron energy of an atom in molecule or crystal, differs but slightly from that in isolated state. Therefore, in most cases it is a good approximation to regard a molecule as a combination of atoms and to account for all interactions as perturbations. From this viewpoint the geometrical structure of substances will be discussed in the next chapters.

Appendix

Supplementary Tables

Experimental data from the *Handbook of Chemistry and Physics*, 88th edn (2007–2008) are presented without reference, otherwise the references to original papers are given.

Table S2.1 'XRD' effective atomic charges in silicates and complex compounds

± A	Be_2SiO_4	Mg_2SiO_4	Mn_2SiO_4	Fe_2SiO_4	Co_2SiO_4
+ M	0.83	1.75	1.35	1.15	1.57
+ Si	2.57	2.11	2.28	2.43	2.21
− O	1.06	1.40	1.25	1.19	1.29
± A	$MgCaSi_2O_6$	$Mg_2Si_2O_6$	$Fe_2Si_2O_6$	$Co_2Si_2O_6$	$LiAlSi_2O_6$
+ M	1.42	1.82	1.12	0.95	1.0, 1.74
+ Si	2.56	2.28	2.19	2.28	1.8
− O	1.33	1.37	1.10	1.08	1.06
± A	$Al_2SiO_4F_2$	$CaAl_2Si_3O_{10}$	$LiFePO_4$	NaH_2PO_4	
+ M	1.53	2, 1.90	1, 1.35	0.2, 0.6	
+ Si	1.75	1.84	0.77	1.8	
− O	1.00	1.14	0.78	0.8	
± A	K_2NiF_4	Cs_2CoCl_4	K_2PdCl_4	K_2PtCl_4	
+ A	1.82	0.7	0.5	1.0	
− X	0.95	0.7	0.6	0.75	
± A	K_2ReCl_6	K_2PdCl_6	K_2OsCl_6	K_2PtCl_6	
+ A	1.6	1.97	2.5	1.88	
− Cl	0.6	0.66	0.75	0.65	
Complex	A	e^*	Complex	A	e^*
$[Co(NH_3)_6]$	N	−0.62	$[Cr(CN)_6]$	C	+0.22
	H	+0.36		N	−0.54
	Co	−0.49		C	−0.38

Table S2.2 Bond metallicity in crystalline halides MX

M^I	F		Cl		Br		I	
	m_1	m_2	m_1	m_2	m_1	m_2	m_1	m_2
Li	0.06	0.07	0.12	0.10	0.14	0.11	0.17	0.12
Na	0.04	0.05	0.11	0.08	0.10	0.09	0.13	0.11
K	0.03	0.03	0.06	0.07	0.08	0.08	0.10	0.09
Rb	0.02	0.03	0.06	0.06	0.08	0.07	0.10	0.09
Cs	0.02	0.02	0.05	0.05	0.06	0.06	0.08	0.08
Cu			0.12	0.08	0.15	0.09	0.18	0.11
Ag	0.06	0.05	0.12	0.09	0.14	0.10	0.18	0.11
Tl	0.06	0.05	0.12	0.10	0.14	0.12	0.18	0.14
M^{II}	O		S		Se		Te	
Be	0.15	0.12	0.24	0.14	0.27	0.17	0.32	0.18
Mg	0.09	0.10	0.15	0.14	0.19	0.16	0.23	0.17
Ca	0.07	0.09	0.13	0.13	0.15	0.15	0.19	0.17

Table S2.2 (continued)

M^I	F		Cl		Br		I	
	m_1	m_2	m_1	m_2	m_1	m_2	m_1	m_2
Sr	0.06	0.07	0.12	0.12	0.13	0.13	0.17	0.16
Ba	0.06	0.06	0.11	0.11	0.13	0.12	0.16	0.14
Zn	0.13	0.11	0.17	0.15	0.20	0.17	0.26	0.19
Cd	0.12	0.11	0.16	0.15	0.19	0.16	0.25	0.19
Hg	0.15		0.20	0.18	0.26	0.20	0.30	0.23
Sn	0.12	0.10	0.16	0.15	0.19	0.16	0.25	0.19
Pb	0.12	0.11	0.16	0.16	0.19	0.18	0.25	0.21
Cr			0.18	0.14	0.21	0.15	0.26	0.18
Mn	0.11	0.09	0.18	0.12	0.20	0.13	0.25	0.17
Fe	0.13	0.10	0.20	0.15	0.23	0.17	0.29	0.20
Co	0.13	0.11	0.21	0.17	0.24	0.18	0.30	0.21
M^{III}	N		P		As		Sb	
B	0.25	0.18	0.38	0.21	0.44	0.23		
Al	0.20	0.18	0.32	0.24	0.38	0.28	0.40	0.26
Ga	0.21	0.18	0.35	0.24	0.42	0.28	0.44	0.28
In	0.21	0.13	0.34	0.19	0.41	0.23	0.44	0.22
Sc	0.15	0.15	0.26	0.22	0.32	0.26	0.33	0.26
Y	0.13	0.13	0.23	0.21	0.29	0.25	0.31	0.25
La	0.12	0.12	0.22	0.20	0.28	0.24	0.29	0.24
U	0.16	0.14	0.28	0.22	0.33	0.26	0.36	0.26

Table S2.3 Dissociation energies of diatomic molecules (kJ/mol) ($E(M_2)$, kJ/mol: Nb_2 513, Tc_2 330, Re_2 432, Os_2 415, Ir_2 361)

M	Molecules					
	MF	MCl	MBr	MI	MH	M_2
H	570	431	366	298	436	436
Li	577	469	419	345	238	105
Na	477	412	363	304	186	74.8
K	489	433	379	322	174	53.2
Rb	494	428	381	319	173	48.6
Cs	517	446	389	338	175	43.9
Cu	427	375	331	289	255	201
Ag	341	311	278	234	202	163
Au	325	302[a]	286[a]	263[a]	292	226
Be	573	434	316	261	221	11.1[b]
Mg	463	312	250	229	155[A]	4.82[b]
Ca	529	409	339	285	163	13.1b[b]
Sr	538	409	365	301	164	12.94[b]
Ba	581	443	402	323	192	19.5[b]
Zn	364	229	180	153	85.8	3.28[c]
Cd	305	208	159	97.2	69.0	3.84[c]
Hg	180	92.0	74.9	34.7	39.8	4.41[c]
B	732	427	391	361	345	290
Al	675	502	429	370	288	133
Ga	584	463	402	334	276	106

Table S2.3 (continued)

M	Molecules					
	MF	MCl	MBr	MI	MH	M_2
In	516	436	384[d]	307	243	82.0
Tl	439	373	331	285	195	59.4
Sc	599	435[e]	365[e]	300[e]	205	163
Y	685	523	481	423		270
La	659	522	446	412		223[E]
C	514	395	318	253	338	618
Si	576	417	358	243	293	320[f]
Ge	523	391	347	238[g]	263	261[f]
Sn	476	350	337	235	264	187
Pb	355	301	248	184	158[h]	83[f]
Ti	569	405	373	262	205	118
Zr	627	530	420	298[i]	312	298
Hf	650			328[j]		328
N[I]	320	321	254	203	331[k]	945
P[I]	459	342	294	243	293[k]	489
As[I]	463	336	280	240	270[k]	386
Sb[I]	430	292	240	183	260[k]	302
Bi[I]	366	285	181	124	212[k]	204
V	590	477	439		209	269
Ta	573	544				390
O[I]	234	269	241	237	428	498
S[I]	344	264	241	194	351	425
Se[I]	317	227	186	158	300	330
Te[I]	326	209	166	134	256	258
Cr	523	378	328	287	190	152
Mo	464		313		211	436
W	597[m]	458[m]	396[m]	328[m]		666
F	159	261	280	272	570	155
Cl	261	243	219	211	431	240
Br	280	217	193	179	366	190
I	272	211	179	151	298	149
Mn	445	338	314	283	251	61.6
Fe	447	330	298[o]	241[p]	148[q]	118
Co	4315	338	326	285[p]	190[q]	163[n]
Ni	437	377	360	293[p]	243[q]	200[n]
Ru	402				234[q]	193
Rh					247[q]	236
Pd					234[q]	> 136
Pt	582				352[q]	307
Th	652	489	364	336		284
U	648	439	377	299		222

[a][2.1], [A][2.2], [b][2.3], [c][2.4, 2.5], [d][2.6], [e][2.7], [E][2.8], [f][2.9], [g][2.10], [h][2.11], [i][2.12], [I][2.13], [j][2.14], [k][2.15, 2.16], [l][2.17], [m][2.18], [n][2.19], [o][2.20], [p][2.21], [q][2.22]

Table S2.4 Dissociation energies of MZ molecules (kJ/mol)

M	Z						
	O	S	Se	Te	N	P	C
Cu	287	274	255	230			
Ag	357	279	210	196			
Au	233	254	251	237			
Be	440[a]	372					
Mg	338[a]	234					
Ca	383[a]	335					
Sr	415[a]	338	251				
Ba	559[a]	418					
Zn	289[a]	225	171	118			
Cd	231[a]	208	128	100			
Hg	269	217	144	89			
B	809[a]	577	462	354	378	347	448
Al	511[a]	332	318	268	278	217	268
Ga	354[a]			265		230	
In	316[a]	288	245	215		198	
Tl	213[a]					209	
Sc	671	477	385	289	464		444
Y	714	528	435	339	477		418
La	799	573	477	381	519		462
C	1076	713	590	564	750	508	607
Si	800	617	534	429	437		452
Ge	660	534	444[b]	409[b]			456
Sn	528	467	401	338			
Pb	374	343	303	250			
Ti	668	491	381	289	476		423
Zr	766	572			565		496
Hf	790				590		540
N	631	467	370		945	617	754
P	599	442	364	298	617	489	507
As	481[c]	389[c]	352[c]	312[c]	489	433	382
Sb	434	379		277	460	357	
Bi	337	315	280	232		282	
V	637	449	347		523		423
Nb	726						524
Ta	839	670			607	611	
O	498	518	465	376	631	589	1076
S	518	425	371	335	464	444	714
Se	430	371	330	293	370	364	590
Te	377	339	2932	258		298	
Cr	461	331			378	378	
Mn	362	301	239				
Th	877				577	372	453
U	755	528			531	293	455
M	Mo	W	Tc	Re	Fe	Co	Ni
D_{MO}	502	720	548	627	407	384	366
M	Pd	Os	Ir	Pt	Ru	Rh	
D_{MO}	381	575	414	415	528	405	

Table S2.4 (continued)

M	Z						
	O	S	Se	Te	N	P	C
M	Fe	Co	Ni	Pt			
D_{MS}	329	331	318	407[d]			
M	Mo	Tc	Fe	Ni	Ru	Rh	
D_{MC}	482	564	376[e]	337	648	580	
M	Pd	Os	Ir	Pt			
D_{MC}	436	608	631	610			

[a][2.23], [b][2.24], [c][2.25], [d][2.26], [e][2.27]

Table S2.5 Dissociation energies D (kJ/mol) of M_2^+ cations

M	D	M	D	M	D	M	D
Ag	168	Cr	129	Li	132	S	522.5
Al	121	Cu	155	Mg	125	Sb	264
Ar	116	Cs	62.5	Mn	129	Se	413
As	364	F	325.5	Mo	449	Si	334
Au	234.5	Fe	272	N	844	Sn	193
B	187	Ga	126	Na	98.5	Sr	108.5
Be	196.5[a]	Ge	274	Nb	577	Ta	666
Bi	199	H	259.5	Ne	125	Te	278
Br	319	He	230	Ni	208	Ti	229
C	602	Hg	134	O	648	Tl	22
Ca	104	I	263	P	481	V	302
Cd	122.5	In	81	Pb	214	Zr	407
Cl	386	K	80	Pd	197	Xe	99.5
Co	269	Kr	84	Pt	318	Y	281
		La	276[b]	Rb	75.5	Zn	60

[a][2.28], [b][2.8]

Table S2.6 Ionization potentials (eV) for M and M^+ atoms

M	I_1(M)	I_2(M)	I_2/I_1
Be	9.32	18.21	1.95
Mg	7.65	15.04	1.97
Ca	6.11	11.87	1.94
Sr	5.69	11.03	1.94
Ba	5.21	10.00	1.92
Zn	9.39	17.96	1.91
Cd	8.99	16.91	1.88
Hg	10.44	18.76	1.80

Table S2.7 Ionization and dissociation energies (kJ/mol) for metal atoms and molecules

A	I(A)	$I(A_2)$	$E(A_2)$	$E(A_2^+)$	E_a(A)	0.9 Φ(A)
Li	520	493	105	132	159	207
Na	496	472	75	99	107	204
K	419	392	53	80	89	193
Rb	403	376	49	76	81	188
Cs	376	357	44	63	76.5	157

Table S2.8 Dissociation energies *D* (kJ/mol) of MH^+ and MO^+ cations

MH^+	*D*	MH^+	*D*	MO^+	*D*	MO^+	*D*
CuH	93	CrH	136	LiO	39	VO	582
AgH	43.5	MoH	176	NaO	37	NbO	688
AuH	144	WH	222	KO	13	TaO	787
BeH	307	OH	488	RbO	29	NO	115
MgH	191	SH	348	CsO	59	PO	791
CaH	284	SeH	304	CuO	134	AsO	495
SrH	209	TeH	305	AgO	123	BiO	174
ZnH	216	MnH	202	BeO	368	CrO	276
HgH	207	TcH	198	MgO	245	MoO	496
ScH	235	ReH	225	CaO	348	WO	695
YH	260	HH	259	SrO	299	SO	524
LaH	243	ClH	453	BaO	441	TeO	339
BH	198	BrH	379	ZnO	161	Re	435
TiH	227	IH	305	ScO	689	FO	335
ZrH	219	FeH	211	YO	698	ClO	468
CH	398	CoH	195	LaO	875	BrO	366
SiH	317	NiH	158	BO	326	IO	316
GeH	377	RuH	160	AlO	146	FeO	343
VH	202	RhH	165	GaO	46	CoO	317
NbH	220	PdH	208	TiO	667	NiO	276
TaH	230	OsH	239	ZrO	753	RuO	372
NH	≥ 436	IrH	306	HfO	685[a]	RhO	295
PH	275	UH	284	CO	811	PdO	145
AsH	291			SiO	478	OsO	418
				GeO	344	IrO	247
				SnO	281	PtO	318
				PbO	247	ThO	848[a]

[a] [2.29]

Table S2.9 Average bond energies (kJ/mol) (Subscripts *p*, *s* and *t* indicate primary, secondary and tertiary carbon atoms; 1 and 2 indicate the number of atoms of a given type, connected to the atom under consideration; superscript indicate the element bonded to the polyvalent atom)

A–B	*E*(A–B)	A–B	*E*(A–B)	A–B	*E*(A–B)
Li–Be	87.4	P=S	441	${}^{F}(C{-}H)_{t}$	398
Li–B	101	P–F	483	${}^{Cl}(C{-}H)_{p}$	405
Li–C	126	P–C	331	${}^{Cl}(C{-}H)_{s}$	403
Li–N	243	O–O	192	${}^{Cl}(C{-}H)_{t}$	401
Li–O	406	S–S	266	${}^{Br}(C{-}H)_{p}$	406
Be–Be	119	P=O	643	${}^{Br}(C{-}H)_{s}$	404
Be–B	186	${}^{Li}(Be{-}H)$	297	${}^{Br}(C{-}H)_{t}$	405
Be–C	232	${}^{BeC}(Be{-}H)$	298	${}^{I}(C{-}H)_{p}$	409
Be–N	340	${}^{CF}(Be{-}H)_{1}$	299	${}^{I}(C{-}H)_{s,t}$	406
Be–O	488	${}^{N}(Be{-}H)$	302	${}^{Li}(N{-}H)_{1}$	384
Be–F	653	${}^{O}(Be{-}H)$	311	${}^{Li}(N{-}H)_{2}$	395
B–B	286	${}^{Li}(B{-}H)_{1}$	383	${}^{Be}(N{-}H)_{1}$	400
B–C	323	${}^{Li}(B{-}H)_{2}$	386	${}^{Be}(N{-}H)_{2}$	401
B–N	443	${}^{Be}(B{-}H)_{1}$	375	${}^{B}(N{-}H)_{1}$	395
B–O	544	${}^{Be}(B{-}H)_{2}$	378	${}^{B}(N{-}H)_{2}$	400
B–F	659	${}^{B}(B{-}H)_{1}$	382	${}^{C}(N{-}H)_{1}$	380

Table S2.9 (continued)

A–B	E(A–B)	A–B	E(A–B)	A–B	E(A–B)
B–Cl	489	$^{B}(B-H)_2$	381	$^{C}(N-H)_2$	383
B–Br	414	$^{C}(B-H)_1$	376	$^{N}(N-H)_1$	373
B–I	334	$^{C}(B-H)_2$	375	$^{N}(N-H)_2$	378
C–C	357	$^{C}(B-H)_3$	375	$^{P}(N-H)_1$	380
C=C	579	$^{N}(B-H)_1$	379	$^{P}(N-H)_2$	390
C≡C	786	$^{N}(B-H)_2$	386	$^{O}(N-H)_1$	371
C–N	319	$^{O}(B-H)_1$	378	$^{O}(N-H)_2$	375
C=N	571	$^{O}(B-H)_2$	374	$^{S}(N-H)_1$	390
C≡N	872	$^{F}(B-H)_1$	372	$^{S}(N-H)_2$	391
C–P	271	$^{Li}(C-H)_p$	433	$^{F}(N-H)_1$	369
C=P	448	$^{Li}(C-H)_s$	428	$^{F}(N-H)_2$	370
C–O	383	$^{Li}(C-H)_t$	426	$^{C}(P-H)_1$	314
C=O	744	$^{Be}(C-H)_p$	431	$^{C}(P-H)_2$	318
C–S	301	$^{Be}(C-H)_s$	428	$^{N}(P-H)_1$	309
C–F	486	$^{Be}(C-H)_t$	426	$^{N}(P-H)_2$	311
C–Cl	359	$^{B}(C-H)_p$	425	$^{P}(P-H)_1$	320
C–Br	300	$^{B}(C-H)_s$	424	$^{P}(P-H)_2$	317
C–I	234	$^{B}(C-H)_t$	423	$^{O}(P-H)_1$	303
Si–C	295	$^{C}(C-H)_p$	411	$^{O}(P-H)_2$	305
Si–F	606	$^{C}(C-H)_s$	408	$^{S}(P-H)_1$	310
Si–Cl	414	$^{C}(C-H)_t$	405	$^{S}(P-H)_2$	312
Si–Br	343	$^{N}(C-H)_p$	406	$^{F}(P-H)_1$	298
Si–I	262	$^{N}(C-H)_s$	402	$^{F}(P-H)_2$	301
N–N	212	$^{N}(C-H)_t$	402	$^{Cl}(P-H)_1$	303
N=N	515	$^{P}(C-H)_p$	413	$^{Cl}(P-H)_2$	306
N≡N	945	$^{P}(C-H)_s$	411	$^{Li}(O-H)$	454
N–P	265	$^{P}(C-H)_t$	410	$^{Be}(O-H)$	471
N=P	450	$^{O}(C-H)_p$	401	$^{B}(O-H)$	466
N–O	223	$^{O}(C-H)_s$	399	$^{C}(O-H)$	452
N=O	541	$^{O}(C-H)_t$	397	$^{S}(O-H)$	458
N–S	224	$^{S}(C-H)_p$	409	$^{F}(O-H)$	433
N=S	413	$^{S}(C-H)_s$	407	$^{N}(S-H)$	355
P–P	211	$^{S}(C-H)_t$	404	$^{P}(S-H)$	362
P=P	360	$^{F}(C-H)_p$	399	$^{O}(S-H)$	346
P–O	358	$^{F}(C-H)_s$	398	$^{S}(S-H)$	354

Table S2.10 Additive energies of π-bonds (kJ/mol)

X	Y	*a*	*b*	*c*	X	Y	*a*	*c*
C	C	222	291	272	N	N	303	251
C	Si	57.5	151	159	N	P	185	184
C	N	252	338	264	N	O	317	259
C	P	177	206.5	180	N	S	189	176
C	S	220	233	218	P	P	149	142
Si	Si	36	101	105	P	O	285	222
Si	N	31	155	151	P	S	180	167
Si	P	95	124	121	O	O	306	306
Si	O	240	233.5	209	O	S	249	249
Si	S	168	182.5	209	S	S	159.5	159.5

[a] [2.30–2.33], [b] [2.34], [c] [2.35]

Table S2.11 Reduced Madelung constants

Structure type	k_M	Structure type	k_M	Structure type	k_M
$AlBr_3$	1.199	$BeCl_2$	1.362	MnF_2, TiO_2	1.589
BCl_3	1.226	SiF_4	1.432	PbF_2, SnO_2	1.602
SnI_4	1.236	CdI_2	1.455	CuCl, ZnS	1.638
$AuCl_3$	1.245	SiO_2	1.467	Y_2O_3	1.672
V_2O_5	1.266	Cu_2O	1.481	CaF_2, ZrO_2	1.680
HgI	1.277	$CrCl_2$	1.500	NiAs	1.733
TlF	1.318	BN	1.528	NaCl, MgO	1.748
AsI_3	1.334	BeO	1.560	CsCl	1.763

Table S2.12 Band gaps (eV) in MX_2 type compounds

M	X				M	X			
	F	Cl	Br	I		O	S	Se	Te
Mg	14.5[a]	9.2	8.2		Ti	3.1[c]	2.0	1.6	1.0
Ca	12.5[b]	6.9		6.0	Zr	5.2[d]	2.1		
Sr	11.0[b]	7.5			Hf	5.5[e]	1.9[f]	1.1	0.4[f]
Ba	9.5[b]	7.0			Si	9.0[c]		1.7	1.0
Zn				4.75[g]	Ge	5.4[c]	3.4	2.5	1.2
Cd	8.7[h]	5.7	4.5	3.5	Sn	3.7[c]	2.1[i]	1.02[i]	
Hg		4.4	3.6[j]	2.35[k]	Pb	1.6	1.0		
Sn		3.9	3.4	2.4	Mo		1.9	1.2[l]	0.9
Pb		4.0	3.1[m]	2.3[n]	W		1.8	1.4	0.1
Mn[o]	10.2	8.3	7.7	5.2	Re		1.5[p]	1.35[p]	
Fe[o]		8.3	7.4	6.0	Ru		1.4[p]	0.9[p]	
Co[o]		8.3	7.4	6.0	Pt	≥ 3.5[q]			
Ni[o]	8.8	8.4	7.5	6.0	U	5.5			

[a][2.36], [b][2.37], [c][2.38], [d][2.39], [e][2.40], [f][2.41, 2.42], [g][2.43], [h][2.44], [i][2.45], [j][2.46], [k][2.47, 2.48], [l][2.49], [m][2.50], [n][2.51], [o][2.52], [p][2.53], [q][2.54]

Table S2.13 Band gaps (eV) in M_nX_m type compounds

M_2X_3	E_g	M_2X_3	E_g	M_nX_m	E_g	M_nX_m	E_g
Sc_2O_3	5.7[a]	Tl_2O_3	2.2	Li_3N	2.2[p]	SbI_3	2.3
Sc_2S_3	2.8	Tl_2Te_3	0.7[h]	Li_2O	8.0[q]	$CrCl_3$	9.5[χ]
Y_2O_3	5.6	As_2O_3	4.5	Cu_2O	2.2[r]	$CrBr_3$	8.0[χ]
La_2O_3	5.4	As_2S_3	2.4[i]	Cu_2S	0.34[s]	ZrS_3	2.5[ϕ]
La_2S_3	2.8	As_2Se_3	1.7	Cu_2Se	1.3[t]	$ZrSe_3$	1.85[ϕ]
La_2Se_3	2.3[b]	As_2Te_3	0.8	Cu_2Te	0.67[u]	HfS_3	2.85[ϕ]
La_2Te_3	1.4	Sb_2O_3	3.25[a]	Ag_2S	1.14[v]	$HfSe_3$	2.15[ϕ]
B_2S_3	3.7[c]	Sb_2S_3	1.7[i]	Ag_2Se	1.58[w]	MoO_3	3.8[a]
Al_2O_3	9.5	Sb_2Se_3	1.2	TlS	0.9[x]	WO_3	2.6[λ]
Al_2S_3	4.1	Sb_2Te_3	0.2[j]	Tl_2S_3	1.0[x]	UO_3	2.3[μ]
Al_2Se_3	3.1	Bi_2O_3	2.85[k]	TlS_2	1.4[x]	TeO_2	3.8[a]
Al_2Te_3	2.4	Bi_2S_3	1.6[i]	Tl_2S_5	1.5[x]	MnS_4	3.7[η]
Ga_2O_3	4.9[d]	Bi_2Se_3	0.8[l]	GeS	1.6[y]	$MnSe_4$	3.3[η]
Ga_2S_3	3.2	Bi_2Te_3	0.2[m]	SiC	3.1[z]	$MnTe_4$	3.2[η]

Table S2.13 (continued)

M_2X_3	E_g	M_2X_3	E_g	M_nX_m	E_g	M_nX_m	E_g
Ga_2Se_3	1.75[e]	Cr_2O_3	1.6	GaSe	2.0[α]	$NbCl_5$	2.7[φ]
Ga_2Te_3	1.2[f]	Cr_2S_3	0.9	MgH_2	5.6[β]	$NbBr_5$	2.0[φ]
In_2O_3	3.3[a]	Cr_2Se_3	0.1	YH_3	2.45[γ]	NbI_5	1.0[φ]
In_2S_3	2.6[g]	Fe_2O_3	2.2[n]	LaF_3	9.7[δ]	V_2O_5	2.5[π]
In_2Se_3	1.5	Fe_2Se_3	1.2	GaF_3	9.8[ε]	Nb_2O_5	3.4[a]
In_2Te_3	1.2[f]	Rh_2O_3	3.4[o]	InF_3	8.2[κ]	Ta_2O_5	4.0[a]

[a][2.38, 2.41, 2.42], [b][2.55], [c][2.56], [d][2.57], [e][2.58], [f][2.59], [g][2.60], [h][2.61], [i][2.62], [j][2.63, 2.64], [k]for α-Bi_2O_3 (for β-Bi_2O_3 $E_g = 2.58$ eV) [2.65], [l][2.66], [m][2.67], [n][2.68], [o][2.69], [p][2.70], [q][2.71], [r][2.72, 2.73], [s][2.74], [t][2.75], [u][2.76], [v][2.77], [w]glass [2.78], [x][2.79], [y][2.80, 2.81], [z][2.82], [α][2.83], [β][2.84], [γ][2.85, 2.86], [δ][2.39], [ε][2.87], [κ][2.88], [χ][2.89], [Φ][2.90], [λ][2.91], [μ][2.92], [η][2.93], [φ][2.94], [π][2.95]

Table S2.14 Additive band gaps (eV) of elements

Li	Be	B	C	N	O	F			
0.4	0.8	1.3	5.5	7.0	6.5	10			
Na	Mg	Al	Si	P	S	Cl	Ne[a]		
−0.3	+0.3	−0.1	+1.2	2.6	2.6	5.2	21.7		
K	Ca	Sc	Ti	V	Cr	Mn	Fe	Co	Ni
−0.8	−0.5	−0.4	−0.3	−0.3	−0.3	−0.2	−0.1	−0.1	−0.1
Cu	Zn	Ga	Ge	As	Se	Br	Ar[a]		
−0.2	+0.2	−0.5	+0.7	+1.2	+1.8	+1.9	14.2		
Rb	Sr	Y	Zr	Nb	Mo	Tc	Ru	Rh	Pd
−1.0	−0.8	−0.7	−0.6	−0.6	−0.6	−0.5	−0.5	−0.5	−0.4
Ag	Cd	In	Sn	Sb	Te	I	Kr[a]		
−0.5	−0.3	−0.8	+0.1	+0.1	+0.35	+ 1.3	11.6		
Cs	Ba	La	Hf	Ta	W	Re	Os	Ir	Pt
−1.2	−1.0	−0.9	−0.8	−0.7	−0.6	-0.6	−0.5	−0.5	−0.5
Au	Hg	Tl	Pb	Bi	P	At	Xe[a]		
−0.5	−0.3	−0.9	−0.7	−0.2	0	+0.7	9.3		

[a][2.96]

Table S2.15 Thermochemical electronegativities of elements. From top to bottom: Pauling [2.97], Allred [2.98], Batsanov [2.99], Smith [2.100]

Li	Be	B	C	N	O	F			
1.0	1.5	2.0	2.5	3.0	3.5	4.0			
0.98	1.57	2.04	2.55	3.04	3.44	3.98			
1.0	1.4	2.0	2.6	2.7	3.2	3.7			
		1.80	2.59	3.11	3.44	3.84			
Na	Mg	Al	Si	P	S	Cl			
0.9	1.2	1.5	1.8	2.1	2.5	3.0			
0.93	1.31	1.61	1.90	2.19	2.58	3.16			
0.9	1.3	1.6	2.0	2.15	2.6	3.2			
			1.71	1.98	2.50	3.06			
K	Ca	Sc	Ti	V	Cr	Mn	Fe^{II}	Co^{II}	Ni^{II}
0.8	1.0	1.3	1.5	1.6	1.6	1.5	1.8	1.8	1.9
0.82	1.00	1.36	1.54	1.63	1.66	1.55	1.83	1.88	1.91
0.7	1.0	1.35	1.7	1.8	1.9	1.9	1.6	1.65	1.7
Cu	Zn	Ga	Ge	As	Se	Br			
1.8	1.6	1.6	1.8	2.0	2.4	2.8			
1.90	1.65	1.81	2.01	2.18	2.55	2.96			
1.5	1.6	1.75	2.1	2.1	2.5	3.0			
			1.93	2.06	2.37	2.86			
Rb	Sr	Y	Zr	Nb	Mo	Tc	Ru	Rh	Pd
0.8	1.0	1.2	1.6	1.6	1.8	1.9	2.2	2.2	2.2
0.82	0.95	1.22	1.33		2.16			2.28	2.20
0.7	0.95	1.25	1.6	1.6	2.2	1.9	2.2	2.2	2.2
Ag	Cd	In	Sn	Sb	Te	I			
1.9	1.7	1.7	1.8	1.9	2.1	2.5			
1.93	1.69	1.78	1.96	2.05	2.10	2.66			
1.7	1.7	1.7	2.0	2.0	2.2	2.7			
			1.79	1.91	2.14	2.47			
Cs	Ba	La	Hf	Ta	W	Re	Os	Ir	Pt
0.7	0.9	1.1	1.3	1.5	1.7	1.9	2.2	2.2	2.2
0.79	0.89	1.10			2.36			2.20	2.28
0.5	0.8	1.1	1.6	1.5	2.2	1.9	2.2	2.2	2.2
Au	Hg	Tl	Pb	Bi	Th	U			
2.4	1.9	1.8	1.8	1.9	1.3	1.7			
2.54	2.00	2.04	2.33	2.02		1.38			
1.8	1.8	1.8	2.1	2.0	1.5	1.6			

Table S2.16 Average thermochemical electronegativities of radicals R

R	χ	R	χ	R	χ	R	χ
CH_3	2.6	NH_2	3.1	BH_2	1.9	$[HCO_3]$	3.4
CF_3	2.9	NF_2	3.2	PH_2	2.3	$[HPO_4]$	3.4
SiF_3	2.0	NCS	3.2	$SiCH_3$	1.9	$[NO_3]$	3.7
$CHCH_2$	2.7	NNN	3.3	OCH_3	3.4	$[SO_4]$	3.7
CCH	2.8	NC	3.3	OC_6H_5	3.5	O_2	3.5
CHO	2.9	NO_2	3.4	OH	3.5		

Table S2.17 Ionization electronegativities according to Pearson (upper lines), Allen (middle lines), Politzer (lower lines)

Li	Be	B	C	N	O	F
0.92	1.43	1.31	1.92	2.23	2.31	3.19
0.87	1.51	1.96	2.43	2.93	3.45	4.01
0.99	1.64	2.14	2.63	3.18	3.52	4.00
Na	Mg	Al	Si	P	S	Cl
0.87	1.17	0.98	1.46	1.72	1.91	2.54
0.83	1.24	1.54	1.83	2.15	2.47	2.74
0.96	1.36	1.59	1.86	2.25	2.53	2.86
K	Ca	Ga	Ge	As	Se	Br
0.74	0.94	0.98	1.40	1.62	1.80	2.33
0.70	0.99	1.68	1.91	2.11	2.32	2.57
0.98	1.08	1.67	1.82	2.09	2.31	2.60

First transition series

Sc	Ti	V	Cr	Mn	Fe	Co	Ni	Cu	Zn
1.03	1.06	1.11	1.14	1.14	1.23	1.31	1.35	1.37	1.44
1.14	1.32	1.46	1.58	1.67	1.72	1.76	1.80	1.77	1.52
1.15	1.21	1.27	1.17	1.33	1.31	1.28	1.38	1.48	1.57

Table S2.18 Average ionization electronegativities of radicals R

R	χ	R	χ	R	χ	R	χ	R	χ
CF_3	3.3	CCH	3.1	NO_2	4.0	OH	3.5	$[ClO_4]$	4.9
CCl_3	2.9	CO	3.7	NO	3.8	SH	2.3	$[ClO_3]$	4.8
CBr_3	2.6	CN	3.8	NC	3.7	SCN	2.9	$[SO_4]$	4.6
CI_3	2.5	NF_2	3.7	NCS	3.5	SF_5	2.9	$[PO_4]$	4.4
CH_3	2.3	NCl_2	3.2	OF	4.1	SeH	2.2	$[CO_3]$	4.3
$CHCH_2$	2.5	NH_2	2.7	OCl	3.7	TeH	2.1		

Table S2.19 Ionization electronegativities of atoms with charges of ±1

A^+	χ	A^+	χ	A^+	χ	A^+	χ	A^-	χ
Li	16.7	Ba	2.5	C	6.3	Bi	4.2	F	–0.1
Na	10.6	Zn	4.7	Si	4.2	V	3.6	Cl	0.2
K	7.2	Cd	4.4	Ge	4.1	Nb	3.6	Br	0.2
Rb	6.2	Hg	5.0	Sn	3.8	Ta	4.2	I	0.2
Cs	5.7	B	6.2	Pb	3.8	O	9.5		
Cu	5.2	Al	4.6	Ti	3.4	S	6.5		
Ag	5.4	Ga	5.0	Zr	3.3	Se	6.0		
Au	5.5	In	4.6	Hf	3.8	Te	5.3		
Be	4.7	Tl	5.0	N	7.8	F	10.3		
Mg	3.9	Sc	3.2	P	5.2	Cl	7.2		
Ca	3.0	Y	3.1	As	4.9	Br	6.5		
Sr	2.8	La	2.8	Sb	4.3	I	5.7		

Table S2.20 Short history of the development of the geometrical electronegativity concept (pioneering works are shown in **bold**)

Year	Authors	Equation	Notes
1942	**Liu**	$\chi = a(N^* + b)/r^{2/3}$	N^* is the number of e-shells
1946	**Gordy**	$\chi = a\ (n + b)/r + c$	n is the number of electrons
1957	Wilmshurst		≈
1964	Yuan		≈
1966	Chandra		≈
1968	Phillips		applied to semiconductors
1979	Ray, Samuel, Parr		for multiple bonds
1982	Inamoto, Masuda		for polar bonds
1983	Owada		n^* instead of n
1988	Luo, Benson		reduced to Pauling' scale
1951	**Cottrell, Sutton**	$\chi = a(Z^*/r)^{1/2} + b$	dimensionality of $E^{1/2}$
1989	Zhang, Kohen		theoretical $Z^*$and r
1993	Batsanov		for normal and vdW molecules
1952	**Sanderson**	$\chi = a(N/r^3) + b$	$N = \Sigma e$
1980	**Allen, Huheey**		for rare gases
1955	**Pritchard, Skinner**	$\chi = a(Z^*/r) + b$	Z^* according to Slater
1964	Batsanov		corrected Z^*
1971	Batsanov		Z^* for valence states
1975	Batsanov		for crystals
1980	Allen, Huheey		for rare gases
1956	**Williams**	$\chi = a\ (n/r)^b$	n is the number of valence electrons
1958	**Allred, Rochow**	$\chi = a(Z^*–b)\ r^2 + c$	Z^* of Slater
1964	Batsanov		corrected Z^*
1971	Batsanov		Z^* for valence states
1975	Batsanov		for crystals
1977	Mande		experimental Z^*
1980	Allen, Huhee		for rare gases
1981	Boyd, Marcus		b is calculated by *ab initio*
1982	Zhang		experimental Z^*

Table S2.20 (continued)

Year	Authors	Equation	Notes
1978	**Batsanov**	$\chi = a(N_e^{1/2})/r$	N_e is the number of outer electrons
1986	**Gorbunov, Kaganyuk**		r is calculated by *ab initio*
1990	**Nagle**	$\chi = a(N/\alpha^{1/2}) + b$	α is the polarizability
2006	**Batsanov**	all formulae	Z* and r for valence states

Table S2.21 Crystalline electronegativities according to Li and Xue [2.101, 2.102]

Li 1.01	Be 1.27	B 1.71	C 2.38	N 2.94	O 3.76	F 4.37			
Na 1.02	Mg 1.23	Al 1.51	Si 1.89	P 2.14	S 2.66	Cl 3.01			
K 1.00	Ca 1.16	Sc 1.41	Ti 1.73	V^{III} 1.54	Cr^{III} 1.59	Mn^{IV} 1.91	Fe^{III} 1.65	Co^{III} 1.69	Ni^{III} 1.70
Cu 1.16	Zn 1.34	Ga 1.58	Ge 1.85	As 2.16	Se 2.45	Br 2.74			
Rb 1.00	Sr 1.14	Y 1.34	Zr 1.61	Nb^{III} 1.50	Mo^{IV} 1.81	Tc^{IV} 1.77	Ru^{IV} 1.85	Rh^{IV} 1.86	Pd^{IV} 1.88
Ag 1.33	Cd 1.28	In 1.48	Sn 1.71	Sb 1.97	Te 2.18	I 2.42			
Cs 1.00	Ba 1.13	La 1.33	Hf 1.71	Ta^{III} 1.54	W^{IV} 1.78	Re^{IV} 1.85	Os^{IV} 1.89	Ir^{IV} 1.88	Pt^{IV} 1.90
Au^{I} 1.11	Hg 1.33	Tl^{I} 1.05	Pb 1.75	Bi 1.90	Th 1.40	U 1.44			

Table S2.22 Effective coordination charges of metal atoms

Metal	Compounds	Ω_{cal}	Ω_{exp}	Metal	Compounds	Ω_{cal}	Ω_{exp}
Cr	$CrSO_4 \cdot 7H_2O$	1.8	1.9	Co	$Co(NO_3)_3$	0.6	1.2
	$Cr(NO_3)_3$	1.3	1.2		$Co(C_5H_5)_2$	0.7	0.4
	K_2CrO_4	0.5	0.1		$Co(C_5H_5)_2Cl$	0.9	1.0
	$Cr(C_6H_6)_2$	1.4	1.3	Ni	$Ni(C_5H_5)_2$	0.6	0.7
Mn	$Mn(NO_3)_2 \cdot 4H_2O$	1.8	1.8		$Ni(C_5H_5)_2Cl$	0.8	1.0
	$K_3Mn(CN)_6$	0.6	0.9	Os	OsO_2	0.7	0.8
	$Mn(C_5H_5)_2$	1.3	1.5		K_2OsCl_6	0.5	0.8
Fe	$(NH_4)_2Fe(SO_4)_2 \cdot 6H_2O$	1.7	1.9		K_2OsO_4	0.7	0.8
	$K_3Fe(CN)_6$	0.4	1.0		K_2OsNCl_5	0.8	0.7
	$Fe(C_5H_5)_2$	0.7	0.6		$KOsO_3N$	0.9	1.0
	$Fe(C_5H_5)_2Cl$	0.8	0.7				

Table S2.23 Comparison of high pressure radii (r_P) and crystallographic radii (r_c) of cations

Cation	r_P	r_c	Cation	r_P	r_c	Cation	r_P	r_c
Li^+	0.75	0.76	Mg^{2+}	0.70	0.72	Sc^{3+}	0.74	0.74
Na^+	0.98	1.02	Ca^{2+}	1.03	1.00	Y^{3+}	0.88	0.90
K^+	1.37	1.38	Sr^{2+}	1.15	1.18	Cr^{3+}	0.67	0.62
Rb^+	1.52	1.52	Ba^{2+}	1.38	1.35	Mn^{3+}	0.66	0.64
Cs^+	1.63	1.67	Zn^{2+}	0.76	0.74	Fe^{3+}	0.66	0.64
Cu^+	0.78	0.77	Cd^{2+}	0.95	0.95	Th^{4+}	1.07	1.05
Ag^+	1.17	1.15	Pb^{2+}	1.23	1.19	U^{4+}	0.97	1.00
Tl^+	1.44	1.50	B^{3+}	0.38	0.27	Zr^{4+}	1.06	0.84
Be^{2+}	0.47	0.45	Al^{3+}	0.63	0.54	Hf^{4+}	0.90	0.83

Table S2.24 Change of effective atomic charges under pressures, de^*/dP 10^2GPa

M	Cl		Br		I	
	[2.103]	[2.104]	[2.103]	[2.104]	[2.103]	[2.104]
Li	1.22	0.8	2.15	0.95	2.87	
Na	1.26	1.1	2.20	1.3	2.91	1.9
K	1.32	1.8	2.26	2.1	2.94	2.8
Rb	1.33	2.1	2.26	2.6	2.93	3.3
Cs	1.38		2.35		3.01	

Supplementary References

2.1 Reynard LM, Evans CJ, Gerry MCL (2001) J Mol Spectr 205: 344
2.2 Shayesteh A, Bernath PF (2011) J Chem Phys 135: 094308
2.3 Heaven MC, Bondybey VE, Merritt JM, Kaledin AL (2011) Chem.Phys Lett 506: 1
2.4 Czajkowski M, Krause L, Bobkowski R (1994) Phys Rev A49: 775
2.5 Kedzierski W, Supronowicz J, Czajkowski M et al (1995) J Mol Spectr 173: 510
2.6 Girichev GV, Giricheva NI, Titov VA et al (1992) J Struct Chem 33: 362
2.7 Gurvich LV, Ezhov YuS, Osina EL, Shenyavskaya EA (1999) Russ J Phys Chem 73: 331
2.8 Liu Y, Zhang C-H, Krasnokutski SA, Yang D-S (2011) J Chem Phys 135: 034309
2.9 Ciccioli A, Gigli G, Meloni G, Testani E (2007) J Chem Phys 127: 054303
2.10 Hillel R, Bouix J, Bernard C (1987) Z anorg allgem Chem 552: 221
2.11 Balasubramanian K (1989) Chem.Rev 89: 1801
2.12 Van der Vis MGM, Cordfunke EHP, Konings RJM (1997) Thermochim Acta 302: 93
2.13 Ponomarev D, Takhistov V, Slayden S, Liebman J (2008) J Molec Struct 876: 34
2.14 Goussis A, Besson J (1986) J Less-Common Met 115: 193
2.15 Berkowitz J (1988) J Chem Phys 89: 7065
2.16 Berkowitz J, Ruscic B, Gibson S et al (1989) J Mol Struct 202: 363
2.17 Nizamov B, Setser DW (2001) J Mol Spectr 206: 53
2.18 Dittmer G, Niemann U (1981) Phil J Res 36: 87
2.19 Gutsev GL, Bauschlicher ChW (2003) J Phys Chem A107: 4755
2.20 Blauschlicher CW (1996) Chem Phys 211: 163
2.21 Han Y-K, Hirao K (2000) J Chem Phys 112: 9353
2.22 Simoes JAM, Beauchamp JL (1990) Chem Rev 90: 629

2.23 Lamoreaux R, Hildenbrand DL, Brewer L (1987) J Phys Chem Refer Data 16: 419
2.24 Giuliano BM, Bizzochi L, Sanchez R et al (2011) J Chem Phys 135: 084303
2.25 O'Hare PAG, Lewis B, Surman S, Volin KJ (1990) J Chem Thermodyn 22: 1191
2.26 Cooke SA, Gerry MCL (2004) J Chem Phys 121: 3486
2.27 Brugh DJ, Morse MD (1997) J Chem Phys 107: 9772
2.28 Antonov IO, Barker BJ, Bondybey VE, Heaven MC (2010) J Chem Phys 133: 074309
2.29 Merritt JM, Bondybey VE, Heaven MC (2009) J Chem Phys 130: 144503
2.30 Leroy G, Sana M, Wilante C, van Zieleghem M-J (1991) J Molec Struct 247: 199
2.31 Leroy G, Temsamani DR, Sana M, Wilante C (1993) J Molec Struct 300: 373
2.32 Leroy G, Temsamani DR, Wilante C (1994) J Molec Struct 306: 21
2.33 Leroy G, Temsamani DR, Wilante C, Dewispelaere J-P (1994) J Molec Struct 309: 113
2.34 von Schleyer PR, Kost D (1988) J Am Chem Soc 110: 2105
2.35 Schmidt M, Truong P, Gordon M (1987) J Am Chem Soc 109: 5217
2.36 Scrocco M (1986) Phys Rev B33: 7228
2.37 Scrocco M (1985) Phys Rev B32: 1301
2.38 Dou Y, Egdell RG, Law DSL et al (1998) J Phys Cond Matter 10: 8447,
2.39 Wiemhofer H-D, Harke S, Vohrer U (1990) Solid State Ionics 40–41: 433
2.40 Cisneros-Morales MC, Aita CR (2010) Appl Phys Lett 96: 191904
2.41 Kliche G (1986) Solid State Commun 59: 587
2.42 Dimitrov V, Sakka S (1996) J Appl Phys 79: 1736
2.43 Tyagi P, Vedeshwar AG (2001) Phys Rev B64: 245406
2.44 Julien C, Eddrief M, Samaras I, Balkanski M (1992) Mater Sci Engin B15: 70
2.45 Roubi L, Carlone C (1988) Canad J Phys 66: 633
2.46 Stanciu GA, Oprica MH, Oud JL et al (1999) J Phys D32: 1928
2.47 da Silva AF, Veissid N, An CY et al (1995) J Appl Phys 78: 5822
2.48 Karmakar S, Sharma SM (2004) Solid State Commun 131: 473
2.49 Anand TJS, Sanjeeviraja C (2001) Vacuum 60: 431
2.50 Ren Q, Liu LQ et al (2000) Mater Res Bull 35: 471
2.51 da Silva AF, Veissid N, An CY et al (1996) Appl Phys Lett 69: 1930
2.52 Thomas J, Polini I (1985) Phys Rev B32: 2522
2.53 Ho CH, Liao PC, Huang YS et al (1997) J Appl Phys 81: 6380
2.54 Gottesfeld S, Maia G, Floriano JB et al (1991) J Electrochem Soc 138: 3219
2.55 Prokofiev AV, Shelykh AI, Golubkov AV, Sharenkova NV (1994) Inorg Mater 30: 326
2.56 Sassaki T, Takizawa H, Uheda K et al (2002) J Solid State Chem 166: 164
2.57 Tu B, Cui Q, Xu P et al (2002) J Phys Cond Matter, 14: 10627
2.58 Adachi S, Ozaki S (1993) Japan J Appl Phys (I) 32: 4446
2.59 Ozaki S, Takada K, Adachi S (1994) Japan J Appl Phys (I) 33: 6213
2.60 Choe S-H, Bang T-H, Kim N-O et al (2001) Semicond Sci Technol 16:98
2.61 Hussein SA, Nassary MM, Gamal GA, Nagat AT (1993) Cryst Res Technol 28: 1021
2.62 Yesugade NS, Lokhande CD, Bhosale CH (1995) Thin Solid Films 263:145
2.63 Lostak P, Novotny R, Kroutil J, Stary Z (1987) Phys Stat Solidi A 104:841
2.64 Lefebre I, Lannoo M, Allan G et al (1987) Phys Rev Lett 59:2471
2.65 Leontie CM, Delibas M, Rusu GI (2001) Mater Res Bull 36:1629
2.66 Torane AP, Bhosale CH (2001) Mater Res Bull 36:1915
2.67 Ismail F, Hanafi Z (1986) Z phys Chem 267:667
2.68 Chernyshova IV, Ponnurangam S, Somasundaran P (2010) Phys Chem Chem Phys 12:14045
2.69 Ghose J, Roy A (1995) Optical studies on Rh_2O_3. In: Schmidt SC, Tao WC (eds) Shock compression of condensed matter. AIP Press, New York
2.70 Fowler P, Tole P, Munn R, Hurst M (1989) Mol Phys 67:141
2.71 Ishii Y, Murakami J-i, Itoh M (1999) J Phys Soc Japan 68:696
2.72 Reimann K, Syassen K (1989) Phys Rev B 39:11113
2.73 Joseph KS, Pradeep B (1994) Pramana 42:41
2.74 Mostafa SN, Mourad MY, Soliman SA (1991) Z phys Chem 171:231

2.75 Haram SK, Santhanam KSV (1994) Thin Solid Films 238:21
2.76 Mostafa SN, Selim SR, Soliman SA, Gadalla EG (1993) Electrochim Acta 38:1699
2.77 Dlala H, Amlouk M, Belgacem S et al (1998) Eur Phys J Appl Phys 2:13
2.78 Kumar MCS, Pradeep B (2002) Semicond Sci Technol 17: 261
2.79 Waki H, Kawamura J, Kamiyama T, Nakamura Y (2002) J Non-Cryst Solids 297:26
2.80 Gauthier M, Polian A, Besson JM, Chevy A (1989) Phys Rev B 40:3837
2.81 Elkorashy A (1990) J Phys Cond Matter 2:6195
2.82 Herve P, Vandamme L (1994) Infrared Phys Technol 35:609
2.83 Gauthier M, Polian A, Besson J, Chevy A (1989) Phys Rev B 40:3837
2.84 Isidorsson J, Giebels IAME, Arwin H, Griessen R (2003) Phys Rev B 68:115112
2.85 Lee MW, Shin WP (1999) J Appl Phys 86:6798
2.86 Wijngaarden RJ, Huiberts JN, Nagengast D et al (2000) J Alloys Compd 308:44
2.87 Varekomp PR, Simpson WC, Shuh DK et al (1994) Phys Rev B 50:14267
2.88 Barrieri A, Counturier G, Elfain A et al (1992) Thin Solid Films 209:38
2.89 Pollini I, Thomas J, Carricarburu B, Mamy R (1989) J Phys Cond Matter 1:7695
2.90 El Ramnani H, Gagnon R, Aubin M (1991) Solid State Commun 77:307
2.91 Kaneko H, Nagao F, Miyake K (1988) J Appl Phys 63:510
2.92 Khila M, Rofail N (1986) Radiochim Acta 40:155
2.93 Goede O, Heimbrodt W, Lamla M, Weinhold V (1988) Phys Stat Solidi B 146:K65
2.94 Hoenle W, Furuseth F, von Schnering HG (1990) Z Naturforsch B 45:952
2.95 Parker JC, Lam DJ, Xu Y-N, Ching WY (1990) Phys Rev B 42:5289
2.96 Sonntag B (1976) Dielectric and optical properties. In: Klein ML, Venables JA (eds) Rare gas solids, vol 1. Acad Press, London
2.97 Pauling L (1960) The nature of the chemical bond, 3rd edn. Cornell Univ Press, Ithaca
2.98 Allred AL (1961) J Inorg Nucl Chem 17:215
2.99 Batsanov SS (2000) Russ J Phys Chem 74:267
2.100 Smith DW (2007) Polyhedron 26:519
2.101 Li K, Xue D (2006) J Phys Chem A110:11332
2.102 Li K, Wang X, Zhang F, Xue D (2008) Phys Rev Lett 100:235504
2.103 Batsanov SS (1997) J Phys Chem Solids 58:527
2.104 Kucharczyk W (1991) J Phys Chem Solids 52:435

References

1. Kossel W (1916) Molecular formation as an issue of the atomic construction. Ann Phys 49:229–362
2. Lewis GN (1916) The atom and the molecule. J Am Chem Soc 38:762–785
3. Langmuir I (1919) The arrangement of electrons in atoms and molecules. J Am Chem Soc 41:868–934
4. Langmuir I (1919) Isomorphism, isosterism and covalence. J Am Chem Soc 41:1543–1559
5. Langmuir I (1920) The octet theory of valence and its applications with special reference to organic nitrogen compounds. J Am Chem Soc 42: 274–292
6. Schwartz WHE (2006) Measuring orbitals: provocation or reality? Angew Chem Int Ed 45:1508–1517
7. Mulliken RS (1978) Chemical bonding. Ann Rev Phys Chem 29:1–30
8. Bader RFW (1990) Atoms in molecules: a quantum theory. Oxford University Press, Oxford
9. Parr RG, Ayers PW, Nalewajski RF (2005) What is an atom in a molecule? J Phys Chem A 109:3957–3959
10. Batsanov SS (1957) On the interrelation between the theory of polarization and the concept of electronegativity. Zh Neorg Khim 2:1482–1487
11. Batsanov SS (2004) Molecular refractions of crystalline inorganic compounds. Russ J Inorg Chem 49:560–568

12. Goldschmidt VM (1954) Geochemistry. Clarendon Press, Oxford
13. O'Keeffe M (1977) On the arrangement of ions in crystals. Acta Cryst A 33:924–927
14. Batsanov SS (1983) On some crystal-chemical peculiarities of inorganic halogenides. Zh Neorg Khim 28:830–836
15. Madden PA, Wilson M (1996) Covalent effects in 'ionic' systems. Chem Soc Rev 25:339–350
16. Gillespie RJ, Silvi B (2002) The octet rule and hypervalence: two misunderstood concepts. Coord Chem Rev 233:53–62
17. von Antropoff A (1924) Die Wertigkeit der Edelgase und ihre Stellung im periodischen System. Angew Chem 37:217–218, 695–696
18. Pauling L (1932) The nature of the chemical bond: the energy of single bonds and the relative electronegativity of atoms. J Am Chem Soc 54:3570–3582
19. Bartlett N (1962) Xenon hexafluoroplatinate Xe^+ $[PtF6]^-$. Proc Chem Soc London 218
20. Bartlett N (1963) New compounds of noble gases: the fluorides of xenon and radon. Amer Scientist 51:114–118
21. Claassen HH, Selig H, Malm JG (1962) Xenon tetrafluoride. J Am Chem Soc 84:3593
22. Chernick CL, Claassen HH, Fields PR et al (1962) Fluorine compounds of xenon and radon. Science 138:136–138.
23. Tramšek M, Žemva B (2006) Synthesis, properties and chemistry of xenon(II) fluoride. Acta Chim Slov 53:105–116
24. Grochala W (2007) Atypical compounds of gases, which have been called 'noble'. Chem Soc Rev 36:1632–1655
25. Goettel KA, Eggert JH, Silvera IF, Moss WC (1989) Optical evidence for the metallization of xenon at 132(5) GPa. Phys Rev Lett 62:665–668
26. Reichlin R, Brister KE, McMahan AK, et al (1989) Evidence for the insulator-metal transition in xenon from optical, X-ray, and band-structure studies to 170 GPa. Phys Rev Lett 62:669–672
27. Batsanov SS (1998) H_2: an archetypal molecule or an odd exception? Struct Chem 9:65–68
28. Pauling L (1960) The nature of the chemical bond, 3rd edn. Cornell Univ Press, Ithaca, New York
29. Gillespie RJ, Robinson EA (1996) Electron domains and the VSEPR model of molecular geometry. Angew Chem Int Ed 35:495–514
30. Morse PM (1929) Diatomic molecules according to the wave mechanics: vibrational levels. Phys Rev 34:57–64
31. Bürgi H-B, Dunitz J (1987) Fractional bonds: relations among their lengths, strengths, and stretching force constants. J Am Chem Soc 109:2924–2926
32. Parr RG, Borkman RF (1968) Simple bond-charge model for potential-energy curves of homonuclear diatomic molecules. J Chem Phys 49:1055–1058
33. Harrison WA (1980) Electronic structure the properties of solids. Freeman, San Francisco
34. Zavitsas AA (2003) The relation between bond lengths and dissociation energies of carbon-carbon bonds. J Phys Chem A 107:897–898
35. Krygowski TM, Cyrański MK (2001) Structural aspects of aromaticity. Chem Rev 101:1385–1420
36. Pauling L, Sherman J (1933) The nature of the chemical bond: the calculation from thermochemlcal data of the energy of resonance of molecules among several electronic structures. J Chem Phys 1:606–617
37. Hückel E (1931) Quantum contributions to the benzene problem. Z Physik 70:204–286
38. Wiberg KB (2001) Aromaticity in monocyclic conjugated carbon rings. Chem Rev 101:1317–1332
39. Hoffmann R, Shaik S, Hiberty PC (2003) A conversation on VB vs MO theory: a never-ending rivalry? Acc Chem Res 36:750–756
40. Pierrefixe SCAH, Bickelhaupt FM (2007) Aromaticity: molecular-orbital picture of an intuitive concept. Chem Eur J 13:6321–6328
41. Gomes JANF, Mallion RB (2001) Aromaticity and ring currents. Chem Rev 101:1349–1384
42. Mitchell RH (2001) Measuring aromaticity by NMR. Chem Rev 101:1301–1316

43. Bühl M, Hirsh A (2001) Spherical aromaticity of fullerenes. Chem Rev 101:1153–1184
44. Chen Z, King RB (2005) Spherical aromaticity: recent work on fullerenes, polyhedral boranes and related structures. Chem Rev 105:3613–3642
45. King RB (2001) Three-dimensional aromaticity in polyhedral boranes and related molecules. Chem Rev 101:1119–1152
46. Boldyrev AI, Wang L-S (2005) All-metal aromaticity and antiaromaticity. Chem Rev 105:3716–3757
47. Hirsh A, Chen Z, Jiao H (2000) Spherical aromaticity in I_h symmetrical fullerenes: the $2(N+1)^2$ rule. Angew Chem Int Ed 39:3915–3917
48. Schleyer P von R (2001) Introduction: aromaticity. Chem Rev 101:1115–1118
49. Schleyer P von R (2005) Introduction: delocalization π and σ. Chem Rev 105:3433–3435
50. Meister J, Schwartz WHE (1994) Principal components of ionicity. J Phys Chem 98:8245–8252
51. Gussoni M, Castiglioni C, Zerbi G (1983) Experimental atomic charges from infrared intensities: comparison with "ab initio" values. Chem Phys Lett 95:483–485
52. Galabov B, Dudev T, Ilieva S (1995) Effective bond charges from experimental IR intensities. Spectrochim Acta A 51:739–754
53. Ilieva S, Galabov B, Dudev T et al (2001) Effective bond charges from infrared intensities in CH_4, SiH_4, GeH_4 and SnH_4. J Mol Struct 565–566:395–398
54. Barinskii RL (1960) Determination of the effective charges of atoms in complexes from the X-ray absorption spectra. J Struct Chem 1:183–190
55. Szigeti B (1949) Polarizability and dielectric constant of ionic crystals. Trans Faraday Soc 45:155–166
56. Batsanov SS (1982) Dielectric methods of studying the chemical bond and the concept of electronegativity. Russ Chem Rev 51:684–697
57. Wagner V, Gundel S, Geurts J et al (1998) Optical and acoustical phonon properties of BeTe. J Cryst Growth 184–185:1067–1071
58. Julien C, Eddrief M, Samaras I, Balkanski M (1992) Optical and electrical characterizations of SnSe, SnS_2 and $SnSe_2$ single crystals. Mater Sci Engin B 15:70–72
59. Schoenes J, Borgschulte A, Carsteanu A-M et al (2003) Structure and bonding in YH_x as derived from elastic and inelastic light scattering. J Alloys Compd 356–357:211–217
60. Jones GO, Martin DH, Mawer PA, Perry CH (1961) Spectroscopy at extreme infra-red wavelengths. II. The lattice resonances of ionic crystals. Proc Roy Soc London A 261:10–27
61. Bosomworth D (1967) Far-infrared optical properties of CaF_2, SrF_2, BaF_2, and CdF_2. Phys Rev 157:709–715
62. Denham P, Field GR, Morse PLR, Wilkinson GR (1970) Optical and dielectric properties and lattice dynamics of some fluorite structure ionic crystals. Proc Roy Soc London A 317:55–77
63. Batana A, Bruno JAO (1990) Volume dependence of the effective charge of zinc-blende-type crystals. J Phys Chem Solids 51:1237–1238
64. Van Vechten J (1969) Quantum dielectric theory of electronegativity in covalent systems. Phys Rev 187:1007–1020
65. Phillips JC, van Vechten J (1970) Spectroscopic analysis of cohesive energies and heats of formation of tetrahedrally coordinated semiconductors. Phys Rev B 2:2147–2160
66. Phillips JC (1970) Ionicity of the chemical bond in crystals. Rev Modern Phys 42:317–356
67. Phillips JC (1974) Electronegativity and tetragonal distortions in $A^{II}B^{IV}C^{V}_2$ semiconductors. J Phys Chem Solids 35:1205–1209
68. Levine BF (1973) d-Electron effects on bond susceptibilities and ionicities. Phys Rev B 7:2591–2600
69. Levine BF (1973) Bond-charge calculation of nonlinear optical susceptibilities for various crystal structures. Phys Rev B 7:2600–2626
70. Levine BF (1973) Bond susceptibilities and ionicities in complex crystal structures. J Chem Phys 59:1463–1486
71. Srivastava VK (1984) Ionic and covalent energy gaps of CsCl crystals. Phys Letters A 102:127–129

72. Al-Douri Y, Aourag H (2002) The effect of pressure on the ionicity of In–V compounds. Physica B 324:173–178
73. Singth BP, Ojha AK, Tripti S (2004) Analysis of ionicity parameters and photoelastic behaviour of A^NB^{8-N} type crystals. Physica B 350:338–347
74. Hertz W (1927) Dielektrizitätskonstante und Brechungsquotient. Z anorg allgem Chem 161:217–220
75. Linke R (1941) On the refraction exponents of PF_5 and OsO_4 and the dielectric constants of OsO_4, SF_6, SeF_6 and TeF_6. Z phys Chem B 48:193–196
76. Sumbaev OI (1970) The effect of the chemical shift of the X-ray K_α lines in heavy atoms. Sov Phys JETP 30:927–933
77. Batsanov SS, Ovsyannikova IA (1966). X-ray spectroscopy and effective charges of atoms in compounds of Mn. In: Chemical bond in semiconductors thermodynamics. Nauka, Minsk (in Russian)
78. Pantelouris A, Kueper G, Hormes J et al (1995) Anionic gold in Cs_3AuO and Rb_3AuO established by X-ray absorption spectroscopy. J Am Chem Soc 117:11749–11753
79. Saltykov V, Nuss J, Konuma M, Jansen M (2009) Investigation of the quasi binary system BaAu–BaPt. Z allgem anorg Chem 635:70–75
80. Saltykov V, Nuss J, Konuma M, Jansen M (2010) $SrAu_{0.5}Pt_{0.5}$ and $CaAu_{0.5}Pt_{0.5}$, analogues to the respective Ba compounds, but featuring purely intermetallic behaviour. Solid State Sci 12:1615–1619.
81. Nefedov VI, Yarzhemsky VG, Chuvaev AV, Tishkina EM (1988) Determination of effective atomic charge, extra-atomic relaxation and Madelung energy in chemical compounds on the basis of X-ray photoelectron and auger transition energies. J Electron Spectr Relat Phenom 46:381–404
82. Jollet F, Noguera C, Thromat N et al (1990) Electronic structure of yttrium oxide. Phys Rev B 42:7587–7595
83. Larsson R, Folkesson B (1991) Atomic charges in some copper compounds derived from XPS data. Acta Chem Scand 45:567–571
84. Larsson R, Folkesson B (1996) Polarity of Cu_3Si. Acta Chem Scand 50:1060–1061
85. Gutenev MS, Makarov LL (1992) Comparison of the static and dynamic atomic charges in ionic-covalent solids. J Phys Chem Solids 53:137–140
86. Dolenko GN (1993) X-ray determination of effective charges on sulphur, phosphorus, silicon and chlorine atoms. J Molec Struct 291:23–57
87. Dolenko GN, Voronkov MG, Elin VP, Yumatov VD (1993) X-ray investigation of the electron structure of organic compounds containing SiS and SiO bonds. J Molec Struct 295:113–120
88. Jolly WL, Perry WB (1974) Calculation of atomic charges by an electronegativity equalization procedure. Inorg Chem 13:2686–2692
89. Debye P (1915) Zerstreuung von Röntgenstrahlen. Ann Phys 46:809–823
90. Tsirel'son VG, Ozerov RP (1996) Electron density and bonding in crystals. Institute of Physics Publishing, Bristol
91. Coppens P (1997) X-ray charge densities and chemical bonding. Oxford University Press, Oxford
92. Koritsanszky TS, Coppens P (2001) Chemical applications of X-ray charge-density analysis. Chem Rev 101:1583–1628
93. Belokoneva EL (1999) Electron density and traditional structural chemistry of silicates. Russ Chem Rev 68:299–316
94. Dunitz JD, Gavezzotti A (2005) Molecular recognition in organic crystals: directed intermolecular bonds or nonlocalized bonding? Angew Chem Int Ed 44: 1766–1787
95. Yufit DS, Mallinson PR, Muir KW, Kozhushkov SI, De Meijere A (1996) Experimental charge density study of dispiro heptane carboxylic acid. Acta Cryst B 52:668–676
96. Jauch W, Reehuis M, Schultz AJ (2004) γ-Ray and neutron diffraction studies of CoF_2: magnetostriction, electron density and magnetic moments. Acta Cryst A 60:51–57
97. Vidal-Valat G, Vidal J-P, Kurki-Suonio K (1978) X-ray study of the atomic charge densities in MgO, CaO, SrO and BaO. Acta Cryst A 34:594–602

98. Sasaki S, Fujino K, Takeuchi Y, Sadanaga R (1980) On the estimation of atomic charges by the X-ray method for some oxides and silicates. Acta Cryst A 36:904–915
99. Kirfel A, Will G (1981) Charge density in anhydrite, $CaSO_4$. Acta Cryst B 37:525–532
100. Sasaki S, Fujino K, Takeuchi Y, Sadanaga R (1982) On the estimation of atomic charges in $Mg_2Si_2O_6$, $Co_2Si_2O_6$, and $Fe_2Si_2O_6$. Z Krist 158:279–297
101. Gonschorek W (1982) X-ray charge density study of rutile. Z Krist 160:187–203
102. Kirfel A, Josten B, Will G (1984) Formal atomic charges in cubic boron nitride BN. Acta Cryst A 40:C178–C179
103. Will G, Kirfel A, Josten B (1986) Charge density and chemical bonding in cubic boron nitride. J Less-Common Met 117:61–71
104. Zorkii PM, Masunov AE (1990) X-ray diffraction studies on electron density in organic crystals. Russ Chem Rev 59:592–606
105. Vidal-Valat G, Vidal J-P, Kurki-Suonio K, Kurki-Suonio R (1992) Evidence on the breakdown of the Born-Oppenheimer approximation in the charge density of crystalline LiH/D. Acta Cryst A 48:46–60
106. Sasaki S (1997) Radial distribution of electron density in magnetite, Fe_3O_4. Acta Cryst B 53:762–766
107. Hill R, Newton M, Gibbs GV (1998) A crystal chemical study of stishovite. J Solid State Chem 47:185–200
108. Tsirel'son VG, Avilov AS, Abramov YuA et al (1998) X-ray and electron diffraction study of MgO. Acta Cryst B 54:8–17
109. Noritake T, Towata S, Aoki M et al (2003) Charge density measurement in MgH_2 by synchrotron X-ray diffraction. J Alloys Comp 356–357:84–86
110. Schoenes J, Borgschulte A, Carsteanu A-M et al (2003) Structure and bonding in YH_x as derived from elastic and inelastic light scattering. J Alloys Comp 356–357:211–217
111. Belokoneva EL, Shcherbakova YuK (2003) Electron density in synthetic escolaite Cr_2O_3 with a corundum structure and its relation to antiferromagnetic properties. Russ J Inorg Chem 48:861–869
112. Whitten AE, Ditrich B, Spackman MA et al (2004) Charge density analysis of two polymorphs of antimony(III) oxide. Dalton Trans 23–29
113. Saravanan R, Jainulabdeen S, Srinivasan N, Kannan YB (2008) X-ray determination of charge transfer in solar grade GaAs. J Phys Chem Solids 69:83–86
114. Noritake T, Aok M, Towata S et al (2002) Chemical bonding of hydrogen in MgH_2. Appl Phys Lett 81:2008–2010
115. Isidorsson J, Giebels IAME, Arwin H, Griessen R (2003) Optical properties of MgH_2 measured in situ by ellipsometry and spectrophotometry. Phys Rev B 68:115112
116. Ichikawa M, Gustafsson T, Olovsson I (1998) Experimental electron density study of NaH_2PO_4 at 30 K. Acta Cryst B 54:29–34
117. Johnson O (1973) Ionic radii for spherical potential ion. Inorg Chem 12:780–785
118. Johnson O (1981) Electron density and electron redistribution in alloys: electron density in elemental metals. J Phys Chem Solids 42:65–76
119. Batsanov SS (2006) Mechanism of metallization of ionic crystals by pressure. Russ J Phys Chem 80:135–138
120. Brechignac C, Broyer M, Cahuzac Ph et al (1988) Probing the transition from van der Waals to metallic mercury clusters. Phys Rev Letters 60:275–278
121. Pastor GM, Stampell P, Bennemann KH (1988) Theory for the transition from van der Waals to covalent to metallic mercury clusters. Europhys Lett 7:419–424
122. Thomas OC, Zheng W, Xu S, Bowen KH Jr (2002) Onset of metallic behavior in magnesium clusters. Phys Rev Lett 89:213–403
123. Batsanov SS (1971) Quantitative characteristics of bond metallicity in crystals. J Struct Chem 12:809–813
124. Batsanov SS (1979) Band gaps of inorganic compounds of the AB type. Russ J Inorg Chem 24:155–157

125. Vegas A, Jansen M (2002) Structural relationships between cations and alloys; an equivalence between oxidation and pressure. Acta Cryst B 58:38–51
126. Liebau F (1999) Silicates and provskites: two themes with variations. Angew Chem Int Ed 38:1733–1737
127. Goldhammer D (1913) Dispersion und Absorption des Lichtes. Tubner-Ferlag, Leipzig
128. Herzfeld K (1927) On atomic properties which make an element a metal. Phys Rev 29:701–705
129. Duffy JA (1986) Chemical bonding in the oxides of the elements: a new appraisal. J Solid State Chem 62:145–157
130. Dimitrov V, Sakka S (1996) Electronic oxide polarizability and optical basicity of simple oxides. J Appl Phys 79:1736–1740
131. Sun L, Ruoff AL, Zha C-S, Stupian G (2006) Optical properties of methane to 288 GPa at 300 K. J Phys Chem Solids 67:2603–2608
132. Sun L, Ruoff AL, Zha C-S, Stupian G (2006) High pressure studies on silane to 210 GPa at 300 K: optical evidence of an insulator–semiconductor transition. J Phys Cond Matter 18:8573–8580
133. Burdett JK (1997) Chemical bond: a dialog. Wiley, Chichester
134. Hemley RJ, Dera P (2000) Molecular crystals. Rev Mineral Geochem 41:335–419
135. Brewer L (1981) The role and significance of empirical and semiempirical correlations. In: O'Keefe M, Navrotsky A (eds) Structure and bonding in crystals, v 1. Acad Press, San Francisco
136. Trömel M (2000) Metallic radii, ionic radii, and valences of solid metallic elements. Z Naturforsch B 55:243–247
137. Jules JL, Lombardi JR (2003) Transition metal dimer internuclear distance from measured force constant. J Phys Chem A 107:1268–1273
138. Batsanov SS, Batsanov AS (2010) Valent states of Cu, Ag, and Au atoms in molecules and solids are the same. Russ J Inorg Chem 55:913–914
139. Batsanov SS (1980) The features of chemical bonding in b-subgroup metals. Zh Neorg Khim 25:615–623
140. Lawaetz P (1971) Effective charges and ionicity. Phys Rev Letters 26:697–700
141. Lucovsky G, Martin RM, Burstein E (1971) Localized effective charges in diatomic molecules. Phys Rev B 4:1367–1374
142. Robertson J (1978) Tight binding band structure of PbI_2 using scaled parameters. Solid State Commun 26:791–794
143. Robertson J (1979) Electronic structure of SnS_2, $SnSe_2$, CdI_2 and PbI_2. J Phys C 12:4753–4766
144. Robertson J (1979) Electronic structure of SnO_2, GeO_2, PbO_2, TeO_2 and MgF_2. J Phys C 12:4767–4776
145. Robertson J (1979) Electronic structure of GaSe, GaS, InSe and GaTe. J Phys C 12:4777–4790
146. Wakamura K, Arai T (1981) Empirical relationship between effective ionic charges and optical dielectric constants in binary and ternary cubic compounds. Phys Rev B 24:7371–7379
147. Liebau F, Wang X (2005) Stoichiometric valence *versus* structural valence: Conclusions drawn from a study of the influence of polyhedron distortion on bond valence sums. Z Krist 220:589–591
148. Liebau F, Wang X, Liebau W (2009) Stoichiometric valence and structural valence—two different sides of the same coin: "bonding power". Chem Eur J 15:2728–2737
149. Frankland E (1853) Ueber eine neue Reihe organischer Körper, welche Metalle enthalten. Liebigs Ann Chem 85:329–373
150. Moelwyn-Hughes EA (1961) Physical chemistry, 2nd edn. Pergamon Press, London
151. Batsanov SS (2008) Dependence of energies on the bond lengths in molecules and crystals. J Struct Chem 49:296–303
152. Batsanov SS (1998) Estimation of the van der Waals radii of elements with the use of the Morse equation. Russ J General Chem 66:495–500
153. Ceccherini S, Moraldi M (2001) Interatomic potentials of group IIB atoms (ground state). Chem Phys Lett 337:386–390

154. Housecroft CE, Wade K, Smith BC (1978) Bond strengths in metal carbonyl clusters. Chem Commun 765–766
155. Hughes AK, Peat KL, Wade K (1996) Structural and bonding trends in osmium carbonyl cluster chemistry: metal–metal and metal–ligand bond lengths and calculated strengths, relative stabilities and enthalpies of formation of some binary osmium carbonyls. Dalton Trans 4639–4647
156. Hughes AK, Wade K (2000) Metal–metal and metal–ligand bond strengths in metal carbonyl clusters. Coord Chem Rev 197:191–229
157. Batsanov SS (2007) Ionization, atomization, and bond energies as functions of distances in inorganic molecules and crystals. Russ J Inorg Chem 52:1223–1229
158. Harrison WA (1980) Electronic structure and the properties of solids. Freeman, San Francisco
159. Fuentealba P, Preuss H, Stoll H, von Szentpály L (1982) A proper account of core-polarization with pseudopotentials: single valence-electron alkali compounds. Chem Phys Lett 89:418–422
160. Von Szentpály L, Fuentealba P, Preuss H, Stoll H (1982) Pseudopotential calculations on $Rb^{+}{}_{2}$, $Cs^{+}{}_{2}$, RbH^{+}, CsH^{+} and the mixed alkali dimer ions. Chem Phys Lett 93:555–559
161. Müller W, Meyer W (1984) Ground-state properties of alkali dimers and their cations (including the elements Li, Na, and K) from ab initio calculations with effective core polarization potentials. J Chem Phys 80:3311–3320
162. Szentpály L von (1995) Valence states and a universal potential energy curve for covalent and ionic bonds. Chem Phys Lett 245:209–214
163. Batsanov SS (2010) Simple semi-empirical method for evaluating bond polarity in molecular and crystalline halides. J Mol Struct 980:225–229
164. Ho J, Polak ML, Lineberger WC (1992) Photoelectron spectroscopy of group IV heavy metal dimers: $Sn^{-}{}_{2}$, $Pb^{-}{}_{2}$, and $SnPb^{-}$. J Chem Phys 96:144–154
165. Ho J, Polak ML, Ervin KM Lineberger WC (1993) Photoelectron spectroscopy of nickel group dimers: $Ni^{-}{}_{2}$, $Pd^{-}{}_{2}$, and $Pt^{-}{}_{2}$. J Chem Phys 99:8542–8551
166. Lippa TP, Xu S-J, Lyapushina SA et al (1998) Photoelectron spectroscopy of As^{-}, As^{2-}, As^{3-}, As^{4-}, and As^{5-}. J Chem Phys 109:10727–10731
167. Nau WM (1997) An electronegativity model for polar ground-state effects on bond dissociation energies. J Phys Organ Chem 10:445–455
168. Erwin KM, Gronert S, Barlow SE et al (1990) Bond strengths of ethylene and acetylene. J Am Chem Soc 112:5750–5759
169. Blanksby SJ, Ellison GB (2003) Bond dissociation energies of organic molecules. Acc Chem Res 36:255–263
170. Ponomarev D, Takhistov V, Slayden S, Liebman J (2008) Enthalpies of formation for free radicals of main group elements' halogenides. J Mol Struct 876:15–33
171. Ponomarev D, Takhistov V, Slayden S, Liebman J (2008) Enthalpies of formation for bi- and triradicals of main group elements' halogenides. J Mol Struct 876:34–55
172. Nikitin MI, Kosinova NM, Tsirelnikov VI (1992) Mass-spectrometric study of the thermodynamic properties of gaseous lowest titanium iodides. High Temp Sci 30:564–572
173. Giricheva NI, Lapshin SB, Girichev GV (1996) Structural, vibrational, and energy characteristics of halide molecules of group II–V elements. J Struct Chem 37:733–746
174. Nikitin MI, Tsirelnikov VI (1992) Determination of enthalpy of formation of gaseous uranium pentafluoride. High Temp Sci 30:730–735
175. Van der Vis MGM, Cordfunke EHP, Konings RJM (1997) Thermochemical properties of zirconium halides: a review. Thermochim Acta 302:93–108
176. Hildenbrand DL, Lau KH, Baglio JW, Struck CW (2001) Thermochemistry of gaseous OSiI, $OSiI_2$, SiI, and SiI_2. J Phys Chem A 105:4114–4117
177. Hildenbrand DL, Lau KH, Sanjurjo A (2003) Experimental thermochemistry of the SiCl and SiBr radicals; enthalpies of formation of species in the Si–Cl and Si–Br systems. J Phys Chem A 107:5448–5451
178. Giricheva NI, Girichev GV, Shlykov SA et al (1995) The joint gas electron diffraction and mass spectrometric study of $GeI_4(g) + Ge(s)$ system: molecular structure of germanium diiodide. J Mol Struct 344:127–134

179. Haaland A, Hammel A, Martinsen K-G et al (1992) Molecular structures of monomeric gallium trichloride, indium trichloride and lead tetrachloride by gas electron diffraction. Dalton Trans 2209–2214
180. Hildenbrand DL, Lau KH, Perez-Mariano J, Sanjurjo A (2008) Thermochemistry of the gaseous vanadium chlorides VCl, VCl_2, VCl_3, and VCl_4. J Phys Chem A 112:9978–9982
181. Grant DJ, Matus MH, Switzer JR, Dixon DA (2008) Bond dissociation energies in second-row compounds. J Phys Chem A 112:3145–3156
182. Hildenbrand DL (1995) Dissociation energies of the monochlorides and dichlorides of Cr, Mn, Fe, Co, and Ni. J Chem Phys 103:2634–2641
183. Hildenbrand DL, Lau KH (1992) Trends and anomalies in the thermodynamics of gaseous thorium and uranium halides. Pure Appl Chem 64:87–92
184. Hildenbrand DL (1996) Dissociation energies of the molecules BCl and BCl^-. J Chem Phys 105:10507–10510
185. Ezhov YuS (1992) Force constants and characteristics of the structure of trihalides. Russ J Phys Chem 66:748–751
186. Gurvich LV, Ezhov YuS, Osina EL, Shenyavskaya EA (1999) The structure of molecules and the thermodynamic properties of scandium halides. Russ J Phys Chem 73:331–344
187. Hildenbrand DL, Lau KH (1995) Thermochemical properties of the gaseous scandium, yttrium, and lanthanum fluorides. J Chem Phys 102:3769–3775
188. Struck C, Baglio J (1991) Estimates for the enthalpies of formation of rare-earth solid and gaseous trihalides. High Temp Sci 31:209–237
189. Ezhov YuS (1995) Variations of molecular constants in metal halide series XY_n and estimates for bismuth trihalide constants. Russ J Phys Chem 69:1805–1809
190. Ezhov YuS (1993) Systems of force constants, coriolis coupling constants, and structure peculiarities of XY_4 tetrahalides. Russ J Phys Chem 67:901–904
191. Giricheva NI, Girichev GV (1999) Mean bond dissociation energies in molecules and the enthalpies of formation of gaseous niobium tetrahalides and oxytrihalides. Russ J Phys Chem 73:372–374
192. Berkowitz J, Ruscic B, Gibson S et al (1989) Bonding and structure in the hydrides of groups III-VI deduced from photoionization studies. J Mol Struct Theochem 202:363–373
193. Jones MN, Pilcher G (1987) Thermochemistry. Ann Rep Progr Chem C84:65–104
194. Craciun R, Long RT, Dixon DA, Christe KO (2010) Electron affinities, fluoride affinities, and heats of formation of the second row transition metal hexafluorides: MF_6 (M = Mo, Tc, Ru, Rh, Pd, Ag). J Phys Chem A114;7571–7582
195. Craciun R, Picone D, Long RT et al (2010) Third row transition metal hexafluorides, extraordinary oxidizers, and Lewis acids: electron affinities, fluoride affinities, and heats of formation of WF_6, ReF_6, OsF_6, IrF_6, PtF_6, and AuF_6. Inorg Chem 49:1056–1070
196. Batsanov SS (2002) Bond polarity as a function of the valence of the central atom. Russ J Inorg Chem 47:663–665
197. Leroy G, Temsamani DR, Sana M, Wilante C (1993) Refinement and extension of the table of standard energies for bonds involving hydrogen and various atoms of groups IV to VII of the Periodic Table. J Molec Struct 300:373–383
198. Batsanov SS (1994) Crystal-chemical estimates of bond energies in metals. Inorg Mater 30:926–927
199. Murphy LR, Meek TL, Allred AL, Allen LC (2000) Evaluation and test of Pauling's electronegativity scale. J Phys Chem A 104:5867–5871
200. Mulliken RS (1952) Magic formula, structure of bond energies and isovalent hybridization. J Phys Chem 56:295–311
201. Mulliken RS, Rieke CA, Orloff D, Orloff H (1949) Formulas and numerical tables for overlap integrals. J Chem Phys 17:1248–1267
202. Mulliken RS (1950) Overlap integrals and chemical binding. J Am Chem Soc 72:4493–4503
203. Jaffe HH (1953) Some overlap integrals involving d orbitals. J Chem Phys 21:258–263
204. Jaffe HH (1954) Studies in molecular orbital theory of valence: multiple bonds involving d-orbitals. J Phys Chem 58:185–190

205. Cotton FA, Leto J (1959) Acceptor properties, reorganization energies, and π bonding in the boron and aluminum halides. J Chem Phys 30:993–998
206. Batsanov SS, Zvyagina RA (1966) Overlap integrals and problem of effective charges, vol 1. Nauka, Novosibirsk (in Russian)
207. Batsanov SS, Kozhevina LI (1969) Overlap integrals, vol 2. Nauka, Novosibirsk (in Russian)
208. Ferreira R (1963) Principle of elecronegativity equalization: bond-dissociation energies. Trans Faraday Soc 59:1075–1079
209. Sanderson RT (1975) Interrelation of bond dissociation energies and contributing bond energies. J Am Chem Soc 97:1367–1372
210. Sanderson RT (1983) Electronegativity and bond energy. J Am Chem Soc 105:2259–2261
211. Sanderson RT (1986) The inert-pair effect on electronegativity. Inorg Chem 25:1856–1858
212. Matcha RL (1983) Theory of the chemical bond: accurate relationship between bond energies and electronegativity differences. J Am Chem Soc 105:4859–4862
213. Bratsch SG (1984) Electronegativity equalization with Pauling units. J Chem Educat 61:588–589
214. Bratsch SG (1985) A group electronegativity method with Pauling units. J Chem Educat 62:101–103
215. Reddy RR, Rao TVR, Viswanath R (1989) Correlation between electronegativity differences and bond energies. J Am Chem Soc 111:2914–2915
216. Smith DW (2002) Comment on "Evaluation and test of Pauling's electronegativity rule". J Phys Chem A 106:5951–5952
217. Smith DW (2004) Effects of exchange energy and spin-orbit coupling on bond energies. J Chem Educat 81:886–890
218. Nasar A, Shamsuddin M (1990) Thermodynamic properties of cadmium selenide. J Less-Common Met 158:131–135
219. Nasar A, Shamsuddin M (1990) Thermodynamic investigations of HgTe. J Less-Common Met 161:87–92
220. Nasar A, Shamsuddin M (1990) Thermodynamic properties of ZnTe. J Less-Common Met 161:93–99
221. Nasar A, Shamsuddin M (1992) Investigations of the thermodynamic properties of zinc chalcogenides. Thermochimica Acta 205:157–169
222. O'Hare PAG, Curtis LA (1995) Thermochemistry of (germanium + sulfur): IV. Critical evaluation of the thermodynamic properties of solid and gaseous germanium sulfide GeS and germanium disulfide GeS_2, and digermanium disulfide Ge_2S_2(g). Enthalpies of dissociation of bonds in GeS(g), GeS_2(g), and Ge_2S_2(g). J Chem Thermodyn 27:643–662
223. Tomaszkiewicz P, Hoppe GA, O'Hare PAG (1995) Thermochemistry of (germanium + tellurium): I. Standard molar enthalpy of formation $\Delta^f H_m{}^\circ$ at the temperature 298.15 K of crystalline germanium monotelluride GeTe by fluorine-bomb calorimetry. A critical assessment of the thermodynamic properties of GeTe(cr and g) and $GeTe_2$(g). J Chem Thermodyn 27:901–919
224. Boone S, Kleppa OJ (1992) Enthalpies of formation for group IV selenides ($GeSe_2$, $GeSe_2$(am), SnSe, $SnSe_2$, PbSe). Thermochim Acta 197:109–121
225. Kotchi A, Gilbert M, Castanet R (1988) Thermodynamic behaviour of the Sn + Te, Pb + Te, Sn + Se and Pb + Se melts according to the associated model. J Less-Common Met 143:L1-L6
226. O'Hare PAG, Lewis BM, Susman S, Volin KJ (1990) Standard molar enthalpies of formation and transition and other thermodynamic properties of the crystalline and vitreous forms of arsenic sesquiselenide (As_2Se_3). Dissociation enthalpies of As-Se bonds. J Chem Thermodyn 22:1191–1206
227. O'Hare PAG (1987) Inorganic chalcogenides: high-tech materials, low-tech thermodynamics. J Chem Thermodyn 19:675–701
228. Cemič L, Kleppa O (1988) High temperature calorimetry of sulfide systems. Phys Chem Miner 16:172–179
229. Goncharuk LV, Lukashenko GM (1986) Thermodynamic properties of chromium selenides, Cr_2Se_3. Russ J Phys Chem 60:1089–1089

230. Dittmer G, Niemann U (1981) Heterogeneous reactions and chemical-transport of tungsten with halogens and oxygen under steady-state conditions of incandescent lamps. Philips J Res 36:87–111
231. Robie RA, Hemingway BS (1985) Low-temperature molar heat capacities and entropies of MnO_2 (pyrolusite), Mn_3O_4 (hausmanite), and Mn_2O_3 (bixbyite). J Chem Thermodyn 17:165–181
232. Johnson GK, Murray WT, Van Deventer EH, Flotow HE (1985) The thermodynamic properties of zirconium ditelluride $ZrTe_2$–1500 K. J Chem Thermodyn 17:751–760
233. Fuger J (1992) Transuranium-element thermochemistry: a look into the past—a glimpse into the future. J Chem Thermodyn 24:337–358
234. O'Hare PAG, Lewis BM, Parkinson BA (1988) Standard molar enthalpy of formation of tungsten diselenide; thermodynamics of the high-temperature vaporization of WSe_2; revised value of the standard molar enthalpy of formation of molybdenite (MoS_2). J Chem Thermodyn 20:681–691
235. Leonidov VYa, Timofeev IV, Lazarev VB, Bozhko AB (1988) Enthalpy of formation of the wurtzite form of boron nitride. Russ J Inorg Chem 33:906–908
236. Ranade MR, Tessier F, Navrotsky A et al (2000) Enthalpy of formation of gallium nitride. J Phys Chem B 104:4060–4063
237. Kulikov IS (1988) Thermodynamics of carbides and nitrides. Metallurgiya, Chelyabinsk (in Russian)
238. Knacke O, Kubaschewski O, Hesselmann K (eds) (1991) Thermochemical properties of inorganic substances, 2nd edn. Springer-Verlag, Berlin
239. Yamaguchi K, Takeda Y, Kameda K, Itagaki K (1994) Measurements of heat of formation of GaP, InP, GaAs, InAs, GaSb and InSb. Materials Transactions JIM 35:596–602
240. Gordienko SP, Fenochka BF, Viksman GSh (1979) Thermodynamics of the lanthanide compounds. Naukova Dumka, Kiev (in Russian)
241. Yamaguchi K, Yoshizawa M, Takeda Y et al (1995) Measurement of thermodynamic properties of Al-Sb system by calorimeters. Materials Transactions JIM 36:432–437
242. Waddington TC (1959) Lattice energies and their significance in inorganic chemistry. Adv Inorg Chem Radiochem 1:157–221
243. Ratkey CD, Harrison BK (1992) Prediction of enthalpies of formation for ionic compounds. Ind Eng Chem Res 31:2362–2369
244. Born M, Lande A (1918) Kristallgitter und bohrsches Atommodell. Verh Dtsch Physik Ges 20:202–209
245. Born M, Lande A (1918) Über Berechnung der Compressibilität regulärer Kristalle aus der Gittertheorie. Verh Dtsch Physik Ges 20:21–216
246. Born M (1919) Eine thermochemischemische Anwendung der Gittertheorie. Verh Dtsch Physik Ges 21:13–24
247. Born M, Mayer JE (1932) Zur Gittertheorie der Ionenkristalle. Z Physik 75:1–18
248. Royer DJ (1968) Bonding theory. McGraw-Hill, New York
249. Bucher M (1990) Cohesive properties of silver halides. J Imaging Sci 34:89–95
250. Johnson QC, Templeton DH (1961) Madelung constants for several structures. J Chem Phys 34:2004–2007
251. Hoppe R (1970) Madelung constants as a new guide in the structural chemistry of solids. Adv Fluorine Chem 6:387–438
252. Alcock NW, Jenkins HDB (1974) Crystal structure and lattice energy of thallium (I) fluoride: inert-pair distortions. Dalton Trans 1907–1911
253. Zucker IJ (1991) Madelung constants and lattice sums for hexagonal crystals. J Phys A 24:873–879
254. Zemann J (1991) Madelung numbers for the theoretical structure type with mutual trigonal prismatic coordination. Acta Cryst A 47:851–852
255. Keshishi A (1996) Calculation of Madelung constant of various ionic structures based on the semisimple Lie algebras. Modern Phys Lett 10:475–485

256. Gaio M, Silvestrelli PL (2009) Efficient calculation of Madelung constants for cubic crystals. Phys Rev B 79:012102
257. Baker AD, Baker MD (2010) Rapid calculation of individual ion Madelung constants and their convergence to bulk value. Am J Phys 78:102–105
258. Izgorodina EI, Bernard UL, Dean PM et al (2009) The Madelung constant of organic salts. Cryst Growth Des 9:4834–4839
259. Kapustinskii A (1933) On the second principle of crystal chemistry. Z Krist 86:359–369
260. Kapustinskii A (1943) Lattice energy of ionic crystals. Acta Physicochim URSS 18:370–377
261. Yatsimirskii KB (1951) Thermochemistry of coordination compounds. Akad Nauk, Moscow (in Russian)
262. Jenkins HDB, Pratt KF (1977) On basic radii of simple and complex ions and repulsion energy of ionic crystals. Proc Roy Soc A 356:115–134
263. Jenkins HDB, Thakur KP (1979) Reappraisal of thermochemical radii for complex ions. J Chem Educat 56:576–577
264. Jenkins HDB, Roobottom HK, Passmore J, Glasser L (1999) Relationships among ionic lattice energies, molecular (formula unit) volumes, and thermochemical radii. Inorg Chem 38:360–3620
265. Glasser L, Jenkins HDB (2000) Lattice energies and unit cell volumes of complex ionic solids. J Am Chem Soc 122:632–638
266. Sorokin NL (2001) Calculations of the lattice energy of fluoride solid solutions with fluorite structure. Russ J Phys Chem 75:1010–1011
267. Aleixo AI, Oliveira PH, Diogo HP, da Piedade MEM (2005) Enthalpies of formation and lattice enthalpies of alkaline metal acetates. Thermochim Acta 428:131–136
268. Yoder CH, Flora NJ (2005) Geochemical applications of the simple salt approximation to the lattice energies of complex materials. Amer Miner 90:488–515
269. Glasser L, Jenkins HDB (2005) Predictive thermodynamics for condensed phases. Chem Soc Rev 34:866–874
270. Glasser L, von Szentpály L (2006) Born-Haber-Fajans cycle generalized: linear energy relation between molecules, crystals and metals. J Am Chem Soc 128:12314–12321
271. Dimitrov V, Sakka S (1996) Electronic oxide polarizability and optical basicity of simple oxides. J Appl Phys 79:1736–1740
272. Xu Y-N, Ching WY (1993) Electronic, optical, and structural properties of some wurtzite crystals. Phys Rev B 48:4335–4351
273. Pandey R, Lepak P, Jaffe JE (1992) Electronic structure of alkaline-earth selenides. Phys Rev B 46:4976–4977
274. KanekoY, Koda T (1988) New developments in IIa–VIb (alkaline-earth chalcogenide) binary semiconductors. J Cryst Growth 86:72–78
275. Rocquefelte X, Whangbo M-H, Jobic S (2005) Structural and electronic factors controlling the refractive indices of the chalcogenides ZnQ and CdQ (Q = O, S, Se, Te). Inorg Chem 44:3594–3598
276. Hanafi ZM, Ismail FM (1988) Colour problem of mercuric oxide photoconductivity and electrical conductivity of mercuric oxide. Z phys Chem 158:8–86
277. Boldish SI, White WB (1998) Optical band gaps of selected ternary sulfide minerals. Amer Miner 83:865–871
278. Sohila S, Rajalakshmi M, Ghosh C et al (2011) Optical and Raman scattering studies on SnS nanoparticles. J Alloys Comp 509:5843–5847
279. Gawlik K-U, Kipp L, Skibowski M et al (1997) HgSe: metal or semiconductor? Phys Rev Lett 78:3165–3168
280. Di Quarto F, Sunseri C, Piazza S, Romano MC (1997) Semiempirical correlation between optical band gap values of oxides and the difference of electronegativity of the elements. J Phys Chem B 101:2519–2525
281. Di Quarto F, Sunseri C, Piazza S, Romano MC (2000) A semiempirical correlation between the optical band gap of hydroxides and the electronegativity of their constituents. Russ J Elektrochem 36:1203–1208

282. Julien C, Eddrief M, Samaras I, Balkanski M (1992) Optical and electrical characterizations of SnSe, SnS_2 and $SnSe_2$ single crystals. Mater Sci Engin B 15:70–72
283. Majumdar A, Xu HZ, Zhao F et al (2004) Bandgap energies and refractive indices of $Pb_{1-x}Sr_xSe$. J Appl Phys 95:939–942
284. Lokhande CD (1992) Chemical deposition of CoS films. Indian J Pure Appl Phys 30:245–247
285. Bai X, Kordesch ME (2001) Structure and optical properties of ScN thin films. Appl Surf Sci 175–176:499–504
286. Hulliger F (1979) Rare earth pnictides. In: Gschneidner KA Jr, Eyring L (eds) Handbook on the physics and chemistry of rare earths, vol 4. Amsterdam: North-Holland
287. Meng J, Ren Y (1991) Investigation of the photoelectronic properties of rare earth monophosphide. Solid State Commun 80: 485–488
288. Miyuata N, Moriki K, Mishima O et al (1989) Optical constants of cubic boron nitride. Phys Rev B 40:12028–12029
289. Onodera A, Nakatani M, Kobayashi M et al (1993) Pressure dependence of the optical-absorption edge of cubic boron nitride. Phys Rev B 48:2777–2780
290. Tarrio C, Schnatterly S (1989) Interband transitions, plasmons, and dispersion in hexagonal boron nitride. Phys Rev B 40:7852–7859
291. Stenzel O, Hahn J, Röder M et al (1996) The optical constants of cubic and hexagonal boron nitride thin films and their relation to the bulk optical constants. Phys Status Solidi 158a:281–287
292. Prasad C, Sahay M (1989) Electronic structure and properties of boron phosphide and boron arsenide. Phys Status Solidi 154b:201–207
293. Vurgaftman I, Meyer JR, Ram-Mohan LR (2001) Band parameters for III–V compound semiconductors and their alloys. J Appl Phys 89:581–5875
294. McBride JR, Hass KC, Weber WH (1991) Resonance-Raman and lattice-dynamics studies of single-crystal PdO. Phys Rev B 44:5016–5028
295. Dey S, Jain VK (2004) Platinum group metal chalcogenides. Platinum Metals Rev 48:16–29
296. Welker H (1952) Uber neue halbleitende Verbindungen. Z Naturforsch 7a:744–749
297. Manca P (1961) A relation between the binding energy and the band-gap energy in semiconductors of diamond or zinc-blende structure. J Phys Chem Solids 20:268–273
298. Vijh AK (1969) Correlation between bond energies and forbidden gaps of inorganic binary compounds. J Phys Chem Solids 30:1999–2005
299. Reddy RR, Ahammed YN (1995) Relationship between refractive index, optical electronegativities and electronic polarizability in alkali halides, III–V, II–VI group semiconductors. Cryst Res Technol 30:263–266
300. Gong X, Gao F, Yamaguchi T et al (1992) Dependence of energy band gap and lattice constant of III-V semiconductors on electronegativity difference of the constituent elements. Cryst Res Technol 27:1087–1096
301. Duffy JA (1977) Variable electronegativity of oxygen in binary oxides: possible relevance to molten fluorides. J Chem Phys 67:2930–2931
302. Duffy JA (1980) Trends in energy gaps of binary compounds: an approach based upon electron transfer parameters from optical spectroscopy. J Phys C 13:2979–2990
303. Mooser E, Pearson WB (1959) On the crystal chemistry of normal valence compounds. Acta Cryst 12:1015–1022
304. Makino Y (1994) Interpretation of band gap, heat of formation and structural mapping for sp-bonded binary compounds on the basis of bond orbital model and orbital electronegativity. Intermetallics 2:55–56
305. Villars P (1983) A three-dimensional structural stability diagram for 998 binary AB intermetallic compounds. J Less-Common Met 92:215–238
306. Villars P (1983) A three-dimensional structural stability diagram for 1011 binary AB_2 intermetallic compounds. J Less-Common Met 99:33–43
307. Phillips JC (1973) Bonds and bands in semiconductors. Academic Press, New York
308. Hooge FN(1960) Relation between electronegativity and energy bandgap. Z phys Chem 24:27–282

309. Shimakawa K (1981) On the compositional dependence of the optical gap in amorphous semiconducting alloys. J Non-Cryst Solids 43:229–244
310. Batsanov SS (1965) A new method of calculating the width of the forbidden zone. J Struct Chem 5:862–864
311. Batsanov SS (1972) Quantitative characteristics of bond metallicity in crystals. J Struct Chem 12:809–813
312. Harrison WA (1980) Electronic structure and properties of solids. Freeman, San Feancisco; Christensen NE, Satpathy S, Pawlowska Z (1987) Bonding and ionicity in semiconductor. Phys Rev 36:1032–1050
313. Veal TD, Mahboob I, McConville CF (2004) Negative band gaps in dilute InN_xSb_{1-x} alloys. Phys Rev Lett 92:136–801
314. Suchet J (1965) Chemical physics of semiconductors. Van Nostrand, Princeton
315. Phillips JC (1970) Ionicity of the chemical bond in crystals. Rev Modern Phys 42:31–356
316. Nethercot AH Jr (1974) Prediction of Fermi energies and photoelectric thresholds based on electronegativity concepts. Phys Rev Lett 33:1088–1091
317. Poole RT, Williams D, Riley J et al (1975) Electronegativity as a unifying concept in the determination of Fermi energies and photoelectric thresholds. Chem Phys Lett 36:401–403
318. Chen ECM, Wentworth WE, Ayala JA (1977) The relationship between the Mulliken electronegativities of the elements and the work functions of metals and nonmetals. J Chem Phys 67:2642–2647
319. Nethercot AH (1981) Electronegativity and a model Hamiltonian for chemical applications. Chem Phys 59:297–313
320. Lonfat M, Marsen B, Sattler K (1999) The energy gap of carbon clusters studied by scanning tunneling spectroscopy. Chem Phys Lett 313:539–543
321. Banerjee R, Jayakrishnan R, Ayub P (2000) Effect of the size-induced structural transformation on the band gap in CdS nanoparticles. J Phys Cond Matt 12:10647–10654
322. Sarangi SN, Sahu SN (2004) CdSe nanocrystalline thin films: composition, structure and optical properties. Physica E 23:159–167
323. Vidal J, Lany S, d'Avezac M et al (2012) Band-structure, optical properties, and defect physics of the photovoltaic semiconductor SnS. Appl Phys Lett 100:032104
324. Franzman MA, Schlenker CW, Thompson ME, Brutchey RL (2010) Solution-phase synthesis of SnSe nanocrystals for use in solar cells. J Am Chem Soc 132:4060–4061
325. Wang Y, Suna A, Mahler W, Kasowski R (1987) PbS in polymers: from molecules to bulk solids. J Chem Phys 87:7315–7322
326. Salem AM, Selim MS, Salem AM (2001) Structure and optical properties of chemically deposited Sb_2S_3 thin film. J Phys D 34:12–17
327. Tyagi P, Vedeshwar AG (2001) Thickness dependent optical properties of CdI_2 films. Physica B 304:166–174
328. Ma DDD, Leo CS, Au FCK et al (2003) Small-diameter silicon nanowire surface. Science 299:1874–1877
329. Wang H, He Y, Chen W et al (2010) High-pressure behavior of β-Ga_2O_3 nanocrystal. J Appl Phys 107:033520
330. Liu B, Li Q, Du X et al (2011) Facile hydrothermal synthesis of CeO_2 nanosheets with high reactive exposure surface. J Alloys Comp 509:6720–6724
331. Ramana CV, Vemuri RS, Fernandez I, Campbell AL (2009) Size-effect on the optical properties of zirconium oxide thin films. Appl Phys Lett 95:231905
332. He Y, Liu JF, Chen W et al (2005) High-pressure behavior of SnO_2 nanocrystals. Phys Rev B 72:212102
333. Gullapalli SK, Vemuri RS, Ramana CV (2010) Structural transformation induced changes in the optical properties of nanocrystalline tungsten oxide thin films. Appl Phys Lett 96:171903
334. Cisneros-Morales MC, Aita CR (2010) The effect of nanocrystallite size in monoclinic HfO_2 films on lattice expansion and near-edge optical absorption. Appl Phys Lett 96:191904
335. Hirai H, Terauchi V, Tanaka M, Kondo K (1999) Band gap of essentially fourfold-coordinated amorphous diamond synthesized from C_{60} fullerene. Phys Rev B 60:6357–6361

336. Alexenskii AE, Osipov VYu, Vul' AYa et al (2001) Optical properties of nanodiamond layers. Phys Solid State 43:14–150
337. Housecroft CE, Constable EC (2010) Chemistry, 4th edn. Pearson, Edinburgh
338. Fajans K (1951) General Chemistry by Linus Pauling. J Phys Chem 55:1107–1108
339. Hückel W (1957) Die chemische Bindung. Kritische Betrachtung der Systematik, der Ausdrucksweisen und der formelmäßigen Darstellung. J prakt Chem 5:105–174
340. Batsanov SS (1960) Comments on Hückel's book. Zh Fiz Khim 34:937–938 (in Russian)
341. Syrkin YaK (1962) Effective charges and electronegativity. Russ Chem Rev 31:197–207
342. Spiridonov VP, Tatevskii VM (1963) On the concept of electronegativity of atoms: content and definitions of electronegativity used by various authors. Zh Fiz Khim 37:994–1000 (in Russian)
343. Spiridonov VP, Tatevskii VM (1963) Analysis of Pauling's scale of electronegativity. Zh Fiz Khim 37:1236–1242 (in Russian)
344. Spiridonov VP, Tatevskii VM (1963) A review of empirical methods of calculating electronegativity by various authors. Zh Fiz Khim 37:1583–1586 (in Russian)
345. Spiridonov VP, Tatevskii VM (1963) A review of semi-empirical and theoretical methods of calculating electronegativities. Zh Fiz Khim 37:1973–1978 (in Russian)
346. Bykov GV (1965) On the electronegativity of atoms (atomic cores) in molecules. Zh Fiz Khim 39:1289–1291 (in Russian)
347. Batsanov SS (1963) On the article "Effective charges and electronegativites" by Ya.K. Syrkin. Zh Fiz Khim 37:1418–1422 (in Russian)
348. Bratsch G (1988) Revised Mulliken electronegativities: calculation and conversion to Pauling units. J Chem Educat 65:34–41
349. Bratsch G (1988) Revised Mulliken electronegativities: applications and limitations. J Chem Educat 65:223–227
350. Batsanov SS (1967) On the articles by V.P. Spiridonov and V.M. Tatevskii criticizing the concept of electronegativity. Zh Fiz Khim 41:2402–2406 (in Russian)
351. Hinze J (1968) Elektronegativität der Valenzzustände. Fortschr chem Forschung 9:448–485
352. Komorowski L (1987) Chemical hardness and Pauling's scale of electronegativities. Z Naturforsch A 42:767–773
353. Komorowski L, Lipinski J (1991) Quantumchemical electronegativity and hardness indices for bonded atoms. Chem Phys 157:45–60
354. Allen LC (1989) Electronegativity is the average one-electron energy of the valence-shell electrons in ground-state free atoms. J Am Chem Soc 111:9003–9014
355. Cherkasov AR, Galkin VI, Zueva EM, Cherkasov RA (1998) The concept of electronegativity: the current state of the problem. Russ Chem Rev 67:375–392
356. Batsanov SS (1968) The concept of electronegativity: conclusions and prospects. Russ Chem Rev 37:332–350
357. Allred AL (1961) Electronegativity values from thermochemical data. J Inorg Nucl Chem 17:215–221
358. Leroy G (1983) Stability of chemical species. Int J Quantum Chem 23:271–308
359. Leroy G, Sana M, Wilante C, van Zieleghem M-J (1991) Revaluation of the bond energy terms for bonds between atoms of the first rows of the Periodic Table, including lithium, beryllium and boron. J Molec Struct 247:199–215
360. Leroy G, Temsamani DR, Wilante C (1994) Refinement and extension of the table of standard energies for bonds containing atoms of the fourth group of the Periodic Table. J Molec Struct 306:21–39
361. Leroy G, Temsamani DR, Wilante C, Dewispelaere J-P (1994) Determination of bond energy terms in phosphorus containing compounds. J Molec Struct 309:113–119
362. Leroy G, Dewispelaere J-P, Benkadour H (1995) Theoretical approach to the thermochemistry of geminal interactions in XY_2H_n compounds ($X = C$, N, O, Si, P, S; $Y = NH_2$, OH, F, SiH_3, PH_2, SH). J Molec Struct Theochem 334:137–143
363. Ochterski JW, Peterson GA, Wiberg KB (1995) A comparison of model chemistries. J Am Chem Soc 117:11299–11308

364. Smith DW (1998) Group electronegativities from electronegativity equilibration. Faraday Trans 94:201–205
365. Smith DW (2007) A new approach to the relationship between bond energy and electronegativity. Polyhedron 26:519–523
366. Matsunaga N, Rogers DW, Zavitsas AA (2003) Pauling's electronegativity equation and a new corollary accurately predict bond dissociation enthalpies and enhance current understanding of the nature of the chemical bond. J Org Chem 68:3158–3172
367. Ionov SP, Alikhanyan AS, Orlovskii VP (1992) On the determination of the electronegativity of both the chemical bond and atom in molecule. Doklady Phys Chem 325:455–456
368. Batsanov SS (2000) Thermochemical electronegativities of metals. Russ J Phys Chem 74:267–270
369. Howard JAK, Hoy VJ, O'Hagan D, Smith GTS (1996) How good is fluorine as a hydrogen bond acceptor? Tetrahedron 52:12613–12622
370. Dunitz JD, Taylor R (1997) Organic fluorine hardly ever accepts hydrogen bonds. Chem Eur J 3:89–98
371. Bykov GV, Dobrotin RB (1968) Calculation of the electronegativity of fluorine from thermochemical data. Russ Chem Bull 17:226–2271
372. Batsanov SS (1989) Structure and properties of fluorine, oxygen, and nitrogen atoms in covalent bonds. Russ Chem Bull 38:410–412
373. Finemann MA (1958) Correlation of bond dissociation energies of polyatomic molecules using Pauling's electronegativity concept. J Phys Chem 62:947–951
374. Datta D, Singh SN (1990) Evaluation of group electronegativity by Pauling's thermochemical method. J Phys Chem 94:2187–2190
375. Batsanov SS (1962) Electronegativity of elements and chemical bond. Nauka, Novosibirsk (in Russian)
376. Batsanov SS (1990) The concept of electronegativity and structural chemistry. Sov Sci Rev B Chem Rev 15(4):3
377. Batsanov SS (1975) System of electronegativities and effective atomic charges in crystalline compounds. Russian J Inorg Chem 20:1437–1440
378. Batsanov SS (1990) Polar component of the atomization energy and electronegativity of atoms in crystals. Inorg Mater 26:569–572
379. Batsanov SS (2001) Electronegativities of metal atoms in crystalline solids. Inorg Mater 37:23–30
380. Vieillard P, TardyY (1988) Une nouvelle échelle d'électronégativité des ions. Compt Rend Ser II 308:1539–1545
381. Ionov SP, Sevast'yanov DV (1994) Relative chemical potential and structural-thermochemical model of metallic bonds. Zh Neorg Khim 39:2061–2067 (in Russian)
382. Mulliken RS (1934) A new electroaffinity scale; together with data on valence states and on valence ionization potentials and electron affinities. J Chem Phys 2:782–793
383. Mulliken RS (1935) Electroaffinity, molecular orbitals and dipole moments. J Chem Phys 3:573–585
384. Mulliken RS (1937) Discussion of the papers presented at the symposium on molecular structure. J Phys Chem 41:318–320
385. Pritchard HO (1953) The determination of electron affinities. Chem Rev 52:529–563
386. Pritchard HO, Skinner HA (1955) The concept of electronegativity. Chem Rev 55:745–786
387. Skinner HA, Sumner FH (1957) The valence states of the elements V, Cr, Mn, Fe, Co, Ni, and Cu. J Inorg Nucl Chem 4:245–263
388. Pilcher G, Skinner HA (1962) Valence-states of boron, carbon, nitrogen and oxygen. J Inorg Nucl Chem 24:93–952
389. Batsanov SS (1960) Structural-chemical problems of the electronegativity concept. Proc Sibir Branch Acad Sci USSR 1:68–83 (in Russian)
390. Iczkowski RP, Margrave JL (1961) Electronegativity. J Am Chem Soc 83:3547–3551
391. Hinze J, Jaffe HH (1962) Orbital electronegativity of neutral atoms. J Am Chem Soc 84:540–546

392. Hinze J, Whitehead MA, Jaffe HH (1963) Bond and orbital electronegativities. J Am Chem Soc 85:148–154
393. Hinze J, Jaffe HH (1963) Orbital electronegativities of the neutral atoms of the period three and four and of positive ions of period one and two. J Phys Chem 67:1501–1506
394. Parr RG, Donnelly RA, Levy M, Palke WE (1978) Electronegativity: the density functional approach. J Chem Phys 68:3801–3807
395. Parr RG, Yang W (1989) Density-functional theory of atoms molecules. Oxford University Press, New York
396. Parr RG, Bartolotti L (1982) On the geometric mean principle for electronegativity equalization. J Am Chem Soc 104:3801–3803
397. Polizer P, Murray JS (2006) A link between the ionization energy ratios of an atom and its electronegativity and hardness. Chem Phys Lett 431:195–198
398. Sen KD, Jørgensen CK (eds) (1987) Structure and Bonding, vol 66. Springer-Verlag, Berlin
399. Allen LC (1994) Chemistry and electronegativity. Int J Quantum Chem 49:253–277
400. Bergmann D, Hinze J (1996) Electronegativity and molecular properties. Angew Chem Int Ed 35:150–163
401. Sacher E, Currie JF (1988) A comparison of electronegativity series. J Electr Spectr Relat Phenom 46:173–177
402. Valone SM (2011) Quantum mechanical origins of the Iczkowski–Margrave model of chemical potential. J Chem Theory Comput 7:2253–2261
403. Pearson RG (1988) Absolute electronegativity and hardness: application to inorganic chemistry. Inorg Chem 27:734–740
404. Pearson RG (1990) Electronegativity scales. Acc Chem Res 23:1–2
405. Politzer P, Shields ZP-I, Bulat FA, Murray JS (2011) Average local ionization energies as a route to intrinsic atomic electronegativities. J Chem Theory Comput 7:377–384
406. Allen LC (1994) Chemistry and electronegativity. Int J Quant Chem 49:253–277
407. Mann JB, Meek TL, Allen LC (2000) Configuration energies of the main group elements. J Am Chem Soc 122:2780–2783
408. Mann JB, Meek TL, Knight ET et al (2000) Configuration energies of the d-block elements. J Am Chem Soc 122:5132–5137
409. Brown ID, Skowron A (1990) Electronegativity and Lewis acid strength. J Am Chem Soc 112:3401–3403
410. Martynov AI, Batsanov SS (1980) New approach to calculating atomic electronegativities. Russ J Inorg Chem 25:1737–1740
411. Giemza J, Ptak WS (1984) An empirical chemical potential of the atomic core for non-transition element. Chem Phys Lett 104:115–119
412. Bergmann D, Hinze J (1987) Electronegativity and charge distribution. In: Sen KD, Jörgensen CK (eds) Structure and Bonding, vol. 66. Springer-Verlag, Berlin
413. True JE, Thomas TD, Winter RW, Gard GL (2003) Electronegativities from core ionization energies: electronegativities of SF_5 and CF_3. Inorg Chem 42:4437–4441
414. Stevenson DP (1955) Heat of chemisorption of hydrogen in metals. J Chem Phys 23:203–203
415. Trasatti S (1972) Electronegativity, work function, and heat of adsorption of hydrogen on metals. J Chem Soc Faraday Trans I 68:229–236
416. Trasatti S (1972) Work function, electronegativity, and electrochemical behaviour of metals. J Electroanalytical Chem 39:163–184
417. Dritz ME (2003) Properties of elements. Metals, Moscow (in Russian)
418. Miedema AR, De Boer FR, De Chatel PF (1973) Empirical description of the role of electronegativity in alloy formation. J Phys F 3:1558–1576
419. Miedema AR, De Chatel PF, De Boer FR (1980) Cohesion in alloys—fundamentals of a semi-empirical model. Physica B 100:1–28
420. Ray PK, Akinc M, Kramer MJ (2010) Applications of an extended Miedema's model for ternary alloys. J Alloys Comp 489:357–361
421. Parr RG, Pearson RG (1983) Absolute hardness: companion parameter to absolute electronegativity. J Am Chem Soc 105:7512–7516

422. Pearson RG (1993) Chemical hardness—a historical introduction. Structure and Bonding 80:1–10
423. Batsanov S.S. (1986) Experimental foundations of structural chemistry. Standarty, Moscow (in Russian)
424. Cottrell TL, Sutton LE (1951) Covalency, electrovalency and electronegativity. Proc Roy Soc London A 207:49–63
425. Pritchard HO (1953) The determination of electron affinities. Chem Rev 52:529–563
426. Allred AL, Rochow EG (1958) A scale of electronegativity based on electrostatic force. J Inorg Nucl Chem 5:264–268
427. Batsanov SS (1993) A new scale of atomic electronegativities. Russ Chem Bull 42:24–29
428. Batsanov SS (2004) Geometrical electronegativity scale for elements taking into account their valence and physical state. Russ J Inorg Chem 49:1695–1701
429. Sanderson RT (1951) An interpretation of bond lengths and a classification of bonds. Science 114:670–672
430. Sanderson RT (1982) Radical reorganization and bond energies in organic molecules. J Org Chem 47:3835–3839
431. Sanderson RT (1988) Principles of electronegativity: general nature. J Chem Educat 65:112–118
432. Sanderson RT (1988) Principles of electronegativity: applications. J Chem Educat 65:227–231
433. Batsanov SS (1988) Refinement of the Sanderson procedure for calculating electronegativities of atoms. J Struct Chem 29:631–635
434. Batsanov SS (1978) A new approach to the geometric determination of the electronegativities of atoms in crystals. J Struct Chem 19:826–829
435. Gorbunov AI, Kaganyuk DS (1986) A new method for the calculation of electronegativity of atoms. Russ J Phys Chem 60:1406–1407
436. Batsanov SS (1971) Electronegativity and effective charges of atoms. Znanie, Moscow (in Russian)
437. Ray N, Samuels L, Parr RG (1979) Studies of electronegativity equalization. J Chem Phys 70:3680–3684
438. Batsanov SS (1994) Equalization of interatomic distances in polymorphous transformations under pressure. J Struct Chem 35:391–393
439. Batsanov SS (1975) Electronegativity of elements and chemical bond (in Russian). Nauka, Novosibirsk
440. Phillips JC (1968) Covalent bond in crystals: partially ionic binding. Phys Rev 168:905–911
441. Phillips JC (1968) Covalent bond in crystals: anisotropy and quadrupole moments. Phys Rev 168:912–917
442. Phillips JC (1968) Covalent bond in crystals: lattice deformation energies. Phys Rev 168:917–921
443. Phillips JC (1974) Chemical bonding at metal-semiconductor interfaces. J Vacuum Sci Technol 11:947–950
444. Phillips JC, Lucovsky G. (2009) Bonds and bands in semiconductors 2nd ed. Momentum Press
445. Li K, Xue D (2006) Estimation of electronegativity values of elements in different valence states. J Phys Chem A110;11332–11337
446. Li K, Wang X, Zhang F, Xue D (2008) Electronegativity identification of novel superhard materials. Phys Rev Lett 100:235504
447. Pettifor DG (1984) A chemical scale for crystal-structure maps. Solid State Commun 51:31–34
448. Pettifor DG (1985) Phenomenological and microscopic theories of structural stability. J Less-Common Metals 114:7–15
449. Pettifor DG (2003) Structure maps revisited. J Phys Cond Matter 15:V13–V16
450. Campet G, Portiera J, Subramanian MA (2004) Electronegativity versus Fermi energy in oxides: the role of formal oxidation state. Mater Lett 58:437–438
451. Carver JC, Gray RC, Hercules DM (1974) Remote inductive effects evaluated by X-ray photoelectron spectroscopy. J Am Chem Soc 96:6851–6856

452. Gray R, Carver J, Hercules D (1976) An ESCA study of organosilicon compounds. J Electron Spectr Relat Phenom 8:343–357
453. Gray R, Hercules D (1977) Correlations between ESCA chemical shifts and modified Sanderson electronegativity calculations. J Electron Spectr Relat Phenom 12:37–53
454. Ray NK, Samuels L, Parr RG (1979) Studies of electronegativity equalization. J Chem Phys 70:3680–3684
455. Parr RG, Bartolotti LJ, Gadre SR (1980) Electronegativity of the elements from simple $\chi\alpha$ theory. J Am Chem Soc 102:2945–2948
456. Pearson RG (1985) Absolute electronegativity and absolute hardness of Lewis acids and bases. J Am Chem Soc 107:6801–6806
457. Fuentealba P, Parr RG (1991) Higher-order derivatives in density-functional theory, especially the hardness derivative $\partial\eta/\partial N$. J Chem Phys 94:5559–5564
458. Politzer P, Weinstein H (1979) Some relations between electronic distribution and electronegativity. J Chem Phys 71:4218–4220
459. Ferreira R, Amorim AO (1981) Electronegatlvity and the bonding character of molecular orbitals.Theor Chim Acta 58:131–136
460. Amorim AO de, Ferreira R (1981) Electronegativities and the bonding character of molecular orbitals: A remark. Theor Chim Acta 59:551–553
461. Pearson RG (1989) Absolute electronegativity and hardness: applications to organic chemistry. J Org Chem 54:1423–1430
462. Mortier WJ, van Genechten K, Gasteiger J (1985) Electronegativity equalization: application and parametrization. J Am Chem Soc 107:829–835
463. Mortier WJ, Ghosh SK, Shankar S (1986) Electronegativity-equalization method for the calculation of atomic charges in molecules. J Am Chem Soc 108:4315–4320
464. Van Genechten KA, Mortier WJ, Geerlings P (1987) Intrinsic framework electronegativity: a novel concept in solid state chemistry. J Chem Phys 86:5063–5071
465. Uytterhoeven L, Mortier WJ, Geerlings P (1989) Charge distribution and effective electronegativity of aluminophosphate frameworks. J Phys Chem Solids 50:479–486
466. De Proft F, Langenaeker W, Geerlings P (1995) A non-empirical electronegativity equalization scheme: theory and applications using isolated atom properties. J Mol Struct Theochem 339:45–55
467. Bultinck P, Langenaeker W, Lahorte P et al (2002) The electronegativity equalization method: parametrization and validation for atomic charge calculations. J Phys Chem A 106:7887–7894
468. Bultinck P, Langenaeker W, Lahorte P et al (2002) The electronegativity equalization method: applicability of different atomic charge schemes. J Phys Chem A 106:7895–7901
469. von Szentpály L (1991) Studies on electronegativity equalization: consistent diatomic partial charges. J Mol Struct Theochem 233:71–81
470. Donald KJ, Mulder WH, von Szentpály L (2004) Valence-state atoms in molecules: influence of polarization and bond-charge on spectroscopic constants of diatomic molecules. J Phys Chem A 108:595–606
471. Speranza G, Minati L, Anderle M (2006) Covalent interaction and semiempirical modeling of small molecules. J Phys Chem A 110:13857–13863
472. Boudreaux EA (2011) Calculations of bond dissociation energies: new select applications of an old method. J Phys Chem A 115:1713–1720
473. Islam N, Ghosh DC (2010) Evaluation of global hardness of atoms based on the commonality in the basic philosophy of the origin and the operational significance of the electronegativity and the hardness. Eur J Chem 1:83–89
474. Urusov VS (1961) On the calculation of bond ionicity in binary compounds. Zh Neorg Khim 6:2436–2439 (in Russian)
475. Batsanov SS (1964) Calculating the degree of bond ionicity in complex ions by electronegativity method. Zh Neorg Khim 9:1323–1327 (in Russian)
476. Waber JT, Cromer DT (1965) Orbital radii of atoms and ions. J Chem Phys 42:4116–4123
477. Batsanov SS (2011) Calculating atomic charges in molecules and crystals by a new electronegativity equalization method. J Mol Struct 1006:223–226

478. Batsanov SS (2011) Thermodynamic determination of van der Waals radii of metals. J Mol Struct 990:63–66
479. Pauling L (1952) Interatomic distances and bond character in the oxygen acids and related substances. J Phys Chem 56:361–365
480. Pauling L (1929) The principles determining the structure of complex ionic crystals. J Am Chem Soc 51:1010–1026
481. Pauling L (1948) The modern theory of valency. J Chem Soc 1461–1467
482. Sanderson RT (1954) Electronegativities in inorganic chemistry. J Chem Educat 31:238–245
483. Reed JL (2003) Electronegativity: coordination compounds. J Phys Chem A 107:8714–8722
484. Suchet JP (1965) Chemical physics of semiconductors. Van Nostrand, London
485. Suchet JP (1977) Electronegativity, ionicity, and effective atomic charge. J Electrochem Soc 124:30C–35C
486. Noda Y, Ohba S, Sato S, Saito Y (1983) Charge distribution and atomic thermal vibration in lead chalcogenide crystals. Acta Cryst B 39:312–317
487. Feldmann C, Jansen M (1993) Cs_3AuO, the first ternary oxide with anionic gold. Angew Chem Int Ed 32:1049–1050
488. Pantelouris A, Kueper G, Hormes J et al (1995) Anionic gold in Cs_3AuO and Rb_3AuO established by X-ray absorption spectroscopy. J Am Chem Soc 117:11749–11753
489. Feldmann C, Jansen M (1995) Zur kristallchemischen Ähnlichkeit von Aurid- und Halogenid-Ionen. Z anorg allgem Chem 621:1907–1912
490. Mudring A-V, Jansen M (2000) Base-induced disproportionation of elemental gold. Angew Chem Int Ed 39:3066–3067
491. Nuss J, Jansen M (2009) BaAuP and BaAuAs, synthesis via disproportionation of gold upon interaction with pnictides as bases. Z allgem anorg Chem 635:1514–1516
492. Karpov A, Nuss J, Wedig U, Jansen M (2003) Cs_2Pt: a platinide(-II) exhibiting complete charge separation. Angew Chem Int Ed 42:4818–4821
493. Batsanov SS, Ruchkin ED (1959) Mixed halogenides of tetravalent platinum. Zh Neorg Khim 4:1728–1733 (in Russian)
494. Batsanov SS, Ruchkin ED (1965) On the isomerism of mixed halogenides of platinum. Zh Neorg Khim 10:2602–2605 (in Russian)
495. Batsanov SS, Sokolova MN, Ruchkin ED (1971) Mixed halides of gold. Russ Chem Bull 20:1757–1759
496. Batsanov SS (1986) Experimental foundations of structural chemistry. Standarty, Moscow (in Russian)
497. Batsanov SS, Rigin VI (1966) Isomerism of thallium selenobromides. Doklady Akad Nauk SSSR 167:89–90
498. Dehnicke K (1965) Synthesis of oxide halides. Angew Chem Int Ed 4:22–29
499. Custelcean R, Jackson JE (1998) Topochemical control of covalent bond formation by dihydrogen bonding. J Am Chem Soc 120:12935–12941
500. Batana A, Faour J (1984) Pressure dependence of the effective charge of ionic crystals. J Phys Chem Solids 45:571–574
501. Ves S, Strössner K, Cardona M (1986) Pressure dependence of the optical phonon frequencies and the transverse effective charge in AlSb. Solid State Commun 57:483–486
502. Katayama Y, Tsuji K, Oyanagi H, Shimomura O (1998) Extended X-ray absorption fine structure study on liquid selenium under pressure. J Non-Cryst Solids 232–234:93–98
503. Gauthier M, Polian A, Besson J, Chevy A (1989) Optical properties of gallium selenide under high pressure. Phys Rev B 40:3837–3854
504. Talwar DN, Vandevyver M (1990) Pressure-dependent phonon properties of III-V compound semiconductors. Phys Rev B 41:12129–12139
505. Kucharczyk W (1991) Pressure dependence of effective ionic charges in alkali halides. J Phys Chem Solids 52:435–436
506. Errandonea D, Segura A, Muoz V, Chevy A (1999) Effects of pressure and temperature on the dielectric constant of GaS, GaSe, and InSe: role of the electronic contribution. Phys Rev B 60:15866–15874

507. Goi AR, Siegle H, Syassen K et al (2001) Effect of pressure on optical phonon modes and transverse effective charges in GaN and AlN. Phys Rev B 64:035205
508. Yamanaka T, Fukuda T, Mimaki J (2002) Bonding character of SiO_2 stishovite under high pressures up to 30 GPa. Phys Chem Miner 29:633–641
509. Batsanov SS (1999) Pressure effect on the heat of formation of condensed substances. Russ J Phys Chem 73:1–6
510. Ferrante J, Schlosser H, Smith JR (1991) Global expression for representing diatomic potential-energy curves. Phys Rev A 43:3487–3494
511. Ghandehari K, Luo H, Ruoff AL et al (1995) Crystal structure and band gap of rubidium hydride to 120 GPa. Mod Phys Lett B 9:1133–1140
512. Ghandehari K, Luo H, Ruoff AL et al (1995) New high pressure crystal structure and equation of state of cesium hydride to 253 GPa. Phys Rev Lett 74:2264–2267
513. Batsanov SS (2005) Chemical bonding evolution on compression of crystals. J Struct Chem 46:306–314
514. Batsanov SS, Gogulya MF, Brazhnikov MA et al (1994) Behavior of the reacting system Sn + S in shock waves. Comb Explosion Shock Waves 30:361–365
515. Batsanov SS (2006) Mechanism of metallization of ionic crystals by pressure. Russ J Phys Chem 80:135–138
516. Reparaz JS, Muniz LR, Wagner MR et al (2010) Reduction of the transverse effective charge of optical phonons in ZnO under pressure. Appl Phys Lett 96:231906
517. Batsanov SS (2004) Determination of ionic radii from metal compressibilities. J Struct Chem 45:896–899
518. Batsanov SS (1994) Pressure dependence of bond polarities in crystalline materials. Inorg Mater 30:1090–1096
519. Batsanov SS (1997) Effect of high pressure on crystal electronegativities of elements. J Phys Chem Solids 58:527–532
520. Fujii Y, Hase K, Ohishi Y et al (1989) Evidence for molecular dissociation in bromine near 80 GPa. Phys Rev Lett 63:536–539
521. Takemura K, Minomura S, Shimomura O et al (1982) Structural aspects of solid iodine associated with metallization and molecular dissociation under high pressure. Phys Rev B 26:998–1004
522. Brown D, Klages P, Skowron A (2003) Influence of pressure on the lengths of chemical bonds. Acta Cryst B 59:439–448
523. Bokii GB (1948) Bond ionicity from atomic polarization and refraction. Moscow Univ Chem Bull 11:155–160

Chapter 3
"Small" Molecule

3.1 Introduction

A molecule can be defined as a limited group of atoms linked by chemical bonds, bearing no net electric charge, which requires the valences of the component atoms to be balanced. In the gas phase, intermolecular distances vastly exceed the real sizes of molecules, thus the latter are effectively isolated. Some elementary non-metals (N, O, halogens) and many compounds with covalent type of bonding exist as molecules, separated by 'nonbonding' distances and having essentially the same structure, in any aggregate state. Certain substances, particularly metals and strongly ionic compounds (e.g. alkali halides) can form molecules only in the gas phase. Their condensed phases contain continuous networks of strongly bonded atoms (often called, rather inconsistently, 'atomic structures') or ions.[1] Thus, $AuCl_2$ has molecular structure in the gas phase but its crystal comprises Au^+ and $[AuCl_4]^-$ ions. The difference is largely due to the difference between isotropic nature of the Coulomb forces and highly directional covalent bonding.

Geometries of small molecules in gases can be determined by microwave and vibration-rotation spectroscopy or by gas electron diffraction (GED). For diatomic molecules, bond distances can be measured with the precision up to 10^{-5} Å (spectroscopy) or 10^{-3} Å (GED), but the uncertainty steeply increases with the number of atoms (n), molecules with $n > 5$ can be satisfactorily characterized only if the structure is exceptionally symmetrical and rigid, or some *a priori* assumptions can be made. A number of reviews contain compilations of molecular structures determined by spectroscopic methods [1–5] and GED [6–17], which are henceforth cited without further references and updated according to new original publications. A monograph on the structures of sulfur-containing compounds, both organic and inorganic, by all gas-phase methods is available [18].

The interactions between predominantly covalent molecules (e.g. organic) in a condensed phase being two orders of magnitude smaller than their atomization

[1]With certain exceptions. Thus, RuO_4 and OsO_4 form typically molecular crystals, low-melting and volatile, consisting of discrete molecules with tetrahedral coordination of metal and mean bond lengths Ru–O 1.698 Å [3.29] and Os–O 1.70 Å [3.30].

S. S. Batsanov, A. S. Batsanov, *Introduction to Structural Chemistry*,
DOI 10.1007/978-94-007-4771-5_3, © Springer Science+Business Media Dordrecht 2012

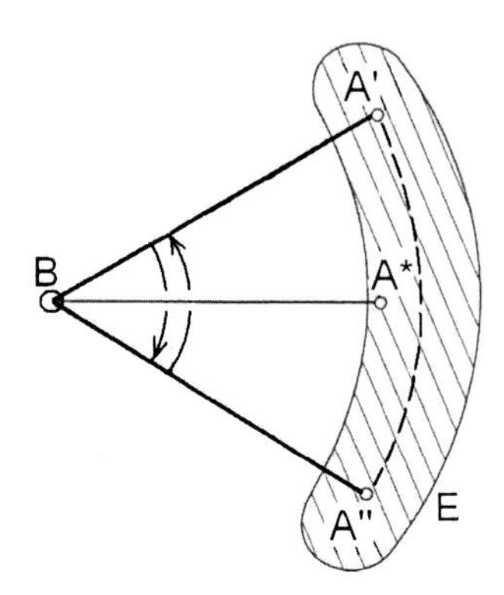

Fig. 3.1 Apparent shortening of bond length as determined by X-ray crystallography (A*–B) compared with actual length A–B: atom A is moving along the arch A′ . . . A″, its observed electron density is smeared over the area E with the time-averaged centroid at A*

energies, it is not surprising that these interactions do not distort the molecular structures substantially. Hence the paradox that reliable *molecular* geometries are known mostly [19] from *crystal* structures, which are mainly studied by X-ray diffraction (XRD) from single crystals or polycrystalline powders. X-rays being scattered by electrons, this method is least efficient in locating hydrogen atoms with their low and easily perturbed electron density. Neutron diffraction (ND) is much better for this purpose, but the sources of radiation (nuclear reactors or spallation sources) are very few. The known crystal structures are vastly more numerous than gas-phase determinations, and are available from several databases [20]. Cambridge Structural Database (CSD) [21, 22] stores all structures containing at least one "organic" carbon atom, except organic biopolymers. Of the latter, protein structures are deposited in the (formerly Brookhaven) Protein Data Bank (PDB) [23] and nucleic acids in the Nucleic Acid Database (NDP) [24]. Comprehensive tables of bond distances in organic [25] and organometallic [26] crystal structures have been compiled using CSD and reproduced *verbatim* in [27, 28].

Bond distance means simply the distance by the straight line between the centers of chemically bonded atoms, although in some cases the maximum concentration of the bonding electron density is known to lie off this line (see Sect. 3.3) In most cases, the observed differences between gas-phase and crystallographic bond distances are due not to any real change of the structure, but to the fact that different methods measure different distances: between atomic centers of mass (spectroscopy), between nuclei (ND) or between maxima of electrostatic potential (GED) or those of electron density (XRD). The differences are mostly small, but the X-ray measurements of A–H bond lengths are up to 0.1 Å shorter than the neutron ones, due to the shift of the hydrogen's only electron into the bond area. Even more importantly, gas-phase measurements give the average interatomic distance, while crystal diffraction yields the distance between average atomic positions. The latter is always systematically shorter than the former (Fig. 3.1). The difference Δd depends on the amplitude of libration ω,

$$\Delta d = d - d\cos\omega = d\omega^2/2 \tag{3.1}$$

Table 3.1 Experimental bond distances (Å) in A_2 type molecules

A	d(A–A)	A	d(A–A)	A	d(A–A)	A	d(A–A)
Li	2.673[a]	Cd	3.76[e]	N	1.098	Te	2.557
Na	3.079[a]	Hg	3.60[e]	P	1.893	H	0.7414
K	3.924[a]	B	1.458[f]	As	2.103	F	1.412
Rb	4.170[a]	Al	2.701[g]	Sb	2.342	Cl	1.988
Cs	4.648[a]	Ga	2.75[h]	Bi	2.656[n]	Br	2.281
Cu	2.220[b]	Tl	3.0[i]	V	1.77[b]	I	2.666
Ag	2.530[b]	Y	2.65[j]	Nb	2.078[b]	Fe	2.02[b]
Au	2.472[b]	C	1.242	Cr	1.679[b]	Ni	2.154[b]
Be	2.452[c]	Si	2.246	Mo	1.929[b]	Pt	2.333[p]
Mg	3.890[d]	Ge	2.368	W	2.048[o]	He	2.967
Ca	4.277[d]	Sn	2.746[k]	Mn	3.4[b]	Ne	3.087
Sr	4.446[d]	Pb	2.927[k]	O	1.208	Ar	3.759
Zn	4.19[b]	Ti	1.942[l]	S	1.889	Kr	4.012
		Zr	2.241[m]	Se	2.166	Xe	4.362[p]

[a][36], [b][37], [c][38], [d][39], [e][40], [f][41], [g][42], [h][43], [i]the best estimated value is 3.11 Å [44], [j][45], [k][46], [l][47], [m][48], [n][49], [o][50], [p][51], [q][52]

hence it increases with temperature, is inversely related to the atomic mass and depends on the structure and perfection of the crystal; for organic molecules it can vary from thousandths to hundredths of Å [31].

In all the abovementioned methods the structure is determined as a set of atomic positions, practically regarding atoms as constant and indivisible entities. High-precision X-ray diffraction studies can be used also to determine the actual distribution of electron (charge) density in crystals and hence the bond character, atomic charges, *etc.*; several hundred such studies of varying quality have been published [32, 33]. More recent techniques of X-ray absorption spectroscopy (XAS), particularly the extended X-ray absorption fine structure (EXAFS), extract information about the nearest environment of an absorbing atom (usually a heavier metal) from the shape of the X-ray absorption curve near the absorption band edge [34]. The latter method can determine the coordination number (within ±20 %), bond distances (with the precision of 0.01–0.05 Å) in *any* aggregate state, including biological systems *in vivo* (see references in [35]).

Tables S3.1–S3.5 present experimental data on the structures of molecules of the MX and MX_n types, where X=H, C, or an element of Groups 15–17.

3.2 Inorganic Molecules and Radicals

3.2.1 *Bond Distances*

Bond lengths in homo-diatomic molecules (A_2) [36–52] are compiled in Table 3.1. For Groups 1 and 17 they correspond to single covalent bonds, for Group 16 to double bonds and for Group 15 to triple bonds. To find out the bond order in other

Table 3.2 Bond distances (Å) in hydride and halide molecules AX. (For MX (except AuX) from [53])

M/X	H	F	Cl	Br	I
H	0.741	0.917	1.275	1.414	1.609
Li	1.596	1.564	2.021	2.170	2.392
Na	1.887	1.926	2.361	2.502	2.711
K	2.240	2.171	2.667	2.821	3.048
Rb	2.367	2.270	2.787	2.945	3.177
Cs	2.494	2.345	2.906	3.072	3.315
Cu	1.462[a]	1.745	2.051	2.173	2.338
Ag	1.618[a]	1.983	2.281	2.393	2.545
Au	1.524[a]	1.918[b]	2.199[b]	2.318[b]	2.471[b]
Be	1.343	1.361	1.797		
Mg	1.678[β]	1.750[c]	2.199		
Ca	1.987[ε]	1.952[c]	2.437[ε]	2.594[ε]	2.829[ε]
Sr	2.146	2.074[c]	2.576[ε]	2.710[e]	2.974[ε]
Ba	2.232	2.159[c]	2.683[ε]	2.845[ε]	
Zn	1.594	1.768[d]	2.130[e]		
Cd	1.781				
Hg	1.740				
B	1.232	1.263[g]	1.716		
Al	1.647[f]	1.654[g]	2.130[g]	2.295[g]	2.537[g]
Ga	1.662[f]	1.774[g]	2.202[h]	2.349[h]	2.576[h]
In	1.838[f]	1.985[g]	2.402[i]	2.543[i]	2.742[i]
Tl	1.873[f]	2.084[g]	2.485	2.618	2.814
Sc	1.775[j]	1.787[k]	2.230[k]	2.381[k]	2.608[k]
Y	1.923[j]	1.928	2.384[l]		2.764[m]
La	2.032[j]	2.023[n]	2.498[n]	2.652[n]	2.879[n]
C	1.120[o]	1.272	1.651		
Si	1.520[o]	1.601			
Ge	1.587[o]	1.745[f]	2.164[p]		
Sn	1.769[o]	1.944[f]			
Pb	1.839[q]	2.08[f]			
Ti	1.779[hh]	1.834[r]	2.260[s]		
Zr		1.854[α]	2.367[t]		
Hf	1.831[u]				
N	1.037[v]	1.370[w]	1.611[x]	1.778[x]	1.965[x]
P	1.422[v]	1.589[w]			2.381[y]
As	1.523[v]	1.740[w]			2.53
Sb	1.711[v]	1.920[w]	2.335[z]		
Bi	1.809[v]	2.034[w]	2.472	2.610	2.800
V		1.776[aa]	2.214[bb]		
O	0.970[cc]	1.354[dd]	1.569[ee]	1.717[ee]	1.868[ff]
S	1.341[cc]	1.601			
Se	1.464[cc]	1.742			
Te	1.656[cc]	1.910[gg]	2.321[gg]		
Cr	1.782[γ]	1.788[ii]	2.206[ii]		
Mn		1.843[jj]			
Re	1.82[kk]	1.843[ll]			
F	0.917	1.412	1.628	1.759	1.910
Cl	1.275	1.628	1.988	2.136	2.321
Br	1.414	1.759	2.136	2.281	2.469
I	1.609	1.910	2.321	2.469	2.666

Table 3.2 (continued)

M/X	H	F	Cl	Br	I
Fe	1.588[mm]	1.780[c]	2.176[c]		
Co	1.513[nn]	1.736[nn]	2.066[oo]		
Ni	1.476	1.740[pp]	2.064[pp]	2.196[pp]	2.348[pp]
Rh	1.59				
F	Ru	Ir	Pt		
M–F	1.916[qq]	1.851[rr]	1.874[ss]		
OH	Cu	Ag	K	Rb	
M–O	2.017[tt]	1.689[tt]	1.828	2.301	
OH	Ca	In	Cl	Br	
M–O	1.985	2.017[uu]	1.689	1.828	
SH	Li	Na	Cu		
M–S[vv]	2.146	2.479	2.139[ww]		
SH	Mg	Ca	Sr	Ba	
M–S[vv]	2.316	2.564	2.706	2.807	

[a][54], [b][55], [β][56, 57], [c][58], [d][59], [ε][60, 61], [e][62], [f][63, 64], [g][65, 66], [h][67], [i][68], [j][69–71], [k][72], [l][73], [m][74], [n][75], [o][76], [p][77], [q][78], [r][79], [s][80], [t][81], [u][82], [v][83], [w][84], [x][85–87], [y][88], [z][89], [aa][90], [bb][91], [cc][92], [dd][93], [ee][94], [ff][95], [gg][96], [hh][97], [ii][98], [jj][99], [kk][100], [ll][101], [mm][102], [nn][103], [oo][104], [pp][105], [qq][106], [rr][107], [ss][108], [tt][109], [uu][110], [vv][111], [ww][112, 113]

Table 3.3 Bond distances d(Å) in MM′ molecules

MM′	LiNa	LiK	LiCs	NaK	NaRb	NaCs
d_{exp}	2.885[a]	3.319[b]	3.668[c]	3.499[d]	3.643[a]	3.850[a]
d_{add}	2.876	3.298	3.662	3.502	3.624	3.863
MM′	KRb	KCs	RbCs	LiCu	LiAg	KAg
d_{exp}	4.034[a]	4.284[e]	4.37[a]	2.26[f]	2.41[f]	2.40[g]
d_{add}	4.047	4.286	4.409	2.446	2.601	3.227

[a][36, 114], [b][115], [c][116], [d][117], [e][118], [f][119], [g][120]

cases, it is necessary to determine the valence state of atoms in these molecules. The bond lengths in MX molecules (where X is hydrogen or a halogen) [53–113] are listed in Table 3.2, those in MX_n molecules are given in Table S3.1, those in MX and MX_2 molecules (where X = C, N, O or a chalcogen) in Tables S3.2–S3.4.

Note the additive character of these distances: for any pair of metals, M and M′, the average difference of bond distances $\Delta d = d(\text{M–X}) - d(\text{M}'\text{–X})$ in hydrides and halides is approximately the same for each X: NaX–LiX = 0.33(2), KX–NaX = 0.31(2), RbX–KX = 0.115(10), CsX–RbX = 0.12(2), AgX–CuX = 0.21(3), Å. Similarly, the M–Cl bond for every alkali metal is longer than the M–F by 0.49(4) Å, while the bonds in sulfides of divalent elements (MS) are longer than those in oxides (MO) by 0.45(5) Å. The bond distances in hetero-atomic molecules of alkali metals show good agreement with the values calculated by the additive rule (Table 3.3).

However, the additive rule works only for atoms of similar electronic structure; it breaks down in we compare, for example, transition and non-transition elements (see LiCu, LiAg, KAg in Table 3.3). Likewise, Na–H and Cu–H bonds differ by 0.424 Å,

Table 3.4 Bond distances d (Å) in AH_n radicals and $AH_{n+1,2}$ molecules

A	AH	AH_2
Be	1.343	1.334
Mg	1.730	1.703
Zn	1.594	1.535
Cd	1.762	1.683
Hg	1.740	1.646
Al	1.645	1.59
O	0.970	0.958
S	1.341	1.336
Se	1.464	1.460
Te	1.656	1.651
A	AH	AH_3
B	1.232	1.193
N	1.038	1.012
P	1.430	1.415
As	1.522	1.511
Sb	1.711	1.700
Bi	1.808	1.778
A	AH_2	AH_4
C	1.107	1.085
Si	1.514	1.480
Ge	1.591	1.514

Na–F and Cu–F by 0.181 Å, while Cu–Cl and Cu–F by 0.306 Å and Ag–Cl and Ag–F by 0.298 Å. Most importantly, a polar (hetero-atomic) bond (A–X) is always shorter than the mean length of the corresponding homo-atomic bonds (A–A and X–X) of the same order, just as its energy (see Sect. 2.3) is always higher than the additive value, both differences increasing with the bond polarity (see below).

The higher the oxidation state of a metal atom, the shorter bonds it forms. Thus, M–Cl bonds in MCl_4 (where M = Ti, Si, Ge, Sn, Pb, V, Cr) are on average 0.075 Å shorter than in MCl_2 (Table S3.2). As the electronegativity of atoms increases on oxidation (see Table 2.15) the shortening can be attributed to a resulting decrease of bond polarity. However, the bond energy is *lower* for higher oxidation states (see Table 2.9). Thus, the shorter M^{IV}–Cl bonds are actually weaker than M^{II}–Cl, contrary to the common notion that 'shorter means stronger'. In the series $CX_2 \rightarrow CX_4$ and CZ $\rightarrow CZ_2$ the formal increase of the carbon valence leads to bond lengthening (Tables S3.3 and S3.4) while the bond energies decrease similarly to those in Si- or Ge-analogues. In hydrides (Table 3.4), accumulation of hydride ligands always reduces bond distances [121] (see also references to Table S3.1). In molecules containing only nonmetals, where the bonding is of nearly covalent type anyway, a (formal) change of oxidation state has little effect on the interatomic distances. Thus, the equatorial Cl–F bonds in ClF, ClF_3 and ClF_5 molecules are similar (1.628, 1.597 and 1.571 Å, respectively), as are axial bonds in ClF_3, FClO, $FClO_2$ and ClF_5, viz. 1.697, 1.697, 1.691 and 1.669 Å [122]. The bonds in the unstable neutral molecules BH_4 (two of 1.182 and two of 1.289 Å, mean 1.235 Å) [123], NH_4 (1.051 Å) and ND_4 (1.048 Å) [124] are similar to those in the corresponding mono-hydrides, or in the cations NH_4^+ (1.029 Å) and ND_4^+ (1.025 Å). However, when bonds are really

Table 3.5 Bond distances d (Å) in A_2, AH molecules and A_2^+, AH^+ cations

A_2	A–A	$(A–A)^+$	AH	A–H	$(A–H)^+$	AH	A–H	$(A–H)^+$
Xe_2	4.362	3.087	BeH	1.343	1.312	PH	1.433	1.404
H_2	0.7414	1.052	MgH	1.730	1.649	CH	1.120	1.131
F_2	1.412	1.322	ZnH	1.594	1.515	NH	1.038	1.045
Cl_2	1.988	1.892	CdH	1.762	1.667	OH	0.970	1.029
O_2	1.208	1.116	HgH	1.740	1.594	SH	1.341	1.338
N_2	1.098	1.116	BH	1.232	1.215	FH	0.917	1.001
P_2	1.893	1.986	AlH	1.648	1.602	ClH	1.275	1.315
Be_2[a]	2.452	2.211	SiH	1.520	1.492	BrH	1.414	1.448

[a][38, 127]

Table 3.6 Bond distances d (Å) in neutral molecules and their anions

MX	M–X	$(M–X)^-$	MX_n	(M–X)	$(M–X)^-$	M_2[g]	(M–M)	$(M–M)^-$
CuH[a]	1.463	1.567	PbO	1.922	1.995	Cu_2	2.220	2.343
CrH	1.668	1.75	PbS	2.287	2.390[e]	Ag_2	2.530	2.654
MnH	1.72	1.82	MoO	1.70	1.72	Au_2	2.472	2.582
CoH	1.526	1.67	FeS	2.04	2.18	Sn_2	2.746	2.659
NiH	1.48	1.61	NiO	1.627	1.66	Pb_2	2.927	2.814
AuO	1.849[b]	1.899[c]	AuO_2[f]	1.793	1.866	Cr_2	1.679	1.705
MgCl	2.199	2.37	RhO_2[f]	1.699	1.735	O_2	1.207	1.26
MgO[d]	1.749	1.794	IrO_2[f]	1.717	1.738	Fe_2	2.02	2.10
ZnO[d]	1.719	1.767	PtO_2[f]	1.719	1.790	Ni_2	2.155	2.257
						Pt_2	2.333	2.407

[a][128], [b][129], [c][130], [d][131], [e][132], [f][133], [g][37]

polar, the oxidation state does make a difference: compare O–X bonds in molecules ClOCl and BrOBr (1.696 and 1.843 Å) with those in OClO and OBrO (1.470 and 1.649 Å) [125].

The geometry of a molecule is substantially affected if it acquires a net electric charge. Bond distances in neutral diatomic molecules and radicals (from Tables 3.1 and 3.2) and the corresponding cations with the charge +1 [1, 126] are compared in Table 3.5, those in neutral molecules and their –1 anions in Table 3.6. Negative charge almost always produces a lengthening of the bonds (relative to the neutral species), provided the bond order remains the same. However, if the addition of an electron alters the electron configuration of the molecule, as in Sn_2 and Pb_2 which experience a $\sigma \rightarrow \pi$ transition, the bond can strengthen and shorten. Positive ionization causes lengthening or contraction, depending on whether the electron is removed from a bonding, non- or anti-bonding orbital. Thus, X–H bonds in hydrides lengthen if X atom is more electronegative than H, and shorten if otherwise.

The oxygen molecule is a special case. A simple Lewis (valence-bond) scheme suggests that this molecule has a bond order of 2, with two lone electron pair on each oxygen atom ($:\ddot{O}=\ddot{O}:$). In fact, it is paramagnetic (triplet) in the ground state, alone among simple diatomic molecules. Rationalization of this fact was one of the early successes of the molecular orbital (MO) theory, according to which the molecule has five fully occupied orbitals (bonding $\sigma 2s$, $\sigma 2p_z$, $\pi 2p_x$, $\pi 2p_y$ and anti-bonding $\sigma^* 2s$)

and two unpaired electrons in the anti-bonding π^*2p_x and π^*2p_y orbitals. Thus the bond order, defined as half of the difference between the number of bonding and anti-bonding electrons, indeed equals 2 in the ground state, increases to 2.5 in the O_2^+ cation and decreases to 1.5 in the O_2^- anion, as one electron is removed from or added to a semi-occupied *anti-bonding* orbital. In agreement with this scheme, the bond energy equals 498 kJ/mol in neutral O_2, 642 kJ/mol in O_2^+ and 408 kJ/mol in O_2^-, while the bond distance is 1.2074, 1.1227 and 1.26 Å, respectively. Molecular oxygen also has two excited singlet states, which are of great importance in oxidation processes. In the first of them, the unpaired electrons remain in different π^* orbitals but have opposite spins, in the second they are paired in the same π^* orbital. These states are 95 and 158 kJ/mol higher in energy than the ground state, and the O–O bond is elongated to 1.2155 and 1.2277 Å, respectively.

Molecular geometries in gaseous *vs* condensed phases are compared in Table S3.5, showing that the effect of the aggregate state is indeed small, because energy of the intermolecular (van der Waals) interaction is small as compared with energy of the chemical bond. Isolated molecules can be studied not only in the gas phase, but also by freezing them at low temperatures into a matrix of inert solid, e.g. nitrogen, methane, argon or other rare gas. Such experiments [134] gave the following M–Cl distances: 2.081 or 2.053 Å for $ZnCl_2$ (in N_2 or Ar matrices), 2.257 Å for $CrCl_2$ (in Ar), 2.207 or 2.156 Å for $FeCl_2$ (in N_2 or CH_4), 2.145 or 2.123 Å for $NiCl_2$ (in N_2 or CH_4). These distances are only slightly (by *ca.* 0.03 Å) longer than in the free molecules (Table S3.1), indicating weak influence of the matrix. Bond distances in the molecules of alkali metals, isolated in Ar matrices (K_2 3.869, Rb_2 4.091, Cs_2 4.547 Å), have been found somewhat shorter than in free molecules (Table 3.2), probably because the latter were studied at much higher temperature [135].

N–H bonds in ammonia (1.045 Å) are shortened to 1.022 and 1.008 Å, respectively, if one H atom (electronegativity $\chi = 2.2$) is replaced with Li ($\chi = 1.0$) or Na ($\chi = 0.9$) and the negative charge on N increases [136]. The effect of bond polarity on interatomic distances is manifest also in XOOX molecules [137–139], where the O–O distance increases as the electronegativity of X decreases and the negative charge accumulates on O:

X	F	Cl	CF_3	H
d(O–O), Å	1.216	1.426	1.437	1.460
χ(X)	4.0	3.1	2.9	2.2

The structure of FOOF is quite exceptional, with the O–O bond practically as short as in O_2 (1.207 Å) and the O–F bonds unusually long (1.586 Å), and can be qualitatively rationalized thus. In the O_2 molecule, which is paramagnetic, oxygen atoms carry one unpaired electron each (in mutually perpendicular *p* orbitals), which is attracted by the more electronegative fluorine to form a polar $^-$F–O=O$^+$–F ↔ F–O$^+$=O–F$^-$ type structure or (in MO terms) a three-center (O_2 F) bond, necessarily a rather weak one. The oxygen-oxygen bonding is left practically unaffected. However, the actual picture may be more complicated, as all attempts since 1962 to calculate the structure and properties of FOOF quantitatively by a variety of quantum-chemical methods,

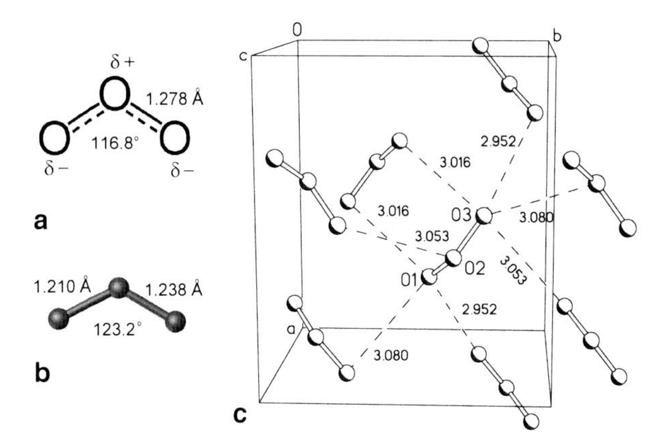

Fig. 3.2 Molecular structure of ozone in gas **a** and crystal **b**, and crystal packing **c**

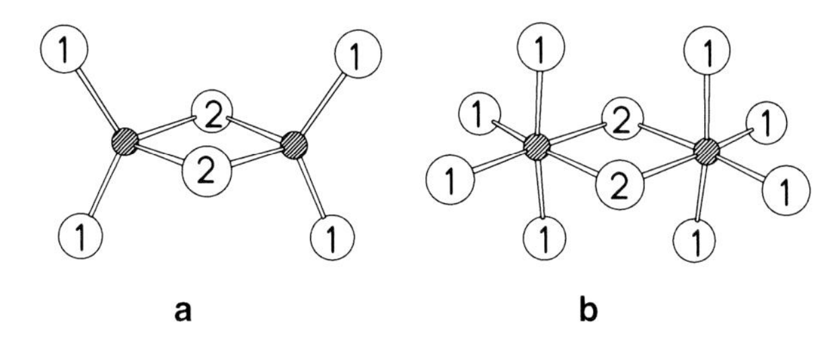

Fig. 3.3 Dimeric molecules A_2X_6 **a** and A_2X_{10} **b**, showing terminal (1) and bridging (2) atoms X

have been notoriously unsuccessful [122]. At the same time, simple crystal-chemical considerations give a rather accurate prediction. From Tables 1.12–1.14 and S1.12 it follows that for the –O = atom, $r(O) = ½(0.72 + 0.60) = 0.66$ Å, $r(O^+) = r(O) - \frac{1}{6} \cdot [r(O) - r(O^{6+})] = 0.56$ Å, $r(F) = 0.71$ and $r(F^-) = 1.09$ Å. Then $d(O–O) = r(O^+) + r(O) = 1.22$, and $d(O–F) = ½\,[r(F) + r(F^-) + r(O) + r(O^+)] = 1.51$ Å.

Ozone is another elementary molecule defying simple Lewis schemes. Since its identification as O_3 in 1865, ozone was supposed to have a cyclic (equilateral triangle) structure with three single bonds. Instead, gas-phase spectroscopic studies (finalised [141] after many inconclusive experiments) and later a crystal structure determination [142] revealed a bent, open structure (Fig. 3.2) with two bonds of the order of 1.5. Note that O–O distances in ozone fit the same relation to electronegativity of the 'substituent' (the third O in this case, $\chi = 3.5$) as in XOOX molecules (see above).

Table S3.6 comprises the results of gas-phase studies of $(AX_n)_2$ dimers (Fig. 3.3) formed by a combination of normal covalent AX_n molecules and having the $X_{n-1}\,A(\mu\text{-}X)_2AX_{n-1}$ structure, with two A–X–A bridges and $2n-2$ terminal X ligands. In each case, the bridging A–X bonds are longer than the terminal ones by *ca.* 10 %. The difference is larger when A is an electron-rich atom (Se, Te or I) in a less-than-maximum oxidation state (+3 or +5), due to strong repulsion between non-bonding valence electrons of the A atom and the bonding ones. In the $(AX_5)_2$ dimers there are two types of terminal A–X bonds: axial and equatorial, with different (by 0.02–0.03 Å) lengths. Ligands always compete for the electron density of the central atom, hence a higher electron-pulling ability (measured by electronegativity, χ) of one ligand inevitably hinders electron withrawal by other ligands, whose bonds with the central atom become weaker and longer. Thus, in X–B=O molecules the boron-oxygen bond lengthens in the succession X=H, Br, Cl, F (1.2021, 1.2047, 1.2062 and 1.2072 Å, respectively) [143], i.e. as the electronegativity of X increases. Similarly, A–H bonds in $H_3Si{-}SiH_3$ (1.492 Å) and $H_3Ge{-}GeH_3$ (1.544 Å) are longer than in SiH_4 (1.480 Å) and GeH_4 (1.517 Å): as one hydride ligand is replaced by a less electronegative Si or Ge, electron-withdrawing by the remaining hydrogen atoms is made easier and their negative charges are enhanced. Further illustrations of this effect are given in Table S3.8, which lists A–X bond distances in molecules containing constant and varying ligands. Similar compensation also takes place between homonuclear bonds of different order. This effect is most evident in organic compounds (see Sect. 3.3), but inorganic molecules also present some examples, e.g. the terminal (1.898 Å) and central (2.155 Å) bonds in the S=S–S=S molecule [144], compared to 1.917 Å in thiozone S_3 [145], the analogue of ozone (see above).

3.2.2 *Bond Angles. VSEPR Concept*

Besides bond distances, the most important parameters of molecular geometry are the angles between bonds (interatomic vectors) and the "coordination polyhedra", formed by ligands around a central atom. Pauling [146] and Slater [147] were the first to rationalize these in terms of different hybridization of the atomic orbitals (s, p and sometimes d) of the central atom, giving rise to such configurations of the ligands as a dumb-bell (sp, dp hybridization), a triangle (sp^2, dp^2, sd^2, pd^2, d^3), a tetrahedron (sp^3, sd^3), a square (dsp^2, d^2p^2), a trigonal pyramid (d^2sp, dp^3, d^3p), a tetragonal pyramid (d^2sp^2, d^4s, d^4p), an octahedron (d^2sp^3), a trigonal bi-pyramid and a square pyramid (dsp^3), and a dodecahedron (d^4sp^3).

Empirical justification of the concept of hybridization is provided by the stereochemistry of carbon, which in saturated compounds has tetrahedral coordination and, in homoleptic cases, all four bonds of equal length, although in the ground state it has the $2s^2 2p_x p_y$ electron configuration and thus can be expected to be divalent. On formation of a compound, one of the s electrons is promoted to the p level and the carbon atom adopts the $2s2p_x p_y p_z$ configuration, which enables it to be tetravalent. Mixing of s and p electrons, which results in bond strengthening, does not affect their spectroscopic properties, therefore electron spectra of a sp^3 carbon atom in its compounds, contain absorption bands corresponding to two types of electrons. Complete identity of bond distances, e.g. in diamond or in methane, justifies the concept of hybridization as a real result of chemical bonding, not just a way of describing a structure.

Equalization of different orbitals through hybridization resembles equalization of valences (oxidation states) of ligands, when they are coordinated to the same atom. Thus, NO and NO_3 radicals entering a coordination sphere, convert into identical NO_2 ligands. Closely related effects include the distribution of two negative charges

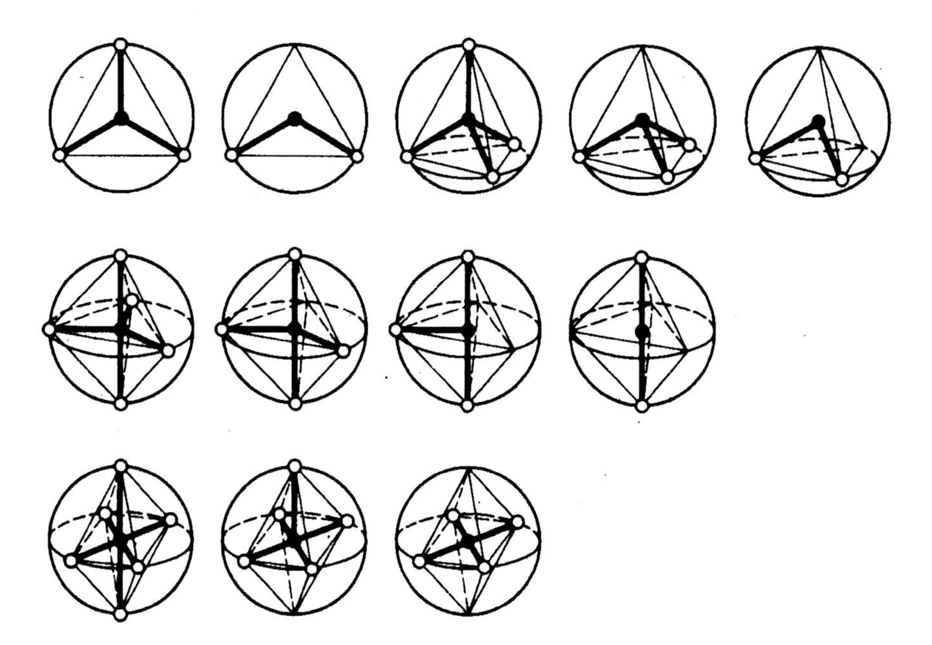

Fig. 3.4 Configurations of molecules according to VSEPR theory (see Table 3.7)

over four or six halogen atoms in MX_4^{2-} and MX_8^{2-}, described in Chap. 1. Similar equalization of electronegativities and chemical electron potentials was discussed in Chap. 2.

Sidgwick and Powell [148] were the first to rationalize the spatial distribution of bonds in molecules by Coulombic interactions of electron pairs in the outer shells of atoms. This approach was further developed by Gillespie et al. into a comprehensive theory of Valence Shell Electron Pair Repulsion (VSEPR) [149–153], based on the Pauli principle. According to VSEPR, covalent bonds (i.e. their electron pairs) *and* lone electron pairs (E) are situated around an atom A in such a way as to minimize their mutual repulsion. In the first approximation, they form a number of regular configurations (polyhedra), shown in Fig. 3.4 and listed in Table 3.7. Because a bonding electron pair is pulled towards the ligand, it occupies a smaller area in the vicinity of A (electron domain, D) than a lone pair. Hence, if the polyhedron has non-equivalent vertices, lone pairs occupy those more distant from the rest, e.g. equatorial positions in a trigonal bipyramid (for molecules of the AX_4E, AX_3E_2, and AX_2E_3 types). If all vertices are equivalent (e.g. in a tetrahedron), the E–A–E angles are widened and the X–A–X angles diminished, compared to the ideal values, as the number of E increases. Thus, in the molecules of CH_4, NH_3 and H_2O which have the configurations AX_4, AX_3E and AX_2E_2 respectively, the H–A–H angles decrease in the succession 109.5°, 107.3° and 104.5°. The distortion becomes stronger when the electronegativity of X increases relative to that of A, as indicated by a comparison of X–A–X angles in hydride and halide molecules (Table S3.8). Multiple bonds

Table 3.7 Molecular configurations according to VSEPR theory

No. of electron pairs total	lone	Molecular type	Molecular shape	Examples
2	0	AX_2	Dumbbell	$BeCl_2$
3	0	AX_3	Trigonal	BF_3
	1	AX_2E	V-shaped	$SnCl_2$
4	0	AX_4	Tetrahedral	CH_4
	1	AX_3E	Trigon. pyramidal	NH_3
	2	AX_2E_2	V-shaped	H_2O
5	0	AX_5	Trigon. bipyramid.	PCl_5
	1	AX_4E	Bisphenoidal	SF_4
	2	AX_3E_2	T-shaped	ClF_3
	3	AX_2E_3	Linear	XeF_2
6	0	AX_6	Octahedral	SF_6
	1	AX_5E	Square pyramidal	IF_5
	2	AX_4E_2	Square	XeF_4

have larger electron domains than single bonds and behave similarly to lone pair, occupying more distant vertices, *etc*. The VSEPR approach is very simple yet efficient in rationalizing stereo-chemical data.

However, molecular structures are also affected by factors other than the electronic and steric repulsion producing many exceptions to the VSEPR rules (see [154–156]). For example, the molecules of alkali earth halides MX_2 are mostly linear (as predicted by VSEPR), but those of CaF_2, SrF_2, $SrCl_2$, $SrBr_2$ and BaX_2 (X = F, Cl, Br, I) have 'bent' configurations with the X–M–X angle different from 180°. The bending can be explained by participation of the *d* orbitals of the previous electron shell, i.e. *sd* hybridization, in the case of heavier metals. Indeed, it should increase with the increase of the principal quantum number and of the bond polarity, because both increases tend to reduce the energy gap between the outer and the penultimate electron shells. Donald and Hoffmann [157] have established the structural preferences in the monomers, dimers, and solid-state structures of MX_2 where M is a Group 2 element. Significant links between bending in the MX_2 monomers and the M_2X_4 dimer structures of D_{2h} or C_{3v} symmetry have been identified (Table S3.10). Quasi-linear or floppy monomers show, in general, only a weak preference for either the D_{2h} or the C_{3v} dimer structure. There is also a relationship between the structural trends in these two (MX_2 and M_2X_4) series of molecular structures and the prevalent structure types in the Group 2 dihalide solids. The most bent monomers tend to crystallize in the CaF_2 and $PbCl_2$ structural types. The rigidly linear monomers condense to form extended solids with low N_c, 4 or 6. The structural preferences in the dimers have been rationalized partly by an MX_2 frontier orbital MO analysis based on sp^x hybridization. The strong preference for the D_{2h} structure in the Be_2X_4 and Mg_2X_4 molecules is explained by the large energy cost of deforming the linear MX_2 fragments to fit them for optimal bonding in the C_{3v} geometry. The C_{3v} dimer structure is favored only when the cation is very large, especially when the minimum energy structure of the MX_2 monomer is bent. The bending in the MX_2 monomer has been explained previously by core polarization interactions between M and X sites and the availability of *d*-orbitals on M for hybridization with the valence *s*-orbitals. This geometry for the $(CaX_2)_2$, $(SrX_2)_2$ and $(BaX_2)_2$ can be rationalized also both by core-polarization effects, as the metal polarizability increases from Be to Ba, and the availability of *d*-orbitals in Ca, Sr and Ba. On the other hand, the lone pairs are sterically inactive in ClF_6^- and BrF_6^-, which are octahedral, or XeF_8^{2-} and IF_8^-, which are regular square-antiprismatic. This can be due to steric overcrowding; indeed, with larger central atom (e.g. IF_6^-) or fewer ligands (XeF_7^-) the lone pairs are active, if rather weakly.

To summarize, VSEPR model has been remarkably successful in predicting the structures of *main-group* compounds, but the predictions remain qualitative. All attempts to quantify them were so far unsuccessful, as well as attempts to extend this approach to transition metals ("extended VSEPR model") taking into account such factors as ligand-ligand repulsion and polarization of the core electron shells of the central atom [158–160]. In the latter case, many other factors must be considered, such as *d*-electron configuration of the central atom, competition between σ and π bonding, etc. (see discussion in [154]).

3.2.3 Non-Stoichiometric and Unusual Molecules

To conclude this section, let us take a closer view of the configurations of gaseous free radicals and complex ions of non-stoichiometric composition, which are intensely studied by Wang, Boldyrev, Kuznetsov et al. Structural chemistry of these species is rather peculiar. Thus, radicals Al_nO ($n = 2, 3$), Al_nS ($n = 3$–9) and Al_nN ($n = 3, 4$) violate the octet rule. In Al_4C and Al_5C radicals, as well as Al_4C^- and Al_5C^- anions, the carbon atom is surrounded in a square-planar fashion by four Al atoms (which are univalent, using only the *p* electron) whereas the fifth Al atom is attached to one of the corners of the square [161, 162]. Other examples of planar species with the central carbon atom are CAl_3Si and CAl_3Si^-, CAl_3Si and CAl_3Si^- [163]. Al_4^{2-}, Ga_4^{2-} and In_4^{2-} anions have planar structures and aromatic bonds due to 2π electrons; the same is true for MAl_3^{2-} anions, where M = Si, Ge, Sn and Pb [164]. The B_5^- and Al_5^- anions are 5-membered rings; in B_5^- the π orbitals are delocalized between all five atoms [165]. Aromatic bond character has also been found in P_5^-, As_5^-, Sb_5^- and Bi_5^- anions, iso-electronic with $C_5H_5^-$ [166]. B_3 and B_3^- are aromatic systems with D_{3h} symmetry, while B_4^- has a slightly distorted square geometry [167]. B_6 and B_6^- are planar but not aromatic [168]. Recently, this group has reported experimental and theoretical evidence that $SiAu_n$ molecules ($n = 2$–4) are structurally and electronically similar to SiH_n [169] and in the linear AuBO molecule the gold atom does mimic hydrogen. The results of photoelectron spectroscopy in the abovementioned works have been supported by quantum chemical calculations.

Molecules of ZnHCl [170] and $ZnHCH_3$ [171] show an example of the competition of Cl and CH_3 ligands for the valence electrons of Zn. Because EN(Cl) > EN(CH_3), the Zn–H bond in the former is more covalent than in the latter, and shorter: Zn–H 1.505 Å and 1.521 Å, respectively.

Another curious case is the HPSi molecule [172] which contains an asymmetrically bridging hydrogen atom (PH 1.488 and Si-H 1.843 Å), in drastic contrast with its lighter analogues HCN, HNC, HNSi and HCP, all of which have linear geometry with terminal hydrogen.

3.3 Organic Molecules

Organic compounds contain very few elements: mainly C, H, N and O, sometimes also S, Se, Te, P, As and halogens, with rather inflexible valence and coordination numbers. Nevertheless, these compounds display a staggering variety of structures, due to the ability of carbon atoms to form up to four covalent bonds with *different* atoms and groups and also to form long chains and cycles of homonuclear (carbon-carbon) bonds. The geometry of organic molecules remains practically unaltered in all aggregate states which can be reached without chemical decomposition. The lengths of element-carbon single bonds in various methyl and ethynyl derivatives, compiled from the review [173] and original papers, are listed in Tables S3.10 and S3.11. This great structural variety obtained from few building blocks is propitious for finding empirical correlations between structural parameters, of which dozens have been derived for specific fragments or classes of compounds (see compilations [174, 175], see also Eq. 2.10).

The main factors affecting bond distances are (i) changes of the bond order, (ii) steric strain, (iii) changes of hybridization and (iv) changes of bond polarity. An increase of bond order (i.e. the number of electron pairs per bond) always increases bond energy and reduces bond length, other things being equal. However, in inorganic species they are seldom equal: usually bond order and bond polarity change simultaneously and their effects are hard to separate. Nonpolar carbon-carbon bonds, e.g. in ethane, ethylene, benzene and acetylene, show an almost "pure" effect of the bond order (Table 3.8). Furthermore, numerous fused-ring aromatic systems display a spectrum of C—C bond orders intermediate between 1 and 2; on this basis Pauling was able to describe the bond length/bond order dependence as a smooth and continuous function (see below).

Steric strain can result from mutual repulsion between bulky atoms (groups), adjacent to the bond in question. Table 3.8 illustrates the progressive lengthening of the C—C single bond in ethane as an increasing number of H atoms replaced with hydrocarbon groups. The effect is much smaller in ethylene and benzene, where the C—C—C angles are wider, and practically absent in the linear acetylene derivatives. Other sources of strain are small rings, which force the atoms to adopt bond angles unnatural for their electron state. Thus, the C—C—C bond angle in cyclopropane (60°) is much smaller than the normal for sp^3 hybridized carbon (109.5°). Indeed, no hybrids of s and p orbitals can form an angle of less than 90°. This results in shortening of the C—C distances, which has been interpreted as an outward bending of these bonds [177] relieving the strain. Studies of charge density distribution in

Table 3.8 Typical carbon-carbon bond lengths (Å) for various bond orders *n*

Molecule/Group	*n*	Gas	Crystal[a]	Molecule/Group	*n*	Gas	Crystal
$H_3C{-}CH_3$	1	1.535		$\equiv C{-}C=$	1	1.425	1.431
$RH_2C{-}CH_2R$	1	1.533[b]	1.524	$\equiv C{-}C\equiv$	1	1.389	1.377
$R_2HC{-}CHR_2$	1	1.545[c]	1.542	$C(CH_3)_4$	1		1.537[f]
$R_3C{-}CHR_2$	1		1.556	benzene	1.5	1.399[f]	1.397
$R_3C{-}CR_3$	1	1.583[c]	1.588	$C_6(CH_3)_6$	1.5		1.411
cyclopropane	1	1.514	1.510	$H_2C=CH_2$	2	1.339[f]	1.313
cyclobutane	1	1.554	1.554	*cis*-RHC=CHR	2	1.346	1.317[a]
cyclopentane	1	1.546	1.543	$R_2C=CR_2$	2	1.351	1.331[a]
cyclohexane	1	1.536	1.535	$H_2C=C=CH_2$	2	1.308[f]	1.294[a]
$=C{-}C(sp^3)$	1	1.506	1.507	$H_2C=C=C=CH_2$	2	1.280[g]	1.269
$=C{-}C=$[d]	1	1.475	1.478	$H_2C=C=C=CH_2$	2	1.320[h]	1.324
$=C{-}C=$[e]	1	1.463	1.455	$HC\equiv CH$	3	1.203[f]	1.186
$\equiv C{-}C(sp^3)$	1	1.459	1.466	$RC\equiv CR$	3	1.212[c]	1.190[a]

[a]average for all derivatives, where R = any substitute, bonded through an sp^3-C atom, [b]average for *n*-alkanes C_mH_{2m+2}, $m = 3$–16; [c]R = Me, [d]non-conjugated, perpendicular or *gauche* conformation, [e]conjugated, planar conformation, [f][176], [g] central bond C=C [176],[h] terminal bond C=C.

Table 3.9 Bond distances C–X (Å) in $C\equiv C{-}(CH_2)_n{-}X$ moieties [179].

X	$n=0$	$n=1$	$n=2$	$n=3$
$C(sp^3)$	1.46	1.49	1.52	
F	1.274	1.383	1.387	1.390
Cl	1.631	1.782	1.786	1.792
Br	1.789	1.901	1.946	1.957
I	1.987	2.117	2.132	2.139

several substituted cyclo-propanes [178] confirmed this model: the peaks of bonding electron density are shifted outward from the inter-nuclear lines (see Fig. 2.3).

A single bond C–C is substantially shortened when it is adjacent to a multiple bond, i.e. involves an sp^2 or sp hybridized carbon atom (see Table 3.8). As the valence electrons of this atom concentrate in the area of the multiple bond, the screening of its nucleus in other directions is diminished and the effective atomic number Z^* correspondingly increases (see Sect. 1.2), hence it attracts the single-bond electron pair stronger and the bond contracts without an increase of the bond order. Thus, an *acetylenic* carbon acts in its *single* bonds in the same way as an electron-withdrawing atom (see below), and causes perturbations further along the chain, e.g. shortening of carbon-carbon or carbon-halogen bonds. As shown in Table 3.9, these perturbations decrease as we move further from the C≡C bond [179] and practically vanish beyond the third atom of the chain. The effect of an *olefinic* carbon is similar but weaker. Cumulated C=C bonds, e.g. in allene $H_2C=C=CH_2$ (Table 3.8), are shorter than those between sp^2 hybridized C atoms, for similar reasons. The effect increases with the number of cumulated double bonds; the C=C bonds in O=C=C=C=O (1.253 Å) [180] and $(i\text{-}Pr_3P)_2ClIr=C=C=C=C=CPh_2$ (1.24 Å) [181] are close to the length of a normal triple bond.

Table 3.10 Bond lengths (Å) affected by a carbo-cationic center

X	C–X	X–H	C–X	X–H
	$H_3C{-}XH_2$		$H_2C^+{-}XH_2$	
N	1.465	1.017	1.282	1.02
P	1.858	1.416	1.638	1.392
As	1.984	1.526	1.746	1.489
Sb	2.182	1.718	1.945	1.670
	$H_3C{-}XH$		$H_2C^+{-}XH$	
O	1.429	0.972	1.257	0.993
S	1.815	1.341	1.618	1.348
Se	1.965	1.472	1.756	1.478
Te	2.159	1.665	1.946	1.668
	$H_3C{-}X$		$H_2C^+{-}X$	
F	1.405		1.244	
Cl	1.778		1.588	
Br	1.950		1.744	
I	2.169		1.945	

The effects of bond polarity, induced by an electronegative atom or a cationic center, are complex. Three possibilities can be considered. If the ionic component of a bond increases with a corresponding decrease of the covalent component, the bond lengthens but its energy increases. Thus, in a CF_4 molecule the fluorine atoms, having higher electronegativity (χ) than carbon, withdraw electrons from the latter and produce polar $C^{\delta+}{-}F^{\delta-}$ bonds. If one F atom is replaced by hydrogen ($\chi_H < \chi_C << \chi_F$) the remaining fluorines will attract electrons stronger, for the lack of competition. Thus the polarity, and hence the length, of the C–F bonds increase in the succession $CF_4 < CHF_3 < CH_2F_2 < CH_3F$; the quantum-chemical calculations confirming this [182]. A similar pattern is observed for fluoride derivatives of B, Si, P and halides of germanium (Table S3.7). For other (less electronegative) halogens the effect is correspondingly weaker. The polarizing effect of a carbo-cationic center is illustrated in Table 3.10 [183], comparing the structures of neutral methyl hydrides of the Group 15–18 elements (X) and their carbo-cations. The positive charge causes a substantial contraction of the C–X bond in all cases, and of the adjacent X–H bond—only if the X atom is not very electronegative. In other words, only a substantially covalent C–X bond can conduct this effect.

Secondly, electrostatic attraction can be superimposed upon a constant covalent interaction, making the bond both shorter and stronger. Thus, in the ethane molecule carbon atoms bear partial negative charges, as $\chi(H) < \chi(C)$. Addition of one fluorine atom reverses the charge on the adjacent C atom, and the resulting $C^{\delta+}{-}C^{\delta-}$ attraction reduces the C–C distance from 1.533 Å in ethane to 1.517 Å in C_2H_5F. A similar contraction of C=C bonds is observed in the succession $H_2C{=}CH_2$, $H_2C{=}CHF$, $H_2C{=}CF_2$ (1.337 → 1.333 → 1.316 Å). Likewise, C≡C bonds in both $CH_3C{\equiv}CF$ (1.200 Å) and $CF_3C{\equiv}CH$ (1.202 Å) are shorter than in $CH_3C{\equiv}CH$ (1.207 Å) [184].

Thirdly, a withdrawal of electron density can, by reducing the screening (see above) enhance the attraction of the bonding electrons to the nuclei and thus

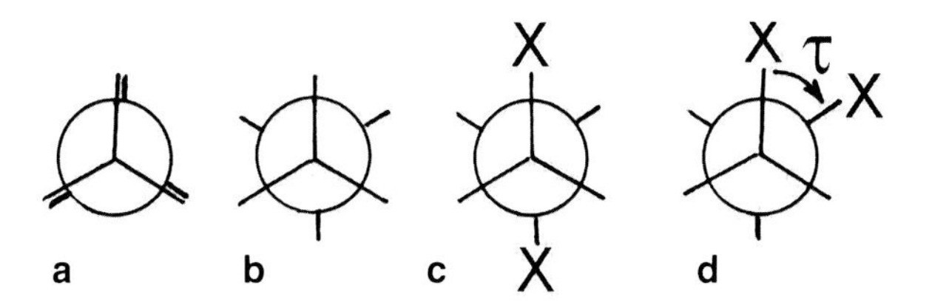

Fig. 3.5 Newman's projections for the eclipsed **a** and staggered **b** conformations of ethane, *trans* **c** and *gauche* **d** conformations of di-substituted ethane, XCH_2CH_2X. The torsion angle X–C–C–X is measured as the rotation τ of the front C–X bond (clockwise for positive) until it eclipses the rear C–X

strengthen and shorten the bond without making it polar, e.g. the C=C bond in $F_2C{=}CF_2$ (1.311 Å, cf. 1.337 Å in ethylene). Note that the C=C bonds in $Cl_2C{=}CCl_2$ (1.355), $Br_2C{=}CBr_2$ (1.363 Å) and $I_2C{=}CI_2$ (1.364 Å) are longer than in ethylene, because the diminishing electron-withdrawing effect of the halogens is more than compensated by the increase of steric repulsion between bulkier halogen atoms. Thus, different effects can be at work simultaneously, and the relation between bond distance and bond strength is by no means straightforward.

Usually, a polyatomic molecule can adopt various *conformations*, i.e. structures differing only by a rotation of one part of molecule relative to another, the axis of rotation coinciding with one of the chemical bonds. Graphically, a conformation can be depicted by a projection of the rotating groups onto a plane, perpendicular to this bond (Newman's projection, Fig. 3.5).

The simplest example is the rotation of CH_3 groups in ethane around the C–C bond. A conformation is called *eclipsed* if the projections (in the direction of the C–C bond) of both CH_3 coincide, and *staggered* if the C–H bonds of one group project onto the middle of the H–C–H angles of the other. The latter is the actual (stable) conformation of ethane, corresponding to the potential energy minimum, because it maximizes the distances between non-bonded atoms. If the substituents are not equivalent, as in symmetrical ethane derivatives XCH_2CH_2X, two different staggered conformations are possible, viz. the *trans* (or *anti*) and *gauche*. According to GED data, the former conformation is thermodynamically more stable for X=Cl, Br and the latter for X=F. Quantitatively, a conformation can be described by a torsion angle τ(X–C–C–X), i.e. a rotation around the C–C axis required to make C–X bonds eclipsed. In this case, $\tau = 180°$ for the *trans* and 60° for the *gauche* conformations.

Conformational behavior of a molecule depends on the rotation barrier, i.e. the potential barrier which must be overcome during a rotation around a given bond. If the barrier is substantially higher than the energy of thermal motion or of inter-molecular interactions, molecules of different conformations behave as distinct chemical entities and are called *isomers*, otherwise they are called *conformers*. For a C=C bond, the barrier is very high (170 ± 40 kJ/mol), because the rotation requires a disruption of a covalent bond. Thus *cis* and *trans* substituted ethylene derivatives are distinct compounds (isomers) rather than different conformers. A barrier of rotation (BR)

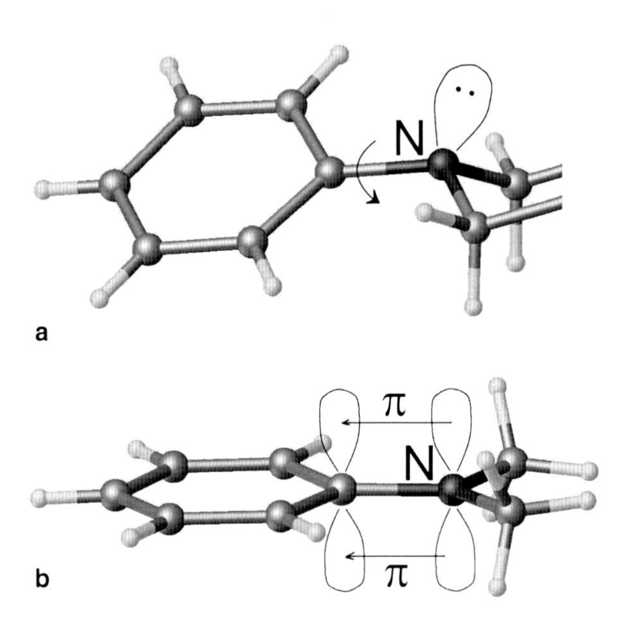

Fig. 3.6 Unconjugated **a** and conjugated **b** aryl-amino moiety

around a single bond between two tetrahedral (saturated) carbon atoms is due mainly to steric repulsion between non-bonded atoms, when they pass through an eclipsed position. In substituted ethanes BR varies from 12 to 75 kJ/mol, depending on sizes of the substitutes. For the same reason, BR generally decreases with the increase of the bond length, e.g. in $A(CH_3)_4$ molecules with A = C, Si, Ge and Sn the barriers around the A–C bonds equals 18, 8.4, 5.4 and 3.3 kJ/mol, respectively. The barrier around the single $C(sp^2)$–$C(sp^3)$ bond in propylene is *ca.* 8 kJ/mol (the most stable conformation being eclipsed) and in carbonyl compounds about half this value. For a single C–C bond adjacent to a C≡C bond, BR is practically nil, i.e. the rotation is free.

When a single C–C bond is enclosed between two double bonds, the C=C–C=C fragment can be either planar (whether *cis* or *trans* about the central bond) or non-planar (perpendicular or *gauche* conformation). The former (but not the latter) conformation permits a π conjugation between the double bonds, increasing the effective order of the central bond, which in conjugated systems is by *ca.* 0.02 Å shorter than in non-conjugated. A similar situation occurs when a double bond is adjacent to an atom with a lone electron pair, as in a C=C–NR_2 moiety where R is an alkyl group. If the lone pair of the N atom can lie parallel to the $p\pi$ orbital of C, conjugation occurs, the N atom adopts a planar-trigonal geometry, with the C=C bond lying in the same plane. If such conformation is impossible because of bulky substitutes, the N atom retains a tetrahedral geometry with one of the vertices occupied by the lone pair. The average C–N distances are 1.355 Å in conjugated systems and 1.416 Å in non-conjugated ones [25] (Fig. 3.6).

Ring systems consisting entirely or mainly of single bonds, also can adopt different conformations. The stable conformation of cyclo-butane is a puckered one;

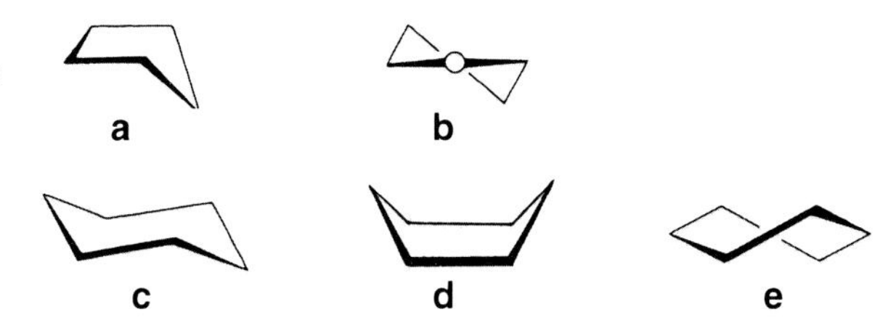

Fig. 3.7 Conformations of a 5-membered ring: envelope **a** and half-chair **b**; those of a 6-membered ring: chair **c**, boat **d** and twist-boat **e**

cyclo-pentane has two stable conformations, known as envelope and half-chair. The most stable conformation of cyclo-hexane, the only one observed in the gas phase at room temperature, is the chair form (Fig. 3.7) predicted in 1890 [185]. The boat and twist-boat conformations have higher energies (by 21–25 kJ/mol). A mathematical formalism permitting an unequivocal quantitative description of a ring conformation has been developed by Cremer and Pople [186].

As already mentioned, carbon forms a rich variety of allotropes, i.e. structurally different elementary substances, which are often conveniently related to different types of organic compounds, e.g. diamond to saturated hydrocarbons, graphite to fused-ring aromatic compounds, etc. In reality, the largest molecules are separated from the smallest pieces of 'bulk' solids by a wide spectrum of intermediate species, of which nano-particles are the most important (see Chap. 8). Thus, the recently synthesized cyclohexamantan $C_{26}H_{30}$ can be formally regarded as a fragment of the diamond structure (see below, Sect. 5.1.2, Fig. 5.4), with tetrahedral bond geometry around each carbon atom and the C–C bonds of 1.568 Å in the central part of the molecule and 1.538 Å on its periphery (cf. 1.544 in diamond) [187], but physically it is more meaningful to regard it as a border case between molecules and nano-particles.

Recently reviewed [188, 189] structures of compounds with polyyne chains, $R(-C{\equiv}C-)_nR'$, are particularly interesting, because in this case the corresponding allotrope of carbon, the one-dimensional carbyne $(-C{\equiv}C-)_\infty$ [190], still remains elusive. It was found that as the length of the chain increases, the lengths of the triple and single bonds converge somewhat, from 1.210 and 1.371 Å in butadiyne to 1.25 and 1.32–1.33 Å, respectively, in the longest chains. However, they converge to *different* asymptotic limits, rather than to one uniform intermediate bond distance. Another remarkable feature of polyynes is the flexibility of the chain, which in this case results not from rotations around single bonds (see above) but from deviations of C–C≡C bond angles from the ideal 180° prescribed by the *sp* hybridization of carbon atoms. The chains can adopt zigzag, S-shaped and bow-type conformations, and one bow-shaped molecule has 37 % of the curvature of a semicircle (Fig. 3.8).

3.4 Organometallic Compounds

Organometallic compounds are often loosely defined as those containing both a metal and an organic group. About half of the structures in the CSD fall into this category; a compilation of metal–element bond distances for which is available [26–28]. In

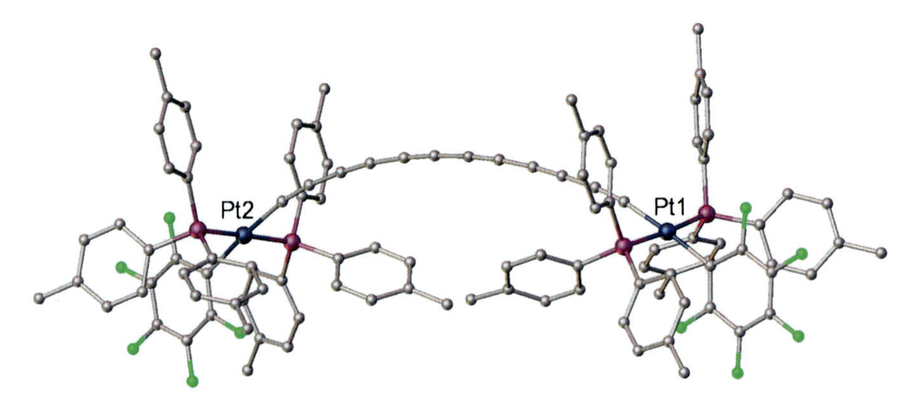

Fig. 3.8 Bending of the dodecahexayne chain in the crystal structure of $[(C_6F_5)(Ptol_3)_2Pt]-(C{\equiv}C)_6-[Pt(Ptol_3)_2(C_6F_5)]$. The terminal Pt–C bonds of the chain form an angle of 116° [191]

Table 3.11 Bond distances (Å) in Co(II) and Ni(II) complexes with organic ligands, for various coordination polyhedra

Bond	Co(II) coordination			Ni(II) coordination		
	Square	Tetrahedral	Octahedral	Square	Tetrahedral	Octahedral
M–O	1.86	1.96	2.11	1.88		2.03
M–N(tetragon)	1.85	2.01	2.16	1.92		2.10
M–N(trigon)				1.87		
M–S	2.17	2.30	2.53	2.20	2.30	2.46
M–Cl		2.26	2.45	2.19		2.43
M–Br		2.39		2.30	2.37	2.54

most of them, however, organic ligands are coordinated through such atoms as O, N, S, P, etc., and the bonding of the metal atom is similar to that in inorganic complexes. Nevertheless, these structures helped to clarify some interesting questions of structural chemistry, i.e. the dependence of bond distances on the type of the coordination polyhedron (for the same coordination number). As shown in Table 3.11, the bonds in tetrahedral Co(II) and Ni(II) complexes are longer than in square-planar ones, due to dsp^2 hybridization involving electrons from the penultimate shell with higher Z^*.

In the strict sense of the term, organometallic compounds must contain direct metal-carbon bonds. These may be of the usual 2-center-2-electron kind (σ bonds), as in alkyl and alkynyl metal derivatives (see Tables S3.10 and S3.11). Alkyls of main group metals (other than Group 14), while monomeric in gas phase, are prone to oligomerization in solutions and to oligo- or polymerization in solids [192]. Thus, gaseous $Be(CH_3)_2$ is a monomer with linear C–Be–C configuration and Be–C bond of 1.698 Å; in crystal it forms polymeric chains ... $Be(\mu\text{-}CH_3)_2Be(\mu\text{-}CH_3)_2Be$... (Be–C 1.93 Å) [193]. Gaseous $Al(CH_3)_3$ at 488 K is a monomer with planar-trigonal coordination (Al–C 1.957 Å) but at 333 K forms dimer $(CH_3)_2Al(\mu\text{-}CH_3)_2Al(CH_3)_2$ with tetrahedral coordination of Al and Al–C distances 1.953 (terminal) and 2.140 Å (bridging) [192]. The latter structure also exists in the solid state [194]. Each bridging

(μ) CH_3 group contributes only one electron to form two M–C bonds, the lengths of which are consistent with bond orders < 1.

Another type of organometallic compounds are π complexes, where an unsaturated organic ligand is coordinated through a group of $n \geq 2$ adjacent atoms (usually C but occasionally B, P or other); such ligands are designated η^n or h^n (*hapto*). The first π complex, Zeise's salt $K[Pt(\eta^2\text{-}H_2C{=}CH_2)Cl_3] \cdot H_2O$, had been synthesized in 1827 [195] but remained enigmatic until its first XRD investigation in 1954 [196]. In it the ethylene ligand is coordinated to Pt symmetrically by both C atoms, the vector between Pt and the ligand center is normal to the latter's plane, but the hydrogen atoms are bent back from this plane (away from Pt) and the C=C bond is lengthened to 1.37 Å [197]. Chatt and Duncanson [198] explained such geometry by simultaneous (synergic) donation of a π electron pair of the C=C bond to Pt and back-donation of *d*-electrons from Pt to the anti-bonding (π^*) orbitals of the C=C bond, which is thereby weakened. This model was later extended, with modifications, to all π complexes. Zeise's salt retains a classical 'inorganic' outlook, and is structurally similar to FeS_2, where an S_2 moiety is also η^2-coordinated. Very different in bonding pattern are compounds in which a metal is coordinated only, or mainly, with π ligands. The seminal π complex therefore is ferrocene, $(\eta^5\text{-}C_5H_5)_2Fe$, which was discovered in 1951 and appeared astonishingly stable. The cyclo-pentadienyl (Cp) anion $C_5H_5^-$, readily formed by elimination of a proton from cyclo-pentadiene C_5H_6 [199], can exist as an uncoordinated anion in crystals [200]. It is a planar regular pentagon with the C–C bond lengths (average 1.397 Å) similar to those in benzene. Ferrocene molecule is a 'sandwich' (Fig. 3.9) with the iron atom lying between parallel Cp rings. A great variety of sandwich complexes, with main group (see review [201]), as well as transition metals, is now known. In most cases, the metal atom is coordinated symmetrically, i.e. it lies on a fivefold axis of the ring. In a coordinated Cp ring, C–C bonds lengthen to *ca.* 1.43–1.44 Å and the C–H bonds tilt out of the ring plane by several degrees toward the metal atom. However, substituents at the Cp ring usually tilt in the outward direction, thus in a $\eta^2\text{-}C_5(CH_3)_5$ ligand (common abbreviation Cp*) the tilt of the C(Cp)–CH_3 bonds is *ca.* 5–6°. Cp being an anion, only divalent metals can form neutral bis-Cp sandwiches suitable for gas phase study. The compilation of M–C distances in cyclopentadienyl complexes [201, 202] is given in Table S3.12.

Although Cp is coordinated by all five carbon atoms, at first it was described by 3-dentate resonance forms, each coordinated by two C=C bonds donating electron pairs and one σ-bonded C atom. This illustrates the difficulties of applying the terms of classical structural chemistry to π complexes, where multi-center bonding and generally low bond polarity makes the concepts of coordination number, valence and oxidation state rather vague. As an alternative, Sidgwick introduced the rule of effective atomic number (EAN, also known as the 'inert gas rule') as the "organometallic" counterpart to the octet rule of Lewis. It states that a π complex is most stable when the transition metal atom has in its outer *ns*, *np* and $(n-1)d$ orbitals the total N_e of 18 electrons. This rule works well for metals of the middle of transition rows; to cover Groups 10 and 11, it was modified as

$$N_e = 12 + 2M \tag{3.2}$$

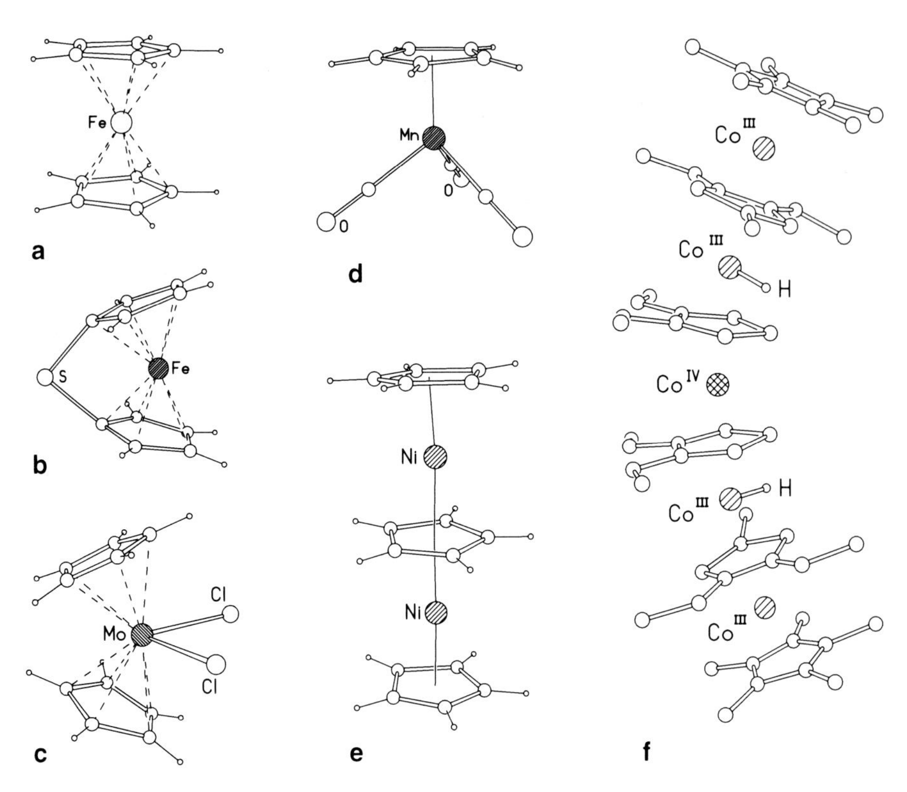

Fig. 3.9 Sandwich complexes: **a** ferrocene, **b** ferrocenophane $S(C_5H_4)_2Fe$, **c** *clino*-sandwich Cp_2MoCl_2, **d** semi-sandwich $CpMn(CO)_3$ with a 'piano stool' configuration, **e** triple-decker sandwich $[CpNi(\mu\text{-}Cp)NiCp]BF_4$, **f** 6-decker $[Cp^*Co(Et_2C_2B_3H_2Me)Co(Et_2C_2B_3H_3)]_2H_2Co$, note that terminal rings are C_5 and bridging ones C_2B_3

where M is the 'dimensionality' of the coordination polyhedron, which equals 3 for a spatial (e.g. tetrahedral or octahedral) coordination, 2 for planar (trigonal or square) and 1 for linear coordination. N_e includes all the valence electrons of the metal atom (presuming it neutral), plus two electrons contributed by each donor-acceptor bond and one by each covalently bonded ligand, corrected for the overall charge of the complex. Thus, for ferrocene $N_e = 10\,(\text{Fe}) + 2 \times 5\ (\text{Cp}) = 18$, which explains its stability. The most widespread π-ligands are η^2-ethylene and η^2-acetylene (contributing 2*e*), η^3-allyl $H_2C \overset{\dots}{\underline{\quad}} CH \overset{\dots}{\underline{\quad}} CH_2$ and η^3-cyclopropenyl C_3H_3 (3*e*), η^4-cis-butadiene $H_2C{=}CHCH{=}CH_2$ and η^4-cyclobutadiene C_4H_4 (3*e*), η^5-Cp (5*e*), η^6-benzene (6*e*), η^7-cyclohepta-trienyl C_7H_7 (7*e*), η^8-cyclooctatetraene C_8H_8 (8*e*), and their derivatives.

The EAN rule is usually not applicable to complexes of early transition metals (Groups 3 and 4), *f*-elements and main-group metals, which have more ionic type of bonding. π-Complexes of 'typical' *d*-metals violating the EAN rule can exist but are less stable and have relatively longer M–C bond distances than those which conform to it. For sandwich complexes, this approach has been developed into the rule of 'electron imbalance' [203, 204]: the M–C distances lengthen proportionally to the sum (δ) of the number of vacancies in the bonding MO (a'_1 and e'_2) and the number of electrons in the anti-bonding MO (e''_1), as illustrated in Table 3.12. It is noteworthy that for all 3*d* metallocenes, the diagonal M–Cp force constant and the mean ionic dissociation energy decrease linearly with the increase of M–C distances, and hence of δ [204]. The imbalance is absent ($\delta = 0$) for neutral ferrocene and the cation of cobaltocene (Cp_2Co^+), while $\delta = 1$ for both the ferrocenium cation Cp_2Fe^+ and neutral cobaltocene Cp_2Co. Hence the same increase of the metal oxidation state (+2 to +3) results in expansion of the sandwich in the case of Fe and its contraction in the case of Co.

The structures of manganocene derivatives depend strongly on whether the metal is in a high-spin ($\delta = 5$) or low-spin ($\delta = 1$) state. Titanocene ($\delta = 4$) is altogether unstable and prone to dimerization, but its analogues with bulky substitutes in the Cp rings have been obtained. This leads naturally to the second condition for stability of organometallic complexes: sufficiently close packing of ligands within the coordination sphere [205]. In this respect also, the geometry of ferrocene is most favourable: the separation between Cp-ring planes (3.3 Å) is close to the normal distance between closely packed aromatic molecules, hence the Fe atom is effectively 'sealed off'. As the M–Cp separations increase, the structure becomes unstable and finally converts into a 'bent sandwich', or *clino* sandwich, with nonparallel Cp rings (Fig. 3.9). This opens up additional coordination sites, which can be filled by auxiliary (usually monodentate) ligands, the number of which depends on both the size *and* electron requirements of the metal atom.

According to the CSD, by now up to 1600 *clino* sandwiches of Ti with one or two auxiliary ligands have been structurally characterised, one-tenth of this number for V, about 50 for Sc, a few for Cr and Mn, and none for Co and Ni. There are numerous derivatives of ferrocene with a covalent bridge between the Cp rings, known as ferrocenophanes. When the bridge is short, the angle α between Cp rings can be as high as 31–32°, as in $S(C_5H_4)_2Fe$, $S(C_5H_3Me)_2Fe$ [206] or $(Me_3Si)_2NB(C_5H_4)_2Fe$ [207], without inducing any auxiliary coordination (Fig. 3.9b). The only example of the latter is the Hg–Fe bonding in the $[(Me_2CC_5H_4)_2Fe–Hg–Fe(C_5H_4CMe_2)](BF_4)_2$ complex, with a –C–C – bridge between the Cp rings and $\alpha = 34°$ [208]. On the other hand, for 2nd- and 3rd-row transition metals, no sandwiches without auxiliary ligands are known except neutral Cp_2Ru [209] and Cp_2Os [210] and their substituted analogues, which have the same stable electron structure as ferrocene. The common configuration is a *clino* sandwich, e.g. Cp_2MX_2 with two mono-dentate ligands and the centers of two Cp-rings forming a (more or less) distorted tetrahedron. It is noteworthy that in *clino* sandwiches, non-bonding electron pairs of the metal atom are usually located in the equatorial plane. Therefore, oxidation of such sandwiches causes M–X bonds to shorten, while the M–Cp distances remain unchanged or even lengthen (Table 3.13) [211, 212].

Only the larger atoms of the early 2nd and 3rd row transition metals, lanthanides and actinides can coordinate three η^5 cyclopentadienyl ligands simultaneously. Complexes of Cp_3M type are known for Y, Zr and most lanthanides, and their analogues

Table 3.12 Bond distances M–C (d, Å), and electron imbalance δ in sandwich complexes

Sandwich	N_e	δ	d, gas[a]	d, cryst[b]	Sandwich	N_e	δ	d, gas	d, cryst
$[Cp^*_2Ti]^+$	13	5		2.310	$[Cp_2Fe]^+$	17	1		2.096
$(C_5Me_4R)_2Ti$[c]	14	4		2.352	Cp_2Fe	18	0	2.064	2.055
Cp_2V	15	3	2.280	2.275	$[Cp_2Co]^+$	18	0		2.031
$[Cp_2Cr]^+$	15	3		2.193	Cp_2Co	19	1	2.119	2.085[d]
Cp_2Cr	16	2	2.169	2.151	$[Cp_2Ni]^+$	19	1		2.075
$[Cp^*_2Mn]^+$	16	2		2.132	Cp_2Ni	20	2	2.196	
Cp_2Mn[e]	17	5	2.380		Cp_2Ru	18	0		
Cp^*_2Mn[f]	17	1		2.112	Cp_2Os	18	0		

[a]GED, [b]CSD, [c]R = $SiMe_3$, [d]Cp-ligand disordered, $d = 2.105$ Å in Cp^*_2Co, [e]high-spin state, [f]low-spin

Table 3.13 Bond distances (Å) in *clino*-sandwiches Cp_2MX_2

Complex	N_e	d(Mo–Cl)	d(Mo–Ω)[a]
Cp_2MoCl_2	18	2.470	1.98
$[Cp_2MoCl_2]^+$	17	2.383	1.99
$[Cp_2MoCl_2]^{2+}$	16	2.284	2.03
Cp_2NbCl_2	17	2.469	2.09
$[Cp_2NbCl_2]^+$	16	2.340	2.084

[a]Ω is the center of the Cp-ring

with bulky substituents at the Cp-rings have been obtained also for Th and U. However, in both known polymorphs of Cp_3La, the lanthanum atom forms additional η^2 [213] or η^1 [214] coordination with Cp-rings of an adjacent molecule; the latter form being iso-structural with Cp_3Pr, Cp_3Lu and Cp_3Sc. For bulkier, substituted η^5 ligands, such as $C_5H_3(SiMe_3)_2$, C_5Me_4R (R = Me, Et, isopropyl, $SiMe_3$) [215], no such additional coordination occurs, and the metal atom lies in the same plane as the centers of the three Cp-rings. This pseudo trigonal coordination is typical also for unsubstituted Cp_3M complexes of the late lanthanides, while it can be completed to pseudo trigonal-bipyramidal by two axial monodentate ligands, as in $Cp_3M(NCMe)_2$ (M = La, Ce, Pr, Nd) [216].

The only tetra-η^5-cyclopentadienyl complexes are Cp_4U and Cp_4Th. In Cp_4Zr and Cp_4Hf, only three and two Cp-rings, respectively, are η^5-coordinated, the rest forming only σ-bonds via a single C atom. These examples illustrate the importance of the ligand packing density in the coordination sphere. As a quantitative measure of it, Tolman [217, 218] suggested to use the apex angle of a cone, having its apex at the metal atom and enclosing within it the van der Waals surface of the entire ligand. Tolman's technique (Fig. 3.11) permitted to rationalize chemical and physical properties of some classes of complexes, e.g. with phosphine ligands. Lobkovsky [219] developed this approach for complicated ligand shapes, suggesting to use solid angles of the ligands, S/r^2, as the measure of their bulkiness. A complex is sterically stable if the sum total of these angles for all ligands lies between 0.85 and 1.00 of the full spherical angle (4π), otherwise it is prone to structural rearrangements. Among other things, this model has successfully rationalized the ability of metal atoms to accommodate different numbers of Cp-ligands.

A remarkable achievement of organometallic chemistry was the preparation of multi-decker sandwich complexes, the simplest of which is $[CpNi(\mu\text{-}Cp)NiCp]^+BF_4^-$ with the mean Ni–C distances of 2.100 Å for the terminal Cp-ligands and 2.164 Å for the central (bridging) one [220, 221]. More often, only the terminal rings are Cp or its derivatives, while bridging rings are boron-carbon heterocycles, e.g. B_3C_2 [222], or consist entirely of hetero-atoms like the planar P_6 ring in the complexes $Cp^*W(\mu\text{-}P_6)WCp^*$ and $(\eta^5\text{-}C_5Me_4Et)M(\mu\text{-}P_6)M(\eta^5\text{-}C_5Me_4Et)$, where M = V or Nb [223, 224]. Such systems can be therefore also regarded as clusters (see Chap. 3.5). The largest one structurally characterized to-date is the 6-decker $[Cp^*Co(Et_2C_2B_3H_2Me)Co(Et_2C_2B_3H_3)]_2H_2Co$ system [225].

The structures of cyclopentadienyl complexes of main group elements also show much structural diversity; they have been comprehensively reviewed in [226]. Univalent metals form 'half-sandwich' CpM molecules (M = Li, Na, K, Ga, In, Tl) in the gas phase, but in the solid state these compounds form infinite polymeric chains, either linear (for Li and Na), or zig-zag with clino sandwich coordination (for larger metal atoms). These polymers can be regarded as multi-decker sandwiches with infinite number of 'decks'. Quasi-monomeric cations $[Cp^*M]^+$ exist in the solid state for M = Ge, Sn and Pb, participating in weak additional interactions with the outer-sphere anions. Genuine (parallel) sandwich structures have been observed in crystals for $[Cp_2Li]^+$ and $[Cp_2Na]^+$, and for Cp_2Mg in both crystalline and gas phase [192]. Clino-sandwich molecules of Cp_2M, where M = Ca, Sr, Ba, Si, Ge, Sn and Pb, are monomeric in the gas phase, but in crystal can remain so only if bulky substitutes are attached to the Cp-rings, otherwise they either acquire auxiliary equatorial ligands or form intricate oligo- or polymeric motifs, particularly so in the case of M = Pb (Fig. 3.10). 'Beryllocene' Cp_2Be [192] and $(C_5HMe_4)_2Be$ [226] in fact have one η^5 and one η^1 ligands, the fact obscured in the former structure by disorder, while Cp^*_2Be has a symmetrical sandwich structure with two η^5 coordinated rings lying at 1.655 Å from the Be atom, with the mean Be–C distance of 2.05 Å.

Sandwich complexes can be formed also with other aromatic ring ligands, particularly with benzene. The seminal compound of this class is dibenzeno-chromium $(\eta^6\text{-}C_6H_6)_2Cr$, which can be regarded as an ideal π complex, Coulomb forces playing no part in the bonding. The measurements of dipole moments of MBz_2 (Bz = benzene) revealed that their structures are symmetric when M is an early transition metal (Sc, Ti, V, Nb, Ta, Zr) but asymmetric in $CoBz_2$ and $NiBz_2$ [227]. M–C distances in benzene complexes are somewhat longer than in cyclopentadienyl ones. Sandwiches with both Cp and benzene ligands are also known.

A very important class of organometallic compounds is complexes of transition metals with carbonyl (CO) ligands (see examples in Tables S3.13) [228–231]. The coordination number in these complexes is determined by the EAN rule, hence the stable complexes are tetrahedral $Ni(CO)_4$ and $Pd(CO)_4$, trigonal-bipyramidal $[Mn(CO)_5]^-$ and $Fe(CO)_5$, octahedral $Cr(CO)_6$, $Mo(CO)_6$ and $W(CO)_6$. Formally CO is an inorganic ligand, the bond in it is shorter (1.128 Å in the gas phase) than in CO_2 (1.160 Å) and is nearly triple (C≡O) in character. Usually CO coordinates with metal in a linear η^1 fashion which can be formally described by the valence-bond scheme M=C=O. However, metal carbonyls are typical π complexes in their

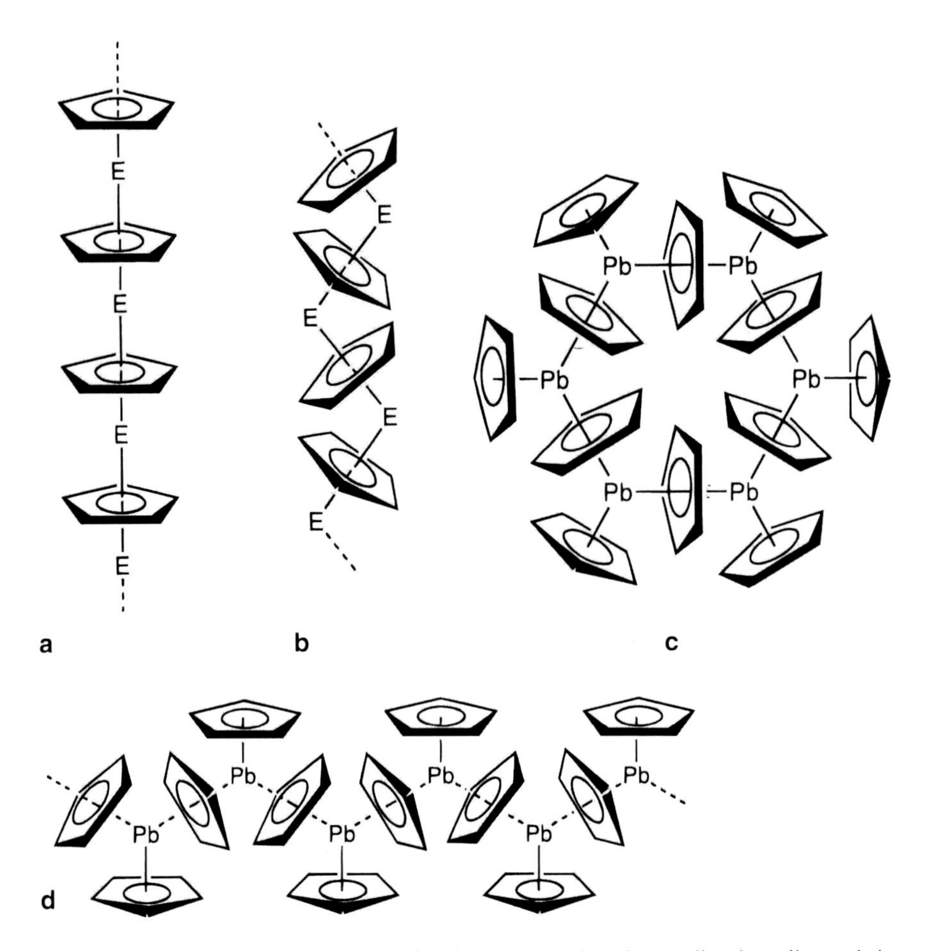

Fig. 3.10 Solid-state structural patterns of main-group metal cyclopentadienyls: **a** linear chain (E = Li, Na), **b** zig-zag chain (E = In, Tl), **c** oligomeric and **d** polymeric forms of plumbocene. (Adapted with premission from [201]. Copyright 1999 American Chemical Society)

Fig. 3.11 Steric parameters of a ligand: **a** Tolman's conical angle $\theta = 2(\theta_1+\theta_2+\theta_3)/3$, **b** Lobkovsky's solid angle S/r^2 (S = spherical surface occupied by the ligand). The latter is more applicable to ligands of arbitrary shape

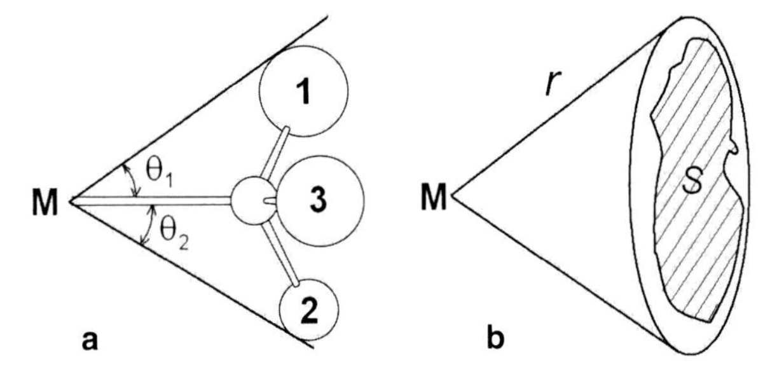

Table 3.14 Mutual influence of CO and η ligands

Complex	d(M–CO), Å	d(C–O), Å	d(M–C-η), Å
Cp^*_2Mn			2.112
$Cp^*Mn(CO)_3$	1.729	1.174	2.126
$ClMn(CO)_5$	1.893, 1.808[a]	1.122, 1.109[a]	
Cp_2Fe			2.055
$CpFe(CO)_2Cl$	1.771	1.125	2.070
$[Fe(CO)_6]^{2+}$	1.910	1.106	
$(C_6H_6)_2Cr$			2.142
$(C_6H_6)Cr(CO)_3$	1.845	1.158	2.233
$Cr(CO)_6$	1.918	1.142	

[a] *trans* to Cl

physical and chemical properties, hence the model of synergic σ donation of a lone electron pair, M⟵:CO, and back π donation of *d*-electrons from the metal atom into the anti-bonding ($p\pi^*$) orbital of the C≡O bond is more appropriate. Thus, although an essentially double (in its length) M=C bond is formed, the actual shift of electron density is very small and the oxidation state of metal is 0. In agreement with this scheme, the C–O bond in the complexes is weaker than in the free CO radical, as indicated both by its lengthening and by lowering of the ν(CO) stretching frequency, compared to 2143 cm^{-1} in the free CO. The strengths of M–C and C–O bonds are inversely related. Adding an electron-donor ligand to the metal center, or replacing an existing ligand by a stronger donor, facilitates the back π-donation and results in a stronger M–C and weaker C–O bond. When more CO ligands are coordinated to the same metal atom, their competition for the same *d* electrons reduces the back donation and results in weaker M–C and stronger C–O bonds (Table S3.13). CO can act not only as a terminal, but also as a bridging ligand in a poly-nuclear complex (cluster), bonding two or three (very rarely four) metal atoms. The C–O distances for the terminal, μ_2 and μ_3 coordination average 1.145, 1.171 and 1.190 Å, respectively [26]. In any of these functions, a CO ligand contributes two electrons to the EAN count. Metal-carbonyl bonding is affected by competition between different ligands for the *d* electrons of metal. Thus, replacing one CO group for another neutral ligand L with lower π acceptor ability, strengthens M–CO bonds with the remaining carbonyls. The effect is stronger in the *trans* than in *cis* positions with respect to L. In some cases, metal-carbonyl coordination is possible only in *trans* positions to non-carbonyl ligands. This feature (similar to the *trans*-effect in inorganic complexes) was reviewed by Compton [232]. However, the abovementioned structural correlations are true only if there is a substantial back donation. If it is weak due to high positive charge on the metal atom, to strong interligand competition, or if the carbonyl is bonded to a main-group atom without any *d* electrons (e.g. B), then CO acts as a pure σ donor and its carbon atom acquires a large positive charge. Since the O atom is negatively charged, the electrostatic attraction can make the C–O bond even *shorter* than in the free molecule. These so-called 'nonclassical' metal carbonyls are rather numerous; they are presented in Table 3.14.

Coordination of the same metal center with carbonyl and cyclopentadienyl (or similar η^n) ligands usually results in a 'piano stool' configuration. The aromatic rings are poorer acceptors of the metal *d* electrons than carbonyls, therefore in such semi-sandwich complexes the M–C(O) bonds are stronger than in purely carbonyl complexes of the same coordination number, and C–O bonds correspondingly weaker. On the contrary, the metal atom is always more distant from the ring plane than in the corresponding sandwich complex (Table 3.14).

Although the majority of η^n ligands are organic groups and molecules, there is an increasing awareness that inorganic molecules or groups with no carbon atoms can play a similar role, e.g. the pseudo-benzene ligand P_6 mentioned above. An analogue of titanocene has been studied, in which the titanium atom is sandwiched between two P_5 rings [233]. Recently studied $[A_7M(CO)_3]^{3-}$ complexes (M = Cr, Mo, W) contain quasi-norbornane ligands A_7, where A = P, As or Sb [234]. In the chromium complexes, the lengths of the P–P, As–As and Sb–Sb bonds (2.121, 2.345 and 2.704 Å), as well as the Cr–P, Cr–As and Cr–Sb bonds (2.514, 2.664 and 2.827 Å, respectively) are usual for covalent bonds. Such inorganic molecules as O_2 and, most remarkably, H_2, also can act as η^2 ligands in π-complexes of transition metals of Groups 6–10 (while N_2 tends to coordinate by *one* atom in a linear fashion). The O–O or H–H bond are weakened in comparison with the free molecules, increasingly so as the bonding with metal is strengthened. Thus, the H–H bond length increases from 0.82 to 1.65 Å as the distances between the metal atom and the center of the H_2 ligand decreases from 1.89 to 1.64 Å [235].

3.5 Clusters

Cluster is commonly defined as a molecule with three or more metal atoms forming a compact group or skeleton (cage) with a substantial direct metal-metal bonding, hence the defining feature of clusters is an apparent discrepancy between the stoichiometry of the compound and the formal valences of its component elements. The first cluster, $Ta_6Cl_{14}{\cdot}7H_2O$, was synthesized in 1907 [236], but not recognized as such until in 1950 Pauling deduced from X-ray diffraction pattern of its solution [237] that the molecule contains a Ta_6 octahedron – an astonishing feat of intuition.

3.5.1 Boron Clusters

The best (and historically the first) understood are boron clusters, which class comprises boranes (poly-nuclear boron hydrides), carboranes (boranes with one or two B atoms replaces with C), and metallo-carboranes with a metal atom incorporated into the cluster. Their structures have been rationalized by Wade in a set of rules [238]. The structure of a boron cluster depends on the numbers of its skeleton atoms (a) and of skeleton electron pairs (SEPs) (p) available for bonding these atoms. If $p = a+1$, the cluster is a closed deltahedron, i.e. a polyhedron having only triangular faces (*closo* structure). If $p = a + 2$, it adopts the form of a higher-order polyhedron with one

Table 3.15 Polyhedra and numbers of SEPs of *closo*-borane clusters $B_nH_n^{2-}$

n	p	Polyhedron, symmetry	n	p	Polyhedron, symmetry
4	5	Tetrahedron, T_d	9	10	Tri-capped trigonal prism, D_{3h}
5	6	Trigonal bipyramid, D_{3h}	10	11	Bi-capped Archimedean antiprism, D_{4d}
6	7	Octahedron, O_h			
7	8	Pentagonal bipyramid, D_{5h}	11	12	Octadecahedron, C_{2v}
8	9	Dodecahedron, D_{2d}	12	13	Icosahedron, I_h

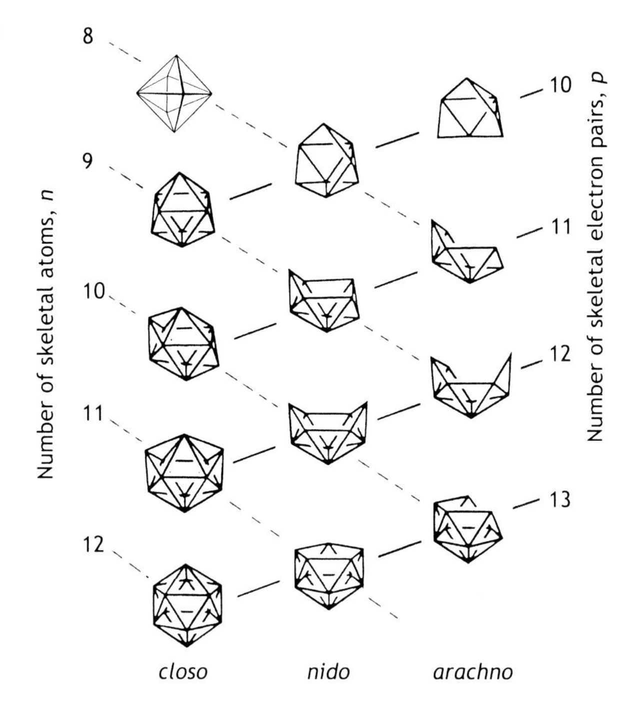

Fig. 3.12 Relations between various boron clusters

vertex missing (*nido* structure), if $p = a + 3$, that of the next larger polyhedron with two vertices missing (*arachno* structure). The polyhedra found in boron clusters are listed in Table 3.15 and Fig. 3.12.

More open structures, e.g. *hypho* (without 3 vertices) for $p = a + 4$, and *klado* (without 4 vertices) for $p = a + 5$, are less common. Each skeletal B or C atom uses one of its valence electrons to form an external bond with H or other univalent substitute, directed away from the center of the polyhedron. The remaining two (B) or three (C) electrons are contributed to skeletal bonding. A skeletal transition metal atom with v valence electrons, receiving x electrons from its external ligands, contributes either $(v+x-12)$ or $(v+x-10)$ skeletal electrons, depending on whether it adopts an 18-electron or 16-electron shell (see Sect. 3.4). Boron clusters do not form polyhedra with $n > 12$, but metallo-carboranes present a few such examples, e.g. $CpCoC_2B_{10}H_{12}$ with $n = 13$ and $Cp_2Co_2C_2B_{10}H_{12}$ with $n = 14$. These polyhedra can be derived from the icosahedron by expanding one or both "tropical" pentagons

Table 3.16 Carbonyl clusters of transition metals

Cluster	Cluster	Cluster
$Os_3(CO)_{12}$	$[Rh_7(CO)_{16}]^{3-}$	$[Rh_{13}(CO)_{24}H_3]^{2-}$
$Ir_4(CO)_{12}$	$Ni_8(CO)_8(PR)_6$	$[Rh_{14}(CO)_{25}]^{4-}$
$Os_5(CO)_{15}$	$[Co_8C(CO)_{18}]^{2-}$	$[Pt_{15}(CO)_{30}]^{2-}$
$[Co_6(CO)_{14}]^{4-}$	$[Rh_9P(CO)_{21}]^{2-}$	$[Rh_{17}(S)_2(CO)_{32}]^{3-}$
$Ru_6C(CO)_{17}$	$[Rh_9P(CO)_{21}]^{2-}$	$[Pt_{19}(CO)_{22}]^{4-}$
$[Rh_6N(CO)_{15}]^{2-}$	$[Fe_4Pt_6(CO)_{22}]^{2-}$	$[Pt_{26}(CO)_{32}]^{2-}$
	$[Pt_{12}(CO)_{24}]^{2-}$	$[Pt_{38}(CO)_{44}]^{4-}$

to hexagons [238]. Nevertheless, larger metallocarborane systems can be formed from two polyhedra with a shared vertex (usually a metal occupied one) or edge.

Boron clusters are electron-deficient i.e. contain bonds with orders lower than 1, which are uncharacteristic for "normal" organic and inorganic molecules. Thus, a *closo* polyhedron $B_nH_n^{2-}$ has $3n-6$ edges, i.e. bonding contacts, but only $2n+2$ electrons. Hence for the clusters with $n=6$ and 12, in which all edges are geometrically equivalent, the bond orders are 0.58 and 0.43, respectively. These orders agree with the observed bond distances, which also vary with the coordination numbers N_c [238]; this provides important data for structural chemistry, because in usual organic molecules $N_c \leq 4$.

Average coordination numbers of B in carborane skeletons varies from 4 to 5.5, and the B–B distances (edges of the polyhedron) from 1.64 to 1.95 Å. The carbon atoms in the same cages have similar N_c (4–6) but form shorter C–C bonds (1.42–1.65 Å). Usually, each B or C atom of the cage bears one hydrogen atom or a single-bonded substituent (halogen, organic or organo-metallic group) pointing outward. A transition metal atom can also occupy one of the skeletal positions. M–B bond distances in such metallocarboranes [239–243] vary, according to the structure, but in most cases are close to the lengths of covalent bonds (adjusted for the bond order) and therefore can be described by the corresponding covalent radii.

Obviously, Wade's rules cannot explain all the structural variety of boron-containing clusters, because a change of composition can alter the bond character as well. Thus, recently synthesized clusters B_nX_n, where $X = Cl, Br, I$, contain n SEPs instead of the $n+1$, normal for closo-boranes (as well as carboranes, metalloboranes, metallocarbonyl complexes). B_9H_9 type clusters have the structure of a 3-capped trigonal prism, while B_9 is a closo borane with $p=n$. Significantly, B–B distances in B_nX_n clusters do not change with X, indicating the similarity between the structures of the cluster skeleton and elementary boron [244]. Stereo-chemical effects on the geometry of carboranes has been reviewed [245]. These can be illustrated by comparing the structure of B_8F_{12} and B_8H_{12}. Notwithstanding the same electron count, the two have very different connectivities (Fig. 3.13) and bond distances, which can be attributed to numerous intramolecular $B\cdots F$ interactions [246].

3.5.2 Transition Metal Clusters

Another important class of clusters are those of transition metals with carbonyl and other ligands, of which osmium-carbonyl clusters are the best understood [247].

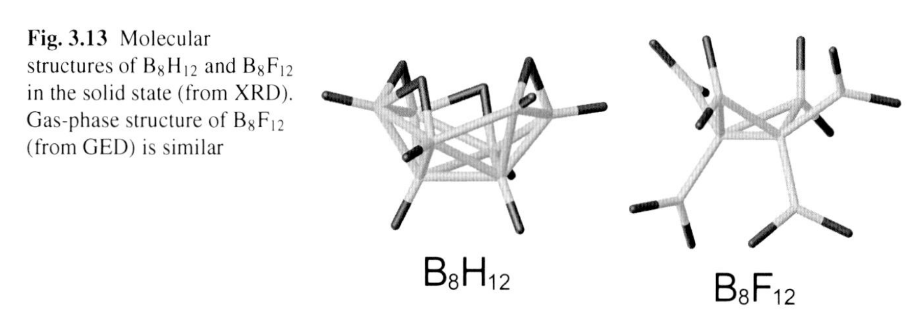

Fig. 3.13 Molecular structures of B_8H_{12} and B_8F_{12} in the solid state (from XRD). Gas-phase structure of B_8F_{12} (from GED) is similar

The smallest of these clusters can be described in terms of 2-electron valence bonds and the EAN rule, e.g. $Os_3(CO)_{12}$ and $Os_4(CO)_{16}$, which contain a triangle and a tetrahedron of Os atoms, respectively, each Os having four CO ligands. However, this approach already breaks down for $Os_6(CO)_{18}$, which instead of the predicted Os_6 octahedron contains a trigonal bipyramid with one face capped by the sixths Os atom (see Fig. 3.14). To explain the structures of these clusters, Wade's theory of boron clusters has been developed by himself [238, 247] and Mingos et al. [248, 249] into a more general Polyhedron Skeletal Electron Pair Theory (PSEPT). According to it, transition-metal skeletons in transition metal clusters correspond to the same types

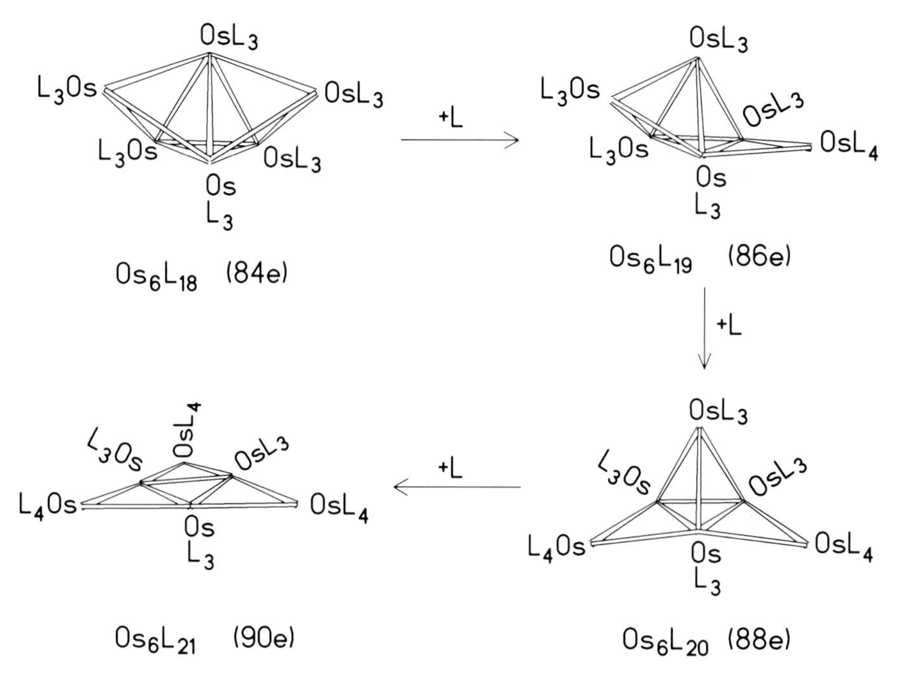

Fig. 3.14 Metal skeletons of some osmium carbonyl clusters (electron numbers in parentheses). L is a CO group or an equivalent 2-electron donor ligand, e.g. phosphine PR_3

of deltahedra as boron clusters, but unlike the latter, can have also additional metal atoms 'capping' their M_3 faces. For an n-vertex *closo* deltahedron with m capping atoms the number of electrons must be

$$2p = 14n + 12m + 2 \tag{3.3}$$

i.e. $14t$ electrons for a mono-capped deltahedron, and $14t$–2 for a bi-capped one, where t is the total number of metal atoms. In such a cluster each metal atom has three CO (or equivalent) ligands and is bonded directly to three or more metal atoms. Adding more SEPs converts *closo* structures into progressively more open structures, where some vertices are occupied by $Os(CO)_4$ groups, linked with only two other Os atoms. Thus $Os_6(CO)_{18}$ has 84 valence electrons (8 from each Os and 2 from each CO), its *closo* structure is TBP, hence $n = 5$ and $m = 1$, and Eq. 3.3 is satisfied: $14 \times 5 + 12 \times 1 + 2 = 84$. Adding six electrons converts the Os_6 skeleton into a planar 'raft' in $Os_6(CO)_{21}$ [250] (Fig. 3.14). However, these rules are much less straightforward than Wade's rules for boron clusters, particularly because there are alternative ways of capping a deltahedron. Hence isomeric structures are possible, especially for larger polyhedra (e.g. for M_{19} a double icosahedron or a six-capped M_{13} cubo-octahedron [251]), between which the rules often cannot give a clear choice.

The interatomic distances in these clusters are longer than in the corresponding bulk metals, but the difference diminishes as the size of the cluster increases, as the effect of the ligand shell of the cluster becomes relatively weaker and the electron structure approaches that in metals [247]; large clusters having broad energy bands instead of separate electron levels [249]. Obviously, meaningful similarity with metal bonding only begins with the clusters containing some 'interior' metal atoms (surrounded only by other metal atoms) and thus approximating fragments of the cubic or hexagonal close packing [252]. The smallest such cluster is $[Rh_{13}(CO)_{24}H_3]^{2-}$, in which the central Rh atom is surrounded by 12 others, while $Pd_{59}(CO)_{32}(PMe_3)_{21}$ contains 11 interior atoms [250]. Clusters also can enclose various non-metal atoms as interior. The typical configurations for clusters with the number of skeletal metal atom varying from 3 to 38, with or without internal atoms, are listed in Table 3.16. Although the majority of clusters contain only one type of metal, those comprising two, three and even 4 different metals, as in $FeCoMoWS(AsMe)_2Cp_2(CO)_7$, are also known.

The borderline between a 'large' cluster molecule and a 'small' particle of a bulk metal can be approached from the latter side as well. Thus, a study of copper particles of *ca.* 10 Å size revealed the $N_c = 6 \pm 1$ and the bond distance of 2.54 Å. For silver, the Ag–Ag distances change from 2.47 Å for an Ag_2 molecule trapped in a rare gas matrix, through 2.86 Å in a particle of 17 Å size (with the mean $N_c = 7.6$) to 2.87 Å in the bulk metal ($N_c = 12$) [253, 254]. Most thorough studies, both experimental and *ab initio*, have been carried out for mercury [249] and showed that the bonding in Hg_n clusters is covalent for $n = 2$–6, with $n > 19$ a transition to van der Waals type occurs [255], and in the range of $n = 20 - 70$ the 6s and 6p orbitals are widening and overlapping, which leads toward the metallic state.

We have already mentioned that structures of large transition metal clusters are usually similar to those of bulk metals. However, some completely novel motifs have also been observed, namely mixed *fcc/chp* packing types, patterns with local 5-fold symmetry (which is forbidden in a crystal lattice with translational symmetry) and strongly perturbed 'amorphous' packing of metal atoms. Belyakova and Slovokhotov [256] reviewed 72 structures of large transition metal clusters containing up to 145 metal atoms, among them some compounds with the metal core shaped as a body-centered icosahedron or a pentagonal prism with capped pentagon faces. They classified large clusters with 5-fold symmetry into two structural classes, viz. (i) atomic skeletons which combine several fused or mutually interpenetrating icosahedra; and (ii) multilayer 'onion' structures, where the central atom is surrounded by successive icosahedral shells.

The structural differences of cluster cores from bulk metals can be attributed [257] to the influence of ligands, which can affect substantially the configuration of a cluster seeking the potential energy minimum, up to producing atomic arrangements unthinkable in regular lattices. Cluster structures are also adopted by halides, chalcogenides and pnictides of transition metals. Their structural chemistry is based on the following simple principles [258]: a reduction of the metal's formal valence increases the tendency to form larger clusters, while an increase of the size of a ligand results in an increase of the M–M distance in clusters of the same type. The former rule can be illustrated by the succession: $NbCl_5$ (mononuclear molecule), Nb_2I_8 (dumbbell), Nb_3I_8 (triangle), $[Nb_6I_8]I_3$ (octahedron), the latter by a comparison of Nb–X distances in Nb_3X_8, viz. 2.81 (X=Cl), 2.88 (X=Br) and 3.00 Å (X=I). Very widespread structural units in such clusters are octahedral groups M_6X_8 and M_6X_{12}, which can be joined through shared vertices, edges or faces, forming infinite chains, ribbons or 3-dimensional networks. The motif with shared vertices results in the $M_{2/2}M_4X_{8/2}$, or M_5X_4 composition, e.g. Ti_5Te_4, V_5S_4, V_5Se_4, V_5Sb_4, Nb_5Se_4, Nb_5Te_4, Nb_5Sb_4, Ta_5Sb_4, and Mo_5As_4. Edge-sharing gives the $M_{4/2}M_4X_{8/4}$, or M_2X composition, as in Ti_2S, Ti_2Se, Zr_2S, Zr_2Se, Hf_2P, Hf_2As, Nb_2Se, Ta_2P and Ta_2As. If M_6X_8 octahedra are linked into a 3-dimensional network, the formula will be $M_{6/2}X_{8/8}$, or M_3X, e.g. in Cu_3Au and U_3Si. Face-sharing M_6X_8 clusters give MX type compounds, while similar linkage of M_6X_{12} results in the MX_2 composition. Often different types of "building blocks" coexist in the same structure, while some of the atomic positions may remain void in either the M or the X-arrays, resulting in the most diverse motifs. Thus, the structure of $Nb_{21}S_8$ contains isolated (M_5X_4) and tetramer ($4M_5X_4$) octahedra; the structure of Ti_8S_3 contains double chains ($2M_5X_4$) linked to networks of $4M_5X_4$ chains.

Very important from the chemical viewpoint are cluster halides of lanthanide metals (Ln) with the formal valence $v < 3$ (Table 3.17). Thus, the structure and physical properties of LnI_2 type compounds correspond in fact to the formula $Ln^{3+}(I^-)_2e^-$, while the halides with $v < 2$ have typically cluster structures based on chains (single or double) of M_6 octahedra, surrounded by halogen atoms. The LnX type compounds constitute a peculiar class of '2-dimensional metals': the Ln–Ln bond distance in them depends on the size of the anion (compare Tb–Tb distances of 3.79

Table 3.17 Types of clusters of lanthanide chalcogenides

Compound	Type	Structure[a]	Examples	Compound	Type	Structure[a]	Examples
M_7X_{12}	M_6X_{12}	discrete M_6	Sc_7Cl_{12}	M_4X_5	M_6X_{12}	IC	Er_4I_5
			Ln_7I_{12}	M_7X_{10}	M_6X_{12}	DC	Er_7I_{10}
M_2X_3	M_6X_8	IC	Ln_2Cl_3				Sc_7Cl_{10}
			Ln_2Br_3				
M_2X_3	M_6X_{12}	IC	Tb_2Br_3	M_6X_7	M_6X_{12}	DC	Tb_6Br_7
M_5X_8	M_6X_{12}	IC	Sc_5Cl_8				Er_6I_7
			Gd_5Br_8	MX	M_6X_{12}	layers	(Sc, Ln) Cl

[a]IC – isolated chains, DC – double chains

Table 3.18 Orders (n) of metal-metal bonds in clusters

Cluster	n	Cluster	n	Cluster	n	Cluster	n
ZrCl	0.81	Nb_6F_{15}	1.10	ZrBr	0.66	Ta_6Cl_{15}	0.69
Zr_6Cl_{12}	0.88	Mo_6Cl_8	1.05	Zr_6I_{12}	0.68	Mo_6Br_8	0.94
		Ti_2S	0.77			Ti_2Se	0.55

Å in TbCl and 3.84 Å in TbBr). Note that mono-iodides of lanthanides do not exist at all, probably because Ln–Ln bond would be too long to stabilize the structure.

M_5 cluster units, although less common than M_6, exist in compounds of the M_2X type, such as Fe_2P, while M_4 clusters can be found in Sc_3P_2, Zr_3As_2, Cr_3C_2. Whilst $Mo_4I_7{}^{2+}$ provides an example of an isolated M_4 cluster, the Ni_2Si structure consists of a network of M_4X_6 clusters, linked through shared vertices. As has been mentioned above, the center of the cluster can be occupied not only by a metal, but also by a nonmetal atom, e.g. hydrogen, as in the structure of HNb_6I_{11}. If the center of a M_6X_8 cluster is occupied, the result is a perovskite-type structure, as in Mn_3GeC and Fe_3GeN. The centers of all octahedra are occupied in Nb_5Ge_3B and Hf_5Sn_4, which have the structure of the Ti_5Ga_4 type.

Naturally, in the structural chemistry of clusters much attention is paid to the peculiarities of metal-metal bonds [243, 259, 260]. It has been noticed that homonuclear bonds are typically formed by metals in the lowest valence state, and that heavier transition metals give stronger M–M bond than the lighter ones. Thus, Cr–Cr, Mo–Mo and W–W bond lengths in iso-structural di-nuclear halide complexes are 3.12, 2.66 and 2.41 Å, respectively. The shortening is due to an increase of the bond order. This permitted to establish the normal Re–Re bond distances for the whole range of bond orders (n): 2.90 ($n = 1$), 2.47 ($n = 2$), 2.30 ($n = 3$) and 2.22 Å ($n = 4$). However, usually in metallic clusters $v \leq 1$ and steric interactions affect the length (and hence the order) of the bonds very substantially, as shown in Table 3.18 for some clusters with ligands of different sizes [261, 262]. Although metal-metal distances in a cluster skeleton depend on the valence of the metal atoms, the polarity of their bonds with ligands, the size and structure of the latter, mostly these distances vary around certain average values, listed in Table 3.19 [239, 263–265].

Table 3.19 Homonuclear metal-metal bond lengths (Å) in clusters

M	d(M—M)	M	d(M—M)	M	d(M—M)	M	d(M—M)
Cu	2.65	Nb	2.88	W	2.75	Rh	2.79
Au	2.84	Ta	2.80	Mn	2.84	Pd	2.72
Al	2.77	Sb	2.82	Fe	2.63	Os	2.87
Zr	3.17	Bi	3.04	Co	2.50	Ir	2.81
V	2.91	Cr	2.77	Ni	2.61	Pt	2.69
		Mo	2.76	Ru	2.84		

Table 3.20 Cluster ions of main-group metals $M_n^{x-/+}$

n	Cluster configuration	Examples
4	Square	Hg_4^{6-}, Bi_4^{2-}, Te_4^{2+}
	Tetrahedron	Tl_4^{8-}, Si_4^{4-}, Ge_4^{4-}, Pb_4^{4-}
5	Planar	Si_5^{0}, Ge_5^{0}
	Trigonal bipyramide	Sn_5^{2-}, Pb_5^{2-}, Bi_5^{3+}
6	Trigonal prism	Te_6^{4+}
7	Mono-capped octahedron	Pb_7^{4-}, P_7^{3-}, As_7^{3-}, Sb_7^{3-}
8	Square antiprism	Bi_8^{2+}
9	Tricapped trigonal prism	Ge_9^{2-}, Bi_9^{5+}
	Mono-capped square antiprism	Ge_9^{4-}, Sn_9^{4-}, Pb_9^{4-}

3.5.3 *Clusters of Main Group Elements*

All these examples refer to the compounds of transition metals, and for a long time it was believed that only these metals (and boron) can form clusters. However, today clusters (in the form of poly-anions, poly-cations or neutral molecules) are known practically for all elements. Von Schnering [266] has compared the frequency of occurrence of cluster structures for elements with the atomization energies of the corresponding elemental solids and found that these distributions are similar, with the maxima for heavier Groups 5–7 metals and for Groups 13–15. These elements give the most stable clusters. The structures of some zinc, cadmium, mercury, and indium compounds contain poly-cations, viz. M_2^{2+} in ZnP_2 and CdP_2, In_2^{4+} and In_3^{5+} in In_6Se_7 and In_4Se_3. Crystal structures of certain halides, chalcogenides and pnictides of main-group metals contain cluster moieties with different charges [252, 266], listed in Table 3.20.

Anions E_9^{4-}, where $E = $ Si, Ge, Sn, Pb, have $2n + 4$ skeletal electrons and hence, according to PSEPT, should adopt *nido* structures, which has been confirmed experimentally, while their derivative clusters $[(OC)_3M(\eta^4\text{-}E_9)]^{4-}$ (where M = Cr, Mo, W), have closo structures (bi-capped Archimedian antiprism) with $2n + 2$ electrons [267]. Corbett [268] reviewed from the viewpoint of Wade's theory, the structures of these and many other mixed-metal clusters, a brief listing of which is given in Table 3.21 (compare with Table 3.20).

Sub-halides of bismuth, obtained though partial oxidation of inter-metallic phases by halogens, show interesting structural possibilities by changing the dimensionality of the structure with the change of composition. Thus, $Bi_{5.6}Ni_5I$ and $Bi_{12}Rh_3Br_2$ have 3-dimensional metallic networks, $Bi_{13}Pt_3I_7$, $Bi_{12}Ni_4I_3$ and $Bi_{13}Ni_4X_6$ have 2-dimensional, while $Bi_{6.8}Ni_2Br_5$, $Bi_9Rh_2X_3$ and Bi_4RuX_2 have 1D chains, with

Table 3.21 Number of electron pairs p and the structures of clusters

Composition and charge of a cluster	Structure	p
E_4^{2-} (E = Sb, Bi), E_4^{2+} (E = Se, Te), Bi_8^{2+}	*arachno*	$n+3$
$Sn_2Bi_2^{2-}$, $Pb_2Sb_2^{2-}$, $InBi_3^{2-}$, E_9^{4-} (E = Ge, Sn), $In_4Bi_5^{3-}$ In_5^{9-},	*nido*	$n+2$
E_5^{2-}(E = Ge, Sn, Pb), Ge_9^{2-}, $TlSn_8^{3-}$, Sn_9^{3-}, Bi_5^{3+}, Tl_5^{7-}, Ga_6^{8-}	*closo*	$n+1$
Tl_6^{6-}, Tl_7^{7-}, Tl_9^{9-}, E_{11}^{7-} (E = Ga, In, Tl)	*hypo*	n

corresponding changes of electro-physical properties [269]. Cluster structures are also known for non-metals, such as phosphorus [270], which forms P_5^-, P_6^{4-}, P_6^{6-}, P_7^-, P_7^{3-}, P_{10}^{6-}, P_{11}^-, P_{11}^{3-} and P_{15}^- anions in salts with alkali metals, as well as various tube-like structures. The P—P bond distances in the anions vary from 2.197 Å in P_7^{3-} to 2.233 Å in P_{11}^{3-}, comparable to those in the white (2.209 Å), violet (2.215 Å) and black (2.228 Å) polymorphs of elementary phosphorus. Very similar P—P distances are found in polycyclic phosphanes, such as $(PPh)_6$ (2.235 Å), $(PPh)_5$ (2.211 Å), $(PCF_3)_5$ (2.223 Å), $(PCF_3)_4$ (2.213 Å), $(Phex)_4$ (2.224 Å), $(Pbu)_4$ (2.212 Å) and $(Pbu)_3$ (2.203 Å) [271], in 4-membered MP_3 rings found in the complexes $Cp_2M(PR)_3$ with M = Zr, Hf and R = Ph, Cy (mean 2.186 Å) [272], or in halido-phosphides of mercury, $Hg_2P_3X_2$ (mean 2.196(1) Å) [273]. Such stability of the P—P bond distances justifies the description of these clusters as comprising rigid building blocks: dumbbells and triangles with 4 or 5 free valences, respectively [274].

3.5.4 *Fullerenes*

The molecules of fullerenes, discovered since 1986, represent convex polyhedra with carbon atoms in their vertices, which have only penta- and hexagonal faces, i.e., the carbocycles characterized by the minimum steric strains. According to the Euler theorem, such polyhedra have exactly 12 pentagonal and any number of hexagonal faces. Carbon atoms in these closed polyhedra (cages) have C—C distances intermediate between single- and double bond lengths. Thus fullerenes belong simultaneously to the realms of inorganic (as an allotropic modification of carbon), organic (as aromatic 'hydrocarbons' without hydrogen) and cluster chemistry. Structures of fullerenes have been reviewed in [275–278], but in this fast-developing area all surveys tend to become outdated very quickly.

The most common (buckminster-) fullerene is a C_{60} polyhedron (Fig. 3.15) comprising pentagonal (C_5) and hexagonal (C_6) faces. All carbon atoms are topologically equivalent, but there are two different types of edges (bonds): those separating C_5 from C_6, or C_6 from C_6 faces. These edges, known as 6:5 and 6:6, have the lengths of 1.45 and 1.38 Å, respectively. Hence, the latter bonds are of substantially multiple character, and suitable for π-coordination with transition metals. Quasi-sperical form of fullerene molecules and isotropic, van der Waals nature of interactions between them, results in close packing motif in the solid state. The crystal structure of pure C60 is essentially a close packing of spheres, either face-centered cubic or

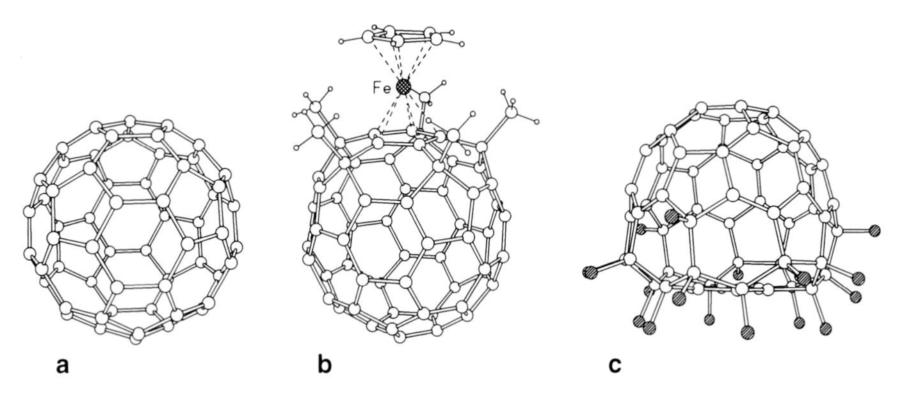

Fig. 3.15 Molecular structures of fullerene C_{60} **a**, $CpFeC_{60}Me_5$ **b** and $C_{60}F_{18}$ **c**

hexagonal, complicated by rotational disorder of the nearly spherical molecule. Pure crystalline buckmisterfullerene prepared by re-sublimation, has an *fcc* lattice. On cooling down to 258 K, a phase transition occurs, which is accompanied by partial ordering of the molecules and lowering of the symmetry. The high temperature or pressure treatment leads to formation of the polymer phases of C_{60} of orthorhombic, rhombohedral, or tetragonal symmetries [279].

Other common fullerenes are C_{70}, C_{78} and C_{84}, although fullerenes up to C_{124} have been studied energetically [280, 281]. The ellipsoidal cage of C_{70}, which can be described as two hemispheres of C_{60} reconnected together through a belt of 10 additional C atoms, contains 5 different types of carbon atoms and 8 types of C–C bonds. Both C_{60} and C_{70} have unique structures, but for larger cages different structural isomers are possible. Thus, C_{84} can have 24 isomers (of which 5 have been prepared so far) whereas C_{96} can have 187 [278]. Over 1100 other fullerene-containing structures currently known can be divided into six classes:

(i) Molecular complexes and co-crystals with various organic or inorganic molecules, in which a fullerene molecule remains electro-neutral.
(ii) Fullerides, i.e. salts in which the fullerene acquires a negative charge, which can be delocalized about the cage. The counter-ions can be organic, organometallic, or alkali (alkali earth) metal cations. The latter type of fullerides M_xC_n is also known as intercalation compounds [275, 277]. They are metallic-type conductors (and in some cases superconductors) and can contain very large number of metal atoms per fullerene unit, e.g. in $Li_{12}C_{60}$, Ba_6C_{60}, Sr_3C_{60} or $K_3Ba_3C_{60}$ [282]. The crystal structure of the superconducting $K_3Ba_3C_{60}$ fulleride ($T_c = 5.4$ K) has been studied by synchrotron X-ray and neutron powder diffraction between 10 and 295 K. It is body-centered cubic at all temperatures with an essentially half-full t_{1g}-derived band. Close contacts between Ba^{2+} and K^+ ions and neighboring C_{60} units imply a strong orbital hybridization, which leads to broadening of the conduction band. Ba^{2+} and K^+ cations are disordered in the same distorted tetrahedral interstitial sites, but they are displaced by a different distance from the center of the site. The resulting local distortions

may be responsible for the relationship between T_c and cubic lattice parameter in $A_3Ba_3C_{60}$ (A = Na, K, Rb, Cs) fullerides. Anions C_n^- and C_n^{2-} are stable to $n = 124$ [279, 283].

(iii) Functionalized fullerenes, in which covalently bonded atoms or groups are attached to the fullerene carbons. Of most interest are halogenated fullerenes, such as $C_{60}F_{18}$, $C_{60}F_{36}$, $C_{60}F_{48}$, $C_{60}Cl_6$, $C_{60}Br_6$, $C_{60}Br_8$, $C_{60}Br_{24}$, or $C_{70}Cl_{10}$ and $C_{70}Br_{10}$. In the latter structure, the C–C bonds adjacent to bromine atoms, are elongated to 1.51 Å, compared to other C–C (1.44 Å) and C=C (1.39 Å) bonds in the cage [284]. Note that while the aromaticity of non-functionalized fullerene is rather low, conversion of some cage atoms from 3- to 4-coordinated facilitates delocalization in the rest of the cage. Thus, in $C_{60}F_{18}$ all F atoms are attached to one hemisphere of the cage and create a belt of 4-coordinated carbon atoms surrounding one (un-fluorinated) C_6 face which requires geometry of a fully aromatic benzene ring [285]. The cage thus acquires a peculiar hemispherical 'turtle carapace' shape (Fig. 3.15c).

(iv) Organometallic compounds, in which fullerene acts as a π ligand, the metal atom bonding to the cage on its outer side. Unsubstituted fullerene coordinates usually in a η^2 fashion, by its 6:6 bond. In polynuclear (cluster) compounds, η^1 coordination also occurs alongside η^2. However, in a substituted fullerene a C_5 face can acquire sufficient aromaticity (see iii) to act as a η^5 ligand, analogous to a cyclopentadienyl ring. Such coordination is observed in $CpFeC_{60}Me_5$ (Fig. 3.15b), an analogue of ferrocene, with the average distances Fe–C(Cp) 2.033 and Fe–C (fullerene) 2.089 Å [286]. Note that all five cage atoms adjacent to the η^5 coordinated ring, are bonded to methyl groups and are 4-coordinated (sp^3). Similar η^5 coordination has been observed in complexes of functionalized C_{60} with thallium [287].

(v) Endohedral complexes designated $M@C_n$, with metal atom(s) residing inside the fullerene cage. Most of these compounds have one incorporated metal atom (M = Ca, Sr, Ba, Sc, Y, La, Ti or Fe) [276], but there is a number of cases with two or three, particularly a small class of $M_3N@C_n$ derivatives, where the cage contains three metal atoms bridged by a planar-trigonal N atom. These are known for M = Sc, Y, Pm, Gd, Tb, Tm and Lu, usually with $n = 80$ but also with $n = 68$, 84 and 92 [288–290,330, 331]. Each metal atom can be described as coordinated with a C_6 and an adjacent C_5 faces (a pseudo-indenyl moiety).

(vi) Oligo- and polymeric fullerenes, the cages of which are linked either through one C–C bond ([1 + 1] type bridging), or through two such bonds formed between adjacent pairs of atoms ([2 + 2] bridge, involving a 4-membered carbon ring). Polymerization can occur at high temperature or under pressure. The first $(C_{60}^-)_2$ dimers had been found in metastable phases of the MC_{60} composition, where M = K, Rb, Cs; later the same type of polymerization was proven in many other metal fullerides, such as Na_2RbC_{60}, Li_3CsC_{60} [277].

Since boron nitride is a structural analogue of carbon both in its graphite (hexagonal, *h*-BN) and diamond (cubic, *c*-BN) forms, one may ask whether it is not possible to prepare a BN analogue of fullerene? The problem is that while in other forms of

BN, boron atoms are surrounded only by nitrogen ones and *vice versa*, the presence of 5-membered rings in a fullerene cage makes unavoidable the energetically unfavorable B–B or N–N bonds. Although in the heterofullerene $(C_{59}N)_2$, in which one carbon atom of a C_{60} cage is replaced with nitrogen, the resulting electron defect (one per cage) is satisfactorily absorbed by the structure, the same seems unlikely if the concentration of defects is much higher. Recent reports [291, 292] on preparation of solid solutions between graphite and *h*-BN or between diamond and *c*-BN, although apparently encouraging, require verification, because distinguishing such solid solutions from fine microcrystalline mixtures is impossible by conventional XRD. This can be done either by total X-ray scattering, which can elucidate local structure, or a crystal-optical study (neither is yet available). Nevertheless, the formation of a BN-based fullerene is possible, if it incorporates a certain amount of oxygen atoms. In fact, admixtures of oxygen are present in all BN nanotubes prepared to the present day. Indeed, such an oxygen-doped BN fullerene has been incidentally synthesized in 1965 from turbostratic BN subjected to shock wave compression by an explosion [294]. The product, called the E-phase of BN, has been reproduced later by many other methods [295–301], its structure models are considered in [302]. The lengths of single and multiple homonuclear bonds are given in Table S3.15.

3.6 Coordination Compounds

The majority of inorganic substances contain bonded atoms of three or more different elements. Salts with mixed anions (LaOF, PbFCl, BiOX, etc) have been discussed in Sect. 5.2, those with mixed cations are discussed here, as well as complex compounds which are combinations of intrinsically stable molecules. In complex compounds the bonding is essentially covalent within the coordination polyhedron but essentially ionic outside it. Complex compounds where the central atom is metal (especially transition metal) are known as coordination compounds. Thus, KNO_3 and $BaSO_4$ are complex compounds, but not coordination compounds. The amount of structural information on such compounds is immense, therefore we shall concentrate on a few problems.

In compounds with mixed ligands, the central atom can adopt a valence state which is not stable otherwise. Stabilization of such states is caused by introduction into the structure of a component which reduces the concentration of structural tension. For example, CuI_2 and AuI_3 do not exist, whereas CuICl, CuIBr, and AuI_2Cl, AuI_2Br and AuIBrCl are quite stable [303, 304]. The stabilization in these cases is caused by a shift of the electron density from metals to F, Cl or Br ligands, preventing the reduction of the metal cation by the electrons of the M–I bond.

In M_nAX_6 compounds where M is an alkali metal, A is a polyvalent metal and X is hydrogen or a halogen, the coordination polyhedron around A is usually an octahedron, see Table S3.16. In square-planar complexes of bivalent platinum, M_2PtCl_4 [305, 306], the Pt–X distances are close to those in octahedral $[PtX_6]^{2-}$ complexes, viz. for $M = H$, K, NH_4 and Cs they equal 2.32, 2.312, 2.305 and 2.300 Å, respectively. Similar situation is found in Rb_2PtBr_4 and K_2PtBr_6 with Pt–Br distances of 2.435 and 2.464 Å, respectively. Similar distances Pt^{II}–X in the square and Pt^{IV}–X in the octahedron coordination can be explained by regarding Pt^{II} as $Pt^{IV}E$, where E is a lone electron pair. Actually, in the crystals of M_2PtX_4, the alkali metal cations are located above the centers of PtX_4 squares, along the axis of the lone electron pair of platinum. In the crystalline Pt^{II} compounds, electrons are shifted towards the ligands, this increases the electron density of the Pt–X bond in Pt^{II} in comparison with Pt^{IV} and therefore the Pt^{II}–X bond is shorter than Pt^{IV}–X, viz. the average Pt^{II}–I 2.620 Å and Pt^{IV}–I 2.690 Å [307]. It is noteworthy that Cs_2PtI_6, a classical complex of tetravalent platinum with uniform Pt–I distances (2.697 Å), has an isomer in the form of $Cs_2PtI_4{\cdot}I_2$ with square-planar coordination of Pt^{II} and Pt–I bonds of 2.621 Å. However, the environment of the Pt^{II} atom is complemented to a tetragonal bipyramid (an octahedron stretched along one of its fourfold axes) by two atoms of iodine molecules, at the distances of 3.233 Å. Under pressure of $P \geq 2$ GPa, $Cs_2PtI_4{\cdot}I_2$ transforms into Cs_2PtI_6 [308].

In coordination compounds with diverse ligands there is a competition between the ligands for attracting electrons from the central atom. This competition takes place mainly between *trans*-ligands, i.e. along the same X–A–Y coordinate. Generally, the higher the electronegativity of X, i.e. the more polar the X–A, the more covalent is the opposite bond A–Y. This effect is common for transition metals in octahedral and especially in square-planar coordination, but is most conspicuous for square complexes of Pt, where it was first discovered. In 1926 Chernyaev [309] (for English-language account see [310]), while studying substitution reactions in platinum complexes, observed the ability of ligands to labilize (i.e. increase the rate of substitution of) their *trans*-partners in the succession:

$$CO, CN, C_2H_4 > NO_2, I, SCN > Br, Cl > OH, F > NH_3, H_2O,$$

which he named *trans*-influence and predicted (correctly) that it would be found in complexes of other metals with mixed ligands [311]. Today, this set of phenomena is usually called '*trans*-effect'. The kinetic *trans*-effect (actually observed by Chernyaev) can be distinguished from the thermodynamic *trans*-effect (the change of the ground state of the A–Y bond) or the structural *trans*-effect (the change of the bond distance), although of course all three are different manifestations of the same electronic interactions within the coordination polyhedron, namely the redistribution of electron density which causes the complementary changes of the bond lengths and strengths, spectroscopic frequencies, effective charges, etc. Table S3.17 illustrates how the Pt–Cl distances increase with the degree of covalency of the bond in the *trans*-position to chlorine [312–318]. A similar series of the *trans*-effect were observed for Pt–C [319] and Pt–P bonds [320], and also in complexes of the Group 8, 9, 10 metals [321]. Today, the following succession of *trans*-effect is generally accepted: C_2H_4, CN^-, CO, NO, H > CH_3^-, SR_2, AsR_3, PR_3, > SO_3^{2-} > Ph^-, $SC(NH_2)_2$, NO_2^-, SCN^-, I^- > Br^- > Cl^- > pyridine > NH_3 > OH^-, H_2O, F^-.

From Tables S3.17 and S3.18 it is evident that the bond is always longer if in the *trans*-position there is an atom with lower EN or with higher bond order (multiplicity). In a distorted coordination polyhedron, the longest bond always lies opposite to the shortest one, whereas bonds of intermediate lengths are *trans* to one another. A good illustration is the crystal structure of $SnWO_4$ which comprises WO_6 and SnO_6 octahedra linked through common vertices. Two Sn–O *trans*-bonds measure 2.18 and 2.82 Å, two W–O *trans*-bonds 1.80 and 2.14 Å, while other Sn–O distances equal 2.39 Å and W–O distances, 1.89 Å. In the similar structure of $HgWO_4$, the O–W–O dimensions are 1.733/2.197 and 1.953/1.953 Å, the O–Hg–O dimensions 2.044/2.743 and 2.633/2.633 Å [322]. This illustrates the directed character of atomic interactions in linear systems of covalent bonds, channeled through certain bonding orbitals, whereas *cis*-ligands exercise much smaller influence.

Available models of *trans*-effect describe it qualitatively correctly [198, 323–326]. One should keep in mind that the *trans*-effect can be masked by steric effects [314, 327] or π interactions [328]. Besides, as the bond polarity in a complex compound increases, the electrostatic interactions of the nearest ligands, i.e. those in *cis*-positions to one another, also increases. By present estimates, this *cis*-effect (not involving the central atom!) can become comparable with the *trans*-effect if bond ionicity exceeds 0.6. Thus, the structures of ionic substances are governed principally by the *cis*-effects, and those of mostly covalent transition-metal complexes, by the *trans*-effect. It has been proven experimentally that the successions of *trans*- and *cis*-effects mirror one another, so that the most active *trans*-ligands show the minimum *cis*-effect, and *vice versa* [329]. It is necessary to emphasize (as Chernyaev emphasized from the start) that the essence of the *trans*-effect is *ionization* of a bond, not its *weakening* in the sense of lower bond energy. In fact, a more polar bond can be stronger. Therefore, the chemical consequences of *trans*-effect depend on the solvent used: ionization of a bond labilizes it in a polar medium, but in non-polar organic solvents the succession of *trans*-effect can be altered or even reversed.

Averaged lengths of the M–X σ bonds (X=terminal atoms of hydrogen, halogens, chalcogens, CO, H_2O and NH_3) in coordination compounds are listed in Table S3.19, together with the distances for bridging (μ) atoms of oxygen and chalcogens. As one can see, the bridging metal-halogen bonds are, on average, 10 % longer than the terminal bonds, whereas the M=O or M=(Chalcogen) bonds are ca 10 % shorter than corresponding single bonds.

Appendix

Supplementary Table

Table S3.1 Bond distances (Å) in hydride and halide molecules AX_n

M/X	H	F	Cl	Br	I
Molecules AX_2					
Cu		1.700[a]	2.035[b]	2.22	
Be	1.334[c]	1.374	1.791	1.932	2.10
Mg	1.703[c]	1.746	2.179[e]	2.308[f]	2.52
Ca	2.04[B]	2.10	2.483[e]	2.592	2.822
Sr	2.177[B]	2.20	2.607[d]	2.748	2.990
Ba	2.274[B]	2.32	2.768	2.899[g]	3.130
Zn	1.524[c]	1.743[a]	2.064	2.194	2.389
Cd	1.683[c]	1.97	2.266	2.386	2.570
Hg	1.633[c]	1.96	2.240	2.374	2.558
Ti		1.73	2.30	2.46	2.66
Zr		1.93			
C	1.107[h]	1.303[h]	1.716[h]	1.74[h]	2.085[h]
Si	1.514[h]	1.590[h]	2.076[h]	2.227[h]	
Ge	1.591[h]	1.732[h]	2.169[h]	2.359[h]	2.540[h]
Sn		2.06	2.338[h]	2.501[h]	2.699[h]
Pb		2.036[h]	2.447[h]	2.597[h]	2.804[h]
V		1.76	2.172	2.43	2.62
P	1.434[i]	1.579[l]			
As	1.516				
Cr		1.796[a]	2.207	2.39	2.58
O	0.958	1.405	1.696[k]	1.838[k]	
S	1.336[l]	1.587[l]	2.015[l]		
Se	1.460[l]		2.157[l]		
Te	1.651[l]		2.329[l]		
Mn		1.812[a]	2.184[A]	2.328[A]	2.538
Fe	1.648[m]	1.770[a]	2.128[A]	2.272[A]	2.51
Co		1.756[a]	2.090[A]	2.223[A]	2.475
Ni		1.730[a]	2.056[A]	2.177[A]	
Molecules AX_3					
Au		1.906[s]			
B	1.187	1.313	1.742	1.893	2.112
Al		1.628	2.052[n]	2.210[n]	2.449[n]
Ga		1.716	2.092	2.243	2.458
In			2.273	2.462	2.641
Sc		1.808[o]	2.285[p]	2.44	2.62[q]
Y		2.04	2.422[e]	2.63	2.817
La		2.077[r]	2.589[r]	2.739[r]	2.961[r]
Ti		1.90	2.205	2.40	2.568
Zr		1.93	2.30	2.39	2.59
V		1.721[o]	2.10		
N	1.012[t]	1.365	1.759		

Table S3.1 (continued)

M/X	H	F	Cl	Br	I
P	1.415[t]	1.563[u]	2.040	2.216	2.43
As	1.511[B]	1.706	2.162	2.329	2.557
Sb	1.700[B]	1.876[v]	2.328	2.490	2.710[w]
Bi	1.778[y]	1.979[v]	2.423	2.63	2.791[w]
Cr		1.720[a]			
Mn		1.739[y]			
Fe		1.763[z]	2.145	2.245	2.50
Molecules AX_4					
C	1.085[α]	1.315[h]	1.767[h]	1.942[h]	2.166[h]
Si	1.480[α]	1.554[h]	2.019[h]	2.183[h]	2.43
Ge	1.517[β]	1.670[o]	2.108[o]	2.264[o]	2.507[o]
Sn	1.711[α]	1.97	2.281	2.423[γ]	2.651[γ]
Pb	1.741[δ]	2.08	2.369	2.50	2.74
Ti		1.754	2.170	2.339	2.546
Zr		1.886[ε]	2.319[ε]	2.452[ε]	2.65[ε]
Hf		1.909	2.320	2.450	2.662
V			2.138	2.276	
Nb			2.249[ε]	2.394[ε]	2.604[ε]
Cr		1.706	2.130		
Mo		1.828[Φ]	2.32[Φ]	2.39[Φ]	
W			2.248	2.40	
Th		2.124	2.567	2.72	
U		2.059	2.506	2.681	2.975
Molecules AX_5					
P		1.551[π]	2.065[π]		
As		1.678	2.151[λ]		
Sb			2.295[λ]		
V		1.718			
Nb		1.88	2.280	2.455	
Ta		1.86	2.285[μ]	2.441	2.66
Mo		1.815[Φ]	2.267		
W			2.261[μ]	2.40	
Re			2.248[μ]		
Molecules AF_6					
S	Se	Te	Mo	W	Tc
1.557	1.685[ν]	1.815[ν]	1.817[σ]	1.829[ν]	1.812[σ]
Re	Ru	Rh	Os	Ir	
1.834[σ]	1.818[σ]	1.824[σ]	1.827[σ]	1.833[σ]	
Pt	Ar	Kr	Xe		
1.849[σ]	1.842[ω]	1.868[ω]	2.004[ω]		

[a][3.1], [b][3.2], [c][3.3], [d][3.4], [e][3.5], [f][3.6], [g][3.7], [h][3.8], [i][3.9], [I][3.216], [k][3.10], [l][3.11], [m][3.12], [n][3.13], [o][3.14], [p][3.15], [q][3.16], [r][3.17], [s][3.18], [t][3.19], [u][3.20], [v][3.21], [w][3.22], [x][3.23], [y][3.24], [z][3.25], [α][3.26], [β][3.27], [γ])[3.28], [δ][3.29], [ε][3.30], [Φ][3.31], [λ][3.32], [μ][3.33], [ν][3.34], [π][3.35], [σ][3.36], [ω][3.37], [A][3.38], [B][3.39], [C]Structures (octahedral) of MX_6, d(M−X): UCl_6 2.460 Å [3.40], WCl_6 2.281 Å [3.41], WBr_6 2.454 Å [3.42]; Structures (pentagonal-bipyramidal) of AF_7: five $d_{equat} = 1.851$ Å and two $d_{axial} = 1.823$ Å in ReF_7 [3.43], $5d_{equat} = 1.849$ Å and $2d_{axial} = 1.795$ Å in IF_7 [3.44]

Table S3.2 Bond distances (Å) in oxide, sulfide, nitride and carbide molecules MX

M	X				M	X			
	O	S	N	C		O	S	N	C
Li	1.688[a]				V	1.580[m]	2.053[y]	1.566[u]	
Na	2.051[a]			2.232[b]	Nb	1.690[m]	2.16	1.662[u]	1.700[p]
K	2.167[a]	2.817[c]		2.528[b]	Ta	1.702[z]	2.105[y]	1.683[z]	
Rb	2.254[a]				N	1.151		1.098	1.172[α]
Cs	2.300[a]				As	1.624	2.017		1.680[α]
Cu	1.791[d]	2.055[c]			Sb	1.826		1.836[β]	
Ag	2.077[d]				Bi	1.934	2.319	2.262	
Au	1.849[d]				Cr	1.621[γ]	2.078[δ]	1.563[ε]	1.619
Be	1.331	1.742			Mo			1.636[u]	1.676[p]
Mg	1.749[f]	2.144[g]			W	1.672[λ]		1.667[μ]	1.747[ν]
Ca	1.822	2.320[c]			S	1.500[s]	1.889		
Sr	1.920	2.441[h]			Se	1.639			
Ba	1.940	2.507[i]			Mn	1.648[γ]	2.068[c]		
Zn	1.705[f]	2.046[c]			Re	1.640[π]		1.638[π]	
B	1.204[j]	1.609	1.274[k]		F	1.354[ρ]	1.601		
Al	1.618[l]	2.029[m]	1.686		Cl	1.569[σ]			
Sc	1.668[m]	2.139[c]	1.687[n]		Br	1.717[σ]			
Y	1.790[m]	2.28[m]	1.804[o]	2.050[p]	I	1.868[σ]			
La	1.826	2.352[q]			Fe	1.619[γ]	2.04[τ]	1.580[ν]	1.593η
C	1.138[r]	1.541[r]	1.172	1.242	Co	1.631[γ]	1.978[θ]		
Si	1.510[s]	1.929[s]		1.802[t]	Ni	1.631[γ]	1.962[φ]		1.627[χ]
Ge	1.647[r]	2.024[r]			Ru			1.571[ψ]	1.607[p]
Sn	1.832	2.209			Rh	1.717[ξ]			1.613[p]
Pb	1.922	2.287			Pd				1.712[p]
Ti	1.620[m]	2.082[c]	1.583[u]		Ir	1.772[ω]		1.607[ω]	1.609[ω]
Zr	1.714[v]	2.156[v]	1.697[u]	1.807[w]	Pt	1.727[ϕ]	2.040[ϕ]	1.682[ϕ]	1.679[ϕ]
Hf	1.725[x]	2.156[x]	1.69		Th	1.840[x]			

[a][3.45], [b][3.46], [c][3.47], [d][3.48], [e][3.49], [f][3.50], [g][3.51], [h][3.52], [i][3.53], [j][3.54], [k][3.55], [l][3.56], [m][3.57], [n][3.58], [o][3.59], [p][3.60], [q][3.61], [r][3.62], [s][3.63], [t][3.64], [u][3.65], [v][3.66], [w][3.67], [x][3.68], [y][3.69], [z][3.70], [α][3.71], [β][3.72], [γ][3.73], [δ][3.74], [ε][3.75], [λ][3.76], [μ][3.77], [ν][3.78], [π][3.79], [ρ][3.80], [σ][3.81], [τ][3.82], [3.83], [η][3.84], [θ][3.85], [φ][3.86], [χ][3.87], [ψ][3.88], [ξ][3.89], [ω][3.90], ϕ[3.91], Phosphides (Å): CP 1.562 Å [3.71], NP 1.491, PP 1.893, AsP 1.999, SbP 2.205, BiP 2.293 Å [3.92], CrP 2.117 Å [3.93], RhP 1.86 Å [3.94]

Table S3.3 Bond distances (Å) in selenide and telluride molecules MA

M	Se	Te	M	Se	Te
Li		2.490[a]	Ge	2.135[d]	2.340[d]
Cu	2.108[b]	2.349[b]	Sn	2.326	2.523
C	1.676		Pb	2.402	2.595
Si	2.058[c]	2.274[c]	Bi	2.477[e]	

[a][3.95], [b][3.96], [c][3.97], [d][3.98], [e][3.99]

Table S3.4 Bond distances (Å) in oxide and chalcogenide molecules

MX_2	O	S	Se	MO_2		M_2O	
C[a]	1.168	1.561	1.700	S	1.431	Li	1.606[e]
Si[a]	1.521	1.924		Se	1.612	Be	1.396[f]
Ge[a]	1.628	2.005	2.139	Rh	1.699[d]	Ga	1.822
Ti	1.704[b]			Ir	1.717[d]	In	2.017
Hf	1.176[c]			Pt	1.719[d]	Tl	2.102
Nb	1.713			Au	1.793[d]		

[a][3.100], [b][3.101], [c][3.102], [d][3.103], [e][3.104], [f][3.105]

Table S3.5 Bond distances (Å) in molecules in different aggregate states

Molecule	Gas	Liquid	Crystal	Molecule	Gas	Crystal
F_2	1.412	1.36[a]	1.49[b]	CI_4	2.166[k]	2.155[l]
Cl_2	1.988	1.95[a]	1.994[c]	SiF_4	1.554[k]	1.521[m]
Br_2	2.288[d]	2.29[a]	2.314[d]	$SiCl_4$	2.019[k]	2.008[n]
I_2	2.677[e]	2.701[e]	2.715[b]	GeF_4	1.673[k]	1.689[o]
KrF_2	1.889[f]		1.894[f]	$GeCl_4$	2.113[p]	2.096[p]
PCl_3	2.040[g]		2.034[g]	$GeBr_4$	2.272[k]	2.272[q]
PBr_3	2.216[g]	2.24[h]	2.213[g]	GeI_4	2.507	2.498[r]
PI_3	2.43[g]		2.463[g]	$SnCl_4$	2.280[s]	2.279[s]
AsF_5	1.678		1.688[i]	$PbCl_4$	2.373[t]	2.362[t]
$AsCl_3$	2.161[g]		2.167[g]	$TiCl_4$	2.170[u]	2.163[u]
$AsBr_3$	2.325		2.364[g]	$TiBr_4$	2.31[u]	2.29[u]
AsI_3	2.557[g]		2.597[g]	SF_6	1.557	1.556[v]
$SbCl_3$	2.328		2.36[g]	TeF_6	1.815	1.785[w]
$SbBr_3$	2.483		2.50	MoF_6	1.825	1.817[x]
SbI_3	2.710		2.87[g]	WF_6	1.829	1.826[x]
CF_4	1.323		1.314	ReF_6	1.829	1.823[x]
CCl_4	1.767	1.767	1.773	OsF_6	1.816	1.827[x]
CBr_4	1.935	1.913[j]	1.912	PtF_6	1.839	1.849[x]

[a][3.106], [b][3.107], [c][3.108], [d][3.109], [e][3.110], [f][3.111], [g][3.112], [h][3.113], [i][3.114], [j][3.115], [k][3.8], [l][3.116], [m][3.117], [n][3.118], [o][3.119], [p][3.120], [q]for liquid $GeBr_4$ [3.121], [r][3.122], [s][3.123], [t][3.124], [u][3.125], [v][3.126], [w][3.127], [x][3.36]

Table S3.6 Bridging and terminal bond distances M–X (Å) in the $(MX_n)_2$ molecules

MX_n	d_{bridg}	d_{term}	MX_n	d_{bridg}	d_{term}	MX_n	d_{bridg}	d_{term}
LiF	1.740	1.564[a]	$AuCl_3$[f]	2.355	2.236	$FeBr_3$	2.537	2.294
NaF	2.081	1.926[a]	BH_3[g]	1.339	1.196	UCl_5	2.70	2.43
NaCl	2.584	2.361[a]	AlH_3[h]	1.694	1.498	UBr_5	2.88	2.65
NaBr	2.740	2.502[a]	AlF_3[i]	1.72	1.60	$NbCl_5$	2.56	2.27
KF	2.347	2.172[a]	$AlCl_3$[i]	2.254	2.066	$NbBr_5$[p]	2.715	2.435
KCl	2.950	2.667[a]	$AlBr_3$[i]	2.417	2.223	NbI_5[q]	2.935	2.670
KBr	3.202	2.829[a]	AlI_3[j]	2.656	2.460	$TaCl_5$	2.56	2.28
RbF	2.448	2.270[a]	GaH_3[g]	1.723	1.550	TaI_5[q]	2.932	2.662
RbCl	3.008	2.787[a]	$GaCl_3$[j]	2.297	2.103	$SbCl_5$[r]	2.580	2.298
RbBr	3.181	2.945[a]	$GaBr_3$[j]	2.463	2.260	SbI_5[s]	3.079	2.796
CsF	2.696	2.345[a]	GaI_3[j]	2.682	2.492	BiF_5	2.11	1.90
CsCl	3.017	2.906[a]	$InCl_3$[k]	2.470	2.260	$BiCl_5$[s]	2.902	2.562

Table S3.6 (continued)

MX_n	d_{bridg}	d_{term}	MX_n	d_{bridg}	d_{term}	MX_n	d_{bridg}	d_{term}
CsBr	3.356	3.072[a]	InI_3[k]	2.80	2.614	$MoCl_5$[t]	2.541	2.265
CuCl[b]	2.166	2.051[a]	$ScCl_3$[l]	2.475	2.260	WCl_5	2.52	2.25
BeF_2[c]	1.553	1.375	ScI_3[m]	2.741	2.615	$SeBr_5$[u]	2.87	2.48
$BeCl_2$[c]	1.968	1.828	YCl_3[n]	2.657	2.438	$TeCl_5$[u]	2.85	2.44
MgF_2[c]	1.880	1.730	YI_3	3.023	2.806	$TeBr_5$[u]	2.96	2.61
$MgCl_2$[c]	2.362	2.188	$SeCl_3$[n]	2.836	2.420	RuF_5[v]	2.001	1.808
$MgBr_2$[d]	2.526	2.334	$SeBr_3$[n]	2.836	2.420	$RhBr_5$[w]	2.601	2.459
MnI_2	2.746	2.548	ICl_3[o]	2.70	2.39	OsF_5[v]	2.043	1.831
AuF_3[e]	2.033	1.876	$FeCl_3$[c]	2.329	2.129	CoO[x]	1.765	1.631

[a]In the isolated MX molecule (the dimer has no terminal M–X bonds), [b][3.128], [c][3.129], [d][3.6], [e][3.18], [f][3.130], [g][3.131], [h][3.132], [i][3.11], [j][3.133], [k][3.134], [l][3.5], [m][3.135], [n][3.136], [o][3.137], [p][3.138], [q][3.139], [r][3.32], [s][3.140], [t][3.141], [u][3.142], [v][3.143], [w][3.144], [x]for the bridging bond from [3.145], for the terminal bond from [3.73]

Table S3.7 Bond distances M–X (Å) in the homologous series of molecules

	d(B–F)[a]		d(C–F)[b]		d(C–F)[c]		d(C–Cl)[b]
BF_3	1.307	CF_4	1.319	FC(O)F	1.316	CF_4	1.767
BHF_2	1.311	CHF_3	1.333	FC(O)Cl	1.334	$CHCl_3$	1.758
BH_2F	1.321	CH_2F_2	1.357	FC(O)Br	1.326	CH_2Cl_2	1.767
		CH_3F	1.391	FC(O)I	1.343	CH_3Cl	1.787
				FC(O)H	1.346		
	d(Si–F)		d(P–F)[d]		d(Ge–F)[e]		d(Ge–Cl)[e]
SiF_4	1.556	PF_5	1.577	GeF_4	1.670	$GeCl_4$	2.112
$SiHF_3$	1.565	PF_4Cl	1.581	$MeGeF_3$	1.714	$MeGeCl_3$	2.132
SiH_2F_2	1.577	PF_3Cl_2	1.593	Me_2GeF_2	1.739	Me_2GeCl_2	2.143
SiH_3F	1.593	PF_2Cl_3	1.596			Me_3GeCl	2.173
		$PFCl_4$	1.597				
	d(P–F)		d(As–F)[f]		d(Ge–Br)[e]		d(Sn–Cl)[g]
PF_5	1.577	AsF_3	1.706	$GeBr_4$	2.272	$SnCl_4$	2.281
$PMeF_4$	1.612	$MeAsF_2$	1.734	$MeGeBr_3$	2.276	$MeSnCl_3$	2.304
PMe_2F_3	1.643	Me_2AsF	1.758	Me_2GeBr_2	2.303	Me_2SnCl_2	2.327
PMe_3F_2	1.685			Me_3GeBr	2.325	Me_3SnCl	2.351

[a][3.146], [b][3.147], [c][3.148], [d][3.149], [e][3.150], [f][3.151], [g][3.152]

Table S3.8 Bond angles X–A–X (ω°) in hydride and halide molecules [3.153]

AX_3	ω	AX_3	ω	AX_2	ω
NH_3	107.3	PI_3	102.0	OH_2	104.5
PH_3	93.8	PBr_3	101.1	SH_2	92.1
AsH_3	91.8	PCl_3	100.3	SeH_2	90.6
SbH_3	91.7	PF_3	97.8	TeH_2	90.3
NF_3	102.2	AsI_3	100.2	OF_2	103.1
PF_3	97.8	$AsBr_3$	99.8	SF_2	98.2
AsF_3	96.1	$AsCl_3$	98.6	SeF_2	94
SbF_3	87.3	AsF_3	96.1	OCl_2	111.2
NCl_3	107.1	SbI_3	99.1	SCl_2	102.8
PCl_3	100.3	$SbBr_3$	98.2	$SeCl_2$	99.6
$AsCl_3$	98.6	$SbCl_3$	97.2	$TeCl_2$	97.0
$SbCl_3$	97.2	SbF_3	87.3		

Table S3.9 Correlation between the MX_2 and $(MX_2)_2$ molecular symmetry and the structural type of the MX_2 crystals

X	MX_2	Be	Mg	Ca	Sr	Ba
F	monomer	linear	linear	bent	bent	bent
	dimer	D_{2h}	D_{2h}	C_{3v}	C_{3v}	C_{3v}
	structural type	SiO_2	TiO_2	CaF_2	CaF_2	CaF_2
Cl	monomer	linear	linear	linear	bent	bent
	dimer	D_{2h}	D_{2h}	D_{2h}	C_{3v}	C_{3v}
	structural type	SiS_2	$CdCl_2$	$CaCl_2$	CaF_2	CaF_2
Br	monomer	linear	linear	linear	q-linear	bent
	dimer	D_{2h}	D_{2h}	D_{2h}	C_{3v}	C_{3v}
	structural type	SiS_2	CdI_2	$CaCl_2$	$SrBr_2$	$PbCl_2$
I	monomer	linear	linear	linear	linear	bent
	dimer	D_{2h}	D_{2h}	D_{2h}	C_{3v}	C_{3v}
	structural type	SiS_2	CdI_2	CdI_2	SrI_2	$PbCl_2$

Table S3.10 Bond lengths A–C (d) in $A(CH_3)_n$ molecules in the gas phase

A	n	d, Å	A	n	d, Å	A	n	d, Å	A	n	d, Å
H	1	1.092	I	1	2.132	Al	3	1.957	Sn	4	2.144
Li	1	1.959	Be	2	1.698	Ga	3	1.967	Pb	4	2.238
Na	1	2.299	Mg	2	2.126	In	3	2.161	Te	4	2.171[c]
K	1	2.634	Zn	2	1.930	Tl	3	2.206	Cr	4	2.038
Cu	1	1.881[a]	Cd	2	2.112	N	3	1.454	Ta	5	2.138
Mg	1	2.105	Hg	2	2.083	P	3	1.847	As	5	2.01
Ca	1	2.348[b]	O	2	1.416	As	3	1.968	Sb	5	2.173[f]
Sr	1	2.487[b]	S	2	1.807	Sb	3	2.163	Bi	5	2.28[g]
Ba	1	2.564	Se	2	1.943	Bi	3	2.267	Mo	6	2.150[d]
F	1	1.389	Te	2	2.142	C	4	1.537	W	6	2.144[e]
Cl	1	1.785	Mn	2	2.01	Si	4	1.875	Te	6	2.193
Br	1	1.933	B	3	1.578	Ge	4	1.945	Re	6	2.128[e]

[a][3.154], [b][3.155], [c][3.156], [d][3.157], [e][3.158], [f][3.159], [g][3.160]

Table S3.11 Bond lengths X–C (d) in $X(C{\equiv}CH)_n$ molecules in the gas phase [3.161–3.165]

X	n	d, Å	X	n	d, Å	X	n	d, Å
Li	1	1.888	F	1	1.27	CH_3[b]	1	1.459
Na	1	2.221	Cl	1	1.63	CF_3[b]	1	1.438
K	1	2.540	Br	1	1.80	Ge[c]	1	1.896
Cu	1	1.818[a]	I	1	1.99	Sn	4	2.067

[a][3.166], [b][3.167], [c]in $H_3GeC{\equiv}CH$

Table S3.12 Distances M–C (d, Å) in cyclopentadienyl complexes (metallocenes)

M	*d*	M	*d*	M	*d*	M	*d*
Li	2.26	B	1.68	Sn	2.68	Tc	2.29
Na	2.68	Al	2.39	Pb	2.79	Re	2.30
K	3.03	Ga	2.40	V	2.33	Fe	2.12
Rb	3.14	In	2.59	Nb	2.41	Co	2.10
Cs	3.35	Tl	2.66	Ta	2.44	Ni	2.10
Cu	2.27	Sc	2.53	As	2.47	Ru	2.22
Be	1.93	Y	2.65	Sb	2.40	Rh	2.20
Mg	2.34	La	2.84	Bi	2.69	Pd	2.34
Ca	2.61	Ti	2.39	Cr	2.21	Os	2.17
Sr	2.75	Zr	2.51	Mo	2.36	Ir	2.16
Ba	2.90	Hf	2.48	W	2.33	Pt	2.33
Zn	2.28	Si	2.42	Mn	2.15	Th	2.82
		Ge	2.53			U	2.75

Table S3.13 Bond distances (Å) in metal-carbonyl molecules, (Crystal structures (source: CSD), except where specified)

Complex	Type[a]	d(M–C)	d(C–O)	Complex	Type[a]	d(M–C)	d(C–O)
free CO			1.128	$Fe(CO)_5$[c]	TBp	1.803 eq	1.117
$[Ti(CO)_6]^{2-}$	Oc	2.038	1.168			1.811 ax	1.133
$[Zr(CO)_6]^{2-}$	Oc	2.210	1.162	$[Fe(CO)_6]^{2+}$	Oc	1.910	1.106
$[Hf(CO)_6]^{2-}$	Oc	2.178	1.163	$[Os(CO)_6]^{2+}$	Oc	2.027	1.102
$V(CO)_6$	Oc	2.001	1.128	$[Co(CO)_4]^-$	T	1.778	1.156
$[V(CO)_6]^-$	Oc	1.913	1.169	$[Co(CO)_4]^{2-}$	T	1.73	1.16
$[Nb(CO)_6]^-$	Oc	2.089	1.160	$[Rh(CO)_4]^+$	SQ	1.951	1.118
$[Ta(CO)_6]^-$	Oc	2.083	1.149	$[Ir(CO)_6]^{3+}$	O	2.029	1.090
$Cr(CO)_6$[d]	Oc	1.916	1.171	NiCO[e]		1.641	1.193
$Cr(CO)_6$	Oc	1.918	1.142	$Ni(CO)_3$[e]	T	1.839	1.121
$Mo(CO)_6$[d]	Oc	2.063	1.145	$Ni(CO)_4$	T	1.815	1.128
$Mo(CO)_6$	Oc	2.057	1.128	PdCO[c]		1.844	1.138
$W(CO)_6$[d]	Oc	2.059	1.148	$[Pd(CO)_4]^{2+}$	SQ	1.992	1.106
$W(CO)_6$	Oc	2.049	1.143	PtCO[c]		1.760	1.148
$[Mn(CO)_5]^-$	TBp	1.798 eq	1.156	$[Pd(CO)_4]^{2+}$	SQ	1.982	1.110
		1.820 ax	1.147	$[Cu(CO)_2]^+$	L	1.916	1.112
$[Mn(CO)_5]^-$	SqP	1.810 bs	1.159	$[Cu(CO)_4]^+$	T	1.965	1.110
		1.794 ap	1.155	$[Ag(CO)_2]^+$	L	2.14	1.08
$[Re(CO)_6]^+$	Oc	2.01	1.13	$[Au(CO)_2]^+$	L	1.972	1.11
FeCO[c]		1.727	1.160	$[Hg(CO)_2]^+$	L	2.083	1.104
$[Fe(CO)_4]^{2-}$	T	1.742	1.169	$B_{12}H_{10}(CO)_2$	B	1.543	1.119
$Fe(CO)_5$[c]	TBp	1.842 eq	1.149	Me_2HC^+–CO	T	1.458	1.101
		1.810 ax	1.142				

[a]Metal coordination: Oc – octahedral, TBp – trigonal bipyramidal (ax – axial, eq – equatorial ligand), SqP – square pyramidal (ap – apical, bs – basal ligand), T – tetrahedral, SQ – square-planar, L – linear, C –borane cage atom with 5 cage neighbors + CO, [b][3.168], [c]gas phase from [3.169], [d]GED [3.170], [e][3.171]

Table S3.14 Bond distances (Å) in carbonyl-halides X–MCO, from gas-phase MW spectra [3.169]

M	X = F		X = Cl		X = Br	
	d(M–X)	d(C–O)	d(M–X)	d(C–O)	d(M–X)	d(C–O)
Cu	1.765	1.131	1.795	1.129	1.803	1.128
Ag	1.965	1.126	2.013	1.124	2.028	1.124
Au	1.847	1.134	1.883	1.132	1.892	1.132

Table S3.15 Lengths of homonuclear bonds A–A (Å) of different orders

A	Bond order, n			A	Bond order, n			A	Bond order, n		
	1	2	3		1	2	3		1	2	3
B	1.76	1.52	1.36	As	2.42	2.28	2.10	Cr	2.64		2.22
Al	2.64	2.50		Sb	2.86	2.66	2.34	Mo	2.76		2.24
Ga	2.52	2.39	2.32	Bi	3.08	2.83	2.66	W	2.70	2.50	2.32
C	1.54	1.34	1.20	Nb	2.84	2.72		Cl	1.98	1.78	
Si	2.35	2.14	1.97	Ta	3.00	2.74		Br	2.28	2.08	
Ge	2.44	2.24	2.13	O	1.42	1.21		I	2.66	2.46	
Sn	2.81	2.62		S	2.06	1.89	1.74	Re	2.62	2.39	2.26
N	1.46	1.25	1.10	Se	2.37	2.17		Ru	2.56	2.40	
P	2.22	2.00	1.86	Te	2.83	2.56		Os	2.82		2.17
								Pt	2.59	2.20	

Quadruple bonds:

A	Mo	W	Re
$n = 4$	2.12	2.22	2.22

Table S3.16 Bond lengths (Å) in octahedral structures of the M_nAX_6 coordination compounds

M_nAX_6	d_{A-X}	M_nAX_6	d_{A-X}	M_nAX_6	d_{A-X}	M_nAX_6	d_{A-X}
K_3ReH_6	1.707[a]	Li_2MoF_6	1.936[l]	$LiPdF_6$	1.899[l]	Rb_2TeCl_6	2.538[r]
Na_3RhH_6	1.67[b]	K_2ReF_6	1.953[l]	$KOsF_6$	1.882[n]	Na_2UCl_6	2.641[v]
Na_3IrH_6	1.68[b]	Li_2RuF_6	1.92[l]	$LiIrF_6$	1.875[m]	K_2ReCl_6	2.354[w]
K_2PtH_6	1.640[c]	Li_2RhF_6	1.903[m]	$KPtF_6$	1.886[n]	K_2PdCl_6	2.309[x]
Sr_2MgH_6	2.03[d]	Cs_2PdF_6	1.89[f]	K_3IrCl_6	2.357[o]	K_2OsCl_6	2.334[w]
Mg_2FeH_6	1.56[b]	K_2OsF_6	1.927[n]	Cs_3CrCl_6	2.324[p]	K_2IrCl_6	2.371[y]
Mg_2RuH_6	1.67[b]	K_2IrF_6	1.928[n]	Am_3BiCl_6	2.712[q]	Cs_2IrCl_6	2.332[z]
Mg_2OsH_6	1.68[b]	K_2PtF_6	1.921[n]	K_2SnCl_6	2.404[q]	K_2PtCl_6	2.316[w]
Li_3ScF_6	2.018[e]	$KAuF_6$	1.882[l]	Am_2SnCl_6	2.418[r]	Rb_3IrBr_6	2.508[y]
Na_3CuF_6	1.89[f]	$LiNbF_6$	1.863[l]	Am_2SnCl_6	2.446[s]	Am_2SnBr_6	2.622[s]
Cs_2CuF_6	1.757[f]	$KTaF_6$	1.860[l]	$(NEt_4)_2ZrCl_6$	2.463[t]	K_2TeBr_6	2.694[r]
Am_2SiF_6	1.688[g]	$KAsF_6$	1.719[l]	$(NEt_4)_2HfCl_6$	2.456[t]	Am_2TeBr_6	2.701[r]
Na_2SnF_6	1.958[h]	$LiSbF_6$	1.88[l]	$RNbCl_6$	2.35[u]	Rb_2TeBr_6	2.701[r]
K_2TiF_6	1.860[i]	$KReF_6$	1.863[l]	$RTaCl_6$	2.35[u]	$(PPh_4)_2UBr_6$	2.664[α]
Tl_2TiF_6	1.91[k]	$LiRuF_6$	1.851[l]	$RWCl_6$	2.32[u]	$(H_3O)_2TcBr_6$	2.506[β]
K_2HfF_6	1.991[l]	$LiRhF_6$	1.854[l]	Am_2TeCl_6	2.538[r]		

Note: Am = NH_4, R = Alkyl, [a][3.172], [b][3.173], [c][3.174], [d][3.175], [e][3.176], [f][3.177], [g][3.178], [h][3.179], [i][3.180], [k][3.181], [l][3.182], [m][3.183], [n][3.184], [o][3.185], [p][3.186], [q][3.187], [r][3.188], [s][3.189], [t][3.190], [u][3.191], [v][3.192], [w][3.193], [x][3.194], [y][3.195], [z][3.196], [α][3.197], [β][3.198], [γ][3.199], [δ][3.200], [ε][3.201]

Table S3.17 *Trans*-effect in platinum(II) complexes: the effect of EN of *trans*-atom [3.202–3.208]

Complex	d(Pt—Cl), Å	*trans*-atom	EN of *trans*-atom
$[Pt(Acac)_2Cl]^-$	2.276	O	3.4
$[Pt(PEt_3)_2Cl_2]$-*trans*	2.294	Cl	3.1
$[PtCl_2(C_2H_6OS)Py]$	2.316	S	2.6
$[Pt(PMe_3)_2Cl_2]$-*cis*	2.376	P	2.2
$[Pt(PPh_2Et)_2HCl]$-*trans*	2.422	P	2.2
$[Pt(PPhMe_2)_2Cl(SiPh_3)]$	2.465	Si	1.9

Table S3.18 *Trans*-effect in complexes of transition metals: the effect of multiple bonds [3.197, 3.209–3.211]

Compounds	X—M—Y	d(M—Y),Å
K_2NbOF_5	O=Nb—F	2.06
	F—Nb—F	1.84
K_2MoOCl_5	O=Mo—Cl	2.63
	Cl—Mo—Cl	2.40
$(NH_4)_2MoOBr_5$	O=Mo—Br	2.83
	Br—Mo—Br	2.55
K_2ReOCl_5	O=Re—Cl	2.47
	Cl—Re—Cl	2.39
K_2OsNCl_5	N≡Os—Cl	2.605
	Cl—Os—Cl	2.36

Table S3.19 Mean bond lengths (Å) in coordination compounds [3.194], [3.210], [3.212–3.215]

M	H	F	Cl	Br	I	O	S	Se	Te	H_2O	NH_3	CO
Cu	1.6	1.9	2.3	2.4	2.6	2.0	2.3			2.0	1.0	1.8
Ag			2.3	2.45	2.6		2.35	2.6		2.1	2.1	2.0
Au			2.3	2.4	2.55		2.25			2.1	2.0	
Zn	1.6		2.25	2.4	2.6	2.1	2.3			2.1	2.0	
Cd			2.5	2.6	2.75	2.3	2.45	2.65		2.3	2.2	
Hg			2.4	2.55	2.7	2.2	2.4	2.6	2.7	2.3	2.3	
Ti	2.0	1.9	2.3	2.45	2.65	1.85	2.4		2.7	2.1		2.0
Zr	2.0	2.05	2.4	2.6	2.8	2.0	2.5	2.7	2.85		2.3	2.2
Hf	2.0	2.0	2.45		2.85	2.0	2.5		2.9		2.3	
V	1.7	1.8	2.3	2.5	2.65	1.9	2.3			2.1	2.2	1.95
Nb	1.7	1.85	2.35	2.6	2.8	1.95	2.45	2.65		2.2		2.05
Ta	1.8	1.85	2.35	2.6	2.8	1.9	2.5	2.65		2.1		2.0
Cr		1.8	2.35	2.5	2.7	1.8	2.4	2.5	2.8	2.0	2.1	1.9
Mo	1.75	1.95	2.4	2.6	2.8	1.9	2.4	2.6	2.9	2.2	2.2	2.0
W	1.75	1.95	2.4	2.6	2.8	1.9	2.4	2.6	2.9	2.2	2.2	2.0
Mn	1.6	1.9	2.35	2.55	2.75	1.9	2.3	2.45	2.7	2.2	2.2	1.8
Tc	1.7		2.3	2.6		1.95	2.4	2.55		2.1	2.1	1.9
Re	1.7	1.95	2.4	2.5	2.7	2.0	2.35	2.6	2.65	2.2	2.2	1.95
Fe	1.6	1.9	2.35	2.4	2.6	1.9	2.25	2.55	2.6	2.2	1.8	
Co	155	1.9	2.35	2.4	2.6	1.9	2.25	2.5	2.55	2.1	1.8	
Ni	1.5	1.9	2.35	2.4	2.6	1.9	2.2	2.45	2.5	2.1	1.8	
Ru	1.7	1.9	2.4	2.5	2.7		2.2			2.1	2.1	1.9
Rh	1.7	1.9	2.4	2.5	2.7		2.3			2.1	2.1	1.85
Pd	1.6	1.9	2.3	2.45		1.9	2.35	2.4		2.2	2.1	1.9
Os	1.7	1.9	2.35	2.55	2.8	1.9	2.4			2.2	2.1	1.9
Ir	1.7	1.9	2.35	2.5	2.75	1.9	2.35	2.2	2.2	2.2	2.0	2.0
Pt	1.65	1.9	2.3	2.45	2.7	2.0	2.3	2.3	2.3			2.05

Supplementary References

3.1 Spiridonov VP, Gershikov AG, Lyutsarev VS (1990) J Mol Struct 221:79
3.2 Beattie IR, Brown JB, Crozet P et al (1997) Inorg Chem 36:3207
3.3 Shayesteh A, Yu S, Bernath PF (2005) Chem Eur J 11:4709
3.4 Varga Z, Lanza G, Minichino C, Hargittai M (2006) Chem Eur J 12:8345
3.5 Reffy B, Marsden CJ, Hargittai M (2003) J Phys Chem A107:1840
3.6 Reffy B, Kolonits M, Hargittai M (2005) J Phys Chem A109:8379
3.7 Hargittai M, Kolonits M, Schulz G (2001) J Mol Struct 241:567–568
3.8 Hargittai M, Schulz G, Hargittai I (2001) Russ Chem Bul 50:1903
3.9 Hirao T, Hayakashi S-I, Yamamoto S, Saito S (1998) J Mol Spectr 187:153
3.10 Müller HSP, Cohen EA (1997) J Chem Phys 106:8344
3.11 Gillespie RJ, Hargittai I (1991) The VSEPR model of molecular geometry. Allyn and Bacon, Boston
3.12 Körsgen H, Urban W, Brown JM (1999) J Chem Phys 110:3861
3.13 Aarset K, Hayakashi S-I, Yamamoto S et al (1999) J Phys Chem A103:1644
3.14 Giricheva NI Girichev GV, Shlykov SA et al (1995) J Mol Struct 344:127
3.15 Haaland A, Martinsen K-G, Shorokhov DI et al (1998) J Chem Soc Dalton Trans 2787
3.16 Ezhov Yu S, K omarov S A, Sevast'yanov VG (1995) Russ J Phys Chem 69:1910
3.17 Kovacs A, Konings RJ (2004) J Phys Chem Refer Data 33:377; Giricheva NI, Shlykov SA, Girichev GV, Galanin IE (2006) J Struct Chem 47:850
3.18 Reffy B, Kolonits M, Schulz A, KlapötkeThM (2000) J Am Chem Soc 122:3127
3.19 do Varella MTN, Bettega MHF, da Silva AJR (1999) J Chem Phys 110:2452
3.20 Naib H, Sari-Zizi N, Bürger H et al (1993) J Mol Spectr 159:249
3.21 Molnar J, Kolonits M, Hargittai M (1997) J Mol Struct 441:413–414
3.22 Molnar J, Kolonits M, Hargittai M et al (1996) Inorg Chem 35:7639
3.23 Jerzembeck W, Bürger H, Constantin FKL et al (2004) J Mol Spectr 226:24
3.24 Hargittai M, Reffy B, Kolonits M et al (1997) J Am Chem Soc 119:9042
3.25 Hargittai M, Kolonits M, Tremmel J et al (1990) Struct Chem 1:75
3.26 Bettega MHF, Natalense APP, Lima MAP, Ferreira LG (1995) J Chem Phys 103:10566
3.27 Pierre G, Boudon VM, Kadmi EB et al (2002) J Mol Spectr 216:408
3.28 Reuter H, Pawlak R (2001) Z Krist 216:34
3.29 Wang X, Andrews L (2003) J Am Chem Soc 125:6581
3.30 Giricheva NI, Girichev GV (1999) Russ J Phys Chem 73:401
3.31 Girichev GV, Giricheva NI, Krasnova OG (2001) J Mol Struct 203:567–568
3.32 Haupt S, Seppelt K (2002) Z anorg allgem Chem 628:729
3.33 Faegri K, Haaland A, Martinsen K-G et al (1997) J Chem Soc Dalton Trans 1013
3.34 Richardson AD, Hedberg K, Lucier GM (2000) Inorg Chem 39:2787
3.35 Mache C, Boughdiri S, Barthelat J-C (1986) Inorg Chem 25:2828
3.36 Drews Th, Supel J, Hagenbach A, Seppelt K (2006) Inorg Chem 45:3782
3.37 Pilme J, Robinson EA, Gillespie RJ (2006) Inorg Chem 45:6198
3.38 Hodges PJ, Brown JM, Ashworth SH (2000) J Mol Spectr 237:205; Hargittai M, Subbotina NYu, Kolonits M, Gershikov AG (1991) J Chem Phys 94:7278
3.39 Aldridge S, Downs AJ (2001) Chem Rev 101:3305
3.40 Ezhov YuS, Komarov SA, Sevast'yanov VG, Bazhanov VI (1993) J Struct Chem 34:473
3.41 Haaland A, Martinsen K-G, Shlykov S (1992) Acta Chem Scand 46:1208
3.42 Willing W, Mueller U (1987) Acta Cryst C43:1425
3.43 Vogt T, Fitch A, Cockcroft JK (1994) Science 263:1265
3.44 Marx R, Mahjoub AR, Seppelt K, Ibberson RM (1994) J Chem Phys 101:585
3.45 Hirota E (1995) Bull Chem Soc. Jpn 68:1

3.46 Sheridan PM, Xin J, Ziurys LM et al (2002) J Chem Phys 116:5544
3.47 Thompsen JM, Ziurys LM (2001) Chem Phys Lett 344:75
3.48 Okabayashi T, Koto F, Tsukamoto K et al (2005) Chem Phys Lett 403:223
3.49 O'Brien LC, Oberlink AE, Roos BO (2006) J Phys Chem A110:11954
3.50 Bauschlicher CW, Partridge H (2001) Chem Phys Lett 342:441
3.51 Walker KA, Gerry MCL (1997) J Mol Spectr 182:178
3.52 Halfen DT, Apponi AJ, Thompsen JM, Ziuris LM (2001) J Chem Phys 115:11131
3.53 Morbi Z, Bernath PF (1995) J Mol Spectr 171:210
3.54 Osiac M, Pöpske J, Davies PB (2001) Chem Phys Lett 344:92
3.55 Ram RS, Bernath PF (1996) J Mol Spectr 180:414
3.56 Launila O, Jonsson J (1994) J Mol Struct 168:1
3.57 Launila O, Jonsson J (1994) J Mol Struct 168:483
3.58 Ram RS, Bernath PF (1992) J Chem Phys 96:6344
3.59 Jakubek ZJ, Nakhate SG, Simard B (2003) J Mol Spectr 219:145
3.60 DaBell RS, Meyer RG, Morse MD (2001) J Chem Phys 114:2938; Balfour WJ (1999) J Mol Spectr 198:393
3.61 Winkel RJ, Davis SP, Abrams MC (1996) Appl Opt 35:2874
3.62 Harrison JF (2006) J Phys Chem A110:10848
3.63 Sanz ME, McCarthy MC, Thaddeus P (2003) J Chem Phys 119:11715
3.64 Deo MN, Kawaguchi K (2004) J Mol Spectr 228:76
3.65 Peter SL, Dunn TM (1989) J Chem Phys 90:5333
3.66 Beaton SA, Gerry MCL (1999) J Chem Phys 110:10715
3.67 Rixon SJ, Chowdhury PK, Merer AJ (2004) J Mol Spectr 228:554
3.68 Coocke SA, Gerry MCL (2002) J Mol Spectr 216:122; Merritt JM, Bondybey VE, Heaven MC (2009) J Chem Phys 130:144503
3.69 Ran Q, Tam WS, Cheung AS-C (2003) J Mol Spectr 220:87
3.70 Ram RS, Bernath PF (2003) J Mol Spectr 221:7
3.71 Yang J, Clouthier D J (2011) J Chem Phys 135:054309
3.72 Cooke AA, Gerry MCL (2004) PCCP 6:4579
3.73 Adam AG, Azuma Y, Barry JA et al (1987) J Chem Phys 86:5231
3.74 Shi Q, Ran Q, Tam WS et al (2001) Chem Phys Lett 339:154
3.75 Balfour WJ, Qian C, Zhou C (1997) J Chem Phys 106:4383
3.76 Ram RS, Levin J, Li G et al (2001) Chem Phys Lett 343:437
3.77 Ram RS, Bernath PF (1994) J Opt Soc Amer B11:225
3.78 Sickafoose SM, Smith AW, Morse MD (2002) J Chem Phys 116:993
3.79 Ram RS, Bernath PF, Balfour WJ et al (1994) J Mol Spectr 168:350
3.80 Miller CE, Drouin BJ (2001) J Mol Spectr 205:312
3.81 Peterson KA, Shepler BC, Figgen D, Stoll H (2006) J Phys Chem A110:13877
3.82 Zhai H-J, Kiran B, Wang LS (2003) J Phys Chem A107:2821
3.83 Aiuchi K, Shibuya K (2000) J Mol Spectr 204:235
3.84 Allen MD, Ziurys LM (1997) J Chem Phys 106:3494
3.85 Flory MA, McLamarrah SK, Ziurys LM (2005) J Chem Phys 123:164312
3.86 Yamamoto T, Tanimoto M, Okabayashi T (2007) PCCP 9:3744
3.87 Borin AC (2001) Chem Phys 274:99
3.88 Ram RS, Bernath PF (2002) J Mol Spectr 213:170
3.89 Aldener M, Hansson A, Petterson A et al (2002) J Mol Spectr 216:131
3.90 Ram RS, Bernath PF (1999) J Mol Spectr 193:363
3.91 Cooke SA, Gerry MCL (2004) J Chem Phys 121:3486
3.92 Leung F, Cooke SA, Gerry MCL (2006) J Mol Spectr 238:36
3.93 Adam AG, Slaney ME, Tokaryk DW, Balfour WJ (2007) Chem Phys Lett 450:25
3.94 Li R, and Balfour WJ (2004) J Phys Chem A108:8145
3.95 Setzer KD, Fink EH, Alekseyev AB et al (2001) J Mol Spectr 206:181

3.96 Okabayashi T, Koto F, Tsukamoto K et al (2005) Chem Phys Lett 403:223
3.97 Giuliano BM, Bizzochi L, Grabow J-U (2008) J Mol Spectr 251:261
3.98 Giuliano BM, Bizzochi L, Sanchez R et al (2011) J Chem Phys 135:084303
3.99 Setzer KD, Breidohr R, Mainecke F, Fink EH (2009) J Mol Spectr 258:50
3.100 Deakyne CA, Li L, Zheng W, Xu D (2002) J Chem Thermodyn 34:185
3.101 Wang H, Steimle TC, Apetrei C, Maier JP (2009) PCCP 11:2649
3.102 Lessari A, Suenram RD, Brugh D (2002) J Chem Phys 117:9651
3.103 Gong Y, Zhou M (2009) J Phys Chem A113:4990
3.104 Bellert D, Breckenridge WH (2001) J Chem Phys 114:2871
3.105 Merritt JM, Bondybey VE, Heaven MC (2009) J Phys Chem A113:13300
3.106 Misawa M (1989) J Chem Phys 91: 2575; Andreani C, Cilloco F, Osae E (1986) Mol Phys 57:931
3.107 Donohue J (1982) The structure of the elements. RE Krieger Publ Co, Malabar Fl
3.108 Powell BM, Heal KM, Torrie BH (1984) Mol Phys 53:929
3.109 Filipponi A, Ottaviano L, Passacantando M et al (1993) Phys Rev E48:4575
3.110 Buontempo U, Filipponi A, Postorino P, Zaccari R (1998) J Chem Phys 108:4131
3.111 Lehmann JF, Dixon DA, Schrobilgen GJ (2001) Inorg Chem 40:3002
3.112 Galy J, Enjalbert R (1982) J Solid State Chem 44:1
3.113 Misawa M, Fukunaga T, Suzuki K (1990) J Chem Phys 92:5486
3.114 Köhler J, Simon A, Hoppe R (1989) Z anorg allgem Chem 575:55
3.115 Bako I, Dore JC, Huxley DW (1997) Chem Phys 216:119
3.116 Pohl S (1982) Z Krist 159:211
3.117 Yang OB, Andersson S (1987) Acta Cryst B43:1
3.118 Zakharov LN, Antipin MYu, Struchkov YuT et al (1986) Sov Phys Cryst 31:99
3.119 Yang OB, Andersson S (1987) Acta Cryst B43:1
3.120 Merz K, Driess M (2002) Acta Cryst C58:i101
3.121 Ludwig KF (1987) J Chem Phys 87:613
3.122 Walz L, Thiery D, Peters E-M et al (1993) Z Krist 208:207
3.123 Reuter H, Pawlak R (2000) Z anorg allgem Chem 626:925
3.124 Maley IJ, Parsons S, Pulham CR (2002) Acta Cryst C58:i79
3.125 Troyanov SI, Snigereva EM (2000) Russ J Inorg Chem 45:580
3.126 Kiefte H, Penney R, Clouter MJ (1988) J Chem Phys 88:5846
3.127 Bartell LS, Powell BM (1992) Mol Phys 75:689
3.128 Hargittai M (2003) Chem Eur J 9:327
3.129 Hargittai M (2000) Chem Rev 100:2233
3.130 Hargittai M, Schulz A, Reffy B, Kolonits M (2001) J Am Chem Soc 123:1449
3.131 Pulham CR, Downs A, Goode M et al (1991) J Am Chem Soc 113:5149; Mitzel NW (2003) Angew Chem Int Ed 42:3856
3.132 Wehmschulte RJ, Power PP (1994) Inorg Chem 33:5611
3.133 Troyanov SI, Krahl T, Kemnitz E (2004) Z Krist 219:88
3.134 Girichev GV, Giricheva NI et al (1992) J Struct Chem 33:362 838
3.135 Ezhov YuS, Komarov SA, Sevast'yanov VG (1997) J Struct Chem 38:403
3.136 Hauge S, Janickis V, Marøy K (1998) Acta Chem Scand 52:435
3.137 Boswijk KH, Wiebenga EH (1954) Acta Cryst 7:417
3.138 Hönle W, Furuseth S, von Schnering HG (1990) Z Naturforsch 45b:952
3.139 Krebs B, Sinram D (1980) Z Naturforsch 35b:12
3.140 Breunig H, Denker M, Schulz RE, Lork E (1998) Z anorg allgem Chem 624:81
3.141 Beck J, Wolf F (1997) Acta Cryst B53:895
3.142 Hauge S, Marøy K (1998) Acta Chem Scand 52:445
3.143 Page EM, Rice D, Almond M et al (1993) Inorg Chem 32:4311
3.144 Boyd SE, Field LD, Hambley TW (1994) Acta Cryst C50:1019

3.145 Danset D, Manaron L (2005) Phys Chem Chem Phys 7:583
3.146 Takeo H, Sugie M, Matsumura C (1993) J Mol Spectr 158:201
3.147 Villamanan RM, Chen WD, Wlodarczak G et al (1995) J Mol Spectr 171:223
3.148 Chiappero MS, Argüello GA, Garcia P et al (2004) Chem Eur J 10:917
3.149 Macho C, Minkwitz R, Rohmann J et al (1986) Inorg Chem 25:2828
3.150 Aarset K, Page EM (2004) J Phys Chem A108:5474
3.151 Downs A, Greene TM, McGrady GS et al (1996) Inorg Chem 35:6952
3.152 Fujii H, Kimura M (1971) Bull Chem Soc Japan 44:2643
3.153 Gillespie RJ, Robinson EA (1996) Angew Chem Int Ed 35:495
3.154 Grotjahn DB, Halfen DWT, Ziurys LM, Cooksy A (2004) J Am Chem Soc 126:12621
3.155 Sheridan PM, Dick MJ, Wang J-G, Bernath PF (2005) J Phys Chem A109:10547
3.156 Liang B, Andrews L, Li J, Bursten BE (2004) Inorg Chem 43:882
3.157 Roesler B, Seppelt K (2000) Angew Chem Int Ed 39:1259
3.158 Kleinhenz V, Pfennig S, Seppelt K (1998) Chem Eur J 4:1687
3.159 Haaland A, Hammel A, Rypdal K et al (1992) Angew Chem Int Ed 31:1464
3.160 Wallenhauer S, Seppelt K (1995) Inorg Chem 34:116
3.161 Mastryukov VS, Simonsen SH (1996) Adv Molec Struct Res 2:163
3.162 Grotjahn DB, Schade C, El-Nahasa A et al (1998) Angew Chem Int Ed 37:2678
3.163 Green TM, Downs AJ, Pulham CR et al (1998) Organometallics 17:5287
3.164 Apponi AJ, Brewster MA, Ziurys LM (1998) Chem Phys Lett 298:161
3.165 Grotjahn DB, Pesch TC, Brewster MA, Ziurys LM (2000) J Am Chem Soc 122:4735
3.166 Sun M, Halfen DT, Min J et al (2010) J Chem Phys 133:174301
3.167 Blanco S, Sanz ME, Mata S et al (2003) Chem Phys Lett 375:355
3.168 McClelland BW, Robiette AG, Hedberg L, Hedberg K (2001) Inorg Chem 40:1358
3.169 Walker NR, Hui JK-H, Gerry MCL (2002) J Phys Chem A106:5803
3.170 Arnesen SP, Seip HM (1996) Acta Chem Scand 20:2711
3.171 Martinez A, Morse MD (2006) J Chem Phys 124:124316
3.172 Bronger W, Auffermann G, Schilder H (1998) Z anorg allgem Chem 624:497
3.173 Bronger W (1991) Angew Chem Int Ed 30:759
3.174 Bronger W (1994) Angew Chem Int Ed 33:1112
3.175 Bertheville B, Yvon K (1995) J Alloys Compd 228:197
3.176 Tyagi AK, Köhler J, Balog P, Weber J (2005) J Solid State Chem 178:2620
3.177 Müller BG (1987) Angew Chem 99:1120
3.178 Schlemper EO, Hamilton WC, Rush JJ (1966) J Chem Phys 44:2499
3.179 Benner G, Hoppe R (1990) J Fluor Chem 48:219
3.180 Göbel O (2000) Acta Cryst C56:521
3.181 Chang J-H, Köhler J (2000) Mater Res Bull 35:25
3.182 Graudejus O, Wilkinson AP, Chacón LC, Bartlett N (2000) Inorg Chem 39:2794
3.183 Fitz H, Müller BG, Graudejus O, Bartlett N (2002) Z anorg allgem Chem 628:133
3.184 Brisden A, Holloway J, Hope E et al (1992) J Chem Soc Dalton Trans 139
3.185 Coll R, Fergusson J, Penfold B et al (1987) Austral J Chem 40:2115
3.186 Sassmannshausen M, Lutz HD (2001) Z anorg allgem Chem 627:1071
3.187 Belkyyal I, Mohklisse R, Tanouti B et al (1997) Eur J Solid State Inorg Chem 34:1085
3.188 Abriel W, du Bois A (1989) Z Naturforsch 44b:1187
3.189 Reutov OA, Aslanov LA, Petrosyan VS (1988) J Struct Chem 29:918
3.190 Ruhlandt-Senge K, Bacher A-D, Müller U (1990) Acta Cryst C46:1925
3.191 Beck J, Schlörb T (1999) Z Krist 214:780
3.192 Bendall P, Fitch A, Fender B (1983) J Appl Cryst 16:164
3.193 Takazawa H, Ohba S, SaitoY (1990) Acta Cryst B46:166
3.194 Takazawa H, Ohba S, SaitoY (1988) Acta Cryst B44:580
3.195 Rankin DWH, Penfold B, Fergusson J (1983) Austral J Chem 36:871

3.196 Coll RK, Fergusson SE, Penfold BR et al (1990) Inorg Chim Acta 177:107
3.197 Bohrer R, Conradi E, Müller U (1988) Z anorg allgem Chem 558:119
3.198 Spitzin VI, Kryutchkov SV, Grigoriev MS, Kuzina AF (1988) Z anorg allgem Chem 563:136
3.199 Zipp A (1988) Coord Chem Rev 84:47
3.200 Maletka K, Fischer P, Murasik A, Szczepanik W (1992) J Appl Cryst 25:1
3.201 Thiele G, Mrozek C, Kammerer D, Wittmann K (1983) Z Naturforsch 38b:905
3.202 Mason R, Robertson GB, Pauling PJ (1969) J Chem Soc A 485
3.203 Messmer GG, Amma EL (1966) Inorg Chem 5:1775
3.204 Hartley FR (1973) Chem Soc Rev 2:163
3.205 Messmer GG, Amma EL, Ibers JA (1967) Inorg Chem 6:725
3.206 Eisenberg R, Ibers JA (1965) Inorg Chem 4:773
3.207 Belsky VK, Konovalov VE, Kukushkin VYu (1991) Acta Cryst C47:292
3.208 Kapoor P, Lövqvist K, Oskarsson Å (1995) Acta Cryst C51:611
3.209 Schüpp B, Heines P, Savin A, Keller H-L (2000) Inorg Chem 39 732
3.210 Manojlovic-Muir L, Muir K (1974) Inorg Chim Acta 10:47
3.211 Poraij-Koshits MA, Atovmyan LO (1974) Crystal chemistry and stereochemistry of oxide compounds of molybdenum. Nauka, Moscow (in Russian)
3.212 Thiele G, Weigl W, Wochner H (1986) Z anorg allgem Chem 539:141
3.213 Aurivillius K, Stolhandske C (1980) Z Krist 153:121
3.214 Bircsak Z, Harrison WTA (1998) Acta Cryst C54:1554
3.215 El-Bali B, Bolte M, Boukhari A et al (1999) Acta Cryst C55:701
3.216 Saito S, Endo Y, Hirota R (1986) J Chem Phys 85:1778

References

1. Huber KP, Herzberg G (1979) Molecular spectra and molecular structure. 4, Constants of diatomic molecules. Van Nostrand, New York
2. Lide DR (ed) (2007–2008) Handbook of chemistry and physics, 88nd edn. CRC Press, New York
3. Harmony MD, Laurie VW, Kuczkowski RL et al (1979) Molecular structures of gas-phase polyatomic molecules determined by spectroscopic methods. J Phys Chem Ref Data 8:619–721
4. Demaison J, Dubrulle A, Hüttner W, Tiemann E (1982) Molecular constants: diamagnetic molecules. In: Hellwege K-H, Hellwege AM (eds) Landolt-Börnstein, vol II/14a. Springer, Berlin
5. Brown JM, Demaison J, Dubrulle A et al (1983) Molecular Constants. In: Hellwege K-H, Hellwege AM (eds) Landolt-Börnstein, vol II/14b. Springer, Berlin
6. (a) Girichev GV, Giricheva NI, Titov VA, Chusova TP (1992) Structural, vibrational and energetic characteristics of gallium and indium halide molecules. J Struct Chem 33: 362–372; (b) Girichev GV, Giricheva NI, Krasnova OG et al (1992) Electron-diffraction investigation of the molecular structure of CoF_3. J Struct Chem 33:838–843
7. Spiridonov VP, Gershikov AG, Lyutsarev VS (1990) Electron diffraction analysis of XY_2 and XY_3 molecules with large-amplitude motion. J Mol Struct 221:79–94
8. Ezhov YuS (1992) Force constants and characteristics of the structure of trihalides. Russ J Phys Chem 66:748–751
9. Ezhov YuS (1993) Systems of force constants, catiolis coupling constants, and structure peculiarities of XY_4 tetrahalides. Russ J Phys Chem 67:901–904
10. Ezhov YuS (1995) Variations of molecular constants in metal halide series XY_n and estimates for bismuth trihalide constants. Russ J Phys Chem 69:1805–1809
11. Giricheva NI, Lapshin SB, Girichev GV (1996) Structural, vibrational and energy characteristics of halide molecules of group II-V elements. J Struct Chem 37:733–746

12. Hargittai M (2000) Molecular structure of metal halides. Chem Rev 100:2233–2301
13. Spiridonov VP, Vogt N, Vogt J (2001) Determination of molecular structure in terms of potential energy functions from gas-phase electron diffraction supplemented by other experimental and computational data. Struct Chem 12:349–376
14. Callomon JH, Hirota E, Kuchitsu K et al (1976) Structure data of free polyatomic molecules. In: Hellwege K-H, Hellwege AM (eds) Landolt-Börnstein, vol II/7. Springer, Berlin 1976
15. Callomon JH, Hirota E, Iijima T et al (1987) Structure data of free polyatomic molecules. In: Hellwege K-H, Hellwege AM (eds) Landolt-Börnstein, vol II/15. Springer, Berlin 1976
16. Hargittai I, Hargittai M (eds) (1988) Stereochemical applications of gas-phase electron diffraction. Part B: Structural information for selected classes of compounds. VCH, New York
17. Vilkov LV, Mastryukov VC, Sadova NI (1978) Determination of the geometrical structure of free molecules. Khimiya, Leningrad (in Russian)
18. Hargittai I (1985) The structure of volatile sulfur compounds. Reidel, Dordrecht
19. Hargittai I, Hargittai M (1987) Gas-solid molecular structure differences. Phys Chem Minerals 14:413–425
20. Allen FH (1998) The development, status and scientific impact of crystallographic databases. Acta Cryst A54:758–771
21. Allen FH (2002) The Cambridge Structural Database: a quarter of a million crystal structures and rising. Acta Cryst B58:380–388
22. Allen FH, Taylor R (2004) Research applications of the Cambrideg Structural Database. Chem Soc Rev 33:463–475
23. Berman HM (2008) The Protein Data Bank: a historical perspective. Acta Cryst A64:88–95
24. Berman HM, Olson WK, Beveridge DL et al (1992) The Nucleic-Acid Database. Biophys J 63:751–759
25. Allen FH, Kennard O, Watson DG et al (1987) Tables of bond lengths determined by X-ray and neutron diffraction: bond lengths in organic compounds. J Chem Soc Perkin Trans 2 Supplement S1-S19
26. Orpen A. Brammer L. Allen FH et al (1989) Tables of bond lengths determined by X-ray and neutron diffraction: organometallic compounds and co-ordination complexes of the d- and f-block metals. J Chem Soc Dalton Trans Supplement S1-S83
27. Bürgi H-B, Dunitz JD (eds) (1994) Structure correlations, vol 2. VCH, Weinheim
28. Wilson AJC (ed) (1992) International tables for crystallography, vol C. Kluwer, Dordrecht
29. Pley M, Wickleder MS (2005) Two crystalline modifications of RuO_4. J Solid State Chem 178:3206–3209
30. Krebs B, Hasse K (1976) Refinements of crystal structures of $KTcO_4$, $KReO_4$ and OsO_4. Acta Cryst B32:1334–1337
31. Trueblood KN (1992) Diffraction studies of molecular motion in crystals. In: Domenicano A, Hargittai I (eds) Accurate molecular structures. Oxford University Press, Oxford
32. Coppens P (1997) X-Ray charge density and chemical bonding. Oxford Univ Press, Oxford
33. Koritsanszky TS, Coppens P (2001) Chemical applications of X-ray charge-density analysis. Chem Rev 101:1583–1627
34. Teo BK (1986) EXAFS: basic principles of data analysis. Springer, Berlin
35. Zubavichus YaV, Slovokhotov YuL (2001) X-ray synchrotron radiation in physicochemical studies. Russ Chem Rev 70:373–404
36. Lombardi E, Jansen L (1986) Model analysis of ground-state dissociation energies and equilibrium separations in alkali-metal diatomic compounds. Phys Rev A33:2907–2912
37. Jules JL, Lombardi JR (2003) Transition metal dimer internuclear distances from measured force constants. J Phys Chem A107:1268–1273
38. Merritt JM, Kaledin AL, Bondybey VE, Heaven MC (2008) The ionization energy of Be_2. Phys Chem Chem Phys 10:4006–4013
39. Allard O, Pashov A, Knöckel H, Tiemann E (2002) Ground-state potential of the Ca dimer from Fourier-transform spectroscopy. Phys Rev A66:042503

40. Strojecki M, Ruszczak M, Łukomsky M, Kaperski J (2007) Is Cd_2 truly a van der Waals molecule? Analysis of rotational profiles. Chem Phys 340:171–180
41. Brazier CR, Carrick PG (1992) Observation of several new electronic transitions of the B_2 molecule. J Chem Phys 96:8684–8690
42. Fu Z, Lemire GW, Bishea GA, Morse MD (1990) Spectroscopy and electronic structure of jet-cooled Al_2. J Chem Phys 93:8420–8441
43. Tan X, Dagdigian PJ (2003) Electronic spectrum of the gallium dimer. J Phys Chem A107:2642–2649
44. Han Y-K, Hirao K (2000) On the ground-state spectroscopic constants of Tl_2. J Chem Phys 112:9353–9355
45. Fang L, Chen X, Shen X et al (2000) Spectroscopy of yttrium dimers in argon matrices. Low Temp Phys 26:752–755
46. Ho J, Polak ML, Lineberger WC (1992) Photoelectron spectroscopy of group IV heavy metal dimers: $Sn^-{}_2$, $Pb^-{}_2$, and $SnPb^-$. J Chem Phys 96:144–154
47. Doverstal M, Lindgren B, Sassenberg U et al (1992) The band system of jet-cooled Ti_2. J Chem Phys 97:7087–7092
48. Doverstal M, Karlsson L, Lindgren B, Sassenberg U (1998) Resonant two-photon ionization spectroscopy studies of jet-cooled Zr_2. J Phys B31:795–804
49. Barrow RF, Taher F, D'incan J et al (1996) Electronic states of Bi_2. Mol Phys 87:725–733
50. Kraus D, Lorentz M, Bondybey VE (2001) On the dimers of the VIB group: a new NIR electronic state of Mo_2. Phys Chem Comm 4:44–48
51. Airola MB, Morse MD (2002) Rotationally resolved spectroscopy of Pt_2. J Chem Phys 116:1313–1317
52. Tsukiyama K, Kasuya T (1992) Vacuum ultraviolet laser spectroscopy of Xe_2. J Mol Spectr 151:312–321
53. Donald KJ, Mulder WH, von Szentpaly L (2004) Valence-state atoms in molecules: influence of polarization and bond-charge on spectroscopic constants of diatomic molecules. J Phys Chem A108:595–606
54. Seto JY, Morbi Z, Harron FC et al (1999) Vibration-rotation emission spectra and combined isotopomer analyses for the coinage metal hydrides: CuH & CuD, AgH & AgD, and AuH & AuD. J Chem Phys 110:11756–11767
55. Reynard LM, Evans CJ, Gerry MCL (2001) The pure rotational spectrum of AuI. J Mol Spectr 205:344–346
56. Shayesteh A, Bernath PF (2011) Rotational analysis and deperturbation of the emission spectra of MgH. J Chem Phys 135:094308
57. Ram RS, Tereszchuk K, Gordon E et al (2011) Fourier transform emission spectroscopy of CaH and CaD. J Mol Spectr 266:86–91
58. Allen MD, Ziurys LM (1997) Millimeter-wave spectroscopy of FeF: rotational analysis and bonding study. J Chem Phys 106:3494–3503
59. Flory MA, McLamarrah SK, Ziurys LM (2006) The pure rotational spectrum of ZnF. J Chem Phys 125:194304
60. Törring T, Ernst W, Kändler J (1989) Energies and electric dipole moments of the low lying electronic states of the alkaline earth monohalides from an electrostatic polarization model. J Chem Phys 90:4927–4932
61. Dickinson CS, Coxon JA (2003) Deperturbation analysis of the $A^2\Pi$-$B^2\Sigma^+$ interaction of SrBr. J Mol Spectr 221:269–278
62. Tenenbaum ED, Flory MA, Pulliam RL, Ziurys LM (2007) The pure rotational spectrum of ZnCl: variations in zinc halide bonding. J Mol Spectr 244:153–159
63. Balasubramanian K (1989) Spectroscopic properties and potential energy curves for heavy p-block diatomic hydrides, halides, and chalconides. Chem Rev 89:1801–1840
64. Urban R-D, Jones H (1992) The ground-state infrared spectra of aluminium monodeuteride. Chem Phys Lett 190:609–613
65. Ogilvie J, Uehara H, Horiai K (1995) Vibration-rotational spectra of GaF and molecular properties of diatomic fluorides of elements in group 13. J Chem Soc Faraday Trans 91:3007–3013

66. Hargittai M, Varga Z (2007) Molecular constants of aluminum monohalides: caveats for computations of simple inorganic molecules. J Phys Chem A111:6–8
67. Singh VB (2005) Spectroscopic studies of diatomic gallium halides. J Phys Chem Ref Data 34:23–37
68. Mishra SK, Yadav RKS, Singh VB, Rai SB (2004) Spectroscopic studies of diatomic indium halides. J Phys Chem Ref Data 33:453–470
69. Ram RS, Bernath PF (1994) High-resolution Fourier-transform emission spectroscopy of YH. J Chem Phys 101:9283–9288
70. Ram RS, Bernath PF (1996) Fourier transform emission spectroscopy of new infrared systems of LaH and LaD. J Chem Phys 104:6444–6451
71. Ram RS, Bernath PF (1996) Fourier transform emission spectroscopy of ScH and ScD. J Chem Phys 105:2668–2674
72. Xia ZH, Xia Y, Chan M-C, Cheung ASC (2011) Laser spectroscopy of ScI. J Mol Spectr 268:3–6
73. Simard B, James AM, Hackett PA (1992) Molecular beam Stark spectroscopy of yttrium monochloride. J Chem Phys 96:2565–2572
74. Wannous G, Effantin C, Bernard A et al (1999) Laser-excited fluorescence spectra of yttrium monoiodide. J Mol Spectr 198:10–17
75. Rubinoff DS, Evans CJ, Gerry MCL (2003) The pure rotational spectra of the lanthanum monohalides, LaF, LaCl, LaBr, LaI. J Mol Spectr 218:169–179
76. Towle JP, Brown JM (1993) The infrared spectrum of the GeH radical. Mol Phys 78:249–261
77. Tanaka K, Honjou H, Tsuchiya MJ, Tanaka T (2008) Microwave spectrum of GeCl radical. J Mol Spectr 251:369–373
78. Setzer KD, Borkowska-Burnecka J, Ziurys LM (2008) High-resolution Fourier-transform study of the fine structure transitions of PbH and PbD. J Mol Spectr 252:176–184
79. Sheridan PM, McLamrrah SK, Ziurys LM (2003) The pure rotational spectrum of TiF. J Chem Phys 119:9496–9503
80. Ram RS, Bernath PF (2004) Infrared emission spectroscopy of TiCl. J Mol Spectr 227:43–49
81. Ram RS, Adam AG, Sha W et al (2001) The electronic structure of ZrCl. J Chem Phys 114:3977–3987
82. Ram RS, Bernath PF (1994) Fourier-transform emission spectroscopy of HFH and HFD. J Chem Phys 101:74–79
83. Hensel K, Hughes R, Brown J (1995) IR spectrum of the AsH radical, recorded by laser magnetic resonance. J Chem Soc Faraday Trans 91:2999–3004
84. Fink EH, Setzer KD, Ramsay DA et al (1996) High-resolution study of the fine-structure transition of BiF. J Mol Spectr 178:143–156
85. Kobayashi K, Saito S (1998) The microwave spectrum of the NF radical in the second electronically excited state. J Chem Phys 108:6606–6610
86. Sakamaki T, Okabayashi T, Tanimoto M (1998) Microwave spectroscopy of the NBr radical. J Chem Phys 109:7169–7165
87. Shestakov O, Gielen R, Setzer KD, Fink EH (1998) Gas phase LIF study of the $b^1\Sigma^+$ $(b0^+)$ $\ll X^3\Sigma-(X_10^+, X_21)$ transition of NI. J Mol Spectr 192:139–147
88. Setzer KD, Beutel M, Fink EH (2003) High-resolution study of the $b^1\Sigma^+$ $(b0^+)\rightarrow X^3\Sigma-(X_10^+)$ transition of PI. J Mol Spectr 221:19–22
89. Cooke SA, Gerry MCL (2005) Born-Oppenheimer breakdown effects and hyperfine structure in the rotational spectra of SbF and SbCl. J Mol Spectr 234:195–203
90. Ram RS, Bernath PF, Davis SP (2002) Infrared emission spectroscopy of VF. J Chem Phys 116:7035–7039
91. Ram RS, Bernath PF, Davis SP (2001) Fourier transform infrared emission spectroscopy of VCl. J Chem Phys 114:4457–4460
92. Gillet DA, Towle JP, Islam M, Brown JM (1994) The infrared spectrum of isotopomers of the TeH radical. J Mol Spectr 163:459–482
93. Miller CE, Drouin BJ (2001) The potential energy surfaces of FO. J Mol Spectr 205:312–318

94. Peterson KA, Shepler BC, Figgen D, Stoll H (2006) On the spectroscopic and thermochemical properties of ClO, BrO, IO, and their anions. J Phys Chem A110:13877–13833
95. Miller CE, Cohen EA (2001) Rotational spectroscopy of IO. J Chem Phys 115:6459–6470
96. Ziebarth K, Setzer KD, Fink EH (1995) High-resolution study of the fine-structure transitions of ^{130}TeF and ^{130}Te^{35}Cl. J Mol Spectr 173:488–498
97. Ram RS, Bernath PF (1995) High-resolution Fourier transform emission spectroscopy of CrD. J Mol Spectr 172:91–101
98. Bencheikh M, Koivisto R, Launila O, Flament JP (1997) The low-lying electronic states of CrF and CrCl. J Chem Phys 106:6231–6239
99. Launila O, Simard B, James AM (1993) Spectroscopy of MnF: rotational analysis in the near-ultraviolet region. J Mol Spectr 159:161–174
100. Dai DG, Balasubramanian K (1993) Spectroscopic properties and potential energy curves for 30 electronic states of ReH. J Mol Spectr 158:455–467
101. Launila O, James AM, Simard B (1994) Molecular beam laser spectroscopy of ReF. J Mol Spectr 164:559–569
102. Balfour WJ, Brown JM, Wallace L (2004) Electronic spectra of iron monohydride in the infrared near. J Chem Phys 121:7735–7742
103. Wang H, Zhuang X, Steimle TC (2009) The permanent electric dipole moments of cobalt monofluoride, CoF, and monohydride, CoH. J Chem Phys 131:114315
104. Ram RS, Gordon I, Hirao T et al (2007) Fourier transform emission spectroscopy of CoCl. J Mol Spectr 243:69–77
105. Tam WS, Leung JW-H, Hu S-M, Cheung AS-C (2003) Laser spectroscopy of NiI. J Chem Phys 119:12245–12250
106. Steimle TC, Virgo WL, Ma T (2006) The permanent electric dipole moment and hyperfine interaction in ruthenium monoflouride (RuF). J Chem Phys 124:024309
107. Zhuang X, Steimle TC, Linton C (2010) The electric dipole moment of iridium monofluoride, IrF. J Chem Phys 133:164310
108. Handler KG, Harris RA, O'Brien LC, O'Brien JJ (2011) Intracavity laser absorption spectroscopy of platinum fluoride, PtF. J Mol Spectr 265:39–46
109. Whiteham CJ, Ozeki H, Saito S (2000) Microwave spectra of CuOD and AgOD: molecular structure and harmonic force field of CuOH and AgOH. J Chem Phys 112:641–646
110. Lakin NM, Varberg TD, Brown JM (1997) The detection of lines in the microwave spectrum of indium hydroxide, InOH, and its isotopomers. J Mol Spectr 183:34–41
111. Janczyk A, Walter SK, Ziurys LM (2005) Examining the transition metal hydrosulfides: the pure rotational spectrum of CuSH. Chem Phys Lett 401:211–216
112. Martinez A, Morse MD (2011) Spectroscopy of diatomic ZrF and ZrCl: 760–555 nm. J Chem Phys 135:024308
113. Kokkin DL, Reilly NJ, McCarthy MC, Stanton JF (2011) Experimental and theoretical investigation of the electronic transition of CuSH. J Mol Spectr 265:23–27
114. Pashov A, Docenko O, Tamanis M et al (2005) Potentials for modeling cold collisions between Na (3 S) and Rb (5 S) atoms. Phys Rev A72:062505
115. Bednarska V, Jackowska I, Jastrzebski W, Kowalczyk P (1998) The molecular constants and potential energy curve of the ground state in KLi. J Mol Spectr 189:244–248
116. Staanum P, Pashov A, Knöckel H, Tiemann E (2007) $X^1\Sigma^+$ and $a^3\Sigma^+$ states of LiCs studied by Fourier-transform spectroscopy. Phys Rev A75:042513
117. Krou-Adohi A, Giraud-Cotton S (1998) The ground state of NaK revisited. J Mol Spectr 190:171–188
118. Ferber R, Klincare I, Nikolayeva O et al (2008) The ground electronic state of KCs studied by Fourier transform spectroscopy. J Chem Phys 128:244–316
119. Brock LR, Knight AM, Reddic JE et al (1997) Photoionization spectroscopy of ionic metal dimers: LiCu and LiAg. J Chem Phys 106:6268–6278
120. Yeh CS, Robbins DL, Pilgrim JS, Duncan MA (1993) Photoionization electronic spectroscopy of AgK. Chem Phys Lett 206:509–514
121. Aldridge S, Downs AJ (2001) Hydrides of the main-group metals. Chem Rev 101:3305–3366

122. Müller HSP (2001) The rotational spectrum of chlorine trifluoride, ClF_3. Phys Chem Chem Phys 3:1570–1575
123. Andrews L, Wang X (2002) Infrared spectrum of the novel electron-deficient BH_4 radical in solid neon. J Am Chem Soc 124:7280–7281
124. Signorell R, Palm H, Merkt F (1997) Structure of the ammonium radical from a rotationally resolved photoelectron spectrum. J Chem Phys 106:6523–6533
125. Müller HSP, Miller CE, Cohen EA (1996) Dibromine monoxide, Br_2O, and bromine dioxide, OBrO: spectroscopic properties, molecular structures, and harmonic force fields. Angew Chem Int Ed 35:2129–2131
126. Pople JA, Curtiss LA (1987) Ionization energies and proton affinities of AH_n species ($A = C$ to F and Si to Cl); heats of formation of their cations. J Phys Chem 91:155–162
127. Antonov IO, Barker BJ, Bondybey VE, Heaven MC (2010) Spectroscopic characterization of Be_2^+ and the ionization energy of Be_2. J Chem Phys 133:074309
128. Calvi RMD, Andrews DH, Lineberger WC (2007) Negative ion photoelectron spectroscopy of copper hydrides. Chem Phys Lett 442:12–16
129. Okabayashi T, Koto F, Tsukamoto K et al (2005) Pure rotational spectrum of gold monoxide (AuO). Chem Phys Lett 403:223–227
130. Ichino T, Gianola AJ, Andrews DH, Lineberger WC (2004) Photoelectron spectroscopy of AuO^- and AuS^-. J Phys Chem A108:11307–11313
131. Kim JH, Li X, Wang L-S et al (2001) Vibrationally resolved photoelectron spectroscopy of MgO^- and ZnO^- and the low-lying electronic states of MgO, MgO^-, and ZnO. J Phys Chem A105:5709–5718
132. Fancher CA, de Clercq HL, Bowen KH (2002) Photoelectron spectroscopy of PbS^-. Chem Phys Lett 366:197–199
133. Gong Y, Zhou M (2009) Infrared spectra of transition-metal dioxide anions: MO^{2-} ($M = Rh$, Ir, Pt, Au) in solid argon. J Phys Chem A113:4990–4995
134. Beattie IR, Spicer MD, Young NA (1994) Interatomic distances for some first row transition element dichlorides isolated in cryogenic matrices. J Chem Phys 100:8700–8705
135. Kornath A, Zoermer A, Ludwig R (1999) Raman spectroscopic investigation of matrix isolated Rb_2, Rb_3, Cs_2, and Cs_3. Inorg Chem 38:4696–4699
136. Grotjahn DB, Sheridan PM, Al Jihad I, Ziurys LM (2001) First synthesis and structural determination of a monomeric, unsolvated lithium amide, $LiNH_2$. J Am Chem Soc 123:5489–5494
137. Nikitin IV (2002) Oxygen compounds of halogens X_2O_2. Russ Chem Rev 71:85–98
138. Pernice H, Berkei M, Henkel G et al (2004) Bis(fluoroformyl)trioxide, FC(O)OOOC(O)F. Angew Chem Int Ed 43:2843–2846
139. Savariault JM, Lehmann MS (1980) Experimental determination of the deformation electron density in hydrogen peroxide. J Am Chem Soc 102:1298–1303
140. Kraka E, He Y, Cremer D (2001) Quantum chemical descriptions of FOOF: the unsolved problem of predicting its equilibrium geometry. J Phys Chem A105:3269–3276
141. Tanaka T, Morino Y (1970) Coriolis interaction and anharmonic potential function of ozone from the microwave spectra in the excited vibrational states. J Mol Spectr 33:538–551
142. Marx R, Ibberson RM (2001) Powder diffraction study on solid ozone. Solid State Sciences 3:195–202
143. Kasuya T, Okabayashi T, Watanabe S et al (1998) Microwave spectroscopy of BrBO. J Mol Spectr 191:374–380
144. Thorwirth S, McCarthy MC, Gottlieb CA et al (2005) Rotational spectroscopy and equilibrium structures of S_3 and S_4. J Chem Phys 123:054326
145. McCarthy MC, Thorwirth S, Gottlieb CA, Thaddeus P (2004) The rotational spectrum and geometrical structure of thiozone, S_3. J Am Chem Soc 126:4096–4097
146. Pauling L (1931) The nature of the chemical bond: application of results obtained from the quantum mechanics and from a theory of paramagnetic susceptibility to the structure of molecules. J Am Chem Soc 53:1367–1400
147. Slater JC (1931) Directed valence in polyatomic molecules. Phys Rev 37:481–489

148. Sidgwick NV, Powell HE (1940) Stereochemical types and valency groups. Proc Roy Soc A176:153–180
149. Gillespie RJ, Hargittai I (1991) The VSEPR model of molecular geometry. Allyn and Bacon, Boston
150. Gillespie RJ, Nyholm RS (1957) Inorganic stereochemistry. Quarterly Rev 11:339–380
151. Gillespie RJ (1972) Molecular geometry. Van Nostrand Reinhold, London
152. Gillespie RJ, Robinson EA (1996) Electron domains and the VSEPR model of molecular geometry. Angew Chem Int Ed 35:495–514
153. Pilme J, Robinson EA, Gillespie RJ (2006) A topological study of the geometry of AF_6E molecules: Weak and inactive lone pairs. Inorg Chem 45:6198–6204
154. Kaupp M (2001) Non-VSEPR structures and bonding in d^0 systems. Angew Chem Int Ed 40:3535–3565
155. Bytheway I, Gillespie RJ, Tang T-H et al (1995) Core distortions and geometries of the difluorides and dihydrides of Ca, Sr, and Ba. Inorg Chem 34:2407–2414
156. Gillespie RJ, Bytheway I, Tang T-H, Bader RWF (1996) Geometry of the fluorides, oxofluorides, hydrides, and methanides of vanadium(V), chromium(VI), and molybdenum(VI): Understanding the geometry of non-VSEPR molecules in terms of core distortion. Inorg Chem 35:3954–3963
157. Donald KJ, Hoffmann R (2006) Solid memory: structural preferences in group 2 dihalide monomers, dimers, and solids. J Am Chem Soc 128:11236–11249
158. Gillespie RJ, Robinson EA, Heard GL (1998) Bonding and geometry of OCF_3^-, ONF_3, and related molecules in terms of the ligand close packing model. Inorg Chem 37:6884–6889
159. Heard GL, Gillespie RJ, Rankin DWH (2000) Ligand close packing and the geometries of $A(XY)_4$ and some related molecules. J Mol Struct 520:237–248
160. Robinson EA, Gillespie RJ (2003) Ligand close packing and the geometry of the fluorides of the nonmetals of periods 3, 4, and 5. Inorg Chem 42:3865–3872
161. Li X, Wang LS, Boldyrev AI, Simons J (1999) Tetracoordinated planar carbon in the Al_4C^- anion. J Am Chem Soc 121:6033–6038
162. Boldyrev AI, Simons J, Li X, Wang L-S (1999) The electronic structure and chemical bonding of hypermetallic Al_5C by ab initio calculations and anion photoelectron spectroscopy. J Chem Phys 111:4993–4998
163. Wang LS, Boldyrev AI, Li X, Simons J (2000) Experimental observation of pentaatomic tetracoordinate planar carbon-containing molecules. J Am Chem Soc 122:7681–7687
164. Kuznetsov AE, Boldyrev AI, Li X, Wang LS (2001) On the aromaticity of square planar Ga_4^{2-} and In_4^{2-} in gaseous $NaGa_4^-$ and $NaIn_4^-$ clusters. J Am Chem Soc 123:8825–8831
165. Zhai HJ, Wang LS, Alexandrova AN, Boldyrev AI (2002) Electronic structure and chemical bonding of B_5^- and B_5 by photoelectron spectroscopy and ab initio calculations. J Chem Phys 117:7917–7924
166. Zhai HJ, Wang LS, Kuznetsov AE, Boldyrev AI (2002) Probing the electronic structure and aromaticity of pentapnictogen cluster anions Pn_5^- (Pn = P, As, Sb, and Bi) using photoelectron spectroscopy and ab initio calculations. J Phys Chem A106:5600–5606
167. Zhai HJ, Wang LS, Alexandrova AN et al (2003) Photoelectron spectroscopy and *ab initio* study of B_3^- and B_4^- anions and their neutrals. J Phys Chem A107:9319–9328
168. Alexandrova AN, Boldyrev AI, Zhai HJ et al (2003) Structure and bonding in B_6^- and B_6: planarity and antiaromaticity. J Phys Chem A107:1359–1369
169. Kiran B, Li X, Zhai H-J et al (2004) $[SiAu_4]$: aurosilane. Angew Chem Int Ed 43:2125–2129
170. Pulliam RL, Sun M, Flory MA, Ziurys LM (2009) The sub-millimeter and Fourier transform microwave spectrum of HZnCl. J Mol Spectr 257:128–132
171. Flory MA, Apponi AJ, Zack LN, Ziurys LM (2010) Activation of methane by zinc: gas-phase synthesis, structure, and bonding of $HZnCH_3$. J Am Chem Soc 132:17186–17192
172. Lattanzi V, Thorwirth S, Halfen DT et al (2010) Bonding in the heavy analogue of hydrogen cyanide: the curious case of bridged HPSi. Angew Chem Int Ed 49:5661–5664
173. Batsanov SS (1998) Calculation of van der Waals radii of atoms from bond distances. J Molec Struct Theochem 468:151–159

174. Mastryukov VS, Simonsen SH (1996) Empirical correlations in structural chemistry. Adv Molec Struct Res 2:163–189
175. Exner K, von Schleyer P (2001) Theoretical bond energies: a critical evaluation. J Phys Chem A105:3407–3416
176. Leal JP (2006) Additive methods for prediction of thermochemical properties. The Laidler method revisited: hydrocarbons. J Phys Chem Ref Data 35:55–76
177. Coulson CA, Moffit WE (1949) The properties of certain strained hydrocarbons. Phil Mag 40:1–35
178. Koritsanszky T, Buschmann J, Luger P (1996) Topological analysis of experimental electron densities: the different C—C bonds in bullvalene. J Phys Chem 100:10547–10553
179. Stolevik R, Postmyr L (1997) Bond length variations in molecules containing triple bonds and halogen substituents. J Mol Struct 403:207–211
180. Ellern A, Drews Th, Seppelt K (2001) The structure of carbon suboxide, C_3O_2, in the solid state. Z anorg allgem Chem 627:73–76
181. Lass RW, Steinert P, Wolf J, Werner H (1996) Synthesis and molecular structure of the first neutral transition-metal complex containing a linear $M{=}C{=}C{=}C{=}C{=}CR_2$ chain. Chem Eur J 2:19–23
182. Speranza G, Minati L, Anderle M (2006) Covalent interaction and semiempirical modeling of small molecules. J Phys Chem A110:13857–13863
183. Kapp J, Schade C, El-Nahasa A, von Schleyer P (1996) Heavy element π donation is not less effective. Angew Chem Int Ed 35:2236–2238
184. Blanco S, Sanz ME, Mata S et al (2003) Molecular beam pulsed-discharge Fourier transform microwave spectra of $CH_3{-}C{\equiv}C{-}F$, $CH_3{-}(C{\equiv}C)_2{-}F$, and $CH_3{-}(C{\equiv}C)_3{-}F$. Chem Phys Lett 375:355–363
185. Sachse H (1890) Ueber die geometrischen Isomerien der Hexamethylenderivate. Chem Ber 23:1363–1370
186. Cremer D, Pople JA (1975) General definition of ring puckering coordinates. J Am Chem Soc 97:1354–1358
187. Dahl JEP, Moldowan JM, Peakman TM et al (2003) Isolation and structural proof of the large diamond molecule, cyclohexamantane ($C_{26}H_{30}$). Angew Chem Int Ed 42:2040–2044
188. Szafert S, Gladysz JA (2003) Carbon in one dimension: structural analysis of the higher conjugated polyynes. Chem Rev 103:4175–4206
189. Szafert S, Gladysz JA (2006) Update 1 of: carbon in one dimension. Chem Rev 106:PR1-PR33
190. Mel'nichenko VM, Sladkov AM, Nikulin YN (1982) Structure of polymeric carbon. Russ Chem Rev 51:421–438
191. Mohr W, Stahl J, Hampel F, Gladysz JA (2001) Bent and stretched but not yet to the breaking point: the first structurally characterized 1,3,5,7,9,11,13,15-octayne. Inorg Chem 40:3263–3264
192. Haaland A (1988) Organometallic compounds of main group elements. In: Hargittai I, Hargittai M (eds) Stereochemical appilication of gas-phase electron diffraction, Part B. VCH, New York
193. Snow AI, Rundle RE (1951) The structure of dimethylberillium. Acta Cryst 4:348–352
194. Huttman JC, Streib WE (1971) Crystallographic evidence of the three-centre bond in hexamethyldialuminium. J Chem Soc D:911–912
195. Zeise WC (1831) Von der Wirkung zwischen Platinchlorid und Alkohol, und von den dabei entstehenden neuen Substanzen. Annalen der Physik und Chemie 97:497–541
196. Wunderlich JA, Mellor DP (1954) A note on the crystal structure of Zeise salt. Acta Crystallogr 7:130
197. Jarvis JAJ, Kilbourn BT, Owston PG (1971) A redetermination of the crystal and molecular structure of Zeise's salt. Acta Crystallogr B27:366–372
198. Chatt J, Duncanson LA, Venanzi LM (1955) Directing effects in inorganic substitution reactions: a hypothesis to explain the trans-effect. J Chem Soc 4456–4460
199. Thiele J (1901) Ueber Abkömmlinge des Cyclopentadiëns. Chem Ber 34:68–71

200. Harder S (1999) Can C—H··· C(π) bonding be classified as hydrogen bonding? A systematic investigation of C—H··· C(π) bonding to cyclopentadienyl anions. Chem Eur J 5:1852–1861
201. Jutzi P, Burford N (1999) Structurally diverse π-cyclopentadienyl complexes of the main group elements. Chem Rev 99:969–990
202. Batsanov SS (2000) Intramolecular contact radii similar to van der Waals ones. Rus J Inorg Chem 45:892–896
203. Gard E, Haaland A, Novak DP, Seip R (1975) Molecular structures of dicyclopentadienyl-vanadium, $(C_5H_5)_2V$, and dicyclopentadienylchromium, $(C_5H_5)_2Cr$, determined by gas-phase electron-diffraction. J Organomet Chem 88:181–189
204. Haaland A (1979) Molecular structure and bonding in the 3d metallocenes. Acc Chem Res 12:415–422
205. Zakharov LN, Saf'yanov YuN, Domrachev GA (1990) In: Porai-Koshits MA (ed) Problems of crystal chemistry. Nauka, Moscow (in Russian)
206. Rulkens R, Gates DP, Balaishis D et al (1997) Highly strained, ring-tilted [1]ferrocenophanes containing group 16 elements in the bridge. J Am Chem Soc 119:10976–10986
207. Braunschweig H, Dirk R, Müller M et al (1997) Incorporation of a first row element into the bridge of a strained metallocenophane: synthesis of a boron-bridged [1]ferrocenophane. Angew Chem Int Ed 36:2338–2340
208. Watanabe M, Nagasawa A, Sato M et al (1998) Molecular structure of Hg-bridged tetramethyl[2]ferrocenophane salt and related salts. Bull Chem Soc Jpn 71:1071–1079
209. Seiler P, Dunitz JD (1980) Redetermination of the ruthenocene structure. Acta Cryst B36:2946–2950
210. Boeyens JCA, Levendis DC, Bruce MI, Williams ML (1986) Crystal structure of osmocene, $Os(\eta\text{-}C_5H_5)_2$. J Crystallogr Spectrosc Res 16:519–524
211. Prout K, Cameron TS, Forder RA (1974) Crystal and molecular structures of bent bis-π-cyclopentadienyl-metal complexes. Acta Cryst B30:2290–2304
212. Gowik P, Klapotke T, White P (1989) Dications of molybdenocene(VI) and tungstenocene(VI) dichlorides. Chem Ber 122:1649–1650
213. Eggers SH, Kopf J, Fischer RD (1986) The X-ray structure of $(C_5L_5)_3La^{III}$ – a notably stable polymer displaying more than 3 different La... C interactions. Organometallics 5:383–385
214. Rebizant J, Apostolidis C, Spirlet MR, Kanellakopulos B (1988) Structure of a new polymorphic form of tris(cyclopentadienyl)lanthanum(III). Acta Cryst C44:614–616
215. Evans WJ, Davis BL, Ziller JW (2001) Synthesis and structure of tris(alkyl- and silyl-tetramethylcyclopentadienyl) complexes of lanthanum. Inorg Chem 40:6341–6348
216. Spirlet MR, Rebizant J, Apostolidis C, Kanellakopulos B (1987) Structure of tris(η^5-cyclopentadienyl)bis(propiononitrile)lanthanum(III). Acta Cryst C43:2322–2324
217. Tolman CA (1970) Phosphorus ligand exchange equilibriums on zerovalent nickel: dominant role for steric effects. J Am Chem Soc 92:2956–2965
218. Tolman CA (1977) Steric effects of phosphorus ligands in organometallic chemistry and homogeneous catalysis. Chem Rev 77:313–348
219. Lobkovsky EB (1984) Steric factor dependence of the structures of certain Cp-containing compounds. J Organomet Chem 277:53–59
220. Dubler E, Textor M, Oswald H-R, Salzer A (1974) X-ray structure-analysis of triple-decker sandwich complex tris (η-cyclopentadienyl)dinickel tetrafluoroborate. Angew Chem 86:135–136
221. Dubler E, Textor M, Oswald H-R, Jameson GB (1983) The structure of μ-(η-cyclopentadienyl)-bis[(η-cyclopentadienyl)nickel(II)] tetrafluoroborate at 190 and 295 K. Acta Cryst B39:607–612
222. Wang X, Sabat M, Grimes RN (1995) Organotransition-metal metallacarboranes: directed synthesis of carborane-end-capped multidecker sandwiches. J Am Chem Soc 117:12227–12234
223. Scherer O, Schwalb J, Swarowsky H et al (1988) Triple-decker sandwich complexes with cyclo-P_6 as middle deck. Chem Ber 121:443–449

224. Scherer O, Vondung J, Wolmershäuser G (1989) Tetraphosphacyclobutadiene as complex ligand. Angew Chem Int Ed 28:1355–1357
225. Nagao S, Kato A, Nakajima A, Kaya K (2000) Multiple-decker sandwich poly-ferrocene clusters. J Am Chem Soc 122:4221–4222
226. del Mar-Conejo M, Fernández R, Gutiérrrez-Pueble E et al (2000) Synthesis and X-ray structures of $[Be(C_5Me_4H)_2]$ and $[Be(C_5Me_5)_2]$. Angew Chem Int Ed 39:1949–1951
227. Rayane D, Allouche A-R, Antoine R et al (2003) Electric dipole of metal–benzene sandwiches. Chem Phys Lett 375:506–510
228. McClelland BW, Robiette AG, Hedberg L, Hedberg K (2001) Iron pentacarbonyl: are the axial or the equatorial iron-carbon bonds longer in the gaseous molecule? Inorg Chem 40:1358–1362
229. Walker NR, Hui JK-H, Gerry MCL (2002) Microwave spectrum, geometry, and hyperfine constants of PdCO. J Phys Chem A106:5803–5808
230. Arnesen SP, Seip HM (1966) Studies on failure of first Born approximation in electron diffraction: molybdenum- and tungsten hexacarbonyl. Acta Chem Scand 20:2711–2727
231. Martinez A, Morse MD (2006) Infrared diode laser spectroscopy of jet-cooled NiCO, $Ni(CO)_3(^{13}CO)$, and $Ni(CO)_3(C^{18}O)$. J Chem Phys 124:124316
232. Compton N, Errington R, Norman N (1990) Transition metal complexes incorporating atoms of the heavier main-group elements. Adv Organomet Chem 31:91–182
233. Sitzmann H (2002) The decaphosphatitanocene dianion – A new chapter in the chemistry of naked polyphosphorus ligands. Angew Chem Int Ed 41:2723–2724
234. Charles S, Eichhorn BW, Rheingold AL, Bott SG (1994) Synthesis, structure and properties of the $[E_7M(CO)_3]^{3-}$ complexes where E=P, As, Sb and M=Cr, Mo, W. J Am Chem Soc 116: 8077–8086
235. Klooster WT, Koetzie TF, Jia G et al (1994) Single crystal neutron diffraction study of the complex $[Ru(H\cdots H)(C_5Me_5)(dppm)]BF_4$ which contains an elongated dihydrogen ligand. J Am Chem Soc 116:7677–7681
236. Chabrie MC (1907) Sur un nouveau chlorure de tantale. Compt Rend 144:804–806
237. Vaughan PA, Sturdivant JH, Pauling L (1950) The determination of the structures of complex molecules and ions from X-ray diffraction by their solutions: the structures of the groups $PtBr_6^-$, $PtCl_6^-$, $Nb_6Cl_{12}^{++}$, $Ta_6Br_{12}^{++}$, and $Ta_6Cl_{12}^{++}$. J Am Chem Soc 72:5477–5486
238. Wade K (1976) Structural and bonding patterns in cluster chemistry. Adv Inorgan Chem Radiochem 18:1–66
239. Mastryukov VS, Dorofeeva OV, Vilkov LV (1980) Internuclear distances in carbaboranes. Russ Chem Rev 49:1181–1187
240. Ferguson G, Parvez M, MacCutrtain JA et al (1987) Reactions of heteroboranes – synthesis of $[2,2\text{-}(PPh_3)_2\text{-}1,2\text{-}SePtB_{10}H_{10}] \cdot CH_2Cl_2$, its crystal and molecular structure and that of $SeB_{11}H_{11}$. J Chem Soc Dalton Trans 699–704
241. Wynd AJ, McLennan AJ, Reed D, Welch AJ (1987) Gold-boron chemistry: synthetic, structural, and spectroscopic studies on the compounds $[5,6\text{-}\mu\text{-}(AuPR_3)\text{-nido-}B_{10}H_{13}]$ (R=cyclo-C_6H_{11} or C_6H_4Me-2). J Chem Soc Dalton Trans 2761–2768
242. Housecroft C (1991) Boron atoms in transition metal clusters. Adv Organometal Chem 33:1–50
243. Cotton FA, Walton RA (1993) Multiple bonds between metal atoms, 2nd ed. Clarendon Press, Oxford
244. Binder H, Kellner R, Vaas K et al (1999) The closo-cluster triad: B_9X_9, $[B_9X_9]^-$, and $[B_9X_9]^{2-}$ with tricapped trigonal prisms (X=Cl, Br, I). Z anorg allgem Chem 625:1059–1072
245. Welch AJ (2000) Steric effects in metallacarboranes. In: Braunstein P, Oro LA, Raithby PR (eds) Metal clusters in chemistry, vol 1. Wiley-VCH, Weinheim
246. Timms PL, Norman NC, Pardoe JAJ et al (2005) The structures of higher boron halides B_8X_{12} (X=F, Cl, Br and I) by gas-phase electron diffraction. J Chem Soc Dalton Trans 607–616
247. Huges AK, Wade K (2000) Metal-metal and metal-ligand bond strengths in metal carbonyl clusters. Coord Chem Rev 197:191–229

248. Mason R, Thomas KM, Mingos DMP (1973) Stereochemistry of octadecacarbonylhexaosmium(0): novel hexanuclear complex based on a bicapped tetrahedron of metal atoms. J Am Chem Soc 95:3802–3804
249. Mingos DMP, Johnston RL (1987) Theoretical models of cluster bonding. Structure and Bonding 68:29–87
250. Lewis J (2000) Retrospective and prospective considerations in cluster chemistry. In: Braunstein P, Oro LA, Raithby PR (eds) Metal Clusters in Chemistry, vol 2. Wiley-VCH, Weinheim
251. Benfield R (1992) Mean coordination numbers and the nonmetal-metal transition in clusters. J Chem Soc Faraday Trans 88:1107–1110
252. Chini P (1980) Large metal-carbonyl clusters. J Organomet Chem 200:37–61
253. Montano PA, Zhao J, Ramanathan M et al (1986) Structure of copper microclusters isolated in solid argon. Phys Rev Lett 56:2076–2079
254. Montano PA, Zhao J, Ramanathan M et al (1989) Structure of silver microclusters. Chem Phys Lett 164:126–130
255. Santiso E, Müller EA (2002) Dense packing of binary and polydisperse hard spheres. Mol Phys 100:2461–2469
256. Belyakova OA, Slovokhotov YuL (2003) Structures of large transition metal clusters. Russ Chem Bull 52:2299–2327
257. Teo BK, Zhang H (1995) Polyicosahedricity: icosahedron to icosahedron of icosahedra growth pathway for bimetallic (Au-Ag) and trimetallic (Au-Ag-M; M=Pt, Pd, Ni) supraclusters; synthetic strategies, site preference, and stereochemical principles. Coord Chem Rev 143:611–636
258. Simon A (1981) Condensed metal clusters. Angew Chem 93:1–22
259. Templeton J (1979) Metal-metal bonds of order four. Progr Inorg Chem 26:211–300
260. Serre J (1981) Metal-metal bonds. Int J Quantum Chem 19:1171–1183
261. Corbett JD (1981) Correlation of metal-metal bonding in halides and chalcides of the early transition elements with that in the metals. J Solid State Chem 37:335–351
262. Corbett JD (1981) Chevrel phases – an analysis of their metal-metal bonding and crystalchemistry. J Solid State Chem 39:56–74
263. Holloway CE, Melnik M (1985) Tantalum coordination compounds: classification and analysis of crystallographic and structural data. Rev Inorg Chem 7:1–74
264. Holloway CE, Melnik M (1985) Vanadium coordination compounds: classification and analysis of crystallographic and structural data. Rev Inorg Chem 7:75–160
265. Holloway CE, Melnik M (1985) Niobium coordination compounds: classification and analysis of crystallographic and structural data. Rev Inorg Chem 7:161–250
266. Schnering H-G von (1981) Homonucleare Bindungen bei Hauptgruppenelementen. Angew Chem 93:44–63
267. Campbell J, Mercier HPA, Franke H et al (2002) Syntheses, crystal structures, and density functional theory calculations of the *closo*-$[1\text{-}M(CO)_3(\eta^4\text{-}E_9)]^{4-}$ (E=Sn, Pb; M=Mo, W) cluster anions. Inorg Chem 41:86–107
268. Corbett JD (2000) Polyanionic clusters and networks of the early p-element metals in the solid state: beyond the Zintl boundary. Angew Chem Int Ed 39:670–690
269. Ruck M (2001) From metal to the molecule – ternary bismuth subhalides. Angew Chem Int Ed 40:1182–1193
270. von Schnering H-G, Hönle W (1988) Chemistry and structural chemistry of phosphides and polyphosphides. Chem Rev 88:243–273
271. Baudler M, Glinka K (1993) Contributions to the chemistry of phosphorus: monocyclic and polycyclic phosphines. Chem Rev 93:1623–1667
272. Ho J, Breen TL, Ozarowski A, Stephan DW (1994) Early metal mediated P-P bond formation in $Cp_2M(PR)_2$ and $Cp_2M(PR)_3$ complexes. Inorg Chem 33:865–870
273. Shevelkov AV, Dikarev EV, Popovkin BA (1994) Helical chains in the structures of Hg_2P_3Br and Hg_2P_3Cl. Z Krist 209:583–585
274. Häser M (1994) Structural rules of phosphorus. J Am Chem Soc 116:6925–6926

275. Balch AL (2000) Structural inorganic chemistry of fullerenes and fullerene-like compounds. In: Kadish KM, Ruoff RS (eds) Fullerenes: chemistry, physics and technology. Wiley-Interscience, New York
276. Shinohara H (2000) Endohedral metallofullerenes: production, separation and structural properties. In: Kadish KM, Ruoff RS (eds) Fullerenes: chemistry, physics and technology. Wiley-Interscience, New York
277. Prassides K, Margadonna S (2000) Structures of fullerene-based solids. In: Kadish KM, Ruoff RS (eds) Fullerenes: chemistry, physics and technology. Wiley-Interscience, New York
278. Neretin IS, Slovokhotov YuL (2004) Chemical crystallography of fullerenes. Russ Chem Rev 73:455–486
279. Kawasaki S, Yao A, Matsuoka Y et al (2003) Elastic properties of pressure-polymerized fullerenes. Solid State Commun 125:637–640
280. Hampe O, Neumaier M, Blom MN, Kappes MM (2002) On the generation and stability of isolated doubly negatively charged fullerenes. Chem Phys Lett 354:303–309
281. Hampe O, Neumaller M, Blom MN, Kappes MM (2003) Electron attachment to negative fullerene ions: a Fourier transform mass spectrometric study. Int J Mass Spectrom 229:93–98
282. Margadonna S, Aslanis E, Li WZ et al (2000) Crystal structure of superconducting $K_3Ba_3C_{60}$. Chem Mater 12:2736–2740
283. Ehrler OT, Weber JM, Furche F, Kappes MM (2003) Photoelectron spectroscopy of C_{84} dianions. Phys Rev Lett 91:113006
284. Troyanov SI, Popov AA, Denisenko NI et al (2003) The first X-ray crystal structures of halogenated [70]fullerene: $C_{70}Br_{10}$ and $C_{70}Br_{10} \cdot 3Br_2$. Angew Chem Int Ed 42:2395–2398
285. Neretin IS, Lyssenko KA, Antipin MY et al (2000) $C_{60}F_{18}$, a flattened fullerene: alias a hexa-substituted benzene. Angew Chem Int Ed 39:3273–3276
286. Sawamura M, Kuninobu Y, Toganoh M et al (2002) Hybrid of ferrocene and fullerene. J Am Chem Soc 124:9354–9355
287. Sawamura M, Iikura H, Hirai A, Nakamura E (1998) Synthesis of π-indenyl-type fullerene ligand and its metal complexes via quantitative trisarylation of C_{70}. J Am Chem Soc 120:8285–8286
288. Stevenson S, Rice G, Glass T et al (1999) Small-bandgap endohedral metallofullerenes in high yield and purity. Nature 401:55–57
289. Stevenson S, Lee HM, Olmstead MM et al (2002) Preparation and crystallographic characterization of a new endohedral, $Lu_3N@C_{80} \cdot 5$(o-xylene), and comparison with $Sc_3N@C_{80} \cdot 5$(o-xylene). Chem Eur J 8:4528–4535
290. Olmstead MM, de Bettencourt-Dias A, Ducamp JC et al (2000) Isolation and crystallographic characterization of $ErSc_2N@C_{80}$. J Am Chem Soc 122:12220–12226
291. Ming LC, Zinin P, Meng Y et al (2006) A cubic phase of C_3N_4 synthesized in the diamond-anvil cell. J Appl Phys 99:033520
292. Filonenko VP, Davydov VA, Zibrov IP et al (2010) High pressure synthesis of new heterodiamond phase. Diamond Relat Mater 19:541–544
293. Chorpa NG, Zettl A (2000) Boron-nitride containing nanotubes. In: Kadish KM, Ruoff RS (eds) Fullerenes: chemistry, physics and technology. Wiley-Interscience, New York
294. Batsanov SS, Blokhina GV, Deribas AA (1965) The effects of explosions on materials. J Struct Chem 6:209–213
295. Akashi T, Sawaoka A, Saito S, Araki M (1976) Structural changes of boron-nitride caused by multiple shock compressions. Jpn J Appl Phys 15:891–892
296. Sokolowski M (1979) Deposition of wurtzite type boron-nitride layers by reactive pulse plasma crystallization. J Cryst Growth 46:136–138
297. Rusek A, Sokolowski M, Sokolowska A (1981) Formation of E-phase BN layers and shock-wave compressed BN on boron as a result of boron reactive electro-erosion. J Mater Sci 16:2021–2023
298. Fedoseev DV, Varshavskaya IG, Lavrent'ev AV, Deryagin BV (1983) Phase transformations of small-size solid particles under laser heating. Dokl Phys Chem 270:416–418

299. Akashi T, Pak H-R, Sawaoka A (1986) Structural changes of wurtzite-type and zincblende-type boron nitrides by shock treatments. J Mater Sci 21:4060–4066
300. Nameki H, Sekine T, Kobayashi T et al (1996) Rapid quench formation of E-BN from shocked turbostratic BN precursors. J Mater Sci Lett 15:1492–1494
301. Gasgnier M, Szwarc H, Ronez A (2000) Low-energy ball-milling: transformations of boron nitride powders. Crystallographic and chemical characterizations. J Mater Sci 35:3003–3009
302. Batsanov SS (2011) Features of phase transformations in boron nitride. Diamond Relat Mater 20:660–664
303. Batsanov SS, Zalivina EN, Derbeneva SS, Borodaevsky VE (1968) Synthesis and properties of copper bromo- and iodo-chlorides. Doklady Acad Nauk SSSR 181:599–602 (in Russian)
304. Batsanov SS, Sokolova MN, Ruchkin ED (1971) Mixed halides of gold. Russ Chem Bull 20:1757–1759
305. Takazawa H, Ohba S, Saito Y (1990) Electron-density distribution in crystals of $K_2[ReCL_6]$, $K_2[OsCl_6]$, $K_2[PtCl_6]$ and $K_2[PtCl4]$ at 120 K. Acta Cryst B46:166–174
306. Bengtsson LA, Oskarsson A (1992) Intermolecular effects on the geometry of $[PtCl_4]^{2-}$ – X-ray diffraction studies of aqueous H_2PtCl_4 and crystalline $(NH_4)_2PtCl_4$. Acta Chem Scand 46:707–711
307. Thiele G, Weigl W, Wochner H (1986) Platinum iodides PtI_2 and Pt_3I_8. Z anorg allgem Chem 539:141–153
308. Schüpp B, Heines P, Savin A, Keller H-L (2000) Crystal structures and pressure-induced redox reaction of $Cs_2PdI_4 \cdot I_2$ to Cs_2PdI_6. Inorg Chem 39:732–735
309. Chernyaev II (1926) The mononitrites of bivalent platinum. Ann Inst Platine (USSR) 4:243–275
310. Kauffmann GB (1977) Il'ya Il'ich Chernyaev (1893–1966) and the trans-effect. J Chem Educ 54:86–89
311. Chernyaev II (1954) Experimental proof of the law of trans-influence. Izvestiya Sectora Platiny 28:34 (in Russian)
312. Poraij-Koshits MA, Khodasheva TS, Antsyshkina AS (1971) Progress in crystal chemistry of complex compounds. In: Gilinskaya EA (ed) Crystal Chemistry 7:5. Academy of Sciences, Moscow (in Russian)
313. Poraij-Koshits MA (1978) Structural effects of mutual influence of ligands in transition- and non-transition-metal complexes. Koordinatsionnaya Khimiya 4:842–866 (in Russian)
314. Hartley FR (1973) The cis- and trans-effects of ligands. Chem Soc Rev 2:163–179
315. Russell DR, Mazid MA, Tucker PA (1980) Crystal structures of hydrido-tris(triethylphosphine)platinum(II), fluoro-tris(triethylphosphine)platinum(II), and chloro-tris(triethylphosphine)platinum(II) salts. J Chem Soc Dalton Trans 1737–1742
316. Blau R, Espenson J (1986) Correlations of platinum-195-phosphorus-31 coupling constants with platinum-ligand and platinum-platinum bond lengths. Inorg Chem 25:878–880
317. Belsky VK, Konovalov VE, Kukushkin VYu (1991) Structure of cis-dichloro-(dimethylsulfoxide)-(pyridine)platinum(II). Acta Cryst C47:292–294
318. Kapoor P, Lövqvist K, Oskarsson Å (1998) Cis/trans influences in platinum(II) complexes: X-ray crystal structures of cis-dichloro(dimethyl sulfide)(dimethyl sulfoxide)platinum(II) and cis-dichloro(dimethyl sulfide)(dimethyl phenyl phosphine)platinum(II). J Mol Struct 470:39–47
319. Lövqvist KC, Wendt OF, Leipoldt JG (1996) trans-Influence on bond distances in platinum(II) complexes: structures of trans-$[PtPhI(Me_2S)_2]$ and trans-$[PtI_2(Me_2S)_2]$. Acta Chem Scand 50:1069–1073
320. Waddell PG, Slawin AMZ, Woollins JD (2010) Correlating Pt–P bond lengths and Pt–P coupling constants. J Chem Soc Dalton Trans 39:8620–8625
321. Aslanov LA, Mason R, Wheeler AG, Whimp PO (1970) Stereochemistries of and bonding in complexes of third-row transition-metal halides with tertiary phosphines. J Chem Soc Chem Commun 30–31
322. Dahlborg MBÅ, Svensson G (2002) $HgWO_4$ synthesized at high pressure and temperature. Acta Cryst C58:i35–i36

323. Coe BJ, Glenwright SJ (2000) Trans-effects in octahedral transition metal complexes. Coord Chem Rev 203:5–80
324. Orgel LE (1956) An electronic interpretation of the trans effect in platinous complexes. J Inorg Nucl Chem 2:137–140
325. Batsanov SS (1956) On the mechanism of trans-effect. Doklady Academii Nauk SSSR 110:390–392 (in Russian)
326. Batsanov SS (1959) Crystal-chemical characteristics of trans-effect. Zhurnal Neorganicheskoi Khimii 4:1715–1727 (in Russian)
327. Manojlovic-Muir L, Muir K (1974) Trans-influence of ligands in platinum(II) complexes – significance of bond length data. Inorg Chim Acta 10:47–49
328. Poraij-Koshits MA, Atovmyan LO (1974) Crystal chemistry and stereochemistry of oxide compounds of molybdenum. Nauka, Moscow (in Russian)
329. Dixon K, Moss K, Smith M (1975) Trifluoromethylthio-complexes of platinum(II) – measurement of trans-influence by fluorine-19 nuclear magnetic resonance spectroscopy. J Chem Soc Dalton Trans 990–998
330. Lu X, Akasaka T, Nagase S (2011) Chemistry of endohedral metallofullerenes: the role of metals. Chem Commun 47:5942–5957
331. Yang H, Jin H, Hong B et al (2011) Large endohedral fullerenes containing two metal ions, $Sm_2@D_2(35)\text{-}C_{88}$, $Sm_2@C_1(21)\text{-}C_{90}$, and $Sm_2@D_3(85)\text{-}C_{92}$. J Am Chem Soc 133:16911–16919

Chapter 4
Intermolecular Forces

The mutual attraction of neutral molecules or atoms in different molecules is caused by the van der Waals or donor-acceptor interactions, or a formation of hydrogen bonds (H-bonds) which define features of structures and heats of evaporation (sublimation) of molecular substances. In the reviews of Bent [1] and Haaland [2] the reader can find numerous examples of thermodynamic, structural, physical and chemical properties, and features of products of molecular interactions. Many energetic and geometrical characteristics of molecules with the van der Waals (vdW) interaction are presented in the previous chapters of the book. Therefore here the attention will be drawn only to some principal questions and problems of structural chemistry and thermodynamics which have not been considered earlier.

4.1 Van der Waals Interaction

The experimental heats of sublimation in molecular substances (compiled in the supplementary to this chapter) contribute only a few percent or even fractions of a percent of the atomization energies, varying from several kilo joules per moles to tens of kilo joules per moles. In isolated vdW-molecules of the Rg · Rg, Rg · M, and Rg · X types where Rg is a rare gas, M is a metal and X is a nonmetal, the vdW energies are even smaller: they vary from 0.1 up to 2.5 kJ/mol. The origin of the attractive forces between non-polar molecules was explained in the early twentieth century. In 1928 Wang showed that there is a long-range attractive energy between two hydrogen atoms which varies as D^{-6} where D is their separation [3]. Soon afterwards, London [4, 5] presented his 'general theory of molecular forces' and gave an approximate formula relating the interaction energy to the polarizability of the free molecule,

$$E_{vdW} = \frac{3}{2}\frac{I_A I_B}{I_A + I_B}\frac{\alpha_A \alpha_B}{D_{AB}^6} \quad (4.1)$$

where $I_{A,B}$ are the ionization potentials of the atoms A and B, $\alpha_{A,B}$ is their polarizability and D_{AB} is the intermolecular (vdW) distance. London showed that these

S. S. Batsanov, A. S. Batsanov, *Introduction to Structural Chemistry*,
DOI 10.1007/978-94-007-4771-5_4, © Springer Science+Business Media Dordrecht 2012

forces arise from the quantum-mechanical fluctuations in the coordinates of the electrons and called them the dispersion effect; later theoretical aspects of the dispersion (vdW) interaction were considered in reviews [6, 7]. Merging all constants into C_6, Eq. 4.1 can be re-written as

$$kD = \left(\frac{C_6}{E_{vdW}}\right)^{1/6} \tag{4.2}$$

where $k = (2/3)^{1/6} = 0.935$. The experiment gives $k = 1.05$ for vdW molecules of the Rg · A type [8] and the Rg · M type, where M = Zn, Cd, Hg [9], i.e. the discrepancy with the theory is ca. 10 %.

According to the theory of Slater and Kirkwood [10],

$$C_6 = K\frac{\alpha_1\alpha_2}{(\alpha_1/N_1)^{1/2} + (\alpha_2/N_2)^{1/2}}\text{Å}^6 \tag{4.3}$$

where K is a constant and $N_{1,2}$ are the numbers of the interacting electrons in atoms 1 and 2. A simple method of computing $N_{1,2}$ has been suggested [11], resulting in Eq. 4.4 for the vdW energy

$$E_W = 0.72\frac{C_6}{D^6} \tag{4.4}$$

(E_w in meV, if D in Å) which gives good agreement with the experiment. An efficient and rather simple method of calculating C_6, is described in [12]. There are also empirical approximations, agreeing well with experimental data. Thus, replacing D^6 with V^2 and taking into account that the differences of ionization potentials of organic molecules are small, we obtain

$$E_W \sim k\left(\frac{\alpha}{V}\right)^2 = c(F_{LL})^2 \tag{4.5}$$

where F_{LL} is the Lorentz-Lorenz function. Thus, Eq. 4.5 reproduces the dependence of evaporation heats of organic compounds on their refractive indices [13].

Essentially different nature of the chemical bonding and vdW interaction of atoms causes not only a sharp distinction in the bond distances and energies, but also qualitatively different change of these characteristics on transition from homo- to heteroatomic molecules. For the chemical bonds $d_{AB} < 1/2\ (d_{AA} + d_{BB})$ and $E_{AB} > 1/2\ (E_{AA} + E_{BB})$, whereas for the vdW interactions this picture is reversed:

$$D_{AB} \geq \frac{1}{2}(D_{AA} + D_{BB}),\ E^{W}{}_{AB} < \frac{1}{2}\left(E^{W}{}_{AA} + E^{W}{}_{BB}\right) \tag{4.6}$$

The deviations of the observed bond lengths in vdW complexes from additivity (Eq. 4.6) are small and have been recognized only recently [14]. The difference between the additive and experimental dissociation energies of vdW complexes are more pronounced, however, they also went unnoticed until 1996 [15, 16]. At the same time, from the London equation it follows directly that

$$\Delta E_{AB}^{W} = \frac{1}{2}(E_{AA}^{W} + E_{BB}^{W}) - E_{AB}^{W}$$
$$= \frac{3}{8}\left[I_{1}\left(\frac{\alpha_{A}^{2}}{D_{A}^{6}}\right) + I_{B}\left(\frac{\alpha_{B}^{2}}{D_{B}^{6}}\right) - 4\frac{I_{A}I_{B}}{I_{A}+I_{B}}\left(\frac{\alpha_{A}\alpha_{B}}{D_{AB}^{6}}\right)\right] \quad (4.7)$$

Since the ionization potentials of various atoms differ relatively little, for atoms of similar size,

$$\Delta E_{AB}^{W} \approx \frac{3}{8}\frac{I}{D^{6}}(\alpha_{A}^{2} + \alpha_{B}^{2} - 2\alpha_{A}\alpha_{B}) = c(\alpha_{A} - \alpha_{B})^{2} \quad (4.8)$$

i.e. the energy of a heteroatomic vdW bond is always smaller that the additive value [17]. However, this is a rather rough approximation. Since actual interatomic distances in heteroatomic vdW molecules always exceed the additive values, the approximation $D_{AA} \approx D_{BB} \approx D_{AB}$ tends to overestimate the denominator in Eq. 4.1 and effectively reduces the power of α. The study [18] proved that the formula

$$\Delta E_{W} = c\Delta\alpha^{1.2}$$

gives better agreement with the experimental data for different vdW contacts, including rare gas dimers (all possible combinations), rare gas–metal (Mg, Ca, Sr, Zn, Cd, or Hg), vdW metal–metal complexes, and complexes of molecules (H_2, O_2, N_2 or CH_4) with rare gases and metals.

The above mentioned formulae are based on the assumption that a vdW interaction of the atoms at any distance do not affect their polarizabilities. In fact, the observed polarizabilities of rare gases and molecular substances vary, depending on the aggregate state. The relative variations range from 0.3 % for Ar to 16.8 % for I_2 [19]. On condensation of molecules, the polarizability can decrease (e.g., by 3.2 % for CF_4 or $SnBr_4$) or increase (by 3.0 % for Cl_2, 6.6 % for Br_2, 16.8 % for I_2) [20]. Given that the effective molecular volume always decreases on condensation, this increase of polarizability can be caused only by accumulation of electron density between molecules. Some clue can be provided by the so-called Müller's factor, the ratio between the (relative) changes of refraction, *R*, and of volume, *V* [21],

$$\Lambda_{o} = \frac{\Delta R}{R}\frac{V}{\Delta V}$$

Other things being equal, refraction corresponds to volume (see Chap. 1), hence the deviation of Λ_o from 1 usually indicates a qualitative change of bonding mode. It is known that heating or compression of gases affects their refractions very little, by 10^{-2} to 10^{-4} % [20], but on compression of condensed substances the situation is very different. Thus, solid hydrogen under compression at $P = 10$–130 GPa shows $\Lambda_o = 0.31$ [22]. Compression of HCl, CO_2 and glycerin at $P = 1.5$–4.5 GPa gives

Table 4.1 Properties of rare gas molecules and crystals

Value	He	Ne	Ar	Kr	Xe
ZPE, kJ/mol	0.318	0.624	0.870	0.670	0.515
ΔH_s, kJ/mol	0.083	2.085	7.64	10.62	14.72
ΔH_s/ 1/2 E(Rg–Rg)	1.9	12.0	12.9	12.6	12.7
N_c in crystals	12	12	12	12	12
d(Rg–Rg)$_{mol}$, Å	2.967	3.087	3.759	4.012	4.362
d(Rg–Rg)$_{cryst}$, Å	3.664[a]	3.156	3.755	3.992	4.335
d_{cryst}/d_{mol}	1.235	1.022	0.999	0.995	0.994
ΔH_{vap}/ 1/2 E(Rg–Rg)	1.6	10.3	8.8	10.6	11.1
N_c in liquids	4	9.5	8.5	8.5	9.2
d(Rg–Rg)$_{liq}$, Å	3.15	3.11	3.76	4.02	4.38
d_{liqt}/d_{mol}	1.062	1.007	1.000	1.002	1.004

[a] For $P = 29.7$ bar at T = 1.73 K

$\Lambda_o = 0.21$ [23], 0.22 [24] and 0.29 [25, 26], respectively. Thus, the relative change of polarizability makes a substantial fraction (20–30 %) of the change of molar volumes; hence the power of D in Eq. 4.1 must be different.

It has been shown [17, 27, 28] that the ratio $N = 2\Delta H_s/\varepsilon$, where ΔH_s is the sublimation heat of a condensed rare gas or a molecular substance (A_2, CH_4, C_6H_6) and ε is the dissociation energy of the corresponding dimers (Rg_2, $(A_2)_2$, etc.), is close to the coordination number, N_c, of the molecules or atoms in question in the solid state. Thus, vdW energy is an additive quantity, as, indeed, Pauling found long ago [29] by calculating ΔH_s of Xe. Helium is the exception to this rule: it has $N = 2$ although it too crystallizes in a close-packed structure with $N_c = 12$. However, this weakening of vdW interactions in the condensed state correlates well with the super-fluidity of helium.

Let us consider the vdW interaction in helium in detail. Liquid helium is the only substance that does not solidify down to 0 K in the absence of external pressure. This is explained by the quantum character of the substance, whose zero point energy (ZPE) exceeds the crystal lattice energy [30]. At the same time, the macroscopic properties of helium (its crystal structure and thermochemical characteristics) do not differ fundamentally from those of other rare gases; this allows to treat it in classical terms. Table 4.1 lists the structural and thermodynamic properties of rare gas molecules and crystals (see also [31]).

One can see that the interatomic distances in solid helium are much longer than in the vdW molecule, whereas for other rare gases the differences are insignificant. The situation is similar with liquid rare gases. The reason why helium atoms move further apart on condensation requires further study, although the fact of helium expanding on solidification was known since 1934 [32].

The critical amplitudes of atom oscillations in condensed helium are close to those in other rare gases. It follows that the Lindemann criterion (see Chap. 6) cannot be used to explain the special features of helium solidification. In this context, note a remarkable similarity of the changes in the properties of MX and He_2 caused by the transition from molecular to the solid state: along with an increase in N_c, we

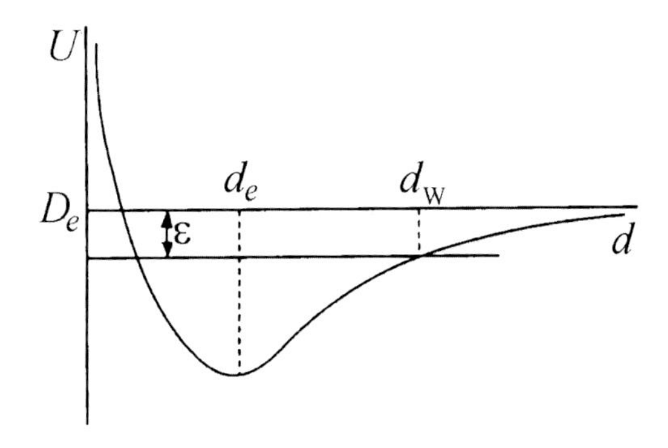

Fig. 4.1 Calculation of van der Waals distances from the Morse potential curve

observe the elongation of interatomic distances by 20–25 % and an increase in the enthalpy of atomization by a factor of 1.5 (Chap. 2). The lengthening of M–X bonds with a simultaneous increase of their energy is known to result from the decrease in the electron density of each bond when their number increases. On the other hand, the London model of van der Waals interactions suggests no similar effect: here the interaction of atoms only depends on their size and polarizability, and the number of contacts is irrelevant. However, there are alternative approaches. Thus, Feynman in his paper [33] on the electrostatic explanation of chemical bonding noted that vdW attractive forces are not the result of (London's) "interactions between oscillating dipoles" but arise from the accumulation of electron density between the atoms; the modern high-level *ab initio* calculations of the He_2 dimer [34] confirmed his prediction. Slater [35] also insisted that there is no fundamental difference between vdW and covalent bonding; later Bader [36–38] showed that the variations of the total, kinetic, and potential energies as functions of the internuclear separation for covalent, polar, and vdW interactions indicate a common underlying quantum mechanism. Recent calculations [39] of the covalent energies (E_c) of several pairs of nonmetal atoms at vdW distances by Mulliken's "magic formula" [40],

$$E_c \sim \bar{I}_{AB} \frac{S_{AB}}{1 + S_{AB}} \tag{4.9}$$

where $\bar{I}_{AB}$ is the mean ionization potential of atoms A and B, and S_{AB} is the overlap integral of the A–B bond, gave the values close to the experimental (from the enthalpy of sublimation) energies of vdW interactions. It has been shown also [41] that for the Morse potential

$$U = D_e[1 - e^{-\beta(d-d_e)}]^2 \tag{4.10}$$

where U is the potential energy, D_e is the bond dissociation energy, β is a constant, d_e and d are the equilibrium and the running bond lengths respectively, the line corresponding to the vdW energy (ε) intersects the potential curve at a point corresponding to the sum of vdW radii, d_w (see Fig. 4.1) Analytically, this reduces Eq. 4.10 to

$$\varepsilon = D_e\left\{1 - [1 - e^{-\beta(d_W-d_e)}]^2\right\} \tag{4.11}$$

Equation 4.11 gives satisfactory agreement between experimental or theoretical values of ε and the vdW distances estimated by other means.

According to the approach of Slater and Kirkwood [10, 11], the energies of interaction between isolated atoms should be calculated taking into account the number of their electrons. Then the main difference between helium and other rare gases is that it has only two electrons, insufficient to provide full fledged interaction with 12 neighbors in the crystal structure. The bonds should then be weakened (elongated), and external pressure is required to stabilize the solid state.

Minemoto et al. [42] measured the longitudinal (α_l) and transverse (α_t) axes of the polarizability ellipsoid for the Ar_2, Kr_2, and Xe_2 molecules and found that the differences $\Delta\alpha = \alpha_l - \alpha_t$ (0.45, 0.72, and 1.23 $Å^3$, respectively) do not agree with the model of a two contacting (hard) spherical atoms. Since, by definition (see also [43]), the cross-section of the polarizability ellipsoid is equal to or larger than the polarizability of an isolated atom (α), and the latter equals 1.64 (Ar), 2.48 (Kr) and 4.04 $Å^3$ (Xe) (see Chap. 11, Table 11.5), the ratio between the Rg···Rg distance and the cross section (γ) can be easily calculated. Obviously,

$$\gamma = \left(\frac{\alpha_l}{\alpha_t}\right)^{1/3} = \left(1 + \frac{\Delta\alpha}{\alpha_t}\right)^{1/3} \leq \left(1 + \frac{\Delta\alpha}{\alpha}\right)^{1/3} \tag{4.12}$$

Substitution of the experimental values of α and $\Delta\alpha$ into Eq. 4.12 gives $\gamma = 1.09$ for all three elements. It follows that, for example, the Xe_2 molecule comprises two flattened (almost twofold) ellipsoids rather than two spheres contacting along the Rg···Rg vector. This indicates a rather strong interatomic interaction.

It is also noteworthy that vdW complexes often possess substantial dipole moments, viz. $Ar \cdot CO_2$ (0.068 D), $Ar \cdot SO_3$ (0.268 D), Ar · HCCH (0.027 D), $N_2 \cdot SO_3$ (0.460 D), $CO_2 \cdot CO$ (0.249 D), which directly indicate a shift of electrons from one atom to another [44]. Indeed, appreciable dipole moments were observed even for heteronuclear complexes of rare gases, e.g. Ne · Kr (0.011 D) or Ne · Xe (0.012 D)!

At the same time, a London-type formula for the dispersion interactions in the endohedral A@B systems (where one subsystem A is inside another subsystem B) has been derived by Pyykkö et al. [45]. It involves the static dipole polarizability, $\alpha(A)$ of the inner system, A, and a new type of dipole polarizability, $\alpha^{-2}(B)$ with an r^{-2} radial operator, for the outer system, B. The second-order correction energy is

$$\Delta E^{(2)} \approx -\frac{3}{4}\frac{I_A I_B}{I_A + I_B}\alpha(A)\alpha^{-2}(B) \tag{4.13}$$

The new formula has no explicit dependence on the radius of B. The predicted interaction energies are compared against MP2 super-molecular calculations for $A@C_{60}$, where A = He–Xe, Zn, Cd, Hg, CH_4. This expression agrees reasonably with the explicit MP2 interaction energies for the super-molecule A@B. The question of how these energies can be measured is an interesting one; however, for an open-structure outer subsystem B, e.g. a bowl- or a ring-shaped one, the energy can easily be measured. Binding energies and interatomic distances in vdW complexes of different types are given in supplementary Tables S4.1–S4.6 and S4.7–S4.10, respectively.

Van der Waals interactions are particularly important for understanding organic solids and biochemical processes, such as enzyme-substrate or enzyme-inhibitor recognition. The cohesion between organic molecules (and hence the vaporization and sublimation enthalpies, melting and boiling points) increases with the molecular size. This is not surprising from any point of view, since both the number of possible atom-atom contacts and the number of electrons responsible for London's attraction increase simultaneously. Thus, the standard vaporization enthalpies of normal alkanes, C_nH_{2n+2}, show an almost perfect linear dependence on n, $\Delta H°(\text{vap}) = n \times 4.9$ kJ mol^{-1} [46]. The total lattice energy of organic solids correlates linearly with the total number of valence electrons per molecule, Z_v [47]

$$E = 1.2\ Z_V + 20\ \text{kJ mol}^{-1}$$

whereas sublimation energies of non-hydrogen bonded oxohydrocarbons are described by the equation

$$\Delta H_{\text{sub}} = 0.841\ Z_V + 39.3\ \text{kJ mol}^{-1}$$

There are two basically different concepts of van der Waals interactions in organic crystals (and condensed phases generally). According to one view, the solid is held together by attraction between immediately contacting atoms of adjacent molecules, more distant interactions are (implicitly) regarded as irrelevant. The alternative approach emphasized the less localized, more diffuse interactions involving entire molecules, or indeed, the entire structure. A separate, but closely related question is: to what point on the potential curve (Fig. 4.1) the closest intermolecular distances correspond? The former view necessary implies that they correspond to the potential minimum, or at least are close to it. The alternative view, expressed particularly by Allinger [48, 49], is that the shortest contacts are shorter than d_e and therefore repulsive, they in fact counterbalance the attractive interactions between more distant atoms (see Fig. 4.2). The problem is complicated by the fact that the precise shape of the potential curve is usually unknown: direct experimental measurements are only available for rare gases and for carbon (from the compressibility of graphite). The fact that the distances and energies (see above) of vdW contacts in Rg_2 molecules and Rg crystals are practically identical [15], suggests that the vdW distance is defined mainly by contacting atoms and is affected by other surrounding atoms relatively little.

Dunitz and Gavezzotti [50–52] have discussed this important question in depth and concluded that the immediate-contact model is satisfactory and indispensable as the first approximation, but on the whole "intermolecular links are not drawn between single atoms". Accordingly, they proposed to replace the old technique of calculating cohesive energy as a sum of pair-wise atom-atom interactions (assuming the atoms spherically-symmetrical and unaffected by their environment) by a more sophisticated approximation, the so-called PIXEL approach. In the latter, the electron density of a separate molecule is calculated by quantum-chemical methods, then intermolecular interactions (coulombic, polarization, dispersion and repulsion components) are calculated between partitions of this electron distributions, much

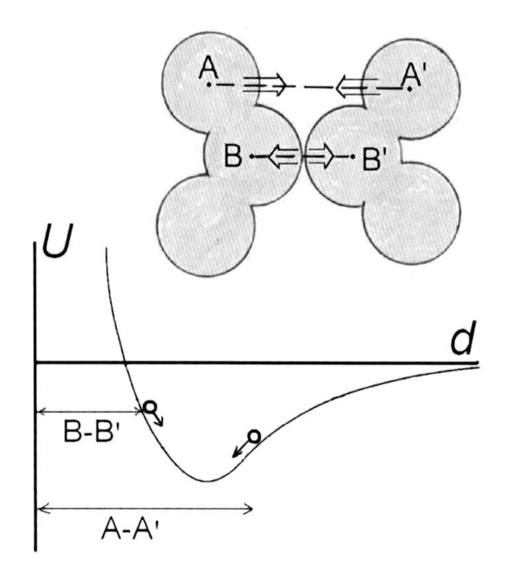

Fig. 4.2 Intermolecular interactions according to Allinger. The shortest contact B—B′ is repulsive, but is counterbalanced (and enforced) by attractive interactions between more distant atoms (A—A′)

smaller than atoms. At present, this method is competitive with thorough quantum calculations in accuracy and much less costly in computer time. Therefore it allowed to calculate, with sufficient precision, what the equilibrium separation between two molecules would be, if unaffected by the rest of the crystal structure. It turned out that only in the minority of cases (ca. 9 %, always in the presence of highly polar functional groups, e.g. CN, NO_2, SO_2) the first coordination shell includes shortened, destabilizing contacts, enforced by the rest of the structure [52]. Thus, although Allinger's effect definitely exists, it is less widespread than is often supposed. Below we shall demonstrate that the contraction of vdW contacts in solids can be explained by effective positive charges on atoms. If similar charges existed on atoms in gas-phase molecules (see [53]), their vdW radii would also decrease compared to those of isolated atoms and would approach the crystallographic radii.

4.2 Interdependence of the Lengths of Covalent and van der Waals Bonds

At the same time, indisputably there are strongly shortened distances in organic compounds which cannot be explained in this way. Experimental studies show that real bond lengths may vary all the way from the sum of covalent radii to the sum of vdW radii [54–56]. Though the anomalous shortening of a I···I distance was observed as early as 1928 [57], the comprehensive theory of this subject is still lacking. Some authors, among others Kitaigorodskii, Porai-Koshits, Zefirov and Zorkii [58–61] considered it from purely geometrical viewpoint, explaining the shortening of the distances by a distortion of vdW spheres (see in detail in Sect. 4.2). Others explained the contraction of vdW contacts by covalent bonding between molecules; the

Table 4.2 Lengths and orders of the covalent and van der Waals I–I bonds

Covalent bond		Van der Waals bond		
d_{exp}, Å	q	$1-q$	D_{cal}, Å	D_{exp}, Å
2.67	1.000	0.000	4.30	4.30
2.70	0.922	0.078	3.61	3.68
2.75	0.805	0.195	3.27	3.30
2.80	0.704	0.296	3.12	3.10
2.85	0.615	0.385	3.02	3.00
2.90	0.537	0.463	2.95	2.93
2.92	0.509	0.491	2.93	2.92

Table 4.3 Lengths and orders of the covalent and van der Waals S–S bonds

Covalent bond		Van der Waals bond		
d_{exp}, Å	q	$1-q$	D_{cal}, Å	D_{exp}, Å
2.06	1.000	0.000	3.60	3.60
2.17	0.742	0.258	2.56	2.57
2.22	0.649	0.351	2.45	2.48
2.26	0.582	0.418	2.38	2.39
2.30	0.523	0.477	2.33	2.36
2.34	0.469	0.531	2.29	2.34

existence of such bonding in crystalline iodine was established long ago by nuclear quadrupole resonance studies [62]. Alternatively, this effect was explained [63–67] by donor-acceptor interactions between molecules. In other works, the shortening of intermolecular distance was interpreted as an evidence of a new type of interaction, variously called 'specific', 'secondary' or the 'interaction that cannot be described by the classical chemical bonding' [7, 56].

Actually, these variations in bond lengths are of crystal-chemical nature and can be understood within the known concepts. As an example, let us consider a linear triatomic system $I_1\cdots I_2\cdots I_3$. Table 4.2 lists the inter-molecular distances D vs. the intra-molecular bond length d obtained by averaging the experimental data [54, 55, 68–73]. Since the D *vs* d curve (Fig. 4.3) shows a hyperbolic behavior [58, 74], it can be described by the equation of O'Keeffe and Brese [75]

$$-\Delta d = 0.37 \ln q \tag{4.14}$$

where Δd is the change of d, and q is the bond order, defined as the valence divided by the coordination number, v/N_c. If the lengthening of the intra-molecular (short) I–I bond is caused by a partial transition of valence electrons to the intermolecular region, then one can determine from Eq. 4.14 the q corresponding to a particular intra-molecular bond length, express the intermolecular bond order as $1-q$, and to calculate, again using Eq. 4.14, the new bond length. The results of such calculations (D_{cal}) are presented in Tables 4.2 and 4.3.

These data prove that the transformation of a vdW bond into a covalent one is the result of electron density transfer. The change of d(I–I) upon the transition from a I_2 molecule to a symmetrical $I_1\cdots I_2\cdots I_3$ system corresponds exactly to the decrease of q from 1 to 0.5 as a result of N_c of the central atom increasing from 1 to 2.

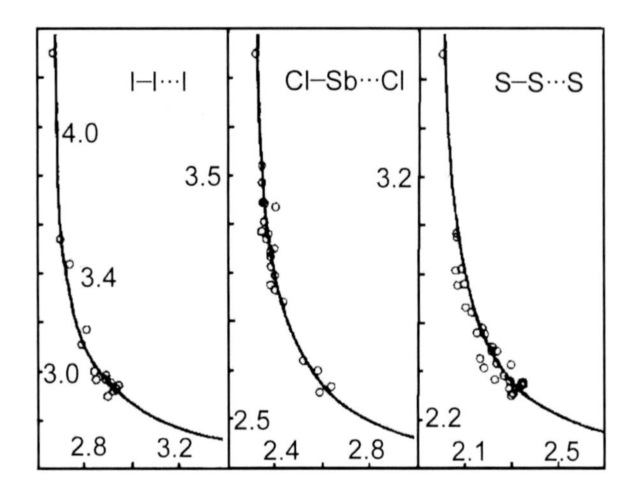

Fig. 4.3 Interdependence of the lengths of covalent and van der Waals bonds (in Å). Reprinted with permission from [55]. Copyright 1975 Wiley

Similar behavior (Fig. 4.3) was observed in the systems X–H· · ·Y [58, 75, 76], Cl–Sb· · ·Cl, X–Cd· · ·X, Br–Br· · ·Br, and S–S· · ·S [77].

Thus, the formation of a symmetrical three-center system from a covalent and a vdW bonds is equivalent to the transformation of a terminal ligand into a bridging one, e.g. on dimerisation $2AX_n \rightarrow A_2X_{2n}$, and all changes in interatomic distances fall within the same range: the covalent bond length increases by *ca.* 0.25 Å, and the vdW distance decreases by *ca.* 1.35 Å. Finally, it has been widely taken for granted that any interatomic distance smaller than Σr_{vdW} is evidence for a bonding interaction, and even a linear correlation between Σr_{cov} and Σr_{vdW} has been set up and used to calculate bond orders [78].

4.3 Van der Waals Radii

4.3.1 Introduction

Van der Waals (vdW) radii describe the distances between atoms with closed outer electron shells (rare gases) or non-bonding distances between atoms in different molecules. Thus vdW radii define the outer size (and shape) of atoms and molecules. In fact, this problem was first encountered in the physics of gases. In 1879 van der Waals [79] realized that deviations of gases (near the critical point) from ideal behavior may be rationalized considering molecules as hard (i.e. mutually repelling on collision) spheres, attracting each other on longer distances. (Note then even such molecule as CO_2 was considered as spherical!). Thus in the real gas equation

$$\left(P + \frac{a}{V^2}\right)(V - b) = RT \tag{4.15}$$

the constant b is linked to the proper volume of the molecule in collision theory, $b = 4V_0$, while a represents the attraction.

In structural chemistry, the first step was made in 1932 by Magat [80] and Mack [81] who introduced the concept of non-valence radius (R) for an atom situated at the periphery of a molecule and called it respectively 'the atomic domain radius' or 'Wirkungradius', implying that this radius determines intermolecular distances. The term 'vdW radius' was coined later by Pauling [82] as the intermolecular forces became known as vdW interactions. *A propos*, van der Waals himself introduced no radii: b in Eq 4.15 has the dimensionality of volume. V_o is smaller than the volume of the same molecule calculated with crystallographic vdW radii (V_m, see below). For not-too-polar organic molecules, $V_o/V_m \approx 0.6$, but this ratio can be much higher for molecules forming strong hydrogen bonds, e.g. methanol (0.80) or acetic acid (0.87).

Today the literature on this subject is immense, and there are many systems of vdW radii, differing not only in their sources and actual values but to some extent in the physical meaning. In the first place, the atom has no clear-cut boundary surface. The vdW forces comprise an attractive component due mainly to dispersion interactions, and a repulsive one due to Pauli exclusion. The repulsive force is significant only at short distances, but then rises quite steeply. The last quality provides the justification for representing atoms in vdW interactions as hard spheres and molecules as agglomerates of (partially overlapping) such spheres, which in a crystal must contact as close as possible but cannot deform or penetrate each other (Kitaigorodskii's principle of close packing). This is achieved when the 'bumps' of one molecule fit into the 'hollows' of another.

4.3.2 *Crystallographic van der Waals Radii*

It was also assumed that vdW radius of an atom depends solely on its atomic number, invariant of the covalently bonded neighbors and of the counterpart in the vdW contact, and is perfectly isotropic. Thus a vdW radius can be determined simply as half the intermolecular contact distance between equivalent atoms, and subtracted from a heteroatomic distance to obtain another one, *etc*. The earliest systems of vdW radii, those of Pauling [82] and later of Bondi [83] and Kitaigorodskii [84] were derived in this way, from the scant and not-too-precise (especially concerning hydrogen atoms) crystallographic data then available. As the experimental data accumulated, it became clear that vdW radii can vary within a few tenths of an angstrom, depending on the structural environment. Thus to derive these radii from the shortest contact distances is to under-estimate them, but finding a meaningful statistical average is not easy. It is, of course, simple to compile a histogram of all distances between certain types of atoms in a crystal structure, but it will be a superposition of two very different distributions: those of (a) immediate contacts and (b) general background. The former distribution is bell-curved but the latter is open-ended and monotonically increasing with distance, hence the sum of both shows an ill-defined maximum, or

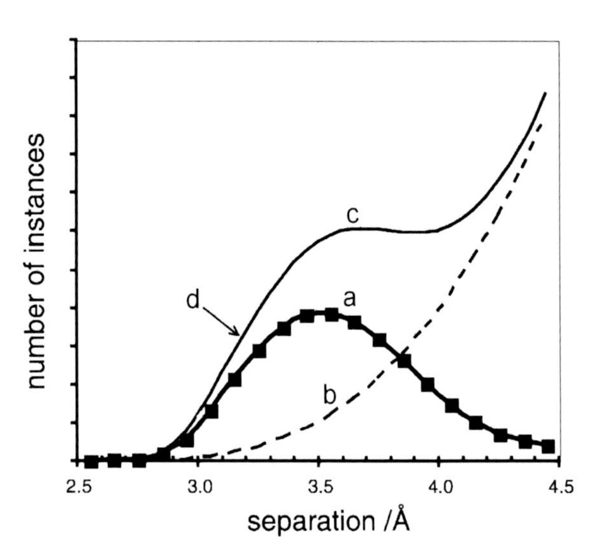

Fig. 4.4 Typical distribution of intermolecular distances (**a**) immediate contacts, (**b**) general background, (**c**) total observed distribution, (**d**) half-height (inflexion) point. (Reproduced from [85], http://dx.doi.org/10.1039/B206867B, by permission of The Royal Society of Chemistry (RSC) for the Centre National de la Recherche Scientifique (CNRS) and the RSC)

only a shoulder [85]. Nevertheless, the left side of the histogram (where immediate contacts dominate) strongly resembles a Gaussian curve and has an inflection point at half-height (Fig. 4.4). Rowland and Taylor [86] suggested using this inflection point to define 'normal' vdW distances for various pairs of atoms and then developed a system of vdW radii to fit these distances, assuming additivity. Unlike the earlier systems, this one is based on vast experimental data from Cambridge Structural Database, but although the procedure is clear mathematically, the physical meaning of the resulting radii is somewhat obscure. Nevertheless, these radii are close to earlier systems, particularly that of Bondi.

In principle, the vdW radius of an atom must depend on its effective charge. However, Pauling [82] found that vdW radius of a nonmetal coincide with its anionic radius, "inasmuch as the bonded atom presents the same face to the outside world in directions away from its bond as the ion does in all direction" and used this rule to suggest vdW radii for some elements. He also noticed that vdW radii exceed the covalent radii (r) of the same elements by $\delta \approx 0.8$ Å; Bondi used this rule in constructing his system of radii (he assumed $\delta = 0.76$ Å), as well as some other physical properties (e.g. critical volumes). This value can be deduced from covalent radii using the equations

$$r_1 - r_n = 0.3 \log n \tag{4.16}$$

$$r_1 - r_n = 0.185 \ln n \tag{4.17}$$

where n is the bond order ($= v/N_c$, see Eq. 4.14), and the general principle that lengthening of an intra-molecular bond leads to shortening of the intermolecular distance, until ultimately the two interactions become indistinguishable [1]. For univalent elements, the transformation of diatomic molecules into continuous solids of *hcp* or *fcc* type means a reduction of n from 1 to 1/12 and hence an increase of r by

0.324 Å (from Eq. 4.16 [82]) or 0.460 Å (from Eq. 4.17 [75, 87]). If upon a 'molecule-to-metal' phase transition R decreases by the same amount, then $R - r_1$ will be twice this, i.e. 0.78 ± 0.15 Å. For compounds with $N_c = 4$ the difference between vdW and covalent radii is 0.70 ± 0.07 Å [89]. The crystallographic vdW radii of nonmetals obtained by different authors from organic or inorganic compounds are very close, see Table S4.11.

Van der Waals radii of metals are difficult to determine directly, because there is only a small (although increasing) number of structures in which metal atoms can be in immediate contact with atoms of another molecule, without forming a strong (ionic, covalent or metallic) bond. Even for the same metal, intermolecular distances strongly depend on the electronegativities of the ligands. Thus, in the crystals of $KAuX_4$ with X = Cl, Br and I, the Au···Au distances equal 4.310 Å [90], 4.515 Å [91] and 4.843 Å [92], respectively. The Pd···Pd distances in K_2PdCl_4 and K_2PdBr_4 are 4.116 Å [93] and 4.309 Å [94], while the Pt···Pt distances in K_2PtCl_4 and K_2PtBr_4 are 4.105 Å [95] and 4.326 Å [96]. Thus the effective size of metal atom increases as the ligands become less electron-withdrawing. One should not suppose that the difference is due to steric repulsion of larger ligands (Cl < Br < I), for the M···M distances are even longer in Na_2PdH_4 (Pd···Pd 5.018 Å) and Na_2PtH_4 (Pt···Pt 5.020 Å) [97], where the ligand (hydride) is the smallest, but the electron density on metal atoms the highest, the electronegativities of Pd, Pt and H being the same. Similar extrapolation gives R(Au) = 2.07 Å. Besides this, M···M distance depends on the conformation, i.e. the torsion angle X–M···M–Y [98]. Due to all these complications, for the majority of metals vdW radii remained undefined or unsatisfactory until recently [99–101]. Table S4.12 lists the vdW radii of metals calculated from the intermolecular distances M···M, M···C or M···H in crystalline organometallic compounds (assuming R = 1.7 Å for C and 1.0 for H) which show large discrepancies. It may be useful to estimate vdW radii of metals indirectly from bond distances, since the latter are much better defined than contact distances. Regarding the formation of an A_2 molecule as an overlap of two spherical atoms of vdW radii (Fig. 4.5), the overlapping volume is

$$\Delta V = \frac{2}{3}\pi(R - r)^2(2R + r) \tag{4.18}$$

This refers to the coordination number $N_c = 1$. If these molecules are fused into a close-packed (*fcc* or *hcp*) metallic structure, N_c increases to 12 while the number of valence electrons remains the same, hence the overlap of each pare of spheres will diminish,

$$\Delta V_1 = 12\Delta V_{12} \; or \; (R_1 - r_1)^2(2R_1 + r_1) = 12(R_{12} - r_{12})^2(2R_{12} + r_{12}) \tag{4.19}$$

Experiments show that a phase transition of molecular solid into metal is indeed accompanied by shortening of intermolecular and a simultaneous elongation of intramolecular bonds, so that each coordination number is characterized by the R and r values of its own. Thus, Eq. 4.19 can be solved to determine the vdW radii of metals, listed in Table S4.13 [102].

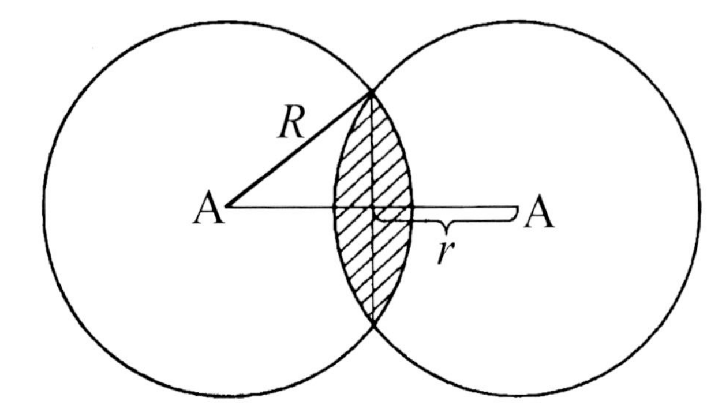

Fig. 4.5 An A_2 molecule modeled as overlapping spheres (R = vdW radius, r = covalent radius)

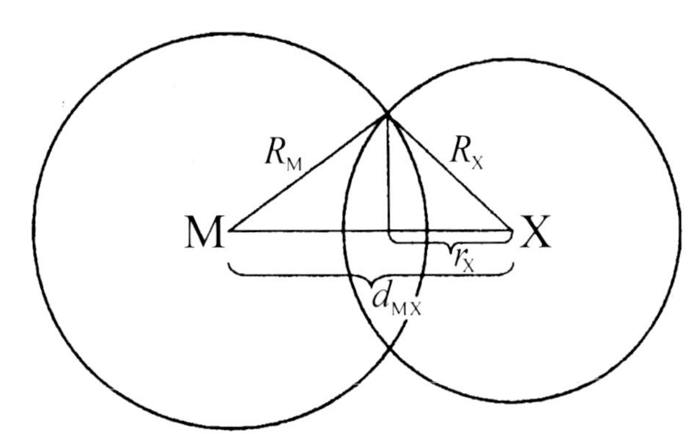

Fig. 4.6 Geometrical calculation of the vdW radius of M in molecules MX

By similar reasoning (Fig. 4.6), vdW radii of metals can be estimated from M–X bond distances d,

$$R_{\mathrm{M}} = \left(R_{\mathrm{X}}^2 + d^2 - 2d\ r_{\mathrm{X}}\right)^{1/2} \tag{4.20}$$

Using an orbital radius for r_X (see Chap. 1, Table 1.8) and knowing R_X (Table S4.11) and d_{MX} from the structures of $M(CH_3)_n$, MCl_n and MBr_n, R_M have been calculated [53], which are also presented in Table S4.13, together with the radii for Au from the recent data on the bond lengths in the AuCl and AuBr molecules), and for V, Th and U from the bond lengths in VX_2, ThX_4, and UX_4.

The intermolecular contact radii R_{IC} in crystals of MX with tetrahedral coordination coincide with the vdW radii of the elements of the fifth period and their compounds. The R_{IC} of metals determined from the $M\cdots C(CH_3)$ distances in $M(C_5Me_5)_n$ molecules, are close to the vdW radii determined by independent methods. The likely reason is that the CH_3 group can tilt away from the plane of cyclopentadienyl ring without significant energy loss and hence the repulsion between M and $C(CH_3)$ is similar to the intermolecular interaction, see Table S4.13 [103].

Nag et al. [104] assumed that the bond valences (v) of atoms at vdW distances have negligible but finite value, and, using Eq. 4.16, found that the vdW radii of d-block elements at $v = 0.01$ are close to those derived in other ways (Table S4.13). Hu et al. [105] have calculated the vdW radii of all elements (including, for the first time, 4*f* elements) from the average volumes of elements in crystals, single covalent radii, Allinger's radii, as well as bond valence parameters for metal–oxygen bonds (Table S4.13). They greatly simplified Nag's calculations by using only d_1(M–O) and R(O) = 1.40 Å in order to extend the study of the vdW radii to all metal atoms.

The special case is the vdW radii of elements in organic compounds. Treating atoms in molecules as rigid spheres, one can calculate the packing factor (ρ) of molecules in crystals, which ranges, according to Kitaigorodskii [84], from 0.65 to 0.77—remarkably similar to $\rho = 0.7405$ found for closely-packed spheres. Later it was shown [106] that large planar or disk-shaped molecules can give much higher ρ, approaching the ideal value (0.909) for packing of perfect disks. Of course, these packing factors should not be confused with those calculated for metals and other inorganic solids—for the latter assume atoms to be spheres not of vdW radius, but of much smaller metallic or covalent radius.

However, one can combine the 'organic' and 'inorganic' concepts by taking into account the covalent and vdW contributions to ρ. As a first approximation, it can be taken that, in the plane defined by the intersection of vdW spheres, ρ is unity (bar packing), whereas in the other directions $\rho = 0.7405$ (close packing of atomic spheres). The area of the base of the spherical segment cut from the vdW sphere distance r from its center is $\pi(R^2 - r^2)$, and the surface area of the vdW sphere is $4\pi R^2$. The ratio of these values is $\sigma = (R^2 - r^2)/4R^2$; for N_c sections, we have $S^* = \sigma N_c$. The total packing factor is then given by

$$\rho^* = S^* + (1 - S^*) \times 0.7405 \quad (4.21)$$

With increasing N_c, $S^* \to 1$, attaining at $N_c = 12$:

$$S^* = \frac{12}{4R^2} \times (R^2 - r^2) = 1$$

and hence

$$R = \left(\frac{3}{2}\right)^{2} r \quad (4.22)$$

Table S4.13 presents the results of calculations using Eq. 4.22 (see [107]).

As mentioned above, we used the Morse function (Eq. 4.10) to calculate vdW radii. This function is adequate for $d = d_e$ and $d = \infty$ and can be used to describe the situation in the vicinity of these points, in particular, of the vdW interaction, when $d = \Sigma R$. Calculating the energy of the vdW interaction in a diatomic system by Eq. 4.4 and using experimental interatomic distances and bond energies in molecules, the vdW radii (Å) of the following nonmetals have been determined [41]: H 1.00, F 1.53, Cl 1.90, Br 2.12, I 2.30, O 1.58, S 1.93, Se 2.07, Te 2.22, N 1.53, P 1.93, and As 2.0, see Table S4.13. Finally, close results were obtained by old Pauling's rule (normal covalent radius plus 0.8 Å) [108, 109] and by the modern system of the single-bond covalent radii according to Alvarez et al. [110], and Pyykkö and Atsumi [111]. The similarity of vdW radii determined by several independent methods is noteworthy; and allows us to compile a comprehensive system of isotropic crystallographic vdW radii (Table 4.4).

Table 4.4 Isotropic vdW radii (Å) of elements: crystallographic (*upper lines*) and equilibrium (*lower lines*)

Li	Be											B	C	N	O	F
2.1	1.8											1.7	1.7	1.6	1.5	1.4
2.6	2.2											2.0	2.0	1.8	1.7	1.6
Na	Mg											Al	Si	P	S	Cl
2.4	2.1											2.0	2.0	1.9	1.8	1.8
2.8	2.4											2.4	2.3	2.1	2.1	2.0
K	Ca	Sc	Ti	V	Cr	Mn	Fe	Co	Ni	Cu	Zn	Ga	Ge	As	Se	Br
2.8	2.4	2.2	2.1	2.0	2.0	2.0	2.0	1.9	1.9	1.9	2.0	2.0	2.0	2.0	1.9	1.9
3.0	2.8	2.6	2.4	2.3	2.2	2.2	2.3	2.2	2.2	2.3	2.2	2.4	2.3	2.2	2.2	2.1
Rb	Sr	Y	Zr	Nb	Mo	Tc	Ru	Rh	Pd	Ag	Cd	In	Sn	Sb	Te	I
3.0	2.6	2.4	2.3	2.1	2.1	2.1	2.0	2.0	2.0	2.0	2.2	2.2	2.2	2.2	2.1	2.1
3.1	2.9	2.7	2.6	2.5	2.4	2.4	2.4	2.3	2.3	2.5	2.4	2.5	2.5	2.4	2.4	2.2
Cs	Ba	La	Hf	Ta	W	Re	Os	Ir	Pt	Au	Hg	Tl	Pb	Bi	Po	At
3.1	2.7	2.5	2.2	2.2	2.1	2.1	2.0	2.0	2.0	2.0	2.1	2.2	2.3	2.3	2.0	2.0
3.3	3.0	2.8	2.5	2.4	2.4	2.3	2.3	2.3	2.4	2.4	2.2	2.5	2.5	2.5		
		Th	U													
		2.4	2.4													
		2.7	2.6													

4.3.3 *Equilibrium Radii of Atoms*

The minimum of the potential curve of vdW interaction of two isolated atoms corresponds to the sum of their equilibrium radii, R_e. Since different interatomic potentials are used to calculate the vdW energy, the reported R_e vary considerably, and are sometimes regarded as pure adjustable parameters in molecular-mechanics calculations, without any strict physical meaning. However, *ab initio* calculations show that the equilibrium vdW radii are physically meaningful, namely, they define the surface enclosing up to 99 % of the electron density of an atom or molecule [110–112]. They also define the distance at which the steric repulsion energy becomes comparable to ambient thermal energy, kT [113],

$$R_{\mathrm{vdW}} = 298k = 0.592\ \mathrm{kcal/mol}$$

Table S4.14 shows R_e from these works for elements of the second and third rows of the Periodical Table. As can see, there is only qualitative accordance between different calculation methods.

The difference between the equilibrium radii of identical elements used by different authors in the molecular mechanics calculations is very large (see Table S4.15) because these values were optimized for narrow ranges of compounds. Allinger et al. [49, 114, 115] have calculated the vdW radii for all of the atoms in the Periodic Table using experimental data for carbon and rare gases; Table S4.16, upper lines.

It is interesting to compare Allinger's results with R_M calculated from the available interatomic distances in of MX_n molecules by Eq. 4.20, depending significantly on the M–X bond polarity. This dependence was described in [53] by an equation, similar to that suggested earlier (but for different purposes) by Blom and Haaland

$$R_{M(X)} = R_M{}^{o} - a\Delta\chi_{MX}{}^{1.4} \tag{4.23}$$

Table 4.5 Experimental and equilibrium radii (Å) of metal atoms in M·Rg molecules

Atom	Na	K	Rb	Mg	Zn	Cd	C
R_{exp}	2.8	2.9	3.0	2.4	2.2	2.3	2.05
R_{eq}	2.68	3.07	3.23	2.41	2.27	2.48	1.97

where $R_{M(X)}$ is the vdW radius of the M atom participating in a M–X bond, $\Delta\chi$ is the difference of electronegativities (in Pauling's scale), $R_M{}^o$ is the vdW radius of the neutral M atom, and a is the constant ($a = 0.13$). The values of $R_M{}^o$ are listed in the middle lines of Table S4.16. In a similar way, $R_M{}^o$ have been calculated from the intra-molecular radii [103], these are listed in the lower lines of Table S4.16.

Tabulated equilibrium radii are close to the experimental radii of atoms in the M_2 molecules of Groups 2 and 12, where bonds are close to the vdW type, or in the vdW complexes Rg · A and M · Rg calculated in [14, 18]. An important point in the latter case is that the distances in heteronuclear vdW molecules turned out to be larger than sums of the vdW radii because of the polarization effects, e.g. the loosening influence of the smaller atoms on the electron polarizability of the larger one:

$$D_{A-B} = \frac{1}{2}(D_{A-A} + D_{B-B}) + \Delta R_{A-B} \tag{4.24}$$

where

$$\Delta R_{A-B} = c\left[\left(\frac{\alpha_A - \alpha_B}{\alpha_B}\right)\right]^{2/3} \tag{4.25}$$

$c = 0.045$, α is the electronic polarizability, and $\alpha_A > \alpha_B$. These relationships were established in [14] and the deviations from the additive rule have been discussed in [18, 106]. Equations 4.24, 4.25 are adapted for all vdW molecules, that permitted to determine the experimental radii in these molecules (Table 4.5). As one can see, these radii are close to the equilibrium values in vdW molecules (Table S4.16). This fact allows to create, by combining the above mentioned data, a comprehensive system of equilibrium vdW radii of elements. These radii are presented in the lower lines of Table 4.4, according to [14, 18, 107] or the calculations using additional experimental data.

4.3.4 Anisotropic van der Waals Radii

Structural studies of crystalline iodine show that the intermolecular distance depends on the crystallographic direction and the difference equals ~0.9 Å [116, 117]. This result at first was also confirmed by Jdanov and Zvonkova [118] for the longitudinal (R_l) and transverse (R_t) vdW radii of Br and H; later this effect has been studied in numerous structural works (see for example [88]).

So, Nyburg [119, 120] determined the anisotropic vdW radii of halogens (Table 4.7) in the structures of X_2 molecules; we calculated [121, 122] the analogous

Table 4.6 Anisotropic vdW radii (Å) of X atoms in the C–X···X bonds according to [122–124]

A	H	F	Cl	Br	I	O	S
R_t	1.26	1.38	1.78	1.84	2.13	1.64	2.03
R_l	1.01	1.30	1.58	1.54	1.76	1.44	1.60
A	Se	Te	N	P	As	Sb	Bi
R_t	2.15	2.33	1.62			2.12	2.25
R_l	1.70	1.84	1.42	1.91	1.85	1.83	1.80

radii for fluorine, hydrogen, oxygen and nitrogen from structures of the corresponding molecules in a solid state. It is known that X_2 molecules can be distributed in the crystal space both in an isotropic manner (through random rotation) and in a strictly ordered orientation. Structures of the first type are realized in α- and β-H_2, which can be described as the *hcp* and *fcc* packing of spheres with the radius $R = R_l + 1/2 \cdot d$(H–H). Knowing the cell parameters of these polymorphs and the length of the H–H bond, we obtain R_l(H) = 1.516 Å for both cases. In the structures of β-F_2 and γ-O_2, there are molecules of two types: 25 % are disordered (to an effective spherical symmetry) and 75 % are ordered. The nearest distances between the ordered (parallel oriented) molecules in these structures are equal to 3.34 Å and 3.42 Å, respectively. Hence, the transverse vdW radii in these structures are R_t(F) = 1.67 Å and R_t(O) = 1.71 Å. The separations between molecules of the two types (ordered and disordered), d_{od}, can be described by Eq. 4.26

$$d_{od}(X\cdots X) = R_t(X) + R_l(X) + \frac{1}{2}d(X\text{–}X) \tag{4.26}$$

Knowing d_{od}(X···X) in the structures of β-F_2 and γ-O_2 (3.73 Å and 3.82 Å, respectively), R_t(X), equal to 1.67 Å and 1.71 Å, respectively, and d(X–X) in these structures (1.48 Å and 1.22 Å, respectively), we obtain R_l(F) = 1.32 Å and R_l(O) = 1.50 Å. From the structures of α- and γ-O_2, the averaged values of R_t(O) = 1.67 and R_l(O) = 1.49 Å have been calculated, from β- and δ-N_2, the values R_t(N) = 1.75 and R_l(N) = 1.55 Å. These radii are given in Table 4.7 as supplements to the data obtained by Nyburg for halogen molecules. Nyburg and Faerman [123, 124] presented some anisotropic R(X) radii in the C–X···X systems; these data were supplemented or corrected in [122] using new structural data (Table 4.6).

In crystal structures of HX, the intermolecular distances H···X were determined for the angles ∠XHX = 180° and X···X for ∠XXX = 90° [125–127]. Using the $\Delta R = R_t - R_l$ for Cl (0.20 Å), Br (0.30 Å) and I (0.37 Å) according to Nyburg [119, 120], we established the anisotropic vdW radii of hydrogen and halogens in these molecules [128] (Table S4.17). In [128] were determined also R_l(H) in CH_4 (1.04 Å) and H_2S (1.02 Å).

The longitudinal vdW radii of H in gas-phase Rg · HX molecules were computed from structural data in [128]. However, in these complexes the HX axis forms a certain angle (θ) with the Rg···H direction and therefore the latter separation corresponds to the intermediate (between R_l and R_t) value of R(H). Using the dependence R(H) = f(180° – θ) [106], the following values of R_l(H) in Rg · HX were obtained

[128]: 0.82 Å for HF, 0.94 Å for OH, 1.04 Å for HCl, 1.12 Å for HBr. Thus, the smaller is $R_l(H)$, the larger is the atomic charge in HX. In the $H_2 \cdot HCl$ complex $R_l(H) = 0.95$ Å (for HCl), and in acetylene $R_l(H) = 0.89$ Å.

The transverse and longitudinal vdW radii have been also determined experimentally in other gas-phase molecules. Thus, in T-shaped vdW complexes $Rg \cdot A_2$ ($Rg = He$, Ne, Ar, Kr, Xe; $A = H$, O, N or a halogen) the radii of A (perpendicular to the A–A bond line) were calculated from structural data [99]; such complexes are rigid and $R_t(A)$ does not depend (within ± 0.05 Å) on the type of Rg. In some $(A_2)_2$ dimers the A_2 molecules contact side-to-side and thus R_t is equal to one-half of the (experimentally determined) separation between the molecular centers of mass. These R_t values are close to the corresponding radii of A in $Rg \cdot A_2$ complexes: the former radii exceed the latter by 0.05 Å on average, due to different modes of molecular packing: projection into hollow in $Rg \cdot A_2$ or projection against projection in $(A_2)_2$ [121, 122].

To determine the longitudinal radii in bonds or molecules, one can use the experimental anisotropy of the bond or molecular electronic polarizability (α_l and α_t). Polarizability being the optical volume of a molecule, the expression $\gamma = (\alpha_l/\alpha_t)^{1/3}$ corresponds to the ratio between the "length" and "thickness" of this molecule:

$$\gamma = \left(\frac{\alpha_l}{\alpha_t}\right)^{1/3} = \frac{L}{T}, \tag{4.27}$$

where for the X_2 molecule

$$L = 2R_l(X) + d(X\text{–}X) \text{ and } T = 2R_t(X) \tag{4.28}$$

For the CO_2 and CS_2 molecules

$$L = 2R_l(X) + 2d(C\text{–}X) \tag{4.29}$$

For C_2H_2

$$L = 2R_l(H) + d(C{\equiv}C) + 2d(C\text{–}H) \tag{4.30}$$

For the BX_3 and MX_4, more complicated procedures were devised, taking into account the deformation of atomic volumes of the central atoms in these molecules and the intra-molecular contact radii (see [129, 130] for details). The results of the calculations of anisotropic vdW radii for the abovementioned molecular structures are listed in Table S4.18; as one can see, the anisotropy of the X radii in molecules depends on the molecular environment. A summary of the averaged anisotropic vdW radii in the X_2 and AX_2 molecules is given in Table 4.7.

Whether the anisotropy of the vdW area of atoms is of intra- or intermolecular nature can be ascertained also by calculating the vdW configuration of isolated molecules. Thus, Bader et al. [131] found that in isolated Li_2, B_2, C_2, N_2, O_2, and F_2 molecules, the vdW radius in the bond direction (longitudinal radius, R_l) is always smaller than the transverse radius R_t (Fig. 4.7). Later, the same was shown in calculations of M_2, MX, CO_2, C_2H_4, C_2H_2 molecules [132, 133].

Table 4.7 Averaged values of the anisotropic vdW radii (Å) in X_2 molecules

Molecule	H_2	F_2	Cl_2	Br_2	I_2	O_2	N_2
$R_t(X)$	1.7	1.6	1.9	2.0	2.2	1.8	1.9
$R_l(X)$	1.5	1.3	1.45	1.45	1.5	1.55	1.6

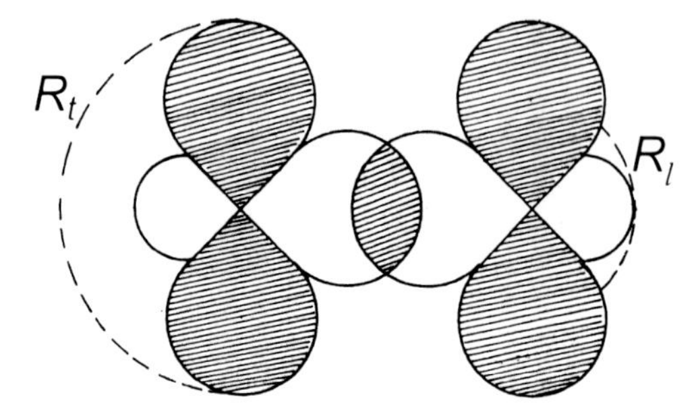

Fig. 4.7 Anisotropic vdW radii: transversal R_t and longitudinal R_l

The calculations by Ishikawa et al. [134] demonstrate that the anisotropy in the R of halogen atoms depends on bond polarity in isolated molecules. The difference $\Delta R = R_t - R_l$ in AX molecules, where A = alkali metals or halogens, change as follows: for X = F from 0.028 (KF) to 0.101 Å (F_2); for X = Cl from 0.038 (KCl) to 0.257 Å (ClF); for X = Br from 0.041 (KBr) to 0.299 Å (FBr). Thus, the larger is the covalency and bond polarizability, the greater is the anisotropy, since the formation of a covalent bond is accompanied by an electron transfer from p_z orbitals to the bonding region, and, accordingly, a decrease in electron density at the opposite lobes of the orbitals. With decreasing bond covalence, the shift of electrons to the overlap region decreases; in ion pairs, there are no bond electrons and, accordingly, very small anisotropy. *Ab initio* calculations for a large number of molecules [113] showed that the anisotropy in the vdW radii depends on the molecular environment.

4.3.5 Concluding Remarks

Experimental data and theoretical estimations show that the vdW radii are not constants—they depend on the molecular structure and atomic charges, and also on the anisotropy of the chemical bonds. They also depend on the aggregate state. As shown above, the vdW radius of an atom is a function of its effective charge. When the net charge on the metal atom is zero, we have the equilibrium vdW radius ($R_e \approx R$), if $e^* > 0$ then $R < R_e$, while a negative charge on the nonmetal atom does not influence the vdW radius greatly because $R \approx r_{X^-}$ (as shown by Pauling). Because the cohesive energies per 'bond' (contact) in the molecular and crystalline states of rare gases are practically identical and the interatomic distances therefore also are similar, the vdW distance is defined only by contacting atoms and depends on other atoms in the molecular or crystal space rather little. Therefore the view [49, 58–61, 114, 115] that the vdW contacts in solids are shorter than the equilibrium distances because of multi-particle interactions, can be disputed: above we have shown that the shortening is caused by the effective positive charges on atoms. If these charges are present in gas-state molecules (see [102, 113]) their vdW radii also become much smaller than for the corresponding isolated atoms and approach the crystallographic radii. Other factors also can be at work. Thus, vdW radii of halogens decrease from CX_4 to SnX_4 because of the stronger vdW interaction between molecules with higher electronic polarizabilities [103]. Similar situation is observed in liquid tetra-chlorides: R_{Cl} is 1.75 Å in CCl_4, 1.63 Å in $SiCl_4$, 1.53 Å in $GeCl_4$ [135]. Finally, the vdW radii of halogen, deduced from the structure of X_2 in the liquid state, are notably larger than in AX_4 molecules, since the polarizabilities of X_2 are smaller than those of the corresponding AX_4: $R(F_2) = 1.54$, $R(Cl_2) = 1.89$, $R(Br_2) = 2.03$, $R(I_2) = 2.23$ Å [136]. Nevertheless, there are some strongly shortened distances in organic condensed phases which cannot be accounted for by these factors.

4.4 Donor–Acceptor Interactions

Donor–acceptor (D $\rightarrow$ A) bond is formed between an atom with a vacant orbital (A, usually metal) and one with a lone electron pair (D, usually non-metal). DA interactions result in the formation of coordination compounds. The paradox of DA bonding is that formally $\chi(D) > \chi(A)$ but the electron transfer goes in the opposite direction. The explanation is that D atoms have closed electron shells, i.e. they are vdW atoms and as such have very low electronegativity, $\chi = 0.2 - 1.3$ (see [14]). Therefore the transfer of lone electron pairs to the polyvalent metals having $\chi \geq 1.5$, is not surprising. Although the lengths of DA bonds are close to those of normal valent bonds, the energies of the former are much lower. Table 4.8 compares the experimental interatomic distances and bond energies in DA complexes [137–143] and in crystalline oxides (from Table 2.11) for the same coordination number. A similar comparison for M–N bonds can be found in Table S4.19.

This sharp distinction can be explained thus. For a chemical bond to be established, molecules must be brought within a short distance, overcoming the vdW repulsion. For HCl and NH_3 this barrier amounts to 94.4 kJ/mol, which is close to the difference (100 kJ/mol) between ΔH_f of iso-structural salts of NH_4 and Rb [144]. Transition of vdW complex, e.g. $NH_3 \cdot HX$ into NH_4X, can occur in the solid state due to the Madelung electrostatic interactions compensating this energy expenditure. Mulliken [145], studying the interaction of NH_3 and HCl, concluded that in the gas phase they can form only a vdW complex held together by dispersion forces, and a transition to ionic NH_4Cl can be due only to electrostatic reasons. *Ab initio* calculations have shown that $NH_x(Me)_{3-x} \cdot HX$ complexes are not ionic pairs: although H–X is somewhat elongated, it is always much shorter than N$\cdots$H [146]. Similar calculations for $NH_3 \cdot HF$ gave the H–N$\cdots$H distances of 1.01 and 1.70 Å, respectively [147]. The changes of DA interactions and of 'partially formed bonds' (intermediate between covalent and vdW) on transition from gas to solid state, have

Table 4.8 Lengths and energies of donor-acceptor and valent metal-oxygen bonds

M	d(M–O), Å		E(M–O), kJ/mol	
	$M(OH_2)_n$	M_nO_m	$M{-}OH_2$	M–O
Li	2.13	2.00	74	145
Na	2.48	2.41	67	110
K	2.79	2.79	56	98
Rb	3.04	2.92	53	94
Cs	3.21	2.86[a]	48	94
Be	1.62	1.65	69	293
Mg	2.08	2.11	67	166
Ca	2.42	2.40	59	177
Sr	2.61	2.58	66	167
Ba	2.77	2.77	62	164
Zn	2.09	1.98, 2.14	58	182
Cd	2.31	2.35	50	103
Y	2.38	2.27		292
La	2.54	2.55	55	243
Al	1.88	1.91	89	257
Ga	1.94	2.00		198
In	2.11	2.16		180
Tl	2.24	2.27	55	125
V	1.99	2.01		250
Bi	2.44	2.40		145
Cr	1.99	1.99		223
Mn^{II}	2.18	2.22	60	153
Mn^{III}	1.99	2.01		190
Fe^{II}	2.12	2.15	63	154
Fe^{III}	2.01	2.03		200
Co^{II}	2.09	2.13	62	156
Co^{III}	1.87			
Ni^{II}	2.06	2.09	63	147

[a] In the Cs_2O, $N_c(Cs) = 3$, in crystallohydrates, $N_c(Cs) = 6–8$

been reviewed by Leopold et al. [148], who concluded that local (condensed) environment can distort such molecules drastically from their gas phase geometry, if the distortion increases the molecular dipole moment. In polar or polarizable media this lowers the total energy and thus offsets the energy cost of the distortion.

Interaction of HX with NH_3, H_2O or BF_3 yields vdW complexes of the $AH_n \cdot HX$ or $BF_3 \cdot XH$ types, where $d(A\cdots H)$ or $d(B\cdots F)$ are by 0.8–1.2 Å longer than the d(A–H) and d(B–F) in normal molecules. To convert $AH_n \cdot HX$ to $AH_{n+1}X$ or $BF_3 \cdot HX$ to HBF_4 requires an expense of energy to shorten the $A\cdots H$ or $B\cdots F$ distances to the lengths of the normal chemical bonds (ΔW_{sh}). In a crystal, this energy can be compensated by the Madelung interaction of NH_4^+ or H_3O^+ ions with X^-, or BF_4^- with H^+ in crystal lattices; in the gaseous state there is no such compensation, hence there are only $NH_3 \cdot HX$ complexes, but no NH_4X salts (see [149]).

As mentioned above, the dissociation energy of DA complexes vary in very wide limits, from 1 to 200 kJ/mol [150–152]. However, these values, strictly speaking, are not the $M\cdots X$ bond energies, they are ΔH_f of two molecules, similarly to the formation heat of NaCl from the Na_2 and Cl_2 molecules. From the viewpoint of structural chemistry, the DA mechanism of bonding two molecules in a complex does not differ from that of the dimer molecule formation, e.g. $BeCl_2 \rightarrow Be_2Cl_4$, $BH_3 \rightarrow B_2H_6$, $NbF_5 \rightarrow Nb_2F_{10}$, etc., or from condensation of NaCl molecules into a crystal.

Schiemenz [153] developed an original criterion for the attractive interaction between nitrogen and its *peri* neighbor in 8-dialkylamino-naphthyl compounds. He noticed that the CNC angles in tertiary amines consistently lie between 109.47° (tetrahedral) and 120° (planar), i.e. the nitrogen atom is partially planarized. Hence, the bond formation and possibly even 'weak bonding interactions' should be identifiable by the decrease of the degree of planarization (γ) as defined in equation:

$$\gamma(\%) = 100\frac{\sum(C-N-C) - 3\cdot 109.47^{\circ}}{3\cdot 120^{\circ} - 3\cdot 109.47^{\circ}} = 100\frac{\sum(C-M-C) - 328.4^{\circ}}{31.6^{\circ}} \quad (4.31)$$

The degree of pyramidalization at the N atom of these compounds permits to distinguish between covalent bond type attraction and other attractive forces. For example, in strained 1,8-disubstituted naphthalene derivatives the distance between *peri*-substituents is always intermediate between the sum of the covalent radii and the sum of vdW radii; if the substituents are potentially a donor and an acceptor (e.g. N or P vs Si), this may be interpreted as a DA bonding. However, Schiemenz et al. [154–156] have shown that the proximity between these substituents is due only to the rigidity of the naphthalene skeleton, indeed, the observed inter-substituent distances substantially longer than 2.50–2.60 Å are evidence *for* steric repulsion and *against* DA interactions.

4.5 Hydrogen Bond

In conclusion, let us consider the hydrogen bond (H-bond). Numerous books and reviews (see a review [157] and references therein) are devoted to the association of polar molecules through a hydrogen atom. According to Pauling [82]: "under certain conditions an atom of hydrogen is attracted by rather strong forces to two atoms, instead of only one, so that it may be considered to be acting as a bond between them. This is called the hydrogen bond". However, today the term "H-bond" is used in a much broader sense. Furthermore, it is often invoked to explain the facts which are not exclusively, or even mainly, due to H-bonding, nor are specific only for it. Therefore, we need first of all to clarify the terminology (see also [158]).

What properties of substances are usually attributed to H-bonds? A textbook example is the melting point, which decreases together with the molecular mass in the succession H_2Te, H_2Se, H_2S (−50, −65, −84 °C) but then jumps to 0 °C for H_2O. The boiling temperature (T_b) of these substances show a parallel trend. The series HI, HBr, HCl, HF is similar in both respects. However, in these cases two opposite factors simultaneously operate: as the polarity of molecules increases, so does the Coulomb interaction, but simultaneously the polarizability decreases, and with it the vdW interaction. Similar breaks in the temperature dependences are observed in substances which contain no hydrogen at all, witness the boiling points of BeI_2 590°, $BeBr_2$ 520°, $BeCl_2$ 500° but BeF_2 1160 °C; AlI_3 180°, $AlBr_3$ 98°, but $AlCl_3$ 193° and AlF_3 1040 °C; or for melting points, ZnI_2 446°, $ZnBr_2$ 392°, $ZnCl_2$ 326°, but ZnF_2 872 °C.

At the same time, it is indisputable that in ice or liquid water the H-bonds do exist, collapsing completely only at very high temperatures or pressures. These bonds prevent denser packing of molecules, therefore H_2O is among the few substances which decrease their density at freezing. To estimate the real influence of H-bonds on the structures and physical or chemical properties of substances, it is necessary to define their energies. The most reliable experimental values of the dissociation heats of gaseous dimers $(H_2O)_2$, $(NH_3)_2$, $(HF)_2$, $(HCl)_2$ according to [159–162] are equal to 15.5, 24, 18, 9.6 kJ/mol, respectively. Comparison of these values with the data from Table 2.11 shows that H-bond energies are two orders of magnitude smaller than the chemical bond energies of H–O, H–N, H–F, and H–Cl. Therefore changes of physical properties due to formation of H-bonds should be insignificant.

However, neutron diffraction (ND) data on different chemical substances with X–H···A hydrogen bonds show clear correlations between decreasing of the H···A distances and lengthening of the covalent X–H bonds. This is shown in Fig. 4.8 which contains data from 58 organic compounds, 18 crystallo-hydrates, NaOH, clinochlore, kalcinite and $M(OH)_2$ oxides (M = Mn, Co and Mg) [163]. Table S4.20 illustrates how on strengthening of the O–H···O hydrogen bond, the O–H and H···O distances converge and ultimately become equal. Today this trend is confirmed by thousands of crystal structure determinations. Average distances O–H and H···A (where A = O, N, F, Cl) in crystalline hydrates of the various structures according to [164] are listed in Table 4.9.

The O–H···O bond is the key to the properties of water. At present, 13 crystalline modifications of water ices are definitely known (Table S4.21). In all of them, each water molecule forms four H-bonds with neighboring molecules, acting as the donor of hydrogen atoms in two of them and as an acceptor in the other two. The four H-bonds are directed towards the vertices of an ideal or slightly distorted tetrahedron.

Of special interest are the changes of the strengths of H-bonds under varying thermodynamic conditions. Thus, a neutron diffraction study of water at 400 °C gave d(O–H) = 1 Å, d(H···H) = 1.55 Å, N_c(H) = 0.98, and N_c(O) = 2.15, thus indicating the absence of H-bonds [165]. Lowering the temperature results in the N_c(O) increasing to 4 because of H-bond formation [166]. Equalization of distances in the H–A···H systems occurs under high pressures: at 42 GPa in HBr, 51 GPa in HCl, 60 GPa in H_2O [167] and 43 GPa in H_2S [168].

Structural data from *in situ* high pressure ND studies are analyzed in [163], they show that the O–H and H···O distances follow the same correlations as have been established at ambient conditions on different compounds. Another pressure effect is the evolution of the double-welled, hydrogen bond potential into a single-well potential. According to *ab initio* calculations, the bulk modulus must have a discontinuity at this point and this can be an indication for hydrogen bond symmetrization; it means that the hydrogen bond symmetrization is a second-order phase transition.

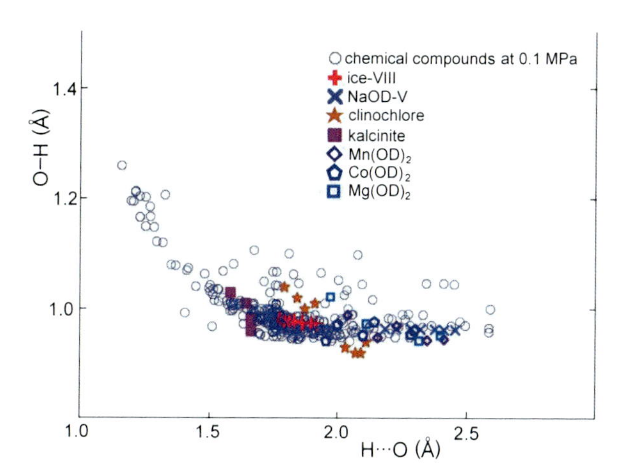

Fig. 4.8 Correlation of the O–H and H···O distances in O–H···O hydrogen bonds. Reprinted from [163] by permission of the publisher (Taylor and Francis Ltd., http://www.tandf.co.uk/journals, Copyright 2007)

Table 4.9 Distances (Å) and angles (°) in O–H···A hydrogen bonds

	O	N	F	Cl
O–H	0.965	0.945	0.967	0.954
H···A	1.857	2.235	1.716	2.254
H–O–H	107.0	103.9	108.1	106.3

The effect of H-bonds on the distances between heavier atoms is often exaggerated, e.g. when all the changes of interatomic distances in crystalline hydroxides are attributed to them. The real role of H-bonds in such changes was analyzed in [169] by comparing the lengths of M–OH and M–F bonds in iso-structural MF_n and $M(OH)_n$ compounds. It appears that the difference $\Delta d = d(\text{M–OH}) - d(\text{M–F}) = 0.08 \pm 0.02$ Å does not depend on the *strength* of H-bonds. The reason for this constancy of Δd is that, provided the changes of valences, coordination numbers and polarities of atoms are equal, the effect of H-bonds on interatomic distances is insignificant.

Dissolution of ionic compounds in water also affects H-bonds. The interactions between water and ions perturbs the long range order of the hydrogen bonding network in the bulk liquid, e.g., by breaking or forming hydrogen bonds. XRAS and Raman scattering studies [170] showed that water molecules in the first coordination shell of strongly hydrated ions have fewer broken hydrogen bonds compared to liquid water, whereas those in the first solvation sphere of weakly hydrated ions have more. The magnitude of the effect follows the Hofmeister series:

$$SO_4{}^{2-} > HPO_4{}^{2-} > F^- > Cl^- > Br^- > I^- > NO_3{}^- > ClO_4{}^-$$

$$NH_4{}^+ < Cs^+ < Rb^+ < K^+ < Na^+ < Ca^{2+} < Mg^{2+} < Al^{3+}$$

Detection of H-bonds by IR-spectroscopy relies on displacement of the absorption band, e.g. of ν(OH), however, this effect actually depends both on the structure of the molecule as a whole, and on the attached mass of vibrating atoms. The most prospective method of studying H-bonds is through the electronic structure of substances, where the effect of mechanical characteristics of molecules is small. One of such methods is a study of the electronic polarizability of the H-bonds considered in Sect. 11.4.

The most important property of the H^+ ion is its extremely small size, resulting in $N_c = 1$ or 2. It is this that makes H-bonds directed and saturated. Although the energy of the H-bond is small, it can alter such features of the structures which are already on, or close to, the threshold of thermodynamic stability, e.g. to interfere with the close packing of molecules (ice), to lower the symmetry (crystalline KF has cubic symmetry, whereas NH_4F is tetragonal), or to influence the composition of substances. Thus, because of the formation of H-bonds between atoms of hydrogen in $NH_4{}^+$ ions and anions, crystalline hydrates of ammonium salts have fewer molecules of water than their K analogues: NH_4F *vs* $KF \cdot 4H_2O$, $(NH_4)_2CO_3$ *vs* $K_2CO_3 \cdot 6H_2O$, $(NH_4)_3PO_4 \cdot 3H_2O$ *vs* $K_3PO_4 \cdot 7H_2O$, $(NH_4)_3AsO_4$ *vs* $K_3AsO_4 \cdot 7H_2O$, $(NH_4)_3Fe(CN)_6 \cdot 1.5H_2O$ *vs* $K_4Fe(CN)_6 \cdot 3H_2O$, *etc*. For the same reason, replacement of H atoms with K in $[H_nAO_m]^{x-}$ anions reduces the number of molecules of water of crystallization, viz.

$K_3AsO_4 \cdot 10H_2O$	$K_3PO_4 \cdot 7H_2O$	$K_2CO_3 \cdot 6H_2O$	$KF \cdot 4H_2O$
$K_2HAsO_4 \cdot H_2O$	$K_2HPO_4 \cdot 3H_2O$	$KHCO_3$	KHF
KH_2AsO_4	KH_2PO_4		

On the contrary, the quantity of molecules of water in hydrates increases with the negative charge on the oxygen atom in the AO_n-anion, viz.

$$KNO_3,\ KClO_4,\ K_2SO_4 < K_2CO_3 \cdot 6H_2O < K_3PO_4 \cdot 7H_2O < K_3AsO_4 \cdot 10H_2O$$

Note, however, that the strength of H_nAO_m acids increases in the opposite succession because of weakening influence of H-bonds on the dissociation of the protons.

So far we have discussed the H-bonds with strongly electronegative acceptors, such as F, O or N, but H-bonds can form with other atoms having $\chi \geq 2$, in particular with metals which can be both donors and acceptors in D–H···M and M–H···A systems [171]. The geometries of D–H···X–M interactions (D = C, N, O; X = halogen) have been examined in detail using many thousands of entries from the CSD, alongside their counterparts in which metal halide groups (X–M) are replaced by halocarbon groups (X–C) or halide ions (X^-). The following succession of relative *strength* of the D–H···X–M hydrogen bonds was established:

$$D\text{–}H\cdots F\text{–}M \gg D\text{–}H\cdots Cl\text{–}M \geq D\text{–}H\cdots Br\text{–}M > D\text{–}H\cdots I\text{–}M,$$

which shows remarkably good qualitative agreement with the trend in the energies for intramolecular N–H···X–Ir hydrogen bonds (X = halogen) [172]. Very interesting are the D–H···H–M interactions (D = N, O), known as 'di-hydrogen bonds' or 'proton–hydride bonds'. They involve an interaction between a protic (δ^+) and a hydridic (δ^-) hydrogen atoms, and occur when the hydridic hydrogen atom is bonded to an electropositive main group element (*e.g*. B, Al, Ga), see in detail in [157, 134] and Fig. 4.9.

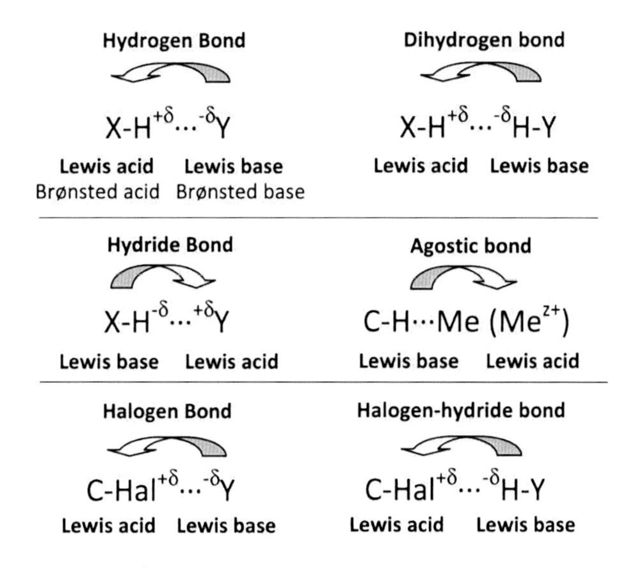

Fig. 4.9 Different types of interactions of the hydrogen atom with atoms of other elements. Reprinted with permission from [173]. Copyright 2006 American Chemical Society

Grabowski [157] concluded that the hydrogen-bonding interaction can be explained in two ways. One is to treat it as a mainly electrostatic interaction. The other approach, while appreciating the electrostatic interaction, holds that other kinds of interaction energy terms are also important, mainly the charge transfer and the dispersive energies. However, as mentioned above, the structural diversity of compounds with hydrogen bonds can be reduced to simple rules whereby bond lengths vary due to changing coordination number, without invoking additional terms.

Halogen atoms, similarly to hydrogen, tend to interact with atoms which possess lone electron pairs. This effect, called 'halogen bond' by analogy with H-bond, is believed to play important role in self-assembly of halogen-containing compounds [174].

Appendix

Supplementary Tables

Table S4.1 Bond energies E (J/mol) and deviations from additive values in molecules of rare gases (From the review [4.1] and original papers, independent measurements are averaged, $\Delta E = 1/2\,[E(\mathrm{Rg \cdot Rg}) + E(\mathrm{Rg' \, Rg'})] - E_{\mathrm{exp}}(\mathrm{Rg \, Rg'})$)

Rg · Rg′	E	ΔE	Rg · Rg′	E	ΔE	Rg · Rg′	E	ΔE
He · He	*90.5*[a]	–	Ne · Ne	*351*	–	Ar · Kr	1386[b,c]	58
He · Ne	177	44	Ne · Ar	548[c]	220	Ar · Xe	1539[b,c]	188
He · Ar	250	388	Ne · Kr	577[c]	450	Kr · Kr	*1702*[c,d]	–
He · Kr	253	643	Ne · Xe	586[c]	724	Kr · Xe	1898[b,c]	88
He · Xe	255	925	Ar · Ar	*1186*	–	Xe · Xe	*2269*[c,f]	–

[a]Average of [4.2–4.4], [b][4.5], [c][4.6], [d][4.7], [e][4.8]

Table S4.2 Bond energies (kJ/mol) in van der Waals molecules Rg · M

M	He	Ne	Ar	Kr	Xe	M	He	Ne	Ar	Kr	Xe
H		0.23	0.40[a]	0.565	0.66	Mg	0.17[m]	0.29[f,m]	0.525[b]		
F	0.205	0.405	0.655	0.695	0.78	Zn		0.30[n]	1.06[j,n]	1.424[n]	1.94[o]
Cl[b]	0.255	0.47	1.24	1.61	1.96	Cd	0.17[p]	0.43[f,q]	1.27[f,n]	1.555[o]	2.105[o]
Br					3.03[c]	Hg		0.55[f,j,r]	1.62[j,n,r,s]	2.14[r,s]	3.04[r,s]
I			1.53[d]	1.95[d]		B		0.25[t]	1.22[t]	1.90[t]	
O	0.20	0.415	0.75	0.90	1.14	Al		0.17[u]	1.68[v,w,x]	2.27[v,x,y]	3.80[v,y]
S		0.563[aa]	1.64[aa]	2.16[aa]		Ga			0.24[v]	0.36[v]	1.56[v]
N			0.75	1.01		In[v,x]			0.92	1.18	2.80
Li		0.11	0.505[e]	0.81	1.22	Tl			0.37[v]	0.42[v]	0.465[v]
Na		0.10[f]	0.50[f]	0.82[g,h]	1.40[i]	M	Rb	Cu	Au	Ca	Sr
K			0.49[j]	0.85	1.28	Ar	0.87	4.88[k]	1.55[k]	0.74[j]	0.81[j]
Cs			0.54	0.885	1.31	M	C	Si	Ge	Sn	Ni
Ag			0.95[j,k]	1.65	3.30[l]	Ar[z]	1.44	2.14	<2.09	<1.65	0.62[ab]

[a][4.5], [b][4.7], [c][4.9,4.10], [d][4.8], [e][4.11], [f][4.12], [g][4.13], [h][4.14], [i][4.15], [j][4.16, 4.17], [k][4.18], [l][4.19], [m][4.20], [n][4.21], [o][4.22], [p][4.23], [q][4.24], [r][4.25], [s][4.26], [t][4.27], [u][4.28], [v][4.29], [w][4.30], [x][4.31], [y][4.32], [z][4.33], [aa][4.196], [ab][4.34]

Table S4.3 Bond energies (kJ/mol) in van der Waals complexes of Atom · Molecule type

Molecule	Atom						
	He	Ne	Ar	Kr	Xe	F	O
H_2	0.13	0.275	0.61	0.705	0.78	0.40	0.45
Cl_2	0.17[a]	0.715[a,b]	2.25[a,b]	2.81[a]	3.425[a]		
Br_2	0.20[c]	0.98[d]					
I_2	0.215[e]	0.90[e]	2.165[f]				
O_2	0.245[g,h]	0.515[h]	1.135[h]	1.295[h,i]	1.51[h,i]		
N_2	0.22[g,h]	0.40[h,j]	0.955[h,k]	1.13[h]	1.385[l,m]		
Ag_2			0.275[n]	0.395[n]	1.23[n]		
NO	0.04[o]	0.346[o]	1.122[o]	1.315[w]	1.45[w]		
N_2O		0.96[p]	1.995[p]	2.30[p]	2.55[p]		
CO_2		0.83[p]	1.78[p]	2.065[p]	2.28[p]		
HF			1.21[q,r]	1.59[q]	2.165[q]		
HCl			1.375[q,r]	1.84[q]	2.46[q]		
CH_4	0.21	0.53[s]	1.355[s]	1.645[s]	1.89[s]	0.75	0.90
C_6H_6	0.985[t]	1.88[t,u]	4.53[t,u]	5.64[t,u]	6.82[t,u]		
C_2H_2			1.20[v]	1.44[v]	1.67[v]		

[a][4.35], [b][4.36], [c][4.37], [d][4.38], [e][4.39], [f]Average of [4.40] and [4.41], [g][4.42], [h][4.1], [i][4.43], [j][4.44], [k][4.45], [l][4.46], [m][4.47], [n][4.48], [o][4.49], [p][4.50], [q][4.51], [r][4.52], [s][4.53], [t][4.54], [u][4.55], [v][4.56], [w][4.197]

Table S4.4 Bond energies (kJ/mol) in van der Waals Molecule · Molecule/Atom complexes

Molecule/atom	Molecule				
	H_2	O_2	N_2	NH_3	CH_4
H_2	0.29	0.57			0.335[a]
O_2	0.57	1.64[b]	1.435[b]		1.34
N_2		1.435[b]	1.135[b]		
NH_3				7.895[c]	
CH_4		1.34			1.65[d]
Li			363		833
Cd	0.395[e]				1.45[f]
Hg	0.44[e]		1.31[h]	2.98[h]	2.14[f]
Al			4.235[g]		

[a][4.57], [b][4.58], [c][4.59], [d][4.60], [e][4.61], [f][4.62], [g][4.63]

Table S4.5 Bond energies (kJ/mol) in Atom · Ion type complexes

Ion	Atom				
He	Ne	Ar	Kr	Xe	
Li^+	7.91[a]	11.9[a]	24.5[a]	29.7[a]	36.9[a]
Na^+	4.34[a]	6.95[a]	16.1[a]	20.2[a]	25.95[a]
K^+	2.12[a]	3.86[a]	10.6[a]	13.7[a]	18.2[a]
Rb^+	1.74[a]	3.18[a]	9.36[a]	12.35[a]	16.7[a]
Cs^+	1.35[a]	2.51[a]	8.20[a]	11.1[a]	15.3[a]
Be^+		3.77[b]	49.2[c]		
Mg^+	0.78[d]	2.58[d]	15.1[c,d,e]	23.1[d,e,f]	34.4[d,e]
Ca^+		1.48	9.04[c,e]	15.6[e,f]	21.7[e]
Sr+		0.92[f]	9.90[c,f]	14.7[f]	
Ba^+			8.13[c]		
Hg^+			22.0[g]	37.9[g]	72.2[g]
B^+			25.7[g]		
Al^+			11.75[h]	18.3[h]	
Zr^+			32.4[i]		
V^+			36.0[j,k,l]	46.3[j,k]	62.7[j]
Nb^+			37.2[j]	55.1[j]	69.6[j]
Fe^+			45.3[m]		
Co^+			49.9[k,n]	65.2[k,n]	82.0[o]
Ni^+			54.7[p]		
F^-	1.83[a]	3.28[a]	9.65[a]	12.6[a]	17.0[a]
Cl^-	1.06[a]	2.12[a]	7.33[a]	9.68[a,q]	14.3[a]
Br^-	0.87[a]	1.83[a]	6.0[a,r]	8.56[a,q]	13.0[a,q]
I^-	0.67[a]	1.54[a]	5.2[a,r]	7.6[a,r]	12.1
S^-		1.093[s]	5.161[s]	7.814[s]	

[a][4.64], [b][4.65], [c][4.66], [d][4.67], [e][4.68], [f][4.69], [g][4.70], [h][4.71], [i][4.72], [j][4.73], [k][4.74], [l][4.75], [m][4.76], [n][4.77], [o][4.78], [p][4.79], [q][4.80], [r][4.6], [s][4.10a]

Table S4.6 Bond energies (kJ/mol) in Ion · Molecule type complexes

Ion	Molecule							
	H_2	N_2	NH_3	H_2O	CO	CH_4	C_2H_4	C_6H_6
Li^+								161[a]
Na^+				95[b]				92.6[a]
K^+				74.9[c]				73.3[a]
Rb^+								68.5[a]
Cs^+								64.6[a]
Cu^+	64.4[d]		237[e]	157[b]	149[b]		176[e]	218[f]
Ag^+			170[g]		89[b]			162[h]
Mg^+				119[b]			67.5[i]	
Ca+		21.0[j]						
Zn^+	15.7[k]							
Cd^+	136[h]							
Al+				104[b]				147[c]
Ti^+	41.8[d]		197[e]	154[b]	118[b]	70[b]	146[f]	258[f]
V^+	42.7[d]		192[e]	147[b]	113[b]		124[f]	233[f]
Cr^+	31.8[d]		183[e]	129[b]	90[b]		95.5[f]	170[f]
Mn^+	8.0[k]		147[e]	119[b]	25[b]		90.7[f]	133[f]
Fe^+	69.0[d]	54.0[l]	184[e]	128[b]	153[l]	57[b]	145[f]	207[f]
Co^+	76.2[d]		219[e]	161[b]	174[b]	90[b]	186[f]	255[f]
Ni^+	72.4[d]	111[b]	238[e]	180[b]	178[b]		182[f]	243[f]
F^-		18.8[m]		110[n]	41.0[m]	28.0[o]		
Cl^-						15.9[o]		64.9[o]
Br^-	4.4[p]					13.0[o]		
I^-	3.0[p]					10.9[o]		

[a][4.81], [b][4.82], [c][4.83], [d][4.84], [e][4.85], [f][4.86], [g][4.87], [h][4.88], [i][4.89], [j][4.90], [k][4.81], [l][4.76], [m][4.82], [n][4.83], [o][4.84], [p][4.85]

Table S4.7 Interatomic distances (Å) in van der Waals molecules A · Rg[a]

A	Rg				
	He	Ne	Ar	Kr	Xe
Li		5.01	4.86	4.84	4.84
Na		5.14	5.02	4.93	5.01
K			5.15	5.15	5.22
Rb				5.29	
Cs			5.50	5.44	5.47
Ag			4.0		
Mg		4.40	4.49		4.56
Zn			4.38[b]	4.20	
Cd		4.27	4.31[b]	4.36	4.55
Hg		3.91	3.99[b]	4.07	4.25
B			3.61		
Al			3.64	3.81	
In			3.86	3.9	
Tl			4.35	4.34	4.27
Si			4.0		
N			3.78	3.86	
O	3.27	3.30	3.60	3.75	3.90
S		3.596[c]	3.722[c]	3.781[c]	
H		3.15	3.58	3.62	3.82

Table S4.7 (continued)

A	Rg				
	He	Ne	Ar	Kr	Xe
F	3.03	3.15	3.50	3.65	3.78
Cl	3.49	3.61	3.88	3.95	4.06
Br			3.89		
I			4.11	4.20	
He	2.97	3.03	3.48	3.70	3.99
Ne	3.03	3.09	3.51	3.66	3.88
Ar	3.48	3.51	3.76	3.88	4.07
Kr	3.70	3.66	3.88	4.01	4.19
Xe	3.99	3.88	4.07	4.19	4.36

[a][4.1, 4.96, 4.97], [b][4.21], [c][4.196]

Table S4.8 Distances (Å) between centers of mass in van der Waals complexes $Rg \cdot A_nB_m$ and $Hg \cdot A_nB_m$

A_nB_m/Rg	He	Ne	Ar	Kr	Xe	Hg
H_2	3.24	3.30	3.58	3.72	3.94	
O_2	3.52	3.62	3.72	3.88	4.02	
N_2	3.69	3.72	3.71	3.84	4.10[a]	4.17[c]
Cl_2	3.67	3.57	3.72			
Br_2	3.84	3.67				
I_2	4.48		4.02			
HF	3.11		3.51	3.61	3.78	
HCl	3.45		3.98	4.08	4.25	
HBr			4.13	4.24		
OH		2.767[d]	2.789[d]	2.860[d]		
SH		2.902[d]	2.900[d]	2.968[d]		
CO		3.645[e]	3.849[e]	3.976[e]	4.194[e]	
CO_2[b]	3.559[h]	3.290[f]	3.504[f]	3.624[f]	3.815[f]	
C_2H_2[b]		4.01[g]	4.04[g]	4.15[g]	4.28[g]	
CH_4[b]	3.85	3.78	3.999[e]	4.097[e]	4.264[r]	4.0[c]
C_5H_5	3.77[i]	3.58[i]				
C_6H_6	3.20[j]	3.30[j]	3.58[j]	3.73[j]	3.89[j]	
SiH_4		4.13[h]	4.043[h]			
SiF_4			3.804[k]	3.942[k]		
H_2O	3.948		3.794[l]	3.91	3.95	3.6[c]
H_2S		3.959[l]	4.044[l]			
NO_2			3.490[m]	3.595[m]	3.774[m]	
N_2O	3.393[h]	3.241[f]	3.469[f]	3.594[f]	3.781[f]	
NH_3[n]		3.723[o]	3.836[o]	3.992[o]	4.067[o]	
BF_3[p]		3.09	3.32	3.45		
OCS	3.827[h]	3.54[f]	3.70[f]	3.81[f]	3.98[f]	
Li_2	6.87[q]	6.90[q]	6.35[q]	6.51[q]	6.39[q]	
Na_2		6.1[q]	7.53[q]			

[a][4.98], [b]Rg···C distances, [c][4.99], [d]Rg···H distances of [4.100], [e][4.101], [f][4.103], [g][4.104], [h][4.105], [i][4.106], [j][4.54], [k][4.107], [l][4.108], [m][4.109], [n]D(Cu···NH_3) = 2.00 Å, D(Ag···NH_3) = 2.42 Å [4.110], [o][4.111], [p]Rg···B distances of [4.112], [q][4.113], [r][4.102]

Table S4.9 Distances Rg···M (Å) in van der Waals complexes Rg · MX ([4.114], d(Ar···NaCl) = 2.89 Å)

Rg	CuF	CuCl	CuBr	Rg	AgF	AgCl	AgBr
Ar	2.22	2.27	2.30	Ar	2.56	2.61	2.64
Kr	2.32	2.36	2.47	Kr	2.59	2.64	2.66
Rg	AuF	AuCl	AuBr	Xe	2.65	2.70	
Ar	2.39	2.47	2.50	$H_2O\cdots M$[a]	CuCl	AgF	AgCl
Kr	2.46	2.52			1.914	2.168[b]	2.198
				$H_2S\cdots M$[a]	CuCl	AgCl	
					2.153	2.384	

[a][4.115], [b][4.198]

Table S4.10 Distances d(Å) in van der Waals complexes comprising two molecules (The distance between molecules' centers of mass, except where specified, X = halogen atom)

Molecule	H_2	F_2	Cl_2	Br_2	O_2	N_2	H_2O	CO_2	N_2O	NH_3
H_2	3.44[a]				3.62[b]		3.16[c]			
H_2O	3.16[c]	2.748[d]	2.848[d]			3.37[e]	2.98[f]			2.02[g]
H_2S		3.20[h]	3.25[h]	3.18[h]						2.32[g]
HCl	3.99[c]						3.227[j]			1.85[g]
HI						3.70[i]	3.745[j]			
CO			3.134[k]	3.105[k]						3.70[l]
CO_2				5.116[m]				3.60[n]		2.99[g]
C_2H_4			3.13[o]	3.07[o]						
C_2H_2			3.163	3.134[p]				3.285[q]	3.305[q]	2.33[g]
NH_3		2.71[r]	2.73[r]	2.72[r]			2.98[s]	2.99[f]		3.27[g]
C_6H_6						3.31[t]	3.33[t]	3.27[t]		
N_2					3.70[u]	3.81[u]	3.37[d]			
N_2O								3.47[n]	3.42[n]	
O_2	3.62[b]				3.56[u]	3.70[u]				

[a][4.2], [b][4.1], [c]H···O [4.116], [d]X···O [4.115], [e]N···O [4.117], [f]O···O, [g]N···X [4.118], [h]X···S [4.119], [i][4.120], [j]X···O [4.121], [k]X···C [4.122], [l][4.123], [m][4.124], [n][4.125], [o]X···center of C_2H_4 [4.126], [p]X···center of C_2H_2 [4.127], [q][4.128], [r]X···N [4.129], [s][4.130], [t][4.131], [u][4.58]

Table S4.11 Van der Waals radii of nonmetals, determined by X-ray analysis from 1939 to 2005

Authors	H	F	Cl	Br	I	O	S	N	C
Pauling[a]	1.2	1.35	1.80	1.95	2.15	1.40	1.85	1.5	1.70
Kitaigorodskii[b]	1.17		1.78	1.95	2.1	1.36		1.57	
Bondi[c]	1.20	1.47	1.75	1.85	1.98	1.50	1.80	1.55	1.70
Alcock[d]		1.47	1.75	1.85	1.98	1.52	1.80	1.55	
Gavezzotti[e]	1.2	1.35	1.8	1.95	2.1	1.4	1.85	1.5	1.7
Zefirov, Zorkii[f]	1.16	1.40	1.90	1.97	2.14	1.29	1.84	1.50	1.72
Tran et al.[g]	1.2	1.4	1.8	2.0	2.1	1.4	1.8	1.6	1.7
Batsanov[h]			1.80	1.90	2.10	1.51[a]	1.80		1.68
Rouland et al.[i]	1.10	1.46	1.76	1.87	2.03	1.58	1.81	1.64	1.71
Zorkii, Stukalin[j]	1.17		1.80			1.34		1.52	

[a][4.132], [b][4.133], [c][4.134], [d][4.135], [e][4.136], [f][4.137], [g][4.138], [h][4.139], [i][4.140], [j][4.141]

Table S4.12 The van der Waals radii (Å) of metals, calculated from structural data

M	*R*	M	*R*	M	*R*
Na	2.76[a]	Al	2.02[b]	Bi	2.22[b], 2.07[ac], 2.15[s]
K	3.13[a]	Ga	(2.05)[b], 2.06[n]	Cr	(2.0)[b]
Rb	3.17[a], 3.03[b]	In	1.98[n]	Mo	(2.2)[b]
Cs	3.3[a], 3.43[b]	Tl	2.02[o], 2.05[p], 2.00[q]	W	(2.2)[b]
Cu	1.7[c]	Ti	(2.0)[b]	Se	1.85[a], 1.92[b], 1.90[s]
Ag	1.9[d]	Zr	(2.3)[b], 1.89[r]	Te	2.02[a], (2.06)[b], 2.06[s]
Au	1.82[e], 2.07[f], 2.06[g], 2.28[h]	Si	1.92[a], 1.95[b], 2.10[s]	Fe	(2.1)[b], 2.10[aa]
Be	1.53[b]	Ge	2.00[a], 2.11[b], 1.95[s]	Co	1.6[b]
Ca	2.17[a], 2.19[i], 2.31[b]	Sn	2.29[a], (2.15)[b]	Ni	1.6[b]
Sr	2.36[a], 2.49[b]		2.10[s], 2.32[t]	Ru	2.16[ab]
Ba	2.51[j], 2.68[b]	Pb	2.03[u], 2.34[v]	Rh	1.94[ac]
Zn	1.95[b], (2.0)[b]	V	(2.0)[b]	Pd	2.1[c], 2.2[ad]
Cd	(2.05)[b]	Nb	2.1[w], 2.25[x]	Os	2.26[a]
Hg	1.8[b], (1.9)[b], 1.95[k], 1.86[l]	Ta	2.23[x]	Ir	1.84[ae]
Y	1.8[m]	P	1.85[b], 1.80[s]	Pt	1.7[c,j], 1.9[cc], 2.2[ag]
B	1.92[b]	As	1.96[b], 1.85[s]	U	2.47[ah]
		Sb	2.06[b], 1.98[y], 2.05[s]		

[a][4.139], [b][4.142], [c][4.143], [d][4.144], [e][4.145], [f][4.146, 4.147], [g][4.148], [h]The average of [4.146, 4.147], [i][4.149], [j][4.150], [k][4.151], [l][4.152], [m][4.153], [n][4.154], [o][4.155], [p][4.156], [q][4.157], [r][4.158], [s][4.135], [t][4.159], [u][4.160], [v][4.161], [w][4.162], [x][4.163], [y][4.164], [z][4.165], [aa][4.166], [ab][4.167], [ac][4.168], [ad][4.169], [ae][4.170], [af][4.171], [ag][4.172], [ah][4.173]

Table S4.13 Crystallographic van der Waals radii (Å) of metals

M	[4.174]	[4.175]	[4.176]	[4.177]	[4.178]	[4.179]	[4.180]	[4.181]	[4.182]
Li	2.24	2.25	2.14			2.14	2.01	2.0	2.14
Na	2.57	2.40	2.39			2.38	2.48	2.7	2.45
K	3.00	2.67	2.67			2.52	3.00	3.1	2.85
Rb	3.12	2.78	2.76			2.61	3.18	3.3	2.92
Cs	3.31	2.90	2.93			2.75	3.46	3.5	3.11
Cu	2.00	2.16	2.12	1.86	1.92	1.96	1.68	1.8	1.92
Ag	2.13	2.25	2.29	2.03	2.10	2.11	1.90	2.1	2.07
Au	2.13	2.18		2.17	2.10	2.14	1.86	2.3	2.04
Be	1.86	2.03	1.92	2.05		1.69	1.47	2.2	1.78
Mg	2.27	2.22	2.18	2.05		2.00	2.13	2.7	2.22
Ca	2.61	2.43	2.27	2.21		2.27	2.60	3.0	2.53
Sr	2.78	2.54	2.40	2.24		2.42	2.83	3.2	2.69
Ba	2.85	2.67	2.55	2.51		2.59	2.95	3.4	2.77
Zn	2.02	2.09	2.14	2.10	1.98	2.01	1.85	1.7	2.03
Cd	2.17	2.18	2.29	2.30	2.17	2.18	2.04	2.0	2.16
Hg	2.17	2.15	2.27	2.09	2.24	2.23	2.00	2.1	2.13
Sc	2.28	2.37	2.18	2.16	2.12	2.15	2.16	2.0	2.24
Y	2.45	2.47	2.38	2.19	2.29	2.32	2.43	2.3	2.42
La	2.51	2.58	2.59	2.40	2.45	2.43	2.54	2.4	2.49
B	1.74	1.87	1.72	1.47		1.68	1.34	1.5	1.65
Al	2.11	2.19	2.07	2.11		1.92	1.94	1.9	2.09
Ga	2.08	2.17	2.13	2.08		2.03	1.88	1.7	2.05
In	2.24	2.28	2.32	2.36		2.21	2.18	1.9	2.25
Tl	2.25	2.29	2.42	2.35		2.27	2.22	2.5	2.28

Table S4.13 (continued)

M	[4.174]	[4.175]	[4.176]	[4.177]	[4.178]	[4.179]	[4.180]	[4.181]	[4.182]
Ti	2.14	2.30	2.22	1.87	2.07	2.11	2.02	1.7	2.15
Zr	2.25	2.38	2.31	1.86	2.19	2.23	2.30	2.0	2.33
Hf	2.24	2.34	2.29	2.12	2.19	2.23	2.25	2.1	2.30
Si	2.06	2.06	2.19	2.07		1.93	1.88	1.9	1.98
Ge	2.13	2.10	2.22	2.15		2.05	1.94	1.9	2.02
Sn	2.29	2.21	2.40	2.33		2.23	2.10	2.2	2.20
Pb	2.36	2.24	2.46	2.32		2.37	2.20	2.2	2.27
V	2.03		2.14	1.79	2.06	2.07	1.96		2.11
Nb	2.13	2.34	2.15	2.07	2.17	2.18	2.04		2.16
Ta	2.13	2.26	2.22	2.17	2.18	2.22	2.12		2.22
As	2.16	2.05	2.14	2.06		2.08	1.90	2.0	2.03
Sb	2.33	2.20	2.32	2.25		2.24	2.19	2.1	2.23
Bi	2.42	2.28	2.40	2.43		2.38	2.25	2.2	2.30
Cr	1.97	2.27	2.05	1.89	2.06	2.06	1.90	1.8	2.12
Mo	2.06	2.29	2.16	2.09	2.16	2.17	2.00		2.13
W	2.07	2.23	2.14	2.10	2.18	2.18	2.04		2.16
Mn	1.96	2.15	2.00	1.97	2.04	2.05	1.92	1.7	2.08
Tc	2.04		2.11	2.09	2.16	2.16	2.01		2.14
Re	2.05	2.21	2.11	2.17	2.16	2.16	1.96		2.11
Fe	1.96	2.19	1.98	1.94	2.02	2.04	1.86	1.7	2.11
Co	1.95	2.16	1.97	1.92	1.91	2.00	1.85	1.7	2.04
Ni	1.92	2.14	1.97	1.84	1.98	1.97	1.83	1.6	2.00
Ru	2.02		2.05	2.07	2.17	2.13	1.95		2.11
Rh	2.02		2.04	1.95	2.04	2.10	1.95		2.07
Pd	2.05		2.14	2.03	2.09	2.10	1.98		2.10
Os	2.03		2.02	2.16	2.17	2.16	1.95		2.09
Ir	2.03		2.01	2.02	2.09	2.13	1.94		2.09
Pt	2.06		2.15	2.09	2.09	2.13	1.96		2.10
Th	2.43	2.54	2.50	2.37		2.45	2.48	2.2	2.45
U	2.17	2.51	2.45	2.40		2.41	2.40	2.5	2.40

Table S4.14 Quantum mechanical vdW equilibrium radii (Å) of selected elements

Reference	Li	Be	B	C	N	O	F	Ne
4.183	2.21	2.21	2.07	1.92	1.77	1.68	1.60	1.53
4.184	2.67	1.64	1.88	1.64	1.36	1.27	1.07	0.94
4.185	2.21	2.20	2.06	1.92	1.79	1.70	1.61	1.52
4.186	2.84	2.22	1.78	1.62	1.43	1.36	1.21	1.22
Reference	Na	Mg	Al	Si	P	S	Cl	Ar
4.183	2.25	2.43	2.44	2.35	2.24	2.15	2.07	1.97
4.184	2.79	2.00	2.49	2.26	1.75	1.71	1.46	1.34
4.185	2.25	2.42	2.41	2.33	2.23	2.14	2.08	1.98
4.186	3.07	2.75	2.30	2.21	2.03	1.91	1.86	1.78

Table S4.15 The molecular mechanics equilibrium radii (Å) of selected elements

Authors	Reference	Year	C	N	O	F	H
Mundt et al.	4.187	1983	1.80		1.36		1.17
Woolf, Roux	4.188	1994	2.06	1.85	1.77		1.32
Cornell et al.	4.189	1995	1.91	1.87	1.66	1.75	1.0 ± 0.4

Table S4.16 The equilibrium vdW radii (Å) of elements according to Allinger [4.186] (upper lines) and Batsanov (middle [4.175] and lower [4.176] lines)

Li	Be											B	C	N	O	F
2.55	2.23											2.15	2.04	1.93	1.82	1.71
2.72	2.32											2.05	1.85	1.70	1.64	1.61
2.51	2.13											1.85				
Na	Mg											Al	Si	P	S	Cl
2.70	2.43											2.36	2.29	2.22	2.15	2.07
2.82	2.45											2.47	2.25	2.09	2.00	1.82
2.78	2.48											2.31	2.34			
K	Ca	Sc	Ti	V	Cr	Mn	Fe	Co	Ni	Cu	Zn	Ga	Ge	As	Se	Br
3.09	2.81	2.61	2.39	2.29	2.25	2.24	2.23	2.23	2.22	2.26	2.29	2.46	2.44	2.36	2.29	2.22
3.08	2.77	2.64	2.52	2.50	2.51	2.43	2.34	2.30	2.26	2.30	2.25	2.38	2.23	2.16	2.10	2.00
3.07	2.61	2.45	2.45	2.31	2.24	2.21	2,21	2.20	2.20	2.31	2.37	2.36	2.35	2.25		
Rb	Sr	Y	Zr	Nb	Mo	Tc	Ru	Rh	Pd	Ag	Cd	In	Sn	Sb	Te	I
3.25	3.00	2.71	2.54	2.43	2.39	2.36	2.34	2.34	2.37	2.43	2.50	2.64	2.59	2.52	2.44	2.36
3.22	2.90	2.73	2.63	2.50	2.40					2.34	2.32	2.44	2.34	2.33	2.30	2.15
3.18	2.75	2.68	2.56	2.38	2.39	2.30	2.26	2.23	2.33	2.54	2.42	2.51	2.59	2.49		
Cs	Ba	La	Hf	Ta	W	Re	Os	Ir	Pt	Au	Hg	Tl	Pb	Bi	Po	At
3.44	3.07	2.78	2.53	2.43	2.39	2.37	2.35	2.36	2.39	2.43	2.53	2.59	2.74	2.66	2.59	2.51
3.38	3.05	2.86	2.54	2.44	2.35	2.38					2.25	2.46	2.34	2.40		
3.36	2.92	2.90	2.54	2.47	2.37	2.28	2.23	2.20	2.32		2.48	2.67	2.67	2.59		
		Th	U													
		2.74	2.52													
		2.75	2.73													
		2.80	2.72													

Table S4.17 The anisotropic vdW radii of H and X in solid-state HX molecules [4.190]

HX	Type	$d(\mathrm{H}\cdots\mathrm{X})$	$d(\mathrm{X}\cdots\mathrm{X})$	$R_l(\mathrm{X})$	$\bar{R}_l(X)$	$R_t(\mathrm{X})$	$\bar{R}_t(X)$	$R_l(\mathrm{H})$	$\bar{R}_l(H)$
HCl	I	2.62	3.87	1.83	1.78	2.03	1.98	0.79	0.75
	II	2.46	3.69	1.74		1.94		0.72	
HBr	I	2.78	4.110	1.900	1.87	1.930	2.17	0.88	0.86
	II	2.675	3.968	1.834		1.864		0.84	
HI	I	3.028	4.459	2.045	2.00	2.082	2.37	0.98	0.92
	II	2.813	4.277	1.953		1.990		0.86	

Table S4.18 The optical anisotropy, the length of bonds and the anisotropic vdW radii of X in X_2/AX_n molecules (in Å) [4.191]

X_2/AX_2	H_2	Cl_2	Br_2	I_2	O_2	N_2	CO_2	CS_2	C_2H_2
γ(A–X)	1.1265	1.192	1.207	1.222	1.223	1.132	1.272	1.372	1.175
d(A–X)	0.7414	1.988	2.281	2.666	1.208	1.098	1.163	1.564	1.061
R_t(X)	1.72	1.94	2.05	2.20	1.78	1.95	1.80	2.03	2.17*
R_l(X)	1.57	1.32	1.33	1.35	1.57	1.66	1.37	1.33	0.89
AX_3/AX_4	BF_3	BCl_3	BBr_3	BI_3	CCl_4	$SiCl_4$	$GeCl_4$	$SnCl_4$	
γ(A–X)	1.084	1.149	1.171	1.178	1.262	1.214	1.288	1.293	
d(A–X)	1.311	1.742	1.893	2.118	1.767	2.017	2.108	2.275	
R_t(X)	1.26	1.90	2.06	2.30	1.89	1.99	1.88	1.85	
R_l(X)	1.10	1.42	1.55	1.73	1.31	1.53	1.65	1.86	
AX_4	SiF_4	CBr_4	$SiBr_4$	$GeBr_4$	$SnBr_4$	SiI_4	GeI_4	SnI_4	
γ(A–X)	1.080	1.285	1.238	1.258	1.278	1.228	1.262	1.267	
d(A–X)	1.553	1.935	2.183	2.264	2.44	2.43	2.507	2.64	
R_t(X)	1.25	2.02	2.21	2.18	2.14	2.39	2.54	2.52	
R_l(X)	1.28	1.45	1.64	1.73	1.93	1.81	1.89	2.02	

*Estimate

Table S4.19 Lengths and energies of the donor-acceptor and normal chemical metal-nitrogen bonds

M	d(M–N), Å		E(M–N), kJ/mol	
	$[M(NH_3)_n]X_m$	$M(NH_2)_n$	$M{-}NH_3$	M–N
Cu	2.03	2.05	63	176
Ag	2.12	2.13	49	149
Au	2.02	2.02	77	
Ca	2.55	2.52	75	159
Zn	2.01	2.03	63	173
Sc	2.29	2.25	55	194
La	2.70	2.65	75	201
Al	1.92	1.89		278
Ga	1.92	1.90	81	215
Zr	2.34	2.37		2.41
Hf	2.29	2.35		2.43
Si	1.90	1.92		215
V^{II}	2.22	2.14		201
Cr^{II}	2.24	2.18	53	
Mn^{II}	2.27	2.21	71	136
Fe^{II}	2.21	2.21	73	147
Co^{II}	2.18	2.14	68	
Ni^{II}	2.12	2.09	78	143
Cr^{III}	2.07	2.10	56	164
Co^{III}	1.96	1.95		143
Ru^{III}	2.10	2.10		
Ir^{III}	2.24	2.13		
Pt	2.04	2.09	84	

Table S4.20 Interatomic distances (Å) in the O–H···O system

Compound	d(O–H)	d(H···O)	Compound	d(O–H)	d(H···O)
$Ca(OH)_2$	0.936	2.397	H_2O	1.01	1.75
$Ni(OH)_2$	0.943	2.183	$B(OH)_3$	1.02	1.70
$Be(OH)_2$[a]	0.956	2.007	$KHCO_3$[c]	1.023	1.587
$H_2C_2O_4 \cdot 2H_2O$	0.960	1.917	KD_2AsO_4	1.03	1.49
$Cu_2(OH)_2CO_3$	0.97	1.84	KH_2AsO_4	1.06	1.46
$NiSO_4 \cdot 6H_2O$[b]	0.975	1.800	KH_2PO_4	1.08	1.41
HIO_3	0.99	1.78	$Ni(DMG)_2$	1.22	1.22

[a][4.192], [b][4.193], [c][4.194]

Table S4.21 Crystallographic parameters of the ice crystal structures [4.195]

Ice	Space group	T (K)	P (bar)	Unit cell parameters (Å)	z[a]	ρ (g/cm³)	Degree of proton ordering
Ih	$P6_3/mmc$	98	1	$a = 4.48, c = 7.31$	4	0.92	Completely disordered
Ic	$Fd\bar{3}m$	98	1	$a = 6.350$	8	0.92	The same
II	$R\bar{3}$	123	1	$a = 7.78, \alpha = 113.1^\circ$	12	1.17	Completely ordered
III	$P4_12_12$	240	2.5	$a = 6.67, c = 6.96$	12	1.286	Disordered
IV	$R\bar{3}c$	110	1	$a = 7.60, \alpha = 70.1^\circ$	16	1.272	Completely disordered
V	$A2/a$	98	1	$a = 9.22, b = 7.54$ $c = 10.35, \beta = 109.2^\circ$	28	1.231	Disordered
VI	$P4_2/nmc$	98	1	$a = 6.27, c = 5.79$	10	1.31	Completely disordered
VII	$Pn\bar{3}m$	295	24	$a = 3.344$	2	1.778	The same
VIII	$I4_1/amd$	10	24	$a = 4.656, c = 6.775$	8	1.810	Completely ordered
IX	$P4_12_12$	110	1	$a = 6.73, c = 6.83$	12	1.13	Ordered
X	$Pn\bar{3}m$				2		Protons reside at midpoints of O···O bonds
XI	$Cmc2_1$	5	1	$a = 4.502, b = 7.798$ $c = 7.328$	8	0.93	Ordered
XII	$I\bar{4}2d$	1.5	1	$a = 8.282, c = 4.036$	12	1.440	Disordered

[a]The number of molecules per unit cell

Supplementary References

4.1 Cambi R, Cappelletti D, Liuti G, Pirani F (1991) J Chem Phys 95:1852
4.2 Chalasinski G, Gutowski M (1988) Chem Rev 88:943
4.3 Aziz RA, Slaman MJ (1991) J Chem Phys 94:8047
4.4 Luo F, McBane GC, Kim G et al (1993) J Chem Phys 98:3564
4.5 Xu Y, Jäger W, Djauhari J, Gerry MCL (1995) J Chem Phys 103:2827
4.6 Zhao Y, Yourshaw I, Reiser G et al (1994) J Chem Phys 101:6538
4.7 Aquilanti V, Cappelletti D, Lorent V et al (1993) J Phys Chem 97:2063
4.8 Tsukiyama K, Kasuya T (1992) J Mol Spectr 151:312
4.9 Clevenger JO (1994) Chem Phys Lett 231:515
4.10 Clevenger JO, Tellinghuisen J (1995) J Chem Phys 103:9611
4.11 Brühl R, Zimmermann D (1995) Chem Phys Lett 233:455; Brühl R, Zimmermann D (2001) J Chem Phys 115:7892
4.12 Wallace I, Breckenridge WH (1993) J Chem Phys 98:2768
4.13 Brühl R, Kapetanakis J, Zimmermann D (1991) J Chem Phys 94:5865

4.14 Zanger E, Schmatloch V, Zimmermann D (1988) J Chem Phys 88:5396
4.15 Baunmann P, Zimmermann D, Brühl R (1992) J Mol Spectr 155:277
4.16 Jouvet C, Lardeux-Dedonder C, Martrenchard S, Solgadi D (1991) J Chem Phys 94:1759
4.17 Bokelmann F, Zimmermann D (1996) J Chem Phys 104:923
4.18 Knight AM, Strangassinger A, Duncan MA (1997) Chem Phys Lett 273:265
4.19 Brock LR, Duncan MA (1995) J Chem Phys 103:9200
4.20 Leung AWK, Julian RR, Breckenridge WH (1999) J Chem Phys 111:4999
4.21 Koperski J, Czajkowski M (2002) J Mol Spectr 212:162; Strojecki M, Koperski J (2009) Chem Phys Lett 479:189
4.22 Wallace I, Kaup JG, Breckenridge WH (1991) J Phys Chem 95:8060; Wallace I, Ryter J, Breckenridge WH (1992) J Chem Phys 96:136
4.23 Koperski J, Czajkowski M (1998) J Chem Phys 109:459
4.24 Wallace I, Funk DJ, Kaup JG, Breckenridge WH (1992) J Chem Phys 97:3135
4.25 Duval M-C, Soep B, Breckenridge WH (1991) J Phys Chem 95:7145
4.26 Collier MA, McCaffrey JG (2003) J Chem Phys 119:11878
4.27 Yang X, Dagdigian PJ (1997) J Phys Chem A101:3509
4.28 Yang X, Dagdigian PJ, Alexer MH (1998) J Chem Phys 108:3522
4.29 Strangassinger A, Knight AM, Duncan MA (1998) J Chem Phys 108:5732
4.30 McQuaid MJ, Gole JL, Heaven MC (1990) J Chem Phys 92:2733; Heidecke S, Fu Z, Colt JR, Morse MD (1992) J Chem Phys 97:1692
4.31 Callender CL, Mitchell SA, Hackett PA (1989) J Chem Phys 90:5252
4.32 Fu Z, Massik S, Kaup JG et al (1992) J Chem Phys 97:1683
4.33 Tao C, Dagdigian PJ (2004) J Chem Phys 120:7512
4.34 Kawamoto Y, Honma K (1998) Chem Phys Lett 298:227
4.35 Bieler C R, Spence KE, Ja KC (1991) J Phys Chem 95:5058
4.36 Rohrbacher A, Ja KC, Beneventi L et al (1997) J Phys Chem A 101:6528
4.37 Boucher DS, Strasfeld DB, Loomis RA et al (2005) J Chem Phys 123:104312
4.38 Cabrera JA, Bieler CR, Olbricht BC et al (2005) J Chem Phys 123:054311
4.39 Cline J et al (1987) In: Weber A (ed) Structure and dynamics of weakly bound molecular complexes. Reidel, Dordrecht, NATO ASI Series 212:533
4.40 Miller AES, Chuang C-C, Fu HC et al (1999) J Chem Phys 111:7844
4.41 Burroughs A, Heaven MC (2001) J Chem Phys 114:7027
4.42 Beneventi L, Casavecchia P, Volpi G (1986) J Chem Phys 85:7011
4.43 Aquilanti V, Ascenzi D, Cappelletti D et al (1998) J Chem Phys 109:3898
4.44 Jäger W, Xu Y, Armstrong G et al (1998) J Chem Phys 109:5420
4.45 Munteanu CR, Cacheiro JL, Fernez B (2004) J Chem Phys 121:10419
4.46 Cappelletti D, Liuti G, Luzzatti E, Pirani F (1994) J Chem Phys 101:1225
4.47 Wen Q, Jäger W (2005) J Chem Phys 122:214310
4.48 Robbins D, Willey KF, Yeh CS, Duncan MA (1992) J Phys Chem 96:4824
4.49 Holmess-Ross HL, Lawrance WD (2011) J Chem Phys 135:014302
4.50 Herrebout WA, Qian HB, Yamaguchi H et al (1998) J Molec Spectr 189:235
4.51 Frazer GT, Pine AS (1986) J Chem Phys 85:2502
4.52 McIntosh, A, Wang, Z, Castillo-Chara, J et al (1999) J Chem Phys 111:5764
4.53 Luiti G, Pirani F, Buck U, Schmidt B (1988) ChemPhys 126:1
4.54 Pirani F, Porrini M, Cavalli S et al (2003) ChemPhys Lett 367:405
4.55 Brupbacher T, Makarewicz J, Bauder A (1994) J Chem Phys 101:9736
4.56 Cappelletti D, Bartholomei M, Carmona-Novillo E et al (2007) J Chem Phys 126:064311
4.57 McKellar ARW, Roth DA, Winnewisser G (1999) J Chem Phys 110:9989
4.58 Aquilanti V, Bartolomei M, Cappelletti D et al (2001) Phys Chem Chem Phys 3:3891
4.59 Case AS, Heid CG, Kable SH, Crim FF (2011) J Chem Phys 135:084312
4.60 Reid BP, O'Loughlin MJ, Sparks RK (1985) J Chem Phys 83:5656
4.61 Wallace I, Funk DJ, Kaup JG, Breckenridge WH (1992) J Chem Phys 97:3135
4.62 Wallace I, Breckenridge WH (1992) J Chem Phys 97:2318
4.63 Yang X, Gerasimov I, Dagdigian PJ (1998) Chem Phys 239:207

4.64 Cappelletti D, Liuti G, Pirani F (1991) Chem Phys Lett183:297
4.65 Frenking G (1989) J Chem Phys 93:3410
4.66 Lüder C, Velegrakis M (1996) J Chem Phys 105:2167
4.67 Burns KL, Bellert D, Leung AW-K, Breckenridge WH (2001) J Chem Phys 114:7877
4.68 Kaup JG, Breckenridge WH (1997) J Chem Phys 107:5283
4.69 Prekas D, Feng B-H, Velegrakis M (1998) J Chem Phys 108:2712
4.70 Breckenridge WH, Jouvet C, Soep B (1995) Adv Metal Semicond Cluster 3:1
4.71 Heidecke SA, Fu Z, Colt JR, Morse MD (1992) J Chem Phys 97:1692
4.72 Scurlock C, Pilgrim J, Duncan MA (1995) J Chem Phys 103:3292
4.73 Bellert D, Buthelezi T, Hayes T, Brucat PJ et al (1997) Chem Phys Lett 277:27
4.74 Lessen DE, Brucat PJ (1989) J Chem Phys 91:4522
4.75 Hayes T, Bellert D, Buthelezi T, Brucat PJ (1998) Chem Phys Lett 287:22
4.76 Tjelta B, Armentrout PB (1997) J Phys Chem A101: 2064
4.77 Buthelezi T, Bellert D, Lewis V, Brucat PJ et al (1995) Chem Phys Lett 242: 627
4.78 Haynes C, Armentrout PB (1996) Chem Phys Lett 249:64
4.79 Asher R, Bellert D, Buthelezi T, Brucat PJ (1994) Chem Phys Lett 228:599
4.80 Yourshaw I, Lenzer T, Reiseer G, Neumark DM (1998) J Chem Phys 109:5247
4.81 Amicangelo JC, Armentrout PB (2000) J Phys Chem A104:11420
4.82 Armentrout PB (1995) Acc Chem Rev 28:430
4.83 Ma JC, Dougherty DA (1997) Chem Rev 97:1303
4.84 Kemper PR, Weis P, Bowers MT (1998) Chem Phys Lett 293:503
4.85 Walter D, Armentrout PB (1998) J Am Chem Soc 120:3176
4.86 Sievers MR, Jarvis LM, Armentrout PB (1998) J Am Chem Soc 120:3176
4.87 Miyawaki J, Sugawara K-I (2003) J Chem Phys 119:6539
4.88 Ho Y-P, Yang Y-C, Klippenstein S, Dunbar R (1997) J Phys Chem A101:3338
4.89 Chen J, Wong TH, Chang YC et al (1998) J Chem Phys 108:2285
4.90 Pullins SH, Reddic JE, Frana MR, Duncan MA (1998) J Chem Phys 108:2725
4.91 Weis P, Kemper PR, Bowers MT (1997) J Phys Chem A 101:2809
4.92 Hiraoka K, Nasu M, Katsuragawa J et al (1998) J Phys Chem A102: 6916
4.93 Weis P, Kemper PR, Bowers MT (1999) J Am Chem Soc 121:3531
4.94 Hiraoka K, Mizuno T, Iino T et al (2001) J Phys Chem A105:4887
4.95 Wild DA, Loh ZM, Wilson RL, Bieske EJ (2002) J Chem Phys 117:3256
4.96 Batsanov SS (1998) J Chem Soc Dalton Trans 1541
4.97 Collier MA, McCaffrey JG (2003) J Chem Phys 119:11878
4.98 Munteanu CR, Cacheiro JL, Fernez B (2004) J Chem Phys 121:10419; Wen Q, Jäger W (2005) J Chem Phys 122:214310
4.99 Duval M-C, Soep B, Breckenridge WH (1991) J Phys Chem 95:7145; van Wijngaarden J, Jäger W (2000) Mol Phys 98:1575
4.100 Carter CC, Castiglioni C (2000) J Mol Struct 525:1
4.101 Liu Y, Jäger W (2004) J Chem Phys 121:6240
4.102 Wen Q, Jäger W (2006) J Chem Phys 124:014301
4.103 Herrebout WA, Qiun H-B, Yamaguchi H, Howard BJ (1998) J Mol Spectr 189:235
4.104 Liu Y, Jäger W (2003) Phys Chem Chem Phys 5:1744; Cappeletti D, Bartolomei M, Carmona-Novillo E, Pirani F (2007) J Chem Phys 126:064311
4.105 McKellar ARW (2006) J Chem Phys 125:114310
4.106 Yu L, Williamson J, Foster SC, Miller TA (1992) J Chem Phys 97:5273
4.107 Urban R-D, Jörissen LG, Matsumoto Y, Takami M (1995) J Chem Phys 103:3960
4.108 Liu Y, Jäger W (2002) Mol Phys 100:611
4.109 Blanco S, Whitham CJ, Qian H, Howard BJ (2001) Phys Chem Chem Phys 3:3895
4.110 Miyawaki J, Sugawara K-i (2003) J Chem Phys 119:6539
4.111 van Wijngaarden J, Jäger W (2001) J Chem Phys 114:3968; van Wijngaarden J, Jäger W (2001) J Chem Phys 115:6504; Wen Q, Jäger W (2008) J Chem Phys 128:204309
4.112 Lee G-H, Matsuo Y, Takami M (1992) J Chem Phys 96:4079

4.113 Rubahn H-G, Toennies J (1988) Chem Phys 126:7; Rubahn H-G (1990) J Chem Phys 92:5384
4.114 Michaud JM, Cooke SA, Gerry MCL (2004) Inorg Chem 43:3871; Yamazaki E, Okabayashi T, Tanimoto M (2004) J Am Chem Soc 126:1028
4.115 Walker NR, Tew DR, Harris SJ et al (2011) J Chem Phys 135:014307
4.116 Anderson DT, Schuder M, Nesbitt DJ (1998) Chem Phys 239:253; Weida MJ, Nesbitt DJ (1999) J Chem Phys 110:156
4.117 Leung HO, Marshall MD, Suenram RD (1989) J Chem Phys 90:700
4.118 Nelson D, Fraser G, Klemperer W (1987) Science, 238:1670
4.119 Legon AC, Thumwood JM (2001) Phys Chem Chem Phys 3:2758
4.120 Jabs W, McIntosh AL, Lucchese RR et al (2000) J Chem Phys 113:249
4.121 Davey JB, Legon AC, Waclawik ER (2000) Phys Chem Chem Phys 2:1659
4.122 Davey JB, Legon AC, Waclawik ER (1999) Phys Chem Chem Phys 1:3097
4.123 Xia C, Walker KA, McKellar ARW (2001) J Chem Phys 114:4824
4.124 Sazonov A, Beaudet RA (1998) J Phys Chem A102:2792
4.125 Dutton CC, Dows DA, Erkey R et al (1998) J Phys Chem A102:6904
4.126 Legon AC, Thumwood JM (2001) Phys Chem Chem Phys 3:1397
4.127 Davey JB, Legon AC (2001) Chem Phys Lett 350:39
4.128 Peebles SA, Kuczkowski (1999) J Phys Chem A103:3884
4.129 Bloemink HI, Evans CM, Holloway JH, Legon AC (1996) Chem Phys Lett 248:260
4.130 Forest S, Kuczkowski R (1996) J Am Chem Soc 118:217
4.131 Sun S, Bernstein E (1996) J Phys Chem 100:13348; Schäfer M, Bauder A (2000) Mol Phys 98:929
4.132 Pauling L (1939, 1960) The nature of the chemical bond, 1st and 3rd edn. Cornell University Press, Ithaca
4.133 Kitaigorodskii AI (1961) Organic chemical crystallography. Consult Bureau, New York
4.134 Bondi A (1964) J Phys Chem 68:441; Bondi A (1968) Physical properties of molecular crystals, liquids, and Glasses. Wiley, New York
4.135 Alcock NW (1972) Adv Inorg Chem Radiochem 15:1
4.136 Gavezzotti A (1983) J Am Chem Soc 105:5220; Filippini G, Gavezzotti A (1993) Acta Cryst B49:868; Dunitz JD, Gavezzotti A (1999) Acc Chem Res 32:677
4.137 Zefirov YuV, Zorkii PM (1989) Russ Chem Rev 58:421; Zefirov YuV, Zorkii PM (1995) 64: 415; Zefirov YuV (1994) Crystallogr Reports 39:939; Zefirov YuV (1997) Crystallogr Reports 42:111
4.138 Tran D, Hunt JP, Wherl S (1992) Inorg Chem 31:2410
4.139 Batsanov SS (1995) Russ Chem Bull 44:18
4.140 Rowl RS, Taylor R (1996) J Phys Chem 100:7384
4.141 Zorkii PM, Stukalin AA (2005) Crystallogr Reports 50:522
4.142 Bondi A (1964) The heat of sublimation of molecular crystals, analysis and molecular structure correlation. In: Rutner E, Goldfinger P, Hirth JP (eds) Condensation and evaporation of solids. Gordon and Breach, New York; Mantina M, Chamberlin AC, Valero R et al (2009) J Phys Chem A113:5806
4.143 Braga D, Grepioni F, Tedesco E, Biradha K, Desiraju GR (1997) Organometallics 16:1846
4.144 Ardizzoia G, La Monica G, Maspero A, Moret M, Masciocchi N (1997) Inorg Chem 36: 2321
4.145 Pathaneni SS, Desiraju GR (1993) J Chem Soc Dalton Trans 319
4.146 Jones PG, Bembenek E (1992) J Cryst Spectr Res 22:397; Schulz LE, Abram U, Straehle J (1997) Z Anorg Allg Chem 623:1791
4.147 Omrani H, Welter R, Vangelisti R (1999) Acta Cryst C55:13
4.148 Helgesson G, Jagner S (1987) Acta Chem Scand A41:556
4.149 Gregory DH, Bowman A, Naher CF, Weston DP (2000) J Mater Chem 10:1635
4.150 Steinbrenner U, Simon A (1998) Z anorg allg Chem 624:228
4.151 Yang X, Knobler CB, Zheng Z, Hawthorne MF (1994) J Am Chem Soc 116:7142
4.152 Lee H, Diaz M, Knobler CB, Hawthorne MF (2000) Angew Chem Int Ed 39:776

4.153 Evans WJ, Boyle TJ, Ziller JW (1993) Organomet 12:3998
4.154 Loos D, Baum E, Ecker A, Schnöckel, Downs AJ (1997) Angew Chem Int Ed 36:860
4.155 Nagle JK, Balch AL, Olmstead MM (1988) J Am Chem Soc 110:319
4.156 Jutzi P, Schnittger J, Hursthouse MB (1991) Chem Ber 124:1693
4.157 Giester G, Lengauer CL, Tillmans E (2002) J Solid State Chem 168:322
4.158 Ho J, Rousseau R, Stephan DW (1994) Organometallics 13:1918
4.159 Atwood JL, Hunter WE, Cowley AH et al (1981) Chem Commun 925
4.160 Hill RJ (1985) Acta Cryst C41:1281
4.161 Campbell J, Dixon D, Mercier H, Schrobilgen G (1995) Inorg Chem 34:5798
4.162 Mawhorter RJ, Rankin DWH, Robertson HE et al (1994) Organometallics 13:2401
4.163 Krebs B, Sinram D (1980) ZNaturforsch 35b:126
4.164 Breunig, HJ, Denker, M, Ebert, KH (1994)Chem Commun 8756
4.165 Silvestru, C, Breunig, HJ, Althaus H (1999) Chem Rev 99:3277
4.166 Cassidy JM, Whitmire KH (1991) Inorg Chem 30:2788
4.167 Capobianchi A, Paoletti AM, Pennesi G et al (1994) Inorg Chem 33:4635
4.168 Bronger W, Müller P, Kowalczyk J, Auffermann G (1991) J Alloys Comp 176:263
4.169 Martin DS, Bonte JL, Rush RM, Jacobson RA (1975) Acta Cryst B31:2538; Hester JR, Maslen EN, Spadaccini N et al (1993) Acta Cryst B49:842; Bronger, W, Auffermann G (1995) JAlloys Comp, 228:119
4.170 Venturelli A, Rauchfuss TB (1994) J Am Chem Soc 116:4824
4.171 Krebs B, Brendel C, Schafer H (1988) Z anorg allgem Chemie, 561:119; Thiele G, Weigl W, Wochner H (1986) Z anorg allgem Chemie 539:141
4.172 Kroening RF, Rush RM, Martin DS, Clardy JC (1974) Inorg Chem 13:1366; Takazawa H, Ohba S, Saito Y, Sano M (1990) Acta Cryst B46:166; Bronger W, Auffermann G, (1995) J Alloys Comp 228:119
4.173 Ryan RR, Penneman RA, Kanellakopulos B (1975) J Am Chem Soc 97:4258
4.174 Batsanov SS (2000) Russ J Phys Chem 74:1144
4.175 Batsanov SS (1999) J Mol Struct (Theochem) 468:151
4.176 Batsanov SS (2000) Russ J Inorg Chem 45:892
4.177 Hu SZ, Zhou ZH, Robertson BE (2009) ZKrist 224:375
4.178 Nag S, Banerjee J, Datta D (2007) New J Chem 31:832
4.179 Kitaigorodskii AI (1973) Molecular crystals and molecules. Academic, New York
4.180 Batsanov SS (2001) Inorgan Mater 37:871
4.181 Batsanov SS (1998) Russ J Gen Chem 68:495
4.182 Batsanov SS (1991) Russ J Inorg Chem 36:1694; Batsanov SS (1998) Russ J Inorg Chem 43:437
4.183 Bader RFW (1994) Atoms in molecules: a quantum theory. Clarendon, Oxford
4.184 Yang Z-Z, Davidson ER (1997) Int J Quantum Chem 62:47
4.185 Mu W-H, Chasse G A, Fang D-C (2008) Intern J Quantum Chem 108:1422
4.186 Badenhoop J K, Weinhold F (1997) J Chem Phys 107:5422
4.187 Mundt O, Rössler GB, Witthauer C (1983) Z anorg allgem Chemie 506:42
4.188 Woolf TB, Roux B (1994) J Am Chem Soc 116:5916
4.189 Cornell WD, Cieplak P, Boyly CI et al (1995) J Am Chem Soc 117:5179
4.190 Batsanov SS (1999) Struct Chem 10:395
4.191 Batsanov SS (1998) Russ J Coord Chem 24:453; Batsanov SS (2002) Russ J General Chem 72: 1153
4.192 Stahl R, Jung C, Lutz HD et al (1998) Z anorg allgem Chem 624:1130
4.193 Ptasiewicz-Bak H, Olovsson I, McIntyre G (1993) Acta Cryst B49:192
4.194 Jeffrey GA, Yeon Y (1986) Acta Cryst, B42:410
4.195 Zheligovskaya EA, Malenkov GG (2006) Russ Chem Rev 75:57
4.196 Gar E, Neumark DM (2011) J Chem Phys 135:024302
4.197 Mack P, Dyke JM, Wright TG (1998) J Chem Soc Faraday, 94:629
4.198 Stephens SL, Tew DR, Walker NR, Legon AC (2011) J Mol Spectr 267:163

References

1. Bent HA (1968) Structural chemistry of donor-acceptor interactions. Chem Rev 68:587–648
2. Haaland A (1989) Covalent versus dative bonds to main group metals. Angew Chem Int Ed 28:992–1007
3. Wang SC (1927) The mutual influence between hydrogen atoms. Phys Z 28:663–666
4. London F (1930) On the theory and systematic of molecular forces. Z Physik 63:245–279
5. London F (1937) The general theory of molecular forces. Trans Faraday Soc 33:8–26
6. Buckingham AD, Fowler PW, Hutson JM (1988) Theoretical studies of van der Waals molecules and intermolecular forces. Chem Rev 88:963–988
7. Pyykkö P (1997) Strong-shell interactions in inorganic chemistry. Chem Rev 97:597–636
8. Ihm G, Cole MW, Toigo F, Scoles G (1987) Systematic trends in van der Waals interactions: atom-atom and atom-surface cases. J Chem Phys 87:3995–3999
9. Czajkowki M, Krause L, Bobkowski R (1994) $D1(5^1P_1) \leftarrow X0^+(5^1S_0)$ spectra of CdNe and CdAr excited in crossed molecular and laser beams. Phys Rev A49:775–786
10. Slater JC, Kirkwood JG (1931) The van der Waals forces in gases. Phys Rev 37:682–697
11. Cambi R, Cappelletti D, Liuti G, Pirani F (1991) Generalized correlations in terms of polarizability for van der Waals interaction potential parameter calculations. J Chem Phys 95:1852–1861
12. Andersson Y, Rydberg H (1999) Dispersion coefficients for van der Waals complexes, including C_{60}–C_{60}. Physica Scripta 60:211–216
13. Huyskens P (1989) Differences in the structures of highly polar and hydrogen-bonded liquids. J Mol Struct 198:123–133
14. Batsanov SS (1994) Van der Waals radii of metals from spectroscopic data. Russ Chem Bull 43:1300–1304
15. Batsanov SS (1996) Thermodynamic peculiarities of the formation of van der Waals molecules. Dokl Chem (Engl Transl) 349:176–178
16. Alkorta I, Rozas I, Elguero J (1998) Charge-transfer complexes between dihalogen compounds and electron donors. J Phys Chem A102:9278–9285
17. Batsanov SS (1998) Some characteristics of van der Waals interaction of atoms. Russ J Phys Chem 72:894–897
18. Batsanov SS (1998) On the additivity of van der Waals radii. Dalton Trans 1541–1545
19. Batsanov SS (1966) Refractometry and chemical structure. Van Nostrand, Princeton
20. Hohm U (1994) Dipole polarizability and bond dissociation energy. J Chem Phys 101:6362–6364
21. Müller H (1935) Theory of the photoelastic effect of cubic crystals. Phys Rev 47:947–957
22. Evans WJ, Silvera IF (1998) Index of refraction, polarizability, and equation of state of solid molecular hydrogen. Phys Rev B57:14105–14109
23. Shimizu H, Kamabuchi K, Kume T, Sasaki S (1999) High-pressure elastic properties of the orientationally disordered and hydrogen-bonded phase of solid HCl. Phys Rev B59:11727–11732
24. Shimizu H, Kitagawa T, Sasaki S (1993) Acoustic velocities, refractive index, and elastic constants of liquid and solid CO_2 at high pressures up to 6 GPa. Phys Rev B47:11567–11570
25. Olinger B (1982) The compression of solid CO_2 at 296 K to 10 GPa. J Chem Phys 77:6255–6258
26. Peterson CF, Rosenberg JT (1969) Index of refraction of ethanol and glycerol under shock. J Appl Phys 40:3044–3046
27. Smirnov BM (1993) Mechanisms of melting of rare gas solids. Physica Scripta 48:483–486
28. Runeberg N, Pyykkö P (1998) Relativistic pseudopotential calculations on Xe_2, RnXe, and Rn_2: the van der Waals properties of radon. Int J Quantum Chem 66:131–140
29. Pauling L (1970) General chemistry. 3rd edn. Freeman & Co, San-Francisco
30. Glyde HR (1976) Solid helium. In: Klein ML, Venables JA (eds) Rare gas solids. Academic, London, 1:121

31. Batsanov SS (2009) The dynamic criteria of melting-crystallization. Russ J Phys Chem A83:1836–1841
32. Simon F (1934) Behaviour of condensed helium near absolute zero. Nature 133:529
33. Feynman RP (1939) Forces in molecules. Phys Rev 56:340–343
34. Allen MJ, Tozer DJ (2002) Helium dimer dispersion forces and correlation potentials in density functional theory. J Chem Phys 117:11113–11120
35. Slater JC (1972) Hellmann-Feynman and virial theorems in the $X\alpha$ method. J Chem Phys 57:2389–2396
36. Bader RFW, Hernandez-Trujillo J, Cortes-Guzman F (2006) Chemical bonding: from Lewis to atoms in molecules. J Comput Chem 28:4–14
37. Bader RFW (2009) Bond paths are not chemical bonds. J Phys Chem A113:10391–10396
38. Bader RFW (2010) Definition of molecular structure: by choice or by appeal to observation? J Phys Chem A114:7431–7444
39. Batsanov SS (2010) The energy of covalent bonds between nonmetal atoms at van der Waals distances. Russ J Inorg Chem 55:1112–1113
40. Mulliken RS (1952) Magic formula, structure of bond energies and isovalent hybridization. J Phys Chem 56:295–311
41. Batsanov SS (1998) Estimation of the van der Waals radii of elements with the use of the Morse equation. Russ J Gen Chem 68:495–500
42. Minemoto S, Sakai H (2011) Measuring polarizability anisotropies of rare gas diatomic molecules. J Chem Phys 134:214305
43. Deiglmayr J, Aymar M, Wester R et al (2008) Calculations of static dipole polarizabilities of alkali dimers. J Chem Phys 129:064309
44. Muenter JS, Bhattacharjee R (1998) The electric dipole moment of the CO_2-CO van der Waals complex. J Mol Spectr 190:290–293
45. Pyykkö P, Wang C, Straka M, Vaara J (2007) A London-type formula for the dispersion interactions of endohedral A@B systems. Phys Chem Chem Phys 9:2954–2958
46. Dunitz JD, Gavezzotti A (1999) Attractions and repulsions in molecular crystals: what can be learned from the crystal structures of condensed ring aromatic hydrocarbons. Acc Chem Res 32:677–684
47. Gavezzotti A (1994) Are crystal structures predictable? Acc Chem Res 27:309–314
48. Allinger NI, Miller MA, Van Catledge FA, Hirsh JA (1967) Conformational analysis.LVII. The calculation of the conformational structures of hydrocarbons by the Westheimer-Hendrickson-Wiberg method. J Am Chem Soc 89:4345–4357
49. Allinger NI, Zhou X, Bergsma J (1994) Molecular mechanics parameters. J Mol Struct Theochem 312:69–83
50. Dunitz JD, Gavezzotti A (2005) Molecular recognition in organic crystals. Angew Chem Int Ed 44:1766–1786
51. Dunitz JD, Gavezzotti A (2009) How molecules stick together in organic crystals. Chem Soc Rev 38:2622–2633
52. Gavezzotti A (2010) The lines-of-force landscape of interactions between molecules in crystals. Acta Cryst B66:396–406
53. Batsanov SS (1999) Calculation of van der Waals radii of atoms from bond distances. J Mol Struct Theochem 468:151–159
54. Bürgi H-B (1975) Zur Beziehung zwischen Struktur und Energie: Bestimmung der Stereochemie von Reaktionswegen aus Kristallstrukturdaten. Angew Chem 87:461–475
55. Batsanov SS (2001) Effect of intermolecular distances on the probability of formation of covalent bonds. Russ J Phys Chem 75:672–674
56. Beckmann J, Dakternieks D, Duthie A, Mitchell C (2005) The utility of hypercoordination and secondary bonding for the synthesis of a binary organoelement oxo cluster. J Chem Soc Dalton Trans 1563–1564
57. Harris PM, Mack E, Blake FC (1928) The atomic arrangement in the crystal of orthorhombic iodine. J Am Chem Soc 50:1583–1600

58. Zefirov YuV, Zorkii PM (1989) Van der Waals radii and their applications in chemistry. Russ Chem Rev 58:421–440
59. Zefirov YuV, Zorkii PM (1995) New applications of van der Waals radii in chemistry. Russ Chem Rev 64:415–428
60. Zefirov YuV (1994) Van der Waals radii and specific interactions in molecular crystals. Crystallogr Rep 39:939–945
61. Zefirov YuV (1997) Comparative analysis of the systems of van der Waals radii. Crystallogr Rep 42:111–117
62. Townes CH, Dailey BP (1952) Nuclear quadrupole effects and electronic structure of molecules in the solid state. J Chem Phys 20:35
63. Hassel O, Rømming C (1962) Direct structural evidence for weak charge-transfer bonds in solids containing chemically saturated molecules. Quart Rev Chem Soc 16:1–18
64. Alcock NW (1972) Secondary bonding to nonmetallic elements. Adv Inorg Chem Radiochem 15:1–58
65. Takemura K, Minomura S, Shimomura O et al (1982) Structural aspects of solid iodine associated with metallization and molecular dissociation under high-pressure. Phys Rev B26:998–1004
66. Masunov AE, Zorkii PM (1992) Donor-acceptor nature of specific nonbonded interactions of sulfur and halogen atoms. Influence on the geometry and packing of molecules. J Struct Chem 33:423–435
67. Porai-Koshits MA, Kukina GA, Shevchenko YuN, Sergienko VS (1996) Secondary bonds in tetraamine complexes of transition metals. Koordinatsionnaya Khimiya 22:83–105 (in Russian)
68. Herbstein FH, Kapon M (1975) I_{16}^{4-} Ions in crystalline (theobromine)$_2$ · H_2I_8. J Chem Soc Chem Commun 677–678
69. Dvorkin AA, Simonov YuA, Malinovskii TI et al (1977) Molecular and crystal structure of bis(α-benzyldioximato)-di-(β-picoline)iron(III) pentaiodide. Doklady Akademii Nauk SSSR (in Russian) 234:1372–1375
70. Passmore J, Taylor P, Widden T, White PS (1979) Preparation and crystal-structure of pentaiodinium hexafluoroantimonate(V) containing I^{153}. Canad J Chem 57:968–973
71. Gray LR, Gulliver DJ, Levason W, Webster M (1983) Coordination chemistry of higher oxidation states. Reaction of palladium(II) iodo complexes with molecular iodine. Inorg Chem 22:2362–2366
72. Pravez M, Wang M, Boorman PM (1996) Tetraphenyphosphonium triiodide. Acta Cryst C52:377–378
73. Svensson PH, Kloo L (2003) Synthesis, structure and bonding in polyiodide and metal iodide—iodine systems. Chem Rev 103:1649–1684
74. Dubler E, Linowski L (1975) Proof of existence of a linear, centrosymmetric polyiodide ion I_{24}^{-} – crystal-structure of $Cu(NH_3)_4I_4$. Helv Chim Acta 58:2604–2609
75. O'Keefe M, Brese NE (1992) Bond-valence parameters for anion-anion bonds in solids. Acta Cryst B48:152–154
76. Gilli P, Bertolasi V, Ferretti V, Gilli G (1994) Evidence for resonance-assisted hydrogen bonding. J Am Chem Soc 116:909–915
77. Einstein FW, Jones RDG (1973) Crystal structure containing an antimony-iron σ-bond. Inorg Chem 12:1690–1696
78. Schiemenz GP (2007) The sum of van der Waals radii—a pitfall in the search for bonding. Z Naturforsch 62b:235–243
79. Van der Waals JD (1881) Die Kontinuität des gasförmingen und flüssingen Zustandes. JA Barth, Leipzig
80. Magat M (1932) Uber die „Wirkungsradien" gebundener Atome und den Orthoeffekt beim Dipolmoment. Z Phys Chem B16:1–18
81. Mack E Jr (1932) The spacing of non-polar molecules in crystal lattices. J Am Chem Soc 54:2141–2165
82. Pauling L (1939) The nature of the chemical bond, 1st edn. Cornell University Press, Ithaca

83. Bondi A (1964) Van der Waals volumes and radii. J Phys Chem 68:441–451
84. Kitaigorodskii AI (1973) Molecular crystals and molecules. Academic, New York
85. Dance I (2003) Distance criteria for crystal packing analysis of supramolecular motifs. New J Chem 27:22–27
86. Rowland RS, Taylor R (1996) Intermolecular nonbonded contact distances in organic crystal structures. J Phys Chem 100:7384–7391
87. O'Keeffe M, Brese NE (1991) Atom sizes and bond lengths in molecules and crystals. J Am Chem Soc 113:3226–3229
88. Brese NE, O'Keefe M (1991) Bond-valence parameters for solids. Acta Cryst B47:192–197
89. Batsanov SS (2001) Relationship between the covalent and van der Waals radii of elements. Russ J Inorg Chem 46:1374–1375
90. Jones PG, Bembenek E (1992) Low-temperature redetermination of the structures of 3 gold compounds. J Cryst Spectr Res 22:397–401
91. Omrani H, Welter R, Vangelisti R (1999) Potassium tetrabromoaurate. Acta Cryst C55:13–14
92. Schulz LE, Abram U, Straehle J (1997) Synthese, Eigenschaften und Struktur von $LiAuI_4$ und $KAuI_4$. Z Anorg Allgem Chem 623:1791–1795
93. Hester JR, Maslen EN, Spadaccini N, Ishizawa N, Satow Y (1993) Electron density in potassium tetrachloropalladate (K_2PdCl_4). Acta Cryst B49:842–846
94. Martin DS, Bonte JL, Rush RM, Jacobson RA (1975) Potassium tetrabromopalladate. Acta Cryst B31:2538–2539
95. Takazawa H, Ohba S, Saito Y, Sano M (1990) Electron density distribution in crystals of $K_2[MCl_6]$ (M = Re, Os, Pt) and $K_2[PtCl_4]$ at 120 K. Acta Cryst B46:166–174
96. Kroening RF, Rush RM, Martin DS, Clardy JC (1974) Polarized crystal absorption spectra and crystal structure for potassium tetrabromoplatinate. Inorg Chem 13:1366–1373
97. Bronger W, Auffermann G (1995) High-pressure synthesis and structure of Na_2PdH_4. J Alloys Compd 228:119–121
98. Pathaneni SS, Desiraju GR (1993) Database analysis of Au···Au interactions. J Chem Soc Dalton Trans 319–322
99. Bondi A (1966) Van der Waals volumes and radii of metals in covalent compounds. J Phys Chem 70:3006–3007
100. Ganty AJ, Deacon GB (1980) The van der Waals radius of mercury. Inorg Chim Acta 45:L225–L227
101. Mingos DMP, Rohl AL (1991) Size and shape characteristics of inorganic molecules and ions and their relevance to molecular packing problems. J Chem Soc Dalton Trans 3419–3425
102. Batsanov SS (2000) The determination of van der Waals radii from the structural characteristics of metals. Russ J Phys Chem 74:1144–1147
103. Batsanov SS (2000) Intramolecular contact radii similar to van der Waals ones. Russ J Inorg Chem 45:892–896
104. Nag S, Banerjee J, Datta D (2007) Estimation of the van der Waals radii of the d-block elements using the concept of block valence. New J Chem 31:832–834
105. Hu SZ, Zhou ZH, Robertson BE (2009) Consistent approaches to van der Waals radii for the metallic elements. Z Krist 224:375–383
106. Batsanov SS (2008) Experimental foundations of structural chemistry. Moscow University Press, Moscow
107. Batsanov SS (2001) Van der Waals radii of elements. Inorg Mater 37:871–885
108. Batsanov SS (1991) Atomic radii of elements. Russ J Inorg Chem 36: 1694–1706
109. Batsanov SS (1998) Covalent metallic radii. Russ J Inorg Chem 43: 437–439
110. Cordero B, Gromez V, Platero-Prats AE et al (2008) Covalent radii revisited. J Chem Soc Dalton Trans 2832–2838
111. Pyykkö P, Atsumi M (2009) Molecular single-bond covalent radii for elements 1–118. Chem Eur J 15:186–197
112. Mu W-H, Chasse GA, Fang D-C (2008) Test and modification of the van der Waals radii employed in the default PCM model. Intern J Quantum Chem 108:1422–1434

113. Badenhoop JK, Weinhold F (1997) Natural steric analysys: *ab initio* van der Waals radii of atoms and ions. J Chem Phys 107:5422–5432
114. Allinger NL (1976) Calculations of molecular structure and energy by force-field methods. Adv Phys Organ Chem 13:1–82
115. Allinger NL, Yuh YH, Lii J-H (1989) Molecular mechanics. The MM3 force field for hydrocarbons. J Am Chem Soc 111:8551–8566
116. K itaigorodskii AI, Khotsyanova TL, Struchkov YuT (1953) On the crystal structure of iodine. Z Fizicheskoi Khimii 27:780–781 (in Russian)
117. Ibberson RM, Moze O, Petrillo C (1992) High resolution neutron powder diffraction studies of the low temperature crystal structure of molecular iodine. Mol Phys 76:395–403
118. Jdanov GS, Zvonkova ZV (1954) Travaux de Institute de Crystallographie: Communications au III Congress International de Crystallographie 10:79
119. Nyburg SC, Szymanski JT (1968) The effective shape of the covalently bound fluorine atom. J Chem Soc Chem Commun 669–671
120. Nyburg SC (1979) 'Polar flattening': non-spherical effective shapes of atoms in crystals. Acta Cryst A35:641–645
121. Batsanov SS (2000) Anisotropy of atomic van der Waals radii in the gas-phase and condensed molecules. Struct Chem 11:177–183
122. Batsanov SS (2001) Anisotropy of the van der Waals configuration of atoms in complex, condensed, and gas-phase molecules. Russ J Coord Chem 27:890–896
123. Nyburg SC, Faerman CH (1985) A revision of van der Waals atomic radii for molecular crystals: N, O, F, S, Cl, Se, Br and I bonded to carbon. Acta Cryst B41:274–279
124. Nyburg SC, Faerman CH, Pracad L (1987) A revision of van der Waals atomic radii for molecular crystals. II: hydrogen bonded to carbon. Acta Cryst B43:106–110
125. Sandor E, Farrow RFC (1967) Crystal structure of solid hydrogen chloride and deuterium chloride. Nature 213:171–172
126. Sandor E, Farrow RFC (1967) Crystal structure of cubic deuterium chloride. Nature 215:1265–1266
127. Ikram A, Torrie BH, Powell BM (1993) Structures of solid deuterium bromide and deuterium iodide. Mol Phys 79:1037–1049
128. Batsanov SS (1999) Van der Waals radii of hydrogen in gas-phase and cond ensed molecules. Struct Chem 10:395–400
129. Batsanov SS (1998) Structural features of van der Waals complexes. Russ J Coord Chem 24:453–456
130. Batsanov SS (2002) Correlation between the anisotropy of the electronic polarizability of molecules and the anisotropy of the van der Waals atomic radii. Russ J General Chem 72:1153–1156
131. Bader RFW, Henneker WH, Cade PE (1967) Molecular charge distribution and chemical binding. J Chem Phys 46:3341–3363
132. Bader RFW, Bandrauk AD (1968) Molecular charge distributions and chemical binding. III. The isoelectronic series N_2, CO, BF, and C_2, BeO, LiF. J Chem Phys 49:1653–1665
133. Bader RFW, Carroll MT, Cheeseman JR, Chang C (1987) Properties of atoms in molecules: atomic volumes. J Am Chem Soc 109:7968–7979
134. Ishikawa M, Ikuta S, Katada M, Sano H (1990) Anisotropy of van der Waals radii of atoms in molecules: alkali-metal and halogen atoms. Acta Cryst B46:592–598
135. Montague DG, Chowdhury MR, Dore JC, Reed J (1983) A RISM analysis of structural data for tetrahedral molecular systems. Mol Phys 50:1–23
136. Misawa M (1989) Molecular orientational correlation in liquid halogens. J Chem Phys 91:2575–2580
137. von Schnering HG, Chang J-H, Peters K et al (2003) Structure and bonding of the hexameric platinum(II) dichloride. Z Anorg Allgem Chem 629:516–522
138. Thiele G, Wegl W, Wochner H (1986) Die Platiniodide PtI_2 und Pt_3I_8. Z Anorg Allgem Chem 539:141–153

139. Thiele G, Steiert M, Wagner D, Wocher H (1984) Darstellung und Kristallstruktur von PtI_3, einem valenzgemischten Platin (II, IV)-iodid. Z Anorg Allgem Chem 516:207–213
140. von Schnering, HG, Chang J-H, Freberg M et al (2004) Structure and bonding of the mixed-valent platinum trihalides, $PtCl_3$ and $PtBr_3$. Z Anorg Allgem Chem 630:109–116
141. Senin MD, Akhachinski VV, Markushin YuE et al (1993) The production, structure, and properties of beryllium hydride. Inorg Mater 29:1416–1420
142. Sampath S, Lantzky KM, Benmore CJ et al (2003) Structural quantum isotope effect in amorphous beryllium hydride. J Chem Phys 119:12499–12502
143. Wright AF, Fitch AN, Wright AC (1988) The preparation and structure of the α- and β-quartz polymorphs of beryllium fluoride. J Solid State Chem 73:298–304
144. Batsanov SS (2002) Donor-acceptor mechanism of complex formation. Russ J Coord Chem 28:1–5
145. Mulliken RS (1952) Molecular compounds and their spectra. J Phys Chem 56:801–822
146. Kurnig IJ, Schneider S (1987) Ab initio investigation of the structure of hydrogen halide-amine complexes in the gas-phase and in a polarizable medium. Int J Quantum Chem 14:47–56
147. Brindle CA, Chaban GM, Gerber RB et al (2005) Anharmonic vibrational spectroscopy calculations for $(NH_3)(HF)$ and $(NH_3)(DF)$. Phys Chem Chem Phys 7:945–954
148. Leopold KR, Canagaratna M, Phillips JA (1997) Partially bonded molecules from the solid state to the stratosphere. Acc Chem Res 30:57–64
149. Takazawa H, Ohba S, Saito Y (1988) Electron-density distribution in crystals of dipotassium tetrachloropalladate(II) and dipotassium hexachloropalladate(IV). Acta Cryst B44:580–585
150. Guryanova EN, Goldstein IP, Romm IP (1975) The donor-acceptor bond. Wiley, New York
151. Timoshkin AY, Suvorov AV, Bettinger HF, Schaefer HF III (1999) Role of the terminal atoms in the donor-acceptor complexes MX_3-D (M=Al, Ga, In; X=F, Cl, Br, I; D=YH_3, YX_3, X^-; Y=N, P, As). J Am Chem Soc 121:5687–5699
152. Bucher M (1990) Cohesive properties of silver-halides. J Imag Science 34:89–95
153. Schiemenz GP (2006) Peri-interactions in naphthalenes—pyramidalization versus planarization at nitrogen as a measure of peri bond formation. Z Naturforsch 61b:535–554
154. Schiemenz GP (2002) Dative N→P/Si interactions—a historical approach. Z Anorg Allgem Chem 628:2597–2604
155. Schiemenz GP, Näther C, Pörksen S (2003) Peri-interactions in naphthalenes—in search of independent criteria for N→P bonding. Z Naturforsch 58b:663–671
156. Schiemenz GP (2004) Peri-Interactions in naphthalenes—the significance of linear and T-shaped arrangements. Z Naturforsch 59b:807–816
157. Grabowski SJ (2011) What is covalency of hydrogen bonding? Chem Rev 111:2597–2625
158. Desiraju GR (2011) A bond by any other name. Angew Chem Int Ed 50:52–59
159. Curtiss LA, Blander M (1988) Thermodynamic properties of gas-phase hydrogen-bonded complexes. Chem Rev 88:827–841
160. Nesbitt D (1988) High-resolution infrared spectroscopy of weakly bound molecular complexes. Chem Rev 88:843–870
161. Hobza P, Zahradnik R (1988) Intermolecular interactions between medium-sized systems. Chem Rev 88:871–897
162. Fiadzomor PAY, Keen AM, Grant RB, Orr-Ewing AJ (2008) Interaction energy of water dimers from pressure broadening of near-IR absorption lines. Chem Phys Lett 462:188–191
163. Sikka SK (2007) On some hydrogen bond correlations at high pressures. High Press Res 27:313–319
164. Lutz HD (1988) Bonding and structure of water molecules in solid hydrates: correlation of spectroscopic and structural data. Struct Bond 69:97–125
165. Postorino P, Tromp R, Ricci M-A et al (1993) The interatomic structure of water at supercritical temperatures. Nature 366:668–670
166. Szornel K, Egelstaff P, McLaurin G, Whalley E (1994) The local bonding in water from −20 to 220 °C. J Phys Cond Matter 6:8373–8382
167. Nälslund L-Å, Edwards DC, Wernet P et al (2005) X-ray absorption spectroscopy study of the hydrogen bond network in the bulk water or aqueous solutions. J Phys Chem A 109:5995–6002

168. Fujihisa H, Yamawaki H, Sakashita M et al (2004) Molecular dissociation and two low-temperature high-pressure phases of H_2S. Phys Rev B69:214102
169. Batsanov SS, Bokii GB (1962) Hydrogen bonds and interatomic distances in hydroxides. J Struct Chem 3:691–692
170. Brammer L (2003) Metals and hydrogen bonds. J Chem Soc Dalton Trans 3145–3157
171. Peris E, Lee JC, Rambo JR et al (1995) Factors affecting the strength of N-H···H-Ir hydrogen bonds. J Am Chem Soc 117:3485–3491
172. Sproul G (2001) Electronegativity and bond type: predicting bond type. J Chem Educat 78:387–390
173. Lipkowski P, Grabowski SJ, Leszczynski J (2006) Properties of the halogen-hydride interaction. J Phys Chem A110:10296–10302
174. Metrangolo P, Resnati G (2001) Halogen bonding: a paradigm in supramolecular chemistry. Chem Eur J 7:2511–2519

Chapter 5
Crystal Structure – Idealised

The concept of crystal structure is the cornerstone of materials science. The atomic arrangement is governed by the nature of interactions between atoms. Therefore information on the crystal structure is very important for understanding of the properties of matter and the nature of chemical bond.

5.1 Structures of Elements

Crystal structures of elements are basically different for metals and non-metals. The former usually have close-packed structures, viz. body-centered cubic (*bcc*), face-centered cubic (*fcc*) or close-packed hexagonal (*hcp*). Non-metals form molecular, chain-type, layered or a diamond-like ('open') crystal structures. The coordination numbers of atoms (N_c) in structures of condensed molecules and simple elemental bodies with ordinary bonds are equal to the valences of atoms, but in metal structures, the N_c (8 in *bcc*, 12 in *fcc* and *hcp*) considerably exceeds the atomic valences. The structural types of elements at ambient conditions are distributed in the Periodic Table as follows: metals of Group 1 through 13 and elements of the 6-th period in Groups 14 through 16, have 3-dimensional lattices with closely packed atoms; other elements of Groups 14–17 have molecular, chainlike or framework structures where atoms are connected by directed covalent bonds. Group 12 metals' structures present a borderline case: although formally *hcp*, the gap in bond lengths between the nearest six and the next six atoms is so large (0.3–0.5 Å) that it is possible to regard only the former as 'bonded'. Atoms of inert (rare) gases in the solid state are bound by isotropic van der Waals forces and therefore have highly-symmetrical, close-packed structures (like metals) but their physical properties are typical of molecular solids.

Thanks to their simplicity, the structures of most elements at ambient conditions have been established in early days of X-ray crystallography and are excellently reviewed in a number of books [1–5]. However, changes in thermodynamic conditions can drastically affect the structure of solids, e.g. transforming an atomic lattice into gaseous molecules at high temperature or, on the contrary, converting a molecular crystal into a metal under high pressure. Increase of coordination number (N_c) of

S. S. Batsanov, A. S. Batsanov, *Introduction to Structural Chemistry*,
DOI 10.1007/978-94-007-4771-5_5, © Springer Science+Business Media Dordrecht 2012

Table 5.1 Interatomic distances (Å) in structures of metals

Li[a] 3.039	Be[b] 2.226 2.286								
Na[a] 3.716	Mg[b] 3.197 3.209	Al[c] 2.863	Si[f] 2.352	P[h] 2.48					
K[a] 4.608	Ca[c] 3.947	Sc[b] 3.256 3.309	Ti[b] 2.896 2.951	V[a] 2.620	Cr[a] 2.498	Mn[c] 2.731	Fe[a] 2.482	Co[b] 2.501 2.507	Ni[c] 2.492
Cu[c] 2.556	Zn[b] 2.665 2.913	Ga[g] A11	Ge[f] 2.450	As[h] 2.76	Se[h] 2.74	Br[a] 2.72			
Rb[a] 4.939	Sr[c] 4.302	Y[b] 3.551 3.647	Zr[b] 3.179 3.231	Nb[a] 2.858	Mo[a] 2.726	Tc[b] 2.703 2.735	Ru[b] 2.650 2.706	Rh[c] 2.690	Pd[c] 2.751
Ag[c] 2.889	Cd[b] 2.979 3.293	In[d] 3.251 3.373	Sn[e] 3.022 3.181	Sb[h] 3.12	Te[h] 3.14	I[d] 3.24			
Cs[a] 5.318	Ba[a] 4.347	La[c] 3.737	Hf[b] 3.127 3.195	Ta[a] 2.860	W[a] 2.741	Re[b] 2.741 2.760	Os[b] 2.675 2.735	Ir[c] 2.713	Pt[c] 2.775
Au[c] 2.884	Hg[b] 3.000 3.466	Tl[b] 3.408 3.457	Pb[c] 3.500	Bi[h] 3.28	α-Po[h] 3.352				
Fr	Ra[a] 4.458	Ac[c] 3.755	Th[c] 3.595	Pa[i] 3.212 3.238	U[j] A20	Np	Pu[c] 3.279	Am[c] 3.461	

High-pressure phases of Si, Ge, Sn (α-Po structure, $N_c = 6$) have $d = 2.58$, 2.68, 3.08 Å, respectively; Si and I in structures with $N_c = 12$ have $d = 2.75$ and 3.24 Å; Sb, Bi, Se, Te, Br in *bcc* structure, $d = 3.26$, 3.46, 2.98, 3.40, 2.72 Å, respectively [6].
[a]A2, $N_c = 8$, [b]A3, $N_c = 6 + 6$, [c]A1, $N_c = 12$, [d]A6, $N_c = 4 + 8$, [e]A5, $N_c = 4 + 2$, [f]A4, $N_c = 4$, [g]$2.442 + 2.712 \times 2 + 2.742 \times 2 + 2.801 \times 2$, [h]α-Po, $N_c = 6$, [i]*bct*, $N_c = 8 + 2$, [j]$2.77 \times 2 + 2.86 \times 2 + 3.28 \times 4 + 3.37 \times 4$

atoms always increases the packing density, measured as the fraction (ρ) of the crystal space occupied by the (supposedly spherical and contacting) atoms. Thus, for spheres of equal size,

$$\begin{array}{llllllll} N_c = & 3 & 4 & 6 & 8 & 10 & 11 & 12 \\ \rho = & 0.169 & 0.340 & 0.564 & 0.680 & 0.698 & 0.712 & 0.7405 \end{array}$$

In the present section we discuss the structures of elements under normal and high pressures. A brief review of the high pressure techniques and equations of state for solids will be given in Chap. 10, values of pressures are henceforth expressed in gigapascals, GPa (1 GPa = 10 kbar ≈ 9869.2 atm).

5.1.1 Structures of Metals

Interatomic distances in metallic structures at normal pressure are presented in Table 5.1. Usually in a metallic crystal all atoms are equivalent, but there are exceptions, particularly the α- and β-phases of manganese. Thus, in the unit cell of

Table 5.2 Properties of the Groups 1 and 11 elements in the crystalline and molecular states

M	T_m	ρ	B_o	$E_a(M)$	$E(M_2)$	k_E	$d_c(M)$	$d(M_2)$	k_d
	°C	g/cm^3	GPa	kJ/mol			Å		
K	63.4	0.86	3.0	89.0	53.2	1.673	4.608	3.924	1.174
Cu	1085	8.93	142	337.4	201	1.679	2.556	2.220	1.151
Rb	39.3	1.53	2.3	80.9	48.6	1.665	4.939	4.170	1.184
Ag	961	10.5	106	284.6	163	1.746	2.889	2.530	1.142
Cs	28.4	1.90	1.8	76.5	43.9	1.742	5.318	4.648	1.144
Au	1064	19.3	171	368.4	221	1.667	2.884	2.472	1.167

α-Mn one can distinguish four groups consisting of 2, 8, 24 and 24 atoms, having different interatomic distances and unrelated by any symmetric transformation. Probably, these atoms have different electronic states. The situation is similar in the β-Mn structure where there are atoms with the same coordination numbers, but with different inter-nuclear distances. The most striking example is represented by structure of gallium where distances in the coordination polyhedron are equal to: 1×2.442 Å $+ 2 \times 2.712$ Å $+ 2 \times 2.742$ Å $+ 2 \times 2.801$ Å; here it is possible even to speak about a Ga_2 molecule. Coordination numbers in structures with distorted polyhedra will be considered in detail in Sect. 5.3.

Densities of elements in *bcc* and *fcc*/*hcp* structures differ not so strongly (≤ 2 %) as one may expect, given the increase of N_c from 8 to 12, because in the *bcc*-structure, besides 8 atoms at the nearest distance d, there are 6 more atoms at $1.155d$, i.e. in this case the real N_c is $8+6$. Nevertheless, physical properties of metals in Group 1 (*bcc*) and Group 11 (*fcc*) differ sharply, as shown in Table 5.2 where the melting temperatures (T_m), densities (ρ), bulk moduli (B_o), interatomic distances (d) and the atomization energies (E_a) are compared for metals of the same periods. Differences in properties are so great that Pauling [7, 8], taking into account the magnetic properties of d-elements, attributed them to increased 'metallic valences' (v), for example, attributing $v = 5.56$ to Cu, Ag and Au. Brewer [9] and Trömel [10, 11] also explained the higher ρ and T_m of Group 11 metals by invoking higher effective valences (4 or 3, respectively).

However, the ratios (k) of energies and interatomic distances in solid M *vs* gaseous molecules M_2 of the Groups 1 and 11 elements are close: average $k_E = E(M)/E(M_2)$ for elements of these groups are equal to 1.69 and 1.70, respectively, while $k_d = d(M)/d(M_2) = 1.17$ and 1.15, respectively (Table 5.2). If Cu, Ag and Au in the solid state had valences several times higher than in molecules, this would have resulted in a relative change of bond lengths and energies under transitions from the molecular elements (where $v = 1$ by definition) to the solid ones. However, this does not occur. Hence, properties of metals are due only to different electronic structure of the isolated atoms.

At the same time, according to the Drude theory, a structure of a metal consists of cations surrounded by electronic gas, where the cation charge is equal to the atomic valence and its valence electrons belong to the whole crystal. However, atoms in the solid metals cannot be ionized to more than +1, because of high magnitudes of the consequent ionization potentials. This limit is evident from the work functions, which for metals are close to half the first ionization potential (see Sect. 1.6). Besides, atomic radii of metals r_M (defined as half the shortest internuclear distance in a structure) are larger than cation radii, and as it will be shown later, r_M can be calculated from covalent radii by applying a simple correction for the difference in coordination numbers. Thus, the metallic bonding can be regarded as an undirected covalent bond. Thus, according to solid-state structural data, atoms in metals only partly lose their valence electrons. The physical interpretation of the electronic structure of elements can be found in the books by Coulson, Harrison, Kittel and Burdett [12–15].

Studies of structural transformations in metals under pressure reveal important features of the electronic structure of atoms. Isotropic compression of a body shortens the interatomic distances and increases the bond energies (which are proportional to d^{-1} or d^{-2}), but simultaneously the repulsive energy (due to the overlap between the electron shell of approaching atoms) will grow more sharply (as d^{-12}). At some stage of compression, the repulsion exceeds the attraction and the given substance can no longer retain the initial structure under new thermodynamic conditions. Then the structure undergoes an abrupt rearrangement to diminish the electron repulsion. So, alkali metals with a *bcc* lattice convert under pressure into a *fcc* phase. Further compaction of a solid causes a compression of atoms, whose electron shells are forced nearer to the nucleus. This effect is the stronger, the larger the nucleus-electron distance and, hence, the weaker the bonding. As a result, the energy gaps between *s*, *p* and *d* electrons diminish, facilitating a transition of some of the outer *s* electrons in Li and Na into *p* states, and $s \rightarrow d$ valence transitions in K, Rb and Cs. These orbitals mix with the radial-symmetric *s* functions to form hybrids which show a tendency to form bonds with an increased directional character. These changes lead to a deviation from dense-packed structures and to formation of allotropes with a smaller number of nearest neighbors. The high-pressure behavior of metals has been reviewed by McMahon and Nelmes [16] and recently by Degtyareva [301].

Phase transitions under pressure are best researched for Cs. Here the following succession of transformations has been established (the pressures of transitions in GPa are shown above the arrows, the volume changes in per cent under the arrows and N_c in parentheses):

$$\text{Cs-I, } bcc\ (8) \xrightarrow[1.0]{2.2} \text{Cs-II, } fcc\ (12) \xrightarrow[9.1]{4.2} \text{Cs-III } fcc\ (12) \xrightarrow[4.3]{4.3} \text{Cs-IV (8)}$$

$$\xrightarrow{\geq 12} \text{Cs-V (6)} \xrightarrow[34.9]{72} \text{Cs-VI } hcp\ (12)$$

The isomorphous transformation at $P = 4.2$ GPa with a reduction in volume is caused by the $6s \rightarrow 5d$ transition, but a decrease of N_c at 4.3 GPa is due to the $5p \rightarrow 5d$ electronic transition [17–19]. These values pressures correspond to the completion of the transition, but in fact a partial transition of the external *s* electrons to the *p* or *d* shells occurs continuously from the very beginning of the compression.

Pressures of the first $bcc \rightarrow fcc$ phase transformations in Li, Na, K, Rb, and Cs are equal to 7.5, 65, 11.6, 7, and 2.2 GPa, respectively. Structures following the *fcc* phases in these metals have very complex coordination polyhedra and reduced coordination numbers [20]. Structures of Cu, Ag, and Au (*fcc*) remain stable up to

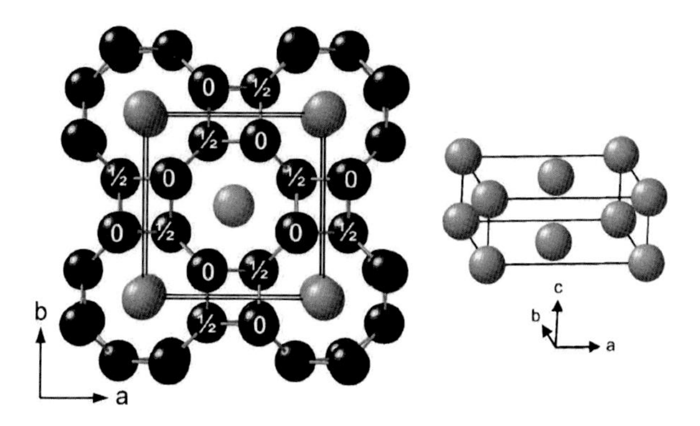

Fig. 5.1 Crystal structure of Sr-V, showing host (*dark*, labels are *z* coordinates) and guest (*light*) atoms. *Right*: guest substructure alone. Reprinted (Fig. 2) with permission from [22]. Copyright 2000 by the American Physical Society, http://link.aps.org/doi/10.1103/PhysRevB.61.3135

very high (>100 GPa) pressures, due to high hardness (B_0) of these elements (see Table 5.2) preventing a reduction in their sizes under pressure and, hence, a formation of hybrid orbitals.

Numerous polymorphic transformations under pressure are observed in alkaline-earth metals. Thus, Be with a distorted *hcp* array at ambient conditions, transforms into a *bcc* type at 1,540 K and normal pressure [21], but at a room temperature it remains stable up to $P = 66$ GPa. Mg undergoes a pressure-driven *hcp* → *bcc* transformation at *ca.* 50 GPa. In Ca a *fcc* phase transforms first into a *bcc* type at 19.5 GPa and then at 32 GPa into a primitive cubic (*sc*) lattice with $N_c = 6$; the stability of this atomic array has been attributed to the *sd* valence band hybridization. Sr shows rich structural transformations under pressure: at 3.5 GPa the polymorphic transition *fcc* → *bcc*, at 24 GPa the *bcc* → β-Sn ($N_c = 4 + 2$) transformation, then at 35 GPa occurs a new transition into a very complex structure. Further compression at $P > 46$ GPa induces a formation of another complex structure (Sr-V) containing two types of atoms: one forming a framework and the other occupying the resulting channels (Fig. 5.1), so that Sr forms a host-guest structure with itself! Barium at 5.5 GPa goes from a *bcc* ambient phase to an *hcp*, then at 12.4 GPa it transforms into an atomic structure similar to Sr-V. However, although the hosts of the Ba and Sr modifications are strictly isotypic, the average N_c of host and guest atoms corresponds to 9 and 10, respectively. Finally, at 48 GPa barium re-enters an *hcp* arrangement [22–25]. Interestingly, the *c/a* ratio of the cell parameters in this phase remains very close to the ideal value 1.633 between 48 and 90 GPa, which contrasts sharply to the ambient pressure *hcp* modification with a substantially lower and strongly pressure-dependent axial ratio (from 1.576 at 5.9 GPa to 1.498 at 11.4 GPa). Zn and Cd have no phase transitions under high pressures, but Hg has four modifications: a rhombohedral phase (α) is stable from 0 to 3.7 GPa, at this pressure α-Hg transforms into β-Hg (a tetragonal phase, *bct*), then at 12 GPa there is a β-Hg → γ-Hg (an orthorhombic form) transition, and γ-Hg at 37 GPa transforms into δ-Hg (*hcp* type with an axial ratio 1.76) [26].

Boron is known both in the amorphous form ($\overline{d}$ = 1.80 Å, $N_c = 6.3$), and in three crystalline modifications, viz. α (orthorhombic, $\overline{d}$ = 1.802 Å, $N_c = 6.5$), β (orthorhombic, $\overline{d}$ = 1.803 Å, $N_c = 6.6$) and γ (tetragonal, $\overline{d}$ = 1.802 Å, $N_c = 6.5$) [27]. These bond lengths and coordination numbers are the average, effective values.

The atomic lattice of γ-B comprises icosahedra linked by two additional atoms having tetrahedral coordination, and 5 bonds of each atom in icosahedra on edges of a pentagonal pyramids; the shortest distance B–B is 1.601 Å and the average 1.802 Å. Aluminum with its ambient-pressure *fcc* arrangement reveals a pressure-induced phase transition to a *hcp* structure at 217 GPa [28]. α-Gallium transforms at 2.5 GPa into a *bcc* structure and at 14 GPa into a *bct* phase with a $c/a \approx 1.57$ at pressures around 20 GPa. The lattice distortion leads to an increased *sp* hybridization of the valence orbitals. Upon further increase of pressure, the axial ratio decreases almost linearly until $c/a = 1.414$ is reached at 120 GPa, which corresponds to a transition from a *bct* to *fcc* structures [29]. For In, the ambient-pressure structure remains stable up to 56 GPa but at higher pressure undergoes an orthorhombic distortion [30]. Tl transforms from *hcp* to *fcc* structures already at 4 GPa and this array remains a stable phase up to 68 GPa [31].

α-Ti at normal pressure and 1,155 K converts into the *bcc* form, i.e. heat influences it in the same direction as a very high pressure. Zr and Hf undergo under pressure similar transformations α (*hcp*) → ω → β(*bcc*) at 35 and 71 GPa, respectively, and Ti goes from ω to a distorted-*hcp* γ-phase at 116 GPa. This is caused by electronic transitions from a wide *sp* band to a narrow *d* one. The increase of the number of *d* electrons at high pressure is caused by closing of *d* and *s* bands, i.e. by formation upon compression of atoms of a more compact electronic structure [32]. Silicon transforms at 10.3 GPa from its ambient diamond-type structure ($N_c = 4$) to the metallic phase, Si-II, with a β-Sn type structure ($N_c = 4 + 2$). Upon further compression, silicon changes to an orthorhombic phase (Si-III) with an intermediate crystal structure between a β-Sn and a primitive hexagonal (*hp*) lattice of atoms with a 4 + 4 coordination of Si. The *hp* phase is the stable arrangement of Si-V ($N_c = 8$) at $P > 16$ GPa. Around 40 GPa, it transforms into the orthorhombic Si-VI ($N_c = 10$ and $N_c = 11$), which is isotypic to Cs-V. Above 42 GPa, Si becomes an *hcp* Si-VII, which is followed by an *fcc* phase Si-VIII above 78 GPa. Upon decompression, an unstable phase (γ-silicon) is formed which has a distorted tetrahedral coordination and pressurizing this modification produces Si-VIII, which also comprises four-bonded silicon [20]. The phase transition of germanium from a diamond-like to a β-Sn type structure occurs at 10.6 GPa, but the change into an orthorhombic *Imma* phase at $P \geq 75$ GPa. At 81 GPa, a hexagonal primitive array forms and at 100 GPa a structural transition has been observed into the orthorhombic *Cmca* structure isotypic with Si-VI, which remains stable up to 160 GPa. At 180 GPa germanium transforms into an *hcp* network [33]. Tin has smaller number of pressure-induced phase reorganizations compared Si and Ge: a β-Sn phase transforms at 45 GPa into a *bcc* arrangement, which is stable up to 120 GPa [34].

Arsenic, antimony and bismuth under ambient conditions form layered rhombohedral lattices (the structural type A7) in which each atom is connected with three nearest neighbors at distances respectively of 2.516, 2.908 and 3.071 Å (internal bonds), and with three atoms of a following layer (external bonds) at much greater distances 3.121 (As), 3.355 (Sb), 3.529 (Bi) Å. These features allow to speak of $N_c = 3$, i.e. of a molecular structure. Phase transformations in these elements at high pressures lead to shortening of external bonds and lengthening of internal ones, until

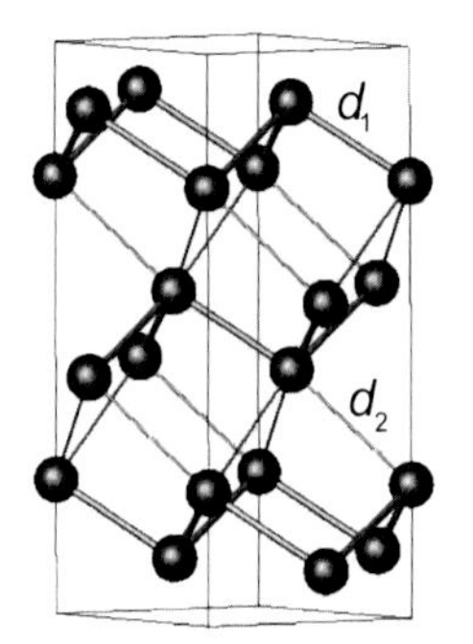

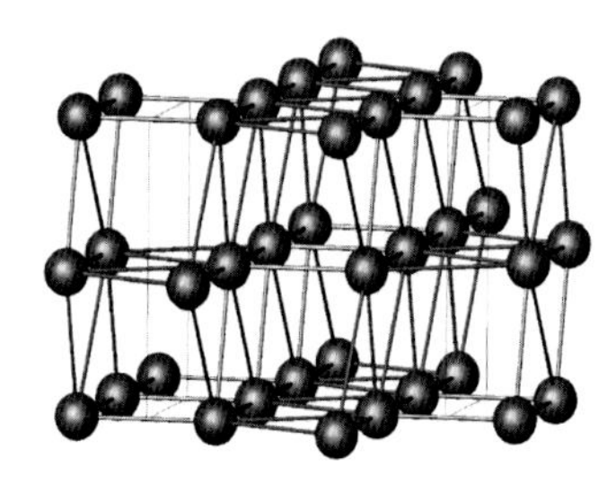

Fig. 5.2 *Left* rhombohedral As-type structure of As, Sb and Bi at ambient conditions, shown on hexagonal crystal axes. The intra-layer interatomic distances (d_1) are indicated by thick lines and the interlayer ones (d_2) by thinner lines. *Right* the Bi-II structure, where each atom has 7 nearest neighbors at distances from 3.147–3.706 Å. Reprinted from [35] by permission of the publisher (Taylor and Francis Ltd., http://www.tandf.co.uk/journals, Copyright 2004)

all bonds become equal at the final point of structural evolution, in a *bcc* structure with $N_c = 8$. Arsenic undergoes a discontinuous transition from rhombohedral (α, As-I) to a primitive cubic phase (*sc*, β-Po type, As-II) with metallic properties at 25 GPa (Fig. 5.2). The pressure-induced transformations of a semiconductor into an elemental metal are due to the enhanced overlap of the *sp* orbitals originating from the reduction of interatomic distances by compression. At 48 GPa a structural transition into a new phase (As-III) has been established. Further compression leads to a *bcc* phase at $P \geq 100$ GPa.

Elemental antimony under normal conditions has a structure of A7 and at 8.5–12 GPa transforms into a modification with a complex atomic lattice (Sb-II) [20], which at 28 GPa converts to a *bcc* structure. The heavier analogue Bi has the same phase transitions as Sb, at generally lower pressures: Bi-I (A7 structure) transforms at 2.5 GPa to a strongly distorted *sc* structure (Bi-II). Already at 2.8 GPa there occurs a structural reorganization into the Bi-III phase, which has an incommensurate crystal structure with $N_c \approx 9$. This arrangement is very similar to that found for high-pressure allotropes of As and Sb. Upon further compression, up to $P \geq 8$ GPa, bismuth transforms to a *bcc* solid [20].

Note also that all the structures discussed above, are determined for bulk materials. If normal thermodynamic conditions are close to the limit of stability of a given phase, then even small external influences (e.g. grinding of the crystals) can change its structure. The effect of particle size on stability of crystal structure of pure cobalt has been studied [36], showing that while common *hcp* allotrope of Co is stable in the bulk, particles with diameters of 100–200 Å have a *fcc* structure and those of 20–50 Å have a *bcc* lattice. In these structures, the particles the minimum of surface free energy, which yields a stable equilibrium configuration of atoms with minimal internal energy.

5.1.2 Structures of Non-Metals

Rare gas solids can be considered as the simplest nonmetal systems. However, the high pressure polymorphism in solid helium is surprisingly complex for a simple system. Helium crystallizes only under pressure in a *fcc* structure and has a small range of stability of a *bcc* structure near $\sim 3 \times 10^{-3}$ GPa and 1.5 K. X-ray diffraction measurement showed that He crystallizes at 11.5 GPa and 298 K in a *hcp* structure. This modification is the only solid phase observed at room temperature to 23 GPa. Single-crystal X-ray studies of Ne and Ar showed the stability of the *fcc* phases of these elements up to 110 and 80 GPa, respectively [37, 38]. However, in Ar at $P \geq 50$ GPa a *hcp* modification appears and its concentration increases as pressure grows; the *fcc* $\rightarrow$ *hcp* transition completes at $P \approx 300$ GPa. Kr starts a similar transition at 3.2 and completes it at *ca.* 170 GPa. Xe also starts a *fcc* $\rightarrow$ *hcp* transformation at 3 GPa and completes it by 70 GPa [39]. Optical and electric-resistivity measurements provided evidence for the metallization of Xe at 136–155 GPa, but below 120 GPa a semiconductor behavior was observed [37, 38]. Interatomic distances in the solid He at 29.7 bar and Ne, Ar, Kr, Xe under ambient pressure are equal to 3.664, 3.156, 3.755, 3.992, and 4.335 Å, respectively, i.e. are very close to the interatomic distances in the corresponding molecules Rg_2 (the average difference is ± 0.8 %, see Table 3.1), while for metals this increase is of an order of magnitude larger.

Solid hydrogen has two modifications: α (A1) and β (A3), which consist of orientationally disordered, freely rotating H_2 molecules forming a close-packed motif. The cubic structure of α-hydrogen *(fcc)* has the cell parameter $a = 5.338$ Å, in hexagonal structure β-H_2 (*hcp*) $a = 3.776$, $c = 6.162$ Å; the radius of a 'spherical' molecule of hydrogen in both structures is the same, 1.887 Å. The axial ratio $c/a = 1.63$ at low pressures (and in the zero-pressure solid at low temperatures), decreases to *ca.* 1.58 at the highest pressures, showing that the material becomes more anisotropic because of the hampered "rotation" of H_2 molecules. Intra-molecular distance in the solid hydrogen is not determined, but taking into account the smallness of its heat of sublimation (see Sect. 1.3) it is accepted to be the same as the bond length in a gas-state molecule. On compression to $P \leq 30$ GPa, solid hydrogen shows only a shortening of intermolecular contacts, but above this pressure spectroscopic data indicate some lengthening of the H–H bond. Extrapolation of these changes to the full equalization of all distances in the crystal structure (which means a transformation of a molecular into a metallic solid) suggested $P = 280$ GPa [40]. However, experiments have shown that at 320 GPa the metallization still does not occur. Extrapolation of more recent data shifted the expected threshold of metallization to 450 GPa [41].

Evolution of theoretical estimations of the metallization pressure in hydrogen is interesting. The first work in this direction belongs to Wigner and Huntington [42] who calculated the energy of a *bcc* lattice for hydrogen and concluded that it is difficult to achieve at high pressure, but the intermediate, layered structure may be attainable. Min et al. [43] calculated the pressure of two transitions: (i) an insulator to metal, due to a collapse of the valence band and the conduction band in a molecular structure at 170 ± 20 GPa, and (ii) a molecular solid to a monatomic metal at 400 ± 100 GPa. In 1988 Hemley and Mao [44] observed the former transition at 150 GPa, but the latter transition is not reached to the present day. The literature on experimental and theoretical studies of the structure of solid hydrogen at high pressures has been reviewed in detail by Henley and Dera [45].

Fluorine, F_2, crystallizes in two modifications: α (monoclinic with parameters $a = 5.50$ Å, $b = 3.28$ Å, $c = 7.28$ Å, $\beta = 102.17°$) and β (cubic phase with $a = 6.67$ Å). The bond lengths F–F in solid, gaseous and liquid fluorine are close and presented in Table S3.5. At ambient conditions, iodine is solid with the orthorhombic (space group *Cmca*) structure, but Cl_2 and Br_2 have the same structure at ambient temperature and under low pressures. This structure is layered, with the molecules lying in planes which are directed perpendicular to the *a*-axis. The distances between the closest halogen atoms lying in neighboring planes, are comparable with the van der Waals diameter. On the contrary, inside the molecule planes (the *bc* planes), the shortest intermolecular bond is significantly shorter than the van der Waals diameter. At heating of condensed molecules Cl_2 and Br_2, a reduction of intra-molecular (d) and an increase of intermolecular distances (D) have been observed [46] (Table 5.15).

Under compression of diatomic molecular solids, the intermolecular distances shrink and the intra-molecular distance increases (to a lesser extent) or remains constant. During the first stages of compression, the predominant effect on the structural arrangement is the decrease of the interlayer distances. Both distances become progressively comparable, leading at a given pressure to molecular dissociation. At this pressure both the intermolecular and intra-molecular distances are the same. However, it is only the general scheme. Thus, the intra-molecular distance in bromine initially increases up to 25 GPa as expected; but then this distance suddenly begins to decrease. This sudden change in sign of the pressure induced bond distance is attributed to a new phase transition. A maximum reduction of the inter-molecular distance is observed at 65 ± 5 GPa where again a phase transition occurs [47].

At 21 GPa iodine transforms into the metallic structure of a *bco* type (iodine-II, $N_c = 4 + 8$) [48], which on further compression continuously approaches *bct*, reaching this phase at 43 GPa (iodine-III), with the further equalization of interatomic distances. The axial ratio *c/a* of a *bct*-phase continuously approaches 1 and at 55 GPa there occurs a first-order phase transition into a *fcc* structure (iodine-IV) which is stable up to 276 GPa [49, 50]. Bromine at 80 GPa transforms into a *bcc* lattice. However, the onset of the metallic behavior takes place in the direction perpendicular to the layers via a progressive gap closure and at a pressure lower than the dissociation pressure (13 GPa for I_2 and 25 GPa for Br_2). In the layered direction the onset of a metallic behavior was also observed, but at a higher pressure [51]. Takemura et al. [52] discovered that between I and II phases there exists a new intermediate phase (iodine-V), using He as the pressurizing medium to obtain the pure hydrostatic compression and specify the pressure of the first transition (23.2–24.6 GPa). They characterized the new phase (a *fct* type, iodine-V), and the phase II (at $P = 25.6 - 30.4$ GPa). Reduction of volume at transitions I → V is equal to 2.0 % and at V → II only 0.2 %. Recently a new phase of solid bromine was revealed [53] at $P > 80$ GPa by Raman scattering experiments. This phase was found to be the same as the iodine-V with an incommensurate structure, discovered by Takemura et al. In

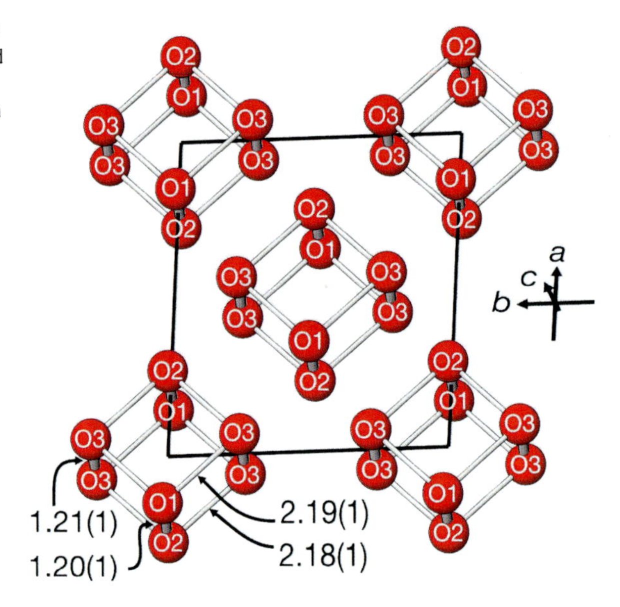

Fig. 5.3 The crystal structure of ε-oxygen at 17.6 GPa (bond distances in Å). Reprinted from [56] by permission from Macmillan Publishers Ltd, Copyright 2006

the incommensurate phases of both bromine and iodine, Raman-active soft modes were clearly found in the low frequency region. The data suggest that the monatomic phase II appears above 30 and 115 GPa for iodine and bromine, respectively.

Oxygen crystallizes on cooling, forming a unique solid, an elemental molecular magnet; this magnetism arises from the electronic ground state of the O_2 molecule. Solid oxygen has molecular structures of a few types: the low-temperature α phase is monoclinic (*C*2/*m*). The $\alpha \rightarrow \delta$ transition occurs at ~ 3 GPa and low temperatures and $\delta \rightarrow \varepsilon$ at 8 GPa. Other researchers discovered the phase transitions $\beta \rightarrow \delta$ at 9.6 GPa and $\delta \rightarrow \varepsilon$ at 9.9 GP. These phases have colours of pink, orange, and red, respectively. The red colour of the ε phase becomes darker with increasing pressure. ND experiments, which directly probed the electronic spin structure of ε-oxygen, showed that magnetism indeed collapses as this phase forms. The ε-oxygen structure is the unique arrangement in which diatomic O_2 molecules remain intact, but with additional linkages giving rise to $(O_2)_4$ clusters (Fig. 5.3). The new structure is also consistent with the observed non-magnetic state of ε-oxygen. There may be parallels between the behavior of ε-oxygen and the charge-transfer activity observed in hydrogen at much higher pressure. At 96 GPa the $\varepsilon \rightarrow \zeta$ transition has been discovered, corresponding to the 'insulator → molecular conductor' transformation; this phase is stable up to 116 GPa; the authors regard it as a semimetal [54–56]. A new dark, amorphous, apparently non-molecular phase of oxygen has been found above 180 GPa at 80 K and then at room and elevated temperatures; it is stable up to 270 GPa at 10–510 K [57]; at $P > 100$ GPa superconductivity has been observed [58]. Powder XRD [59] proved that ε-oxygen has a structure consisting of an O_8 cluster with 4 molecules.

Crystal modifications of sulfur, orthorhombic (*o*), monoclinic (*m*), and rhombohedral (*r*), have in the structures rings S_8 (*o* and *m*, $d = 2.060$ Å) or S_6 (*r*, $d = 2.057$ Å). On heating, rings are broken and turn into chains of various length, reaching the maximum at $T = 473$ K, and on further heating break into shorter pieces. *o*-S at room temperature and $P = 37.5$ GPa transforms into the S-II form (of *bct* type) with parallel chains of atoms. Further transition into the S-III phase begins at 75 and is completed at 100 GPa. In the structure of this phase (similar to Se-IV and Te-III, see below) $N_c = 6$ with following distances: $2 \times 2.208 + 4 \times 2.225$ Å. Finally, at 162 GPa sulfur transforms into a structure of β-Po type with an octahedral coordination of atoms [60].

Selenium crystallizes in the monoclinic, rhombohedral or orthorhombic molecular structures containing the Se_8, Se_7 and Se_6 rings, respectively ($d = 2.32$, 2.33 Å), and the trigonal structure consisting of spiral-like infinite chains, with 2.37 Å distances between atoms in the chain [61]. For tellurium under ambient conditions only a trigonal structure is known, with bond lengths of 2.83 Å. At normal conditions, in these elements the distances (d) are shorter in chains than between them (D), therefore on compression D diminishes faster than d and the anisotropy factor $f_a = D/d$ approaches 1 [62]. Using experimental data, the authors have obtained by extrapolation for $f_a = 1$ the equalization pressures 14 and 4 GPa for Se and Te, respectively. However, in fact Se still has $f_a = 1.3$ at $P = 14$ GPa and Te has $f_a = 1.2$ at $P = 4$ GPa, i.e. f_a is not an ideal parameter and one should take into account the difference in strengths of the intra- and inter-chain bonds, as will be shown later. The trigonal selenium (Se-I) was studied [63] under $P \leq 150$ GPa revealing the following phases: at 14 GPa (Se-II), at 23 GPa (Se-III), at 28 GPa (Se-IV), at 60 GPa (Se-V) and at 140 GPa (Se-VI). Se-I has the structure with $N_c = 2$, Se-III a layered *bcm* structure with $N_c = 4$, Se-IV a layered *bco* structure with the same N_c, Se-V has a lattice of the β-Po type ($N_c = 6$) and Se-VI crystallizes in a *bcc* structure with $N_c = 8$. On the Se-V → Se-VI transition, the Se–Se distance between the chains diminishes, and that in the layer increases, until at $P = 150$ GPa both converge to 2.42 Å, which is longer than in the initial structure at normal pressure (2.37 Å). This increase of bond length upon a reduction of atomic volume from 81.8 $Å^3$ to 22.5 $Å^3$ on the phase transition Se-I → Se-VI is caused by the increase of N_c from 2 (Se-I) to 8 (Se-VI). The monoclinic Se at $P = 12$ GPa transforms into Se-III (for trigonal selenium it requires 23 GPa) and at 33 GPa into Se-IV. The rhombohedral selenium smoothly transforms into a trigonal phase in the interval from 4 to 15 GPa and at 16 GPa converts into a metal state. Kawamura et al. [64] found an interesting dependence of the band structure on the geometrical structures of the selenium modifications. If we compare the average intermolecular distances with the band gaps and the pressures of metallization, we obtain an unusual result: the longer the intermolecular distances, the smaller is the pressure necessary for a transition into a metal state,

Modifications	Trigonal	Rhombohedral	Monoclinic
D (Se···Se), Å	3.44	3.54	3.80
E_g, eV	1.8	1.9	2.1
P_M, GPa	23	16	12

Structural changes in tellurium under pressure are similar [65], but the phase transition pressures are essentially lower than those in selenium and sulfur, viz. for the Group 16 elements (in GPa) [66]:

Structures	*bco*	β-*Po*	*bcc*
S	84	162	700 (by extrapolation of c/a to 1)
Se	28	60	140
Te	6	11	27

Nitrogen at low temperature and pressure forms a molecular crystal where strong N≡N bonds co-exist with weak intermolecular interactions. The cubic *Pa*3 structure (α phase) has the lowest free energy well at low temperatures and pressures. With increasing pressure, the system transforms to β and γ phases. A further increase in pressure leads to δ-N_2 and, at higher pressure, ε-N_2 [37–39]. Numerous optical experiments at $P > 100$ GPa show a formation of a non-molecular phase, which at 300 K and *ca.* 140 GPa transforms to a semi-conductor phase. An allotropic form was reported [67, 68] where each atom participates in three single bonds; it was synthesized from molecular nitrogen at T > 2,000 K and $P > 110$ GPa, but at T = 1,400 K it required 140 GPa. This polymeric nitrogen has cubic *gauche* structure (cg-N) and very large bulk modulus (300 GPa).

Since numerous attempts to realize the transition of a molecule into a monatomic metal in F_2, Cl_2, O_2, N_2 have not succeeded, it makes sense to use for these transformations the ability of some solids to absorb 100- and even 1,000-fold volumes of such gases. Another approach was suggested by Ashcroft [69], who noted that CH_4, SiH_4, and GeH_4 might become metallic at lower pressure since hydrogen is 'chemically'pre-compressed by the Group 14 atoms.

At ambient conditions phosphorus has many modifications [70]. White phosphorus, composed of P_4 molecules, exists as three allotropes at different temperatures and pressures. In the room temperature α-modification, P_4 molecules dynamically rotate around their centers of mass. One of the two low temperature modifications is β-P_4, with $d(\mathrm{P{-}P}) = 2.204$ Å [71]. In its crystal structure the molecules have fixed orientations. The other low temperature modification, γ-P_4, can be obtained from the α-phase upon slow warming, but on further heating γ-P_4 converts to β-P_4, which at *ca.* 193 K reversibly transforms back to α-P_4. The phosphorous atoms in the γ-P_4 form almost perfect tetrahedron where each atom is connected with three other at the distances of 2.17 Å. Then, there is the black modification of phosphorus, whose behavior under pressure is most thoroughly studied. Black phosphorous with 3 covalent bonds plus two additional interlayer contacts (A17 structure) is reported to become metallic at 1.7 GPa without changing its crystal structure. At 4.2 GPa, a semiconducting phase with a rhombohedral structure (A7, 3 + 3 coordination) is formed which transforms to a β-Po ($N_c = 6$) structure with metallic properties at 10.2 GPa. At 103 GPa this phase turns into another allotrope which is followed at P > 137 GPa by a hexagonal primitive array ($N_c = 2 + 6$) finally changing to a *bcc* ($N_c = 8$) at 262 GPa [72]. This structural sequence was attributed to a $s \rightarrow d$ transition associated with hybridization at elevated pressures [73]. Whereas in the initial

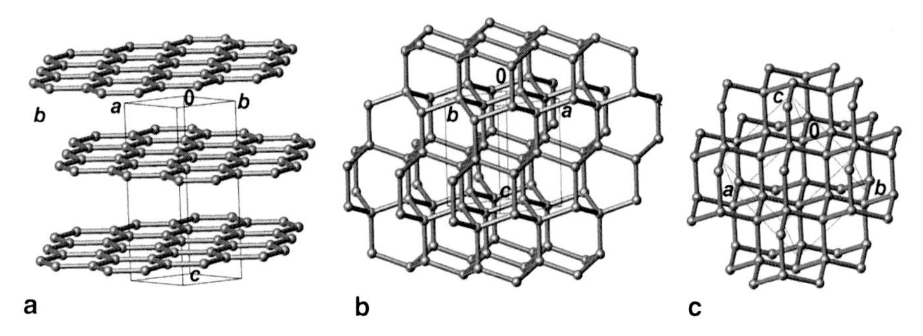

Fig. 5.4 Crystal structures of graphite **a**, hexagonal diamond, or lonsdalite **b** and cubic diamond **c**, drawn to the same scale and showing unit cells

phase the P atom forms covalent bonds with three neighbors ($N_c = 3$) and under ambient conditions the length of the P–P bonds inside the skeleton is 2.250 Å and that between skeletons 3.665 Å, in the same phase at 5.5 GPa the internal distance is decreased by 0.065 Å, while the external by an average of 0.215 Å. After the first phase transition these distances become, respectively, 2.20 and 2.81 Å (at 5.5 GPa), then at $P = 9.7$ GPa they change to 2.22 and 2.66 Å, and after the second transition at 10.3 GPa all bonds are equalized and in a primitive cubic cell become 2.39 Å [74, 75].

Carbon has many modifications, besides fullerenes discussed in Chap. 3, the most important of them being diamond and graphite (Fig. 5.4). The structural relations between carbon polymorphs are described in [76] as reconstructive phase transitions with displacive mechanisms occurring *via* a common substructure. The polymorphs are shown to correspond to limit states resulting from critical fractional displacements and critical strains. Diamond is the example of the structure type A4, in which all Group 14 elements crystallize with strictly tetrahedral coordination of atoms and the distances C–C 1.5445, Si–Si 2.3517, Ge–Ge 2.4408, Sn–Sn (grey tin) 2.8099 Å. There is another, hexagonal modification of diamond (lonsdalite, or w-diamond[1]) with a structure where each successive layer of tetrahedra is rotated by 60° in relation to previous one. Although quite stable at ambient conditions, this form, unlike cubic diamond, does not occur naturally. It was obtained in 1967 during the quest to synthesize diamond from graphite. Another hexagonal modification of carbon is graphite, which has the structure (A9) consisting of infinite layers of regular hexagons. The C–C distances within a layer are all equivalent, at 1.362 Å, while the shortest contacts between the layers are 3.334 Å, i.e. it is possible to present the coordination polyhedron as a distorted tetrahedron: $3 \times 1.362 + 1 \times 3.334$ Å. Graphite transforms to diamond at pressures above 10 GPa with the formation of the ideal tetrahedral coordination with distances 1.544 Å. It is interesting that the thermal expansion coefficient of graphite is much greater in the direction parallel

[1]From 'wurtzite', because this form stands in the same relation to cubic diamond as wurtzite does to zinc blende.

to the layers (4×10^{-6} K^{-1}) than perpendicular to them (3×10^{-7} K^{-1}) while for compressibility it is the other way round. There is also a rhombohedral modification of graphite, which also consists of hexagonal layers but shifted with respect to each other. The density and interatomic distances in both modifications of graphite are the same. Finally, one can mention *carbine* which is a product of chemical synthesis with the structure consisting of infinite chains ≡C–C≡C–C≡ [77, 78], however, unlike other allotropes, its existence is still disputed.

Note that the increase of coordination numbers (N_c) in the structures of elements is accompanied by an increase in the bond distances ($\overline{d}$) and the packing density of atoms (ρ):

N_c	3		4		6		8		12
Structure	A9	→	A4	→	A5	→	A2	→	A1
$\overline{d}$	1.00		1.02		1.09		1.11		1.14
ρ	0.17		0.34		0.56		0.68		0.74

This apparent contradiction is explained in Sect. 1.4.2, see also [79].

Transition from solid to molecular nonmetals is accompanied by a change in interatomic distances due to change in bond order: for S, Se, Te, $d(X\text{–}X)/d(X{=}X) = 1.094 \pm 0.9\ \%$, for P, As, Sb, Bi, $d(X\text{–}X)/d(X{\equiv}X) = 1.176 \pm 2.3\ \%$. For O and N these ratios are equal to 1.17 and 1.33, respectively, i.e. the relative shortening of multiple bonds diminishes on going down in Groups 15 and 16. The ratios of single A–A bond length to the shortest A=A distances in $R_2A{=}AR_2$ compounds where A=C, Si, Ge and Sn, are 1.15, 1.10, 1.10 and 1.01, respectively [80]. Ratios of the dissociation energy in A_2 molecules (Table S2.3) to the energy of single bond E(A–A) from Table 2.10 show similar successions:

$$\mathrm{O}\,(2.6) \rightarrow \mathrm{Se}\,(1.8) \rightarrow \mathrm{Te}\,(1.55)$$

$$\mathrm{N}\,(4.5) \rightarrow \mathrm{As}\,(2.3) \rightarrow \mathrm{Sb}\,(2.0) \rightarrow \mathrm{Bi}\,(2.0)$$

$$\mathrm{C}\,(1.7) \rightarrow \mathrm{Ge}\,(1.25) \rightarrow \mathrm{Sn}\,(1.25) \rightarrow \mathrm{Pb}\,(1.2)$$

A survey of homonuclear multiple bonds between main group elements based on experimental data of the Group 14 and 15 elements double-bond systems, $R_nA{=}AR_n$ (A=C → Pb, $n = 2$; A=N → Bi, $n = 1$) [80] shows that for heavier elements (from the 3-rd and certainly from the 4-th Period), such systems often do not behave as expected for compounds with multiple bonds. They have *trans* bent structures and show enormous variation in their bond lengths. It was argued [80] that the classical multiple bond indicators—bond lengths and bond strengths—have no meaning for multiple bonds in which elements from the higher periods are involved, and are useful only for bonds involving at least one atom of the 2-nd Period.

5.2 Binary Inorganic Crystalline Compounds

In this section the structures of binary crystal compounds are considered, including those which contain metal or non-metal atoms of different types. The limited volume of the book does not allow to pay attention to structures of inter-metallic compounds (see reviews [81, 82]) and of metal-rich compounds, which are in detail discussed by Rao [83] and Franzen [84].

In structures of inorganic crystalline compounds, coordination numbers (N_c) vary from 2 up to 12. Here and further by N_c we mean the number of atoms coordinated around the central metal atom; the N_c of non-metals in M_nX_m structures is equal to $N_c(M)n/m$. In solid compounds with a framework lattice, the N_c multiplied by the ligand valence (v_X) always exceeds the valence of the metal:

$$v_M < N_c v_X \tag{5.1}$$

which leads to metal atoms combining into giant polymers where all bonds are bridging and consequently longer than those in isolated molecules. The physical reason of lengthening of internuclear distances in crystals consists in the reduction of the electron density per bond because of higher ligancy of the central atom.

5.2.1 *Crystal Structures of Halides, Oxides, Chalcogenides, Pnictides*

The lowest coordination number of atoms in structures of binary solids is 2. Such structures (zigzag chains of atoms) are found in AuCl ($d = 2.36$ Å), AuBr (2.42 Å), AuI (2.60 Å), HgO (2.03 Å) and HgS (2.37 Å). Compounds with the tetrahedral coordination of atoms are much more numerous; Table 5.3 shows the bond lengths in crystals MX of structural type ZnS (sphalerite, B3). As one can see, these structures are formed of Groups 11/17, 12/16 and 13/15 elements (sp^3) with a small difference in electronegativities (ENs), forming essentially covalent bonds. Therefore tetrahedral structures are usually considered as typically covalent configurations. Comparison of Tables S2.1 and 5.3 shows that transition from $N_c = 1$ (molecule) to $N_c = 4$ (crystal) is accompanied by an increase in bond lengths, on average by 0.290(3) Å, or by a factor of 1.133(9).

In compounds MX with the wurtzite structures (B4) there are 3 basal and one apical M–X bonds with distances: CuH $3 \times 1.765 + 1.729$ Å [85], AgI $3 \times 2.819 + 2.798$ Å, BeO $3 \times 1.646 + 1.657$ Å, ZnO $3 \times 1.974 + 1.988$ Å, CdS $3 \times 2.526 + 2.532$ Å, CdSe $3 \times 2.630 + 2.635$ Å, AlN $3 \times 1.889 + 1.903$ Å, GaN $3 \times 1.949 + 1.956$ Å [86]. If we compare these data and the values in Table 5.3, the average distances in compounds with structure B4 are very close to distances of the corresponding solids of the structural type B3.

Square coordination of atoms exists in CuO, SnO, PdO and PtO, where M–O bond distances equal 1.954, 2.22, 2.01 and 2.02 Å, respectively. In PbO the lead

Table 5.3 Interatomic distances (Å) in compounds of the B3 structural type

M(I)	F	Cl	Br	I
Li				2.74
NH_4	2.708			
Cu		2.345	2.464	2.624
Ag				2.812
M(II)	O	S	Se	Te
Be	1.649	2.107	2.225	2.436
Mg		2.45[a]	2.53	2.780[b]
Zn	1.978	2.342	2.454	2.637
Cd		2.528	2.620	2.806
Hg		2.535	2.635	2.797
Mn		2.431[c]	2.546[c]	2.744[c]
M(III)	N	P	As	Sb
B	1.566[d]	1.965[d]	2.068[d]	
Al	1.896[e]	2.367[e]	2.451[e]	2.657[e]
Ga	1.948[e]	2.360[e]	2.448[e]	2.640[e]
In	2.156[e]	2.542[e]	2.623[e]	2.806[e]
Nb	2.09	2.35		
Cr	2.36[f]	2.45[f]	2.54[f]	2.62[f]
Fe	1.865[g]			

[a][88], [b][89], [c][90], [d][91], [e][92], [f][93], [g][94]; bond lengths are equal to: BePo 2.528, ZnPo 2.732, CdPo 2.886 Å [95]

atom is displaced out of the plane of the oxygen square, and d(Pb–O) = 2.30 Å. In PdS the atom of metal is surrounded by four sulfur atoms in a square at distances of 2.33 Å [87]. Bonds in these structures correspond to the dsp^2 or d^2p^2 hybridization and are also of essentially covalent character.

In Table 5.4 there are listed interatomic distances in MX solids of the structure type NaCl (B1), where atoms have octahedral coordination. These compounds contain metal atoms with low ENs, forming bonds of an essentially ionic character. Therefore compounds with the B1 structure are usually considered as typically ionic. Comparison of interatomic distances in structures of the B3 and B1 types (Tables 5.3 and 5.4) shows that such increase in coordination of atoms is accompanied by an increase of the bond lengths by a factor of 1.080(9). Interatomic distances in crystal of this type are additive. Differences of the bond lengths in halides MX are: $\Delta d_{\text{Na-Li}} = d(\text{Na–X}) - d(\text{Li–X}) = 0.28$ Å, $\Delta d_{\text{K-Na}} = 0.34$ Å, $\Delta d_{\text{Rb-K}} = 0.015$ Å, $\Delta d_{\text{Cs-Rb}} = 0.18$ Å, $\Delta d_{\text{Cs-NH}_4} = 0.17$ Å, $\Delta d_{\text{NH}_4\text{-Ag}} = 0.52$ Å, $\Delta d_{\text{Tl-Ag}} = 0.41$ Å. This principle works very well because of similar character of chemical bonds (for example, for halides K, Rb and Cs the deviation is *ca.* 5 %). On the contrary, if we compare hydrides and fluorides of alkali metals where the bond character is different, we get $\Delta d = d(\text{M–H}) - d(\text{M–F}) = 0.15$ Å ± 35 %. In the case of oxides and chalcogenides of the MX type, the additive principle is correct within 8 %. Comparison of data in Tables 3.2, S3.1 and 5.4 shows, that the ratio of the bond lengths for $N_c = 1$ to those for $N_c = 6$ is equal to 1.22 ± 3.8 %. The six-fold coordination (a trigonal prism) exists in the structure types NiAs, TiP and MnP, as listed in Table S5.2. The differences of ENs of atoms in these substances are smaller than in typical representatives of the B1 class, so the bonds here have intermediate, polar-covalent character.

Table 5.4 Interatomic distances (Å) in compounds of the NaCl structural type (LiH 2.042, NaH 2.445, KH 2.856, RbH 3.025, CsH 3.195 Å; PtN 2.402 Å [96]; AuCl 3.16 Å [97])

M(I)	F	Cl	Br	I	M(II)	O	S	Se	Te
Li	2.009	2.566	2.747	3.025	Tm		2.71	2.82	3.00
Na	2.307	2.814	2.981	3.231	Yb	2.44	2.84	2.94	3.265
K	2.664	3.139	3.293	3.526	Th	2.60	2.842[e]	2.945[e]	
Rb	2.815	3.285	3.434	3.663	U	2.447	2.744	2.878[e]	3.076[e]
Cs	3.005	3.47	3.615	3.83	Np		2.766[e]	2.903[e]	3.101[e]
NH_4	2.885	3.300	3.437	3.630	Pu	2.48	2.772[e]	2.900[e]	3.089[e]
Ag	2.465	2.774	2.887	3.035	Am		2.80	2.91	3.088[f]
Tl	2.88	3.16	3.297	3.47	Cm		2.79	2.90	3.075
M(II)	O	S	Se	Te	M(III)	N	P	As	Sb
Mg	2.106	2.596	2.732		Sc	2.25	2.66	2.74	2.92[g]
Ca	2.405	2.842	2.962	3.174	Y	2.44	2.83	2.89	3.085[g]
Sr	2.580	3.012	3.116	3.330	La[h]	2.648	3.018	3.068	3.245
Ba	2.770	3.193	3.296	3.500	Ce	2.606	2.95	3.03	3.20
Ra		3.29	3.40		Pr	2.568	2.946	3.009	3.182
Zn	2.140[a]	2.530[b]	2.670[c]		Nd	2.562	2.913	2.979	3.15
Cd	2.348	2.72	2.84	3.051	Sm	2.518	2.875	2.955	3.13
Y		2.733	2.879	3.048	Gd	2.487	2.854	2.932	3.110
La[D]	2.57	2.926	3.034	3..218	Tb	2.461	2.838	2.908	3.085
Ce	2.54	2.89	2.99	3.18	Dy	2.448	2.822	2.895	3.07
Pr	2.52	2.87	2.97	3.16	Ho	2.432	2.808	2.88	3.06
Nd	2.51	2.85	2.954	3.130	Er	2.418	2.798	2.867	3.048
Sm	2.57	2.985	3.100	3.297	Tm	2.40	2.78	2.86	3.04
Eu	2.57	2.984	3.092	3.292	Yb	2.388	2.772	2.845	3.034
Gd		2.78	2.89	3.07	Lu	2.383	2.766	2.84	3.028
Tb		2.76	2.87	3.05	Th	2.583[e]	2.914[e]	2.989[e]	3.159[e]
Dy	2.66	2.75	2.85	3.04	U	2.444	2.792	2.888[e]	3.102[e]
Ho		2.73	2.84	3.02	Np		2.804[f]	2.918[f]	
Er	2.54	2.72	2.83	3.01	Pu	2.45	2.832[f]	2.928	
Ti	2.088[d]				Am	2.50	2.855	2.94	
	2.30	2.58			In		2.762[i]	2.88[j]	
Sn			3.01	3.16	AlN	GaN	InN	TiN	ZrN
Pb		2.968	3.061	3.226	2.022[k]	2.076[k]	2.344[k]	2.12	2.289[k]
Mn	2.222	2.610	2.725	3.013	HfN	VN	TaN	CrN	WN
Fe	2.154				2.263[l]	2.072[e]	2.168[m]	2.074	2.06
Co	2.130				LaBi	CeBi	PrBi	NdBi	SmBi
Ni	2.088				3.28	3.24	3.22	3.21	3.18
SiC	TiC	ZrC	HfC	VC	TbBi	HoBi	NpBi	PuBi	AmBi
2.02[n]	2.163	2.344	2.323	2.091	3.14	3.11	3.185[f]	3.179[f]	3.163[f]
NbC	TaC	ThC	UC	PtC	CaPo	SrPo	BaPo	HgPo	PbPo
2.233	2.228	2.66	2.48	2.407[o]	3.257[p]	3.398[p]	3.560[p]	3.125[p]	3.295[p]

[a] [98, 99], [b] for all LnX from [100], [c] [101], [d] [102], [D] [103], [e] [104], [f] [105], [g] [106], [h] [107], [i] [108], [j] [109], [k] [110]; [l] [111], [m] [112], [n] [113], [o] [114], [p] [95], HgO 2.524 Å [115]

Table 5.5 Interatomic distances (Å) in compounds of the CsCl structural type (RbAu 3.55, CsAu 3.69 Å [116, 117]; CoSi 2.439 Å [118])

M(I)	H	F	Cl	Br	I
Na		2.36[a]	3.00[b]		
K	2.96[c]	2.78[d]	3.28[e,f]	3.46[e,f]	3.76[e,f]
Rb	3.16[c]	2.87	3.41[e,f]	3.57[e,f]	3.84[e,f]
Cs	3.312[c]	3.09	3.566	3.720	3.956
Ag		2.595[g]			
Tl			3.327	3.443	3.64
NH_4			3.350	3.515	3.784
M(II)	**O**	**S**	**Se**	**Te**	**Sb**
Ca	2.52[h]	3.00[i]	3.13[i]	3.30[j]	
Sr	2.65[k]	3.125[k]	3.26[k]	3.475[k]	
Ba	2.96[l]	3.37[l]	3.42[m]	3.697[n]	
Cd	2.48[o]				
La		2.988[p]	3.123[p]	3.364[p]	
Th			3.12[q]	3.31[q]	3.33[q]
U				3.24[q]	3.23[q]

[a][119], [b][120], [c][121], [d][122], [e][123], [f][124], [g][125], [h][126], [i][127], [j][128], [k][129], [l][130], [m][131], [n][132], [o][133], [p][103], [q][134]

Crystals of the structure type CsCl (B2) have cubic coordination of atoms (Table 5.5). Atoms of metals in these compounds are large, with low ENs and hence they form the most ionic bonds. The average ratios $d(N_c = 8)/d(N_c = 1)$ for the substances listed in Table 5.5, is equal to 1.26(5).

Alkali halides, having at ambient conditions the B1 structure, undergo under pressure a phase transition with an increase of N_c. N_c can be reduced either by epitaxial growth of the given substance on a substrate of corresponding structure, or by the crystal growth of solid phase encapsulated within single walled carbon nanotubes. These tubes have cylindrical cavities with a strictly limited diameter range, typically 10–20 Å. The effect of these ultra-thin capillaries on the crystallization of encapsulated molten binary species is to produce reduced or modified coordination structures. So, in KI crystals formed within 14–16 Å diameter nanotubes, the atoms with three separate coordination numbers 6, 5 and 4 are exhibited at the center, face and corner of the crystal, respectively. The interatomic distances K–I are anisotropic: along the nanotube on the surface $d(\text{K–I}) = 2.37$, and in the volume 2.58 Å, perpendicularly to the nanotube axis respectively 2.46 and 2.75 Å [135, 136]. Phase transformations with a decrease of coordination numbers have been observed also

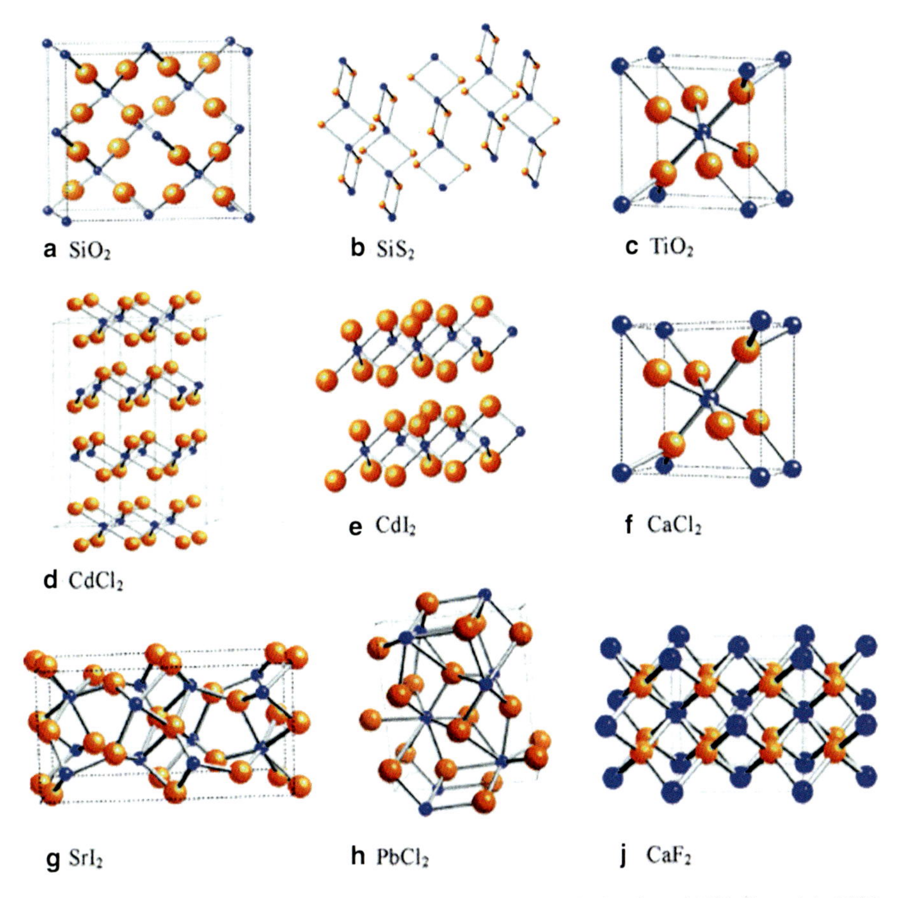

Fig. 5.5 Structural types of MX_2 crystals. Reprinted with permission from [158]. Copyright 2006 American Chemical Society

in other crystalline compounds: AgX ($N_c = 4 \rightarrow 3$), SrI_2 ($7 \rightarrow 6, 4$), BaI_2 ($9 \rightarrow 6, 5$), PbI_2 ($6 \rightarrow 5$), $LnCl_3$ ($6 \rightarrow 5$). At the same time, molecular substances (Al_2Cl_6, SnI_4,WCl_6) or crystals with a chain-like structure ($ZrCl_4$ and $HfCl_4$) incorporated within nanotubes have the same N_c as in bulk (see [137–139]). Ionic bonds prove more labile than covalent ones: interatomic distances, for example, in fullerenes (from C_{60} to C_{84}) when placed into nanotubes, are only a few per cent shorter than in bulk samples [140], while for Sb_2O_3 the difference exceeds 10 % [141].

Crystals of the MX_2 type (Fig. 5.5) have coordination numbers from 2 up to 12. Usually as examples of crystals with $N_c = 2$, halides of mercury are chosen which are considered as molecular substances. However, actually, in the structure type $HgCl_2$ the array of atoms resembles the structure of $PbCl_2$ but with shorter distances between Hg and two nearest neighbors. Therefore $N_c = 2$ in this case only formally; the problem of distorted coordination polyhedra and effective coordination numbers will be discussed later. Twofold coordination of the metal atoms is found in structures

Table 5.6 Interatomic distances (Å) in compounds of the MX_2 tetrahedral structures (*plp* is the low-pressure phase, *php* is the high-pressure phase, *pht* is the high-temperature phase, *mc* is the monoclinic modification, tetra is the tetrahedral one)

MX_2	d(M–X)	MX_2	d(M–X)	MX_2	d(M–X)	MX_2	d(M–X)
BeH_2	1.45[a]	SiO_2		GeO_2		$Be(NH_2)_2$[e]	1.746
BeF_2	1.540[b]	quartz	1.607	quartz	1.739[h]	$Mg(NH_2)_2$[e]	2.084
$BeCl_2$	2.026[c]	tridymite	1.606	cristobalite	1.75	$Zn(NH_2)_2$[e]	2.028
$BeBr_2$	2.185[c]	cristobalite	1.604	GeS_2		$Mn(NH_2)_2$[e]	2.121
BeI_2	2.417[c]	coesite	1.613	plp	2.19	$Be(CN)_2$[i]	1.718
$ZnCl_2$	2.346	SiS_2		pht	2.217	$Mg(CN)_2$[i]	2.108
$ZnBr_2$	2.415[d]	plp	2.14	$GeSe_2$[g]		$Be(OH)_2$[j]	1.632
ZnI_2	2.645[e]	php	2.13	mc	2.356	$Zn(OH)_2$[j]	1.956
HgI_2	2.788[f]	$SiSe_2$	2.275	tetra	2.359		

[a][147, 148], [b][149], [c][150], [d][151], [e][152], [f][153], [g][154], [h][155], [i][156], [j][157]

of the cuprite type, the bond lengths in which are equal to: Cu_2O 1.84, Ag_2O 2.05, Ag_2S 2.12, Ag_2Se 2.16, and Au_2S 2.17 Å [142].

Sn is tri-coordinate in the structures of SnF_2, $KSnF_3$ and NH_4SnF_3. The average length of the Sn–F bond is 2.14 Å in SnF_2, 2.11 Å in $KSnF_3$, 2.12 Å in NH_4SnF_3. Square coordination is found in structures of PdX_2 and PtX_2, where interatomic distances (Å) are equal to:

Pd–Cl	Pd–Br	Pd–I	Pt–Cl	Pt–Br	Pt–I
2.31	2.46	2.60	2.315 [143]	2.48	2.597 [144]

Structures of PtX_3 (X=Cl, Br, I) are a combination of the PtX_2 and PtX_4 motives: a cubic closest packing of X-anions forms the basis of an optimized arrangement of cubo-octahedral $[Pt_6X_{12}]$ cluster molecules with Pt^{II} and enantiomers of helical chains of edge-sharing $[PtX_2X_{4/2}]$ octahedra with Pt^{IV} in *cis*-configurations. Average distances are: $\bar{d}(Pt^{II}–Cl) = 2.314$, $\bar{d}(Pt^{II}–Br) = 2.445$, $\bar{d}(Pt^{II}–I) = 2.618$ Å, $\bar{d}(Pt^{IV}–Cl) = 2.342$, $\bar{d}(Pt^{IV}–Br) = 2.487$, $\bar{d}(Pt^{IV}–I) = 2.685$ Å. In coordination octahedra of these compounds, shorter bonds are in *trans*-position to longer ones: $d(Pt^{IV}–Cl) = 2.286$ *vs* $d(Pt^{IV}–Cl) = 2.376$ Å, $d(Pt^{IV}–Br) = 2.474$ *vs* $d(Pt^{IV}–Br) = 2.563$ Å, $d(Pt^{IV}–I) = 2.652$ *vs* $d(Pt^{IV}–I) = 2.745$ Å [145, 146].

More numerous tetrahedral structures of MX_2 solids differ in the way in which the tetrahedra are linked together. If in SiO_2 and BeF_2 there is a three-dimensional (frame) linkage, in SiS_2 and $BeCl_2$ the tetrahedra are connected into infinite chains. However, this distinction affects the lengths of chemical bonds rather little. Therefore in Table 5.6 all compounds MX_2 in the tetrahedral structures are presented without indications of the ways of tetrahedra connecting.

The Be–O and Zn–O distances in $M(OH)_2$ are slightly shorter than in MO: 1.632 against 1.649 Å, and 1.956 against 1.978 Å, respectively, due to reduction of coordination numbers of anions in $M(OH)_2$ in comparison with MO (2 against 4). Comparison of Tables S3.1 and 5.6 shows that the ratio of the bond lengths in the crystals ($N_c = 4$) to those in the molecules ($N_c = 2$) of BeX_2, ZnX_2 and HgX_2 is 1.11(3).

Table 5.7 Interatomic distances (Å) in structures of the rutile type [159–163]

MX_2	4[a]d(M–X)	2d(M–X)	MO_2	4d(M–O)	2d(M–O)
CuF_2	1.93	2.27	SiO_2	1.757	1.808
AgF_2	2.071	2.584	GeO_2	1.874	1.906
MgF_2	1.994	1.984	SnO_2	2.058	2.047
MgH_2	1.955	1.935	PbO_2	2.167	2.154
$CaCl_2$	2.740	2.743	TiO_2	1.949	1.980
$CaBr_2$	2.901	2.914	VO_2	1.921	1.933
ZnF_2	2.040	2.019	NbO_2	2.079	2.005
VF_2	2.092	2.074	TaO_2	2.030	1.998
CrF_2	2.01	2.43	TeO_2	2.321	2.032
$CrCl_2$	2.393	2.903	CrO_2	1.911	1.891
$CrBr_2$	2.54	3.00	MoO_2	1.959	2.062
CrI_2	2.74	3.24	WO_2	1.951	2.062
MnF_2	2.131	2.104	MnO_2	1.882	1.894
FeF_2	2.118	2.002	RuO_2	1.985	1.941
CoF_2	2.057	2.015	OsO_2	2.006	1.962
NiF_2	2.021	1.981	IrO_2	1.998	1.958
PdF_2	2.16	2.17	PtO_2	2.003	1.989

[a]Figures indicate the number of such bonds (ditto in Tables 5.9, 5.10, and 5.11)

Higher coordination of atoms obtains in the structure of rutile (TiO_2), where the atom of metal is located in the center of a slightly distorted octahedron with two Ti–O distances differing from the other four. Representatives of this structural type are listed in Table 5.7. The average internuclear distances for $N_c = 6$ are longer than those for $N_c = 2$, by a factor of 1.15(3).

Compounds with CdX_2 structures, where atoms of metals have $N_c = 6$ (trigonal prism), are listed in Table S5.3. In these structures atoms of halogens form slightly deformed close packing motif, where cations occupy one-half of octahedral voids. The structure type MoS_2 is similar to the CdI_2 structure, therefore Table S5.3 gives also the bond lengths in chalcogenides of some metals crystallizing in this structure. The ratio $d(N_c = 6)/d(N_c = 2)$ is equal to 1.16(2). Cs_2O, Tl_2S [164] and Ag_2F [165] also crystallize in the structural type CdI_2 but with reversed positions of cations and anions; such structures are known as anti-CdI_2. The Cs–O, Tl–S and Ag–F distances are equal to 2.86, 2.91 and 2.451 Å, respectively. Comparison of Tables 5.7 and S5.3 shows that, as in the case of tetrahedral MX_2 structures, d(M–OH) $\leq$ d(M–O) for the same reason. Metal atoms have $N_c = 6$ in the structures of pyrite and marcasite (FeS_2), where *di*atomic anions have −1 charge (Table S5.4). Note that the lengths of the S–S and P–P bonds in these substances, as a rule, increases with the decrease of M–S or M–P distances, due to a shift of the valence electrons from X–X to M–X bonds. In the structural type SrI_2 the metal atom is in the center of the polyhedron consisting of a square and a triangle ($N_c = 7$, mean Sr–I distance 3.35 Å). EuI_2 and $YbCl_2$ have similar structures with the average MX distances of 3.34 and 2.84 Å, respectively.

Very important is the structural type of CaF_2, where the metal atom is surrounded by 8 atoms of fluorine occupying the vertices of a cube. Examples of this type are presented in Table 5.8. The last column of this table lists the oxides and chalcogenides of alkali metals, which have structures of the anti-fluorite type, i.e. with the metal

Table 5.8 Interatomic distances (Å) in structures of the CaF_2 type

MX_2	d(M–X)	MO_2	d(M–O)	M_2X	d(M–X)
AgF_2	2.36	CeO_2	2.34	Li_2O	2.00
CaF_2	2.365	PrO_2	2.32	Li_2S	2.470
SrF_2	2.511	TbO_2	2.26	Li_2Se	2.600
$SrCl_2$	3.021	ThO_2	2.424	Li_2Te	2.815
$SrBr_2$[a]	3.17	PaO_2	2.38	Na_2O	2.408
BaF_2	2.683	UO_2	2.368	Na_2S	2.83
$BaCl_2$	3.17	NpO_2	2.35	Na_2Se	2.95
RaF_2	2.757	PuO_2	2.33	Na_2Te	3.17
CdF_2	2.333	AmO_2	2.33	K_2O	2.792
HgF_2	2.398	CmO_2	2.32	K_2S	3.20
LaF_2	2.527[f]	TiO_2	2.109	K_2Se	3.324
CeF_2	2.488[f]	ZrO_2	2.276	K_2Te	3.530
PrF_2	2.478[f]	HfO_2	2.22	Rb_2O	2.925
NdF_2	2.459[f]	SnO_2	2.132	Rb_2S	3.35
SmF_2	2.542[f]	PbO_2	2.316	Rb_2Se	3.47[c]
EuF_2	2.530[f]	RuO_2	2.106[b]	Rb_2Te	3.676[d]
YbF_2	2.424[f]	PdO_2	2.43	Cu_2S	2.409
ErF_2	2.382[f]	ScOF	2.414	Cu_2Se	2.529
PbF_2	2.570	YOF	2.322	Ag_2Te	2.846
MnF_2	2.25	LaOF	2.492	Be_2C	1.880[e]
CoF_2	2.13	CeOF	2.474	Mg_2Si	2.74
NiF_2	2.10	PrOF	2.444	Mg_2Ge	2.762
PdF_2	2.30	NdOF	2.423	Mg_2Sn	2.928
YH_2	2.255[g]	SmOF	2.390	Mg_2Pb	2.95
LaH_2	2.451[g]	HoOF	2.391	Al_2Au	2.59
TiH_2	1.927[h]	AcOF	2.573	Al_2Pt	2.56
ZrH_2	2.076[g]	PuOF	2.473		

[a]Distorted CaF_2, [b][169], [c][170], [d][171], [e][172], [f][173, 174], [g][175], [h][176]

in the place of fluorine and non-metal atoms in the place of calcium. The average $d(N_c = 8)/d(N_c = 2)$ ratio is 1.18(4). Chemical bonds in these structures have a high degree of ionicity. Atoms of uranium have the same coordination in the structures of US_2 and USe_2 where d(U–X) are 2.84 and 2.957 Å, respectively [166].

In the structural type of $PbCl_2$, chlorine atoms form strongly deformed hexagonal close-packing. The lead atom lies in the center of the common face of two joined octahedra and is surrounded by 9 atoms of chlorine: six form a trigonal prism, three more lie in one plane (perpendicular to the faces of the prism) and form a triangle with the Pb atom in its center. In this structure crystallize CaH_2 and SrH_2 with the mean M–H distances of 2.41 and 2.59 Å, respectively [167], $BaCl_2$ ($\bar{d} = 3.24$ Å), $BaBr_2$ ($\bar{d} = 3.38$ Å), and BaI_2 ($\bar{d} = 3.67$ Å), as well as PbF_2, $PbCl_2$ and $SnCl_2$ with the average distances of 2.65 Å, 3.14 Å and 3.24 Å. Cs_2S, Cs_2Se and Cs_2Te belong to the anti-$PbCl_2$ structural type [168].

Rb_2Te adopts the CaF_2, $PbCl_2$, and at high temperatures, the Ni_2In structures. $PbCl_2$ and Ni_2In structures can be transformed into each other by a displacive rearrangement mechanism. Recently synthesized Cs_2Pt also has Ni_2In structure in the form of a tricapped trigonal prism with three cesium atoms at 3.28 six more at 4.04 Å

Table 5.9 Interatomic distances (Å) in structures of the $ZrCl_4$ type

MX_4	$2d$(M–X)	$2d$(M–X)	$2d$(M–X)	MX_4	$2d$(M–X)	$2d$(M–X)	$2d$(M–X)
TiF_4[a]	1.716	1.932	1.970	NbI_4	2.676	2.755	2.905
$ZrCl_4$	2.307	2.498	2.655	CrF_4[a]	1.677	1.870	1.983
$ZrBr_4$	2.461	2.649	2.806	$TeCl_4$	3 × 2.311		3 × 2.929
α-ZrI_4[b]	2.696	2.876	3.026	TeI_4	2.769	3.108	3.232
β-ZrI_4[b]	2.694	2.871	3.026	MnF_4[a]	1.700	1.850	1.946
γ-ZrI_4[b]	2.693	2.874	3.027	$TcCl_4$	2.242	2.383	2.492
$HfCl_4$	2.295	2.482	2.635	$ReCl_4$	2.260	2.361–2.414	2.424–2.442
HfI_4[b]	2.677	2.852	3.002	RuF_4[e]	1.85	1.98	2.01
SnF_4[c]	1.874	2.025	2.025	PdF_4[f]	1.91	1.94	2.00
PbF_4[c]	1.944	2.124	2.124	$OsCl_4$	2.261	2.378	2.378
VF_4[a]	1.696	1.918	1.923	$PtCl_4$	2.297	2.331	2.396
NbF_4[d]	1.856	2.042	2.042	$PtBr_4$	2.456	2.469	2.538
$NbCl_4$	2.291	2.425	2.523	PtI_4	2.652	2.654	2.716

[a][190], [b][191], [c][192], [d][193], [e][194], [f][124]

from the Pt [177]. In this sense the Pt anion can be regarded as a homologue of chalcogenides, taking into account that $\chi(\mathrm{Pt}) \approx \chi(\mathrm{Te})$, see Table 2.15 which indicates the ionic nature of Cs_2Pt.

In the structure PbFCl, the Pb atom also has $N_c = 9$, it is surrounded by 4 atoms of F at the distance of 2.52 Å, four Cl at 3.07 Å and one Cl at 3.21 Å. In this structural type, many compounds of MXY composition crystallize [178–180], which will be considered later.

To conclude the survey of MX_2 structural types, let us consider the AlB_2 family where the atom of metal is connected with 12 atoms of boron, being situated between two B_6 hexagons. This or similar structures are typical for the compounds of the metals of Groups 3–8. Internuclear distances in this structural type AlB_2 are listed in Table S5.5 according to [181].

Some crystalline halides, MX_3, have molecular structures (e.g. M_2X_6, considered in Chap. 3), a square coordination is found in the structure of AuF_3 ($d = 1.92$ and 2.04 Å [182]), other halides have structures of the ScF_3 and FeF_3 types (Table S5.6) or $FeCl_3$ and $AlCl_3$ types (Table S5.7) in which the metal atom is located in the center of an octahedron formed by halogen atoms. The ratio $d(N_c = 6)/d(N_c = 3)$ averages 1.10(1).

Halides of lanthanides and actinides (except fluorides) crystallize in the structural types UI_3 and UCl_3. In the former, halogens are located in the vertices of a distorted trigonal prism, with two more atoms capping two side-faces of the prism; in Table S5.8 are given the bond lengths in compounds of the structural type UI_3. In the structure UCl_3,the atom of metal is in the center of a trigonal prism with *three* extra atoms located at slightly longer distances above the sides of the prism; the lengths of bonds in representatives of this structural type are presented in Table S5.9. Fluorides of the rare-earth metals have structures of two types, YF_3 and LaF_3. In the former, the metal atom is coordinated by six fluorines at the vertices of a trigonal prism and three more completing a 9-vertex coordination polyhedron. The same structure is adopted by TlF_3 and BiF_3, however, here the 9-th atom of fluorine lies at a greater distance and hence $N_c = 8$. In Table S5.10 are presented the averaged interatomic distances M–F for these coordination numbers in YF_3 and its analogues. Coordination polyhedron in the structure of tysonite, LaF_3, is a superposition of a trigonal prism and a trigonal bipyramid. La has five nearest fluoro-ligands, six other at a greater distance and two further still, this the structure is usually regarded as 9-coordinate. In Table S5.11 are listed the M–F distances and the ligancies in the coordination polyhedra of LaF_3, CeF_3 and UF_3.

The solid halides of the Group 15 elements have peculiar structures in which three nearest ligands form an AX_3 molecule, while 3–6 more distant atoms complement this molecule to a polyhedron with $N_c = 6$, 8 or 9. Comparison of the condensed and gaseous molecules of this type shows that the difference of the bond lengths d(M–X) in their structures does not exceed the experimental error, hence, additional layers of atoms at distances exceeding the M–X bond length by 1 Å or more, do not affect the chemical bonds and geometrical parameters of molecules any more.

Compounds MX_4 crystallize in different structural types: compounds of the Group 14 elements (except fluorides of Ti, Sn and Pb) have molecular structures but compounds of other metals crystallize in the structural type $ZrCl_4$. In this structure, octahedra of the chlorine atoms are linked by vertices with two nearest octahedra, forming infinite zigzag chains. Two terminal Cl atoms are placed in a *cis*-position, opposite to them there are two longest bridging bonds, the third pair of bridging bonds has intermediate length. Such system of bonds when the shortest bond lies opposite to the longest one, is the general rule for crystal structures (see below). Similar 'conservation' of the total length in a three-center system was observed also in molecular structures with mixed ligands (Chap. 2). In Table 5.9 are listed the experimental data on structures of the $ZrCl_4$ and related types, where octahedra are linked in different ways. Tetra-halides of actinides, except fluorides, crystallize in the structure type UCl_4 where the U atom is surrounded by four Cl forming a flattened tetrahedron, with four more (from surrounding tetrahedra) lying at greater distances. In Table 5.10 the M–X distances in these structures are listed. In the structures of HfF_4 and ThF_4, belonging to this type, $\bar{d}$(M–F) = 2.094 and 2.325 Å, respectively [183]. In the UF_4 structure there are two types of uranium atoms: U(1) is coordinated by eight F atoms at distances ranging from 2.25–2.35 Å, U(2) is connected with eight F at 2.23–2.32 Å; $\bar{d}$(U–F) = 2.28 Å. The ratio $d(N_c = 8)/d(N_c = 4)$ averages 1.10(1).

In structures of all the studied MX_5 (except β-UF_5) metal atoms are in the center of octahedra connected by two vertices. However, the mode of linking can differ: in UCl_5 two octahedra are connected to form an U_2Cl_{10} molecule, in NbF_5 four octahedra are combined into a Nb_4F_{20} molecule, in VF_5 the octahedra form an infinite chain. Since the dimeric molecules have been described earlier (Table S5.7), Table 5.11 gives the data only for tetrameric MF_5 and VF_5. The maximum coordination ($N_c = 8$) in MX_5 is found in the structure of β-UF_5, where two terminal U–F bonds (1.96 Å) are much shorter than six bridging ones (2.27 Å).

All the AX_6 compounds studied have the molecular structures considered above. IF_7 and ReF_7 also have molecular structures with pentagonal bipyramidal coordination, where the mean lengths of the equatorial bonds equal to 1.849 and 1.851 Å, and the axial ones, 1.795 and 1.823 Å, respectively [184, 185].

Table 5.10 Interatomic distances (Å) in structures of the UCl_4 type

MX_4	$4d$(M–X)	$4d$(M–X)	MX_4	$4d$(M–X)	$4d$(M–X)
α-$ThCl_4$	2.85	2.89	$PaCl_4$	2.64	2.95
β-$ThCl_4$	2.72	2.90	$PaBr_4$	2.77	3.07
α-$ThBr_4$	2.909	3.020	UCl_4	2.638	2.869
β-$ThBr_4$	2.85	3.12	$NpCl_4$	2.60	2.93

Table 5.11 Interatomic distances (Å) in structures of pentafluorides

MF_5	$4d$(M–F)	$2d$(M–F)	MF_5	$4d$(M–F)	$2d$(M–F)
VF_5	1.69	1.96	MoF_5	1.78	2.06
NbF_5	1.77	2.06	RuF_5	1.90	2.08
TaF_5	1.77	2.06	OsF_5	1.84	2.03

Oxides and chalcogenides of the M_nX_m type are studied less than halides because of the difficulties of obtaining single crystals, although today the powder materials can be studied by diffraction methods with sufficient accuracy. Oxides of trivalent metals crystallize mainly in the structural types of Al_2O_3 (corundum) and Mn_2O_3. In the former, the Al atom is distorted-octahedrally coordinated by six O, in the latter the Mn atom is in the center of a cube which has two vertices vacant; thus, $N_c(M) = 6$ in both cases (Table 5.12). In the structure of Au_2O_3, the Au atom has $N_c = 4$ [186], in β-Ga_2O_3 there are two types of the Ga atoms: one is surrounded by a tetrahedron of four oxygens, the other by an octahedron of six O, with d(Ga–O) = 1.841 and 1.993 Å, respectively [187]. In the structural type La_2O_3 the La atom is placed in the center of a deformed oxygen octahedron and one more O lies above a face of the octahedron. The average distances of M–O in La_2O_3, Ce_2O_3, Pr_2O_3, Nd_2O_3 and Ac_2O_3 are 2.55 Å, 2.505 Å [188], 2.523 Å [189], 2.50 Å and 2.61 Å, respectively. Structures of the Sb_2O_3 modifications (senarmontite and valentinite) contain tetrahedra where only three vertices are occupied by O atoms and the 4-th by a lone electron pair; d(Sb–O) in these phases are 1.977 Å and 2.006 Å, respectively. In the α-Bi_2O_3 structure there are two geometrically different Bi atoms with $N_c = 5$ and 6; however in both cases the Bi atom has three short bonds with oxygen ($\bar{d} = 2.187$ Å and 2.205 Å, respectively), and other much longer bonds: 2.546, 2.629 Å, and 2.422, 2.559, 2.787 Å, respectively for $N_c = 5$ and $N_c = 6$.

M_2S_3 compounds are remarkable for great variety of structural types, often having strongly distorted coordination polyhedra in which it is impossible to establish any

Table 5.12 Interatomic distances (Å) in structures of the Al_2O_3 and Mn_2O_3 types

M_2O_3[a]	d(M–O)	M_2O_3[a]	d(M–O)	M_2O_3	d(M–O)	M_2O_3	d(M–O)
Sc	2.12	Dy	2.27	Al[b]	1.91[c]	V[a]	2.01
Y	2.27	Ho	2.26	Ga[b]	2.00	Cr[b]	1.99
Sm	2.32	Er	2.25	In[b]	2.16	Mn[a]	2.01
Eu	2.32	Tm	2.25	Tl[a]	2.27[d]	Fe[b]	2.03
Gd	2.30	Yb	2.22	Ti[b]	2.05	Rh[b]	2.03
Tb	2.29	Lu	2.22				

[a] Structural type of Mn_2O_3, [b] Structural type of Al_2O_3, [c] [195], [d] [196]

definite coordination number. This problem will be discussed later; presently we consider the M_2X_3 chalcogenides crystallizing in the structure of ZnS, where every third position in the metal sub-lattice is vacant. The M–X distances in these structures are: Al_2S_3 2.233 Å [197], α-Ga_2S_3 2.244 Å, β-Ga_2S_3 2.258 Å, Ga_2Se_3 2.351 Å, Ga_2Te_3 2.549 Å, and α-In_2S_3 2.326 Å. A similar structure is found for Al_2Te_3 [198] with d(Al–Te) = 2.628 Å. In the Sc_2S_3 structure the atom of Sc is surrounded by 6 atoms of S on an average distance of 2.587 Å [199].

In the structural type of Th_3P_4 ($N_c = 8$), besides the seminal compound, there crystallize U_3P_4, U_3As_4, U_3Sb_4, Zr_3N_4 (4 × 2.19 + 4 × 2.49 Å), Hf_3N_4 (4 × 2.17 + 4 × 2.47 Å) [200], and also M_2X_3 compounds (with vacancies in the non-metal sub-lattice), where M = Ln, Ac, and X = chalcogens; Table S5.12. The atom of metal is surrounded in this structure by 8 chalcogen atoms in a cubic arrangement.

Chalcogenides of the Group 15 elements have structures where there are 3 short bonds and 3 or 4 longer ones. For example, in As_2S_3 there are 3 × 2.24 Å + 1 × 3.49 Å + 2 × 3.59 Å distances, in As_2Se_3 three bonds with $d = 2.427$ Å and 4 more distances: 3.373 Å, 3.513 Å, 3.725 Å and 4.143 Å [201], in the As_2Te_3 structure there are 3 × 2.708 Å + 4 × 3.660 Å distances [202], in the Sb_2Se_3 there are two types of the Sb atoms, each of which forms three short bonds (2.674 and 2.732 Å) and three or four longer ones (3.226 and 3.245 Å) according to [203]; in Bi_2Te_3 also there are three short (3.066 Å) and three longer (3.258 Å) bonds [204]. Thus, in these structures, the elements form 3 covalent bonds with the nearest neighbors and 3 intermolecular bonds with the next layer of atoms.

In conclusion, we can note the general rules of interatomic distances in binary compounds. Within each structural type the bond lengths obey the additivity principle. An increase in the coordination number increases interatomic distances, and this change is the same for identical relative changes of N_c, for all ratios $N_c/N'_c = n$, where $n = 1 \div 8$. It allows to describe the dependence of d(M–X) on N_c by the following average values:

$N_c \rightarrow N'_c$:	$1 \rightarrow 2$	$1 \rightarrow 3$	$1 \rightarrow 4$	$1 \rightarrow 6$	$1 \rightarrow 8$
$d(N_c)/d(N'_c)$:	1.10	1.15	1.17	1.22	1.26

Note that, as was shown above, the variations in the *relative* changes of the interatomic distances upon a change of coordination numbers are several times smaller than the variations of the absolute values of the bond lengths. The relative change in interatomic distances on the $N_c(3) \rightarrow N_c(8)$ transition in structures of binary compounds equals 1.26/1.15 = 1.10, i.e. is close to the value of 1.11 found for similar transitions in elemental solids (see Sect. 5.1.2). It means that in the case of spherically-symmetric valence electrons (in metals and ionic compounds) the interatomic distances are governed by geometrical factors.

The increase in the atomic valence noticeably reduces the bond length. Thus, for oxides and fluorides an increase of the valence of metals from 2 to 3 at the same N_c reduces the bond length by 9.3 %, whereas an increase of v from 3 up to 4 decreases d by 5.4 %. The greatest variation of valences was observed in the structures containing $[FeO_4]^{n-}$ where d(Fe–O) increase in the succession 1.647 → 1.720 → 1.807 → 1.889 Å at the progressive reduction of v(Fe): 6 → 5 → 4 → 3 [205]. Therefore in

Table 5.13 Interatomic distances (Å) in compounds with mixed ligands

$M_kX_lY_m$	d(M–X)	$M_kX_lY_m$	d(M–X)	$M_kX_lY_m$	d(M–X)	$M_kX_lY_m$	d(M–X)
$Cu_2(OH)_3Cl$	1.93	$Hg_3Cl_2O_2$	2.84	ZrSiO	2.77	NbPS	2.58
$Cu_2(OH)_3Br$	1.97	$Hg_3Cl_2S_2$	8B2.87	ZrSiS	2.82	NbPSe	2.63
$Cu_2(OH)I$	2.02	$Hg_3Cl_2Se_2$	2.90	ZrSiSe	2.83	POF_3	1.436
CaHCl	2.17	$Hg_3Cl_2Te_2$	2.99	ZrSiTe	2.87	PO_2F	1.464
CaHBr	2.22	NdFO	2.37	ZrNCl	2.140[f]	PO_3F^{2-}	1.500
CaHI	2.49	NdFS	2.53	ZrNBr	2.152[f]	PO_4^{3-}	1.533
Ca_2IN	3.280[a]	NdFSe	2.57	ZrNI	2.175[f]	SbF_5	1.89
Ca_2IP	3.389[a]	NdFTe	2.70	SbIS	3.11[d]	SbF_4Cl	1.95
Ca_2IAs	3.408[a]	ThAsS	2.46	SbISe	3.14[d]	SbF_2Cl_3	2.02
CaMgSi	3.39	ThAsSe	2.50	SbITe	3.22[d]	$SbFCl_4$	2.12
CaMgGe	3.42	ThAsTe	2.61	BiOF	2.28	SO_2F_2	1.386[e]
CaMgSn	3.56	ThOS	2.416[b]	BiOCl	2.31	SO_2FCl	1.408[e]
BaFCl	2.649	ThOTe	2.434[c]	BiOBr	2.32	SO_2Cl_2	1.418[e]
BaFBr	2.665	UOS	2.34	BiOI	2.33	Cr_2S_2S	2.40
BaFI	2.694	UOSe	2.36	PbClCl	2.98	Cr_2S_2Se	2.48
$BaMg_2Si_2$	3.62	UOTe	2.39	PbClBr	3.00	Cr_2S_2Te	2.69
$BaMg_2Ge_2$	3.67	USbS	2.42	PbClI	3.03	FeOF	2.00
$BaMg_2Sn_2$	3.86	USbSe	2.53			FeOCl	2.03
$BaMg_2Pb_2$	3.88	USbTe	2.65				

[a][206], [b][207], [c][208], [d][209], [e][210], [f][111, 211]

additive calculations it is necessary to consider not only coordination numbers, but also the valence states of the atoms.

5.2.2 *Structures of Compounds with Diverse Bonds*

The length of a given bond in a crystal structure depends on others ligands present. As shown in Table 5.13, a substitution in the AX_n polyhedron of a X ligand by another atom with lower EN increases d(A–X) of the remaining ligands, because of the increase of polarity of their bonds.

Solid solutions are essentially more convenient materials for studying the mutual influence of atoms because of the possible monotonic change in composition with retention of the structure of a whole series of compounds [212]. It is well-known that between the cell parameters of a mixed crystal and the concentration of the components there is a linear relation (Vegard's law). If the KBr–KI system, which gives a continuous series of solid solutions, is taken as an example, then the picture of gradual change in the K–X bond ionicity as the composition of the solution is varied becomes clear. Let's begin by considering pure KBr into which KI will be introduced gradually. It is evident that as the iodine concentration is increased, the competition with the bromine for withdrawing electrons from K (due to the lower electronegativity of iodine) will be reduced, and the ionicity of the K–Br bond and the lattice parameter of the solid solution will increase. Since the halogen atoms in it are distributed statistically, hence the K–Br length will be increased. The K–I bond ionicity in the same system will be changed in the opposite sense: owing to the competition with the bromine the polarity of the K–I bond (and its length) will decrease as the bromine concentration in the coordination polyhedron KX_6 increases. Thus, in solid solutions MX^I–MX^{II} the bond length M–X^I increases with the growth of the covalent character in the M–X^{II} bond. Therefore, it is in solid solutions that the highest effective charges of atoms and the ultimate values of ionic radii can be found.

Similar changes in bond lengths occur in a structure of the same substance if its bonds are in the *trans*-position to different atoms (Table S5.13). The increase in bond lengths which occurs in the *trans*-position to shorter bonds, is observed in the three-center systems X^I–M–X^{II} in some oxides and halides (Table S5.14). As can be seen, $\Delta d = d(\mathrm{M{-}X^{II}}) - d(\mathrm{M{-}X^{I}})$ increases with the decrease in the M–X^I distance, since for the formation of a shorter bond (of greater multiplicity or covalency) it is necessary to transfer electrons from the bond in the *trans*-position, with a corresponding increase in its length. We shall encounter this rule later but presently we can mention hydrogen bonds of the O–H···O type where hydrogen plays a role of the central atom in this coordinate.

5.3 Interconversions of Crystal Structures

An important section of crystal chemistry is the interrelation and genesis of crystal structures. The elementary geometrical operations transforming one structural type to another one are not only of theoretical but also, as will be shown below, of practical value. Thus, the structure of NaCl can be converted into the CsCl type by an interplanar movement and an anti-parallel displacement of atoms in adjacent (100) NaCl layers (see [213]), a CsCl structure turns into the CaF_2 type if the center of every other cube is occupied, or in ZnS if the vertices are occupied in every other cube. Stretching an octahedron along the 4-fold axis or removing two axial atoms altogether, transforms it into a square; a 180° rotation of one face of an octahedron transforms it into a trigonal prism (Fig. 5.6). More complex geometrical transformations are described in articles [214–218].

According to O'Keefe, the crystal structures of compounds are formed on the basis of the metal structure in which the non-metal atoms are notionally inserted. This problem has been reviewed comprehensively by Vegas et al. [219, 220] who have shown that the structures of about a hundred compounds are related to the structures of their parent metals or alloys, in which non-metal atoms simply occupy the interstitial voids. It is interesting that oxides of metals often reproduce the structures of the high-pressure phase of initial alloys, i.e. from the structural point of view an oxidation

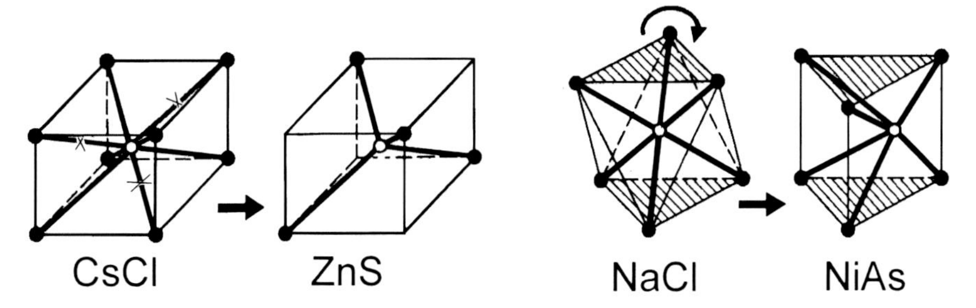

Fig. 5.6 Geometrical transformations of structures by elimination of atoms (*left*) or shear (*right*)

is equivalent to compression. Since properties of a substance are in many respects defined by its volume, it is obvious that the result does not depend on the means by which the volume is altered. This conclusion, to a certain extent, applies to chemical ways of compressing materials (see [221]). It has been shown [222] that properties of the solid solutions of alkali halides change in the same way upon replacement of a larger anion by a smaller one (e.g., Cl for Br in KBr) as upon physical compression of KBr up to the same volume.

5.4 Effective Coordination Number

The concept of the coordination number as the number of nearest neighbors situated at equal distances from the central atom, has been introduced by Werner in 1893 [223], and for the NaCl structure by Pfeifer in 1915 [224]. The above mentioned structural types ZnS, NaCl, CsCl, CaF_2, $CdCl_2$, FeS_2 are characterized by regular coordination polyhedra, but other structures often contain distorted polyhedra. Moreover, as the accuracy in the structure determinations improved, it became clear that regular coordination is an exception rather than the rule. Therefore the study of reasons of distortion in coordination polyhedra and their quantitative description became an important problem of theoretical crystal chemistry. Some factors influencing the distortion are of electronic nature. Thus, the presence in the metal atom of a lone electron pair (Tl^{I}, Sn^{II}, Sb^{III}) leads to a difference in the interatomic distances in the coordination sphere of the central atom. Presence of an unfilled electron level also leads to nonequivalent interaction of *d*-electrons with *p* orbitals in *x*-, *y*- and *z*-directions (the Jahn-Teller effect). The strong deviation from the ideal octahedron in FeF_2 comes from the Fe^{2+} ion having the d^6 configuration in the high-spin state, which would require a singly occupied *d* orbital and direct Fe–Fe interactions mainly along the *c*-axis [225]. However, in many cases this effect cannot explain a distortion of the coordination polyhedra. Apparently, there are more general reasons of such phenomenon being widespread in the world of crystals.

In 1961 Bauer [226], within the framework of the ionic model, demonstrated that in structures of the TiO_2 type the minimum energy of a crystal lattice is reached at unequal bond lengths, namely four bonds in an octahedron should be longer than the other two. Though only fluorides conform to this rule, it is essential that a distortion of the coordination polyhedron is caused by the action of spherically symmetric forces. From a geometrical viewpoint, it corresponds to the fact that unequal spheres can achieve higher packing density than identical spheres, as small spheres can fit between larger ones.

How we can define N_c if ligands are situated at different distances? For the first time this question was addressed by Frank and Casper [227] who defined the effective coordination number (N_c*) as the number of planes in the Voronoi polyhedron. This approach assumes that all atoms have identical volume, all planes in a polyhedron are equivalent, i.e. there are no distinctions in the character of the bonds formed by different ligands. The metal structures generally conform to these criteria [228] but application of this approach to binary inorganic and complex compounds [229, 230] is not straightforward and requires some accounting for different atom sizes.

The first crystal chemical method of defining N_c* in binary compounds belongs to Witting [231]. He assumed that the contribution of the nearest atom to N_c* equals 1, an atom at double that distance contributes nothing [232], and for intermediate distances he used a linear interpolation between 0 and 1. Later Hoppe [233] suggested another method: the lines are drawn from the center of the coordination polyhedron to all the neighboring atoms, and at the points of contact of the central atom and ligands are drawn the secant planes. The ratio of the areas of all planes to the nearest (greatest) plane gives contributions of all ligands to N_c*. Later Melhorn and Hoppe [234] suggested to define N_c* by comparison of the real interatomic distance in the coordination polyhedron (d_{exp}) with the sum of effective ionic radii (d_{teor}) by the formula:

$$N_c{}^* = \sum_i n_i \exp\left[1 - \left(\frac{d_{\text{exp}}}{d_{\text{teor}}}\right)^6\right] \tag{5.2}$$

where n_i is the ligancy with the identical bond length. The review of the results obtained by this method is given in [235]. In the work of Brown and Wu [236] the effective coordination numbers of atoms were calculated from the bond valences (V_b):

$$N_c{}^* = \frac{Z}{V_{\text{bmax}}} \tag{5.3}$$

where Z is the formal valence and V_{bmax} is the maximum bond valence (this characteristic will be considered later). In 1977 Batsanov [237] proposed the energetic criteria for the determination of N_c*, according to which the contribution (fraction) of each ligand is determined by the ratio $E(\text{M}-\text{X}_i)/E(\text{M}-\text{X}_o)$ where X_o denotes the nearest ligand and X_i other ligands. In the framework of the ionic approximation,

$$N_c{}^* = \sum_i \frac{E(M - X_i)}{E(M - X_o)} = \sum_i \frac{d(M - X_o)}{d(M - X_i)} \tag{5.4}$$

Simultaneously, a similar formula was proposed by Brunner [238], and later by Beck [239]. In [237] N_c* was calculated in terms of the covalent theory. Since the overlap integrals (S) of the bonds are directly proportional to the bond energies (see Sect. 2.3), then

$$N_c^* = \sum_i \frac{E(M - X_i)}{E(M - X_o)} = \sum_i \left(\frac{S(M - X_i)}{1 + S(M - X_i)} : \frac{S(M - X_o)}{1 + S(M - X_o)} \right) \tag{5.5}$$

where S is the overlap integral for the shortest ($M-X_o$) any other ($M-X_i$) bond. The described ionic and covalent approaches to the calculation of N_c*give similar results, which confirms the objectivity of these values. Interestingly, in the case of halides and oxides of some metals an increase in the separations of ligands from the central atom and hence a decrease in the energy of Coulomb interaction is accompanied by an increase of the overlap integrals for the s–$p\sigma$ and $p\sigma$–$p\sigma$ bonds, i.e. of the covalent energy. Thus, the actual distances are such as to optimize the *sum* of the covalent and ionic energies. In other words, the variable distances in the coordination polyhedron correspond to several minima on the potential surface of a given compound.

In 1978 Carter [240] proposed to calculate N_c* under the general formula:

$$\frac{1}{N_c^*} = \sum_1^N \left(\frac{A_i}{A_t} \right)^2 \tag{5.6}$$

where A_i are the force, bond energy, force constant, or overlap integral, and $A_t = \Sigma A_i$. If all A_i are the same, $N_c^* = N_c$, if not, then $N_c^* < N_c$. However, difficulties in the experimental determination or the theoretical account of these parameters, and also of the hybridization and valence states of atoms, stimulated development of procedures for the purely geometrical estimation of N_c*. Thus, it has been proposed [241] to define the contribution of ligands to the coordination numbers in proportion to the square of distance from the central atom:

$$N_c^* = \Sigma \left(\frac{d_a}{d_i} \right)^2 \tag{5.7}$$

where d_a is the sum of atomic radii and d_i is the actual bond length. In Table 5.14 the results of estimations N_c* by Eq. 5.7 are listed.

Besides the distortion of the coordination polyhedron, deviations of N_c from integer values can be caused by point defects (vacancy and interstitial) in a crystal structure or in the first coordination sphere of an amorphous solid. These values of N_c* are defined experimentally by XRD (see Chap. 7) and by optical methods. Thus, Wemple [242] estimated by spectroscopy that in the structure of As_2S_3 the $N_c^* = 3.4 \pm 0.2$ whereas Eq. 5.7 gives 3.7. Optical estimates of N_c* of Se and Te are 2.8 and 3.0, respectively, whereas the calculated values are 2.8 and 3.1. When the composition of chalcogenide glasses (Ge-S, Ge-Se, As-Se, Ge-As-Se) is altered, there occur phase transformations with changing structural, mechanical and electric properties

Table 5.14 Bond lengths (Å) and effective coordination numbers

M_nX_m	Bond length (number of such bonds in parentheses)						N_c*
Hg_2F_2	2.133 (2)	2.715 (4)					4.1
Hg_2Cl_2	2.43 (2)	3.21 (4)					4.5
Hg_2Br_2	2.71 (2)	3.32 (4)					4.5
TlF	2.539	2.623	2.792 (2)	3.254	3.496		4.4
	2.251	2.521	2.665 (2)	3.069	3.905		4.8
TlI	3.34	3.50 (4)	3.87 (2)				6.0
CuF_2	1.910 (2)	1.929 (2)	2.305 (2)				5.3
$CuCl_2$	2.30 (4)	2.95 (2)					5.4
$CuBr_2$	2.40 (4)	3.18 (2)					5.8
α-SnF_2	2.057	2.102	2.156	2.671	2.834	3.221	4.2
2.048	2.197	2.276	2.386	2.494	3.309		4.3
β-SnF_2	1.89	2.26	2.40	2.41	2.49		3.9
γ-SnF_2	2.13 (2)	2.32 (2)					3.2
$SnCl_2$	2.66	2.78	3.06	3.22	3.30	3.86	3.8
$SnBr_2$	2.81	2.90 (2)	3.11 (2)	3.41 (2)			5.0
SnI_2	3.000	3.198 (2)	3.251 (2)	3.718 (2)			5.2
	3.174 (2)	3.147 (2)					4.9
PbF_2	2.41 (2)	2.45	2.53	2.64	2.69 (2)	3.03 (2)	7.2
$PbCl_2$	2.86	2.90 (2)	3.06	3.08 (3)	3.64 (2)		7.3
$PbBr_2$	2.967 (2)	2.995	3.223 (2)	3.259	3.353	3.846 (2)	7.4
CuO[a]	1.886	1.956	1.958	2.041	2.774	2.801	4.9
HgO	2.033 (2)	2.79 (2)	2.90 (2)				4.3
HgS	2.359 (2)	3.10 (2)	3.30 (2)				4.7

[a][244]

at $N_c* = 2.67$, this means a transition from 2D to 3D lattice [243]. Non-integer coordination numbers are characteristic of structures of liquids and amorphous solids which will be considered later.

5.5 Bond Valence (Bond Strength, Bond Order)

Knowing the coordination numbers, one can solve a number of crystal chemical problems. Thus, 80 years ago Goldschmidt discovered the dependence of the chemical bond length on the coordination number of a cation [245]. At the same time Pauling introduced the concept of bond strength [246], according to which the valence of an atom *i*, which is linked to atoms *j*, obeys the relationship

$$V_i = \Sigma_j v_{ij} \tag{5.8}$$

where v_{ij} is the valence of the bond between the two atoms *i* and *j*. This relationship is the expression of the bond-valence sum rule and derives from Pauling's electroneutrality principle (see Chap. 2). In 1947 Pauling also formulated a relationship between bond length and bond order [247]:

$$d = d_1 - A \log V \tag{5.9}$$

where d is the length of a chemical bond, d_1 is the length of the same type of bond for $v = 1$, and the parameter A equals 0.71 for covalent bonds and 0.60 for metallic structures. Then, especially in the 1970s, this equation was applied to interpret experimental bond lengths and to predict bond lengths in crystals. From the beginning of the 1980s until now the formulae almost exclusively used in the literature correlating bond valence and bond length have been the following:

Brown and Shannon [248]:

$$v = \left(\frac{d}{d_1}\right)^{-N} \tag{5.10}$$

Zachariasen [249]:

$$v = \exp\left[\frac{(d_1 - d)}{B}\right] \tag{5.11}$$

Brown and Altermatt [250]:

$$v = \exp\left[\frac{(d_1 - d)}{0.37}\right] \tag{5.12}$$

where N is an empirical parameter which in many cases has values in the range 4–7, d_1 is the constant for a given type of A–X bond, whereas B is assumed to equal 0.314 Å [249], 0.37 Å [250] or 0.305 Å [251]. Values of N and d_1 for various chemical bonds are reported [252–270]. The history, theory and development of the bond valence model can be found in the book by Brown [271], where many bond valence parameters for Eq. 5.12 are presented. As one can see from presented data, the values of these parameters depend on the bond types and the oxidation state of atoms [272], and recently the parameter B has been found to be linearly dependent on the coordination number of the central atom in two series of polyhedra, LaO_n ($n = 7$–12) and MoO_n ($n = 5$–7) [251].

A statistical study carried out by Zocchi on the applicability of various techniques to chemical bonds showed that, despite fluctuations, the Brown-Altermatt equation does not provide satisfactory results because the parameter B cannot be constant for different bonds. This conclusion seems to corroborate the doubts already reported by Efremov [258] and Urusov [273] regarding the assumed 'universality' of the constant $B = 0.37$ Å.

At the same time, one can envisage an alternative dependence of bond distances on the coordination number. As reported in [274], the ratios of interatomic distances in the crystalline monohalides MX for $N_c = 2$, 4, 6 and 8 to the bond lengths in corresponding molecules, $k_{mn} = d(N_c = m)/d(N_c = n)$ are equal to: $k_{21} = 1.056 \pm 1.0$ %, $k_{41} = 1.126 \pm 1.3$ %, $k_{61} = 1.220 \pm 3.7$ %, $k_{81} = 1.264 \pm 3.8$ %, $k_{41} = 1.174 \pm 3.6$ %, $k_{61} = 1.291 \pm 5.8$ %, while the absolute differences, $\Delta d_{mn} =$

Table 5.15 Corrections of bond lengths for changing coordination numbers

$N_c(m) \rightarrow N_c(n)$	$8 \rightarrow 1$	$6 \rightarrow 1$	$4 \rightarrow 1$	$2 \rightarrow 1$
k_{mn}	1.26	1.25	1.16	1.06
Δd_{mn}(Å)	0.70	0.54	0.29	0.12
$N_c(m) \rightarrow N_c(n)$	$8 \rightarrow 2$	$6 \rightarrow 2$	$4 \rightarrow 2$	
k_{mn}	1.19	1.17	1.11	
Δd_{mn}(Å)	0.40	0.36	0.21	
$N_c(m) \rightarrow N_c(n)$	$9 \rightarrow 3$	$6 \rightarrow 3$		
k_{mn}	1.17	1.11		
Δd_{mn}(Å)	0.44	0.21		
$N_c(m) \rightarrow N_c(n)$	$8 \rightarrow 6$	$6 \rightarrow 4$		
k_{mn}	1.04	1.08		
Δd_{mn}(Å)	0.13	0.19		

d(crystal)–d(mol) are equal to: $\Delta d_{21} = 0.131$ Å $\pm$ 16.5 %, $\Delta d_{41} = 0.284$ Å $\pm$ 3.7 %, $\Delta d_{61} = 0.537$ Å $\pm$ 15.1 %, $\Delta d_{81} = 0.698$ Å $\pm$ 12.5 %, $\Delta d_{41} = 0.293$, Å $\pm$ 16.9 %, $\Delta d_{61} = 0.544$ Å $\pm$ 24.1 %. For MX_2 (X is a halogen or oxygen): $k_{42} = 1.108 \pm 2.3$ %, $k_{62} = 1.171 \pm 2.8$ %, $k_{82} = 1.194 \pm 3.8$ %; $\Delta d_{42} = 0.213$ Å $\pm$ 30.0 %, $\Delta d_{62} =$ 0.363 Å $\pm$ 18.8 %, $\Delta d_{82} = 0.395$ Å $\pm$ 15.7 %. For trihalides $k_{63} = 1.107 \pm 1.4$ %, $k_{93} = 1.174 \pm 1.4$ %, $\Delta d_{63} = 0.212$ Å $\pm$ 18.5 %, $\Delta d_{93} = 0.437$ Å $\pm$ 6.7 %. For polymorphic transformation, $k_{64} = 1.086 \pm 0.9$ %, $\Delta d_{64} = 0.170$ Å $\pm$ 14.5 %, $k_{86} =$ 1.050 ± 1.3 %; $\Delta d_{86} = 0.124$ Å $\pm$ 28.9 %. Hence, the average relative changes in the bond lengths in binary compounds with changing N_C can be described as follows:

N_c (m) $\rightarrow$ N_c (n)	$1 \rightarrow 1.33$	$1 \rightarrow 1.5$	$1 \rightarrow 2$	$1 \rightarrow 3$	$1 \rightarrow 4$	$1 \rightarrow 6$	$1 \rightarrow 8$
k_{mn}	1.05	1.08	1.10	1.17	1.19	1.27	1.30

These changes are similar to some changes in the ionic radii when N_c changes in the structure. Thus, according to Pauling [7], for K^+ or Cl^- the nature of the change in the ionic radii with changing N_c is as follows:

N_c (m) $\rightarrow$ N_c (n)	$1 \rightarrow 2$	$1 \rightarrow 3$	$1 \rightarrow 4$	$1 \rightarrow 6$	$1 \rightarrow 8$
k_{mn}	1.10	1.17	1.19	1.27	1.30

Thus the deviations from the mean relative coefficients (k_{mn}) are always many times smaller than from the mean absolute coefficients (Δd_{mn}), which indicates the role of the atomic size. Generally, the changes of interatomic distances in crystals at changing coordination numbers are specific for each transition and valence state. It is necessary to take this fact into account when using the coordination corrections. In Table 5.15 the averaged values of these corrections are given, calculated from the above mentioned data. As seen from this table, the interatomic distances in crystal structures naturally change with variation of the bond order, which can be used to analyse chemical bonding.

5.6 Ternary Compounds

The majority of inorganic substances contain atoms that are simultaneously connected with elements of two or more kinds. Compounds with mixed anions (LaOF, PbFCl, $BiOX_2$, etc) are considered above, those with mixed cations, known as ternary compounds, are discussed here. In compounds with mixed ligands there exist the valence states of atoms which are unstable in the case of homoleptic bonds. Stabilization of such states is caused by introduction into the structure of a substance, of a fragment reducing the 'concentration of tension'. For example, CuI_2 and AuI_3 do not exist, but mixed halides CuIX and $AuIX_2$ are quite stable [275, 276].

Table S5.15 shows the interatomic distances in ternary compounds. In all cases, compounds with identical N_c are compared to exclude its influence on the bond lengths. It is obvious that on variation of M in compounds $M_nAO(F)_m$ the increase of χ(M) or reduction of d(M–X) decrease the electron density in the A–O(F) bond that increases its length.

Structures of nitrates $M(NO_3)_n$ demonstrate the effect of variation in d(M–O) on the d(N–O). Thus, in Cu $(NO_3)_2$ the d(N–O) varies from 1.13 to 1.35 Å depending on d(Cu–O): if $d(Cu–O_I) = 2.0$ Å then $d(O_I–N) = 1.32$ Å, if $d(Cu–O_{II}) = 2.5$ Å then $d(O_{II}–N) = 1.15$ Å. In the structure of $Co(NO_3)_3$, nitrate groups are connected with metal via two oxygens and the third oxygen remains free, the former have d(N–O) of 1.28 Å, and the latter of 1.19 Å. The picture is similar in Tl_2CO_3: $d(C–O) = 1.24$ *vs* $d(Tl–O) = 2.82$ Å, and 1.28 *vs* $d(Tl–O) = 2.68$ Å; in $CdSO_4$ and $HgSO_4$, the longest bond S–O *vs* the shortest bonds Cd–O (2.228 Å) and Hg–O (2.221 Å) in system M–S–O [277].

In the structures of $BaCo_2(PO_4)_2$ or $BaNi_2(PO_4)_2$ one can see the influence of χ(M) on the P–O lengths: $d(P–O) = 1.510$ Å *vs* Ba–O bond and 1.555 Å *vs* Co–O [278], or $d(P–O) = 1.502$ Å *vs* Ba–O and 1.570 Å *vs* Ni–O [279]. In the structures of two phases of Na_2CO_3 a dependence of d(C–O) on d(Na–O) was observed: $d(C–O) = 1.187$ *vs* $d(Na–O) = 2.441$ Å in α-phase and $d(Na–O) = 1.293$ *vs* $d(Na–O) = 2.333$ Å in β-phase [280]. A hydrogen atom in the outer sphere of the complex makes the maximum effect. Thus, in the structure of $KHCO_3$ where the carbonate-ion is connected simultaneously to K and H which strongly differ in their EN, the longest and shortest C–O bonds are realized simultaneously [281]. A similar situation is observed in the structure of $Ba(HSO_4)_2$ [282]. In this context, the change in the structure of $BaSO_4$ at heating is noteworthy: d(Ba–O) increases from 2.941 to 3.023 Å (at 1010 °C) while d(S–O) decreases from 1.475 to 1.448 Å [283]. The distances within and between the coordination polyhedra in K_2PtCl_6 similarly change on heating [284]. The natural changes of the bond separations are established in the $W(NO)_3Cl_3$ structure where d(W–Cl) decreases as d(W–NO) increases along the Cl–W–NO coordinate [285]. The maximum influence of external atoms on d(A–O) in complex ions is observed in acids: H_3PO_4, H_2SO_4, HNO_3 [286–288] (Table S5.15).

In $KMnF_3$, $NaMnF_3$ and $TlMnF_3$, d(Mn–F) increases with the increase in EN of the univalent metal, in the succession 2.093, 2.114, 2.125 Å. However, in the case

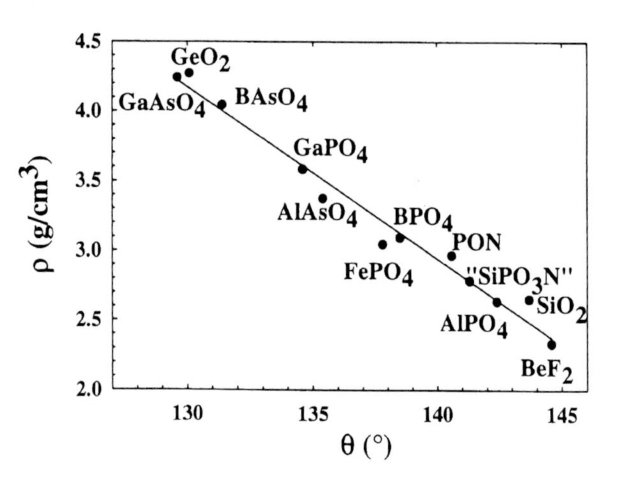

Fig. 5.7 Evolution of density as a function of the inter-tetrahedral bridging angle Θ for a series of a α-quartz homeotypes. Reprinted with permission from [290], ©2004 Oldenbourg Wissenschaftsverlag GmbH

of more ionic compounds the steric factor may play the key role. Thus, d(Mg–F) in K_2MgF_4 and Rb_2MgF_4 are, accordingly, equal to 1.99 Å and 2.01 Å, i.e. the larger Rb^+ ion simply pushes F^- anions apart in a solid solution MF–MgF_2. Similar situation is observed in ternary hydrides K_2MgD_4 and Cs_2MgD_4 where the Mg–D distance increases from 2.01 to 2.11 Å on increase of the size of alkaline metal [289]. Distances Sn–O in $CaSnO_3$, $SrSnO_3$, $BaSnO_3$ increase as 1.960, 2.016, 2.136 Å, i.e. together with the sizes of alkali earth metals. Similarly, in the series $CaTiO_3$, $SrTiO_3$, $BaTiO_3$, Ti–O distances increase as 1.90, 1.95, 2.00 Å.

Mutual influence of atoms at variations of M in isotypic compounds MP(As)O_4 consisting of MO_4 and P(As)O_4 tetrahedra, is described in [290]. The densities of α-quartz-type material depend directly on the structural distortion (inter-tetrahedral bridging angle Θ) present (Fig. 5.7)

Difference of distances A–X in the structures of M_kAX_m and AX_n with the same N_c(A) is caused only by the EN difference, $\Delta\chi = \chi(A) - \chi(M)$ (Table S5.16). From the observed changes of distances one can conclude that the ionicity of M–X and covalency of A–X bonds in ternary compounds are both higher than in binary oxides and halides. Thus, a change of the cation leads to a definite change of the sizes of complex ions. Because the EN of atoms changes in inverse proportion to their size (see Sect. 2.4), we obtain the compensatory dependence: the smaller is the size of an external cation, the larger are dimensions of the complex anion that dominates the dense packing of ions. Such interdependence is most evident in silicates.

Very interesting are the structures of K_3BrO, K_3IO, K_3AuO [291] where the Au atom plays a role of a halogen and bears a negative charge, due to its high electron affinity (2.31 eV); in these structures d(K–O) = 2.61, 2.64, 2.62 Å, respectively, d(K–Br) = 3.69 Å, d(K–I) = 3.73 Å, d(K–Au) = 3.71 Å.

5.7 Structural Features of Silicates

For a long time the structures of silicates were described as frameworks of close-packed anions, with smaller cations occupying the gaps. In 1959 Belov [292] opened a new era of crystal chemistry of silicates by proving that the basic motif of the structure is defined by large cations. He found that the lengths of edges in the SiO_4 tetrahedron are close to the edges of MO_6 octahedra with M = Mg, Al, Fe, which permits to fill the crystal space more densely by coordination polyhedra SiO_4 and MO_6. In the case of larger cations, the harmony in structural units can be reached only if SiO_4 units are connected into Si_2O_7 radicals, i.e. the cation size controls the anionic structure.

The major results of XRD analysis of silicates are reviewed in Belov's book [292] and need not be reiterated here. The most characteristic feature of silicates is tetrahedral coordination of silicon and approximate constant lengths of d(Si–O) $\approx$ 1.6 Å. However, in the structure of fluoro-zeolite there are both SiO_4 tetrahedra with $\overline{d}$(Si–O) = 1.59 Å and SiO_4F structural units where d(Si–O) = 1.68 Å and d(Si–F) = 1.74 Å [293]. Lengthening of d(Si–O) in the latter case is due to the increase of N_c from 4 to 5. Generally speaking, for Si–F bonds $N_c = 6$ is characteristic, e.g. in structures of the M_2SiF_6 type, but for oxygen compounds for a long time only tetrahedral environment was known. However, after the discovery of a transformation in SiO_2 of a quartz → rutile type with $N_c = 4 \rightarrow N_c = 6$, the tetrahedral coordination of silicon in Si–O bonds ceased to be an axiom and is assumed only for the silicates which are generated under normal thermodynamic conditions. As above mentioned, O'Keefe [294] has shown that the Si$\cdots$Si distance in crystal and glass silicates, in nitrides, nitride-silicides, SiC and even in SiX_4 is practically constant, at 3.08 ± 0.06 Å. According to O'Keefe, at constant lengths of bonds for a given coordination number the system aspires to occupy the maximum volume that corresponds to a minimum of the crystal lattice energy.

Besides the discussed compounds based on normal chemical bonds, molecules can connect by the donor-acceptor (DA) mechanism; the binding energy in such adducts is intermediate between the energies of normal chemical and vdW bonds. The composition and structure of DA-adducts are considered in reviews [295–298]; here we can note the general property uniting them with normal complex compounds, namely, that intramolecular bond lengths increase with the shortening of intermolecular contacts. An important feature of these adducts is the dependence of intermolecular distances on the angle under which molecules connect [299, 300]. The anisotropy of external configuration in molecules are specially considered in Sect. 4.3.4, in connection with vdW radii.

Appendix

Supplementary Tables

Table S5.1 Variation of bond lengths in structures of condensed halogens under heating

X_2	Cl_2				Br_2			
T, K	22	55	100	160	5	80	170	250
d, Å	1.994	1.989	1.985	1.979	2.301	2.294	2.289	2.286
D, Å	3.258	3.274	3.296	3.331	3.286	3.303	3.329	3.368

Table S5.2 M–X distances (Å) in structures of the NiAs, TiP (*) and MnP (**) types

M	S	Se	Te	M	P	As	Sb	Bi
Be			2.66[a]	Ti	2.48*	2.60	2.82	
Ti	2.49	2.58	2.73	V[c]	2.407	2.543**	2.815	
Zr			2.95[b]	Nb		2.62	2.79	
V	2.42	2.55	2.68	Cr	2.38**	2.51**	2.74	
Cr	2.46	2.60	2.77	Mn	2.36*	2.57	2.78	2.91
Mn			2.92	Fe	2.31**	2.43**	2.67	
Fe	2.49	2.57	2.61	Co	2.30**	2.42**	2.58	
Co	2.34	2.48	2.62	Ni		2.43	2.62	2.70
Ni	2.39	2.50	2.65	Ru			2.66	
Rh			2.70	Rh			2.68**	2.76
Pd			2.78	Pd			2.73	
				Ir			2.70	
				Pt			2.75	2.84

[a][5.1], [b][5.2], [c][5.3]

Table S5.3 M–X distances (Å) in structures of the CdX_2 type

M	OH	Cl	Br	I	M	S	Se	Te
Mg	2.099[a]	2.505[d]	2.735[c]	2.918[d]	Cu			2.65[f]
Ca	2.369[a]			3.12	Zn			2.65[f]
Zn			2.74[e]		Ti	2.428[g]	2.554[g]	2.770[f]
Cd	2.314[a]	2.652[b]	2.782[b]	2.989[d]	Zr	2.56	2.66	2.819[f]
Ti		2.499[d]		2.65[d]	Hf	2.56	2.66	2.822[f]
Zr		2.60		2.97	Sn	2.564[g]		
Ge				2.93	V	2.37	2.492[g]	2.67
Pb				3.227[d]	Nb	2.474[g]	2.59[g]	2.82
Mn	2.196[a]	2.593[b]	2.727[c]	2.92	Ta	2.468[g]	2.59	2.82
Fe	2.139[a]	2.483[d]	2.636[d]	2.88	Mo	2.42[g]	2.527[g]	2.716[g]
Co	2.100[a]	2.508[b]	2.63	2.83	W	2.415[g]	2.526[g]	2.76
Ni	2.073[a]	2.474[b]	2.628[b]	2.74	Re	2.41[g]	2.50[g]	
					Co		2.344	2.565[f]
					Ni			2.590[f]
					Rh			2.636[g]
					Pd			2.693 h
					Ir	2.350[i]	2.476[i,j]	2.650[g]
					Pt	2.340[g]	2.513[g]	2.676[g]

[a][5.4], [b][5.5], [c][5.6], [d][5.7], [e][5.8], [f][5.9], [g][5.10], [h][5.11], [i][5.12], [j][5.13]

Table S5.4 M—X distances (Å) in structures of the FeS_2 type [5.14]

M	S		Se		Te	
	d(M—S)	*d*(S—S)	*d*(M—Se)	*d*(Se—Se)	*d*(M—Te)	*d*(Te—Te)
Cu	2.453[a]	2.030[a]	2.557[d]	2.331[d]	2.749[i]	2.746[i]
Mn	2.593[a]	2.091[a]	2.702[d]	2.332[g]	2.907[i]	2.750[i]
Fe	2.262[a]	2.177[a]	2.377[d]	2.535[d]	2.619[i]	2.926[d]
Co	2.325[a]	2.113[a]	2.438[d]	2.446[g]	2.594[d]	2.914[d]
Ni	2.399[a]	2.072[a]	2.488[d]	2.417[d]	2.653[i]	2.650[i]
Ru	2.352[b]	2.171[b]	2.471[b]	2.453[b]	2.648[i]	2.790[i]
Pd	2.30	2.13	2.44	2.36		
Os	2.351[c]	2.216[c]	2.48	2.43	2.647[i]	2.830[i]
Ir	2.368[d]	2.299[d]	2.476[d]	2.555[d]	2.653[i]	2.883[i]
M	P		As		Sb	
	d(M—P)	*d*(P—P)	*d*(M—As)	*d*(As—As)	*d*(M—Sb)	*d*(Sb—Sb)
Au					2.76	2.86
Si[e]	2.400	2.13	2.519	2.38		
Cr					2.72	2.88
Mn					2.718[j]	2.842[j]
Fe	2.24	2.27	2.38	2.49	2.59	2.89
Ni			2.38	2.45	2.56	2.88
Ru[f]	2.363	2.234	2.461	2.475	2.641	2.863
Pd			2.495[h]	2.420[h]	2.676[h]	2.838[h]
Os[f]	2.367	2.248	2.469	2.469	2.641	2.889
Pt	2.39	2.18	2.49	2.41	2.671[h]	2.782[h]

[a][5.15], [b][5.16], [c][5.17], [d][5.10], [e][5.18], [f][5.19], [g][5.20], [h][5.21], [i][5.9], [j][5.22]

Table S5.5 Interatomic distances (Å) in structures of the AlB_2 type

M	*d*(M—B)	*d*(B—B)	*d*(M—M)
Sc	2.53	1.816	3.517
Ti	2.38	1.748	3.228
V	2.30	1.727	3.050
Cr	2.30	1.714	3.066
Mn	2.31	1.736	3.037
Re	2.24	1.821	2.901
Ru		1.864	
Os	2.23	1.873	

Table S5.6 Interatomic distances (Å) in structures of the ScF_3 and FeF_3 types

MF_3	*d*(M—F)	MF_3	*d*(M—F)	MF_3	*d*(M—F)
Sc	2.01	V	1.935[a]	Fe	1.925[a]
Al	1.80[b]	Nb	1.95	Co	1.89
Ga	1.88[b]	Ta	1.95	Ru	1.982
In	2.053	Cr	1.90	Rh	1.98
Ti	1.97	Mo	2.04	Pd	2.04
Zr	1.98	Mn	1.93	Ir	2.01

[a][5.23], [b][5.24]

Table S5.7 Interatomic distances (Å) in structures of the $FeCl_3$ and $AlCl_3$ types [5.25–5.27]

MCl_3	d(M–Cl)	MBr_3	d(M–Br)	MX_3	d(M–X)
$ScCl_3$	2.52	$GdBr_3$	2.868	$TcBr_3$	2.489
YCl_3	2.633	$TbBr_3$	2.855	$RuBr_3$	2.470
$AlCl_3$	2.31	$DyBr_3$	2.836	$RhBr_3$	2.44
$TiCl_3$	2.47	$HoBr_3$	2.825	$IrBr_3$	2.49
$ZrCl_3$	2.54	$ErBr_3$	2.816	TiI_3	2.76
VCl_3	2.45	$TmBr_3$	2.805	ZrI_3	2.910
$CrCl_3$	2.38	$YbBr_3$	2.798	BiI_3	3.09
$MoCl_3$	2.47	$LuBr_3$	2.792	ReI_3	2.74
$ReCl_3$	2.46	$TiBr_3$	2.582		
$FeCl_3$	2.37	$ZrBr_3$	2.676		
$RuCl_3$	2.34	$CrBr_3$	2.57		
$RhCl_3$	2.31	$MoBr_3$	2.555		

Table S5.8 Distances (Å) M–X in structures of the UI_3 type

MX_3	In prism		Over face	MX_3	In prism		Over face
	2d(M–X)	4d(M–X)	2d(M–X)		2d(M–X)	4d(M–X)	2d(M–X)
LaI_3	3.339	3.342	3.396	UI_3	3.165	3.244	3.456
$TbCl_3$	2.70	2.79	2.95	$CmBr_3$	2.865	2.983	3.137
				$CfCl_3$	2.690	2.806	2.940

Table S5.9 Distances (Å) M–X in structures of the UCl_3 type

MX_3	6d(M–X)	3d(M–X)	MX_3	6d(M–X)	3d(M–X)
$LaCl_3$	2.950	2.953	UCl_3	2.931	2.938
$LaBr_3$	3.095[a]	3.156[a]	UBr_3	3.062	3.145
$NdCl_3$	2.886	2.923	$PuCl_3$	2.886	2.919
$EuCl_3$	2.835	2.919	$AmCl_3$	2.874	2.915
$GdCl_3$	2.822	2.918	$CmCl_3$	2.859	2.914
			$CfCl_3$	2.815	2.924

[a] [5.26]

Table S5.10 Averaged distances (Å) M–X in structures of the YF_3 type

MF_3	Y	Sm	Ho	Yb	Tl	Bi
N_c	9	9	9	9	8	8
d(M–F)	2.32	2.39	2.32	2.30	2.28	2.38

Table S5.11 Distances (Å) M–X in structures of the LaF_3 type

MF_3	2 ×	1 ×	2 ×	2 ×	2 ×	2 ×
La	2.421	2.436	2.467	2.482	2.638	2.999
Ce	2.400	2.419	2.445	2.460	2.621	2.974
U	2.41	2.44	2.47	2.48	2.63	3.01

Table S5.12 Distances (Å) M–X in structures of the Th_3P_4 type

M_2X_3	S	Se	M_2X_3	S	Se
La	3.014	3.135	Eu	2.947	
Ce	2.978	3.093	Gd		3.003
Pr	2.967	3.075	Ac	3.096	
Nd	2.942	3.052	Pu	2.912	
Sm	2.954	3.026	d(U–Te) = 3.237 Å in U_2Te_3		

Table S5.13 Distances (Å) in coordinates of the X–M–Y type

Substances	X–M–Y	*d*(M–Y)	Substances	X–M–Y	*d*(M–Y)
$TiNEt_2Cl_3$	NEt_2–Ti–Cl	2.71	$WSCl_4$	S–W–Cl	3.05
	Cl–Ti–Cl	2.48		Cl–W–Cl	2.37
VOF_3	O–V–F	2.34	$WSBr_4$	S–W–Br	3.03
	F–V–F	1.76		Br–W–Br	2.54
$SbCl_4F$	Cl–Sb–Cl	2.32	$ReOF_4$	O–Re–F	2.32
	F–Sb–Cl	2.25		F–Re–F	2.00
$MoOF_4$	O–Mo–F	2.27	FeOCl	Cl–Fe–O	2.10
	F–Mo–F	1.93		O–Fe–O	1.96

Table S5.14 Distances (Å) in coordinates of the X_I–M–X_{II} type

M_nO_m	O_I–M–O_{II}		MX_n	X_I–M–X_{II}	
	d(M–O_I)	d(M–O_{II})		d(M–X_I)	d(M–X_{II})
V_2O_5[a]	1.577	2.791	TlF	2.251	3.905
	1.779	2.017		2.521	3.069
	1.878	1.878		2.665	2.665
CrO_3	1.60	3.10	$TcCl_4$	2.238	2.493
	1.76	1.76		2.246	2.490
MoO_3[b]	1.671	2.332		2.377	2.388
	1.734	2.251	VF_5	1.65	2.00
	1.948	1.948		1.68	1.93
WO_3	1.72	2.16	NbI_4	2.69	2.90
	1.79	2.13		2.76	2.76
	1.89	1.91			

[a][5.28], [b][5.29]

Table S5.15 Changes of bond lengths (Å) in complexes under variation of cations

M_nAO_m	*d*(A–O)	M_nAO_m	*d*(A–O)	M_nAO_m	*d*(A–F)
$RbUO_2(NO_3)_2$	1.58	H_2SO_4	1.426	$(NH_4)_2BeF_4$	1.53
$K_2UO_2F_5$	1.75		1.537[a]	Li_2BeF_4	1.56
$BaUO_4$	1.90	$Ba(HSO_4)_2$	1.443	Tl_2BeF_4	1.61
$CaUO_4$	1.91		1.563[b]	Cs_2ZrF_6	2.035
$MgUO_4$	1.92	H_3PO_4	1.493	Rb_2ZrF_6	2.040
$MnUO_4$	2.13		1.550[c]	K_2ZrF_6	2.126
$GeUO_4$	2.39	$BaNi_2(PO_4)_2$	1.502[d]	$KSbF_6$	1.708
K_2WO_4	1.79		1.570[e]	$NaSbF_6$	1.776
Na_2WO_4	1.88	HNO_3	1.2	$LiSbF_6$	1.876
$CaWO_4$	1.75		1.321[f]	K_2UF_6	2.347
$MgWO_4$	1.95	$Cu(NO_3)_2$	1.15[g]	Na_2UF_6	2.390
$CdWO_4$	1.96		1.32[h]	$CaPdF_6$	1.893
$FeWO_4$	2.06	Na_2CO_3	1.187[i]	$CdPdF_6$	2.075
			1.290[j]		
		$KHCO_3$	1.257		
			1.337[k]		
		Tl_2CO_3	1.24[l]		
			1.28[m]		

[a]S–O(H) [5.30]; [b]S–O(H) [5.31]; [c]P–O(H) [5.32]; [d]P–O(Ba); [e]P–O(Ni) [5.33]; [f]N–O(H) [5.34]; [g]at *d*(Cu–O) = 2.5 Å; [h]at *d*(Cu–O) = 2.0 Å; [i]at *d*(Na–O) = 2.441 Å; [j]at *d*(Na–O) = 2.367 Å [5.35]; [k]C–O(H) [5.36]; [l]at *d*(Tl–O) = 2.82 Å; [m]at *d*(Tl–O) = 2.68 Å

Table S5.16 Bond lengths (Å) in ternary compounds and initial halides and oxides

M_kAX_m	*d*(A–X)	AX_n	*d* (A–X)	MAO_k	*d*(A–O)	A_nO_m	*d* (A–O)
NH_4HgCl_3	2.34	$HgCl_2$	2.53	$KAlO_2$	1.66		
$CsCoCl_3$	2.45	$CoCl_2$	2.51	$YAlO_3$	1.85	Al_2O_3	1.91
$KNiCl_3$	2.40	$NiCl_2$	2.47	$LaAlO_3$	1.89		
$KTeF_5$	1.94	TeF_4	1.99	$YScO_3$	1.97	Sc_2O_3	2.09
$KRuF_6$	1.91	RuF_5	1.96	$LaScO_3$	2.03		
Na_2SnF_6	1.96	SnF_4	2.00	$BaCeO_3$	2.20	Ce_2O_3	2.34
$CsVCl_3$	2.48	VCl_2	2.54	$CaTiO_3$	1.90	TiO_2	1.96
$CsVBr_3$	2.59	VBr_2	2.67	$BaZrO_3$	2.10	ZrO_2	2.26
$CsVI_3$	2.78	VI_2	2.87	$CaSnO_3$	1.96	SnO_2	2.05
$LiVO_2$	1.95	V_2O_3	2.00	$BaPbO_3$	2.14	PbO_2	2.16
$LaVO_3$	1.96			$KMnO_4$	1.60	Mn_2O_7	1.77
$CeCrO_3$	1.93	Cr_2O_3	1.99	$KFeO_2$	1.73	Fe_2O_3	2.03
K_2HgO_2	1.95	HgO	2.03	$YFeO_3$	1.92		

Supplementary References

5.1 Onodera A, Mimasaka M, Sakamoto I et al (1999) J Phys Chem Solids 60:167
5.2 Örlygsson G, Harbrecht B (2001) J Am Chem Soc 123:4168
5.3 Hofmeister AM (1997) Phys Rev B 56:5835
5.4 Lutz H, Möller H, Schmidt M (1994) J Mol Struct 328:121
5.5 Anderson A, Lo Y, Todoeschuck J (1981) Spectrosc Lett 14:105
5.6 Schneider M, Kuske P, Lutz HD (1992) Acta Cryst B 48:761
5.7 Brogan MA, Blake AJ, Wilson C, Gragory DH (2003) Acta Cryst C 59:i136
5.8 Merrill L (1982) J Phys Chem Ref Data 11:1005
5.9 Jobic S, Brec R, Rouxel J (1992) J Alloys Comp 178:253
5.10 Podberezskaya NV, Magarill SA, Pervukhina NV, Borisov SV (2001) J Struct Chem 42:654
5.11 Bronsema KD, de Boer JL, Jellinek F (1986) Z anorg allgem Chem 540/541:15
5.12 Pell MA, Mironov YuV, Ybers JA (1996) Acta Cryst C 52:1331
5.13 Jobic S, Deniard P, Brec R et al (1990) J Solid State Chem 89:315
5.14 Burdett JK, Candell E, Miller GJ (1986) J Am Chem Soc 108:6561
5.15 Chattopadhyay T, von Schnering H-G, Stansfield RFD, McIntyre GJ (1992) Z Krist 199:13
5.16 Lutz HD, Müller B, Schmidt Th, Stingl Th (1990) Acta Cryst C 46:2003
5.17 Williams D, Pleune B, Leinenweber K, Kouvetakis J (2001) J Solid State Chem 159:244
5.18 Donohue PC, Siemons WJ, Gillson JL (1968) J Phys Chem Solids 28:807
5.19 Kjekshus A, Rakke T, Andersen AF (1977) Acta Chem Scand A31:253
5.20 Muller B, Lutz HD (1991) Solid State Commun 78:469
5.21 Brese NE, von Schnering HG (1994) Z anorg allgemChem 620:393
5.22 Takizawa H, Shimada M, Sato Y, Endo T (1993) Mater Lett 18:11
5.23 Daniel P, Bulou A, Leblanc M et al (1990) Mater Res Bull 25:413
5.24 Le Bail A, Jacoboni C, Leblanc M et al (1988) J Solid State Chem 77:96
5.25 Fjellvag H, Karen P (1994) Acta Chem Scand 48:294
5.26 Krämer K, Schleid T, Schulze M et al (1989) Z anorg allgem Chem 575:61

5.27 Poineau F, Rodriguez EE, Forster PM et al (2009) J Am Chem Soc 131:910
5.28 Enjalbert R, Galy J (1986) Acta Cryst C42:1467
5.29 McCarron E, Calabrese J (1991) J Solid State Chem 91:121
5.30 Wasse JC, Howard CA, Thompson H et al (2004) J Chem Phys 121:996
5.31 Näslund J, Persson I, Sandström M (2000) Inorg Chem 39:4012
5.32 Olofsson-Mårtensson M, Häussermann U, Tomkinson J, Noréus D (2000) J Am Chem Soc 122:6960
5.33 Batsanov SS (2002) Russ J Coord Chem 28:1
5.34 Allan DR, Marshall WG, Francis DJ et al (2010) J Chem Soc Dalton Trans 39:3736
5.35 Beattie JK, Best SP, Skelton BW, White AH (1981) J Chem Soc Dalton Trans 2105
5.36 Schmid R, Miah AM, Sapunov VN (2000) Phys Chem Chem Phys 2:97

References

1. Naray-Szabo I (1969) Inorganic crystal chemistry. Akadémiai Kiado, Budapest
2. Donohue J (1982) The structure of the elements. RE Krieger Publ Co, Malabar Fl
3. Young DA (1991) Phase diagrams of the elements. University California Press, Oxford
4. Wells AF (1995) Structural inorganic chemistry, 6th edn. Clarendon Press, Oxford
5. Lide DR (ed) (2007–2008) Handbook of chemistry and physics, 88th edn. CRC Press, New York
6. Batsanov SS (1994) Metallic radii of nonmetals. Russ Chem Bull 43:199–201
7. Pauling L (1960) The nature of the chemical bond, 3rd edn. Cornell Univ Press, Ithaca
8. Pauling L (1989) The nature of metals. Pure Appl Chem 61:2171–2174
9. Brewer L (1981) The role and significance of empirical and semiempirical correlations. O'Keefe M, Navrotsky A (eds) Structure and bonding in crystals, vol 1. Academic Press, San Francisco
10. Trömel M, Alig H, Fink L, Lösel J (1995) Zur kristallchemie der Elemente: Schmelztemperatur, Atomvolumen, formale Wertigkeit und Bindungsvalenzen. Z Krist 210:817–825
11. Trömel M (2000) Metallic radii, ionic radii, and valences of solid metallic elements. Z Naturforsch 55b:243–247
12. Coulson CA (1961) Valence. Oxford Univ Press, Oxford
13. Harrison WA (1980) Electronic structure and the properties of solids. The physics of the chemical bond. Freeman, San Francisco
14. Kittel C (1996) Introduction to solid state physics, 7th edn. Wiley, New York
15. Burdett J (1997) Chemical bond: a dialog. Wiley, Chichester
16. McMahon MI, Nelmes RJ (2006) High-pressure structures and phase transformations in elemental metals. Chem Soc Rev 35:943–963
17. Takemura K, Shimomura O, Fujiihisa H (1991) Cs(VI): a new high-pressure polymorph of cesium above 72 GPa. Phys Rev Lett 66:2014–2017
18. Takemura K, Christensen NE, Novikov DL et al (2000) Phase stability of highly compressed cesium. Phys Rev B61:14399–14404
19. Nelmes RJ, McMahon MI, Loveday JS, Rekhi S (2002) Structure of Rb-III: novel modulated stacking structures in alkali metals. Phys Rev Lett 88:155503
20. Shwarz U (2004) Metallic high-pressure modifications of main group elements. Z Krist 219:376–390
21. Kleykamp H (2000) Thermal properties of beryllium. Thermochim Acta 345:179–1841
22. McMahon MI, Bovornratanaraks T, Allan DR et al (2000) Observation of the incommensurate barium-IV structure in strontium phase V. Phys Rev B61:3135–3138
23. Nelmes RJ, Allan DR, McMahon MI, Belmonte SA (1999) Self-hosting incommensurate structure of barium IV. Phys Rev Lett 83:4081–4084
24. Heine V (2000) Crystal structure: as weird as they come. Nature 403:836–837

25. Takemura K (1994) High-pressure structural study of barium to 90 GPa. Phys Rev B50:16238–16246
26. Schulte O, Holzapfel WB (1993) Phase diagram for mercury up to 67 GPa and 500 K. Phys Rev B48:14009–14012
27. Krishnan S, Anselt S, Felten JJ et al (1998) Structure of liquid boron. Phys Rev Lett 81:586–589
28. Akahama Y, Nishimura M, Kinoshita K, Kawamura H (2006) Evidence of a fcc-hcp transition in aluminum at multimegabar pressure. Phys Rev Lett 96:045505
29. Takemura K, Kobayashi K, Arai M (1998) High-pressure bct-fcc phase transition in Ga. Phys Rev B58:2482–2486
30. Takemura K, Fujihisa H (1993) High-pressure structural phase transition in indium. Phys Rev B47:8465–8470
31. Schulte O, Holzapfel WB (1997) Effect of pressure on the atomic volume of Ga and Tl up to 68 GPa. Phys Rev B55:8122–8128
32. Vohra YK, Spencer PhT (2001) Novel γ-phase of titanium metal at megabar pressures. Phys Rev Lett 86:3068–3071
33. Takemura K, Schwarz U, Syassen K et al (2000) High-pressure Cmca and hcp phases of germanium. Phys Rev B62:R10603-R10606
34. Desgreniers S, Vohra YK, Ruoff AL (1989) Tin at high pressure: an energy-dispersive X-ray-diffraction study to 120 GPa. Phys Rev B39:10359–10361
35. Degtyareva O, McMahon MI, Nelmes RJ (2004) High-pressure structural studies of group 15 elements. High Pressure Research 24:319–356
36. Ram S (2001) Allotropic phase transformations in hcp, fcc and bcc metastable structures in Co-nanoparticles. Mater Sci Engin A304–306:923–927
37. Eremets MI, Hemley RJ, Mao H-K, Gregoryanz E (2001) Semiconducting non-molecular nitrogen up to 240 GPa and its low-pressure stability. Nature 411:170–174
38. Gregoryanz E, Goncharov AF, Hemley RJ, Mao H-K (2002) Raman, infrared, and X-ray evidence for new phases of nitrogen at high pressures and temperatures. Phys Rev B66:224108
39. Errandonea D, Boehler R, Japel S et al (2006) Structural transformation of compressed solid Ar: an X-ray diffraction study to 114 GPa. Phys Rev B73:092106
40. Loubeyre P, Jean-Louis M, Silvera I (1991) Density dependence of the intramolecular distance in solid H_2: spectroscopic determination. Phys Rev B43:10191–10196
41. Loubeyre P, Occelli F, LeToullec R (2002) Optical studies of solid hydrogen to 320 GPa and evidence for black hydrogen. Nature 416:613–617
42. Wigner E, Huntington HB (1935) On the possibility of a metallic modification of hydrogen.J Chem Phys 3:764–770
43. Min BI, Jensen HJF, Freeman AJ (1986) Pressure-induced electronic and structural phase transitions in solid hydrogen. Phys Rev B33:6383–6390
44. Hemley RJ, Mao H-K (1988) Phase transition in solid molecular hydrogen at ultrahigh pressures. Phys Rev Lett 61:857–860
45. Hemley RJ, Dera P (2000) Molecular crystals. In: Hazen RM, Downs RT (eds) High-temperature and high-pressure crystal chemistry, Rev Min Geochem 41:355. MSA, Washington
46. Powell BM, Heal KM, Torrie BH (1984) The temperature dependence of the crystal structures of the solid halogens, bromine and chlorine. Mol Phys 53:929–939
47. San-Miguel A, Libotte H, Gauthier M et al (2007) New phase transition of solid bromine under high pressure. Phys Rev Lett 99:015501
48. Takemura K, Minomura S, Shimomura O (1982) Structural aspects of solid iodine associated with metallization and molecular dissociation under high pressure. Phys Rev B26:998–1004
49. Reichlin R, McMahan AK, Ross M et al (1994) Optical, X-ray, and band-structure studies of iodine at pressures of several megabars. Phys Rev B49:3725–3733
50. Fujihisa H, Fujii Y, Takemura K, Shimomura O (1995) Structural aspects of dense solid halogens under high pressure studied by x-ray diffraction – molecular dissociation and metallization. J Phys Chem Solids 56:1439–1444

51. Miguel AS, Libotte H, Gaspard JP et al (2000) Bromine metallization studied by X-ray absorption spectroscopy. Eur Phys J B17:227–233
52. Takemura K, Sato K, Fujihisa H, Onoda M (2003) Modulated structure of solid iodine during its molecular dissociation under high pressure. Nature 423:971–974
53. Kume T, Hiraoka T, Ohya Y et al (2005) High pressure raman study of bromine and iodine: soft phonon in the incommensurate phase. Phys Rev Lett 94:065506
54. Fujihisa H, Akahama Y, Kowamura H et al (2006) O_8 cluster structure of the epsilon phase of solid oxygen. Phys Rev Lett 97:085503
55. Militzer B, Hemley RJ (2006) Crystallography: solid oxygen takes shape. Nature 443:150–151
56. Lundegaard B, Weck G, McMahon MI et al (2006) Observation of an O_8 molecular lattice in the ε phase of solid oxygen. Nature 443:201–204
57. Shimizu K, Suhara K, Ikumo M et al (1998) Superconductivity in oxygen. Nature 393:767–769
58. Goncharov AF, Gregoryanz E, Hemley RJ, Mao H-K (2003) Molecular character of the metallic high-pressure phase of oxygen. Phys Rev B68:100102 (R)
59. Akahama Y, Kowamura H, Häusermann D et al (1995) New high-pressure structural transition of oxygen at 96 GPa associated with metallization in a molecular solid. Phys Rev Lett 74:4690–4693
60. Heiny C, Lundegaard LF, Falconi S, McMahon MI (2005) Incommensurate sulfur above 100 GPa. Phys Rev B71:020101(R)
61. Nakano K, Akahama Y, Kawamura H et al (2001) Pressure-induced metallization and structural transition of orthorhombic Se. Phys Stat Solidi b223:397–400
62. Keller R, Holzapfel WB, Schulz H (1977) Effect of pressure on the atom positions in Se and Te. Phys Rev B16:4404–4412
63. Akahama Y, Kobayashi M, Kawamura H (1993) Structural studies of pressure-induced phase transitions in selenium up to 150 GPa. Phys Rev B47:20–26
64. Kawamura H, Matsui N, Nakahata I et al (1998) Pressure-induced metallization and structural transition of rhombohedral Se. Solid State Commun 108:677–680
65. Parthasarathy G, Holzapfel WB (1988) High-pressure structural phase transitions in tellurium. Phys Rev B37:8499–8501
66. Luo H, Greene RG, Ruoff AL (1993) β-Po phase of sulfur at 162 GPa: X-ray diffraction study to 212 GPa. Phys Rev Lett 71:2943–2946
67. Eremets MI, Gavriliuk AG, Trojan IA et al (2004) Single-bonded cubic form of nitrogen. Nature Mater 3:558–563
68. Eremets MI, Gavriliuk AG, Trojan IA (2007) Single-crystalline polymeric nitrogen. Appl Phys Lett 90:171904
69. Ashcroft NW (2004) Hydrogen dominant metallic alloys: high-temperature superconductors? Phys Rev Lett 92:187002
70. Okudera H, Dinnebier RE, Simon A (2005) The crystal structure of γ-P_4, a low temperature modification of white phosphorus. Z Krist 220:259–264
71. Simon A, Borrmann H, Horakh J (1997) On the polymorphism of white phosphorus. Chem Ber 130:1235–1240
72. Akahama Y, Kawamura H, Carlson S et al (2000) Structural stability and equation of state of simple-hexagonal phosphorus to 280 GPa: phase transition at 262 GPa. Phys Rev B61:3139–3142
73. Häussermann U (2003) High-pressure structural trends of group 15 elements. Chem Eur J 9:1471–1478
74. Kikegawa T, Iwasaki H (1983) An X-ray diffraction study of lattice compression and phase transition of crystalline phosphorus. Acta Cryst B39:158–164
75. Iwasaki H, Kikegawa T (1984) Simple cubic structure as a stable form of phosphorus under high pressure. In: Homan C, MacCrone RK, Whalley E (eds) High pressure in science and technology. North-Holland, New York
76. Katzke H, Bismayer U, Toledano P (2006) Reconstructive phase transitions between carbon polymorphs: limit states and periodic order-parameters. J Phys Condens Matter 18:5129–5134

77. Goresy AE, Donnay G (1968) A new allotropic form of carbon from the Ries crater. Science 161:363–364
78. Whittaker AG (1978) Carbon: a new view of its high-temperature behavior. Science 200:763–764
79. Batsanov SS (1994) Equalization of interatomic distances in polymorphous transformations under pressure. J Struct Chem 35:391–393
80. Grützmacher H, Fässler TF (2000) Topographical analyses of homonuclear multiple bonds between main group elements. Chem Eur J 6:2317–2325
81. Pearson WB (1972) The crystal chemistry and physics of metals alloys. Wiley, New York
82. Demchyna R, Leoni S, Rosner H, Schwarz U (2006) High-pressure crystal chemistry of binary intermetallic compounds. Z Krist 221:420–434
83. Rao CNR, Pisharody KPR (1976) Transition metal sulfides. Progr Solid State Chem 10:207–270
84. Franzen HF (1978) Structure and bonding in metal-rich compounds: Pnictides, chalcides and halides. Progr Solid State Chem 12:1–39
85. Goedkoop JA, Anderson AF (1955) The crystal structure of copper hydride. Acta Cryst 8:118–119
86. Yoshiasa A, Koto K, Kanamaru F et al (1987) Anharmonic thermal vibrations in wurtzite-type AgI. Acta Cryst B43:434–440
87. Brese NE, Squattrito PJ, Ibers JA (1985) Reinvestigation of the structure of PdS. Acta Cryst C41:1829–1830
88. Konczewicz L, Bigenwald P, Cloitre T et al (1996) MOVPE growth of zincblende magnesium sulphide. J Cryst Growth 159:117–120
89. Dynowska E, Janik E, Bak-Misiuk J et al (1999) Direct measurement of the lattice parameter of thick stable zinc-blende MgTe layer. J Alloys Compd 286:276–278
90. Iwanovski RJ (2001) Comment on the covalent radius of Mn. Chem Phys Lett 350:577–580
91. Wentzcovitch RM, Cohen ML, Lam PK (1987) Theoretical study of BN, BP, and BAs at high pressures. Phys Rev B36:6058–6068
92. Vurgaftman I, Meyer JR, Ram-Mohan LR (2001) Band parameters for III–V compound semiconductors and their alloys. J Appl Phys 89:5815–5875
93. Mavropoulos Th, Galanakis I (2007) A review of the electronic and magnetic properties of tetrahedrally bonded half-metallic ferromagnets. J Phys Cond Matter 19:315221
94. Suzuki K, Morito H, Kaneko T et al (1993) Crystal structure and magnetic properties of the compound FeN. J Alloys Compd 201:11–16
95. Witteman WG, Giorgi AL, Vier DT (1960) The preparation and identification of some intermetallic compounds of polonium. J Phys Chem 64:434–440
96. Gregoryanz E, Sanloup Ch, Somayazulu M et al (2004) Synthesis and characterization of a binary noble metal nitride. Nature Mater 3:294–297
97. Halder A, Kundu P, Ravishankar N, Ramanath G (2009) Directed synthesis of rocksalt AuCl crystals. J Phys Chem C113:5349–5351
98. Recio JM, Blanco MA, Luaa V et al (1998) Compressibility of the high-pressure rocksalt phase of ZnO. Phys Rev B58:8949–8954
99. Desgreniers S (1998) High-density phases of ZnO: structural and compressive parameters. Phys Rev B58:14102–1405
100. Duan C-G, Sabirianov RF, Mei WN et al (2007) Electronic, magnetic and transport properties of rare-earth monopnictides. J Phys Cond Matt 19:315220
101. Pellicer-Porres J, Segura A, Muoz V et al (2001) Cinnabar phase in ZnSe at high pressure. Phys Rev B65:012109
102. Christensen AN (1990) A neutron diffraction investigation on single crystals of titanium oxide, zirconium carbide, and hafnium nitride. Acta Chem Scand 44:851–852
103. Vaitheeswaran G, Kanchana V, Heathman S et al (2007) Elastic constants and high-pressure structural transitions in lanthanum monochalcogenides from experiment and theory. Phys Rev B75:184108

104. Gerward L, Olsen JS, Steenstrup S et al (1990) The pressure-induced transformation B1 to B2 in actinide compounds. J Appl Cryst 23:515–519
105. Wastin F, Spirlet JC, Rebizant J (1995) Progress on solid compounds of actinides. J Alloys Compd 219:232–237
106. Hayashi J, Shirotani I, Hirano K et al (2003) Structural phase transition of ScSb and YSb with a NaCl-type structure at high pressures. Solid State Commun 125:543–546
107. Shirotani I, Yamanashi K, Hayashi J et al (2003) Pressure-induced phase transitions of lanthanide monoarsenides LaAs and LuAs with a NaCl-type structure. Solid State Commun 127:573–576
108. Menoni CS, Spain IL (1987) Equation of state of InP to 19 GPa. Phys Rev B35:7520–7525
109. Vohra YK, Weir ST, Ruoff AL (1985) High-pressure phase transitions and equation of state of the III-V compound InAs up to 27 GPa. Phys Rev B31:7344–7348
110. Uehara S, Masamoto T, Onodera A et al (1997) Equation of state of the rocksalt phase of III–V nitrides to 72 GPa or higher. J Phys Chem Solids 58:2093–2099
111. Chen X, Koiwasaki T, Yamanaka S (2001) High-pressure synthesis and crystal structures of β-MNCl (M=Zr and Hf). J Solid State Chem 159:80–86
112. Mashimo T, Tashiro S, Toya T et al (1993) Synthesis of the B1-type tantalum nitride by shock compression. J Mater Sci 28:3439–3443
113. Yoshida M, Onodera A, Ueno M et al (1993) Pressure-induced phase transition in SiC. Phys Rev B48:10587–10590
114. Ono S, Kikegawa T, Ohishi Y (2005) A high-pressure and high-temperature synthesis of platinum carbide. Solid State Commun 133:55–59
115. Zhou T Schwarz U Hanfland M et al (1998) Effect of pressure on the crystal structure, vibrational modes, and electronic excitations of HgO. Phys Rev B57:153–160
116. Feldmann C, Jansen M (1993) Cs_3AuO, the first ternary oxide with anionic gold. Angew Chem Int Ed 32:1049–1050
117. Zachwieja U (1993) Einkristallzüchtung und strukturverfeinerung von RbAu und CsAu. Z anorg allgem Chem 619:1095–1097
118. Walter D, Karyasa IW (2005) Synthesis and characterization of cobalt monosilicide (CoSi) with CsCl structure stabilized by a β-SiC matrix. Z anorg allgem Chem 631:1285–1288
119. Sato-Sorensen Y (1983) Phase transitions and equations of state for the sodium halides: NaF, NaCl, NaBr, and NaI. J Geophys Res B88:3543–3548
120. Heinz D, Jeanloz R (1984) Compression of the *B2* high-pressure phase of NaCl. Phys Rev B30:6045–6050
121. Hochheimer HD, Strössner K, Hönle V et al (1985) High pressure X-ray investigation of the alkali hydrides NaH, KH, RbH, and CsH. Z phys Chem 143:139–144
122. Yagi T (1978) Experimental determination of thermal expansivity of several alkali halides at high pressures. J Phys Chem Solids 39:563–571
123. Hofmeister AM (1997) IR spectroscopy of alkali halides at very high pressures: Calculation of equations of state and of the response of bulk moduli to the *B*1-*B*2 phase transition. Phys Rev B56:5835–5855
124. Köhler U, Johannsen PG, Holzapfel WB (1997) Equation-of-state data for CsCl-type alkali halides. J Phys Cond Matter 9:5581–5592
125. Hull S, Berastegui P (1998) High-pressure structural behavior of silver(I) fluoride. J Phys Cond Matter 10:7945–7956
126. Richet P, Mao H-K, Bell PM (1988) Bulk moduli of magnesiowüstites from static compression measurements. J Geophys Res B93:15279–15288
127. Luo H, Greene RG, Ghandehari K et al (1994) Structural phase transformations and the equations of state of calcium chalcogenides at high pressure. Phys Rev B50:16232–16237
128. Zimmer H, Winzen H, Syassen K (1985) High-pressure phase transitions in CaTe and SrTe. Phys Rev B32:4066–4070
129. Luo H, Greene RG, Ruoff AL (1994) High-pressure phase transformation and the equation of state of SrSe. Phys Rev B49:15341–15345

130. Weir ST, Vohra YK, Ruoff AL (1986) High-pressure phase transitions and the equations of state of BaS and BaO. Phys Rev B33:4221–4226
131. Grzybowski T, Ruoff AL (1983) High-pressure phase transition in BaSe. Phys Rev B27:6502–6503
132. Grzybowski T, Ruoff AL (1984) Band-overlap metallization of BaTe. Phys Rev Lett 53:489–492
133. Liu H, Mao H-K, Somayazulu M et al (2004) *B*1-to-*B*2 phase transition of transition-metal monoxide CdO under strong compression. Phys Rev B70:094114
134. Gerward L, Olsen JS, Steenstrup S et al (1990) The pressure-induced transformation B1 to B2 in actinide compounds. J Appl Cryst 23:515–519
135. Meyer RR, Sloan J, Dunin-Borkowski RE et al (2000) Discrete atom imaging of one-dimensional crystals formed within single-walled carbon nanotubes. Science 289:1324–1326
136. Sloan J, Novotny MC, Bailey SR et al (2000) Two layer 4:4 co-ordinated KI crystals grown within single walled carbon nanotubes. Chem Phys Lett 329:61–65
137. Sloan J, Grosvenor SJ, Friedrichs S et al (2002) A one-dimensional BaI_2 chain with five- and six-coordination, formed within a single-walled carbon nanotube. Angew Chem Int Ed 41:1156–1159
138. Sloan J, Kirkland AI, Hutchison JL, Green MLH (2002) Structural characterization of atomically regulated nanocrystals formed within single-walled carbon nanotubes using electron microscopy. Acc Chem Res 35:1054–1062
139. Sloan J, Kirkland AI, Hutchison JL, Green MLH (2002) Integral atomic layer architectures of 1D crystals inserted into single walled carbon nanotubes. Chem Commun:1319–1332
140. Hirahara K, Bandow S, Suenaga K et al (2001) Electron diffraction study of one-dimensional crystals of fullerenes. Phys Rev B64:115420
141. Friedrichs S, Sloan J, Green MLH et al (2001) Simultaneous determination of inclusion crystallography and nanotube conformation for a Sb_2O_3/single-walled nanotube composite. Phys Rev B64:045406
142. Ishikawa K, Isonaga T, Wakita S, Suzuki Y (1995) Structure and electrical properties of Au_2S. Solid State Ionics 79:60–66
143. von Schnering HG, Chang J-H, Peters K et al (2003) Structure and bonding of the hexameric platinum(II) dichloride, Pt_6Cl_{12} (β-$PtCl_2$). Z anorg allgem Chem 629:516–522
144. Thiele G, Wegl W, Wochner H (1986) Die Platiniodide PtI_2 und Pt_3I_8. Z anorg allgem Chem 539:141–153
145. Thiele G, Steiert M, Wagner D, Wocher H (1984) Darstellung und Kristallstruktur von PtI_3, einem valenzgemischten Platin (II, IV)-iodid. Z anorg allgem Chem 516:207–213
146. Schnering HG von, Chang J-H, Freberg M et al (2004) Structure and bonding of the mixed-valent platinum trihalides, $PtCl_3$ and $PtBr_3$. Z anorg allgem Chem 630:109–116
147. Senin MD, Akhachinski VV, Markushin YuE et al (1993) The production, structure, and properties of beryllium hydride. Inorg Mater 29:1416–1420
148. Sampath S, Lantzky KM, Benmore CJ et al (2003) Structural quantum isotope effects in amorphous beryllium hydride. J Chem Phys 119:12499–12502
149. Wright AF, Fitch AN, Wright AC (1988) The preparation and structure of the α- and β-quartz polymorphs of beryllium fluoride. J Solid State Chem 73:298–304
150. Troyanov SI (2000) Crystal modifications of beryllium dihalides $BeCl_2$, $BeBr_2$, and BeI_2. Russ J Inorg Chem 45:1481–1486
151. Chieh C, White MA (1984) Crystal structure of anhydrous zinc bromide. Z Krist 166:189–197
152. Fröhling B, Kreiner G, Jacobs H (1999) Synthese und Kristallstruktur von Mangan(II)- und Zinkamid, $Mn(NH_2)_2$ und $Zn(NH_2)_2$. Z anorg allgem Chem 625:211–216
153. Hostettler M, Birkedal H, Schwarzenbach D (2002) The structure of orange HgI_2. I. Polytypic layer structure. Acta Cryst B58:903–913
154. Grande T, Ishii M, Akaishi M et al (1999) Structural properties of $GeSe_2$ at high pressures. J Solid State Chem 145:167–173
155. Micoulaut M, Cormier L, Henderson G S (2006) The structure of amorphous, crystalline and liquid GeO_2. J Phys Cond Matter 18:R753–R784

156. Williams D, Pleune B, Leinenweber K, Kouvetakis J (2001) Synthesis and structural properties of the binary framework C–N compounds of Be, Mg, Al, and Tl. J Solid State Chem 159:244–250
157. Jacobs H, Niemann A, Kockelmann W (2005) Tieftemperaturuntersuchungen von Wasserstoffbrückenbindungen in den Hydroxiden β-Be$(OH)_2$ und ε-Zn$(OH)_2$ mit Raman-Spektroskopie sowie Röntgen- und Neutronenbeugung. Z anorg allgem Chem 631:1247–1254
158. Donald KJ, Hoffmann R (2006) Solid memory: structural preferences in group 2 dihalide monomers, dimers, and solids. J Am Chem Soc 128:11236–11249
159. Baur WH, Khan AA (1971) Rutile-type compounds. IV. SiO_2, GeO_2 and a comparison with other rutile-type structures. Acta Cryst B27:2133–2139
160. Range K-J, Rau F, Klement U, Heys A (1987) β-PtO_2: high pressure synthesis of single crystals and structure refinement. Mater Res Bull 22:1541–1547
161. Bolzan AA, Fong C, Kennedy BJ, Howard CJ (1997) Structural studies of rutile-type metal dioxides. Acta Cryst B53:373–380
162. Bortz M, Bertheville B, Böttger G, Yvon K (1999) Structure of the high pressure phase γ-MgH_2 by neutron powder diffraction. J Alloys Comp 287:L4–L6
163. O'Toole NJ, Streltsov VA (2001) Synchrotron X-ray analysis of the electron density in CoF_2 and ZnF_2. Acta Cryst B57:128–135
164. Giester G, Lengauer CL, Tillmanns E, Zemann J (2002) Tl_2S: Re-determination of crystal structure and stereochemical discussion. J Solid State Chem 168:322–330
165. Müller BG (1987) Fluoride mit kupfer, silber, gold und palladium. Angew Chem 99:1120–1135
166. Beck HP, Dausch W (1989) The refinement of α-USe_2, twinning in a $SrBr_2$-type structure. J Solid State Chem 80:3–39
167. Sichla T, Jacobs H (1996) Single crystal X-ray structure determination on calcium and strontium deuteride. Eur J Solid State Inorg Chem 33:453–461
168. Schewe-Miller I, Bötcher P (1991) Synthesis and crystal structures of K_5Se_3, Cs_5Te_3 and Cs_2Te. Z Krist 196:137–151
169. Haines J, Leger JM (1993) Phase transitions in ruthenium dioxide up to 40 GPa: Mechanism for the rutile-to-fluorite phase transformation and a model for the high-pressure behavior of stishovite SiO_2. Phys Rev B48:13344–13350
170. von Sommer H, Hoppe R (1977) Die Kristallstruktur von Cs_2S mit einer Bemerkung über Cs_2Se, Cs_2Te, Rb_2Se und Rb_2Te. Z anorg allgem Chem 429:118–130
171. Stöwe K, Appel S (2002) Polymorphic forms of rubidium telluride Rb_2Te. Angew Chem Int Ed 41:2725–2730
172. Tzeng CT, Tsuei K-D, Lo W-S (1998) Experimental electronic structure of Be_2C. Phys Rev B58:6837–6843
173. Batsanov SS, Egorov VA, Khvostov YuB (1976) Shock synthesis of difluorides of lantanides. Proc Acad Sci USSR Dokl Chem 227:251–252
174. Egorov VA, Temnitskii IN, Martynov AI, Batsanov SS (1979) Variation of lanthanides' valences under shock compression of reacting systems. Russ J Inorg Chem 24:1881–1882
175. Ito M, Setoyama D, Matsunaga J et al (2006) Effect of electronegativity on the mechanical properties of metal hydrides with a fluorite structure. J Alloys Comp 426:67–71
176. Kalita PE, Sinogeikin SV, Lipinska-Kalita K et al (2010) Equation of state of TiH_2 up to 90 GPa: A synchrotron x-ray diffraction study and ab initio calculations. J Appl Phys 108:043511
177. Karpov A, Nuss J, Wedig U, Jansen M (2003) Cs_2Pt: A platinide (II) exhibiting complete charge separation. Angew Chem Int Ed 42:4818
178. Batsanov SS, Kolomijchuk VN (1968) The crystal chemistry of salts with mixed anions. J Struct Chem 9:282–298
179. Flahaut J (1974) Les structures type PbFCl et type anti-Fe_2As des composés ternaires à deux anions MXY. J Solid State Chem 9:124–131
180. Beck HP (1976) A study on mixed halide compounds MFX (M=Ca, Sr, Eu, Ba; X=Cl, Br, I) J Solid State Chem 17:275–282

181. Burdett JK, Candell E, Miller GJ (1986) Electronic structure of transition-metal borides with the AlB_2 structure. J Am Chem Soc 108:6561–6568
182. Einstein FWB, Rao PR, Trotter J, Bartlett N (1967) The crystal structure of gold trifluoride. J Chem Soc A 478–482
183. Benner G, Müller BG (1990) Zur Kenntnis binärer Fluoride des ZrF_4-Typs: HfF_4 und ThF_4. Z anorg allgem Chem 588:33–42
184. Marx R, Mahjoub A, Seppelt K, Ibberson R (1994) Time-of-flight neutron diffraction study on the low temperature phases of IF_7. J Chem Phys 101:585–593
185. Vogt T, Fitch AN, Cockcroft JK (1994) Crystal and molecular structures of rhenium heptafluoride. Science 263:1265–1267
186. Weiher N, Willnef EA, Figulla-Kroschel C et al (2003) Extended X-ray absorption fine-structure (EXAFS) of a complex oxide structure: a full multiple scattering analysis of the Au L_3-edge EXAFS of Au_2O_3. Solid State Commun 125:317–322
187. Åhman J, Svenssson G, Albertsson J (1996) A reinvestigation of β-gallium oxide. Acta Cryst C52:1336–1338
188. Bärnighausen H, Schiller G (1985) The crystal structure of A-Ce_2O_3. J Less-Common Met 110:385–390
189. Wolf R, Hoppe R (1985) Eine Notiz zum A-typ der Lanthanoidoxide: Über Pr_2O_3. Z anorg allgem Chem 529:61–64
190. Müller BG (1997) Synthese binärer und ternärer Fluoride durch Druckfluorierung. Eur J Solid State Inorg Chem 34:627–643
191. Troyanov SI, Antipin MYu, Struchkov YuT, Simonov MA (1986) Crystal structure of HfI_4. Russ J Inorg Chem 31:1080–1082
192. Bork M, Hoppe R (1996) Zum Aufbau von PbF_4 mit Strukturverfeinerung an SnF_4. Z anorg allgem Chem 622:1557–1563
193. Krämer O, Müller BG (1995) Zur Struktur des Chromtetrafluorids. Z anorg allgem Chem 621:1969–1972
194. Krebs B, Sinram D (1980) Darstellung, Struktur und Eigenschaften einer neuen Modifikation von NbI_5. Z Naturforsch 35b:12–16
195. Maslen EN, Streltsov VA, Streltsova NR et al (1993) Synchrotron X-ray study of the electron density in α-Al_2O_3. Acta Cryst B49:973–980
196. Otto H, Baltrusch R, Brandt H-J (1993) Further evidence for Tl^{3+} in Tl-based superconductors from improved bond strength parameters involving new structural data of cubic Tl_2O_3. Physica C215:205–208
197. Eisenmann B (1992) Crystal structure of α-dialuminium trisulfide, Al_2S_3. Z Krist 198:307–308
198. Conrad O, Schiemann A, Krebs B (1997) Die Kristallstruktur von β-Al_2Te_3. Z anorg allgem Chem 623:1006–1010
199. Dismukes JP, White JG (1964) The preparation, properties, and crystal structures of some scandium sulfides in the range Sc_2S_3-ScS. Inorg Chem 3:1220–1228
200. Zerr A, Miehe G, Riedel R (2003) Synthesis of cubic zirconium and hafnium nitride having Th_3P_4 structure. Nature Mater 2:185–189
201. Stergiou AC, Rentzeperis PJ (1985) The crystal structure of arsenic selenide, As_2Se_3. Z Krist 173:185–191
202. Stergiou AC, Rentzeperis PJ (1985) Hydrothermal growth and the crystal structure of arsenic telluride, As_2Te_3. Z Krist 172:139–145
203. Voutsas GP, Papazoglou AG, Rentzeperis PJ, Siapkas D (1985) The crystal structure of antimony selenide, Sb_2Se_3. Z Krist 171:261–268
204. Feuutelais Y, Legendre B, Rodier N, Agafonov V (1993) A study of the phases in the bismuth – tellurium system. Mater Res Bull 28:591–596
205. Weller MT, Hector AL (2000) The structure of the $Fe^{IV}O_4{}^{4-}$ ion. Angew Chem Int Ed 39:4162–4163
206. Hadenfeldt C, Herdejürgen H (1988) Darstellung und Kristallstruktur der Calciumpnictidiodide Ca_2NI, Ca_2PI und Ca_2AsI. Z anorg allgem Chem 558:35–40

207. Gensini M, Gering E, Benedict U et al (1991) High-pressure X-ray diffraction study of ThOS and UOSe by synchrotron radiation. J Less-Common Met 171:L9-L12
208. Beck HP, Dausch W (1989) Die Verfeinerung der Kristallstruktur von ThOTe. Z anorg allgem Chem 571:162–164
209. Jumas J-C, Olivier-Fourcade J, Ibanez A, Philippot E (1986)^{121}Sb Mössbauer studies on some antimony III chalcogenides and chalcogenohalides. Application to the structural approach of sulfide glasses. Hiperfine Interact 28:777–780
210. Mootz D, Merschenz-Quack A (1988) Structures of sulfuryl halides: SO_2F_2,SO_2ClF and SO_2Cl_2. Acta Cryst C44:924–925
211. Chen X, Fukuoka H, Yamanaka S (2002) High-pressure synthesis and crystal structures of β-MNX (M=Zr, Hf; X=Br, I). J Solid State Chem 163:77–83
212. Batsanov SS (1977) Calculation of the effective charges of atoms in solid solutions. Russ J Inorg Chem 22:941–943
213. Leoni S, Zahn D (2004) Putting the squeeze on NaCl: modelling and simulation of the pressure driven B1-B2 phase transition. Z Krist 219:339–344
214. Belov NV (1986) Essays on structural crystallography. Nauka, Moscow (in Russian)
215. Drobot DV, Pisarev EA (1983) Interrelations between the structures of halogenides and oxides of heavy transition metals. Koordinatsionnaya Khimiya 9:1273–1283 (in Russian)
216. Burdett JK, Lee S (1983) Peierls distortions in two and three dimensions and the structures of AB solids. J Am Chem Soc 105:1079–1083
217. Burdett JK (1993) Some electronic aspects of structural maps. J Alloys Compd 197:281–289
218. Müller U (2002) Kristallpackungen mit linear koordinierten Atomen. Z anorg allgem Chem 628:1269–1278
219. Vegas A, Jansen M (2002) Structural relationships between cations and alloys; an equivalence between oxidation and pressure. Acta Cryst B58:38–51
220. Vegas A, Santamaria-Perez D (2003) The structures of ZrNCl, TiOCl and AlOCl in the light of the Zintl-Klemm concept. Z Krist 218:466–469
221. Smirnov IA, Oskotskii VS (1978) Semiconductor-metal phase transition in rare-earth semiconductors (samarium monochalcogenides). Sov Phys Uspekhi 21:117–140
222. Batsanov SS (1988) Correspondence berween the physical and the chemical compression of substances. Russ J Phys Chem 62:265–266
223. Werner A (1893) Beitrag zur Konstitution anorganischer Verbindungen. Z anorg Chem 3:267–330
224. Pfeifer P (1915) Die Kristalle als Molekülverbindungen. Z anorg allgem Chem 92:376–380
225. Riss A, Blaha P, Schwarz K, Zemann J (2003) Theoretical explanation of the octahedral distortion in FeF_2 and MgF_2. Z Krist 218:585–589
226. Baur WH (1961) Uber die Verfeinerung der Kristallstrukturbestimmung einiger vertreter des Rutiltyps. III. Zur Gittertheories des Rutiltyps. Acta Cryst 14:209–213
227. Frank FC, Kasper JS (1958) Complex alloy structures regarded as sphere packings. I. Definitions and basic principles. Acta Cryst 11:184–190
228. Laves F (1967) Space limitation on the geometry of the crystal structures of metals and intermetallic compounds. Phase transition in metals alloys. In: Rudman PS, Stringer J, Jaffee RI (eds) Phase stability in metals and alloys. McGraw-Hill, NewYork
229. Serezhkin VH, Mikhailov YuN, Buslaev YuA (1997) The method of intersecting spheres for determination of coordination numbers of atoms in crystal structures. Russ J Inorg Chem 42:1871–1910
230. Blatov VA, Kuzmina EE, Serezhkin VN (1998) The size of halogen atoms in the structure of molecular crystals. Russ J Phys Chem 72:1284–1287
231. Wieting J (1967) 69 In: Schulze GER (ed) Metallphysik. Akademie, Berlin
232. Trömel M (1986) The crystal-chemistry of irregular coordinations. Z Krist 174:196–197
233. Hoppe R (1970) Die Koordinationszahl – ein "anorganisches Chamäleon". Angew Chem 82:7–16
234. Mehlhorn B, Hoppe R (1976) Neue hexafluorozirkonate (IV): $BaZrF_6$, $PbZrF_6$, $EuZrF_6$, $SrZrF_6$. Z anorg allgem Chem 425:180–188

235. Hoppe R (1979) Effective coordination numbers and mean active fictive ionic radii. Z Krist 150:23–52
236. Brown ID, Wu KK (1976) Empirical parameters for calculating cation–oxygen bond valences. Acta Cryst B32:1957–1959
237. Batsanov SS (1977) Effective coordination number of atoms in crystals. Russ J Inorg Chem 22:631–634
238. Brunner GO (1977) A definition of coordination and its relevance in the structure types AlB_2 and NiAs. Acta Cryst A33:226–227
239. Beck HP (1981) High-pressure polymorphism of BaI_2. Z Naturforsch 36b:1255–1260
240. Carter FL (1978) Quantifying the concept of coordination number. Acta Cryst B34:2962–2966
241. Batsanov SS (1983) On the meaning and calculation techniques of effective coordination numbers. Koordinatsionnaya Khimiya 9:867 (in Russian)
242. Wemple SH (1973) Refractive-index behavior of amorphous semiconductors and glasses. Phys Rev B7:3767–3777
243. Tanaka K (1989) Structural phase transitions in chalcogenide glasses. Phys Rev B39:1270–1279
244. Asbrink S, Waskowska A (1991) CuO: X-ray single-crystal structure determination at 196 K and room temperature. J Phys Cond Matter 3:8173–8180
245. Goldschmidt VM (1929) Crystal structure and chemical constitution. Trans Faraday Soc 25:253–283
246. Pauling L (1929) The principles determining the structure of complex ionic crystals. J Am Chem Soc 51:1010–1026
247. Pauling L (1947) Atomic radii and interatomic distances in metals. J Am Chem Soc 69:542–553
248. Brown ID, Shannon RD (1973) Empirical bond-strength–bond-length curves for oxides. Acta Cryst A29:266–282
249. Zachariasen WH (1978) Bond lengths in oxygen and halogen compounds of d and f elements. J Less-Common Met 62:1–7
250. Brown ID, Altermatt D (1985) Bond-valence parameters obtained from a systematic analysis of the Inorganic Crystal Structure Database. Acta Cryst B41:244–247
251. Zocchi F (2006) Accurate bond valence parameters for M—O bonds (M=C, N, La, Mo, V). Chem Phys Lett 421:277–280
252. Slupecki O, Brown ID (1982) Bond-valence–bond-length parameters for bonds between cations and sulfur. Acta Cryst B38:1078–1079
253. Bart JCJ, Vitarelli P (1983) Valence balance of magnesium in mixed (Cl, O) environments. Inorg Chim Acta 73:215–220
254. Trömel M (1983) Empirische Beziehungen zu den Bindungslängen in Oxiden. 1. Die Nebengruppenelemente Titan bis Eisen. Acta Cryst B39:664–669
255. Trömel M (1984) Empirische Beziehungen zu den Bindungslängen in Oxiden. 2. Leichtere hauptgruppenelemente sowie kobalt, nickel und kupfer. Acta Cryst B40:338–342
256. Trömel M (1986) Empirische Beziehungen zu den Bindungslängen in Oxiden. 3. Die offenen Koordinationen um Sn, Sb, Te, I und Xe in deren niederen Oxidationsstufen. Acta Cryst B42:138–141
257. O'Keeffe M (1989) The prediction and interpretation of *bond* lengths in crystals. Structure and Bonding 71:162–190
258. Efremov VA (1990) Characteristic features of the crystal chemistry of lanthanide molybdates and tungstates. Russ Chem Rev 59:627–642
259. Abramov YuA, Tsirelson VG, Zavodnik VE et al (1995) The chemical bond and atomic displacements in $SrTiO_3$ from X-ray diffraction analysis. Acta Cryst B51:942–951
260. Naskar JP, Hati S, Datta D (1997) New bond-valence sum model. Acta Cryst B53:885–894
261. Borel MM, Leclaire A, Chardon J et al (1998) A molybdenyl chloromonophosphate with an intersecting tunnel Structure: $Ba_3Li_2Cl_2(MoO)_4(PO_4)_6$. J Solid State Chem 141:587–593
262. Wood RM, Palenik GJ (1999) Bond valence sums in coordination chemistry using new R_0 values. Potassium-oxygen complexes. Inorg Chem 38:1031–1034

263. Wood RM, Abboud KA, Palenik GJ (2000) Bond valence sums in coordination chemistry. Calculation of the oxidation state of chromium in complexes containing only Cr—O bonds and a redetermination of the crystal structure of potassium tetra(peroxo)chromate (V). Inorg Chem 39:2065–2068
264. Garcia-Rodriguez L, Rute-Perez A, Piero JR, González-Silgo C (2000) Bond-valence parameters for ammonium–anion interactions. Acta Cryst B56:565–569
265. Shields GP, Raithby PR, Allen FH, Motherwell WDS (2000) The assignment and validation of metal oxidation states in the Cambridge Structural Database. Acta Cryst B56:455–465
266. Adams S (2001) Relationship between bond valence and bond softness of alkali halides and chalcogenides. Acta Cryst B57:278–287
267. Socchi F (2000) Critical comparison of equations correlating valence and length of a chemical bond. Evaluation of the parameters R_1 and B for the Mo—O bond in MoO_6 octahedra. Solid State Sci 2:385–389
268. Socchi F (2001) Some considerations about equations correlating valence and length of a chemical bond. Solid State Sci 3:383–386
269. Trzescowska A, Kruszynski R, Bartczak TJ (2004) New bond-valence parameters for lanthanides. Acta Cryst B60:174–178
270. Trzescowska A, Kruszynski R, Bartczak TJ (2006) Bond-valence parameters of lanthanides. Acta Cryst B62:745–753
271. Brown ID (2002) The chemical bond in inorganic chemistry: the bond valence model. Oxford University Press, Oxford
272. Brese NE, O'Keeffe M (1991) Bond-valence parameters for solids. Acta Cryst B47:192–197
273. Urusov VS (1995) Semi-empirical groundwork of the bond-valence model. Acta Cryst B51:641–649
274. Batsanov SS (2010) Dependence of the bond length in molecules and crystals on coordination numbers of atoms. J Struct Chem 51:281–287
275. Batsanov SS, Zalivina EN, Derbeneva SS, Borodaevsky VE (1968) Synthesis and properties of CuClBr and CuClI. Doklady Acad Sci USSR 181:599–602 (in Russian)
276. Batsanov SS, Sokolova MN, Ruchkin ED (1971) Mixed halides of gold. Russ Chem Bull 20:1757–1759
277. Aurivillius K, Stolhandske C (1980) A reinvestigation of the crystal structures of $HgSO_4$ and $CdSO_4$. Z Krist 153:121–129
278. Bircsak Z, Harrison WTA (1998) Barium cobalt phosphate, $BaCo_2(PO_4)_2$. Acta Cryst C54:1554–1556
279. El-Bali B, Bolte M, Boukhari A et al (1999) $BaNi_2(PO_4)_2$. Acta Cryst C55:701–702
280. Chapuis G, Dusek M, Meyer M, Petricek V (2003) Sodium carbonate revisited. Acta Cryst B59:337–352
281. Adam A, Ciprus V (1994) Synthese und struktur von $[(Ph_3C_6H_2)Te]_2$, $[(Ph_3C_6H_2)Te(AuPPh_3)_2]PF_6$ und $[(Ph_3C_6H_2)TeAuI_2]_2$. Z anorg allgem Chem 620:1678–1685
282. Troyanov SI, Simonov MA, Kemnitz E et al (1986) Crystal structure of $Ba(HSO_4)_2$. Doklady Akademii Nauk SSSR 288:1376–1379
283. Sawada H, Takeuchi Y (1990) The crystal structure of barite, β-$BaSO_4$, at high temperatures. Z Krist 191:161–171
284. Schefer J, Schwarzenbach D, Fischer P et al (1998) Neutron and X-ray diffraction study of the thermal motion in K_2PtCl_6 as a function of temperature. Acta Cryst B54:121–128
285. Hayton TW, Patrick BO, Legzdins P, McNeil WS (2004) The solid-state molecular structure of $W(NO)_3Cl_3$ and the nature of its W–NO bonding. Can J Chem 82:285–292
286. Kemnitz E, Werner C, Troyanov SI (1996) Reinvestigation of crystalline sulfuric Acid and Oxonium Hydrogensulfate. Acta Cryst C52:2665–2668
287. Souhassou M, Espinosa E, Lecomte C, Blessing RH (1995) Experimental electron density in crystalline H_3PO_4. Acta Cryst B51:661–668
288. Lebrun N, Mahe F, Lanuot J et al (2001) A new crystalline phase of nitric acid dehydrate. Acta Cryst C57:1129–1131

289. Bertheville B, Herrmannsdörfer T, Yvon K (2001) Structure data for K_2MgH_4 and Rb_2CaH_4 and comparison with hydride and fluoride analogues. J Alloys Compd 325:L13–L16
290. Hines J, Cambon O, Astier R et al (2004) Crystal structures of α-quartz homeotypes boron phosphate and boron arsenate: structure-property relationships. Z Krist 219:32–37
291. Feldmann C, Jansen M (1995) Zur kristallchemischen Ähnlichkeit von Aurid- und Halogenid-Ionen. Z anorg allgem Chem 621: 1907–1912
292. Belov NV (1976) Essays on structural mineralogy. Nedra, Moscow (in Russian)
293. Fyfe CA, Brouwer DH, Lewis AR et al (2002) Combined solid state NMR and X-ray diffraction investigation of the local structure of the five-coordinate silicon in fluoride-containing as-synthesized STF zeolite. J Am Chem Soc 124:7770–7778
294. O'Keeffe M (1977) On the arrangements of ions in crystals. Acta Cryst A33:924–927
295. Melnik M, Sramko T, Dunaj-Jurco M et al (1994) Crystal structures of nickel complexes. Reviews of inorganic chemistry. Rev Inorg Chim 14:1–346
296. Tisato F, Refosco F, Bandoli G (1994) Structural survey of technetium complexes. Coord Chem Rev 135/136:325–397
297. Holloway CE, Melnik M (1996) Manganese coordination compounds: classification and analysis of crystallographic and structural data. Rev Inorg Chim 16:101–314
298. Bau R, Drabnis MH (1997) Structures of transition metal hydrides determined by neutron diffraction. Inorg Chim Acta 259:27–50
299. Linert W, Gutmann V (1992) Structural and electronic responses of coordination compounds to changes in the molecule and molecular environment. Coord Chem Rev 117:159–183
300. Dvorak MA, Ford RS, Suenram RD et al (1992) Van der Waals vs. covalent bonding: microwave characterization of a structurally intermediate case. J Am Chem Soc 114:108–115
301. Degtyareva O (2010) Crystal structure of simple metals at high pressures. High Pressure Res 30:343–371

Chapter 6
Crystal Structure: Real

By definition, crystal is a solid substance possessing long-range order, so that knowing the law of this order, one can predict the atomic arrangement at any distance from the starting point. An ideal crystal is described as an infinite repetition of an elementary unit (unit cell) in three dimensions (translational symmetry). Another useful mathematical description is the crystal lattice, a 3-dimensional network of points which are identical in a given crystal structure. In the previous chapter we have discussed crystal structures as if they possessed perfect periodicity, i.e. as if the nature and positions of atoms in every unit cell of a given crystal were exactly the same as in any other, at any given moment. The real structure of even a good-quality crystal is very far from this ideal picture, as proven by the fact that if we calculate the mechanical, electric or other macroscopic properties of solids neglecting the deviations from the idealized crystal structures, the results will be wrong by several orders of magnitude. The factors which perturb the periodicity are (i) the thermal motions of atoms, (ii) the static disorder (local or 'global') in the arrangement of atoms, (iii) chemical impurities and/or violations of stoichiometry of the main component and (iv) defects of various types. In fact, it is not easy to separate (ii) and (iii) from (iv)—often these are the same phenomena viewed from different angles, which can be (with reservations) described as 'chemical' and 'physical'. Those who determine and interpret crystal structures, presume perfect periodicity and therefore see imperfections as atoms *in a unit cell* being disordered between different sites, or some sites being only partly occupied, or chemically different atoms sharing the same site. Defects, on the other hand, are violations of periodicity, recognized and analyzed as such. In this chapter, we shall mainly consider problems of (i) and (iv).

6.1 Thermal Motion

The bond lengths in crystalline compounds (as discussed above) are the distances between the mean atomic positions, but in reality atoms in solids are not stationary but undergo thermal vibrations. In the first approximation, these vibrations are considered as harmonic, i.e. conforming to the Hooke's law with the chemical bond acting as a spring. Hence the force acting on a vibrating atom is proportional to

S. S. Batsanov, A. S. Batsanov, *Introduction to Structural Chemistry*,
DOI 10.1007/978-94-007-4771-5_6, © Springer Science+Business Media Dordrecht 2012

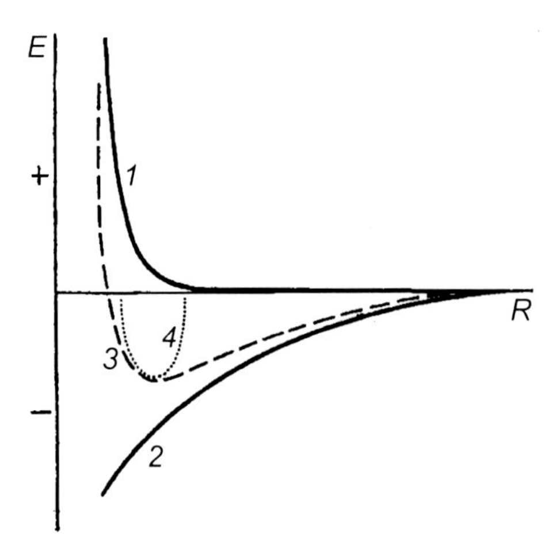

Fig. 6.1 Anharmonicity: *1*—repulsive energy, *2*—attractive energy, *3*—total potential energy, *4*—potential well according to harmonic approximation (the difference between *3* and *4* is exaggerated)

its deviation (ΔR) from the equilibrium position (at a distance r_e), and its potential energy to the square of this deviation:

$$\Delta E = k(\Delta R)^2 \tag{6.1}$$

If this were strictly correct, bodies would not expand on heating. The actual potential well is not described by a symmetrical parabola, as implied by Eq. 6.1, but has a gentler slope on the longer-distance side (Fig. 6.1): the attraction between atoms when the bond is stretched is weaker than their repulsion when the bond is shortened (see Chap. 2). Hence the atom's shifts from the equivalent position (r_e) on the negative and positive sides are also unequal, and always $|-\delta r| < |+\delta r|$. As the amplitude of the atomic vibration (δr) increases on heating, the (usual) results are bond lengthening and bond-bending.

The effect on a crystal structure can be defined by the linear (α) and volume (β) coefficients of thermal expansion (CTE),

$$\alpha = \frac{1}{d}\left(\frac{\Delta d}{\Delta T}\right)_P \tag{6.2}$$

$$\beta = \frac{1}{V}\left(\frac{\Delta V}{\Delta T}\right)_P \tag{6.3}$$

where Δd and ΔV are the changes in length and volume, respectively, and ΔT is the change in temperature; the subscript P signifying constant pressure. For crystals of cubic symmetry (as well as for amorphous materials) the linear CTE is independent of direction, and the volumetric and linear CTEs are simply related as $\beta = 3\alpha$; but for lower symmetry the linear expansion varies with direction. These coefficients themselves vary with temperature. Close to 0 K, there is virtually no change in potential energy of a system with temperature, and hence also almost no thermal expansion.

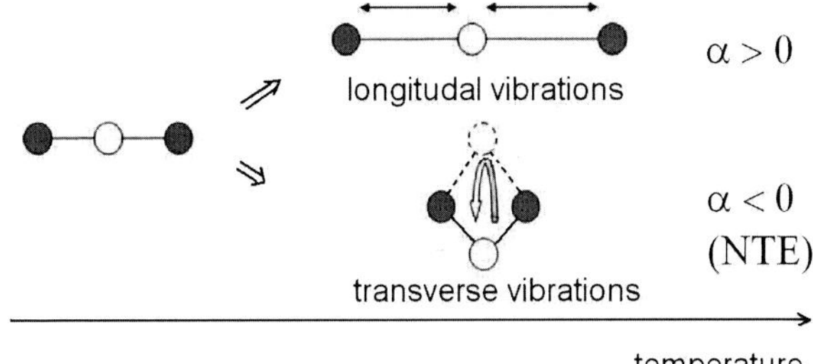

Fig. 6.2 Dependence of thermal expansion coefficient (α) of a solid on the type of vibrations

The usual, positive CTE corresponds to the case when the atoms vibrate along the bond line, that causes an increase in the mean distance between the atoms as the temperature increases, but the relative vibrational motion perpendicular to this line tends to decrease the distance between the mean positions of the two atoms, and so to contract the solid, thus exhibiting negative thermal expansion (NTE); Fig. 6.2.

NTE has been studied experimentally and theoretically for decades, see reviews [1, 2] where several models are considered, explaining the origin of contraction in crystals under heating. What circumstances favor the negative thermal expansion? This question was answered by Blackman [3] thus: "Negative volume expansion coefficients are to be expected from open structures rather than close-packed ones, and those with relatively low shear moduli will be favoured". In fact, because of the tension effect, negative volumetric expansion is most likely to be found in materials of open structure where coordination numbers (N_c) are low. Because N_c of an atom on the surface of a crystal is lower than in its interior (see details in Chap. 7) the vibration amplitude of a surface atoms is approximately twice bigger than that of inner atoms, as have been shown for Ag (111), Pb (111), Pd (100) (111), Pt (100) (110) (111) [3], for α-Ga (010), Si (111), α-Al_2O_3 (0001), H_2O ice (0001) [4], for NaCl [5], and for NaF (100) [6]. Numerous examples of substances with the negative thermal expansion are presented in above mentioned reviews.

In anisotropic materials the volumetric expansion coefficient β can be negative as well as in cubic materials. Even when β is positive, one or two of the three principal linear coefficients may be negative. Relation between the positive and negative coefficient of expansion may be different at different temperatures. Thus, in graphite the CTE perpendicular to the hexagonal axis, $\alpha_\perp$, is negative below room temperature, while the expansion parallel to this axis, $\alpha_\parallel$ is positive, since carbon atoms are tightly bound in planar honeycomb layers, while between the layers there are only weak van der Waals interactions. However, in absolute value NTE is much smaller than CTE because of much greater stiffness within the planes than between them. At high temperatures, additional compressive stress occurs within the planes, so $\alpha_\perp$ becomes positive.

It is well known that the atomic coordinates obtained in a crystal structure analysis are the combined effect of the (equilibrium) atomic arrangement and thermal displacements. Hazen, Downs and Prewitt have shown [7] that the thermal corrections in determining interatomic distances in some structures of silicates can be as large as 5 %, which must be taken into account when using standard values of the atomic or ionic radii. It is significant that bonds in BeO_4 tetrahedra and MgO_6 octahedra in different oxides and silicate structures have similar thermal expansion. Hanzen and Finger [8, 9] have shown that the thermal expansion coefficient depends on such structural parameters as coordination numbers (N_c), cation (z_c) and anion (z_a) valences and an ionic factor (f_i) as

$$\alpha = 4.0\left(\frac{N_c}{f_i z_c z_a}\right) \times 10^{-6} K^{-1} \tag{6.4}$$

where $f_i = 0.50$ for silicates and oxides, 0.75 for halides, 0.40 for chalcogenides, 0.25 for phosphides and arsenides, and 0.20 for nitrides and carbides. This formula permits to predict α for average bond lengths in most coordination groups to within ±20 %, however the strongest and shortest bonds have expansion coefficients smaller than it predicts. For binary and ternary oxides, e.g. when oxygen is the anion, Eq. 6.4 converts to

$$\alpha = 4.0\left(\frac{N_c}{z_c}\right) \times 10^{-6} K^{-1} \tag{6.5}$$

This expression permits to estimate coordination numbers in structures of amorphous or glass compounds by using measurements of CTEs within ca. 10 %.

Accurate studies of thermal expansion of organic crystals under standard conditions are few, while comparing the same crystal structures incidentally determined at different temperatures in different laboratories is meaningless—the effect is completely blurred by systematic errors (see review [11]). In any case, the expansion of molecular solids is much stronger than that of continuous solids, whether covalent or ionic, due to weakness of van der Waals interactions. As a rough guide, one can use the volume thermal expansion coefficient of $0.95(3) \times 10^{-4}$ K^{-1}, obtained in the process of calibrating the atomic volume increments against *all* available organic and organometallic crystal structures [12].

6.2 Lindemann's Hypothesis

Obviously, thermal vibrations of atoms in a solid are strongest on the verge of melting. Sutherland was the first (1891) to suggest that melting occurs when the amplitude of vibrations reaches a certain fraction (equal for all the elements) of the atomic size [13]. In 1910, Lindemann [14] developed this idea and related the critical amplitude to the temperature of melting (T_m) and atomic oscillation frequency ν proportional to the characteristic Debye temperature (Θ). In its modern form [15] the Lindemann's rule states that a material melts at the temperature at which the amplitude of thermal vibration exceeds a certain critical fraction of the interatomic distance, and this fraction depends somewhat on the crystal structure, position in the Periodic Table, and perhaps other unspecified physical quantities. These works initiated numerous studies concerned with empirical and theoretical estimates of critical amplitudes (the history of this question and various versions of the formulae see in [16, 17]) used to describe features of the fusion in crystals of different compositions and structures. According to Lindemann, the mean-square displacement of atoms caused by thermal oscillations in crystals when the melting point is reached is

$$\overline{u}^2 = \frac{3T_m \hbar^2}{kM\Theta^2} \tag{6.6}$$

where $\hbar$ is the Planck constant divided by 2π, k is the Boltzmann constant, Θ_m is the characteristic Debye temperature at the melting point, and M is the absolute mass of the element or compound. The ratio between the displacement ($\bar{u}$) and radius (R) of an atom is

$$\delta_L = \frac{C}{\Theta_m R}\left(\frac{T_m}{M}\right)^{1/2} \tag{6.7}$$

where $C = 12.06$ if R is in Å and T_m in K. However, several authors used different (Einstein or Debye) heat capacity models for calculating δ_L, and approximated R as $V^{1/3}$, not taking into account the geometrical features of different structures. Besides, the X-ray reflection broadening used for the determination of δ_L is caused not only by thermal oscillations of atoms but also by crystal defects and conditions of the experiment. As a result, the values of Lindemann's factor (L-factor) reported in the literature, show a large spread even for the same element. Recently we recalculated δ_L [17] for all studied substances using Eq. 6.7 and obtained a fairly consistent values of δ_L averaging 0.135 ± 0.035.

It seems natural to calculate L-factors directly from the thermochemical characteristics of substance fusion, namely, from the sum of the enthalpy of 'substance preparation for melting' $\Delta H_m = H(T_m) - H(0)$ and the enthalpy of melting proper $\Delta_m H$, e.g. from $H_m = \Delta H_m + \Delta H_m$. Since the energy of a harmonic oscillator is $E = ½ f, \Delta R^2$, where f is the force constant and ΔR is the change in the bond length for $E = H_m$ (when $\Delta R = \delta R$), we obtain

$$\delta_B = \frac{K}{R}\left(\frac{H_m}{f_m}\right)^{1/2} \tag{6.8}$$

where K is a dimensionless constant equal to 0.01822 for f in mdyn/Å and H_m in kilo joules per gram and R in Å. Because the force constant for crystals can be calculated (see [18]) as

$$f_c = 10^{-3}\frac{9B_o V_o}{N_c R^2} \tag{6.9}$$

where B_o is the bulk modulus in GPa, V_o is the molar volume in cm^3, f is the force constant in mdyn/ΔÅ. The force constants at the temperature of melting f_m may be calculated using the obvious equation

$$f_m = f_c \frac{B_m R_m}{B_o V_o} \tag{6.10}$$

Table 6.1 Critical factors for rare gases according to Lindemann and thermodynamics approaches

Rg	R_o, Å	Θ_D, K	T_m, K	δ_L	R_m, Å	B_m, GPa	f_m, mdyn/Å	δ_B
He	1.832	27	0.95	0.12	1.832	0.022	$1.03 \cdot 10^{-4}$	0.13
Ne	1.578	74.4	24.55	0.11	1.592	1.08	$3.32 \cdot 10^{-3}$	0.11
Ar	1.878	89.9	83.80	0.10	1.947	2.83	$5.97 \cdot 10^{-3}$	0.13
Kr	1.996	69.1	115.95	0.10	2.091	3.31	$7.94 \cdot 10^{-3}$	0.12
Xe	2.168	60	161.35	0.10	2.265	3.61	$8.42 \cdot 10^{-3}$	0.13

Using Eqs. 6.7 and 6.8, the values of δ_L and δ_B were calculated [16] for all studied compounds (Table S6.1), giving the averages of $\delta_L = 0.13 \pm 0.04$ and $\delta_B = 0.15 \pm 0.02$; the results for solid rare gases are given in Table 6.1.

What is the physical meaning of the critical amplitude of atomic vibrations in a melting metal? Evidently, bond lengths can be 'stretched' only in a frameworks of a certain stable structure. As shown by Goldschmidt, the relative interatomic distances in the structures of metals with different coordination numbers change as follows:

N_c	4	6	8	12
Structure	A4 →	A5 →	A2 →	A1
$\overline{d}$	1.00	1.07	1.09	1.12

Hence, the change of interatomic distances according to L-factor, exceeds by 13 to 15 % the limits of the stability of the metal structures at any phase transition; that leads to the destruction of the crystalline order, i.e. to amorphization (melting) of the solid.

As seen from Table 6.1, δ_L and δ_B for all rare gases are very similar, but He is the only Rg that does not solidify under any cooling conditions in the absence of external pressure. This is explained usually by the quantum character of the substance, whose zero point energy exceeds the crystal lattice energy. At the same time, the macroscopic properties of helium do not differ fundamentally from those characteristic of other rare gases, which allows us to treat it as an object of classical physics. Thus, it has been noticed [18–20] that the ratio q between the enthalpy of sublimation ΔH_s and half of the bond energy E(Rg–Rg) in Rg_2 molecules, for Ne, Ar, Kr, and Xe averages 12.8 ± 1.4 (on our data [20], $q = 12.5 \pm 0.3$) i.e. is close to the coordination number of Rg atoms in these close-packed crystal structures ($N_c = 12$). This indicates the additive character of van der Waals (vdW) interactions. For helium, however, q is as low as 1.9, although the crystal structure is of the same type. The q values and bond lengths in Rg crystals and Rg_2 molecules are listed in Table 6.2 (ΔH_s and ΔH_v are the sublimation and evaporation heats, respectively).

It is evident that the distances are sharply elongated when helium is crystallized, while they remain constant in the molecules and crystals of other rare gases. The situation is similar in liquid rare gases: $N_c = 8.9$ on average in the liquid state, whereas, in liquid helium, $N_c = 4$, and the bond length is longer by 6 % (cf. 0.3 %

Table 6.2 Properties of rare gas molecules and crystals

Value	He	Ne	Ar	Kr	Xe
$q_{cryst} = \Delta H_s / {}^1\!/_2\, E(Rg–Rg)$	1.9	12.0	12.9	12.6	12.7
$d(Rg–Rg)_{mol}$, Å	2.967	3.087	3.759	4.012	4.362
$d(Rg–Rg)_{cryst}$, Å	3.664	3.156	3.755	3.992	4.335
d_{cryst}/d_{mol}	1.235	1.022	0.999	0.995	0.994
$q_{liq} = \Delta H_v / {}^1\!/_2\, E(Rg–Rg)$	1.6	10.3	8.8	10.6	11.1
N_c (liq)	4	9.5	8.5	8.5	9.2
$d(Rg–Rg)_{liq}$, Å	3.15	3.11	3.76	4.02	4.38
d_{liq}/d_{mol}	1.062	1.007	1.000	1.002	1.004

in other rare gases) on condensation. Thus in the liquid state, $q/N_c = 0.4$ for He, as against the average of 1.1 for other rare gases, i.e. a decrease in the strength of the He–He bond on transition from the gaseous to the liquid state is similar (although smaller) to that observed in solid helium. The reason why helium atoms move apart as a result of condensation requires investigation, although an increase in the volume caused by helium solidification was observed as early as 1934 [21].

Thus, the structural properties and the critical vibration amplitude in condensed He are close to those of other rare gases, but thermodynamic characteristics are sharply different. Hence Lindemann's criterion cannot explain the special features of helium solidification. In this context, note a remarkable analogy in the changes of the properties in ionic MX and He_2 molecules at their condensation to the solid state: in both cases with an increase in N_c, we observe bond length elongation by 20–25 % and an increase in the enthalpy of atomization by a factor of 1.5 (see Chap. 2). In MX it is caused by the Madelung interaction, whereas the vdW interaction of atoms according to London theory depends only on the atomic size and polarizability, but the number of contacts must not influence it. However, by the Slater and Kirkwood approach [22, 23], interaction energies between isolated atoms depend on the number of their electrons and really the main difference between helium and other rare gases lies in their electron densities. Then it can be assumed that two helium electrons do not provide full-fledged interaction with 12 neighbors in its crystal structure, the bond should be weakened (elongated), and external pressure is required to stabilize the solid state. To show that the configuration of Rg_2 molecules does not correspond to a contact between two spherical atoms, we can refer to a recent work [24], where the anisotropy, i.e., the difference between the longitudinal ($\alpha_{||}$) and transverse ($\alpha_{\perp}$) polarizability ellipsoid axes, was measured for the Ar_2, Kr_2, and Xe_2 molecules. The $\Delta\alpha = \alpha_{||} - \alpha_{\perp}$ values for these Rg_2 were found to be 0.5, 0.7, and 1.3 Å^3, respectively. Since by definition (see [25]), the cross-section of the polarizability ellipsoid is equal to or larger than the polarizability of an isolated atom (α_o), and these values for the rare gases specified are 1.64, 2.48, and 4.04 Å^3, respectively (see Chap. 11), the ratio between the Rg_2 length and cross section (γ) is easy to calculate. As has been shown [16], $\gamma \leq 1.09$ for all the three gases. It follows that two flattened (almost by half) ellipsoids rather than two spheres contact each other along the Rg–Rg bond line, which is evidence of very strong interatomic interaction.

This result contradicts the traditional interpretation of the vdW interaction of atoms which is based on London's model (of instantaneous dipoles) and correspondingly on the Lennard-Jones or Buckingham potentials. However, there are alternative approaches. Thus, Feynman in his paper [26] on the electrostatic explanation of chemical bonding noted that vdW attractive forces are not the result of "the interactions between oscillating dipoles", but arise from the accumulation of electron density between the nuclei; the modern *ab initio* calculations of the He_2 dimer [27] confirmed his prediction. Slater [28] also insisted that there is no fundamental distinction between vdW and covalent bonding, later Bader [29, 30] showed that the changes in the total, kinetic, and potential energies with inter-nuclear separation for covalent, polar, and vdW interactions indicate a common underlying quantum mechanism.

It is possible to ask: is there any definite interatomic distance (expressed in relation to the equilibrium distance) at which the chemical bonds are dissociated and the substance decays into free atoms? To answer this question, let us start with the harmonic approximation. Because $B_o = \rho c^2$, where ρ is the density in g/cm^3, c the sound velocity in km/s, and $V_o = A/\rho$ (A is the relative atomic mass) Eq. 6.9 transforms into

$$f_c = 10^{-3} \frac{9Ac^2}{N_c R^2} \tag{6.11}$$

Substituting H_m in Eq. 6.8 by the evaporation heat, ΔH_v (in kJ/g), and combining Eqs. 6.8 and 6.11, we obtain

$$\delta_b = \frac{C}{c_b} \sqrt{\Delta H_v N_c} \tag{6.12}$$

where $C = 0.192$, N_c is the coordination number in the melt, c_b is the sound velocity in it at the boiling point (in km/s), and ΔH_v is expressed in kJ/g. Table S6.2 lists the values of δ_b for metals obtained by equation Eq. 6.12 and the data used. The experimental values of N_c for most metals were taken from [32] or, if unavailable (as for Be, Ta, Mo, W and U), those of solid metals were used instead (italicized in Table S6.2). Direct measurement of sound velocity at, or near, the boiling point are usually unavailable, therefore the values of c_b were obtained by extrapolation, using sound velocities at melting points (c_m), their derivatives $\partial c/\partial T$ [33, 34] and the differences ΔT between boiling and melting temperatures,

$$c_b = c_m - \Delta T \left(\frac{dc}{dt} \right) \tag{6.13}$$

Final results of calculations are given in Table 6.3 under δ_b.

Similarly, it is possible to estimate the critical amplitude of vibrations of atoms in liquid metals at the boiling temperature ($\bar{\delta}_b = 0.50$), exceeding which leads to the rupture of chemical bonds, e.g. to a decay of the melt into free atoms. To verify this conclusion, we can use the universal equation of state (EOS) of metals according to Vinet et al. [35],

$$E(d) = E_o E^*(d^*) \tag{6.14}$$

Table 6.3 Critical amplitudes of atomic vibrations determined by different methods

M	δ_b	δ_P	δ_{pb}	M	δ_b	δ_P	δ_{pb}
Li	0.63	0.75	0.75	Ti		0.54	0.52
Na	0.53	0.63	0.64	Zr		0.58	0.55
K	0.54	0.59	0.56	Hf		0.52	0.52
Rb	0.47	0.57	0.52	Si		0.61	0.65
Cs	0.47	0.58	0.54	Ge	0.40	0.58	0.62
Cu	0.50	0.50	0.46	Sn	0.48		0.49
Ag	0.38	0.45	0.40	Pb	0.37	0.42	0.42
Au	0.48	0.38	0.35	V		0.53	0.53
Be	0.46	0.64	0.60	Nb		0.52	0.54
Mg	0.35	0.45	0.42	Ta	0.36	0.51	0.52
Ca	0.42	0.55	0.50	Sb	0.34		
Sr	0.41	0.53	0.48	Bi	0.31		
Ba	0.47	0.57	0.50	Cr		0.45	0.46
Zn	0.28	0.38	0.33	Mo	0.31	0.45	0.44
Cd	0.26	0.34	0.29	W	0.45	0.46	0.46
Hg	0.24		0.26	Mn	0.54		0.41
Sc			0.52	Tc			0.42
Y		0.58	0.56	Re		0.41	0.40
La	0.52		0.73	Fe	0.40	0.51	0.50
Al	0.53	0.55	0.50	Co	0.43	0.50	0.47
Ga	0.47		0.55	Ni	0.43	0.49	0.49
In	0.46	0.51	0.46	Ru		0.42	0.42
Tl	0.38	0.47	0.40	Rh			0.42
Th		0.60	0.57	Pd	0.37	0.41	0.37
U	0.38		0.50	Os			0.41
				Ir		0.40	0.39
				Pt		0.37	0.39

Here $E(d)$ is the binding energy as a function of the bond length, E_o is the equilibrium binding energy,

$$E^*(d^*) = (1 + d^*)\exp(-d^*), \tag{6.15}$$

and the parameters d^* and l are the scaled lengths,

$$d^* = \frac{(d - d_o)}{l} \tag{6.16}$$

$$l = \left(\frac{E_o}{12\pi B_o V_o{}^{1/3}} \right)^{1/2} \tag{6.17}$$

Using Eqs. 6.14–6.17, one can describe quantitatively the known experimental data and first-principles calculations for *all* classes of metals, covalent and vdW molecules, and determine the values of the *negative* pressure leading to the metal rupture, e.g. to the decay of solids into free atoms, P_R [32]. Taking into account that $E = P\Delta V$ and equaling E to the atomization energy E_a, and P to P_R, one can calculate V_R as

$$\Delta V_R = \frac{E_a}{P_R} \tag{6.18}$$

and, hence, the critical increase of interatomic distances necessary for the full destruction of solids, i.e. their transformation into free atoms, is equal to

$$1 + \delta_p = \left(\frac{V_o + \Delta V_R}{V_o} \right)^{1/3} \tag{6.19}$$

The results of calculations of δ_p, using the experimental values of E_a, B_o, P_R and V_o [36], are presented in Table 6.3, and the necessary experimental data are given in Table S6.3. We see that the average $\bar{\delta}_P = 0.50 \pm 0.08$. The universal EOS allows to estimate the maximum interatomic distances in boiling metals, when steric repulsion energy becomes equal to the thermal energy in boiling points, $E_{Tb} = RT_b$ and condensed bodies transform into free atoms. We have calculated d^* by Eqs. 6.15–6.17 under the condition $E(d) = ET_b$ and then obtained δ_{pb} as

$$\delta_{pb} = \frac{d^* l}{d_o} \tag{6.20}$$

using the values l and d_o from [37]. The values of δ_{pb} for 50 elements are presented in Table 6.3, the average $\bar{\delta}_{pb} = 0.48$; the data required for calculations can be found in Table S6.4.

Thus the averaged values of critical amplitudes of atomic vibrations, $\bar{\delta}_b$, $\bar{\delta}_P$, $\bar{\delta}_{pb}$, vary between 0.48 and 0.50 of the bond lengths. Deviations of δ from the average for different metals result from several causes, mainly (i) experimental errors in measuring the properties of melts at high temperatures and (ii) large discrepancies in N_c measured by different methods (see Chap. 7). Nevertheless, the mean values of δ can be compared with the theoretical estimations using the formula of Gitis and Mikhailov [38],

$$c = (2U)^{1/2} \tag{6.21}$$

where U is the cohesive energy in kJ/g. Blairs showed [30] that this equation for liquid metals can be transformed to

$$c = (2\Delta H_v)^{1/2} \tag{6.22}$$

Combining Eqs. 6.12 and 6.22, we obtain

$$\delta = 0.136\sqrt{N_c} \tag{6.23}$$

Since the coordination numbers in liquid metals are 12 ± 1, the average value of δ_{GMB} according to the Gitis-Mikhailov-Blairs's approach is 0.47(2). If we use Rodean's dependence of the speed velocity on ΔH_v [36], then δ from Eq. 6.23 should be multiplied by 1.1 ± 0.1 and we obtain $\delta_R = 0.52(5)$. Thus, the average value of the critical vibration amplitude determined by Eqs. 6.12, 6.19 and 6.20 is within limits of Gitis-Mikhailov-Blairs's and Rodean's approaches. It is also noteworthy that the stretching limit of the chemical bond (by the factor of 1.5 ± 0.2) corresponds to the sum of the radii of isolated atoms, which are close to the vdW radii of these elements [37]. The results of applying Eq. 6.12 to liquid rare gases are given in Table 6.4 (see [16]). The critical amplitudes for all liquid rare gases have practically the same magnitudes.

Table 6.4 Properties of liquefied rare gases close to boiling points

A	c, km/s	N_c	ΔH_v, kJ/g	δ_b
He	0.207	4	0.018	0.249
Ne	0.593	8.8	0.090	0.288
Ar	0.747	8.5	0.163	0.302
Kr	0.690	8.5	0.108	0.267
Xe	0.631	8.9	0.096	0.281

6.3 Defects in Crystals

6.3.1 Classification of Defects

The defects which disrupt the regular patterns of crystals, can be classified into point defects (zero-dimensional), line defects (1-dimensional), planar (2-dimensional) and bulk defects (3-dimensional). *Point defects* are imperfections of the crystal lattice having dimensions of the order of the atomic size. The formation of point defects in solids was predicted by Frenkel [40]. At high temperatures, the thermal motion of atoms becomes more intensive and some of atoms obtain energies sufficient to leave their lattice sites and occupy interstitial positions. In this case, a vacancy and an interstitial atom, the so-called Frenkel pair, appear simultaneously. A way to create only vacancies has been shown later by Wagner and Schottky [41]: atoms leave their lattice sites and occupy free positions on the surface or at internal imperfections of the crystal (voids, grain boundaries, dislocations). Such vacancies are often called Schottky defects (Fig. 6.3). This mechanism dominates in solids with close-packed lattices where the formation of vacancies requires considerably smaller energies than that of interstitials. In ionic compounds also there are defects of two types, Frenkel and Schottky disorder. In the first case there are equal numbers of cation vacancies

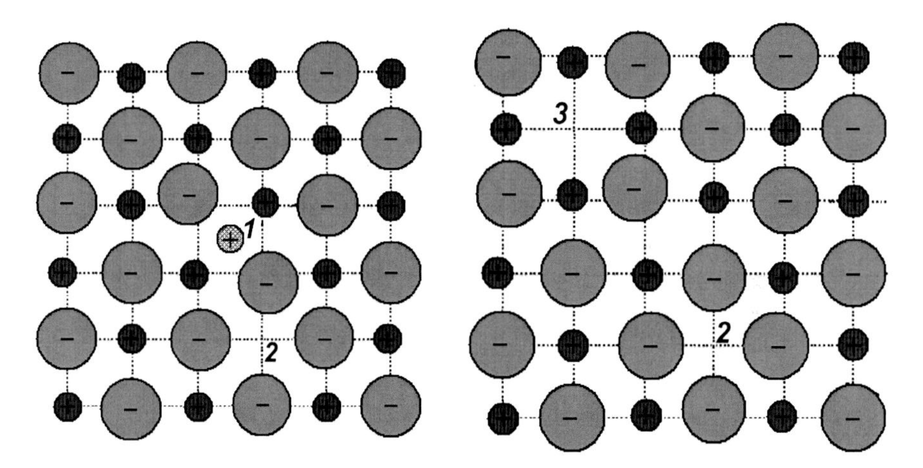

Fig. 6.3 Schematic representation of Frenkel (*left*) and Schottky (*right*) defects: *1*—interstitial cation, *2*—cation vacancy, *3*—anion vacancy

and cation interstitials (the interstitial defects are atoms that occupy a site in the crystal structure at which there is normally no atom at all). In the second case there are equal numbers of cation and anion vacancies. Note that *anion* interstitials are much less common than cation ones, because cations are usually smaller and hence require less distortion to fit in.

Point defects, both vacancies and interstitials, are thermodynamically stable since they lower the Gibbs energy of the crystal. The equilibrium concentrations of point defects rapidly increase with temperature. In metals, vacancies are the predominant point defects in the equilibrium state and their concentrations at high temperatures are much larger than those of interstitials.

Although crystallization is one of the best ways of purifying chemical substances, yet no crystal is free of impurities. An impurity atom is either incorporated in a crystal structure as an interstitial, or replaces an atom of the main element at the same site. In the latter case, known as the substitution defect, the valence (oxidation state) of the replacing and replaced atoms can be the same or different, in which case a charge compensation is required. Finally, there are so-called anti-site defects: for example, in a regular AB structure some of the A atoms occupy the sites properly belonging to B, and *vice versa*. In this case, the periodicity is perturbed with neither a vacancy, nor an interstitial, nor an impurity being present.

Linear defects around which some of the atoms of the crystal lattice are misaligned are known also as dislocations; they are of two basic types, the *edge* dislocation and the *screw* dislocation. 'Mixed' dislocations, combining the aspects of both types, are also common. Edge dislocations are caused by termination of a plane of atoms inside a crystal, while a screw dislocation is a feature in which a helical path is traced around a linear defect by the planes of atoms. The presence of dislocation results in lattice distortion. The direction and magnitude of such distortion is expressed by the Burgers's (b) vector. For an edge type, the b-vector is perpendicular to the dislocation line, whereas in the case of the screw type it is parallel. In metals, b-vector is parallel to close-packed crystallographic directions and its magnitude equals a single interatomic distance. Disclinations are linear defects with "adding" or "subtracting" an angle around a line. Usually they are important only in liquid crystals.

Planar defects also have several varieties. Thus, grain boundaries are surfaces where the crystallographic axes change directions substantially. This usually occurs when two crystals start growing separately and then merge. Anti-phase boundaries occur in ordered alloys with periodically modulated structures: the crystallographic direction remains the same, but on two sides of the boundary surface the modulation has opposite phases. Stacking faults usually occur in close-packed structures. In these, stacking together of *two* layers is always the same: an atom (A) of one layer fits a gap (B) of the other. The third layer can be added in two different ways: an atom of this layer lying directly above an atom (A) or a different gap (C) gap of the first layer. Regular repetition of the former motif (...ABABABAB... sequence) gives the hexagonal close-packing (*hcp*), regular repetition of the latter (...ABCABC...) gives the cubic close-packing (*ccp*) structure, alias face-centred cubic (*fcc*). A widespread enough defect is a stacking fault interrupting either stacking sequence, e.g. producing a ABCABABCAB sequence in an *fcc* structure.

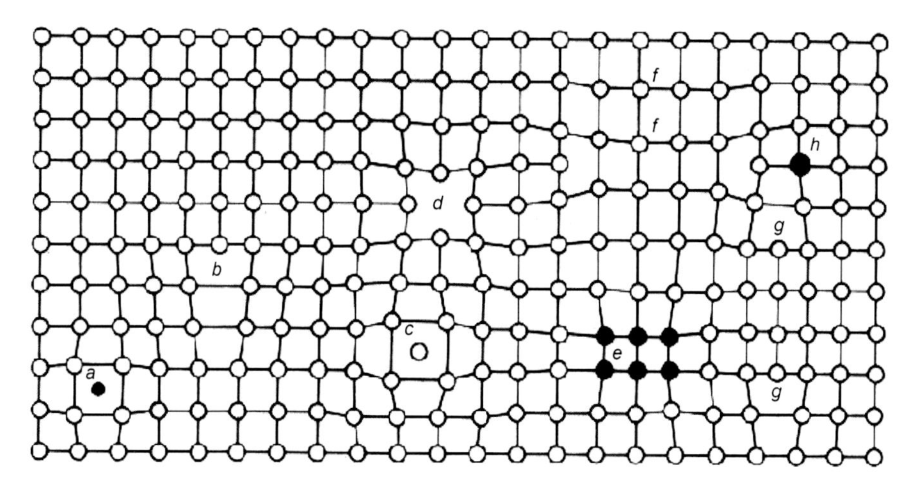

Fig. 6.4 Types of crystal defects: *a* interstitial impurity atom, *b* edge dislocation, *c* self-interstitial atom, *d* vacancy, *e* cluster of impurity atoms, *f* vacancy-type dislocation loop, *g* interstitial type dislocation loop, *h* substitutional impurity atom

Typical *bulk defects* are voids, i.e. small regions without atoms, which can be described as clusters of vacancies. Impurities also can cluster together to form small regions of a different phase. Figure 6.4 gives a schematic representation of the different types of defects in crystals. A brief history of the studies of point defects in metals and a review of the experimental data obtained up to 1998 are available [42].

6.3.2 Defects Induced by Shock Waves

When shock waves are propagated through a crystalline specimen, fragmentation and disorientation of crystallites take place at the boundary between the compressed and unperturbed matter (the shock front) due to steep pressure gradient. Experiments show that the minimum size of grains after the shock compression is *ca.* 10 nm, and a number of atoms in such cluster is *ca.* 10^4 to 10^5. Assuming that dislocations constitute the boundaries of these blocks, the maximum concentration of dislocations must be *ca.* 10^{18} to 10^{19} per cm^3 and the dislocation density in shocked crystals is therefore *ca.* 10^{12} to 10^{13} per cm^2. The residual heat (see Chap. 9) causes some of these dislocations to disappear and the dislocation density is reduced by one or two orders of magnitude. The experimental results for ionic crystals [43] are shown in Table 6.5, the survey of similar works on metals can be found in [44]. The minimum sizes of domains in the shocked polycrystalline materials are listed in Table 6.6.

It has been observed in crystalline alkali halides that after shock treatment the increase of the dislocation density was smallest along the [111] direction and largest along the [112], but during annealing it decreased fastest along the latter direction [44]. The change of microhardness (which, after the passage of shock waves through a crystal, increases by a factor of 1.3 to 2) and of other physical properties is also

Table 6.5 Dislocation density in shocked ionic crystals

Crystal	P, GPa	$\rho_{initial}/cm^2$	$\rho_{shocked}/cm^2$
LiF	8.5	10^5	10^{11}
NaCl	6	10^5	10^{10}
KBr	~6	10^5	10^{10}
CsI	~7	10^5	10^9
MgO	8	10^4–10^5	10^{12}
CaF_2	6	10^5	10^7

Table 6.6 Sizes (D) of crystallites in shocked powders

Material	D, nm	Material	D, nm	Material	D, nm
Diamond, TiO_2	10	CdF_2	18	Al_2SiO_5	25
Al_2O_3	13	CaF_2	19	Ni, LaB_6	27
BN, AlN, Mo	15	BaF_2	22	ZrC	37
LiF, MgO	16	UO_2	23	NaF	44

direction-dependent. The disorientation of the blocks in shock-compressed single-crystals of NaCl and CaF_2 varied in the range from 0.7 to 3° [45, 46]. At the same time, quartz [47] and silicate crystals undergo a complete or partial conversion into the amorphous state under shock compression. The latter phenomenon has been studied in greatest detail by optical and X-ray methods [48–50]. It has been shown that sub-microscopic inclusions of a disordered amorphous phase appear in the bulk of the crystals, and as a result, the anisotropy in properties of the shocked crystals diminishes and their parameters approach those of isotropic solids. Repeated shock compression of crystals affects the anisotropy in a peculiar way. Thus, the first compression of CaF_2 and BaF_2 leads to an anisotropic (because of the different strengths in different directions) crystalline blocks whose size are 18–60 and 22–38 nm in the <110> and <111> directions, respectively [51]. After the second compression, the major axes of the ellipsoids in the CaF_2 and BaF_2 samples are diminished to 38 and 26 nm, respectively, while the shortest axes remain the same; a similar behavior was observed in MgO and Al_2O_3. This tendency towards isometric blocks correlates with the enhancement of the isotropic character in single crystal silicates (see above). Therefore it can be expected that micro-stresses are more likely to occur in larger than in smaller grains, and the dislocation density will increase in direct relation to the grain sizes, rather than in the usually observed inverse relation. Annealing of the defects introduced by shock waves, restores the real structure of a material to the equilibrium state. Table 6.7 presents the results of studying micro-stresses ($<\varepsilon^2>^{1/2} = \Delta d/d$, where d is the inter-planar spacing) in a series of shocked polycrystalline materials ([44] and references therein).

An ordered superlattice of defects has been found [52, 53] in the shocked CeF_4, ThF_4 and UF_4, which transformed into LaF_3-type structures with 25 % of the positions in the cationic sublattice being vacant. Thus it is more correct to present

Table 6.7 Dependence of micro-stresses ($\times 10^3$) on the sizes of crystallites

Crystal	D, nm	$<\varepsilon^2>^{1/2}$	Crystal	D, nm	$<\varepsilon^2>^{1/2}$	Crystal	D, nm	$<\varepsilon^2>^{1/2}$
CaF_2	18	1.6	Al_2O_3	66	2.4	Mo	14.5	1.8
	38	1.8		150	3.6		26.5	3.4

the formulae of the new modifications as $M_{0.75}F_3$. Generally speaking, it is not quite clear why a solid under high pressure undergoes a phase transition with an increase of density with a simultaneous formation of a system of vacancies in the metallic sub-lattice. However, in the above mentioned case it is so: studies of the shocked M_2O_5 (M = V, Nb, Ta) revealed that under high dynamic pressure their structures had been rearranged with a change of the stoichiometry from M_2O_5 to M_xO_2, where x varied from 1 to 0.8 [54–57]. In other words, a cation-deficient rutile structure was formed.

To conclude this section, we should mention a curious case of a pseudo-phase transition in TiO_2 under shock compression [58]. The initial phase belonged to the monoclinic system ($a = 5.85$, $b = 9.34$, $c = 4.14$ Å, $\beta = 107.5°$), but after the shock loading all the weak lines on XRD disappeared because of high concentration of defects and remaining strong ones corresponded to a cubic lattice with $a = 4.176$ Å, i.e. this phase transition is due to the generation of numerous defects, destroying anisotropy of the crystal.

In addition to the changes in physical properties as a result of shock compression of crystals, of alkali halides in particular, the chemical behavior of the substances is altered slightly. In particular, a charge transfer from anions to cations induces basicity in these compounds: alkalinity appears. The concentration of such chemical defects is small (a fraction of a per cent) and depends on the intensity of the shock waves.

Generally speaking, the effects of explosion on materials are very complicated. They include the effects of high pressure and high temperature, plastic deformations, strong straining tensions and electric potentials created on the boundary between the compressed and the uncompressed matter. Thus, strong shock waves can lead to deep structural transformations, although later the high residual temperature may cause a complete annealing of them and recover the initial state. An interesting insight into the phenomena is provided by the separate action of pressure and temperature, in the case of rather low pressure which precludes the generation of high residual temperature.

As the concentration of defects increases under shock loading, we cannot exceed a certain critical value, after which a fragile destruction takes place. In metals, this usually occurs after the third shock treatment. It has been shown [59] that a thermal treatment at 350 °C leads to annealing of the defects and micro-cracks, which allows to repeat the shock hardening. Six cycles of such shock/temperature treatment (STT) of Steel 45 (containing C 0.45, Si 0.27, Cr 0.25, Mn 0.65, Ni 0.25 %, and the balance of Fe) increased its Vickers hardness from $H_V = 156$ for the initial state up to $H_V = 418$. It is very important technologically that low-temperature heating does increase not only the hardness but also the plasticity.

6.3.3 *Real Structure and Melting of Solids*

Because melting is the break-up of the crystal structure, then any preliminary partial destruction of crystal grains must decrease the enthalpy of fusion and probably

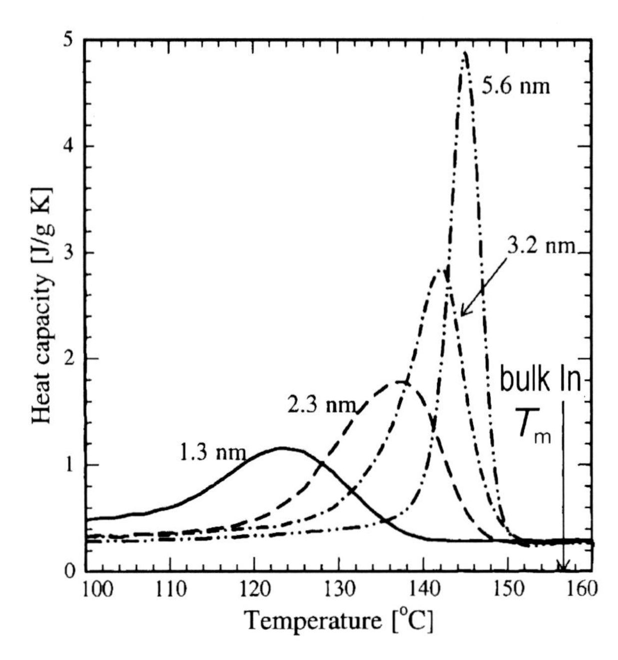

Fig. 6.5 Melting point depression of indium nano-particles of various size (normalized calorimetric curves). Reprinted (Fig. 2) with permission from [71], Copyright 2000 by the American Physical Society, http://link.aps.org/doi/10.1103/PhysRevB.62.10548

Table 6.8 Depression of the melting point, ΔT_m, in nano-materials compared to bulk species

Substance	Na	Au[a]	Cd[b]	Al	In[g]
D, nm	32	2.5	14	22	4
ΔT_m, K	83	600	9	17	110
Substance	Si[c]	Sn[d]	Pb[e]	CdS	TaC[f]
D, nm	4	10	2	2	1.78
ΔT_m, K	500	82	480	1200	3757

According to [64], except where specified, [a]For $D = 5$ nm $\Delta T_m = 300°$ [66], [b][67], [c][68], [d][69], [e]For $D = 4.2$ nm $\Delta T_m = 331°$, for $D = 11.2$ nm $\Delta T_m = 123°$ [70], [f]For $D = 10.0$ nm $\Delta T_m = 1703°$ [70], [g][71]

depress the melting point. The depression of melting temperatures (T_m) in nano-materials in comparison with bulk materials have been proven long ago and confirmed in recent nano-material studies (Fig. 6.5, Table 6.8). Lord Kelvin was first to predict (1871) that T_m of small particles will decrease with their sizes [60]. The first experimental confirmation of Kelvin's idea was obtained by Pawlow in 1909 [61]. Lindemann's criterion leads to the same conclusion. As mentioned above, the vibration amplitude of the surface atoms is approximately double that of the inner atoms [3–6]. The smaller a specimen, the bigger is the proportion of surface atoms and the more important this effect becomes. In nano-particles the surface atoms become predominant, hence it would require much lower temperature to increase δ_B to the critical value of 0.15 than in the case of bulk materials. This directly follows from Eqs. 6.8–6.10, as in the surface layer the structure is altered and the N_c is lower than inside a crystal [62]. Experimental data confirm this conclusion (see reviews [63–65]).

For atomic clusters of sufficiently small size, the melting conditions can be met even at ambient temperature; this size can be estimated by purely thermodynamic method. Let us imagine a solid with the molar volume V_m (in cm^3) to be fragmented into $n = V_m/D^3$ particles of cubic shape with the edge D. The total surface of these particles will be

$$S = \frac{V_m}{D^3} \times 6D^2 = \frac{6V_a N_a}{D} \tag{6.24}$$

and the number of surface atoms

$$N_s = 6\frac{N_a V_a{}^{1/3}}{D} \tag{6.25}$$

where V_a is the atomic volume (in Å^3) and N_a is the Avogadro number. The proportion of surface atoms in the sample is $f = N_s/N_a = 6V_a{}^{1/3}/D$, and the free energy of the surface atoms,

$$E_s = \frac{1}{3} f \times E_a = 2E_a \frac{V_a^{1/3}}{D} \tag{6.26}$$

where E_a is the atomization energy of the body. The condition $E_s = \Delta H_m$ can be taken as the energy criterion of the amorphization [65], assuming that injecting the energy equal to the melting enthalpy will cause melting irrespective of the method by which the energy is introduced into the solid. If this is done by mechanical crushing (grinding), an interesting corollary is that the smallest achievable crystalline grain size, beneath which amorphization occurs, is

$$D = 2D_a \frac{E_a}{\Delta H_m} \tag{6.27}$$

According to Eq. 6.27 the minimum size of crystal grains in elemental solids of the A1–A4 types vary from 2.4 to 46.4 nm with the average of 17 ± 9 nm; the calculations for all elements are presented in Table S6.5.

Indeed, a sharp decrease of the melting enthalpy after a shock-wave treatment has been shown experimentally [72], a thermodynamic interpretation of these results can be found in [64]. Depression of melting temperatures and enthalpies in organic crystals has been reported [73–76] and different models of these dependences are available [77, 78]. A simple empirical rule has been established: the enthalpy of the defect formation is nearly proportional to the melting temperature, viz. in alkali halides $\Delta H_f/T_m = 2 \cdot 10^{-3}$ (for ΔH_f in kJ/mol and T_m in K); it was also shown that $\partial T_m/\partial P = kV_o$, where V_o is the molar volume and $k = 5.7 \cdot 10^{-3}$ K mol/J [79].

6.4 Isomorphism and Solid Solutions

In 1819 Mitscherlich discovered that certain substances of different composition give similar crystal forms, named this phenomenon *isomorphism* and concluded (correctly, as we know today) that it reflects similarity of the atomic structure.

Isomorphism is very important in mineralogy and in the nineteenth century proved very useful for establishing the atomic masses (equivalents). However, the present-day usage of the term is different from the original one: isomorphism now describes the ability of different atoms to replace one another in a crystal structure without altering it substantially. In this latter sense, isomorphism is relevant to the subject of the present chapter. If the substitution is only partial, its product is a solid solution—a crystal structure where symmetrically equivalent atoms are not always chemically equivalent. The isomorphous substitution, as well as the resulting solid solutions, can be of two types, viz. *isovalent* and *heterovalent*. Isomorphism can be perfect, allowing substitution in any proportion from 0 to 100 %, or possible only within certain solubility limits. These limits and the thermodynamic stability of the solid solution depend on many factors. For isovalent solid solutions, Goldschmidt [80] empirically established the following rules:

i. Atoms can replace each other when the difference of their radii does not exceed 10–15 %; it is easier to substitute a smaller atom for a larger one than the other way round;
ii. The electronic structure, and hence the character of chemical bonding, of the two atoms must be similar;
iii. Heating facilitates isomorphism.

It is instructive to compare the formation of solid solutions with the dynamics of the crystal structure [81]. If the difference in sizes of replacing atoms is less than the amplitude of their thermal vibrations, then the substitutions should not destroy the crystal lattice. At ambient temperatures the amplitudes of thermal vibrations of atoms are of the order of 10 % of the bond length, close to the melting point they increase to 13–15 % (see Sect. 6.2), in agreement with the size-limits of isomorphism as defined by Goldschmidt's first rule. This also explains that the atomic vibrations and propensity for isomorphism increase together (rule iii). Later this approach was used to define the temperatures of solid solution formation [82]. A striking illustration of rule (ii) was provided by high pressure experiments: at $P > 10$ GPa potassium metal forms solid solutions with Ag and Ni (which do not exist at normal pressure) because mixing of 3*d* and 4*s* electron shells gives K the character of a transition metal [83].

Vegard [84, 85] established that in ionic salts there is a linear relation between the lattice parameters of a mixed crystal and the concentrations of the components. This rule applies also to mixed crystals of organic compounds. Thus, low-temperature phases of compounds $CBr_{4-n}Cl_n$ (n = 0 – 4) are isostructural and form continuous series of mixed crystals [86]. It has been, moreover, identified that the fractional occupancy of the halogen atoms fully controls the lattice dimensions. Lattice parameters as a function of the fractional occupancy of chlorine atom slightly deviate from the Vegard's law (Fig. 6.6).

Generally speaking, any replacement of atoms distorts the lattice and must increase the concentration of defects in a solid solution. The accumulating tensions not infrequently result in substantial deviations from Vegard law, which are usually positive and can be described by a simple parabolic function [87]. In the KCl–KBr and TlCl–TlBr solid solutions, this effect is strongest around the 1:1 composition [88]. In the $Ga_{1-x}In_xAs$ solid solution, the interatomic distances Ga/In–As change

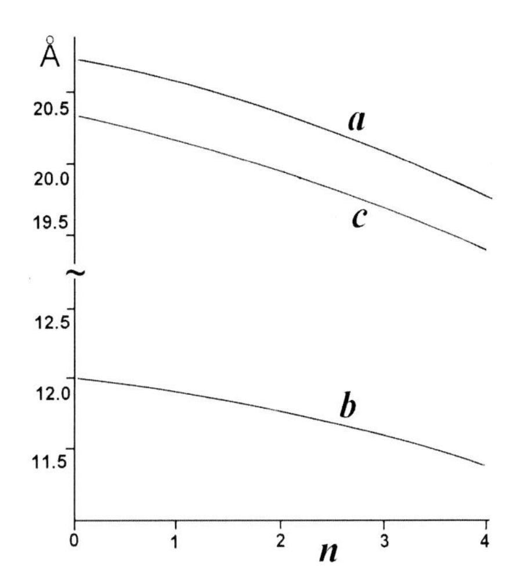

Fig. 6.6 Unit cell parameters of $CBr_{4-n}Cl_n$ solid solutions (monoclinic) as a function of *n*, at *ca* 220 K. (Based on data from [86])

only by 0.044 Å between $x = 0$ to $x = 1$, against the 0.174 Å predicted by Vegard law [89]. Deviations from linearity were also observed in $K_{1-x}Rb_xBr$ and $RbBr_{1-x}I_x$ for x between 0 and 1 [90].

As the solid solution retains the same symmetry as its pure components, atoms of different elements are symmetrically related and hence must occupy the same volume! *A priori*, two (extreme) pictures can be suggested. For example, in a solid solution $MX_{1-x}Y_x$, both X and Y may adopt the same local environments as in pure MX and MY respectively, the rest of the crystal structure absorbing the resulting strain. The observed unit cell parameters will be the statistical average over the whole crystal, not coinciding with any actual local value. Alternatively, we can envisage the whole lattice expanding or contracting uniformly (rigid-lattice model), with the electron shells of individual atoms somehow accommodating to this change. Unfortunately, the standard X-ray diffraction analysis, which is based on coherent (Bragg) scattering, provides no information on the local structure insofar as the latter deviates from the average structure. The information about atoms' positions relative to one another, rather than relative to the crystal lattice system of coordinates, is carried by the non-coherent (diffuse) scattering, which is much more difficult both to measure and to interpret than Bragg reflections; in routine experiments diffuse scattering is simply ignored. Only recently it became possible to extract from it useful structural information in the form of one-dimensional pair distribution function (PDF). This method of 'total' (i.e. Bragg *and* diffuse) scattering [91, 92] is still in an early stage of development and is intrinsically much more difficult and less unequivocal than usual X-ray diffraction. The same can be said of another method of elucidating the geometry of the local environment of a given atom (short-range order), namely the EXAFS spectroscopy (see Sect. 3.1) which can work for any aggregate state, including crystalline. The number of structures studied by both methods remains tiny (compared to the immense mass of standard X-ray crystal structures), usually these are molecules and solids of exceptional technological or biomedical significance, while much less attention is paid to simpler systems. Of these, an EXAFS study of

Table 6.9 Effective atomic charges in alkali halides

M	MCl[a]	MCl–MBr[b]		MBr[a]	X	KX[a]	KX–RbX[b]		RbX[a]
Li	0.74	0.83	0.59	0.71	F	0.93	0.90	1.00	0.95
Na	0.76	0.85	0.60	0.72	Cl	0.80	0.76	0.92	0.85
K	0.80	0.88	0.67	0.77	Br	0.77	0.72	0.88	0.80
Rb	0.85	0.89	0.72	0.80	I	0.74	0.69	0.86	0.78
Cs	0.86	0.92	0.72	0.81					

[a]Pure component, [b]Solid solution (at infinite dilution conditions)

RbBr-RbI solid solutions (and their melts) revealed distinct distributions of Rb–Br and Rb–I distances, practically independent from the concentration [93] and therefore from the observed average structure. Combined EXAFS and neutron scattering studies [94] of cation-substituted solid solutions also found large deviations from the rigid-lattice model; the authors assumed each cation to adopt locally its optimum bonding geometry.

However, there is also evidence in favour of the alternative view [95]. Imagine a solid solution $MX_{1-x}Y_x$ with $r(X) < r(Y)$. For X and Y having similar electronic configurations (according to Goldschmidt's second rule) this implies that X is more electronegative than Y (e.g. X=Cl, Y=I). The competition between X and Y for the electron density of M will result in X having higher, and Y lower, negative charge than in pure MX and MY, respectively. Note that this conclusion has been confirmed by experimental determination of charges by Szigeti's method (Table 6.9) [96]. Obviously, the degree of charge redistribution will increase with x and, since the effective size of an atom is increased by a net negative charge, the sizes of X and Y will converge. Although each atom will adopt its own optimum coordination geometry, this need not be the same as in the corresponding pure salt. Obviously, this issue needs re-investigation on the modern technical level.

There are known cases when formation of a solid solution actually *removes* the valence 'tension' in the structure and stabilizes a state in which an individual compound does not exist. Thus, pure CuF is unstable against decomposition into elemental Cu and CuF_2, but is quite stable as an isomorphic impurity (up to 2 %) in NaF, where $N_c(Cu^+) = 6$ according to spectroscopic data [97], although crystal-chemical rules suggest that CuF should belong to the structural type of ZnS.

Solubility limits in solid solutions can be widened by annealing (tempering) a high-temperature phase or depositing a thin film of a solid solution on a substrate with the necessary structure (epitaxy). Thus, a single-crystal layer (50–60 nm) of the $KCl_{0.5}I_{0.5}$ composition (not known to exist as bulk mixed crystals), have been obtained on a KBr substrate at low temperature [98]. Solid solutions MgO–CaO of all concentrations were prepared as epitaxial layers on the MgO substrate at $T = 300\,°C$ [99], their unit cell parameters conform to Vegard law.

New experimental data have modified Goldschmidt's rules. Thus, the CaO–NiO solution has been prepared by laser heating and subsequent deposition on a substrate of single-crystalline MgO [100]. The solid solution $Mg_{1/3}Fe_{2/3}O$ was prepared by shock compression of the mixture $MgCO_3 + Fe$ [101]. The $Mg_{1-x}Fe_xS$ solutions were found to have the NaCl structure at x < 0.68 and the NiAs structure at x = 0.68 [102]. The following solid solutions with cations of different electronic structures have been prepared: $Be_xZn_{1-x}Se$, $Be_xZn_{1-x}Te$ [103], $Mg_xZn_{1-x}O$ [104], $Mg_xZn_{1-x}Se$ [105], $Sr_xPb_{1-x}Se$ [106], $Sr_xPb_{1-x}(NO_3)_2$ [107], $Sn_{0.5}Ti_{0.5}O_2$ [108], $Mg_xZn_{1-x}TiO_3$ [109], $Mg_xCd_{1-x}CO_3$ [110], $K_xAg_{1-x}NbO_3$ [111]. These examples show that the maximum difference in ionic radii allowing isomorphous substitution, can reach 20 %, and different electronic structure is not an absolute obstacle for the formation of solid solutions.

Heterovalent isomorphism for inorganic substances is rather common, though such solutions are characterized by high concentrations of structural defects and typically *nonstoichiometric* phases. This type of solid solutions is often found in intermetallic compounds [112]. Numerous examples of the heterovalent isomorphism are provided by such systems as LnF_3–MF_2 and LnF_3–Ln_2O_3 where the fluorite-type solid solutions ($M_{1-x}F_2$ and LnOF) are formed; these and other solid solutions with LnF_3 have been reviewed in [113]. An example of substitution between highly dissimilar elements is $Sr_{0.25}Nd_{1.75}NiO_4$ obtained in crystalline state [114].

Goldschmidt [80] and Fersman [115], proceeding from qualitative considerations, formulated the rule of heterovalent isomorphism, according to which isomorphous relations mostly involve ions of the elements positioned in Periodical table on the diagonals (for example, Na^+ and Ca^{2+}, Mg^{2+} and Sc^{3+}, Sc^{3+} and Zr^{4+}, etc) having similar radii (diagonal series of isomorphism).

Appendix

Supplementary Tables

Table S6.1 Critical factors for the threshold of melting in metals, according to Lindemann (δ_L) and Batsanov (δ_B). [6.1]

M	δ_L	δ_B	M	δ_L	δ_B	M	δ_L	δ_B
Li	0.20	0.15	Y	0.11	0.19	Ta	0.16	0.15
Na	0.18	0.15	La	0.16		Cr	0.11	
K	0.19	0.15	Al	0.13	0.16	Mo	0.12	0.14
Rb	0.20	0.15	In	0.14		W	0.11	0.14
Cs	0.10	0.15	Tl	0.07	0.12	Mn	0.12	
Cu	0.14	0.17	Ti	0.19	0.20	Tc	0.12	
Ag	0.14	0.16	Zr	0.15		Re	0.10	0.16
Au	0.14	0.12	Hf	0.12	0.20	Fe	0.13	0.15
Be	0.14	0.18	Si	0.19	0.12	Co	0.13	0.18
Mg	0.12	0.17	Ge	0.11	0.10	Ni	0.13	0.16
Ca	0.16		Sn	0.08		Ru	0.08	0.14
Sr	0.16		Pb	0.12	0.12	Rh	0.09	0.14
Ba	0.17		Th	0.13	0.20	Pd	0.15	0.15
Zn	0.10	0.15	U	0.22		Os	0.10	
Cd	0.09	0.13	V	0.17	0.15	Ir	0.09	0.14
Sc	0.15		Nb	0.18	0.14	Pt	0.15	0.12

Table S6.2 Experimental values of sound velocity at melting and boiling points (c_m and c_b in m/s), evaporation heats (ΔH_v in kJ/g), coordination numbers of atoms in melts and critical amplitudes of atomic vibrations (as fractions of the bond lengths) at boiling temperatures

M	ΔT_{b-m}	c_m	$-dc/dT$	c_b	ΔH_v	N_c	δ_b
Li	1,162	4,554	0.60	3,857	17.04	9.5	0.63
Na	785.1	2,526	0.44	2,180	3.438	10.4	0.53
K	695.5	1,876	0.59	1,465	1.641	10.5	0.54
Rb	648.5	1,251	0.34	1,031	0.681	9.5	0.47
Cs	642.4	983	0.30	790	0.408	9.0	0.47
Cu	1,842	3,440	0.49	2,537	3.870	11.3	0.50
Ag	1,200	2,790	0.39	2,322	1.918	11.3	0.385
Au	1,792	2,568	0.57	1,547	1.378	10.9	0.48
Be	1,182	9,104	see[a]	7,500[b]	27.45	*12*	*0.465*
Mg	440	4,065	0.58	3,810	4.354	10.9	0.35
Ca	642	2,978	0.49	2,663	3.071	11.1	0.42
Sr	605	1,902	0.31	1,714	1.222	11.1	0.41
Ba	1,143	1,331	0.18	1,125	0.712	10.8	0.47
Zn	487	2,850	0.34	2,684	1.484	10.5	0.28
Cd	446	2,256	0.39	2,082	0.749	10.3	0.26
Hg	395.6	1,511	0.48	1,321	0.261	10.0	0.235
La	2,550	2,030	0.08	1,826	2.207	11.1	0.52
Al	1,858	4,561	0.48	3,669	8.970	11.5	0.53
Ga	2,174	2,873	0.26	2,308	3.018	10.4	0.47
In	1,915	2,337	0.29	1,782	1.556	11.6	0.46
Tl	1,169	1,650	0.23	1,381	0.657	11.6	0.38
Sn	2,370	2,464	0.28	1,800	1.862	10.9	0.48
Pb	1,421	1,821	0.28	1,423	0.678	10.9	0.37
Ta	2,441	3,303	0.26	2,668	3.066	*8*	*0.36*
Sb	987	1,900	−0.23	2,127	1.631	8.7	0.34
Bi	1,293	1,640	0.04	1,588	0.745	8.8	0.31
Mo	2,016	4,672	0.47	3,724	4.636	*8*	*0.31*
W	2,250	3,277	0.47	2,220	3.363	*8*	*0.45*
Mn	815	2,442	0.37	2,140	3.355	10.9	0.54
Fe	1,323	4,200	0.50	3,538	5.066	10.6	0.40
Co	1,432	4,031	0.46	3,372	5.075	11.4	0.43
Ni	1,458	4,047	0.39	3,478	5.144	11.6	0.43
Pt	2,057	3,053	0.24	2,559	2.160	11.1	0.37
U	2,795	2,000	≈ 0	2,037[b]	1.384	*12*	*0.38*

[a][6.2]; [b][6.3]

Table S6.3 Experimental atomization energies (kJ/mol), the negative pressures of rupture (GPa), the increase of volume V_R/V_o and the maximum amplitude of vibrations of atoms at the rupture pressure, δ_P

M	E_a	$-P_R$	V_R/V_o	δ_p	M	E_a	P_R	V_R/V_o	δ_p
Li	159.2	2.793	5.378	0.752	Ti	468.9	16.455	3.686	0.545
Na	109.0	1.385	4.329	0.630	Zr	609.8	14.595	3.974	0.584
K	90.8	0.659	4.022	0.590	Hf	612.7	17.695	3.542	0.524
Rb	82.8	0.520	3.847	0.567	Pb	196.8	5.856	2.839	0.416
Cs	79.8	0.393	3.914	0.576	V	511.4	23.55	3.601	0.533
Cu	337.7	19.88	3.389	0.502	Nb	720.7	25.95	3.511	0.520
Ag	285.6	14.13	2.968	0.437	Ta	780.5	29.13	3.458	0.512
Au	364.7	21.64	2.651	0.384	Cr	395.6	26.615	3.056	0.451
Be	321.3	19.45	4.383	0.636	Mo	657.1	34.49	3.025	0.446
Mg	147.6	5.166	3.041	0.449	W	835.6	40.845	3.147	0.465
Ca	176.1	2.432	3.756	0.554	Re	781.5	42.295	2.825	0.414
Sr	162.8	1.872	3.564	0.527	Fe	413.9	24.135	3. 415	0.506
Ba	179.5	1.648	3.872	0.570	Co	423.6	27.27	3.346	0.496
Zn	130.2	8.844	2.607	0.376	Ni	427.9	28.14	3.306	0.490
Cd	111.9	6.21	2.396	0.338	Ru	638.7	40.975	2.882	0.423
Y	423.6	7.191	3.962	0.582	Pd	379.8	23.46	2.826	0.414
Al	322.3	11.91	3.707	0.548	Ir	668.6	44.45	2.761	0.403
In	250.9	6.496	3.451	0.511	Pt	564.6	34.61	2.790	0.408
Tl	180.4	4.966	3.155	0.467	Th	571.7	9.23	4.122	0.603

Table S6.4 Thermal energy (kJ/mol) and the maximum amplitude of atomic vibrations at boiling points

M	E_{Tb}	δ_{pb}	M	E_{Tb}	δ_{pb}	M	E_{Tb}	δ_{pb}
Li	13.43	0.75	Sc	25.85	0.52	V	30.60	0.53
Na	9.61	0.64	Y	30.00	0.56	Nb	41.71	0.54
K	8.58	0.56	La	31.07	0.73	Ta	47.65	0.52
Rb	7.99	0.52	Al	23.21	0.50	Mn	19.40	0.41
Cs	7.85	0.54	Ga	20.59	0.55	Tc	42.82	0.42
Cu	23.57	0.46	In	19.50	0.46	Re	48.80	0.40
Ag	20.25	0.40	Tl	14.52	0.40	Fe	25.06	0.50
Au	26.02	0.35	Ti	29.60	0.52	Co	26.61	0.47
Be	22.80	0.60	Zr	38.93	0.55	Ni	26.49	0.49
Mg	11.33	0.42	Hf	40.54	0.52	Ru	36.77	0.42
Ca	14.61	0.50	Si	29.42	0.65	Rh	32.99	0.42
Sr	13.76	0.48	Ge	25.82	0.62	Pd	26.90	0.37
Ba	18.04	0.50	Sn	23.90	0.49	Os	43.94	0.41
Zn	9.81	0.33	Pb	16.81	0.42	Ir	39.09	0.39
Cd	8.65	0.29	Th	42.08	0.57	Pt	34.07	0.39
Hg	5.24	0.26	U	36.62	0.50			

Table S6.5 Minimum sizes of crystallites (D_{min}), further reduction of which results in amorphization

M	E_a	H_m	D_a	D_{min}	M	E_a	H_m	D_a	D_{min}
	kJ/mol		nm			kl/mol		nm	
Li	159.3	3.00	0.279	29.6	C	716.1	104.6	0.178	2.4
Na	107.4	2.6	0.340	28.1	Si	450.7	50.0	0.2715	4.9
K	89.0	2.3	0.423	32.7	Ge	375	37.0	0.283	5.7
Rb	80.9	2.2	0.453	33.3	Sn	301.5	7.2	0.300	25.1
Cs	76.6	2.1	0.487	35.5	Pb	195.2	4.8	0.312	25.4
Cu	337.5	13.0	0.228	11.8	V	514.3	20.9	0.2405	11.8
Ag	284.6	11.3	0.257	12.9	Nb	721.7	26.4	0.262	14.3
Au	368.4	12.55	0.257	15.1	Ta	783	31.6	0.262	13.0
Be	324	12.2	0.201	10.7	Cr	397.2	16.9	0.229	10.8
Mg	146.9	8.95	0.285	9.4	Mo	657.7	32.0	0.250	10.3
Ca	177.9	8.5	0.352	14.7	W	852	35.4	0.251	12.1
Sr	162.8	8.3	0.383	15.0	Mn	283.7	12.0	0.230	10.9
Ba	181	7.75	0.398	18.6	Tc	657	24.0	0.243	13.3
Zn	130.4	7.3	0.248	8.9	Re	774.9	33.2	0.245	11.4
Cd	111.8	6.2	0.278	10.0	Fe	416.2	13.8	0.2275	13.7
Hg	61.4	2.3	0.285	15.2	Co	427.2	16.2	0.223	11.8
Sc	378.3	14.1	0.292	15.7	Ni	429.7	17.5	0.222	10.9
Y	424.1	11.4	0.321	23.9	Ru	653	24.0	0.2385	13.0
La	430.5	6.2	0.334	46.4	Rh	555	21.5	0.239	12.3
Al	329.8	10.8	0.255	15.6	Pd	375	17.6	0.245	10.4
In	241.3	3.3	0.297	43.4	Os	788.5	31.8	0.2405	11.9
Tl	181.4	4.1	0.305	27.0	Ir	668	26.1	0.242	12.4
Ti	473.4	15.4	0.260	16.0	Pt	565.4	19.6	0.247	14.2
Zr	606	16.9	0.2855	20.5	Th	590	16.1	0.3205	23.5
Hf	619.4	24.0	0.282	14.5					

Supplementary References

6.1 Batsanov SS (2009) Russ J Phys Chem 83:1836
6.2 Boivineau M, Arlès L, Vermeulen JM, Thévenin T (1993) Intern J Thermophysics 14:427
6.3 Boivineau M, Arlès L, Vermeulen JM, Thévenin T (1993) Physica B190:31

References

1. Barrera GD, Bruno JAO, Barron THK, Allan NL (2005) Negative thermal expansion. J Phys Cond Matt 17:R217–R252
2. Miller W, Smith CW, Mackenzie DS, Evans KE (2009) Negative thermal expansion: a review. J Mater Sci 44:5441–5451
3. Blackman M (1958) On negative volume expansion coefficients. Phil Mag 3:831–838
4. Goodman RM, Farrell HH, Samorjai GA (1968) Mean displacement of surface atoms in palladium and lead single crystals. J Chem Phys 48:1046
5. Van Hove MA (2004) Enhanced vibrations at surfaces with back-bonds nearly parallel to the surface. J Phys Chem B108:14265–14269

6. Vogt J (2007) Tensor LEED study of the temperature dependent dynamics of the NaCl(100) single crystal surface. Phys Rev B75:125–423
7. Härtel S, Vogt J, Weiss H (2010) Relaxation and thermal vibrations at the NaF(100) surface. Surf Sci 604:1996–2001
8. Hazen RM, Downs RT, Prewitt CT (2000) Principles of comparative crystal chemistry. Rev Miner Geochem 41:1–33
9. Hazen RM, Finger IW (1982) Comparative crystal chemistry: temperature, composition and the variation of crystal structure. Wiley, New York
10. Hazen RM, Finger IW (1987) High-temperature crystal chemistry of phenakite and chrysoberyl. Phys Chem Minerals 14:426–434
11. Sun CC (2007) Thermal expansion of organic crystals and precision of calculated crystal density. J Pharm Sci 96:1043–1052
12. Hofmann DWM (2002) Fast estimation of crystal densities. Acta Cryst B57:489–493
13. Sutherland W (1891) A kinetic theory of solids, with an experimental introduction. Philos Mag 32:31–43, 215–225, 524–553
14. Lindemann FA (1910) The calculation of molecular natural frequencies. Phys Z 11:609–612
15. Ledbetter H (1991) Atomic frequency and elastic constants. Z Metallkunde 82:820–822
16. Batsanov SS (2009) The dynamic criteria of melting-crystallization. Russ J Phys Chem A83:1836–1841
17. Batsanov SS (2005) Metal electronegativity calculations from spectroscopic data. Russ J Phys Chem 79:725–731
18. Smirnov BM (1993) Mechanisms of melting of rare gas solids. Physica Scripta 48:483–486
19. Runeberg N, Pyykko P (1998) Relativistic pseudopotential calculations on Xe_2, RnXe, and Rn_2: the van der Waals properties of radon. Int J Quantum Chem 66:131–140
20. Batsanov SS (1998) Some characteristics of van der Waals interaction of atoms. Russ J Phys Chem 72:894–897
21. Simon F (1934) Behaviour of condensed helium near absolute zero. Nature 133:529
22. Slater JC, Kirkwood JG (1931) The van der Waals forces in gases. Phys Rev 37:682–697
23. Cambi R, Cappelletti D, Liuti G, Pirani F (1991) Generalized correlations in terms of polarizability for van der Waals interaction potential parameter calculations. J Chem Phys 95:1852–1861
24. Minemoto S, Tanji H, Sakai H (2003) Polarizability anisotropies of rare gas van der Waals dimers studied by laser-induced molecular alignment. J Chem Phys 119:7737–7740
25. Deiglmayr J, Aymar M, Wester R et al (2008) Calculations of static dipole polarizabilities of alkali dimers. J Chem Phys 129:064–309
26. Feynman RP (1939) Forces in molecules. Phys Rev 56:340–343
27. Allen MJ, Tozer DJ (2002) Helium dimer dispersion forces and correlation potentials in density functional theory. J Chem Phys 117:11113–11120
28. Slater JC (1972) Hellmann-Feynman and virial theorems in the $X\alpha$ method. J Chem Phys 57:2389–2396
29. Bader RFW, Hernandez-Trujillo J, Cortes-Guzman F (2007) Chemical bonding: from Lewis to atoms in molecules. J Comput Chem 28:4–14
30. Bader RFW (2009) Bond paths are not chemical bonds. J Phys Chem A113:10391–10396
31. Bader RFW (2010) Definition of molecular structure: by choice or by appeal to observation? J Phys Chem A114:7431–7444
32. Waseda Y (1980) The structure of non-crystalline materials, McGraw-Hill, New York
33. Blairs S (2006) Correlation between surface tension, density, and sound velocity of liquid metals. J Coll Interface Sci 302:312–314
34. Blairs S (2006) Temperature dependence of sound velocity in liquid metals. Phys Chem Liquids 44:597–606
35. Vinet P, Rose JH, Ferrante J, Smith JR (1989) Universal features of the equation of state of solids. J Phys Cond Matter 1:1941–1963
36. Rose JH, Smith JR, Guinea F, Ferrante J (1984) Universal features of the equation of state of metals. Phys Rev B29:2963–2969

37. Batsanov SS (2011) Thermodynamic determination of van der Waals radii of metals. J Molec Struct 990:63–66
38. Gitis MB, Mikhailov IG (1968) Calculating velocity of sound in liquid metals. Sov Phys-Acoust 13:473–476
39. Rodean HC (1974) Evaluation of relations among stress-wave parameters and cohesive energy of condensed materials. J Chem Phys 61:4848–4859
40. Frenkel JI (1926) Thermal movement in solid and liquid bodies. Z Phys 35:652–669
41. Wagner C, Schottky W (1930) Theory of controlled mixed phases. Z phys Chem 11:163–210
42. Kraftmakher Y (1998) Equilibrium vacancies and thermophysical properties of metals. Phys Rep 299:80–188
43. Batsanov SS (1972) Syntheses under shock-wave pressures. In: Hagenmuller P (ed) Preparative methods in solid state chemistry. Academic, New York
44. Batsanov SS (1994) Effects of explosions on materials. Springer, New York
45. Batsanov SS, Zhdan PA, Kolomiichuk VN (1968) Action of explosion on matter. Dynamic compression of single-crystal NaCl. Comb Expl Shock Waves 4:161–163
46. Batsanov SS, Malyshev EM, Kobets LI, Ivanov VA (1969) Preservation and study of fluorite single-crystals under conditions of dynamic compression. Comb Expl Shock Waves 5:306–308
47. Decarli PS, Jamieson JC (1959) Formation of an amorphous form of quartz under shock conditions. J Chem Phys 31:1675–1676
48. Chao ECT (1967) Shock effects in certain rock-forming minerals. Science 156:192–202
49. Stöffler D (1972) Behavior of minerals under shock compression. Fortschr Miner 49:50–113
50. Stöffler D (1974) Physical properties of shocked minerals. Fortschr Miner 51:256–289
51. Moroz EM, Svinina SV, Batsanov SS (1972) Changes in the real structure of certain fluorides as a result of compressive impact. J Struct Chem 13:314–316
52. Batsanov SS, Kiselev YuM, Kopaneva LI (1979) Polymorphic transformation of ThF_4 in shock compression. Russ J Inorg Chem 24:1573–1573
53. Batsanov SS, Kiselev YuM, Kopaneva LI (1980) Polymorphic transformation of UF_4 and CeF_4 in shock compression. Russ J Inorg Chem 25:1102–1103
54. Adadurov GA, Breusov ON, Dremin AN et al (1971) Phase-transitions of shock-compressed t-Nb_2O_5 and h-Nb_2O_5. Comb Expl Shock Waves 7:503–506
55. Adadurov GA, Breusov ON, Dremin AN et al (1972) Formation of a Nb_xO_2 ($0.8 \leq x \leq 1.0$) phase under shock compression of niobium pentoxide. Doklady Akademii Nauk SSSR 202:864–867
56. Syono Y, Kikuchi M, Goto T, Fukuoka K (1983) Formation of rutile-type Ta(IV)O_2 by shock reduction and cation-deficient $Ta_{0.8}O_2$ by subsequent oxidation. J Solid State Chem 50:133–137
57. Kikuchi M, Kusaba K, Fukuoka K, Syono Y (1986) Formation of rutile-type $Nb_{0.94}O_2$ by shock reduction of Nb_2O_5. J Solid State Chem 63:386–390
58. Batsanov SS, Bokarev VP, Lazarev EV (1989) Influence of shock-wave action on chemical activity. Comb Expl Shock Waves 25:85–86
59. Batsanov SS, Sazonov VE, Sekoyan SS, Shmakov AS (1989) Effect of shock-thermal treatment on mechanical properties of steels. Propell Expl Pyrotechn 14:238–240
60. Thomson W (1871) On the equilibrium of vapour at a curved surface of liquid. Phil Mag 42:448–452
61. Pawlow P (1909) The dependency of the melting point on the surface energy of a solid body. Z phys Chem 65:545–548
62. Ouyang G, Zhu WG, Sun CQ et al (2010) Atomistic origin of lattice strain on stiffness of nanoparticles. Phys Chem Chem Phys 12:1543–1549
63. Roduner E (2006) Size matters: why nanomaterials are different. Chem Soc Rev 35:583–592
64. Sun J, Simon SL (2007) The melting behavior of aluminum nanoparticles. Thermochim Acta 463:32–40
65. Batsanov SS (2011) Size effect in the structure and properties of condensed matter. J Struct Chem 52:602–615

66. Buffat P, Borel J-P (1976) Size effect on melting temperature of gold particles. Phys Rev A13:2287–2298
67. Zhang DL, Hutchinson JL, Cantor B (1994) Melting behavior of cadmium particles embedded in an aluminum-matrix. J Mater Sci 29:2147–2151
68. Goldstein AN (1996) The melting of silicon nanocrystals. Appl Phys A62:33–37
69. Lai SL, Guo JY, Petrova V et al (1996) Size-dependent melting properties of small tin particles. Phys Rev Lett 77:99–102
70. Jiang Q, Aya N, Shi FG (1997) Nanotube size-dependent melting of single crystals in carbon nanotubes. Appl Phys A64:627–629
71. Zhang M, Efremov MY, Schittekatte F et al (2000) Size-dependent melting point depression of nanostructures. Phys Rev B 62:10548
72. Batsanov SS, Zolotova ES (1968) Impact synthesis of divalent chromium chalcogenides. Proc Acad Sci USSR Dokl Chem 180:93–96
73. Jackson CL, McKenna GB (1990) The melting behavior of organic materials confined in porous solids. J Chem Phys 93:9002–9011
74. Jackson CL, McKenna GB (1996) Vitrification and crystallization of organic liquids confined to nanoscale pores. Chem Mater 8:2128–2137
75. Jiang Q, Shi HX, Zhao M (1999) Melting thermodynamics of organic nanocrystals. J Chem Phys 111:2176–2180
76. Lonfat M, Marsen B, Sattler K (1999) The energy gap of carbon clusters studied by scanning tunneling spectroscopy. Chem Phys Lett 313:539–543
77. Guisbiers G, Buchaillot L (2009) Modeling the melting enthalpy of nanomaterials. J Phys Chem C113:3566–3568
78. Zhu YF, Lian JS, Jiang Q (2009) Modeling of the melting point, Debye temperature, thermal expansion coefficient, and the specific heat of nanostructured materials. J Phys Chem C113:16896–16900
79. Ksižek K, Górecki T (2000) Vacancies and a generalised melting curve of alkali halides. High Temp-High Press 32:185–192
80. Goldschmidt VM (1926) Geochemische Verteilungsgesetze der Elemente. Skrifter Norske Videnskaps-Akad, Oslo
81. Batsanov SS (1973) Energetic aspect of isomorphism. J Struct Chem 14:72–75
82. Batsanov SS (1982) Chemical transformation of inorganic substances during shock compression. Zhurnal Neorganicheskoi Khimii 27:1903–1905
83. Parker LJ, Atou T, Badding JV (1996) Transition element-like chemistry for potassium under pressure. Science 273:95–97
84. Vegard L (1921) The constitution of the mixed crystals and the filling of space of the atoms. Z Phys 5:17–26
85. Vegard L (1928) X-rays in the service of research on matter. Z Krist 67:239–259
86. Negrier P, Tamarit JL, Barrio M et al (2007) Monoclinic mixed crystals of halogenomethanes $CBr_{4-n}Cl_n$ (n = 0,. . . , 4). Chem Phys 336:150–156
87. Urusov VS (1992) Geometric model for deviations from Vegard's law. J Struct Chem 33:68–79
88. Batsanov SS (1986) Experimental foundation of structural chemistry. Standarty, Moscow (in Russian)
89. Mikkelsen JC, Boyce JB (1983) Extended X-ray-absorption fine structure study of $Ga_{1-x}In_xAs$ random solid solutions. Phys Rev B28:7130–7140
90. Boyce JB, Mikkelsen JC (1985) Local structure of ionic solid-solutions—extended x-ray absorption fine-structure study. Phys Rev B31:6903–6905
91. Egami T, Billinge SJL (2003) Underneath the Bragg peaks: structural analysis of complex materials. Pergamon, Amsterdam
92. Billinge SJL, Dykhne T, Juhas P et al (2010) Characterisation of amorphous and nanocrystalline molecular materials by total scattering. Cryst Eng Comm 12:1366–1368
93. Di Cicco A, Principi E, Filipponi A (2002) Short-range disorder in pseudobinary ionic alloys. Phys Rev B65:212106

94. Binsted N, Owens C, Weller MT (2007) Local structure in solid solutions revealed by combined XAFS/Neutron PD refinement. AIP Conf Proc 882:64
95. Batsanov SS (1978) Limiting values of ionic-radii. Doklady Acad Sci USSR 238:95–97 (in Russian)
96. Batsanov SS (1977) Calculation of the effective charges of atoms in solid solutions. Russ J Inorg Chem 22:941–943
97. Goldberg A, McClure D, Pedrini C (1982) Optical-absorption and emission-spectra of Cu^+, NaF single-crystals. Chem Phys Lett 87:508–511
98. Yang M, Flynn C (1989) Growth of alkali-halides from molecular-beams. Phys Rev Lett 62:2476–2479
99. Hellman E, Hartford E (1994) Epitaxial solid-solution films of immiscible MgO and CaO. Appl Phys Lett 64:1341–1343
100. Mao X, Perry D, Russo R (1993) $Ca_{1-X}Ni_XO$ catalytic thin-films prepared by pulsed-laser deposition. J Mater Res 8:2400–2403
101. Sekine T (1988) Diamond from shocked magnesite. Naturwissenschaften 75:462–463
102. Farrell SP, Fleet ME (2000) Evolution of local electronic structure in cubic $Mg_{1-x}Fe_xS$ by S *K*-edge XANES spectroscopy. Solid State Comm 113:69–72
103. Pages O, Tite T, Chafi A et al (2006) Raman study of the random ZnTe-BeTe mixed crystal: percolation model plus multimode decomposition. J Appl Phys 99:063–507
104. Ohtomo A, Kawasaki M, Koida T et al (1998) $Mg_xZn_{1-x}O$ as a II-VI widegap semiconductor alloy. Appl Phys Lett 72:2466–2468
105. Paszkowicz W, Szuszkewicz W, Dunowska E et al (2004) High-pressure structural and optical properties of wurtzite-type $Zn_{1-x}Mg_xSe$. J Alloys Compd 371:168–171
106. Shen WZ, Tang HF, Jiang LF et al (2002) Band gaps, effective masses and refractive indices of PbSrSe thin films. J Appl Phys 91:192–198
107. Shtukenberg AG (2005) Metastability of atomic ordering in lead-strontium nitrate solid solutions. J Solid State Chem 178:2608–2612
108. Kong LB, Ma J, Huang H (2002) Preparation of the solid solution $Sn_{0.5}Ti_{0.5}O_2$ from an oxide mixture via a mechanochemical process. J Alloys Compd 336:315–319
109. Liferovich RP, Mitchell RH (2004) Geikielite-ecandrewsite solid solutions: synthesis and crystal structures of the $Mg_{1-x}Zn_xTiO_3$ ($0 \leq x \leq 0.8$) series. Acta Cryst B60:496–501
110. Bromiley FA, Boffa-Ballaran T, Zhang M, Langenhorst F (2004) A macroscopic and microscopic investigation of the $MgCO_3$–$CdCO_3$ solid solution. Geochim Cosmochim Acta 68:A87
111. Fu D, Itoh M, Koshihara S-y (2009) Dielectric, ferroelectric, and piezoelectric behaviors of $AgNbO_3$–$KNbO_3$ solid solution. J Appl Phys 106:104104
112. Pearson WB (1972) The crystal chemistry and physics of metals alloys. Wiley-Interscience, New York
113. Sobolev BP (2000) The rare earth trifluorides. Institut d'Estudis Catalans, Barcelona
114. Liu XQ, Wu YJ, Chen XM (2010) Giant dielectric response and polaronic hopping in charge-ordered $Nd_{1.75}Sr_{0.25}NiO_4$ ceramics. Solid State Commun 150:1794–1797
115. Fersman AE (1936) Polar isomorphism. Comptes Rendus de l'Academie des Science de l'URSS 10:119–122

Chapter 7
Amorphous State

Grinding or any other deformation of a solid causes crushing and disorganization of crystal grains, generation of defects, dislocations, micro-cracks, and ultimately amorphization of the crystalline material. The question arises: to what size it is possible to grind a crystal in order for its properties still to correspond to those of the bulk material, and where is the 'size' border between a crystal and an amorphous body?

7.1 Dispersing Powders

Dispersing polycrystalline substances can produce such grains that their properties will differ from those of bulk materials; Roy was the first to consider this question [1]. He concluded that the minimum size of a particle which shows properties of the bulk crystal, according to various physical methods, must exceed 10 nm. Such materials now are called '*nano*-materials' (1 nm = 10 Å). Roy stated that the same amorphous body can have a different short-range atomic order, for example amorphous SiO_2 prepared by different methods has different properties.

In [2] were estimated the minimum quantities of atoms in clusters (N_{min}) as the border between the nano- and bulk crystals, determined by different methods:

Studying of the crystal	Studying of the cluster	N_{min}
Solvation of ions in solutions	Gas thermochemistry of ions	~ 10
Short range order in the structure	Configuration of clusters	~ 15
Work function of metals	Ionization potential of clusters	$\sim 10^2$
Long range order in the structure	Gas electron diffraction	$\sim 10^3$
Fusion of the solid body	Curve of heating of clusters	$\sim 10^6$

Since reducing the grain sizes leads to broadening of XRD lines, one can compute when these lines merge into a halo characteristic of an amorphous substance. For MgO the critical size is around 1 nm, for SiO_2 twice that. Obviously, the higher the symmetry of the crystal, i.e. the fewer lines are present in the XRD pattern, the broader the diffraction lines must be (i.e. the smaller the grain size) to form a

S. S. Batsanov, A. S. Batsanov, *Introduction to Structural Chemistry*,
DOI 10.1007/978-94-007-4771-5_7, © Springer Science+Business Media Dordrecht 2012

continuous halo. Therefore crystals of cubic symmetry, such as NaCl, till now have not been obtained in the glass form.

The critical value of the grain size corresponding to the transformation of a crystal into amorphous matter can be calculated also by thermodynamic method. Crystalline substances can be transformed into the amorphous (glass) state by different methods, viz. irradiation, heating, mechanical grinding, crushing by shock waves, but in all cases one must spend the same amount of energy on destruction of the lattice (W_{destr}) [3]. It is well known that the fusion of a solid begins when the crystal attains a heat content equal to

$$Q_m = \int_o^{Tm} c_p dT \tag{7.1}$$

where c_p is the specific heat and T_m is the melting temperature. Excluding, for simplicity, those crystals which have a first-order phase transition before melting (otherwise Q_m must be increased by the phase transition heat) and assuming that c_p is independent of temperature, we can take this quantity as the energy criterion of the transition of a crystal into an amorphous phase, irrespective of the way by which the energy is introduced into the solid. If this way consists in mechanical crushing/grinding, then, since the mechanical energy goes into an increase of the total surface energy, σ, of the ground crystals, we can predict the point of amorphization at which σ becomes equal to Q_m. The calculations made using this criterion, and experimental data on the specific heat and surface energies of crystalline substances, have shown that the critical size of particles in halides and oxides of alkali and earth-alkali metals is equal to *ca.* 1 nm, and at metals to *ca.* 2.5 nm, which explains why amorphous metals are easier to obtain than amorphous ionic solids. The same sizes can be taken for 'a nucleus of a crystal lattice' in an amorphous solid.

In polycrystalline materials with micrometer-sized grains the surface comprises a very small fraction of the total number of atoms (N) and its effect on the overall properties of the material can be ignored. In nano-sized particles a considerable fraction of the atoms forms the surface of the grain and in very small objects (few nm) the number of surface-related atoms can exceed the number of the 'bulk' atoms. Although the structure of a nano-crystal is not uniform and should be considered as composed of two distinctive parts: the grain core and the shell phases [4, 5], in general, grinding of crystals reduces the mean coordination number $\overline{N}_c$, since atoms at the surface have fewer direct neighbors than atoms in the bulk. Estimations show [6] that crushing grains to 5–10 nm reduces $\overline{N}_c$ by 1 or 2, and the work of crushing (additional surface energy) is close to the heat of the phase transition $N_c = 6 \rightarrow N_\text{c} = 5(4)$ type. The $\overline{N}_c$ of the major structural types in the MX crystalline compounds as a function of particle size have been calculated [7] and are listed in Table 7.1.

Mean coordination numbers in different structural types of metals can be calculated [8] as

$$\overline{N}_c = (1 - y)N_c + yN_c^{surf} \tag{7.2}$$

Table 7.1 Mean coordination numbers and their relative changes as functions of particle size

Structural type	ZnS ($N_c = 4$)		NaCl ($N_c = 6$)		CsCl ($N_c = 8$)	
Particle size[a]	$\overline{N}_c$	$\Delta N_c/N_c$	$\overline{N}_c$	$\Delta N_c/N_c$	$\overline{N}_c$	$\Delta N_c/N_c$
25	3.77	0.057	5.88	0.020	7.11	0.111
20	3.715	0.071	5.85	0.025	6.91	0.136
15	3.625	0.094	5.805	0.032	6.59	0.176
10	3.455	0.136	5.71	0.048	6.01	0.249
6	3.145	0.214	5.535	0.077	5.05	0.369
5	3.00	0.250	5.45	0.092	4.63	0.421
4	2.805	0.299	5.325	0.112	4.10	0.488
3	2.51	0.372	5.13	0.145	3.38	0.578
2	2.03	0.492	4.76	0.206	2.37	0.704
1	1.14	0.715	3.86	0.357	1.00	0.875

[a] as a number of the unit cell parameters (a)

Table 7.2 The bulk and surface coordination numbers in metallic structures

	N_c	N_c^{surf}
α-Po	6	3.274
fcc	12	7.181
bcc	8	4.523
hcp	12	6.265

where y is the fraction of surface atoms in the crystalline grain, N_c is the coordination number in the bulk of metal, N_c^{surf} is the same on the surface of a grain. In Table 7.2 are listed the values of N_c and N_c^{surf} in various structural types, which allow to calculate $\overline{N}_c$ by Eq. 7.2 for samples of any size.

Size-dependent reduction of N_c affects the interatomic distances and hence the volumes (per atom or formula unit) of solids. Remarkably, these volumes can both increase and decrease with the decrease of particle size, as has been established for some metals and compounds (Table S7.1). The paradox is that a reduction of N_c must lead to a decrease of interatomic distances according to Goldschmidt's principle and hence to a decrease of volume, but at the same time it decreases the packing coefficient and hence increases the volume of the body; the net result depends on the balance of these two factors. Table 7.3 shows the differences of interatomic distances determined experimentally between the surface and the inner layers in crystal structures. The interplanar spacing between the atoms of the first and the second layer from the surface is almost always shorter than in the bulk, i.e. the Δd_{12} difference is negative, to compensate for the reduced coordination number of the surface atoms. However, the d_{12} spacings for the Be(0001) and Mg(0001) surfaces and the 'dimer' M–M bonds in elemental Zn, Cd and Hg have been reported to expand [26, 27]. The reason for increasing interatomic distances in these cases consists in the van der Waals nature of the M–M bonds as can be seen from comparing the bond lengths in solid metals and dimer molecules, Feibelman [28] (Table 7.4). The corresponding difference for the second/third layer separation (Δd_{23}) can be both positive and negative.

Table 7.3 Bond length relaxation in the surface layers of crystals at room temperature

Crystal (face)	Δd_{12} (%)	$+\Delta d_{23}$ (%)	Refs
Cu (211)	– 15	– 11	[9]
(311)	– 11.9	1.8	[10]
(320)	– 24	– 16	[11]
Ag (110)	– 8.0	3.2	[12]
Al (331)	– 11.7	4.1	[13]
(110)	– 16.0	3.4	[14]
Pb (001)	– 8.0	3.1	[15]
(111)	– 3.5	0.5	[16]
Ti (0001)	– 4.9	1.4	[17]
W (110)	– 3.0	0.2	[17]
Fe (310)	– 12.0		[18]
Co (1010)	– 12.8	0.8	[19]
Ni	– 12.0		[20]
Rh (001)	– 1.4	– 0.6	[17]
Pt	– 23	– 12	[21]
NaF (100)	– 1.3		[22]
NaCl	– 3.2		[23]
KI	– 1.6		[24]
Diamond	– 5.0		[25]

7.2 Amorphous Solids, Glasses

Increasing concentration of defects in a solid, leading ultimately to amorphization of the latter, can result not only from mechanical stress but also from chemical reactions. For example, when a chemical transformation occurs under a shock compression, because of its short duration the growth of crystalline grains of the products is often limited to the sizes of 1–5 nm and after unloading the material converts into an amorphous state [29]. Radiation amorphization in many respects is similar to the shock-wave effects [30]. In amorphous bodies thermal effect of fusion is small or non-existent, which characterizes it as a 'frozen melt' produced when the speed of cooling a melt exceeds the speed of its crystallization. Crystallization of ionic substances proceeds very quickly and, consequently, obtaining their amorphous phases is complicated. Only a few solid solutions with ionic components are known, but in fact their formation is a result of some additional factors disorganizing the structure. Thus, the amorphous state is stabilized in the solid solution KCl-CsCl both after a shock compression, and on cooling of a fused sample [31],

Table 7.4 Bond lengths (Å) in molecules and solid metals

M	Li	Na	K	Cu	Ag	Au	Al	Bi
$d(M_2)$	2.67	3.08	3.90	2.22	2.53	2.47	2.70	2.66
d(M)	3.02	3.66	4.52	2.56	2.89	2.88	2.86	3.07
M	Be	Mg	Ca	Sr	Zn	Cd	Hg	Mn
$d(M_2)$	2.46	3.89	4.27[a]	4.44[a]	4.19	3.76	3.63	3.40[a]
d(M)	2.22	3.20	3.95	4.30	2.66	2.98	3.01	2.73

[a] [7]

Table 7.5 Structures of amorphous solids and glasses

M	N_c^*	d, Å	M_nX_m	N_c^*	d, Å	M_nX_m	N_c^*	d, Å
B[a]	6.3	1.80	BeF_2[c]	3.9	1.553	$GeSe_2$[g]	4.2	2.36
C[b]	3.9	1.52	$ZnCl_2$	5.1	2.346	P_2O_5[h]	4	1.432
Si	3.8[α]	2.35	B_2O_3	2.9	1.375	As_2O_3[e]	3	1.775
Ge	3.8[α]	2.47	B_2S_3[d]	3.0	1.81	As_2S_3[i]	3	2.27
P	2.9	2.24	GaSb[α]	3.7		As_4S_4[i]	3	2.25
As	3[α]	2.49	SiO_2[c]	3.9	1.611	As_2Se_3[j]	3.0	2.42
Se	2.3[α]	2.347	GeO_2[f]	3.9	1.739	As_2Te_3[k]	2.1	2.70
			GeS_2[g]	4.1	2.21	Sb_2Se_3	2.6	2.58

[a][35], [α][36], [b][37], [c][4], [d][38, 39], [e][40], [f][41], [g][42], [h]for terminal bonds in corner-sharing tetrahedra; the length of the bridging bond is 1.581 Å [43], [i][44], [j][45], [k][46]

but this happens only because at high temperature this solid solution exists in the NaCl-type structure, but below 432 °C one of the components (CsCl) changes to a B2 structure uncharacteristic for KCl, thus hindering crystallization. Glasses with the compositions LiX-KX-CsX-BaX_2 (X = Cl, Br, I) and LiCl-KCl-CsCl also have been synthesized and found to have the following Li–X bond lengths and effective coordination numbers N_c^*(Li): $d(\text{Li}-\text{Cl}) = 2.37$ Å and $N_c^* = 4.4, d(\text{Li}-\text{Br}) = 2.49$ Å and $N_c^* = 3.8, d(\text{Li}-\text{I}) = 2.70$ Å and $N_c^* = 3.9$ [32]. So, in these systems the disorganizing factor is the big difference in the ionic radii of Li^+ and Cs^+, preventing the formation of regular solid solutions according to Goldschmidt's rules, and also, possibly, the hygroscopic nature of LiX and BaX_2. On cooling of the NH_4I-KI solid solutions, the glass state forms due to the dynamical disorder of NH_4^+ ions [33].

Experimental studies of amorphous bodies have generally involved XRD, ND and spectroscopic techniques such as EXAFS and XANES, and also Raman spectroscopy, but theoretical studies of the structure of these materials have generally been carried out by classical or *ab initio* molecular dynamics calculations. In both approaches, as a result, the nearest-neighbors separations found for systems in the glass state are generally comparable to those observed for the corresponding crystals, but unlike the latter, glasses posses no long-range periodicity. Even the short-range order (coordination polyhedron) is only approximately constant, and is characterized by the averaged, effective coordination number N_c^*. Note that N_c^* refers to atoms in the bulk, it should not be confused with the mean coordination number $\overline{N}_c$ (see above) which takes into account the different coordination of the surface and interior atoms in a particle.

Structural change ('polyamorphism') can occur in glasses under high pressures. Due to the technical difficulties associated with performing *in situ* diffraction experiments, pressure effects are usually studied on glasses after the pressure has been released, in which lasting increase of density is often observed. However, this densification does not always lead to a change of coordination. For example, upon static pressure experiments the six-coordinated Ge was not found in the unloaded GeO_2 glass, but *in situ* measurements by both ND (up to 5 GPa) and XRD (up to 15 GPa) detected changes in the oxygen correlations. Shortening of d(Ge–O) is observed below 6 GPa while it increases again at higher pressure, because GeO_4 tetrahedra are rearranged into GeO_6 octahedra. However, the structure of the high-pressure glass, which is based on edge- and corner-sharing octahedra, is not retained after decompression [34]. Results of structural studies of the glasses consisting of individual substances are summarized in Table 7.5.

It is interesting that structural, mechanical and electronic properties in chalcogenide glasses (As-Se, Ge-S, Ge-Se, Ge-As-S) change monotonically with composition, but at $N_c^* = 2.67$ a sharp change in properties occurs, indicating a phase transition from two- to three-dimensional lattice [47, 48]. This phenomenon can be described as 'amorphous morphotropy'.

7.3 Structure of Melts

Let us now consider the formation and structure of melts. If the crystal contains bonds of unequal strength, on heating the weakest ones will be disrupted first, thus near the fusion temperature the system will have a structure intermediate between the crystal and the liquid states, containing fragments of a crystal structure 10–100 Å in size. Partial fusion of a crystal was observed in AgI which at 146 °C undergoes a transition with the cation delocalization, which begins freely to move in the interatomic voids. As a result, the conductivity of the high-temperature *solid* phase of AgI approaches those of molten ionic salts. Thus, at 146 °C the cationic sublattice melts, whereas the anionic sublattice holds out until 555 °C, the point of physical fusion of the crystal. The radial distribution of electron density in this substance has been studied [49]. Due to chaotic distribution of Ag^+, the XRD picture is a halo and the harmonic analysis gives $N_c^* = 5$, and d(Ag–Ag) = 2.75 Å. The behavior of AgBr was investigated in details on heating up to the melting temperature [50]. The lattice parameter expands from 5.82 to 5.94 Å on heating from 190 to 400 °C, the concentration of free Ag^+ ions increases from 1.8 up to 5.0 % and that of fixed Ag^+ ions decreases accordingly. Other examples of positional disordering of atoms on heating have been reviewed by Ubbelode [51].

Interesting transformations occur on heating with substances which contain anisotropic ions. Thus, heating of $NaNO_3$ from 250 to 275 °C produces a phase transition from the $CaCO_3$ into the NaCl-type structure, as the NO_3 ions begin to rotate freely and acquire quasi-spherical shape. Similarly, at ambient conditions in NH_4Cl, NH_4Br and NH_4I the orientation of the cation is fixed by ion-dipole interactions or hydrogen bonds, but at 184.3, 137.8 and 17.6 °C, respectively, these solids undergo phase transitions with the onset of the cation rotation. NH_4NO_3 shows as many as five polymorphic transitions at the following temperatures (in °C):

$$\text{hexagonal} \xrightarrow{-18} \text{orthorhombic} \xrightarrow{+32} \text{orthorhombic} \xrightarrow{84.2} \text{tetragonal} \xrightarrow{125}$$
$$\text{cubic} \xrightarrow{169.6} \text{liquid}$$

In the highest-temperature (cubic) polymorph, both cations and anions statistically 'rotate' and have quasi-spherical symmetry. Increases of the crystal symmetry before fusion can be seen as a preparation of anisotropic crystals to the transition into isotropic liquid state.

Melts usually have lower densities than the corresponding crystals; e.g. MCl_n ($n = 1, 2, 3$) at fusion increase their volumes by *ca.* 20 % [52]. A few substances, particularly Sb, Ga, Bi and H_2O, reduce volume on melting, by 1, 3.2, 3.4 and 8.3 %, respectively. The structures of melted elements, according to [53–64] are presented in Table 7.6 which lists the ligancy and shortest interatomic distances in the crystal and the corresponding liquid, and the temperature of the experiment. At other temperatures, structural characteristics can differ: for example, the melted silicon at 1400 K has $d = 2.438$ Å and $N_c^* = 5.6$, at 1893 K, $d = 2.445$ Å and $N_c^* = 6.2$ [65, 66]; for tellurium at 956 K, $d = 2.82$ Å and $N_c^* = 2.4$, at 1276 K, $d = 2.91$ Å and $N_c^* = 3.1$ [67]. In the liquid silicon (by 50 K above the T_m) under $P = 4$–23 GPa, the Si–O distance remains equal to 2.44 ± 0.02 Å, whereas N_c^* increases from 6.8 to 9.2 [68].

It is necessary to notice that determining N_c^* is the most difficult problem for the XRD study of melts since it depends on a number of assumptions and even conventions, hence the reported values vary widely. For example, in the case of mercury, N_c^* derived from the area of the peak corresponding to the interatomic distance, is 14.6, whereas other approaches can give half this value. Cahoon [69] proposed to estimate N_c^* in liquids from the geometrical relations between N_c^* and the packing density of atoms,

$$\rho = \pi d^3/6V_a$$

using the values of d from XRD of liquids and V_a from density measurements. The results for liquid elements are presented in Table 7.6, based on the densities and interatomic distances measured at the temperatures of fusion.

As we can see, there is a qualitative agreement between the traditional definition of N_c^* from the radial distribution curves and the new 'volumetric' values, but in most cases the latter correspond better to structural changes in solids at their transitions to the liquid state. This table shows also that the coordination of atoms in the structures of metals at fusion, as a rule, varies slightly. Exceptions are the elements which have molecular or *heterodesmic* structures (Ga, Sb, Bi) in which on fusion the directed bonds are broken, packing becomes more dense and N_c increases, that leads to an increase of the solid's density at fusion.

The problem of dense packing of rigid spheres of the same or different sizes has been considered by many authors both theoretically and experimentally, the latter by studying the behavior of suspensions with different concentrations of a solid phase; the details can be found in a review [70]. It has been established that the density of packing of spheres in a liquid is 0.494, in a mixture of a liquid and a crystal 0.545, while a dense packing of *randomly* distributed isometric particles corresponds to $\rho = 0.64$. Hence $N_c \approx 6$ can be regarded as a threshold value for the structure of a melt: if a crystal has $N_c > 6$, it will decrease on fusion, if $N_c < 6$, it will increase. According to this criterion, N_c of Si and Ge at fusion will increase. The structure of nonmetals does not change immediately on fusion, but on further heating N_c increases and metallic conductivity appears [71].

In liquid metals under pressure, as in solids, the coordination numbers tend to increase. Thus, in melted iron $N_c^* = 10.8$ at normal pressure and $T = 1830$ K,

Table 7.6 Comparison of structures of elements in the solid and liquid states

Element	Solid state		Liquid state				
			XRD			$N_c^* = f(d, V)$	
	N_c	d, Å	T, °C	N_c^*	d, Å	N_c^*	d, Å
Li	8	3.04	197	13.0	2.99	7.2	3.00
Na	8	3.72	100	8.5	3.82	7.3	3.68
K	8	4.54	70	8.5	4.65	7.3	4.56
Rb	8	4.95	40	9.5	4.97	7.7	4.97
Cs	8	5.32	100	10.9	5.33	7.6	5.31
Cu	12	2.56	1090	10.1	2.55	7.1	2.50
Ag	12	2.89	1054	11.5	2.85	7.0	2.82
Au	12	2.88	1100	9–11	2.85	7.0	2.80
Mg	12	3.20	675	10.4	3.20	7.0	3.10
Ca	12	3.95		11.1	3.83		
Sr	12	4.30		11.1	4.23		
Ba	8	4.35		10.8	4.31		
Zn	6 + 6	2.66 + 2.91	425	9.9	2.74	6.8	2.66
Cd	6 + 6	2.98 + 3.29	325	8.3	3.06	6.9	3.00
Hg	6 + 6	3.00 + 3.47	20	8.7	3.07	6.6	3.00
B	6.5	1.80	2600	5.8	1.76		
Al	12	2.86	670	9.9	2.86	6.8	2.78
Ga	1 + 6	2.44 + 2.75	20	9.8	2.92	6.8	2.78
In	4 + 8	3.25 + 3.37	170	10.5	3.30	6.8	3.14
Tl	6 + 6	3.41 + 3.46	350	11.5	3.30	7.1	3.30
Sc	6 + 6	3.26 + 3.31		10.3	2.92		
La	12	3.74		11.1	3.87		
Si	4	2.35	1320	5.6	2.412	4.7	2.40
Ge	4	2.45	940	5.7	2.75	5.0	2.60
Sn	6	3.075	240	8.2	3.20	6.7	3.16
Pb	12	3.50	330	12	3.40	7.3	3.40
Ti	6 + 6	2.90 + 2.95		10.9	3.17		
Zr	6 + 6	3.18 + 3.23	1860	12	3.12		
P	3	2.22		3	2.22		
Sb	3	2.91	640	6.8–9.4	2.85	6.7	3.26
Bi	3	3.07	400	8	3.40	6.4	3.34
V	8	2.62		11.0	2.82		
S	2	2.05	130	1.7–2.4	2.07		
Se	2	2.32	230	2.5	2.35	2.4	2.36
Te	2	2.86	450	2.4	2.95		
Cr	8	2.50	1900	10.9	2.58	7.6	2.58
Mn	12	2.73		10.9	2.67		
Fe	8	2.48	1870	12.3	2.55	7.7	2.56
Co	12	2.52	1800	12.1	2.51	7.2	2.48
Ni	12	2.49	1905	11.2	2.48	7.2	2.46
Rh	12	2.69	~2000	~12	3.3		
Pd	12	2.75		10.9	2.71		
Pt	12	2.775	1800	11.2	2.73	7.5	2.73
He	12	3.644	4.24	8.1	3.72		
Ne	12	3.156	28	8.8	3.17		
Ar	12	3.755	83.8	12.5	3.71		
Kr	12	3.992	125	8.5	4.07		
Xe	12	4.336	170	8.9	4.40		

but increases to 11.4 at $T = 2100$ K and $P = 5$ GPa, while the shortest interatomic distances remain the same (2.58 Å) [72]. In the structure of melted indium, N_c^* increases from 9.8 to 12 as the pressure increases from ambient to 7 GPa [73]. ND and XRD study of the melted elements, carried out by Takeda et al. [74–78], yielded the information on the electronic structure of atoms. The electron density maps reveal the following nearest minima and maxima corresponding to the borders of the atomic cores and to centroids of the valence electron clouds (Å):

	Mg	Zn	Ga	Tl	Sn	Pb	Bi	Te
r_{min}	0.75	0.74	0.70	0.80	0.72	0.80	0.7	0.8
r_{max}	1.10	1.10	1.18	1.10	0.96	1.08	0.94	1.7

Evidently, the minimum sizes of atoms correspond to the cation radii, i.e. the structure of a liquid metal really corresponds to a system of cations surrounded by electron clouds. In the case of tellurium the first maximum corresponds to the isolated electronic pair, and the second one to bonded electrons. Tao [79] showed that the atomic diameter in liquid melts near their melting points is close to the covalent diameter multiplied by 0.92.

Similarly to crystals (see Table 7.3), the free surfaces of liquid metals and alloys have been proven experimentally to be layered. Thus, in liquid Ga the contraction in the surface layer is $\Delta d_{12} = 10\,\%$ [80], in In $\Delta d_{12} = 14\,\%$ [81], in liquid Sn $\Delta d_{12} = 9\,\%$ [82]. Remarkably, in liquid Hg (in accordance with the Hg crystal structure) the surface layer is expanded by 10 % because of the van der Waals interaction of the neighboring atoms [83]. Authors of [84] considered the role of surface tension in the surface layering of liquids and noted that according to recent theoretical works, at low enough temperatures the free surfaces of *all* liquids should become layered, unless preempted by bulk freezing.

Besides elements, detailed studies have been carried out on melted halides, oxides and chalcogenides of metals. Table S7.2 presents structural data on the MX halides, and in Table S7.3 halides, chalcogenides and pnictides of multivalent elements are presented.

At fusion of MX salts, N_c decreases and the volume increases. The possible reasons are (i) greater concentration of vacancies, (ii) formation of tetrahedral complexes $[MX_4]^{3-}$ which are statistically mixed with free cations; (iii) a chaotic mixing of crystallites (15–100 atoms) inside which ions are strongly ordered. As shown in Table S7.3, in structures of many melts coordination numbers of cations are close to 4 and only for very large cations $N_c \geq 6$, i.e. both the coordination and bond lengths of atoms decrease in comparison with the solid state. In liquid complex compounds, as well as in corresponding crystals, the mutual compensation of the bond lengths of the central atom and the external cations was observed. For example, on the $ZnCl_2 \rightarrow Rb_2ZnCl_4$ transition, d(Zn–Cl) decreases from 2.31 to 2.28 Å, d(Rb–Cl) lengthens from 3.28 in RbCl to 3.41 Å in Rb_2ZnCl_4 [85]. The effect of N_c(Al) on the Al–X distance in the MX-AlX_3 melts has been studied [86], viz.

Complex	$[AlF_4]^-$	$[AlF_5]^{2-}$	$[AlF_6]^{3-}$	$[AlCl_4]^-$	$[AlCl_6]^{3-}$
Structure	Tetrahedron	Bipyramid	Octahedron	Tetrahedron	Octahedron
d(Al–X)	1.71	1.78, 1.81	1.90	2.15	2.39 Å

The speed of fusion in solids (under action of laser radiation) was determined by superfast XRD [87]. In crystalline Ge and InSb, the long-range order is lost within the time-span of *ca.* 10^{-13} s [88, 89], close to the duration of phase transitions in solids under shock compression.

7.4 Structure of Aqueous Solutions

The most important subjects of structural chemistry of the liquid state are water and water solutions of salts; for the early history and the fundamental results of these studies see [90]. The central concept of these works, as formulated in 1933 by Bernal and Fowler [91], was that water molecules are hydrogen-bonded into a continuous 3D network with $N_c = 4$, donating two and accepting two hydrogen bonds each. However, the continuous network model failed to explain the well-known anomalies of water (non-monotonic temperature dependences of volume and compressibility, and pressure dependences of viscosity). To explain water anomalies, Bernal and Fowler had to introduce two states in the model: they assumed that a dense (quartz-like) and less dense (tridymite-like) structures coexist in water. According to an alternative concept, only in ice $N_c = 4$, but in liquid water $N_c = 4.4$ because on fusion of ice, some molecules 'drop' into intermolecular voids and thus increase the density. In general, weakening or breaking of H-bonds on heating raises the coordination number and density of water. Accordingly, lowering temperature increases the average distance between molecules of water [92].

Other models and a brief summary of historical background are considered in [93] together with a review of recent studies on the structure and dynamics of water (in liquid and amorphous states) by physical methods and computer experiments. Basic concepts offered for describing the structure and dynamics of water are considered. The structure of water is currently viewed as having a uniform 3D network formed by hydrogen-bonded molecules. This network is dissimilar from that found in any crystal structure; it is dynamically and structurally inhomogeneous. The network is rather labile; water molecules continuously change their neighbors, so that the average lifetime of an H-bond is only a few picoseconds. The combined experimental and theoretical investigation of water under extreme conditions (T from 300 to 1500 K and P to 56 GPa) suggests a super-ionic phase above 47 GPa. This super-ionic phase (termed 'dynamically ionized') consists of very short-lived ($< 10^{-14}$ s) H_2O, H_3O^+, and OH^- species; the mobility of oxygen ions decreases abruptly with pressure, while H^+ ions remain very mobile [94].

Solutions of salts in water represent typical insertion structures where ions occupy intermolecular voids, being oriented by the positive or negative charges to the corresponding poles in a dipole of the H_2O molecule. Samoylov [90], having analyzed the XRD data and using an original thermochemical method to estimate the coordination of ions in water solutions, has found, that in diluted solutions the ions replace a molecule of water each and thus acquire $N_c \approx 4$. In concentrated solutions, ions coordinate water molecules around them; the structures resemble those of crystalline

Table 7.7 Bond lengths d (Å) and coordination numbers (N_c) of ions in water solutions and crystallohydrates (from [97, 106–109] and CSD)

Ion	Solution		Crystal		Ion	Solution		Crystal	
	d(M–O)	N_c^*	d(M–O)	N_c		d(A–O)	N_c^*	d(A–O)	N_c
Li^+	1.95	4	1.91	4	Ni^{2+}	2.055	6	2.06	6
Li^+	2.10	6			Pd^{2+}	2.01	4		
Na^+	2.43	6	2.44	6	Pt^{2+}	2.01	4		
K^+	2.81	7	2.87	6	Al^{3+}	1.89	6	1.88	6
Rb^+	2.98	8	3.05		Ga^{3+}	1.96	6	1.95	6
Cs^+	3.07	8	3.2	9	In^{3+}	2.14	6	2.13	6
Cu^+	2.14	4			Tl^{3+}	2.22	6	2.24	6
Ag^+	2.32	4			Sc^{3+}	2.17	8	2.09	6
H_3O^+	2.74	4			Y^{3+}	2.36	8	2.44	9
NH_4^+	2.94	4	3.19	8	La^{3+}	2.52	9	2.54	9
Be^{2+}	1.615	6	1.62	4	Ti^{3+}	2.03	6	2.03	6
Mg^{2+}	2.10	6	2.07	4	V^{3+}	1.99	6	1.99	6
Ca^{2+}	2.46	8	2.40	6	Bi^{3+}	2.41	8		
Sr^{2+}	2.63	8	2.62	8	Cr^{3+}	1.98	6	1.91	6
Ba^{2+}	2.82	8	2.85	8	Fe^{3+}	2.02	6	2.00	6
Cu^{2+}	1.96	6	1.97	6	Co^{3+}	1.87	6	1.98	6
Zn^{2+}	2.08	6	2.12	6	Rh^{3+}	2.04	6		
Cd^{2+}	2.30	6	2.29	6	Ce^{4+}	2.41	8	2.49	6
Hg^{2+}	2.34	6			Zr^{4+}	2.19	8		
Sn^{2+}	2.29	3			Hf^{4+}	2.16	8		
Pb^{2+}	2.54	6			Th^{4+}	2.45	9	2.45	9
V^{2+}	2.14	6	2.12	6	U^{4+}	2.42	9	2.54	9
Cr^{2+}	1.99	6			F^-	2.66	5	2.76	
Mn^{2+}	2.20	6	2.16	6	Cl^-	3.17	6	3.27	
Fe^{2+}	2.12	6	2.13	6	Br^-	3.32	6	3.46	
Co^{2+}	2.08	6	2.09	6	I^-	3.55	6	3.62	

hydrates and the shortest distances between the ions Li^+, Na^+, K^+, OH^-, or Cl^- and the nearest molecules of water coincide, within 5 %, with the sum of the ionic radius and the van der Waals (= anionic) radius of O (1.40 Å). Further XRD studies of the structure of water solutions confirmed this conclusion [95–103]. The averaged values of d(ion–oxygen) and N_c of ions in water are presented in Table 7.7 according to [97, 102, 103]. Metal–halogen distances and corresponding coordination numbers in water solutions of salts are presented in Table S7.4.

It is generally assumed that the interactions between water and ions affect the long range ordering of the hydrogen bonding network in the bulk liquid, for example, by breaking or forming hydrogen bonds. The magnitude of the effect follows the Hofmeister series

Strongly hydrated anions	Weakly hydrated anions
$SO_4^{2-} > HPO_4^{2-} > F^- > Cl^- > Br^- > I^- > NO_3^- > ClO_4^-$	
$NH_4^+ > Cs^+ > Rb^+ > K^+ > Na^+ > Ca^{2+} > Mg^{2+} > Al^{3+}$	
Weakly hydrated cations	Strongly hydrated cations

Recent research of structures of water solutions of alkali halides has revealed a new feature [100]. In aqueous solutions of NaI (6 M) and CsI (3 M), the Cs^+–O distance is 3.00 Å and the hydration number of Cs^+ is 7.9, similar to those of other Group 1 ions at similar concentrations. The nearest-neighbor Cs^+–I^- distance is 3.84 Å and the upper boundary for the number of ion pairs formed is 2.7. The average I^-–O distance (in NaI and CsI solutions) is 3.79 Å and the hydration number of I^- is 8.8, consistent with the typical values for a halide ion. The closest I^-–Na^+ distance is 3.17 Å and the number of ion pairs measured is 1.6 for a 6 M solution. I^-–H, I^-–O and I^-–Cs^+ distances are 3.00, 3.82 and 3.82 Å, respectively. The presence of M^+–I^- bonds in the diluted solutions was explained by the high polarizability of I^-; higher number of M^+–I^- bonds for CsI compared to NaI confirms the importance of polarizing interaction of ions. XAS and XRS studies of the structure of water in the bulk and the first solvation sphere of ions [104] showed that only the water molecules in the first solvation sphere of ions are, in terms of the hydrogen bond configurations, affected by the presence of the ions. Water molecules in the first solvation sphere of strongly hydrated ions have fewer broken hydrogen bonds compared to liquid water, whereas those in the first solvation sphere of weakly hydrated ions have more. Thus, hydrated ions hardly change the local hydrogen bond configuration in bulk liquid water, but in 1 M solutions the average number of hydrogen bonds in these solutions is close to that of pure water.

Comparison of d(M–O) in concentrated water solutions and in crystalline hydrates of the same cations shows that these distances are practically identical (see Table 7.7). Drakin proved [105] that they are close to d(M–O) in oxides with the same coordination numbers. Hence, there are no geometrical distinctions between the normal and donor-acceptor bonds. Freedman and Lewis [106] developed this approach and considered, in particular, two ways in which water molecules can be oriented by cations: either by the negative end of the dipole or by a lone electron pair, the former producing planar-trigonal and the latter tetrahedral geometry at the oxygen atom. XRD study of crystalline hydrates has shown that molecules of water form tetragonal arrangement when the central cation of an aqua-complex has low charge (+1 or +2) or large size (e.g. lanthanide), whereas for small multi-charged cations, as a rule, the trigonal configuration was observed.

Structural studies of binary and complex compounds in organic solvents [110–113] revealed the d(M–O) and d(M–X) bond lengths very close to those in water solutions, although there are certain differences. Thus, according to EXAFS studies in tetramethylurea solutions the coordination polyhedra of Mn^{2+} and Ni^{2+} are square pyramids, those of Cu^{2+} and Co^{2+} are distorted tetrahedra, of Zn^{2+} a tetrahedron, of Cd^{2+} and In^{3+} octahedra, of Fe^{2+} a square-pyramid or a trigonal bipyramid [110], whereas in water solutions all these ions have octahedral coordination. The reason of such distinction is clear: organic solvents have lower dielectric constants than water, and therefore the dissociation of salts is weaker. Besides, organic molecules can form various complexes with the basic substance: $GaCl_3$ is a monomer in the mesitilene solution, forms a η^6-complex with arene, but in benzene $GaCl_3$ exists in the form of $Cl_2Ga(\mu\text{-}Cl)_2GaCl_2$ [111].

Solutions of metals in ammonia combine the properties of metals and liquids. Structures of pure ammonia and solutions of lithium in it were studied repeatedly; NH_3 molecules in a pure liquid have $N_c^* \approx 12$ and $d(N\cdots N) = 3.48$ Å, $N_c^*(H) = 7.5$

and $d(\mathrm{H}\cdots\mathrm{H})$ 2.9 Å [114]. Dissolution of lithium in ammonia (up to 21 mol %) reduces both $d(\mathrm{N}\cdots\mathrm{N})$ and $N_c(\mathrm{N})$, to 3.42 Å and 6, respectively, while $N_c(\mathrm{H})$ drops to 3. Whereas in pure ammonia each N atom forms two H-bonds, in a 8 % lithium solution the average number drops to 0.7 and in a 21 % solution to zero, i.e. all NH_3 molecules are involved in the $Li\cdot(NH_3)_4$ complex. The shortening of the distances between molecules of ammonia in solutions is caused by formation of such complexes with $d(\mathrm{Li–N}) = 2.01$ Å [115]. XRD study of the crystalline $Li(NH_3)_4X$ revealed the average Li–N distance of 2.10 Å [116]. An ND study of solutions of Li and K in ammonia [117] determined the positions of hydrogen atoms and their coordination to H and N: in pure ammonia the mean number of N–H bonds is 3.2, with $d(\mathrm{N–H}) = 1.01$ Å and $d(\mathrm{H}\cdots\mathrm{H}) = 1.62$ Å. Coordination numbers of lithium and potassium in these solutions are equal to 4 and 6, respectively, $d(\mathrm{K}\cdots\mathrm{N}) = 2.85$ Å. A solution of calcium in liquid ammonia shows $d(\mathrm{Ca–N}) = 2.45$ Å, while $N_c^*(\mathrm{Ca})$ changes from 6.5 in a 10 % solution to 7.1 in a 4 % one. In the crystalline complex $Ca(NH_3)_6$, $d(\mathrm{Ca–N}) = 2.56$ Å [118]. Thus, the short range orders in structures of solutions of salts in water and organic solvents, and metals in ammonia are close to those in crystalline compounds of similar composition.

In conclusion, we return to the structure of liquid water. A detailed history of the structural description of water is given in [93, 119], beginning from 1892 when Röntgen proposed that water contains a mixture of two components, one ice-like and the other unknown [120]. Tokushima et al. [119] presented definite experimental observations based on the XES split lone-pair state, which indicate the existence of two distinct structural motifs in liquid water, viz. the tetrahedral and the strongly distorted hydrogen-bonded species, in a 2:1 ratio. This assignment is based on comparison with ice and gas-phase spectra, temperature dependent measurements, excitation energy dependence and theoretical simulations.

Appendix

Supplementary Tables

Table S7.1 Particle sizes (upper lines, nm) and unit cell volumes per atom or formula unit (in Å^3) for macro (V_1) and nano (V_2) crystals

Substance	V_1	V_2	ΔV (%)^a	Ref.	Substance	V_1	V_2	ΔV (%)^a	Ref.
Ag	macro 68.23	10 67.97	−0.4	7.1	Se	70 81.8	13 82.3	0.6	7.9
Au	macro 67.86	30 67.72	−0.2	7.1	c-CdS	50 186.8	5 198.5	6.3	7.10
Sn	31.8 108.2	9.2 107.8	−0.4	7.2	W_2N	macro 70.2	40 71.6	2.0	7.11
Bi	33.2 212.2	8.9 210.9	−0.6	7.2	GaN	macro 45.65	3.2 46.22	1.2	7.12
Pd	macro 58.8	1.4 53.6	−8.9	7.3	NiO	macro 72.56	nano 73.45	1.2	7.13
Pt	macro 60.1	3.7 59.2	−1.5	7.4	Y_2O_3	100 1192	2.6 1215	1.9	7.14
LiF	9 64.8	3 63.1	−2.6	7.5	c-In_2O_3	macro 64.71	6 64.85	0.2	7.60
NaCl	15 178.5	4.8 176.9	−0.9	7.5	TiO_2 rutile	24 62.5	4 63.1	1.0	7.15
NaBr	15 211.9	3.5 206.7	−2.4	7.5	ZrO_2	41 133.3	7 134.4	0.8	7.16
KCl	15 247.3	2 242.0	−2.1	7.5	HfO_2	45 41.2	5 32.0	2.5	7.17
Al_2O_3	67 493	6 482	−2.2	7.6	$BaTiO_3$	macro 64.3	15 64.9	0.9	7.18
β-Ga_2O_3	macro 209.4	14 207.0	−1.1	7.7	ReO_3	macro 48.6 50.2	12 50.0 [b] 51.0 [c]	 2.9 1.6	7.19
h-In_2O_3	macro 61.92	6 61.49	−0.7	7.60					
CdSe	macro 129.5	3 128.7	−0.6	7.8					

a) $\Delta V = \frac{V_2 - V_1}{V_1}$, b) rhombohedral phase, c) monoclinic phase

Table S7.2 Structures of liquid MX halides [7.20–7.28]

M	F		Cl		Br		I	
	d, Å	N_c	*d*, Å	N_c	*d*, Å	N_c	*d*, Å	N_c
Li	1.95	3.7	2.45	3.8	2.64	4.4	2.82	4.6
Na	2.30	4.1	2.77	3.7	3.05	3.5	3.15	4.0
K	2.65	4.9	3.15	4.0	3.32	3.8	3.52	4.0
Rb			3.28	4.8	3.4	4.1	3.65	4.6
Cs			3.48	4.5	3.66	4.6	3.85	4.5
Cu			2.23	3.0	2.45	3.0	2.53	3.2
Ag			2.62		2.73	4.4	2.90	4.5

Table S7.3 Structures of the liquid halides, oxides, and chalcogenides of the M_nX_m type

MX_2	*d*, Å	N_c	$MX_{3,4}$	d, Å	N_c	M_mX_n	d, Å	N_c
$MgCl_2$[a]	2.42	4.3	$FeCl_3$[k]	2.23	3.8	ZnTe[u]	2.69	4.3
$CaCl_2$[a]	2.78	5.3	UCl_3[l]	2.84	6	CdTe[u]	2.81	3.7
$SrCl_2$[a]	2.90	5.1	CCl_4[k]	1.77	4	HgTe[u]	2.93	6.3
$BaCl_2$[a]	3.10	6.4	$CClF_3$[m]	1.75	4	GeS[v]	2.41	3.7
$ZnCl_2$[a]	2.27	4.3	CF_3Cl[m]	1.33	4	GeSe[v]	2.49	3.7
$ZnBr_2$[b]	2.47	4.0	CBr_4[n]	1.93	4	GeTe[v]	2.72	5.1
ZnI_2[b]	2.63	4.2	$CBrF_3$[m]	1.90	4	SnS[v]	2.54	3.2
$CdCl_2$[c]	2.42	3.9	CF_3Br[m]	1.33	4	SnSe[v]	2.75	2.7
$MnCl_2$[d]	2.50	4.0	$SiCl_4$[o]	2.01	4	SnTe[v]	3.03	5.3
$NiCl_2$[e]	2.31	4.3	$GeCl_4$[o]	2.11	4	NiTe[w]	2.56	4
$NiBr_2$[e]	2.47	4.7	$GeBr_4$[p]	2.27	4	GaAs[x]	2.56	5.5
NiI_2[ae]	2.60	4.2	$SnCl_4$[o]	2.29	4	GaSb[x]	2.95	5.4
$AlCl_3$[f]	2.11	4	SnI_4[q]	2.67	8.3	B_2O_3[y]	1.37	5 3
$AlBr_3$[g]	2.29	4	$TiCl_4$[o]	2.17	4	Al_2O_3[z]	1.78	4.2
$GaBr_3$[g]	2.34	4	VCl_4[o]	2.14	4	Ga_2Te_3[α]	2.60	3.5
GaI_3[g]	2.35	3.75	$TeCl_4$[r]	2.36	3.9	In_2Te_3[a]	2.95	~3
$InCl_3$[h]	2.54	4	$TeBr_4$[r]	2.54	3.9	As_2Se_3[k]	2.43	2.6
$ScCl_3$[h]	2.48	4.8	Cu_2Se[a]	2.52	4	Sb_2Se_3[k]	2.72	2.7
ScI_3[h]	2.76	4.7	Cu_2Te[a]	2.53	3.1	Sb_2Te_3[k]	2.93	4.4
YCl_3[h]	2.67	5.7	Ag_2Se[s]	2.74	3.5	Bi_2Se_3[k]	2.83	4.1
$LaCl_3$[i]	2.93	8.2	Ag_2Te[s]	2.88	3.2	Bi_2Te_3[k]	3.18	5.0
$LaBr_3$[i]	3.01	7.4	Tl_2Se[t]	3.30	9	GeO_2[β]	1.74	4.2
LaI_3[i]	3.18	6.7	Tl_2Te[a]	3.41		GeS_2[γ]	2.21	4.1
PBr_3[j]	2.24	3	CuSe[a]	2.52		$GeSe_2$[γ]	2.36	4.2
$SbCl_3$[k]	2.35	3	CuTe[a]	2.55	2.9	V_2O_5[δ]	1.70	3.6

[a][7.29], [b][7.30], [c][7.31], [d][7.32], [e][7.33], [f][7.34], [g][7.35], [h][7.36], [i][7.37], [j][7.38], [k][7.39], [l][7.40], [m][7.41], [n][7.42], [o][7.43], [p][7.44], [q][7.45], [r][7.46], [s][7.47], [t][7.48], [u][7.49], [v][7.50], [w][7.51], [x][7.52, 7.53], [y][7.54], [z][7.55], [α][7.56], [β][7.57], [γ][7.58], [δ][7.59]

Table S7.4 Bond lengths (Å) of metal halides in aqueous solutions

M–X	d(M–X)	N_c(M)	M–X	d(M–X)	N_c(M)
Cu^{II} – Cl	2.26	4	Tl^{III} – Cl	2.43	4
Cu^{II} – Cl	2.43	6	Tl^{III} – Cl	2.59	6
Cu^{II} – Br	2.43	4.2	Cr^{III} – Cl	2.31	6
Ag – Cl	2.29	3.6	Mn^{II} – Cl	2.49	6
Ag – Br	2.43	3.9	Mn^{II} – Br	2.62	6
Zn – Cl	2.29	4	Fe^{III} – Cl	2.25	4
Zn – Br	2.40	3.9	Fe^{III} – Cl	2.33	6
Zn – I	2.61	4	Fe^{II} – Br	2.61	5.9
Cd – Cl	2.58	4	Co^{II} – Cl	2.41	6.1
Cd – I	2.80	4	Co^{II} – Br	2.58	6.1
Hg – Cl	2.47	4	Ni^{II} – Cl	2.44	6
Hg – Br	2.61	4	Ni^{II} – Br	2.58	6
Hg – I	2.78	4	Rh^{III} – Cl	2.33	6
In – Cl	2.52	6	Pt^{II} – Cl	2.31	4
Tl^{III} – Cl	2.37	2	Pt^{IV} – Cl	2.33	6
Tl^{III} – Cl	2.40	3	Pt^{IV} – Br	2.47	6

Supplementary References

7.1 Gu QF, Krauss G, Steurer W et al (2008) Phys Rev Lett 100:045502
7.2 Yu XF, Liu X, Zhang K, Hu ZQ (1999) J Phys Cond Matt 11:937
7.3 Lamber R, Wetjen S, Jaeger NI (1995) Phys Rev B 51:10968
7.4 Solliard C, Flueli M (1985) Surface Sci 156:487
7.5 Boswell FWC (1951) Proc Phys Soc London A 64:465
7.6 Chen B, Penwell D, Benedetti LR et al (2002) Phys Rev B 66:144101
7.7 Wang H, He Y, Chen W et al (2010) J Appl Phys 107:033520
7.8 Zhang J-Y, Wang X-Y, Xiao M et al (2002) Appl Phys Lett 81:2076
7.9 Zhao YH, Zhang K, Lu K (1997) Phys Rev B 56:14322
7.10 Kozhevnikova NS, Rempel AA, Hergert F, Mager A (2009) Thin Solid Films 517:2586
7.11 Ma Y, Cui Q, Shen L, He Zh (2007) J Appl Phys 102:013525
7.12 Lan YC, Chen XL, Xu YP et al (2000) Mater Res Bull 35:2325
7.13 Anspoks A, Kuzmin A, Kalinko A, Timoshenko J (2010) Solid State Commun 150:2270
7.14 Beck Ch, Ehses KH, Hempelmann R, Bruch Ch (2001) Scripta Mater 44:2127
7.15 Kuznetsov AY, Machado R, Gomes LS et al (2009) Appl Phys Lett 94:193117
7.16 Acuna LM, Lamas DG, Fuentes RO et al (2010) J Appl Cryst 43:227
7.17 Cisneros-Morales MC, Aita CR (2010) Appl Phys Lett 96:191904
7.18 Smith MB, Page K, Siegrist T et al (2008) J Am Chem Soc 130:6955
7.19 Biswas KS, Muthu DV, Sood AK et al (2007) J Phys Cond Matt 19:436214
7.20 Antonov BD (1976) Zh Struct Khim 17:46
7.21 Ohno H, Fuzukawa K, Takagi R et al (1983) J Chem Soc Faraday Trans II 79:463
7.22 Rovere M, Tosi MP (1986) Rep Progr Phys 49:1001
7.23 Li J-C, Titman J, Carr G et al (1989) Physica B 156–157:168
7.24 Shirakawa Y, Saito M, Tamaki S et al (1991) J Phys Soc Japan 60:2678
7.25 Takeda S, Inui M, Tamaki S et al (1994) J Phys Soc Japan 63:1794
7.26 Stolz M, Winter R, Howells WS (1994) J Phys Cond Matter 6:3619
7.27 Di Cicco A, Rosolen MJ, Marassi R et al (1996) J Phys Cond Matter 8:10779
7.28 Drewitt JWE, Salmon PS, Takeda S, Kawakita Y (2009) J Phys Cond Matt 21:755104
7.29 Enderby JE, Barnes AC (1990) Rep Progr Phys 53:85

7.30 Allen DA, Howe RA, Wood ND, Howells WS (1991) J Chem Phys 94:5071
7.31 Takagi Y, Itoh N, NakamuraT (1989) J Chem Soc Faraday Trans I 85:493
7.32 Biggin S, Gray M, Enderby JE (1984) J Phys C 17:977
7.33 Wood N D, Howe RA (1988) J Phys C 21:3177
7.34 Badyal YS, Allen DA, Howe RA (1994) J Phys Cond Matter 6:10193
7.35 Saboungi M-L, Howe MA, Price DL (1993) Mol Phys 79:847
7.36 Wasse JC, Salmon PS (1999) J Phys Cond Matter 11:2171
7.37 Wasse JC, Salmon PS (1999) J Phys Cond Matter 11:1381
7.38 Misawa M, Fukunaga T, Suzuki K (1990) J Chem Phys 92:5486
7.39 Batsanov SS (1986) Experimental foundations of structural chemistry. Standarty Moscow (in Russian)
7.40 Okamoto Y, Kobayashi F, Ogawa T (1998) J Alloys Comp 271–273:355
7.41 Mort KA, Johnson KA, Cooper DL et al (1998) J Chem Soc Faraday Trans 94:765
7.42 Bakó I, Dore JC, Huxley DW (1997) Chem Phys 216:119
7.43 Ensico E, Lombardero M, Dore JC (1986) Mol Phys 59:941
7.44 Ludwig KF, Warburton WK, Wilson L, Bienenstock AI (1987) J Chem Phys 87:604
7.45 Fuchizaki K, Kohara S, Ohishi Y, Hamaya N (2007) J Chem Phys 127:064504
7.46 Le Coq D, Bytchkov A, Honkimäki V, Bytchkov E (2008) J Non-Cryst Solids 354:259
7.47 Price DL, Saboungi M-L, Susman S et al (1993) J Phys Cond Matter 5:3087
7.48 Barnes AC, Guo C (1994) J Phys Cond Matter 6:A229
7.49 Gaspard J-P, Raty J-Y, Céolin R, Bellissent R (1996) J Non-Cryst Solids 205–207:75
7.50 Raty J-Y, Gaspard J-P, Bionducci N et al (1999) J Non-Cryst Solids 250–252:277
7.51 Nguyen VT, Gay M, Enderby JE et al (1982) J Phys C 15:4627
7.52 Hattori T, Taga N, Takasugi Y et al (2002) J Non-Cryst Solids 312–314:26
7.53 Hattori T, Tsuji K, Taga N et al (2003) Phys Rev B 68:224106
7.54 Misawa M (1990) J Non-Cryst Solids 122:33
7.55 Landron C, Hennet L, Jenkins TE et al (2001) Phys Rev Lett 86:4839
7.56 Buchanan P, Barnes AC, Whittle KR et al (2001) Mol Phys 99:767
7.57 Kamiya K, Yoko T, Itoh Y, Sakka S (1986) J Non-Cryst Solids 79:285
7.58 Susman S, Volin KJ, Montague DG, Price DL (1990) J Non-Cryst Solids 125:168
7.59 Takeda S, Inui M, Kawakita Y et al (1995) Physica B 213–214:499
7.60 Qi J, Liu JF, He Y et al (2011) J Appl Phys 109:063520

References

1. Roy R (1970) Classification of non-crystalline solids. J Non-Cryst Solids 3:33–40
2. Stace AJ (2002) Metal ion solvation in the gas phase: the quest for higher oxidation states. J Phys Chem A 106:7993–6005
3. Batsanov SS, Bokarev VP (1980) The limit of crushing of inorganic substances. Inorg Mater 16:1131–1133
4. Palosz B, Grzanka E, Gierlotka S et al (2002) Analysis of short and long range atomic order in nanocrystalline diamonds with application of powder diffractometry. Z Krist 217:497–509
5. Rempel A, Magerl A (2010) Non-periodicity in nanoparticles with close-packed structures. Acta Cryst A 66:479–483
6. Bokarev VP (1986) Geometric estimate of the atomic coordination numbers in defect crystals of inorganic substances. Inorg Mater 22:306–307
7. Batsanov SS (2008) Experimental foundations of structural chemistry. Moscow Univ Press, Moscow
8. Pirkkalainen K, Serimaa R (2009) Non-periodicity in nanoparticles with close-packed structures. J Appl Cryst 42:442–447
9. Seyller Th, Diehl RD, Jona F (1999) Low-energy electron diffraction study of the multilayer relaxation of Cu(211). J Vacuum Sci Techn A 17:1635–1638

10. Parkin SR, Watson PR, McFarlane RA, Mitchell KAR (1991) A revised LEED determination of the relaxations present at the (311) surface of copper. Solid State Commun 78:841–843
11. Tian Y, Quinn J, Lin K-W, Jona F (2000) Cu{320} Structure of stepped surfaces. Phys Rev B 61:4904–4909
12. Nascimento VB, Soares EA, de Carvalho VE et al (2003) Thermal expansion of the Ag(110) surface studied by low-energy electron diffraction and density-functional theory. Phys Rev B 68:245408
13. Davis HL, Hannon JB, Ray KB, Plummer FW (1992) Anomalous interplanar expansion at the (0001) surface of Be. Phys Rev Lett 68:2632–2635
14. Li YS, Quinn J, Jonna F, Marcus PM (1989) Low-energy electron diffraction study of multilayer relaxation on a Pb{110} surface. Phys Rev B 40:8239–8244
15. Lin RF, Li YS, Jonna F, Marcus PM (1990) Low-energy electron diffraction study of multilayer relaxation on a Pb{001} surface. Phys Rev B 42:1150–1155
16. Li YS, Jonna F, Marcus PM (1991) Multilayer relaxation of a Pb{111} surface. Phys Rev B 43:6337–6341
17. Teeter G, Erskine JL (1999) Studies of clean metal surface relaxation experiment-theory discrepancies. Surf Rev Lett 6:813–817
18. Geng WT, Kim M, Freeman AJ (2001) Multilayer relaxation and magnetism of a high-index transition metal surface: Fe(310). Phys Rev B 63:245401
19. Over H, Kleinle G, Ertl G et al (1991) A LEED structural analysis of the Co($10\bar{1}0$) surface. Surf Sci 254:L469–L474
20. Geng WT, Freeman AJ, Wu RQ (2001) Magnetism at high-index transition-metal surfaces and the effect of metalloid impurities: Ni (210). Phys Rev B 63:064427
21. Zhang X-G, Van Hove MA, Somorjai GA et al (1991) Efficient determination of multilayer relaxation in the Pt(210) stepped and densely kinked surface. Phys Rev Lett 67:1298–1301
22. Härtel S, Vogt J, Weiss H (2010) Relaxation and thermal vibrations at the NaF(100) surface. Surf Sci 604:1996–2001
23. Vogt J (2007) Tensor LEED study of the temperature dependent dynamics of the NaCl(100) single crystal surface. Phys Rev B 75:125423
24. Okazawa T, Nishimura T, Kido Y (2002) Surface structure and lattice dynamics of KI(001) studied by high-resolution ion scattering combined with molecular dynamics simulation. Phys Rev B 66:125402
25. Palosz B, Pantea C, Grazanka E et al (2006) Investigation of relaxation of nanodiamond surface in real and reciprocal spaces. Diamond Relat Mater 15:1813–1817
26. Sun CQ, Tay BK, Zeng XT et al (2002) Bond-order–bond-length–bond-strength (bond-OLS) correlation mechanism for the shape-and-size dependence of a nanosolid. J Phys Cond Matter 14:7781–7796
27. Sun CQ (2007) Size dependence of nanostructures: impact of bond order deficiency. Prog Solid State Chem 35:1–159
28. Feibelman PJ (1996) Relaxation of hcp(0001) surfaces: a chemical view. Phys Rev B 53:13740–13746
29. Batsanov SS (1987) Effect of high dynamic pressure on the structure of solids. Propellants Explosives Pyrotechnics 12:206–208
30. Batsanov SS (1994) Effects of explosions on materials. Springer-Verlag, New York
31. Batsanov SS, Bokarev VP, Moroz IK (1982) Solid solution in the KCl-CsCl system. Russ J Inorg Chem 26:1557–1559
32. Kinugawa K, Othori N, Kadono K et al (1993) Pulsed neutron diffraction study on the structures of glassy LiX–KX–CsX–BaX_2 (X=Cl, Br, I). J Chem Phys 99:5345–5351
33. Noda Y, Nakao H, Terauchi H (1995) Neutron incoherent scattering of structural glass NH_4I–KI mixed crystal. Physica B 213–214:564–566
34. Micoulaut M, Cormier L, Henderson GS (2006) The structure of amorphous, crystalline and liquid GeO_2. J Phys Conden Matter 18:R753–R784
35. Delaplane RG, Dahlborg U, Howells WS, Lundström T (1988) A neutron diffraction study of amorphous boron. J Non-Cryst Solids 106:66–69

36. Degtyareva VF, Degtyareva O, Mao H-K, Hemley RJ (2006) High-pressure behavior of CdSb: compound decomposition, phase formation, and amorphization. Phys Rev B 73:214108
37. Marks NA, McKenzie DR, Pailthorpe BA et al (1996) Microscopic structure of tetrahedral amorphous carbon. Phys Rev Lett 76:768–771
38. Sinclair RN, Stone CE, Wright AC et al (2001) The structure of vitreous boron sulphide. J Non-Cryst Solids 293–295:383–388
39. Yao W, Martin SW, Petkov V (2005) Structure determination of low-alkali-content $Na_{2S} + B_2S_3$ glasses using neutron and synchrotron X-ray diffraction. J Non-Cryst Solids 351:1995–2002
40. Boswell FWC (1951) Precise determination of lattice constants by electron diffraction and variations in the lattice constants of very small crystallites. Proc Phys Soc London A 64:465–475
41. Yamanaka T, Sugiyama K, Ogata K (1992) Kinetic study of the GeO_2 transition under high pressures using synchrotron X-radiation. J Appl Cryst 25:11–15
42. Batsanov SS, Bokarev VP (1987) Existence of polymorphic modifications in the amorphous state. Inorg Mater 23:946–947
43. Hoppe U, Walter G, Barz A et al (1998) The P-O bond lengths in vitreous P_2O_5 probed by neutron diffraction with high real-space resolution. J Phys Cond Matter 10:261–270
44. Brazhkin VV, Gavrilyuk AG, Lyapin AG et al (2007) AsS: bulk inorganic molecular-based chalcogenide glass. Appl Phys Lett 91:031912
45. Kajihara Y, Inui M, Matsuda K et al (2007) X-ray diffraction measurement of liquid As_2Se_3 by using third-generation synchrotron radiation. J Non-Cryst Solids 353:1985–1989
46. Dongo M, Gerber Th, Hafiz M et al (2006) On the structure of As_2Te_3 glass. J Phys Cond Matter 18:6213–6224
47. Tanaka K (1988) Layer structures in chalcogenide glasses. J Non-Cryst Solids 103:149–150
48. Tanaka K (1989) Structural phase transitions in chalcogenide glasses. Phys Rev B 39:1270–1279
49. Suzuki M, Okazaki H (1977) The structure of α-AgI. Phys Stat Solidi 42a:133–140
50. Keen DA, Hayes W, McGreevy RL (1990) Structural disorder in AgBr on the approach to melting. J Phys Cond Matter 2:2773–2786
51. Ubbelode AR (1978) The molted state of matter. Wiley, New York
52. Tosi MP (1994) Structure of covalent liquids. J Phys Cond Matter 6:A13–A28
53. Waseda Y, Tamaki S (1975) Structural study of Pt and Cr in the liquid state by X-ray diffraction. High Temp-High Press 7:215–220
54. Waseda Y (1980). The structure of non-crystalline materials: liquids and amorphous solids. McGraw Hill, New York.
55. Vahvaselkä KS (1978) X-ray diffraction analysis of liquid Hg, Sn, Zn, Al and Cu. Phys Scripta 18:266–274
56. Krishnan S, Ansell S, Feleten JJ et al (1998) Structure of liquid boron. Phys Rev Lett 81:586–589
57. Anselm A, Krishnan Sh, Felten JJ, Price DL (1998) Structure of supercooled liquid silicon. J Phys Cond Matter 10:L73–L78
58. Di Cicco A, Aquilanti G, Minicucci M et al (1999) Short-range interaction in liquid rhodium probed by x-ray absorption spectroscopy. J Phys Cond Matter 11:L43–L50
59. Wei S, Oyanagi H, Liu W et al (2000) Local structure of liquid gallium studied by X-ray absorption fine structure. J Non-Cryst Solids 275:160–168
60. Katayama Y, Mizutani T, Utsumi W et al (2000) A first-order liquid–liquid phase transition in phosphorus. Nature 403:170–173
61. Holland-Moritz D, Schenk T, Bellisent R et al (2002) Short-range order in undercooled Co melts. J Non-Cryst Solids 312–314:47–51
62. Schenk T, Holland-Moritz D, Simonet V et al (2002) Icosahedral short-range order in deeply undercooled metallic melts. Phys Rev Lett 89:075507
63. Jakse N, Hennet L, Price DL et al (2003) Structural changes on supercooling liquid silicon. Appl Phys Lett 83:4734–4736

64. Salmon PS, Petri I, de Jong PHK et al (2004) Structure of liquid lithium. J Phys Cond Matter 16:195–222
65. Wasse JC, Salmon PS (1999) Structure of molten lanthanum and cerium tri-halides by the method of isomorphic substitution in neutron diffraction. J Phys Cond Matter 11:1381–1396
66. Iwadate, Suzuki, K, Onda, N et al (2006) Structural changes on supercooling liquid silicon. J Alloys Compd 408–412:248–252
67. Ensico E, Lombardero M, Dore JC (1986) A RISM analysis of neutron diffraction data for $SiCl_4/TiCl_4$ and $SnCl_4/TiCl_4$ liquid mixtures. Mol Phys 59:941–952
68. Funamori N, Tsuji K (2002) Pressure-induced structural change of liquid silicon. Phys Rev Lett 88:255508
69. Cahoon JR (2004) The first coordination number for liquid metal. Canad J Phys 82:291–301
70. Santiso E, Müller EA (2002) Dense packing of binary and polydisperse hard spheres. Mol Phys 100:2461–2469
71. Katayama Y, Tsuji K (2003) X-ray structural studies on elemental liquids under high pressures. J Phys Cond Matter 15:6085–6104
72. Sanloup C, Guyot F, Gillet P et al (2000) Structural changes in liquid Fe at high pressures and high temperatures from synchrotron X-ray diffraction. Europhys Lett 52:151–157
73. Shen G, Sata N, Taberlet N et al (2002) Melting studies of indium: determination of the structure and density of melts at high pressures and high temperatures. J Phys Cond Matter 14:10533–10540
74. Takeda S, Tamaki S, Waseda Y (1985) Electron-ion correlation in liquid metals. J Phys Soc Japan 54:2552–2258
75. Takeda S, Tamaki S, Waseda Y, Harada S (1986) Electron's distribution in liquid zinc and lead. J Phys Soc Japan 55:184–192
76. Takeda S, Harada S, Tamaki S, Waseda Y (1988) Electron charge distribution in liquid metals. Z phys Chem NF 157:459–463
77. Takeda S, Inui M, Tamaki S et al (1993) Electron charge distribution in liquid Te. J Phys Soc Japan 62:4277–4286
78. Takeda S, Inui M, Tamaki S et al (1994) Electron-ion correlation in liquid magnesium. J Phys Soc Japan 63:1794–1802
79. Tao DP (2005) Prediction of the coordination numbers of liquid metals. Metallurg Mater Trans A 36:3495–3497
80. Regan MJ, Kawamoto EH, Lee S et al (1995) Surface layering in liquid gallium: an X-ray reflectivity study. Phys Rev Lett 75:2498–2501
81. Tostmann H, DiMasi E, Pershan PS et al (1999) X-ray studies on liquid indium:Surface structure of liquid metals and the effect of capillary waves. Phys Rev B 59:783–791
82. Shpyrko OG, Grigoriev AY, Steimer Ch et al (2004) Anomalous layering at the liquid Sn surface. Phys Rev B 70:224206
83. Magnussen OM, Ocko BM, Regan MJ et al (1995) X-ray reflectivity measurements of surface layering in liquid mercury. Phys Rev Lett 75:4444–4447
84. Shpyrko O, Fukuto M, Pershan P et al (2004) Surface layering of liquids: the role of surface tension. Phys Rev B 69:245423
85. Stolz M, Winter R, Howells WS (1994) The structural properties of liquid and quenched sulphur II. J Phys Cond Matter 6:3619–3628
86. Raty J-Y, Gaspard J-P, Bionducci N et al (1999) On the structure of liquid IV–VI semiconductors. J Non-Cryst Solids 250–252:277–280
87. Thomas JM (2004) Ultrafast electron crystallography: the dawn of a new era. Angew Chem Int Ed 43:2606–2610
88. Sokolowski-Tinten K, Blome C, Dietrich C et al (2001) Femtosecond X-ray measurement of ultrafast melting and large acoustic transients. Phys Rev Lett 87:225701
89. Rousse A, Rischel C, Fourmaux S et al (2001) Non-thermal melting in semiconductors measured at femtosecond resolution. Nature 410:65–68
90. Samoylov OYa (1957) Structure of water solutions of electrolytes and hydration of ionics. Academy of Science of the USSR, Moscow (in Russian)

91. Bernal JD, Fowler RH (1933) A theory of water and ionic solution. J Chem Phys 1:515–548
92. Hattori T, Taga N, Takasugi Y et al (2002) Structure of liquid GaSb under pressure. J Non-Cryst Solids 312–314:26–29
93. Malenkov GG (2006) Sructure and dynamics of liquid water. J Struct Chem 47:S1–S31
94. Goncharov AF, Goldman N, Fried LE et al (2005) Dynamic ionization of water under extreme conditions. Phys Rev Lett 94:125508
95. Markus Y (1988) Ionic radii in aqueous solutions. Chem Rev 88:1475–1498
96. Johansson G (1992) Structures of complexes in solution derived from X-ray diffraction measurements. Adv Inorg Chem 39:159–232
97. Ohtaki H, Radnai T (1993) Structure and dynamics of hydrated ions. Chem Rev 93:1157–1204
98. Blixt J, Glaster J, Mink J et al (1995) Structure of thallium(III) chloride, bromide, and cyanide complexes in aqueous solution. J Am Chem Soc 117:5089–5104
99. D'Angelo P, Bottari E, Festa MR et a (1997) Structural investigation of copper(II) chloride solutions using X-ray absorption spectroscopy. J Chem Phys 107:2807–2812
100. Ramos S, Neilson GW, Barnes AC, Buchanan P (2005) An anomalous x-ray diffraction study of the hydration structures of Cs^+ and I^- in concentrated solutions. J Chem Phys 123:214501
101. Hofer TS, Randolf BR, Rode BM, Persson I (2009) The hydrated platinum (II) ion in aqueous solution—a combined theoretical and EXAFS spectroscopic study. Dalton Trans 1512–1515
102. Persson I (2010) Hydrated metal ions in aqueous solution: how regular are their structures? Pure Appl Chem 82:1901–1917
103. Mähler J, Persson I (2012) A study of the hydration of the alkali metal ions in aqueous solution. Inorg Chem 51:425–438
104. Näslund L-Å, Edwards DC, Wernet P et al (2005) X-ray absorption spectroscopy study of the hydrogen bond network in the bulk water of aqueous solutions. J Phys Chem A 109:5995–6002
105. Drakin SI (1963) Me–H_2O distances in crystal hydrates and radii of ions in aqueous solution. J Struct Chem 4:472–478
106. Friedman HL, Lewis L (1976) The coordination geometry of water in some salt hydrates. J Solution Chem 5:445–455
107. Persson I, Sandström M, Yokoyama H, Chaudhry M (1995) Structure of the solvated strontium and barium ions in aqueous, dimethyl-sulfoxide and pyridine solution, and crystal-structure of strontium and barium hydroxide octahydrate. Z Naturforsch 50:21–37
108. Schmid R, Miah AM, Sapunov VN (2000) A new table of the thermodynamic quantities of ionic hydration: values and some applications (enthalpy-entropy compensation and Born radii). Phys Chem Chem Phys 2:97–102
109. Torapava N, Persson I, Eriksson L, Lundberg D (2009) Hydration and hydrolysis of thorium(IV) in aqueous solution and the structures of two crystalline thorium(IV) hydrates. Inorg Chem 48:11712–11723
110. Inada Y, Sugimoto K, Ozutsumi K, Funahashi S (1994) Solvation structures of manganese(II), iron(II), cobalt(II), nickel(II), copper(II), zinc(II), cadmium(II), and indium(III) ions in 1,1,3,3-tetramethylurea as studied by EXAFS and electronic spectroscopy. Variation of coordination number. Inorg Chem 33:1875–1880
111. Ulvenlund S, Wheatley A, Bengtsson L (1995) Spectroscopic investigation of concentrated solutions of gallium(III) chloride in mesitylene and benzene. Dalton Trans 255–263
112. Inada Y, Hayashi H, Sugimoto K-I, Funahashi S (1999) Solvation structures of manganese(II), iron(II), cobalt(II), nickel(II), copper(II), zinc(II), and gallium(III) ions in methanol, ethanol, dimethyl sulfoxide, and trimethyl phosphate as studied by EXAFS and electronic spectroscopy. J Phys Chem A 103:1401–1406
113. Lundberg D, Ullström A-S, D'Angelo P, Persson I (2007) A structural study of the hydrated and the dimethylsulfoxide, N, N′-dimethylpropyleneurea, and N, N-dimethylthioformamide solvated iron(II) and iron(III) ions in solution and solid state. Inorg Chim Acta 360:1809–1818
114. Thompson H, Wasse JC, Skipper NT et al (2003) Structural studies of ammonia and metallic lithium–ammonia solutions. J Am Chem Soc 125:2572–2581
115. Hayama S, Skipper NT, Wasse JC, Thompson H (2002) X-ray diffraction studies of solutions of lithium in ammonia: the structure of the metal–nonmetal transition. J Chem Phys 116:2991–2996

116. Jacobs H, Barlage H, Friedriszik M (2004) Vergleich der kristallstrukturen der tetraammoniakate von lithiumhalogeniden, $LiBr \cdot 4NH_3$ und $LiI \cdot 4NH_3$, mit der struktur von tetramethylammoniumiodid, $N(CH_3)_4I$. Z anorg allgem Chem 630:645–648
117. Wasse JC, Hayama S, Skipper NT et al (2000) The structure of saturated lithium- and potassium-ammonia solutions as studied by neutron diffraction. J Chem Phys 112:7147–7151
118. Wasse JC, Howard CA, Thompson H et al (2004) The structure of calcium–ammonia solutions by neutron diffraction. J Chem Phys 121:996–1004
119. Tokushima T, Harada Y, Takahashi O et al (2008) High resolution X-ray emission spectroscopy of liquid water: the observation of two structural motifs. Chem Phys Lett 460:387–400
120. Röntgen WC (1892) Über die Konstitution des flüssigen Wassers. Ann Phys Chem 45:91

Chapter 8
Between Molecule and Solid

Crystallography (see Chap. 5) usually works on the assumption that the crystal is infinite—only in this case the mathematical model of the translational (crystal lattice) symmetry is rigorously correct. For macroscopic crystals this is a very sensible approximation: the crystal structure pattern of NaCl is repeated *ca* 1.77 million times per millimeter, and that of the largest species studied crystallographically (a virus) *ca* 6,250 times per millimeter. This model is still valid, with some adjustments and reservations, down to micrometer-sized crystals of ordinary materials. On the other hand, clusters (see Chap. 3) with tens of metal atoms are still essentially molecules. However, the particles of the intermediate size, on the order of nanometers (nm), commonly known as nano-particles, have peculiar properties, qualitatively different from both molecules and bulk solids and varying dramatically with size (even within the nano-range itself). Nano-particles are ubiquitous in nature, and even some man-made examples had been known for a long time but remained largely a scientific curiosity until the late twentieth century.[1] In the last decades, nano-materials became a major focus of theoretical and practical interest in physics, chemistry and materials engineering. Now they are often regarded as another aggregate state of matter. In previous chapters we already discussed some energetic and structural properties of nano-powders; here we consider general trends of these properties in relation to size.

8.1 Energetic Properties of Clusters and Nanoparticles

As was told above, coordination numbers of atoms on the surface of a crystal are less than in the bulk, and most properties of dispersed materials can be formulated and understood in terms of atomic N_c imperfection and its effect on atomic cohesion. In particular, the difference between the cohesive energy of an atom at the surface and

[1] However, scientific investigation of colloidal gold (known since alchemists) was pioneered in 1857 by Faraday [1] who on this subject (as on practically any other) was well ahead of his time.

S. S. Batsanov, A. S. Batsanov, *Introduction to Structural Chemistry*,

DOI 10.1007/978-94-007-4771-5_8,

Table 8.1 Depression of the fusion temperature

Substance	D, nm	ΔT_m, K
Na	32	83
Au	2.5	406
Al	20	13
Ga[a]	5	212
In	10	105
Si[b]	2	500
Sn[c]	5	82
Pb	20	13
Bi[d]	2	120
CdS	2	1,200

ΔT_m under transition the bulk into nanophases [10], [a][11], [b][12], [c][13], [d][14]

that of an atom inside the solid determines the general trends of the melting behavior of a nano-solid. Other physical properties follow the same trend as T_m that is dictated by the effect of N_c imperfection and corresponding changes in bond energies, as will be shown below.

8.1.1 Melting Temperatures and Heats Under Transition from Bulk to Nanophases

The melting process consists of two stages: the preparation of a solid to the onset of melting and the actual destruction of the crystalline order, the work of which is characterized by the melting heat ΔH_m. The physical essence of the first stage is an increase in the amplitude of atomic vibrations to the value above which the body starts to melt (Lindemann criterion, see in detail in Sect. 6.2). Surface atoms having smaller number of bonds than atoms of the interior, their vibration amplitude is correspondingly larger [2–5], and hence, the mean vibration amplitude in the sample will increase with increasing fraction of surface atoms, thus depressing the melting point (see Table 8.1). The inverse relation between particle size and T_m was predicted by W. Thompson in 1871 [6] and confirmed experimentally by Pawlow in 1909 [7]. Since then the melting behavior of small particles, as well as surface layers of bulk solids attracted increasing research interest both theoretically and experimentally, see reviews [8–10]. In many cases, surface melting and evaporation occur at temperatures lower than the corresponding bulk values. The enthalpy of melting (ΔH_m) also decreases with the particle size, so the peak on a differential scanning calorimetry curve not only shifts to lower temperature but also becomes lower and less distinct (see Fig. 6.5). This can be seen as a gradual withering away of the crystallinity: obviously, the definition of a crystal as a solid possessing long-range order becomes meaningless if there is no long range at all. A drastic decrease of ΔH_m in shock-compressed solids, known for a long time [13], is also due to diminution of the grain size.

In order to explain the observed effects, different thermodynamic models have been developed, which establish the dependence of melting temperatures and enthalpy, Debye temperature, and atomic vibration amplitude on the crystal grain size (different models and the history of the question see in reviews [8, 15, 16]). The melting point depression for small crystals is described in the classical thermodynamic approach by the Gibbs–Thomson equation [17, 18]. For spherical particles,

$$T_{\mathrm{m}}(D) = T_{\mathrm{m}}(\infty) - \frac{T_m(\infty)\sigma_{sl}}{\Delta H_m(\infty)\rho_s D} \tag{8.1}$$

where $T_{\mathrm{m}}(\infty)$, $\Delta H_{\mathrm{m}}(\infty)$, and ρ_{s} are the bulk melting temperature, the bulk latent heat of melting, and the solid phase density, respectively; D represents the diameter of a spherical particle and $T_{\mathrm{m}}(D)$ is the melting point of a particle with the size D; σ_{sl} is the solid-liquid interfacial energy. Similar size behavior of the melting points is observed in a variety of substances [19], including organic compounds [20]. Concerning the size dependence of ΔH_{m}, it is assumed that the heat of fusion decreases with particle size due to the increase in the surface energy

$$\Delta H_{\mathrm{m}}(D) = \Delta H_{\mathrm{m}}(\infty) - \frac{2\sigma_{sl}}{\rho_s D} \tag{8.2}$$

where $\Delta H_{\mathrm{m}}(D)$ is the heat of fusion for a particle with the size D. Equation 8.2 is derived assuming an equilibrium melting process and the Laplace formula [21] for the solid-liquid transition. However, the size-dependent heat of fusion is significantly smaller than that predicted by the effects of the surface tension, indicating that a solid nano-particle is at a higher energy than expected, presumably due to the presence of defects or irregularities in the crystal structure at, or emanating from, the surface. This hypothesis has been tested on (ligand-free) metal clusters and nanoparticles, particularly those of tin. Using the technique of Haberland et al. [22, 23], it has been shown [24] that charged tin clusters with 10–30 atoms remain solid at *ca.* 50 K *above* the melting point of bulk tin. This behavior is possibly related to the fact that the structure of the charged clusters is completely different from that of the bulk element. Calorimetry measurements up to 650 K on unsupported Sn_{18}^+, Sn_{19}^+, Sn_{20}^+, and Sn_{21}^+ clusters [25] revealed some small features (which may be due to localized structural changes) but no clear melting transitions. Hence, for clusters of this size the concept of melting is not applicable.

8.1.2 Energy Variation Under Transition from Bulk to Clusters

Let us begin with energetic properties of polyatomic homoatomic molecules (clusters) of metals (M_n) and nonmetals (A_n). Because clusters lie somewhere between molecules and bulk solids, approaches and models to study them have come from both sides; in most cases, condensed matter theorists have applied solid-state models while quantum chemists have moved from calculations on complex molecules to clusters. It has been shown [26] that the experimental data and the results of this model point to a shell model description of metal clusters, therefore further development of the theoretical explanation of physical properties of the metallic cluster followed this direction.

When atoms combine into molecules or clusters, the atomic orbitals are transformed into molecular orbitals and finally merge into continuous bands (see Fig. 2.5). The density of states (DOS) within a band is roughly equal to the number of atoms in the ensemble per eV. In works [27, 28] at first were established the general dependences: for clusters, the ionization potentials (I_c) decrease and the electron affinities (A_c) increase as the size of the cluster (i.e. the number n of atoms in it) increases. The linear extrapolations of both dependences ultimately converge to the work function (Φ) of the corresponding bulk material, viz.

$$I_c = \Phi + CR_c \tag{8.3}$$

$$A_c = \Phi - DR_c \tag{8.4}$$

where R_c is the radius of a cluster, C and D are constants. R_c is proportional to $n^{1/3}$ for metal clusters, and to $n^{1/2}$ for hollow cages, e.g. fullerenes or (car)borane clusters. Above (Sect. 1.1) were given numerous experimental data for different metallic clusters which some authors consider as a confirmation of Eqs. 8.3 and 8.4 (see also Fig. 1.4). However, these equations can be regarded only as approximations, because (i) the largest experimentally studied clusters ($n = 10^2$–10^3) are still quite small to model the bulk and (ii) even in this range the dependences are far from linear. In fact, both curves show a number of local peaks, corresponding to the closest packing modes of atoms (tetrahedron → octahedron → cubo-octahedron → dodecahedron → icosahedron) and to certain "magic" numbers of valence electrons in the cluster. This applies not only to metals, but also to such semimetals as carbon, germanium, tin, selenium and arsenic, as well as to organometallic clusters. The trends of I_c and A_c toward Φ reflect the structural similarity between the *interior* part of a large cluster and the crystalline metal. Note also that an extrapolation of the currently available electron affinities of Ti clusters to $n = \infty$ gives $\Phi = 3.80$ eV, considerably different from the experimental $\Phi = 4.33$ eV [29], possibly because the Ti clusters have different structures than the bulk metal.

As mentioned above (see Sect. 2.3.4, Table 2.16), the experimental values of the band gaps (E_g) in most substances increase as the grain sizes decrease. Diamond is an exception, its E_g decreasing together with the grain sizes. Here, an important role is played by the curvature of the surface layer, which results in the negative electron affinity of nano-diamond particles [30, 31].

Studies by the X-ray photoelectron spectroscopy of the ionic bonding in free nanoscale NaCl clusters with tens or hundreds of atoms [32] revealed cluster-level energy shifts of around 3 eV toward lower binding energy for Na 2*p* and *ca* 1 eV toward higher binding energy for Cl 2*p* relative to the monomer levels. This means that the difference between the electronegativities of Cl and Na atoms, i.e. the bond ionicity, increases as we go from the NaCl monomer to clusters of increasing size. This can be regarded as the first step of the transition from the molecule to the crystal.

8.2 Changes of the Atomic Structure on Transition from Bulk Solids to Nanophases

This topic will be discussed in two steps, firstly the crystals with 3-dimensional structures (metal nanoparticles and 4-coordinate semiconductors) and then those with a layered structure (Bi, Se, Te, etc). A decrease in the crystal sizes can result in a change of the structure type if the surface energy gain exceeds the enthalpy of the corresponding phase transition. Thus, Co has the structure of the *hcp* type in the bulk, the *fcc* type in 10–20 nm particles and the *bcc* type in 2–5 nm particles [33]. Particles of In with the diameter of ≤ 5 nm have the *fcc* structure, and those from 5 nm upward to the bulk have the *bct*-lattice [34]. AgI adopts the cubic structure in the particles larger than 50 nm and the hexagonal one in smaller crystals [35]. InAs has the wurtzite (w) structure up to 40 nm and the sphalerite (zb) structure in grains of ≥ 80 nm [36]. Nano-CdS has the the zb structure for $D = 4$ nm, while the w-phase is stable for the bulk material [37, 38]. On the contrary, MnSe was obtained in the w-form in nanoparticles, whereas the zb phase is stable for bulk crystals [37, 39].

Nano-phases of a number of oxides also exist in structures not occurring in bulk samples, viz. CoO in the hexagonal modification of the ZnO type [40], Al_2O_3 in the γ-form, Y_2O_3 in the monoclinic modification, $BaTiO_3$ in the cubic structure [41], ZrO_2 in the tetragonal and monoclinic modifications [42–44]. TiO_2 is stable in the anatase form with sizes below 14 nm [45], and ReO_3 in the monoclinic modification instead of cubic that is characteristic of bulk samples [46]. Nanocrystalline WO_3 films 9 nm thick have monoclinic structure, but in 50 nm particles adopt the tetragonal one [37, 47]. For iron oxide Fe_2O_3 nano-particles crystallized in mesoporous SiO_2, all four known polymorphs were found to be stable in different size ranges, viz. γ-form for $D < 8$ nm, ε-form between 8 and 30 nm, β-form from 30–50 nm and the magnetic α-modification for larger sizes [48].

The study of structural parameters of the Ge nanophase [49] shows that 9 nm nanoparticles have $N_c = 4$, similar to that of the bulk, but this parameter decreases (down to 3.3) as grains become smaller. At the same time, the interatomic distance increases and approaches that for bulk amorphous Ge. This contradicts the observations for metallic nanophases where a decrease in the size is accompanied by a contraction of the mean interatomic distance due to capillary pressure. For the present case, the increase in interatomic distance apparent in semiconductor samples results from the increase in amorphous fraction with decreasing grain sizes. At the same time, crushing of Te crystals reduces both the N_c and bond lengths, and increases the force constants (K_B) [50], Fig. 8.1. These results indicate that nanosamples of Te are close to the covalent state.

8.3 Size Effect in the Dielectric Permittivity of Crystals

The dielectric permittivity (ε) of solids often changes with the particle size. The dependence is a result of a complicated interplay of several factors, it can vary widely in magnitude and even have different sense. Let us consider the influence of different factors on ε under transition from bulk solids to nanophases.

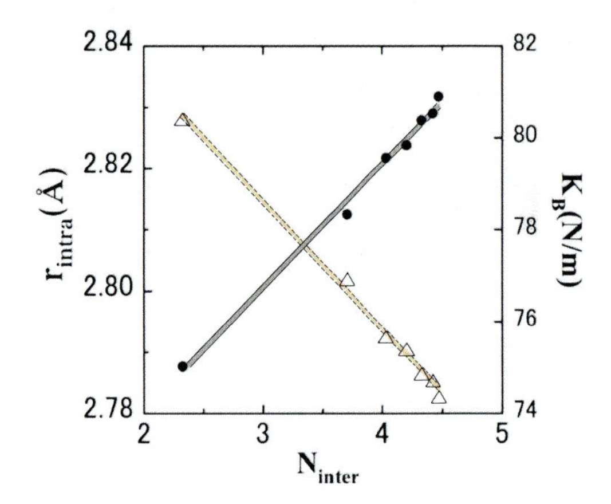

Fig. 8.1 Correlations of the intra-chain bond distance (r_{intra}) with the inter-chain coordination number (N_{inter}) and the force constant (K_B) with N_{inter} for Te nanoparticles. (Adapted with permission from [50]. Copyright 2011 American Chemical Society)

8.3.1 Effect of the Energy Factor

Because ε reflects the ease with which the charges in the substance can be shifted by an external electric field, it is related to the band gap E_g of a solid. A theoretical formula of this dependence was derived by Penn [51] assuming that all the valence electrons of atoms in crystals are delocalized (as in metals) and the laws of the electronic gas are applicable to them,

$$\varepsilon = 1 + (h\omega_p/E_g)^2 \quad (8.5)$$

where ω_p is the plasma frequency of vibrations of the valence electrons,

$$\omega_p = (4\pi e^2 N/mV)^{1/2} \quad (8.6)$$

E_g is the band gap, N is the number of valence electrons, m is their mass, V is the volume of the formula unit of the crystal. Several empirical expression linking ε and E_g are also known, viz [52, 53, 54, 55].

$$\begin{aligned} &\varepsilon^2 \cdot E_g = a, \\ &\varepsilon = b - c\ln E_g, \\ &\varepsilon \cdot E_g = d, \\ &\varepsilon = 1 + \frac{e}{E_g + f} \end{aligned} \quad (8.7)$$

where a, b, c, d, e, f are constants. All these formulae give qualitatively similar results: ε and E_g are inversely correlated. Because (see Sect. 8.1.2) the band gaps in most substances widens as the grain size decreases, ε should decrease. All theoretical works confirm this conclusion. Thus, quantum mechanical calculations of dielectric constant in Si clusters of 100–1,300 atoms [56, 57] predicted ε significantly lower than the bulk value. Other calculations [58, 59] using a self-consistent tight binding approach, have shown that the effective dielectric constant is strongly influenced by the nanostructuring, leading to a strong reduction with respect to the bulk dielectric constant and to the value for a perfectly crystalline layer. These results are explained by the effect of the local fields and by the reduction of the dielectric response near the surfaces of the nanocrystals. Nakamura et al. [60] investigated the dielectric properties of Si(111) ultrathin films using first-principles calculations in external electrostatic fields and found that with increasing thickness of the films, the optical dielectric constant converges to the experimental bulk dielectric constant at the thickness of only eight bilayers. The results show that ε is reduced distinctly in the first few bilayers from the surface, due to penetration of the depolarized charges induced at the surface. Such an effective reduction of the field near the surface is one of the reasons for the decrease of ε for ultrathin films. Density functional theory (DFT) calculations of the dielectric properties of silicon nitride thin films with thickness < 6 nm [61] suggested substantial decrease in the static dielectric constant of crystalline films, as their thickness is reduced. The variation in the response in the proximity of the surface plays a key role in the observed decrease. In addition, amorphization of the films may bring further reduction of dielectric constants. Finally, Kageshima and Fujiwara [62] also studied the dielectric constants of Si(111) nanofilms by using the first-principles calculation and showed that the dielectric constants are significantly smaller than that of the bulk. Furthermore, ε depends on the conductivity type as well as on the film thickness. All these theoretical results correspond to the simple fact that as the size of the system is reduced (ultimately) to a diatomic molecule, the bonding pattern is reduced to a simple covalent bond, so that the Maxwell's law ($\varepsilon = n^2$) becomes valid.

8.3.2 *Effect of the Phase Composition on ε of Barium Titanate*

A good illustration of the size effect is presented by the electric properties of $BaTiO_3$, extensively studied from 1960s onwards, see [63–71] and references therein. $BaTiO_3$ has a perovskite-type structure with the Ti atom situated inside an octahedron formed by six oxygen atoms (Fig. 8.2) [68]. Above the Curie point (130°C for bulk material) the structure has cubic symmetry, the Ti lies exactly at the centre of the O_6 octahedron and the material is para-electric with moderate $\varepsilon \approx 25$ [69]. At ambient conditions, however, the Ti atom is displaced by Δd from its central position along the c axis. This lowers the crystal symmetry to tetragonal (with the unit cell parameters $a = b < c$) and creates a permanent electric dipole. As a result, tetragonal $BaTiO_3$ is ferroelectric with $\varepsilon \approx 2.5 \times 10^3$. It has been found that the Curie temperature and the heat of this phase transition decrease together with the grain size D. Furthermore, in the tetragonal phase itself, the tetragonal distortion (i.e. the c/a ratio) and ε also decrease with D (Table 8.2). As in the case of melting, the phase transition not only shifts to lower temperature for smaller grain sizes, but also becomes increasingly diffuse.

These size-effects can be explained as a combination of the effect of grain boundaries and an intrinsic effect—variation of the tetragonal strain (within the framework

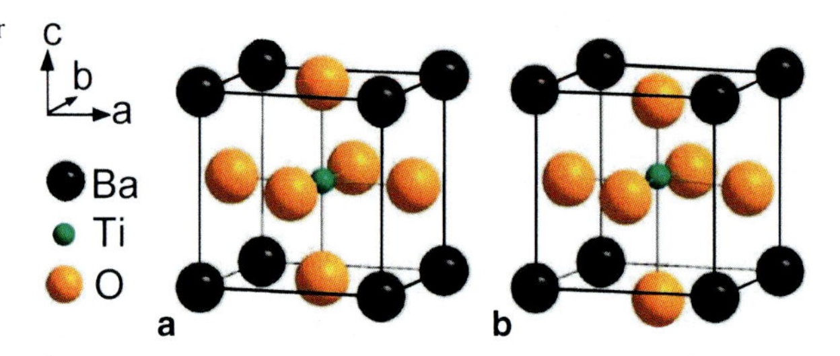

Fig. 8.2 Structural models for $BaTiO_3$ compound: (**a**) cubic and (**b**) tetragonal phase. (Reprinted with permission from [68]. Copyright 2008 American Chemical Society)

Table 8.2 Dependence of the properties of $BaTiO_3$ on the grain size[a]

D, nm	Bulk	1,200	300	100	70	50	45	26
c/a [68]	1.011				1.0065		1.0054	
Δd, Å [68]					0.073		0.097	0.137
T_c, °C [65]	130	125				88		
ε^b		2,520	2,200	1,680		780		
Q, J/mol [65]	210 ± 20	220				44		

[a] a and c are unit cell parameters, Δd is the Ti atom displacement, T_c is the Curie temperature, Q is the heat of the tetragonal/cubic phase transition, ε is the dielectric constant, D is the grain diameter
[b] At 70 °C and 10 kHz [65]

of the Landau-Ginsburg-Devonshire theory) [65]. However, the experimental results are somewhat contradictory and many questions remain open. Thus, high-precision XRD, Raman and EXAFS studies [68] revealed that although the (tetragonal) lattice becomes more pseudo-cubic with decreasing D, the Ti atom displacement actually increases!

It has been suggested that at some critical size (usually estimated at 10–30 nm) the ferroelectricity of $BaTiO_3$ must disappear altogether. However, in the experiments 26 nm particles were found to be still tetragonal [68], although an earlier neutron diffraction study [71] suggested that at 40 nm size this phase becomes pseudo-cubic with $c/a = 1.000(5)$ and partially transforms into a hexagonal polymorph. It seems that the metrics of the unit cell is not a good indicator of the symmetry of the atomic structure. Thus, 50 nm particles appeared cubic for XRD, but their Raman spectra and ferroelectric behavior indicated a tetragonal phase [70], while theoretical calculations suggest that ferroelectric properties can arise from very small (0.005 Å) Ti atom displacements.

8.3.3 *Dielectric Behavior of Ceramic Materials*

Because the high dielectric responses in the tetragonal phase of $BaTiO_3$ are caused by a deformation of the TiO_6 coordination polyhedron and consequently the creation of a permanent electric dipole, it is instructive to consider from this viewpoint the dielectric properties in other ternary oxides. After the size effect in $BaTiO_3$ was clarified, works on the synthesis of materials with high ε were focused on obtaining compounds with the maximum distortion in the crystal structures. For this purpose, numerous solid solutions with the maximum difference in ionic radii were prepared. In this way, crystalline compounds with $\varepsilon > 10^3$ were obtained, viz. $CaCu_3Ti_4O_{11.7}F_{0.3}$, $Ca_{1/4}Cu_{3/4}TiO_3$, $La_{15/8}Sr_{1/8}NiO_4$, $Pr_{3/5}Ca_{2/5}MnO_3$, Sr_2TiMnO_6, $Ba_{0.95}La_{0.05}TiO_{3-x}$ [10]. Later, $\varepsilon > 10^3$ were observed in $Ln_{2/3}Cu_3Ti_4O_{12}$ [72], $LuFe_2O_4$ [73] and in Pr-modified $PbTiO_3$ [74]. In the latter, it was suggested that Pr^{4+}ions may occupy the Ti^{4+}positions, thereby inducing chemical disorder at the Ti sites as well, and so making both the A- and B-sublattices disordered in the perovskite structure. In $Ln_{2-x}Sr_xNiO_4$ [75,76] the dielectric constant increases with an increase of the strontium content or a decrease of the lanthanide ionic radius. The best dielectric properties have been obtained in $Sm_{1.5}Sr_{0.5}NiO_4$ ceramics, where $\varepsilon \approx 10^5$ at 10 kHz., while $La_{1.75}Sr_{0.25}NiO_4$ and $La_{1.75}Sr_{0.25}Ni_{0.7}Al_{0.3}O_4$ ceramics showed $\varepsilon = 2.5 \times 10^4$ and 4.5×10^4, respectively. Hence, the dielectric constant is enhanced by partially substituting nickel with aluminum ions, which have smaller radius [77].

In recent years, materials with the K_2NiF_4 structure and a colossal dielectric permittivity have been widely used, such as $Ba_{1.2}Sr_{0.8}CoO_{4+\delta}$, $LaSrFeO_{4+\delta}$, $Sr_{2-x}La_xMnO_{4+\delta}$ [78–80] and $CaLnAlO_4$ [81–83]. Recently a new type of high permittivity materials $La_{2-x}Ca_xNiO_{4+\delta}$ (x = 0, 0.1, 0.2, 0.3) were reported [84]. All these materials also crystallize in the tetragonal K_2NiF_4 structure, where the smaller Ca^{2+} ion (radius 1.18 Å) has replaced the larger La^{3+} (1.22 Å) with the resulting shrinkage of the unit cell. Finally, Ca-doping makes the structure unstable and the ε values for $La_{2-x}Ca_xNiO_{4+\delta}$ (x = 0, 0.1, 0.2, 0.3) at 1 kHz increase to 5.5×10^2, 3×10^3, 5.8×10^3 and 1.15×10^4, respectively. It has been established that along the c axis, the (La, Ca)O_9 dodecahedra are compressed while the NiO_6 octahedra are stretched. The change of (La, Ca)–O bond lengths within the (La, Ca)O_9 dodecahedra play a role in the enhancement of the dielectric constants in these compounds. Thus, $\varepsilon \approx 10^5$ at room temperature has been reported for single-crystalline $La_{15/8}Sr_{1/8}NiO_4$ [85].

Besides the purely structural (dipole) mechanism of the dielectric enhancement in ternary oxides, there are also other possible reasons for this phenomenon. Thus, Al_2O_3/TiO_2 binary oxides contain no dipoles in both oxide layers which could contribute to polarization, but the external contributions originated from the interfaces between Al_2O_3 and TiO_2 layers result in the giant dielectric constant [86]. This is attributed to the so-called Maxwell–Wagner (MW) relaxation, whereby the surface charge accumulate at the interfaces between the two dielectric media (with different conductivities) which act as micro-capacitors. MW relaxation can be simulated by an equivalent circuit composed of two parallel RC elements connected in series. If we assume that the resistance and capacitance of Al_2O_3 and TiO_2 sublayers are R_1, C_1, and R_2, C_2, respectively, the dielectric permittivity of the Al_2O_3/TiO_2 nanolaminates can be expressed as $\varepsilon = C_1(d/\varepsilon_0 S) = (d/d_1)\,\varepsilon_1$, where d is the sample thickness (150 nm), S is the surface area, and d_1 is the Al_2O_3 sublayer thickness (range from 50–0.2 nm). Based on this expression, the measured dielectric constants could range from 3 to 750 times larger than that of Al_2O_3.

The grain boundary causes a large non-uniform distortion and changes the domain structure in the ceramics without grain growth during sintering. So, the grain boundary plays an important role in dielectric properties of fine-grain $BaTiO_3$ ceramics [87].

In conclusion, one can note that from the very beginning of size effect studies, it has been realized that size effects can be of intrinsic (i.e. related to the changes in atomic polarization) and extrinsic nature. Extrinsic effects can be due to either simple modification of the structures caused by their patterning or processing (e.g. increased contribution of grain boundaries in polycrystalline materials) or more complicated effects which include the influence of inhomogeneous strain, incomplete polarization screening at the surface and defect microstructure. Most of the early studies of the size effects refer to the extrinsic ones; therefore, information is rather contradictory and discrepant even for the same materials prepared by different processing routes. Size effects in ferroelectric thin films are also influenced by extrinsic effects which include grain boundaries, local non-stoichiometry, lack of crystallinity and surface depletion/accumulation. In the case where the extrinsic effects are reduced, it has been found that the ferroelectric properties can still be observed in ferroelectric films of four unit cells (2 nm) thick, thus suggesting a significant difference in the scaling behavior in thin films and nano-powders. It is evident that such a big difference in critical sizes for different samples can be caused by extrinsic effects. We believe that most of the cases reported in the literature are of extrinsic origin [88, 89]. Thus, ε of ceramics can be increased either by ion substitution or by restricting the grain growth. Thus, $BaTiO_3$ doped with small amounts of La and having the grain sizes of 50–250 nm, showed colossal $\varepsilon \approx 10^6$ (at ambient temperature and at 1 kHz) and no Curie transition between -100 and $+150\,°C$, which was attributed to interfacial polarisation involving polarons due to formation of Ti^{3+} cations [90].

8.3.4 Dielectric Properties of Multi-Phase Systems

As we have seen, the highest dielectric permittivity is found in such solids which are heterogeneous on microscopic level. The effect can be even more drastic in multiphase systems, especially if one of the phases is water. In 1934 Smith-Rose [91] discovered that soils, which in dry state have ε of 2–10, increase it by several orders of magnitude if impregnated with water (which has $\varepsilon \approx 80$). Later, similar effects were observed for clays and porous rocks (e.g. sandstone, calcite) [92], i.e. for materials with nano-scale heterogeneous structure and highly developed surface. The observed ε can be as high as 10^4 or 10^5 at low frequencies, whereas the amount of water present can be quite small (several per cent). The mechanism of this effect (which is very important in geological surveying) is one of greatest complexity, still only imperfectly understood (see excellent reviews by Chelidze et al. [93, 94]). Often it is attributed squarely to the Maxwell-Wagner mechanism (see above) but in fact this model comes nowhere near explaining the magnitude of the effect [93]. Several other models have been proposed, none of which explains quantitatively all the observed facts. Nevertheless, it is clear that the effect originates at the interface between the solid and aqueous phases and is closely related to the peculiar properties of the

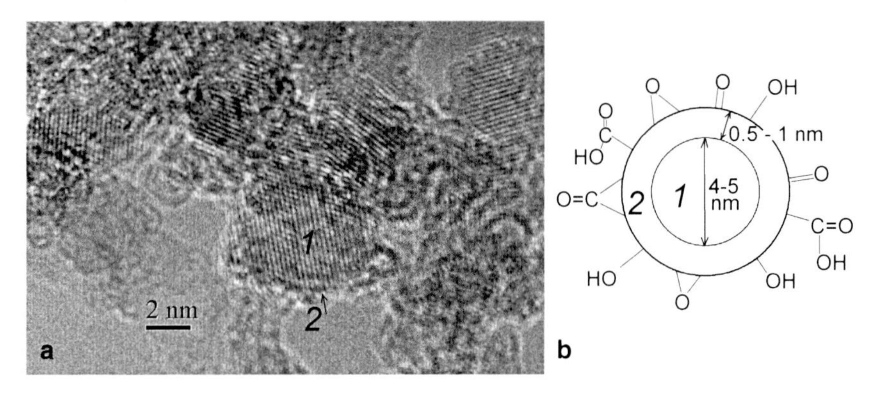

Fig. 8.3 Transmission electron microscopy image (**a**), and schematic structure (**b**) of a DND particle: *1*—diamond core, *2*—non-diamond shell (sp^2 and sp^3 C + admixtures). Sizes of surface functional groups are exaggerated for clarity

Table 8.3 Dielectric permittivity (ε) of detonation diamond nano-powder [103]

Moisture content (%)	a.c. frequency, Hz					
	25	10^2	10^3	10^4	10^5	5×10^5
≈0	90	27	9.3	4.5	3.4	6.0
1.2	3.2×10^4	2.1×10^3	130	25	12	14
2.5	1.1×10^9	2.3×10^7	2.6×10^4	410	46	34
2.8	4.3×10^9	1.9×10^8	2.4×10^5	440	93	34
4.1[a]	3.7×10^{19}	1.1×10^{15}	2.2×10^{11}	10^8	5.9×10^3	510

[a] 6 month exposure to atmospheric air

interfacial water [95] which is presently realized as being ubiquitous and important in many areas [96–98].

Recently, an even more spectacular case of giant ε [99–103] was observed in nano-diamond powders prepared by detonation (see Sect. 10.5) of high explosives with negative oxygen balance, i.e. containing an excess of carbon over oxygen. These so-called detonation nano-diamond (DND) particles (Fig. 8.3) consist of diamond cores of 4–5 nm diameter surrounded by shells of non-diamond carbon terminated by C–H bonds or by oxygen-containing functional groups (hydroxyl, carbonyl, carboxyl, lactone, etc.). Whereas bulk diamond has low static $\varepsilon = 5.67$–5.87, and ordinary synthetic diamond ball-milled to 20–28, 1–2 and 0.1 μm grain sizes gave only modest increases of ε to 7.5, 8.8 and 31.4, respectively, DND displayed colossal ε, exceeding 10^{19} (at 25 Hz) in some cases—by far the highest value known for any system (Table 8.3). However, this giant dielectric permittivity is not intrinsic to DND but is caused by small amounts of water which it spontaneously adsorbs from atmospheric air (DND is hygroscopic, as usual for nanopowders). The water content causing such effect is surprisingly small (≤ 4 %), insufficient to produce even a complete monolayer on the surfaces of DND particles. The effect requires proton-releasing functional groups on the diamond surface interacting with the adsorbed water: it largely disappears if the specially-prepared hydrogen-free DND was used.

On the other hand, even traces of DND drastically affect the physical properties of liquid water, increasing its ε from 80 to over 10^6 and altering some other physical properties (e.g. sound velocity), this effect can be attributed to interfacial water, which is known to differ substantially in its properties from the surrounding 'bulk' water [96–98]. Similar, although much weaker effects were observed in other polar solvents.

The role of adsorbed water on the diamond surface is now recognized as important in other respects as well. Thus, it may be responsible for the surface conductivity of diamond [95, 104]—a highly unusual property, especially as bulk diamond is the best insulator material known.

8.4 Conclusions

As noted in previous chapters of this book, crystal structure parameters (coordination numbers, interatomic distances, unit cell volumes), characteristics of phase transformations, and physical properties of samples can depend on the size of crystal grains; the detailed reviews of the size effect in different mechanical properties of nanomaterials can be found in [9, 105]. In the present chapter it is shown that the size effect is especially significant for electrophysical properties of crystals, including the effect of dispersed powders on the properties of polar liquids contacting with them.

Obviously, the topics discussed in this and other chapters cover only a small part of the structural chemistry of nanomaterials. Besides, this area now experiences such explosive growth that any review becomes outdated very quickly. It is worth recollecting that the first review on this subject (1978) contained 5 references, while in 2008 the number of publications available exceeded 500,000.

The structural features and properties of condensed matter in nanophases open new possibilities of the application of nanocrystals not only in microelectronics and electrotechnics (e.g. for the development of supercapacitors [106] where large ε is required) but also for the solution of fundamentals question of structural chemistry, e.g. how the electronic structure of ionic compounds changes in an environment with colossal dielectric permittivity. This change can result in weakening the electrolytic dissociation even to the point of transition into the molecular state (because the Coulomb interaction is inversely proportional to ε) and for the same reason, to make salts soluble in organic media. Here, we have concentrated on the experimental results, the theoretical explanation of which is still required.

References

1. Faraday M (1857) The Bakerian lecture: experimental relations of gold (and other metals) to light. Philos Trans 147:145–181
2. Hou M, Azzaoui ME, Pattyn H et al (2000) Growth and lattice dynamics of Co nanoparticles embedded in Ag: a combined molecular-dynamics simulation and Mössbauer study. Phys Rev B 62:5117–5128

3. Van Hove MA (2004) Enhanced vibrations at surfaces with back-bonds nearly parallel to the surface. J Phys Chem B 108:14265–14269
4. Vogt J (2007) Tensor LEED study of the temperature dependent dynamics of the NaCl(100) single crystal surface. Phys Rev B 75:125423
5. Härtel S, Vogt J, Weiss H (2010) Relaxation and thermal vibrations at the NaF(100) surface. Surf Sci 604:1996–2001
6. Thomson W (1871) On the equilibrium of vapour at a curved surface of liquid. Phil Mag 42:448–452
7. Pawlow P (1909) The dependency of the melting point on the surface energy of a solid body. Z Phys Chem 65:545–548
8. Roduner E (2006) Size matters: why nanomaterials are different. Chem Soc Rev 35:583
9. Sun CQ (2007) Size dependence of nanostructures: Impact of bond order deficiency. Progr Solid State Chem 35:1–159
10. Batsanov SS (2011) Size effect in the structure and properties of condensed matter. J Struct Chem 52:602–615
11. Parravicini GB, Stella A, Ghigna P et al (2006) Extreme undercooling of liquid metal nanoparticles. Appl Phys Lett 89:033123
12. Goldstein AN (1996) The melting of silicon nanocrystals: Submicron thin-film structures derived from nanocrystal precursors. Appl Phys A 62:33–37
13. Lai SL, Guo JY, Petrova V et al (1996) Size-dependent melting properties of small tin particles. Phys Rev Lett 77:99–102
14. Olson EA, Efremov MYu, Zhang M et al (2005) Size-dependent melting of Bi nanoparticles. J Appl Phys 97:034304
15. Batsanov SS, Zolotova ES (1968) Impact synthesis of divalent chromium chalcogenides. Doklady Akademii Nauk SSSR 180:93–96
16. Guisbiers G, Buchaillot L (2009) Modeling the melting enthalpy of nanomaterials. J Phys Chem C 113:3566–3568
17. Zhu YF, Lian JS, Jiang Q (2009) Modeling of the melting point, Debye temperature, thermal expansion coefficient, and the specific heat of nanostructured materials. J Phys Chem C 113:16896–16900
18. Defay R, Prigogine I (1966) Surface tension and adsorption. Longmans, London
19. Christenson HK (2001) Confinement effects on freezing and melting. J Phys Cond Matter 13:R95-R134
20. Jackson CL, McKenna GB (1990) The melting behaviour of organic materials confined in porous solids. J Chem Phys 93:9002–9011
21. Sun J, Simon SL (2007) The melting behavior of aluminum nanoparticles. Thermochim Acta 463:32–40
22. Schmidt M, Kusche R, Kronmuller W, von Issendorff B, Haberland H (1997) Experimental determination of the melting point and heat capacity for a free cluster of 139 sodium atoms. Phys Rev Lett 79:99–102
23. Schmidt M, Kusche R, von Isserdorff B, Haberland H (1998) Irregular variations in the melting point of size-selected atomic clusters. Nature 393:238–240
24. Shvartsburg AA, Jarrold MF (2000) Solid clusters above the bulk melting point. Phys Rev Lett 85:2530–2532
25. Breaux GA, Neal CM, Cao B, Jarrold MF (2005) Tin clusters that do not melt: calorimetry measurements up to 650 K. Phys Rev B 71:073410
26. Cohen ML, Chou MY, Knight WD, Heer WA de (1987) Physics of metal clusters. J Phys Chem 91:3141–3149
27. Wood DM (1981) Classical size dependence of the work function of small metallic spheres. Phys Rev Lett 46:749–749
28. Rienstra-Kiracole JC, Tschumper GS, Schaefer HF et al (2002) Atomic and molecular electron affinities: photoelectron experiments and theoretical computations. Chem Rev 102:231–282
29. Liu S-R, Zhai H-J, Castro M, Wang L-S (2003) Photoelectron spectroscopy of Ti_n- clusters (n = 1–130). J Chem Phys 118:2108–2115

30. Zhirnov VV, Shenderova OA, Jaeger D L et al (2004) Electron emission properties of detonation nanodiamonds. Phys Solid State 46:657–661
31. Edmonds MT, Pakes CI, Mammadov S et al (2011) Surface band bending and electron affinity as a function of hole accumulation density in surface conducting diamond. Appl Phys Lett 98:102101
32. Zhang C, Andersson T, Svensson S et al (2011) Ionic bonding in free nanoscale NaCl clusters. J Chem Phys 134:124507
33. Ram S (2001) Allotropic phase transformations in hcp, fcc and bcc metastable structures in Co-nanoparticles. Mater Sci Eng A 304–306:923–927
34. Oshima Y, Nangou T, Hirayama H, Takayanagi K (2001) Face centered cubic indium nano-particles studied by UHV-transmission electron microscopy. Surf Sci 476:107–114
35. Gorbunov BZ, Kokutkina NA, Kutsenogii KP, Moroz EM (1979) Influence of the sizes of silver-iodide particles on their crystal-structure. Kristallografiya 24:334–337 (in Russian)
36. Johansson J, Dick KA, Caroff P et al (2010) Diameter dependence of the wurtzite–zinc blende transition in InAs nanowires. J Phys Chem C 114:3837–3842
37. Zhu H, Ma Y, Zhang H et al (2008) Synthesis and compression of nanocrystalline silicon carbide. J Appl Phys 104:123516
38. Jiang JZ, Olsen JS, Gerward L, Morup S (1998) Enhanced bulk modulus and reduced transition pressure in γ-Fe_2O_3 nanocrystals. Europhys Lett 44:620–626
39. Haase M, Alivisatos AP (1992) Arrested solid-solid phase transition in 4-nm-diameter cadmium sulfide nanocrystals. J Phys Chem 96:6756–6762
40. Seo WS, Shim JH, Oh SJ et al (2005) Phase- and size-controlled synthesis of hexagonal and cubic CoO nanocrystals. J Am Chem Soc 127:6188–6189
41. McHale JM, Auroux A, Perrota AJ, Navrotsky A (1997) Surface energies and thermodynamic phase stability in nanocrystalline aluminas. Science 277:788–791
42. Pitcher MW, Ushakov SV, Navrotsky A et al (2005) Energy crossovers in nanocrystalline zirconia. J Am Ceram Soc 88:160–167
43. Ramana CV, Vemuri RS, Fernandez I, Campbell AL (2009) Size-effects on the optical properties of zirconium oxide thin films. Appl Phys Lett 95:231905
44. Lu F, Zhang J, Huang M et al (2011) Phase transformation of nanosized ZrO_2 upon thermal annealing and intense radiation. J Phys Chem C 115:7193–7201
45. Zhang H, Chen B, Banfield JF (2008) Atomic structure of nanometer-sized amorphous TiO_2. Phys Rev B 78:214106
46. Biswas K, Muthu SDV, Sood AK et al (2007) Pressure-induced phase transitions in nanocrystalline ReO_3. J Phys Cond Matter 19:436214
47. Jiang JZ (2004) Phase transformations in nanocrystals. J Mater Sci 39:5103–5110
48. Sakurai S, Namai A, Hashimoto K, Ohkoshi S-I (2009) First observation of phase transformation of all fur Fe_2O_3 phases. J Am Chem Soc 131:18299–18303
49. Araujo LL, Giulian R, Sprouster DJ et al (2008) Size-dependent characterization of embedded Ge nanocrystals: Structural and thermal properties. Phys Rev B78:094112
50. Ikemoto H, Goyo A, Miyanaga T (2011) Size dependence of the local structure and atomic correlations in tellurium nanoparticles. J Phys Chem C115:2931–2937
51. Penn D (1962) Wave-number-dependent dielectric function of semiconductors. Phys Rev 128:2093–2097
52. Moss TS (1950) A relationship between the refractive index and the infra-red threshold of sensitivity for photoconductors. Proc Phys Soc B63:167–175
53. Dionne G, Wooley JC (1972) Optical properties of some $Pb_{1-x}Sn_xTe$ alloys determined from infrared plasma reflectivity measurements. Phys Rev B6:3898–3913
54. Grzybowski TA, Ruoff AL (1984) Band-overlap metallization of BaTe. Phys Rev Lett 53:489–492
55. Herve P, Vandamme LKJ (1994) General relation between refractive index and energy gap in semiconductors. Infrared Phys Technol 35:609–615
56. Wang L-W, Zunger A (1994) Dielectric constants of silicon quantum dots. Phys Rev Lett 73:1039–1042

57. Wang L-W, Zunger A (1996) Pseudopotential calculations of nanoscale CdSe quantum dots. Phys Rev B53:9579–9582
58. Delerue C, Lannoo M, Allan G (2003) Concept of dielectric constant for nanosized systems. Phys Rev B68:115411
59. Delerue C, Allan G (2006) Effective dielectric constant of nanostructured Si layers. Appl Phys Lett 88:173117
60. Nakamura J, Ishihara S, Natori A et al (2006) Dielectric properties of hydrogen-terminated Si(111) ultrathin films. J Appl Phys 99:054309
61. Pham TA, Li T, Shankar S et al (2010) First-principles investigations of the dielectric properties of crystalline and amorphous Si_3N_4 thin films. Appl Phys Lett 96:062902
62. Kageshima H, Fujiwara A (2010) Dielectric constants of atomically thin silicon channels with double gate. Appl Phys Lett 96:193102
63. Arlt G, Hennings D, de With G (1985) Dielectric properties of fine-grained barium titanate ceramic. J Appl Phys 58:1619–1625
64. Frey MH, Payne DA (1996) Grain-size effect on structure and phase transformations for barium titanate. Phys Rev B 54:3158–3168
65. Zhao Z, Buscaglia V, Viviani M et al (2004) Grain-size effects on the ferroelectric behavior of dense nanocrystalline $BaTiO_3$ ceramics. Phys Rev B 70:024107
66. Buscaglia V, Buscaglia MT, Viviani M (2006) Grain size and grain boundary-related effects on the properties of nanocrystalline barium titanate ceramics. J Eur Ceram Soc 26:2889–2898
67. Curecheriu L, Buscaglia MT, Buscaglia V et al (2010) Grain size effect on the nonlinear dielectric properties of barium titanate ceramics. Appl Phys Lett 97:242909
68. Smith MB, Page K, Siegrist Th (2008) Crystal structure and the paraelectric-to-ferroelectric phase transition of nanoscale $BaTiO_3$. J Am Chem Soc 130:6955–6963
69. Golego N, Studenikin SA, Cocivera M (1998) Properties of dielectric $BaTiO_3$ thin films prepared by spray pyrolysis. Chem Mater 10:2000–2005
70. Chavez E, Fuentes S, Zarate RA, Padilla-Campos L (2010) Structural analysis of nanocrystalline $BaTiO_3$. J Mol Struct 984:131–136
71. Yashima M, Hoshina T, Ishimura D et al (2005) Size effect on the crystal structure of barium titanate nanoparticles. J Appl Phys 98:014313
72. Sebald J, Krohns S, Lunkenheimer P et al (2010) Colossal dielectric constants: a common phenomenon in $CaCu_3Ti_4O_{12}$ related materials. Solid State Commun 150:857–860
73. Ren P, Yang Z, Zhu WG et al (2011) Origin of the colossal dielectric permittivity and magnetocapacitance in $LuFe_2O_4$. J Appl Phys 109:074109
74. Kalyani AK, Garg R, Ranjana R (2009) Tendency to promote ferroelectric distortion in Pr-modified $PbTiO_3$. Appl Phys Lett 95:222904
75. Liu XQ, Wu SY, Chen XM, Zhu HY (2008) Giant dielectric response in two-dimensional charge-ordered nickelate ceramics. J Appl Phys 104:054114
76. Liu XQ, Wu SY, Chen XM, Zhu HY (2009) Temperature-stable giant dielectric response in orthorhombic samarium strontium nickelate ceramics. J Appl Phys 105:054104
77. Liu XQ, Wu SY, Chen XM (2010) Enhanced giant dielectric response in Al-substituted $La_{1.75}Sr_{0.25}NiO_4$ ceramics. J Alloys Compd 507:230–235
78. Liping S, Lihua H, Hui Zh et al (2008) La substituted Sr_2MnO_4 as a possible cathode material in SOFC. J Power Sources 179:96–100
79. Jin C, Liu J (2009) Preparation of $Ba_{1.2}Sr_{0.8}CoO_{4+\delta}$ K_2NiF_4-type structure oxide and cathodic behavioral of $Ba_{1.2}Sr_{0.8}CoO_{4+\delta}$-GDC composite cathode for intermediate temperature solid oxide fuel cells. J Alloys Compd 474:573–577
80. Huang J, Jiang X, Li X, Liu A (2009) Preparation and electrochemical properties of $La_{1.0}Sr_{1.0}FeO_{4+\delta}$ as cathode material for intermediate temperature solid oxide fuel cells. J Electroceram 23:67–71
81. Homes CC, Vogt T, Shapiro SM et al (2001) Optical response of high-dielectric-constant perovskite-related oxide. Science 293:673–676
82. Ni L, Chen XM (2007) Dielectric relaxations and formation mechanism of giant dielectric constant step in $CaCu_3Ti_4O_{12}$ ceramics. Appl Phys Lett 91:122905

83. Xiao Y, Chen XM, Liu XQ (2008) Stability and microwave dielectric characteristics of $(Ca_{1-x}Sr_x)LaAlO_4$ ceramics. J Electroceram 21:154–159
84. Shia C-Y, Hu Z-B, Hao Y-M (2011) Structural, magnetic and dielectric properties of $La_{2-x}Ca_xNiO_{4+\delta}$. J Alloys Compd 509:1333–1337
85. Krohns S, Lunkenheimer P, Kant Ch et al (2009) Colossal dielectric constant up to gigahertz at room temperature. Appl Phys Lett 94:122903
86. Li W, Auciello O, Premnath RN, Kabius B (2010) Giant dielectric constant dominated by Maxwell–Wagner relaxation in Al_2O_3/TiO_2 nanolaminates synthesized by atomic layer deposition. Appl Phys Lett 96:162907
87. Hiramatsu T, Tamura T, Wada N (2005) Effects of grain boundary on dielectric properties in fine-grained $BaTiO_3$ ceramics. Mater Sci Eng B 120:55–58
88. Lunkenheimer P, Bobnar V, Pronin AV et al (2002) Origin of apparent colossal dielectric constants. Phys Rev B 66:052105
89. Lunkenheimer P, Fichtl R, Ebbinghaus SG, Loidl A (2004) Nonintrinsic origin of the colossal dielectric constants in $CaCu_3Ti_4O_{12}$. Phys Rev B 70:172102
90. Valdez-Nava Z, Guillemet-Fritsch S, Tenailleau Ch et al (2009) Colossal dielectric permittivity of $BaTiO_3$-based nanocrystalline ceramics sintered by spark plasma sintering. J Electroceram 22:238–244
91. Smith-Rose RL (1934) Electrical measurements on soil with alternating currents. J Instr Electr Eng London 80:379
92. Scott JH, Carroll RD, Cunningham DR (1967) Dielectric constant and electrical conductivity measurements of moist rock: a new laboratory method. J Geophys Res 72:5101–5115
93. Chelidze TL, Gueguen Y (1999) Electrical spectroscopy of porous rocks: a review. I. Theoretical models. Geophys J Int 137:1–15
94. Chelidze TL, Gueguen Y, Ruffet C (1999) Electrical spectroscopy of porous rocks: a review. II. Experimental results and interpretation. Geophys J Int 137:16–34
95. Sommer AP, Zhu D, Bruhne K (2007) Surface conductivity on hydrogen-terminated nanocrystalline diamond: implication of ordered water layers. Cryst Growth Des 7:2298–2301
96. Sommer AP, Zhu D (2008) From microtornadoes to facial rejuvenation: implication of interfacial water layers. Cryst Growth Des 8:3889–3892
97. Brovchenko I, Oleinikova A (2008) Interfacial and confined water. Elsevier, Amsterdam
98. Sommer AP, Hodeck KF, Zhu D et al (2011) Breathing volume into interfacial water with laser light. J Phys Chem Lett 2:562–565
99. Batsanov SS, Poyarkov KB, Gavrilkin SM (2008) Orientational polarization of molecular liquids in contact with diamond crystals. JETP Lett 88:595–596
100. Gavrilkin SM, Poyarkov KB, Matseevich BV, Batsanov SS (2009) Dielectric properties of diamond powder. Inorg Mater 45:980–981
101. Batsanov SS, Poyarkov KB, Gavrilkin SM (2009) The effect of the atomic structure on dielectric properties of nanomaterials. Doklady Phys 54:407–409
102. Batsanov SS, Poyarkov KB, Gavrilkin SM et al (2011) Orientation of water molecules by the diamond surface. Russ J Phys Chem 85:712–715
103. Batsanov SS, Gavrilkin SM, Batsanov AS et al (2012) Giant dielectric permittivity of detonation-produced nanodiamond is caused by water. J Mater Chem 22:11166–11172
104. Nebel CE (2007) Surface-conducting diamond. Science 318:1391–1392
105. Meyers MA, Mishra A, Benson DJ (2006) Mechanical properties of nanocrystalline materials. Progr Mater Sci 51:427–556
106. Zhang LL, Zhao XS (2009) Carbon-based materials as supercapacitor electrodes. Chem Soc Rev 38:2520–2531

Chapter 9
Phase Transition

9.1 Polymorphism

Mitscherlich in 1823 described the phenomenon of polymorphism, i.e. the existence of different crystal modifications of the same chemical substance, and its study went alongside the research of isomorphism (see below). Thus, it was established that MCO_3-type compounds where M = Mg, Zn, Co, Fe, Mn, Cd, have the structure of calcite with $N_c(M) = 6$, while for M = Sr, Pb, Ba, i.e. for cations with larger sizes, they adopt the structure of aragonite with $N_c(M) = 9$. Such regular change of the structure upon variations in the chemical composition within homologous series of compounds is called *morphotropy*. For $CaCO_3$ both modifications are known, which gave the names to these structural classes: the low-temperature form is the mineral calcite, the high-temperature one is aragonite. The calcite → aragonite transformation is a typical case of polymorphism. On the other hand, aragonite is a typical illustration of the importance of the kinetic factor: at ambient conditions this form is thermodynamically unstable with respect to calcite, but the transition is hindered kinetically and only occurs on geological time-scale (10^7–10^8 years).

From the structural point of view, polymorphic transformations can be of two types [1, 2], viz. 'reconstructive transitions' with change in N_c of atoms (Fig. 9.1) and 'displacive transitions', in which the positions of atoms change insignificantly and N_c remains the same (Fig. 9.2). For structural chemistry, the former transformations are most important. As energies of phase transitions amount (at most) to several percent of the atomization energy of solids, exact calculations of the thermodynamic stability of phases are very difficult. At present, the most effective are crystal-chemical approaches to estimating the reasons and results of polymorphism.

Phase transitions can also be classified by their thermodynamic 'order', which is the order of the derivative of the Gibbs free energy. In the present book we are mainly interested in pressure-induced transformations, which involve a discontinuous change of volume, i.e. first-order phase transitions. In second-order phase transitions the second derivative of the Gibbs free energy with respect to volume, which is proportional to the compressibility, is discontinuous. Experimentally it can be very difficult to distinguish between a second-order phase transition and a weakly first-order one.

S. S. Batsanov, A. S. Batsanov, *Introduction to Structural Chemistry*,
DOI 10.1007/978-94-007-4771-5_9, © Springer Science+Business Media Dordrecht 2012

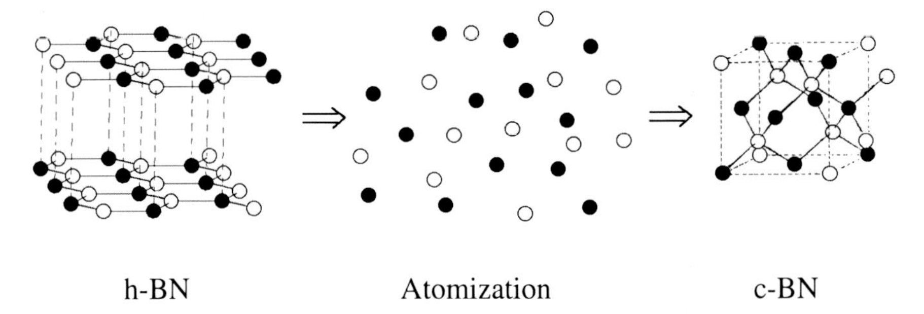

Fig. 9.1 An example of reconstructive phase transition: transformation of hexagonal boron nitride (h-BN) into cubic boron nitride (c-BN) is impossible without complete atomization of the parent solid (here achieved by detonating a mixture of BN and an explosive). The coordination number changes from 3 to 4

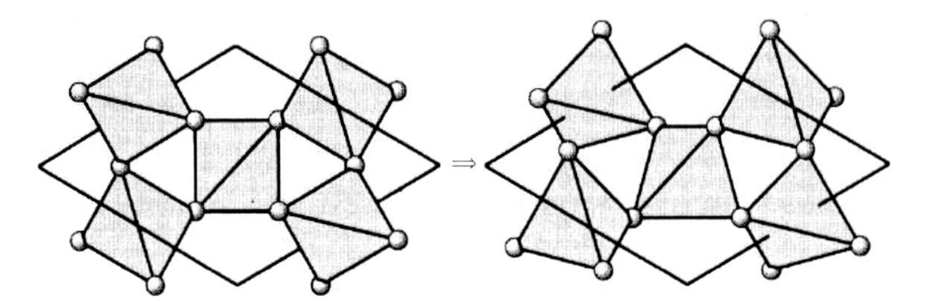

Fig. 9.2 An example of a displacive phase transition: the transformation of β-quartz (high-temperature phase) into α-quartz (low-temperature phase). A view down the hexagonal axis, showing O_4 tetrahedra and the unit cell outline

Presenting the structures of an ionic crystal as a close packing of anions with cations in the voids, it is easy to determine the ratios of cation and anion radii ($k_r = r_+/r_-$) which correspond to various coordination polyhedra. Thus, for a triangle the lower limit is $k_r = 2/\sqrt{3} - 1 = 0.155$, for a tetrahedron $k_r = \sqrt{6}/2 - 1 = 0.255$, for an octahedron $k_r = \sqrt{2} - 1 = 0.414$, for a cube $k_r = \sqrt{3} - 1 = 0.732$. Compression of an ionic compound (which is essentially a compression of anions) increases k_r up to a critical value, whereupon a phase transition occurs with an increase of N_c and of packing density. A similar phenomenon occurs on replacement of a smaller cation by a larger one (the case of morphotropy). However, such simple model does not always correctly predict the coordination of atoms in crystal structures even for oxides, because in reality chemical bonds are not purely ionic and the oxygen ion is sufficiently rigid [3]. If we consider coordination numbers in the structures of halides, the following picture emerges: N_c(MF) as a rule is smaller than N_c(MX) where X = Cl, Br, I; but $N_c(MF_n) \geq N_c(MX_n)$ where $n \geq 3$ [4], although the ionic theory predicts (F^- being smaller) the following succession: N_c(MF) > N_c(MCl) > N_c(MBr) > N_c(MI).

Divergence of the ionic model from the reality is due to repulsion of anions, a destabilizing factor in the structure which depends not only on anion-anion distances but also on effective charges (e^*) of atoms. To relate the effect of repulsion only to the ionic size, it is necessary to make a correction for the differences of the $(e^*)^2$ in defining the Coulombic energy. In alkaline halides the ratios of $(e^*)^2$ for identical cations are equal to: F/Cl = 1.245, Br/Cl = 0.923, I/Cl = 0.826. Multiplying the ideal ionic radii (Table 1.18) by these factors yields certain effective values which can be called the energetic radii of anions:

F^-	Cl^-	Br^-	I^-	O^{2-}	S^{2-}	Se^{2-}	Te^{2-}
2.30	2.25	2.20	2.15	3.05	2.35	2.25	2.15 Å

As can be seen, the energetic radii of anions have the reverse order of magnitudes in comparison with crystal-chemical radii, which explains the reduction of N_c in MF in comparison with others MX halides.

Besides accounting for the effect of repulsion, the geometrical estimation of N_c can be improved further, using the real radii of atoms in polar bonds, and also their hardness. According to the Born–Landé theory, the hardness of an ion is determined by the factor $f_n = n/(n-1)$, where n is the Born coefficient of repulsion. These factors are equal to 1.25 for ions with the He electronic shell, 1.167 for ions of the Ne type, 1.125 for the Ar type and Cu^+, 1.111 for the Kr type and Ag^+, and 1.091 for the Xe type and Au^+. Knowing the bond ionicity, one can calculate the real ionic radii, multiply them by the factor of hardness and so find the real k_r. The coordination numbers thus calculated, agree with the observed ones for some 80 % of structures [5].

The presence of a directed covalent bond (or its appreciable contribution) prevents dense packing of atoms, so such compounds have structures with $N_c \le 4$, while in ionic structures $N_c > 4$. Phillips and Lukowsky [6] used the ionicity factor $f_i = 0.785$ as a criterion to divide structures with $N_c = 4$ and 6. Taking into account that the bond ionicity according to Pauling (i) and the ionicity factor of Phillips are connected by Wemple's equation [7–9]:

$$i^2 = \frac{f_i}{(\varepsilon - 1)} \tag{9.1}$$

where ε is the dielectric constant (≈ 4 for crystalline compounds with polar bonds), the ionicity threshold of Phillips corresponds to $i = 0.5$. Thus, to predict the coordination number it is necessary to estimate the bond ionicity in a substance (see Chap. 2) and then to choose the calculation procedure: if $i > 0.5$ one should apply the geometrical criterion (k_r), if $i < 0.5$ the structure will have $N_c \le 4$. Such an approach gives perfect agreement between the predicted and experimental coordination.

Difficulties in the accurate calculation of bond ionicity and radii of atoms in polar bonds do not permit to predict the parameters of polymorphic transformations by the crystal-chemical method, but a global physical theory does not yet exist. This led to the development of the statistical approach, to structural maps with various coordinates, such as an electronegativity (χ), atomic radius (r), the number of valence electrons, etc. Thus, various structure types were plotted in 2D-maps with the coordinates $\bar{n} - \Delta\chi$, where $\bar{n}$ is the mean principal quantum number (Fig. 9.3) [10–12]; Burdett et al. [13] used as coordinates the Coulombic (C) and homo-polar (E_h)

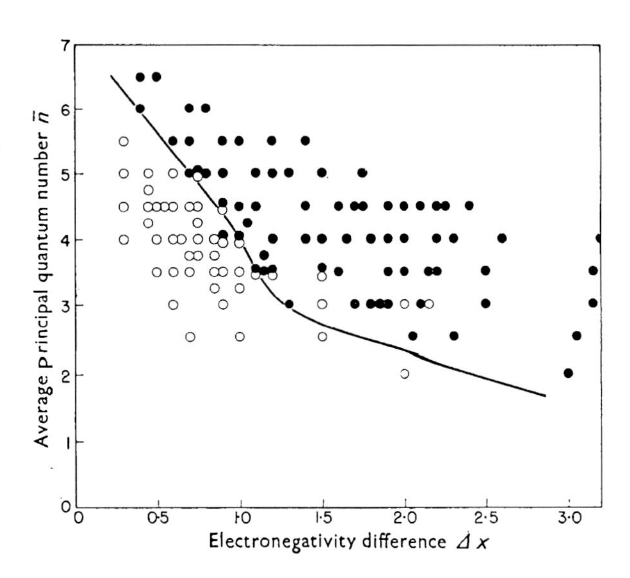

Fig. 9.3 Occurrence of octahedral (*solid circles*) and tetrahedral (*hollow circles*) coordination in AX solids. (Reproduced with permission from [10]; Copyright 1959 International Union of Crystallography)

components of the atomization energy; Pettifor [14–18] described the structures of A_nB_m compounds in terms of 'chemical electronegativities' χ_A and χ_B. For complex compounds, the coordinates Z/r_i and χ have been suggested [19]. Sproul [20] plotted binary compounds on a map with coordinates $\Delta\chi$ and $\bar{\chi}$ and located the regions of existence of covalent, ionic and metal substances. Burdett [21] considered some electronic aspects of structural maps. The structural maps for melted halides of polyvalent and alkaline metals were given by Tosi [22]. Finally, Villars [23–29] used 3D maps with the coordinates Δr_{AB}, $\Delta\chi_{AB}$, Σv_e, where $r_{A,B}$ are the orbital radii of atoms A and B, and Σv_e is the total number of valence electrons in these atoms.

Phase transformations in organic solids occur at milder changes of thermodynamic conditions and consist, as a rule, in changing the orientations of molecules in the crystal space (Fig. 9.4). The geometry of molecules themselves change little and the coordination numbers not at all. Thus the growth of a new phase can be epitaxial, if initial and final molecules have similar structures, or disordered (random) if there is a structural distinction between initial and final states of the system. Crystal-chemical and thermodynamic aspects of polymorphism in organic compounds, as well as issues of terminology are considered in [30–34].

9.1.1 Polyamorphism

In principle, amorphous solids can change their (short-range) atomic structure in the same manner as crystalline solids do. So far, all cases of this so-called polyamorphism have been observed at non-ambient pressures. Thus, the N_c* in amorphous (a-Si) and liquid silicon was found to increase monotonically with pressure [36] and later, a polyamorphous transition was observed [37]. Between the pressures of 3

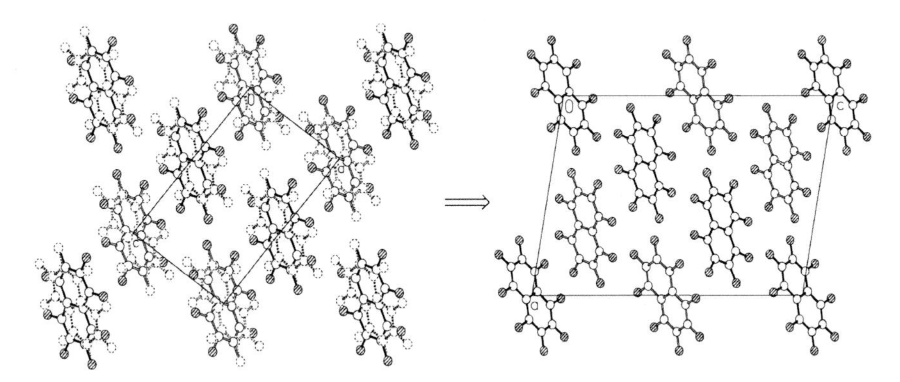

Fig. 9.4 An example of order-disorder phase transition. Molecules are rotating in the high-temperature form of octafluoronaphthalene (*left*) but are stationary in the low-temperature phase (*right*). (Adapted with permission from [35]. Copyright 2010 American Chemical Society)

and 13.5 GPa, the pair distribution function shows primary features at 2.36 Å (the Si–Si bond) and 3.83 Å (the next-nearest neighbor Si··· Si distance) indicative of the local tetrahedral coordination geometry. At $P > 10$ GPa, the latter peak shifts to a shorter separation of *ca.* 3.40 Å and the former to a slightly longer separation, which indicates the formation of a high-coordinate Si species. The reverse change occurs on decompression below 9.5 GPa. Analysis of the MD simulation configurations indicates that the phase transformation may be associated with formation of defect structures within the tetrahedrally-coordinated polyamorphous phase, in which highly coordinated atoms occupy interstitial sites in the amorphous network.

In the structure of the B_2O_3 glass at ambient conditions $N_c(B) = 3$ but it increases with pressure and temperature. According to X-ray diffraction data, at 8 GPa and 300 K the structure contains 15 % of boron atoms with $N_c = 4$, at 9.5 GPa and 300 K, 30 % and at 9.5 GPa and 650 K, 45 % [38], while an earlier solid-state NMR study suggested 5 % at 2 GPa and 27 % at 6 GPa [39].

Some substances show an interplay of polyamorphism and crystalline polymorphism. Thus, an insulator, molecular crystal phase I of SnI_4 transforms into a crystalline metallic phase II above 7.2 GPa, is pressure-amorphized at 15 GPa and re-crystallized at 61 GPa into a non-molecular, polymeric phase III with a *fcc*-lattice [40]. On compression of GeI_4, the Ge–I distance begins to lengthen from $P = 20$ GPa, this precedes the decomposition of this substance [41]. SnO shows strong anisotropy under compression: the *a*-axis is rather incompressible with a linear stiffness coefficient $K_{ao} = 306$ GPa, whereas for the *c*-axis $K_{co} = 43$ GPa. Although SnO is metastable at ambient conditions and decomposes at elevated temperature into Sn and SnO_2, nevertheless the decomposition rate under pressure is higher than at ambient conditions [42, 43].

It is known that when a static pressure is combined with a shear deformation of the sample, the phase transition often takes place at a lower pressure; the same may happen under a shock-wave compression, which always produces shear deformations. These effects can be explained by the shear causing partial amorphization of the sample. The enthalpy of a phase transition ($\Delta H_{tr} = P_{tr} \Delta V_{tr}$) consists of two components—the work of destroying the original order (W_{des}) and the formation work of new atomic arrangement (W_{for}):

$$\Delta H_{tr} = W_{des} + W_{for} \tag{9.2}$$

For a reconstructive phase transition $W_{des} = c\Delta H_m$, where ΔH_m corresponds to complete destruction of the long-range atomic order (as on melting), and c is the extent (fraction) of this destruction of the solid. A melt has a short-range order within the range of 10–30 Å, whereas a mechanical and shock-wave grinding of solids reduces the grain size down to 100 Å at most, hence $c \leq 0.2 \pm 0.1$. From these values of c and Eq. 9.3 it follows that a reduction of the phase transition pressure is

$$\Delta P_{tr} = P - W_{for}/\Delta V_{tr} \approx 0.2\Delta H_m/\Delta V_{tr} \tag{9.3}$$

where ΔH_m is the melting heat (see below) and ΔV_{tr} is the volume difference of the initial and final phases (see Chap. 10). Equation 9.4 gives the following ΔP_{tr} for ionic solids: ZnO 4.1, ZnS 2.9, ZnSe 1.1 GPa; CdS 1.5, CdSe 1.4 GPa; NaF 2.5, NaCl 3.1 GPa; KF 1.9, KCl 1.2, KBr 1.1, KI 1.3 GPa. These estimates indicate how a compression with shear can reduce the phase transition pressure of a solid. In the case of the melt $c = 1$ and the effect is several times larger.

9.2 Energies of Phase Transitions

The energy released during the transitions of a solid into a liquid, a liquid into a gas, and solid into a gas are the melting heat ΔH_m, the evaporation heat ΔH_v, and the sublimation heat ΔH_s, respectively; hence, ideally $\Delta H_s = \Delta H_v + \Delta H_m$. However, because these phase transitions are studied at different temperatures, actually $\Delta H_s \approx \Delta H_v + \Delta H_m$. From the structural viewpoint, ΔH_m in a crystalline solid corresponds to the loss of long-range order, whereas ΔH_v and ΔH_s may correspond to very different processes and be of different orders of magnitude. Evaporation of metals means the disruption of metallic bonds, therefore metals have very high ΔH_s, identical with the atomization energy E_a; hence $\Delta H_s \approx \Delta H_v \gg \Delta H_m$. The same is true for non-metals with polymeric covalent structures (e.g., diamond). Evaporation of solids with molecular structure requires only the disruption of vdW interactions rather than valence bonds, therefore ΔH_v is much lower and can be comparable with ΔH_m.

9.2.1 Melting Heats of Compounds

The melting heats ΔH_m for binary inorganic compounds are presented in Table 9.1, for elements in Table 9.3. Melting heats of organic and organometallic crystals with molecular structures are much smaller. Thus, ΔH_m (kJ/mol) = 5.4 for cyclobutane,

Table 9.1 Melting heats (kJ/mol) of inorganic compounds. Henceforth from [44, 45] except where specified

ν_M	M	F	Cl	Br	I	O	S
I	Li	27.1	19.8	17.7	14.6	35.6	
	Na	33.4	28.2	26.2	23.7	47.7	19.0
	K	27.2	26.3	25.5	24.0	27.0	16.2
	Rb	25.8	24.4	23.3	22.1	21.0	
	Cs	21.7	20.4	23.6	25.6	20.0	
	Cu		7.1	7.2	7.9	65.6	9.6
	Ag	16.7	13.0	9.2	9.4	15.5	7.9
	Tl	14.0	15.6	16.4	14.7	30.3	23.0
II	Cu	55.0	20.4			49	
	Be	4.8	8.7	18.0	21.0	86	
	Mg	58.7	43.1	39.3	26	77	63
	Ca	30.0	28.0	29.1	41.8	80	70
	Sr	29.7	16.2	10.5	19.7	81	63
	Ba	23.4	15.8	32.2	26.5	46	63
	Zn	40	10.3	15.7	17	70	30
	Cd	22.6	48.6	33.4	15.3		43
	Hg	23.0	19.4	17.9	15.6		26
	Sn	10.5	14.5	18.0	18.0	27.7	31.6
	Pb	14.7[c]	21.8[c]	16.4[c]	23.4[c]	25.6	49.4
	Cr	34	45	45	46		25.5
	Mn	30	30.7	33.5	41.8	43.9	26.1
	Fe	50	43	43	39	24	31.5
	Co	58.1	46.0	43	35	50	30
	Ni	69	78	56	48	50.7	30.1
MX[d]	TiN	VN	CrN	MnN	FeN	CoN	NiN
	50	21	25	36	13.5	45	28
MX	ZnSe	ZnTe	CdTe	HgSe	HgTe	SnTe	PbTe
	24[a]	56[ab]	48.5[e]	28[a]	36[a]	45.2	47.4[f]
M_2O_3	B_2O_3	Al_2O_3	Ga_2O_3	In_2O_3	Tl_2O_3	Sc_2O_3	Y_2O_3
	24.5[g]	111[g]	99.8[g]	105[g]	105[g]	127	105
	La_2O_3	Ti_2O_3	As_2O_3	Sb_2O_3	Bi_2O_3	V_2O_3	Cr_2O_3
	125	92	30.1	61.5	28.4	140	125
M_2S_3	B_2S_3	Al_2S_3	Sb_2S_3	Bi_2S_3	Mo_2S_3		
	48.5	66	65.3	37.2	130		
MO_2	SiO_2	GeO_2	SnO_2	TiO_2	ZrO_2	HfO_2	
	9.6[g]	17.2[g]	23.4[g]	68	90	96	
	MoO_2	WO_2	TcO_2	ReO_2	PtO_2	ThO_2	UO_2
	66.9	48.1	75.3	50.2	19.2	90	78

[a][46], [b][47–52], [c][53], [d][54], [e][55], [f][56], [g][57]

5.8 for cyclopentane, 3.6 for cyclohexane, 3.8 for cycloheptane, 3.1 for cyclooctane [58], 17.8 for ferrocene and 19.0 for nickelocene [59].

Knowing the melting energy, one can estimate the extent of the short-range order in liquids. Let us suppose that solid ZnS, NaCl and CsCl (to take representative examples) are being split into cubic blocks of diminishing size and calculate the

Table 9.2 Estimation of the short-range order in melts by the 'grinding' method

Structural type	ZnS		NaCl		CsCl	
Number of bonds						
In vertices	4		24		8	
On edges	$12(N-1)$		$48(N-1)$		$24(N-1)$	
On faces	$12(2N^2-2N+1)$		$30(\bar{I}^2-2N+1)$		$24(N-1)^2$	
In volume	$4(4N^3-6N^2+3N-1)$		$6(4N^3-6N^2+3N-1)$		$8(N-1)^3$	
Σ bonds	$16N^3$		$6(4N^3+4N^2+\bar{n})$		$8N^3$	
Number of atoms						
In vertices	8		8		8	
On edges	$12(N-1)$		$12(N-1)$		$12(N-1)$	
On faces	$6(2N^2-2N+1)$		$6(2N^2-2N+1)$		$6(N-1)^2$	
In volume	$4N^3-6N^2+3N-1$		$4N^3-6N^2+3N-1$		$(N-1)^3$	
Σ atoms	$4N^3+6N^2+3N+1$		$4N^3+21N-11$		$(N+1)^3$	
N	$\overline{N}_c$	$\Delta N_c/N_c$	$\overline{N}_c$	$\Delta N_c/N_c$	$\overline{N}_c$	$\Delta N_c/N_c$
1	1.14	0.72	3.86	0.36	1.00	0.88
2	2.03	0.49	4.76	0.21	2.37	0.70
4	2.80	0.30	5.32	0.11	4.10	0.49
6	3.14	0.21	5.53	0.08	5.05	0.37
10	3.45	0.14	5.71	0.05	6.01	0.25
15	3.62	0.09	5.80	0.03	6.59	0.18
20	3.72	0.07	5.85	0.025	6.91	0.14
25	3.77	0.06	5.88	0.02	7.11	0.11

numbers of atoms and bonds in such blocks on every stage. As objects of study we take crystals of these structural types. Table 9.2 shows numbers of atoms and bonds on the surface and in the interior of the blocks with sizes $N = D/a$, where D is the block edge and a is the unit cell parameter. Ratios of the total number of bonds to the total number of atoms in the blocks give the mean coordination numbers ($\overline{N}_c$) for different grains, whereas the ratio $\Delta N_c/N_c$ defines the number of bonds in the coordination polyhedron which are broken during such 'grinding'. For halides MX of the B3, B1 and B3 types these ratios are equal to 0.09, 0.115 and 0.12, respectively, which corresponds to $\overline{N}_c = 3.6, 5.3$ and 7. These values of $\overline{N}_c$ are close to the experimental N_c* in the corresponding melts (see Chap. 7).

9.2.2 Sublimation Heats of Elements and Compounds

The sublimation (ΔH_s) heats of elements at normal thermodynamic conditions are listed in Table 9.3. ΔH_s vary periodically (Fig. 9.5), similarly to the atomization energies. The $\Delta H_m/\Delta H_s$ ratio for metals is almost constant (0.035 on average), for semiconductor elements with continuous covalent-bond network it is about three times higher, while for molecular structures it is close to 1. Crystalline inorganic compounds with continuous bond network have approximately the same $\Delta H_m/\Delta H_s$ ratios as metals. Values of ΔH_s for oxides halides, and chalcogenides are presented in Tables 9.4–9.6.

Table 9.3 Sublimation (upper lines) and melting heats (lower lines) of elements (kJ/mol)

Li	Be	B	C	N	O	F	Ne	H	He
159.3	324	565	716.7	472.7	249.2	79.4	2.14	1.028	0.09
3.0	7.9	50.2	117	0.71	0.44	0.51	0.33	0.116	0.018
Na	Mg	Al	Si	P	S	Cl	Ar		
107.5	147.1	330.9	450.0	316.5	277.2	121.3	7.70		
2.6	8.5	10.7	50.2	0.66	1.72	6.40	1.18		
K	Ca	Sc	Ti	V	Cr	Mn	Fe	Co	Ni
89.0	177.8	377.8	473.4	515.5	397.5	283.3	415.5	426.7	430.1
2.3	8.55	14.1	14.15	21.5	21.0	12.9	13.8	16.2	17.5
Cu	Zn	Ga	Ge	As	Se	Br	Kr		
337.4	130.4	272.0	372	302.5	227.2	111.9	10.68		
13.3	7.1	5.6	36.9	24.4	6.7	10.6	1.64		
Rb	Sr	Y	Zr	Nb	Mo	Tc	Ru	Rh	Pd
80.9	164.0	424.7	610.0	733.0	659.0	678	650.6	556	376.6
2.2	7.4	11.4	21.0	30	37.5	33.3	38.6	26.6	16.7
Ag	Cd	In	Sn	Sb	Te	I	Xe		
284.9	111.8	243	301.2	264.4	196.6	106.8	14.93		
11.3	6.2	3.3	7.15	19.8	17.4	15.5	2.27		
Cs	Ba	La	Hf	Ta	W	Re	Os	Ir	Pt
76.5	179.1	431.0	618.4	782.0	851.0	774	787	669	565.7
2.1	7.1	9.7	27.2	36.6	52.3	34.1	57.8	41.1	22.2
Au	Hg	Tl	Pb	Bi	Po	At	Th	Pa	U
368.2	61.4	182.2	195.2	209.6	147.0	97.2	602	563	533
12.55	2.3	4.15	4.8	11.1	12.5		13.8	12.3	9.14
	Ra	Ac							
	159	406							
	7.7	12.0							

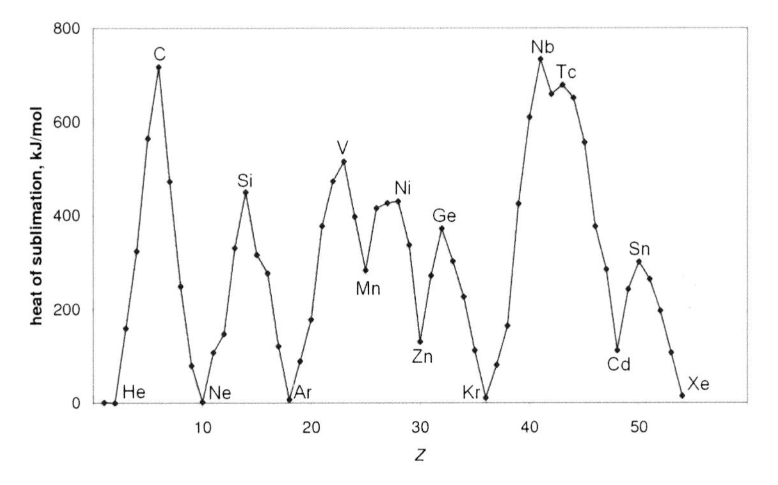

Fig. 9.5 Sublimation heats of elementary solids, Periods 1–5

Table 9.4 Sublimation heats (kJ/mol) of metal oxides

M_2O	Li	Na	K	Rb	Cs	Tl
	425	385	304	272	187	198[a]
MO	Cu	Be	Mg	Ca	Sr	Ba
	462	742[a]	660[a]	679[a]	589[a]	420[a]
	Zn	Cd	Hg	Sn	Pb	Mn
	441[a]	388[a]	184[a]	302[a]	287[a]	509
	Fe	Co	Ni	Pd		
	520	557	505	452		
M_2O_3	Al	Ga	In	As	Sb	Bi
	843	527	439	110	192	290
MO_2	Zr	Hf	Si	Ge	Sn	Pb
	805	872	589[a]	472[a]	536[a]	282[a]

[a] [57]

The coordination numbers of atoms in crystalline compounds exceed their valences, resulting in lower electron density per bond and therefore higher bond polarity. Hence the dependence of ΔH_s on the atomic charges, as has been shown firstly by Fajans [71] (see also [72]) and later by Urusov [73]. This dependence can be elucidated by comparing the sublimation heats of polar compounds with those of the corresponding elements, where the bonds are perfectly covalent. Alkali halides provide the most suitable example, where a crystal → molecule transition changes only the polarity of the chemical bond. Therefore the sublimation heat of a compound exceeds the sum of those of the components.

The sublimation heats of molecular solids, due to weak vdW interactions, are much lower [83, 84]. As a rule, they increase with the size of the molecule, e.g. $\Delta H_s =$ 9.20 kJ/mol for methane, 20.09 for ethane, 27.43 for propane, 49.68 for benzene, 72.63 for naphthalene, 73.42 kJ/mol for ferrocene. For elemental substances, the sublimation heat increases with an increase of polarizability, *viz.* for N_2, Ar and I_2, $\Delta H_s = 6.88$, 7.73 and 62.44 kJ/mol, respectively. Polar molecules, although having lower polarizability, show much higher ΔH_s due to ion-dipole interactions or hydrogen bonding, *cf.* $\Delta H_s = 20.08$ kJ/mol for HCl, 24.15 for N_2O, 26.19 for CO_2, 29.22 for NH_3 and 47.35 for H_2O.

9.2.3 Evaporation Heats of Compounds

As mentioned above, the evaporation heat can be calculated as the difference between the sublimation heat and the melting heat. A survey of direct measurements of ΔH_v from 1910 to 2010 is available [85]; extracts from it are presented in Table 9.7, together with data from the reference book [45] and papers [86]. A comparison with Table 9.6 reveals that ΔH_v are close to ΔH_s for various substances.

Evaporation heats of organic compounds are additive properties, i.e. can be expressed through the sum of increments corresponding to individual atoms, chemical bonds and/or radicals. Thus, Korolev et al. [87] derived the formula ΔH_v (kJ/mol) $= \Delta H_{vX} + \Delta H_{vCH}$ [85], where ΔH_{vX} and ΔH_{vCH} are the increments of

Table 9.5 Sublimation heats of metal chalcogenides (kJ/mol)

Comp.	ΔH_s	Ref.	Comp.	ΔH_s	Ref.	Comp.	ΔH_s	Ref.
Cs_2Te	160	[60]	ZnSe	377	[64]	SnS	220	[68]
Cu_2Se	126	[61]	ZnTe	332	[64]	SnSe	214	[45]
CuS	316	[45]	CdSe	326	[65]	SnTe	222	[45]
CuSe	215	[61]	CdTe	293	[65]	PbS	233	[69], [70]
Ag_2Se	110	[62]	GeS	167	[66]	PbSe	226	[45]
Ag_2Te	160	[63]	GeTe	197	[67]	PbTe	224	[45]

Table 9.6 Sublimation heats of metal halides (kJ/mol)

MX_n	M	X				MX_n	M	X			
		F	Cl	Br	I			F	Cl	Br	I
MX	Li	276	213	195	184	MX_3	Au	364	118	52.8	
	Na	284	230	218	214		B	23.6	38.2		64.0
	K	242	223	216	205		Al	300	119	103	98
	Rb	223	211	203	196		Ga	255	80.3		94.5
	Cs	194	202	202	196		In	335	161	143	119
	Cu		218	226	223		Sc[g]	376	310	276	233
	Ag	214	234	197	221		Y[g]	441	321	293	275
	Au		224	208	224		La[g]	442	326	299	281
	Tl	143	135	133	140		U	470	221	307	
MX_2	Cu	261	220	142			Ti	247	179	176	
	Be	228	134	130	130		V	299	214	188	
	Mg	385	242	226	205		As	46.2	54.4	66.9	95.0
	Ca	436	311	298	279		Sb	102	69.2	78.7	92.5
	Sr	444	348	317	292		Bi	201	116	111	117
	Ba	395	358	343	321		Cr	253	226[h]	237	
	Zn	266	149	144	141		Mn	284			
	Cd	305	196	175	144		Fe	254	141[i]	140[i]	
	Hg	136[a]	79.4	84.1	89.5	MX_4	Ti	99.2	57.1	67.4	98.3
	Sn[b]	168	134	139	149		Zr	238	112[y]	116	128
	Pb	239	186	177	167		Hf	248	102[y]	105	116
	V		246	212	234		C[a]	14.7	43.3	54.5	
	Cr	361	262	239	270		Si	24.7	43.1		66.9
	Mn	319	218	210	208[c]		Ge[a]	31.0	44.6	58.6	87.1
	Fe	316	209	208	173		Sn	146	51.0	57.7	79.1
	Co	312[d]	226	216[e]	192		Th	340	249[f]	197	159
	Ni	317[d]	238	226[e]	214		U	319	213[f]	225	229
	Pt	264					W[j]	465	200	207	218
MCl_4^f	Nb	Ta	Mo	W	Tc	MCl_4^f	Re	Ru	Os	Rh	Ir
	128	144	113	160	112		141	109	147	83	115

[a][74], [b][53], [c][75], [d][76], [e][77], [f][78], [g][79], [h][80], [i][81], [j][82]

evaporation of X and C–H bond, or a more general equation

$$\Delta H_v\,(\text{kJ/mol}) = 4.69\left(n_C - n_Q\right) + 1.3n_Q + a + 3.0$$

where n_C is the sum of all C-atoms, n_Q is the number of quaternary C-atoms, a is the increment of the atomic group. Table 9.8 shows the contributions of different functional groups to the evaporation enthalpy of molecules [88].

Table 9.7 Evaporation heats of inorganic and organo-metallic compounds (kJ/mol)

AH_n	ΔH_v	AF_n	ΔH_v	ACl_n	ΔH_v	ABr_n/AI_n	ΔH_v
HF	25.2	S_2F_2	14.9	S_2Cl_2	41.1	S_2Br_2	53.9
HCl	16.2	BeF_2	213.0	SCl_2	43.8	BBr_3	30.5
HBr	17.6	XeF_2	53.5	$BeCl_2$	105	$AlBr_3$	23.5
HI	19.8	BF_3	19.3	$ZnCl_2$	126	$GaBr_3$	38.9
H_2O_2	48.5	PF_3	16.5	$CdCl_2$	124.3	NBr_3	44.1
H_2O	40.6	AsF_3	20.8	$HgCl_2$	58.9	PBr_3	48.5
H_2S	19.5	ClF_3	27.5	$SnCl_2$	86.8	$AsBr_3$	41.8
H_2Se	19.7	BrF_3	47.6	$PbCl_2$	127	$SbBr_3$	59
H_2Te	19.2	CF_4	13.5	$CrCl_2$	197	$BiBr_3$	75.4
B_2H_6	14.3	SiF_4	15.4	$CoCl_2$	169.0	CBr_4	38.1
NH_3	22.7	SF_4	21.1	$NiCl_2$	147.2	$SiBr_4$	37.9
N_2H_4	41.8	SeF_4	46.4	BCl_3	23.8	$GeBr_4$	46.6
PH_3	14.6	TeF_4	34.3	$GaCl_3$	23.9	$SnBr_4$	43.5
P_2H_4	28.8	XeF_4	60.0	NCl_3	32.9	$TiBr_4$	44.4
AsH_3	16.7	VF_5	44.5	PCl_3	32.6	ZnI_2	117
SbH_3	21.3	NbF_5	52.3	$AsCl_3$	35.0	CdI_2	115
CH_4	8.2	TaF_5	56.9	$SbCl_3$	45.2	HgI_2	64.0
C_2H_2	16.3	PF_5	17.2	$BiCl_3$	72.6	PbI_2	104
C_2H_4	14.0	AsF_5	20.8	CCl_4	29.8	BI_3	40.5
C_2H_6	14.7	MoF_5	51.8	$SiCl_4$	29.9	AlI_3	32.2
SiH_4	12.1	BrF_5	30.6	$GeCl_4$	27.9	GaI_3	56.5
Si_2H_6	21.2	IF_5	41.3	$SnCl_4$	34.9	InI_3	95.4
GeH_4	14.1	ReF_5	58.1	$TiCl_4$	37.5	PI_3	43.9
Ge_2H_6	25.1	OsF_5	65.6	$ZrCl_4$	51.5	AsI_3	56.5
SnH_4	19.0	SF_6	17.1	VCl_4	41.4	SbI_3	68.6
C_4H_8	24.1	SeF_6	18.3	$TeCl_4$	77.0	BiI_3	77.7
C_5H_{10}	28.5	TeF_6	18.8	$NbCl_5$	52.7	SiI_4	50.2
C_6H_{12}	33.0	MoF_6	29.0	$TaCl_5$	54.8	GeI_4	64.2
C_6H_6	33.8	WF_6	25.8	$MoCl_5$	62.8	SnI_4	57.2
C_7H_{14}	38.5	OsF_6	28.1	WCl_6	52.7	TiI_4	58.5
C_8H_{16}	43.4	IrF_6	30.9				
		UF_6	29.5				
$M(CH_3)_2$	ΔH_v	$M(CH_3)_3$	ΔH_v	$M(CH_3)_4$	ΔH_v	$M(C_5H_5)_2$	ΔH_v
Zn	30.0	B	22.0	Si	25.5	Cr	49.5
Cd	37.3	Al	40.0	Ge	27.8	Fe	47.3
Hg	35.6	Tl	39.2	Sn	32.1	Os	56.3
Te	35.1	As	28.3	Pb	36.9	Ru	53.6
		Sb	30.0				
		Bi	35.5				

Vaporization enthalpy appears to be a linear function of the number of carbon atoms not only for alkanes or nitriles but also for alkyl-azides. The dependence of vaporization enthalpy (in kJ/mol) on the number of carbon atoms (N_C = 4–10) for mono-azides $CH_3{-}(CH_2)_n{-}N_3$ obeys the equation $\Delta H_v = 16.6 + 5.8\ N_C$. The dependence of vaporization enthalpy on the number of carbon atoms for di-azides $N_3{-}(CH_2)_n{-}N_3$ with n = (3 to 6) is described by the equation $\Delta H_v = 33.3 + 6.6\ N_C$ [89].

Table 9.8 Increments of the evaporation enthalpy in organic compounds (kJ/mol)

Class	Group	*a*	Class	Group	*a*	Class	Group	*a*
Acid	–COOH	38.8	Amide	–CONH	42.5	Ketone	> CO	10.5
Alcohol	–OH	29.4	Chloride	–Cl	10.8	Nitrile	–CN	16.7
Aldehyde	–CHO	12.9	Bromide	–Br	14.4	Nitro	$-NO_2$	22.8
Amide, pri	$-NH_2$	14.8	Iodide	–I	18.0	Nitrogen	= N–	12.2
Amide, sec	> NH	8.9	Ester	–COO	10.5	Sulfide	> S	13.4
Amide, ter	> N–	6.6	Ether	> O	5.0	Thiol	–SH	13.9

Table 9.9 Enthalpies of phase transitions of elements (kJ/mol)

M	ΔH_{tr}	ΔH_m	M	ΔH_{tr}	ΔH_m	M	ΔH_{tr}	ΔH_m
Ac[a]		10.8	Hf	6.7	27.2	Sc	4.0	14.1
Be[b]	6.1	7.9	La[a]	3.56	9.70	Sm[a]	3.27	12.4
C	2.1	117	Li	0.04	3.00	Sn	2.1	7.15
C_{60}	6.5[c,d]		Na	0.03	2.60	Sr	0.75	7.40
C_{70}	5.9[d]		Ni	0.58	17.5	Tb[a]	4.90	15.8
Ca	0.9	8.55	Mn	6.23	12.9	Th[a]	3.5	17.3
Ce[a]	3.06	8.64	Nd[a]	2.48	9.78	Ti	5.77	14.15
Co	0.45	16.2	Pa[a]	6.6	18.9	Tl	0.4	4.15
Fe	1.11	13.8	Pr[a]	3.12	10.0	U[a]	7.47	15.9
Gd	3.94	13.74	S	0.40	1.72	Y	5.0	11.4
						Zr	5.9	21.0

[a][90], [b][91], [c][92], [d][93], [e][94]

9.2.4 *Enthalpies of Phase Transformations*

Whilst sublimation and melting are first-order phase transitions with a change of both the structure and the aggregate state, polymorphic transformations in crystals change structures within the same aggregate state. Therefore the enthalpy of such transitions (ΔH_{tr}) is much smaller than ΔH_s and is comparable with ΔH_m. Values of ΔH_{tr} for a number of elemental solids and compounds are given in Tables 9.9 and 9.10, respectively.

These values vary strongly, depending on the place of a given species on the phase diagram. Other things being equal, the heat of a phase transition depends on the depth of the structural rearrangement. So, for the sphalerite → wurtzite transition (during which the tetrahedral coordination of metal is preserved) in CdS, CdTe, PbS and PbTe, $\Delta H_{tr} = 0.5$, 2.0, 0.3 and 0.35 kJ/mol, respectively; the transitions of the same compounds into NaCl-type structures require $\Delta H_{tr} = 20.6$, 16.0, 2.3 and 3.0 kJ/mol, respectively [98]. The dependence of thermodynamic properties of oxides on the coordination of cations has been analyzed by Reznitskii [102] who found certain correlations in crystalline oxides and showed a possibility to predict energetic properties of these substances from their structures. Thus the enthalpy of crystals during a phase transformation can change in a range from *ca.* 1 to several hundred kJ/mol, depending on the nature of the structural changes, e.g. a distortion of the structure, the loss of the short- or long-range order, or dissociation into molecules or separate atoms in the gas phase.

Table 9.10 Enthalpies of polymorphic transformations, ΔH_{tr} (kJ/mol) in selected compounds

Compound	N_c	ΔH_{tr}	Comp.	Transformation	ΔH_{tr}
KBr	6 → 8	0.14[a]	SiO_2	Quartz → cristobalite	2.7[h]
KI	6 → 8	0.66[a]		Quartz → coesite	4.9[h]
RbBr	6 → 8	0.43[a]		Quartz → stishovite	49.2[h]
RbI	6 → 8	0.48[a]	GeO_2	Rutile → quartz	22.0[h]
CsF	6 → 8	6.28		Rutile → glass	41.2[h]
CsCl	8 → 6	3.76	TiO_2	Rutile → brookite	0.9[h]
NH_4Cl	8 → 6	5.60		Rutile → anatase	2.9[h]
NH_4Br	8 → 6	3.99	ZrO_2	Monoclinic → tetragonal	5.43[i]
NH_4I	8 → 6	3.39	Al_2O_3	α → γ	18.8[h]
CuCl	α → β	3.97	Sb_2O_3	Cubic → orthorhombic	13.4
CuBr	α → γ	10.4	Cr_2O_3	α → γ	92.3[h]
CuI	α → γ	10.8	Mn_2O_3	α → γ	33.8[h]
AgI	α → β	7.53	Fe_2O_3	α → γ	16.7[h]
BeF_2	α → β	0.31	KNO_3	α → β	5.00[j]
CaF_2	α → β	4.77	NH_4NO_3	T = 305 K	1.59[j]
PbF_2	9 → 8	2.09		T = 398 K	4.22[j]
$PtCl_2$	α → β	0.17	$RbNO_3$	T = 166 K	3.98[j]
AlF_3	α → β	0.56		T = 228 K	2.72[j]
InI_3	5.16[b]		T = 278 K	1.46[j]	
YF_3	A	32.4	$AgNO_3$	T = 433 K	2.55[j]
PbO	B	0.42	Li_2SO_4	T = 848 K	25.5
ZnS	C	5.0[c]	Na_2SO_4	T = 512 K	11.25
CdS	4 → 4	0.50[d]	K_2SO_4	T = 857 K	8.95
HgS	6 → 4	18.0	Rb_2SO_4	T = 928 K	4.18
SnS		6.0[e]	Ag_2SO_4	T = 700 K	18.7
PbS	4 → 4	0.3[d]	$MgSiO_3$	Pyrochlor → garnet	37.0[c]
FeS	α → β	2.88		Pyrochlor → ilmenite	59.2[c]
CdTe	4 → 4	2.0[d]	$MgGeO_3$	Pyrochlor → ilmenite	7.5[c]
SnTe		0.21[e]	$CaGeO_3$	Garnet → perovskite	43.3[c]
PbTe	4 → 4	0.35[d]	$CdTiO_3$	Ilmenite → perovskite	15.0[c]
ZrTe	6 → 6	8.3[f]	Mg_2SiO_4	Olivine → spinel	31.8[c]
Cu_2S	α → β	6.8	Fe_2SiO_4	Olivine → spinel	16.3[c]
$GeSe_2$		1.24[g]	Co_2SiO_4	Olivine → spinel	9.0[c]

A orthorhombic→hexagonal, B tetragonal→orthorhombic, C sphalerite→wurtzite, [a][95], [b][96], [c][97], [d][98], [e][99], [f][100], [g][101], [h][102], [i][103], [j][104]

References

1. Buerger MJ (1951) Phase transformations in solids. Wiley, New York
2. Toledano P, Dmitriev V (1996) Reconstructive phase transitions. World Scientific, Singapore
3. Prewitt CT (1985) Crystal-chemistry—past, present, and future. Amer Miner 70:443–454
4. Batsanov SS (1983) On some crystal-chemical peculiarities of simple inorganic halogenides. Zhurnal Neorganicheskoi Khimii 28:830–836 (in Russian)
5. Batsanov SS (1986) Experimental foundations of structural chemistry. Standarty, Moscow (in Russian)
6. Phillips JC, Lukowsky G (2009) Bonds and bands in semiconductors, 2nd ed Academic, New York
7. Wemple SH (1973) Effective charges and ionicity. Phys Rev B7:4007–4009

8. Wemple SH (1973) Refractive-index behavior of amorphous semiconductors and glasses. Phys Rev B7:3767–3777
9. Revesz AG, Wemple SH, Gibbs GV (1981) Structural ordering related to chemical bonds in random networks. J Physique 42:C4–217–C4–219
10. Mooser E, Pearson WB (1959) On the crystal chemistry of normal valence compounds. Acta Cryst 12:1015–1022
11. Watson RE, Bennet LH (1978) Transition-metals—d-band hybridization, electronegativities and structural stability of intermetallic compounds. Phys Rev B18:6439–6449
12. Watson RE, Bennet LH (1982) Structural maps and parameters important to alloy phase stability. MRS Proceedings 19:99–104
13. Burdett JK, Price GD, Price SL (1981) Factors influencing solid-state structure—an analysis using pseudopotential radii structural maps. Phys. Rev. B 24:2903–2912
14. Pettifor DG (1984) A chemical-scale for crystal-structure maps. Solid State Commun 51:31–34
15. Pettifor DG (1985) Phenomenological and microscopic theories of structural stability. J Less-Comm Met 114:7–15
16. Pettifor DG (1986) The structures of binary compounds: phenomenological structure maps. J Phys C19:285–313
17. Pettifor DG (1992) Theoretical predictions of structure and related properties of intermetallics. Mater Sci Technol 8:345–349
18. Pettifor DG (2003) Structure maps revisited. J Phys Cond Matter 15:V13-V16
19. Dudareva AG, Molodkin AK, Lovetskaya GA (1988) The prediction of compound formation in GaX_3-MX_n and InX_3-MX_n systems, where X = Cl, Br, I. Russ J Inorg Chem 33:916–917
20. Sproul G (1994) Electronegativity and bond type: evaluation of electronegativity scales. J Phys Chem 98:6699–6703
21. Burdett JK (1997) Chemical bond: a dialog. Wiley, Chichester
22. Tosi MP (1994) Melting and liquid structure of polyvalent metal-halides. Z Phys Chem 184:121–138
23. Villars P (1983) A 3-dimensional structural stability diagram for 998 binary AB intermetallic compounds. J Less Comm Met 92:215–238
24. Villars P (1984) A 3-dimensional structural stability diagram for 1011 binary AB_2 intermetallic compounds. J Less Comm Met 99:33–43
25. Villars P (1984) 3-dimensional structural stability diagrams for 648 binary AB_3 and 389 binary A_3B_5 intermetallic compounds. J Less Comm Met 102:199–211
26. Villars P (1985) A semiempirical approach to the prediction of compound formation for 3486 binary alloy systems. J Less Comm Met 109:93–115
27. Villars P (1986) A semiempirical approach to the prediction of compound formation for 96446 ternary alloy systems, II. J Less Comm Met 119:175–188
28. Villars P, Phillips JC (1988) Quantum structural diagrams and high-T_c superconductivity. Phys Rev B37:2345–2348
29. Rabe KM, Phillips JC, Villars P, Brown ID (1992) Global multinary structural chemistry of stable quasicrystals, high-T_c ferroelectrics, and high-T_c superconductors. Phys Rev B45:7650–7676
30. Ubbelode AR (1978) The molted state of matter. Wiley, New York
31. Mnyukh Yu (2001) Fundamentals of solid state phase transitions, ferromagnetism and ferrolectricity. 1st Books Library, Washington DC
32. Bernstein J (2002) Polymorphism in organic crystals. IUCR Monograph on Crystallography, No. 14. Clarendon Press, Oxford
33. Herbstein FH (2006) On the mechanism of some first-order enantiotropic solid-state phase transitions. Acta Cryst B 62:341–383
34. Braga D, Grepioni F (2000) Organometallic polymorphism and phase transitions. Chem Soc Rev 29:229–238
35. Ilott AJ, Palucha S, Batsanov AS et al (2010) Elucidation of structure and dynamics in solid octafluoronaphthalene. J Am Chem Soc 132:5179–5185

36. Tsuji K, Hattori T, Mori T et al (2004) Pressure dependence of the structure of liquid group 14 elements. J Phys Cond Matter 16:S989-S996
37. Daisenberger D, Wilson M, McMillan PF et al (2007) High-pressure X-ray scattering and computer simulation studies of density-induced polyamorphism in silicon. Phys Rev B75:224118
38. Brazhkin VV, Katayama Y, Trachenko K et al (2008) Nature of the structural transformations in B_2O_3 glass under high pressure. Phys Rev Lett 101:035702
39. Lee SK, Mibe K, Fei Y et al (2005) Structure of B_2O_3 glass at high pressure. Phys Rev Lett 94:165507
40. Hamaya N, Sato K, Usui-Watanaba K et al (1997) Amorphization and molecular dissociation of SnI_4 at high pressure. Phys Rev Lett 79:4597–4600
41. Itie JP (1992) X-ray absorption-spectroscopy under high-pressure. Phase Trans 39:81–98
42. Giefers H, Porsch F, Wortmann G (2005) Thermal disproportionation of SnO under high pressure. Solid State Ionics 176:1327–1332
43. Giefers H, Porsch F, Wortmann G (2006) Structural study of SnO at high pressure. Physica B 373:76–81
44. Lide DR (ed) (2007–2008) Handbook of chemistry and physics, 88th edn. CRC Press, New York
45. Glushko VP (ed) (1981) Thermochemical constants of substances. USSR Acad Sci, Moscow (in Russian)
46. Kulakov MP (1990) Change of specific volume of $A^{II}B^{VI}$ compounds on melting. Inorg Mater 26:1947–1950
47. Nasar A, Shamsuddin M (1990) Thermodynamic properties of ZnTe. J Less Comm Met 161:93–99
48. Nasar A, Shamsuddin M (1990) An investigation of thermodynamic properties of cadmium sulphide. Thermochim Acta 197:373–380
49. Nasar A, Shamsuddin M (1990) Thermodynamic properties of cadmium telluride. High Temp Sci 28:245–254
50. Nasar A, Shamsuddin M (1990) Thermodynamic properties of cadmium selenide. J Less Comm Met 158:131–135
51. Nasar A, Shamsuddin M (1990) Thermodynamic investigations of mercury telluride. J Less Comm Met 161:87–92
52. Nasar A, Shamsuddin M (1992) Investigations of the thermodynamic properties of zinc chalcogenides. Thermochim Acta 205:157–169
53. Gurvich LV, Veyts IV, Alcock CB (eds) (1994) Thermodynamical properties of individual substances. CRC Press, Boca Raton, FL
54. Guillermet AF, Frisk K (1994) Thermochemical assessment and systematics of bonding strengths in solid and liquid "MeN" 3d transition-metal nitrides. J Alloys Comp 203:77–89
55. Shamsuddin M, Nasar A (1988/1989) Thermodynamic properties of cadmium telluride. High Temp Sci 28:245–254
56. Huang Y, Brebrick RF (1988) Partial pressures and thermodynamic properties for lead telluride. J Electrochem Soc 135: 486–496
57. Lamoreaux RH, Hildenbrand DL, Brewer L (1987) High-temperature vaporization behavior of oxides of Be, Mg, Ca, Sr, Ba, B, Al, Ga, In, Tl, Si, Ge, Sn, Pb, Zn, Cd, and Hg. J Phys Chem Refer Data 16:419–443
58. Bashir-Hashemi A, Chickos JS, Hanshaw W et al (2004) The enthalpy of sublimation of cubane. Thermochim Acta 424:91–97
59. Rojas A, Vieyra-Eusebio MT (2011) Enthalpies of sublimation of ferrocene and nickelocene. J Chem Thermodyn 43:1738–1747
60. Portman R, Quin M, Sagert N et al (1989) A Knudsen cell mass-spectrometer study of the vaporization of cesium telluride and cesium tellurite. Thermochim Acta 144:21–31
61. Piacente V, Scardala P (1994) A study on the vaporization of copper(II) selenide. J Mater Sci Lett 13:1343–1345
62. Scardala P, Piacente V, Perro D (1990) Standard sublimation enthalpy of solid Ag_2Se. J Less-Comm Met 162:11–21

63. Adami M, Ferro D, Piacente V, Scardala P (1987) Vaporization behavior and sublimation enthalpy of solid Ag_2Te. High Temp Sci 23:173–186
64. Bardi G, Trionfetti G (1990) Vapor-pressure and sublimation enthalpy of zinc selenide and zinc telluride. Thermochim Acta 157:287–294
65. Bardi G, Ieronimakis K, Trionfetti G (1988) Vaporization enthalpy of cadmium selenide and telluride. Thermochim Acta 129:341–343
66. O'Hare PAG, Curtiss LA (1995) Thermochemistry of germanium + sulfur. J Chem Thermodyn 27:643–662
67. Tomaszkiewicz I, Hope GA, O'Hare PAG (1995) Thermochemistry of germanium + tellurium. J Chem Thermodyn 27:901–919
68. Wiedemeier H, Csillag FJ (1979) Equilibrium sublimation and thermodynamic properties of SnS. Thermochim Acta 34:257–265
69. Botor J, Milkowska G, Konieczny J (1989) Vapor-pressure and thermodynamics of PbS(s). Thermochim Acta 137:269–279
70. Konieczny J, Botor J (1990) The application of a thermobalance for determining the vapour pressure and thermodynamic properties. J Therm Analys Calorimetry 36:2015–2019.
71. Fajans K (1967) Degrees of polarity and mutual polarization of ions in the molecules of alkali fluorides, SrO and BaO. Struct Bonding 3:88–105
72. Gopikrishnan CR, Jose D, Datta A (2012) Electronic structure, lattice energies and Born exponents for alkali halides from first principles. AIP Advances 2:012131
73. Urusov VS (1975) Energetic crystal chemistry. Nauka, Moscow (in Russian)
74. Acree W, Chickos JS (2010) Phase transition enthalpy measurements of organic and organometallic compounds. Sublimation, vaporization and fusion enthalpies from 1880 to 2010. J Phys Chem Refer Data 39:043101
75. Nikitin MI, Rakov EG, Tsirel'nikov VI, Khaustov SV (1997) Enthalpies of formation of manganese di- and trifluorides. Russ J Inorg Chem 42:1039–1042
76. Brunetti B, Piacente V (1996) Torsion and Knudsen measurements of cobalt and nickel difluorides and their standard sublimation enthalpies. J Alloys Comp 236:63–69
77. Bardi G, Brunetti B, Ciccariello E, Piacente V (1997) Vapour pressures and sublimation enthalpies of cobalt and nickel dibromides. J Alloys Comp 247:202–205
78. Ionova GV (2002) Thermodynamic properties of halide compounds of tetravalent transactinides. Russ Chem Rev 71:401–416
79. Struck CW, Baglio JA (1991) Estimates for the enthalpies of formation of rate-earth solid and gaseous trihalides. High Temp Sci 31:209–237
80. Hackert A, Plies V (1998) Eine neue methode zur messung von temperaturabhängigen partialdrücken in geschlossenen systemen. Die bestimmung der bildungsenthalpie und -entropie von PtI_2(s). Z anorg allgem Chem 624:74–80
81. Parker VB, Khodakovskii IL (1995) Thermodynamic properties of the aqueous ions (2+ and 3+) of iron and the key compounds of iron. J Phys Chem Refer Data 24:1699–1745
82. Dittmer G, Niemann U (1981) Heterogeneous reactions and chemical-transport of tungsten with halogens and oxygen under steady-state conditions of incandescent lamps. Philips J Res 36:87–111
83. Rojas-Aguilar A, Orozco-Guareo E, Martinez-Herrera M (2001) An experimental system for measurement of enthalpies of sublimation by d.s.c. J Chem Thermodyn 33:1405–1418
84. Lobo LQ, Ferreira AGM (2001) Phase equilibria from the exactly integrated Clapeyron equation. J Chem Thermodyn 33:1597–1617
85. Chickos JS, Acree WE (2003) Enthalpies of vaporization of organic and organometallic compounds, 1880–2002. J Phys Chem Refer Data 32:519–878
86. Huron M-J, Claverie P (1972) Calculation of interaction energy of one molecule with its whole surrounding. J Phys Chem 76:2123–2133
87. Korolev GV, Il'in AA, Sizov EA et al (2000) Increments of enthalpy of vaporization of organic compounds. Russ J General Chem 70:1020–1022
88. Chickos JS, Zhao H, Nichols G (2004) The vaporization enthalpies and vapor pressures of fatty acid methyl esters. Thermochim Acta 424:111–121

89. Verevkin SP, Emel'yanenko VN, Algarra M et al (2011) Vapor pressure and enthalpies of vaporization of azides. J Chem Thermodyn 43:1652–1659
90. Konings RJM, Beneš O (2010) Thermodynamic properties of the f-elements and their compounds: the lanthanide and actinide metals. J Phys Chem Refer Data 39:043102
91. Kleykamp H (2000) Thermal properties of beryllium. Thermochim Acta 345:179–184
92. Digonskii VV, Digonskii SV (1992) Laws of the diamond formation. Nedra, St. Peterburg (in Russian)
93. Diky VV, Kabo GJ (2000) Thermodynamic properties of C60 and C70 fullerenes. Russ Chem Rev 69:95–104
94. Peletskii VE, Petrova II, Samsonov BN (2001) Investigation of the heat of polymorphous transformation in zirconium. High Temp Sci 39:666–669
95. Pistorius CWFT (1965) Polymorphic transitions of alkali bromides and iodides at high pressures to 200°C. J Phys Chem Solids 26:1003–1011
96. Titov VA, Chusova TP, Stepin Yu G (1999) On thermodynamic characteristics of In-I system compounds. Z Anorg Allgem Chem 625:1013–1018
97. Balyakina IV, Gartman VK, Kulakov MP, Peresada GI (1990) Phase transition in cadmium selenide. Inorg Mater 26:2147–2149
98. Leute V, Schmidt R (1991) The quasiternary system (Cd_kPb_{1-k})(S_LTe_{1-L}). Z Phys Chem 172:81–103
99. Leute V, Brinkmann S, Linnenbrink J, Schmidtke HM (1995) The phase diagram of the quasi-ternary system (Sn, Pb)(S, Te). Z Naturforsch 50a:459–467
100. Örlygsson G, Harbrecht B (2001) Structure, properties, and bonding of ZrTe (MnP type), a low-symmetry, high-temperature modification of ZrTe (WC type). J Am Chem Soc 123:4168–4173
101. Stølen S, Johnsen H-B, Abe R et al (1999) Heat capacity and thermodynamic properties of $GeSe_2$. J Chem Thermodyn 31:465–477
102. Reznitskii LA (2000) Energetics of crystalline oxides. Moscow Univ Press, Moscow (in Russian)
103. Moriya Y, Navrotsky A (2006) High-temperature calorimetry of zirconia: heat capacity and thermodynamics of the monoclinic-tetragonal phase transition. J Chem Thermodyn 38:211–223
104. Breuer K-H, Eysel W (1982) The calorimetric calibration of differential scanning calorimetry cells. Thermochim Acta 57:317–329

Chapter 10
Extreme Conditions

Of more than half a million crystal structures known, the overwhelming majority have been determined at ambient conditions or at moderately low temperatures achievable with cooling devices that use liquid-nitrogen (b.p. 77 K). However, the development of science and technology increasingly relies on technologies involving extreme conditions. It should be noted that for many years, variation of temperature has been the main way of changing thermodynamical conditions in structural studies, for example lowering temperature to reduce atomic thermal vibrations and obtain more precise structural data, or to investigate phase transitions and low-temperature phases, or to crystallize and study substances which are liquid, gaseous or unstable at normal conditions. Another approach is the structural investigations at elevated pressure. Pressure is highly efficient for generating phase transitions and new phases, for conformational and structural transformations of molecules, polymerization, and structure-property relations, which are of interest to chemists and physicists. The knowledge of stability of substances at extreme conditions is still far from complete: it is difficult generally to predict pressure-generated reactions and transformations and therefore in this chapter we will consider mainly the corresponding experimental data.

During the last decades, high-pressure crystallography has evolved into a powerful technique which can be routinely applied in laboratories and dedicated synchrotron and neutron facilities. The variation of pressure adds a new thermodynamic dimension to crystal-structure analyses, and extends the understanding of the solid state and materials in general. New areas of a thermodynamic exploration of phase diagrams, polymorphism, transformations between different phases and cohesion forces, structure-property relations, and a deeper understanding of matter at the atomic scale in general are accessible with the high-pressure techniques in hand. Thus, chemical behavior of substances under high pressures sharply differs from that at ambient conditions: the gaseous mixture $O_2 + H_2$ can react with detonation at the normal pressure, while this mixture under 7.6 GPa is absolutely inert [1]; the $NH_3 + H_2O$ mixture forms the ammonium salt NH_4OH at the ambient pressure, but at 6.5 GPa and 300 K these molecules exist as an alloy [2]; on the contrary, the inert gaseous mixture $N_2 + O_2$ (air) at P > 5 GPa transforms into the ionic compound $NO^+NO_3^-$ [3]. A history, guidelines and requirements for performing high-pressure structural studies are outlined in [4].

S. S. Batsanov, A. S. Batsanov, *Introduction to Structural Chemistry*,
DOI 10.1007/978-94-007-4771-5_10, © Springer Science+Business Media Dordrecht 2012

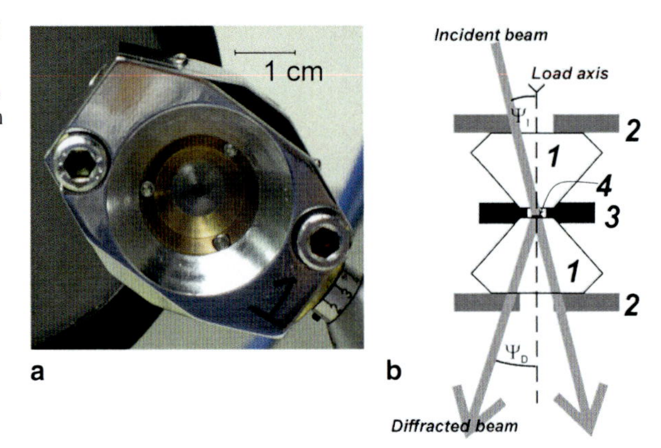

Fig. 10.1 Diamond anvil cell (*left*, viewed down the load axis) and its schematic (*right*, adapted with permission from [16]): *1* anvil (diamond), *2* backing plate (beryllium), *3* gasket (e.g. tungsten), *4* sample chamber with pressure medium (liquid or gas), calibration standard (ruby) and the crystal. Note the smallness of the working volume

High pressures are the most effective means of changing the structure of matter: if heating solids up to the melting point changes their volumes by a few per cent, pressures of even a few kbars reduce the volumes of typical solids or liquids significantly more. Application of XRD at high pressures began in 1935 [5] with $P = 1$ kbar in a quartz cell, in 1938 alkali halides were investigated at 5 kbar in a steel bomb [6]. When tungsten carbide (WC) began to be used as the cell material, pressures up to tens of GPa were obtained. In 1950 Lawson and Tang for the first time applied in XRD researches a diamond anvil cell (DAC) [7] and the limit of pressure reached 100 GPa. Present-day DACs (Fig. 10.1) can give up to 300 GPa [8, 9]. Unfortunately, the higher is the pressure, the smaller the working volume of the cell and the size of the compressed sample: at $P \geq 100$ GPa it is as small *ca* 10^{-3} mm^3, therefore diffraction studies require special experimental settings. Nevertheless, with the powerful sources of radiation now available (e.g. synchrotrons for X-rays) this is enough to obtain structural information, although not the chemically meaningful quantities of products. Different applications of the high pressure technique in chemistry have been reviewed in [10–15].

10.1 Polymorphic Transformations Under High Static Pressures

The qualitative picture of a phase transition of a substance under pressure is very simple. Compression shortens interatomic distances (d) and the bonding energy increases as d^{-1} or d^{-2}, but simultaneously the repulsive energy (due to overlap of electron clouds of nearest atoms) grows more sharply, as d^{-12}. On some stage of compression the repulsion becomes bigger than the attraction and the given substance with the initial structure can no longer exist under the new thermodynamic conditions. Then the structure undergoes an abrupt rearrangement in order to diminish the electron repulsion. This can be achieved in one way only, namely, by increasing the N_c in the structure leading to an increase of the packing density (but also, paradoxically, of the *nearest* interatomic distances, see Sect. 1.4.2), in accordance with the Le Chatelier's principle.

As already stated, it is not easy to calculate the stability of a structure under variable thermodynamic conditions, because the phase transition enthalpy is usually of the order of *ca.* 10^{-3} of the atomization enthalpy. However, for simpler cases the *ab initio* calculations of the high-pressure phases are now possible. In some cases energy differences as small as a few hundreds of joules per moles must be resolved to determine whether a particular structure is stable. On the whole the agreement between theory and experiment is good. There are discrepancies, some of which surely result from inadequacies in the calculations, but overall it is not an exaggeration to say that this field represents a triumph for density-functional theory [17]. Nevertheless, the most interesting results, such as the acquisition of transition-metal properties and complex incommensurate structures by Group 1 and 2 metals under high pressure (see Sect. 5.1.1), although theoretically explicable *post factum*, came totally unexpected. Thus the main source of the information about pressure induced phase transitions is still the experiment. The structural study at high pressures by XRD can be carried out by two methods: either on the unloaded sample if the transition is irreversible, or *in situ* if it is reversible. In the further discussion we will not distinguish these approaches since from crystal-chemical point of view the reversibility of a phase transition does not influence the structural parameters.

Table 10.1 lists the pressures of phase transitions in MX compounds from the ZnS structure type ($N_c = 4$) to the NaCl or NiAs ($N_c = 6$), and into the CsCl ($N_c = 8$) types. This table shows that the pressures of phase transitions (P_{tr}) fall with the increase of ionic radii of both the cations (which agrees with the ionic model) and anions (which contradicts it). This contradiction disappears, if we take into account the polar character of M–X bonding: the increase of the bond covalency in the successions MF $\rightarrow$ MI or MO $\rightarrow$ MTe rises $r(M^+)$ and reduces $r(X^-)$, thus increasing the $k_r = r_+/r_-$ ratio and diminishes P_{tr}. In alkali hydrides [18, 19] the pressure of the structural transformation from the NaCl into the CsCl type decreases with the increase of the cation size, viz. 29.3 GPa in NaH, 4.0 in KH, 2.7 in RbH and 1.2 in CsH. In CsH one more phase transition was discovered at 17.5 GPa [70] whereupon the CsCl structure converts into the CrB type; this phase is stable up to 253 GPa and its compressibility at the this pressure is the highest for the MX compounds, $V/V_o = 0.26$.

Polymorphic transformations into the structure with $N_c > 8$ were observed in cesium halides [71–74] at following pressures: CsCl 65, CsBr 53, CsI 39 GPa; besides, CsI at $P > 200$ GPa undergoes a smooth second-order transition into a *hcp* structure with $N_c = 12$. Very peculiar transitions were observed in samarium chalcogenides SmX: under high pressures the electron transition Sm(II) $\rightarrow$ Sm(III) takes place with a sharp isomorphic decrease in the lattice parameters. Further compression leads to the transformation of the NaCl $\rightarrow$ CsCl type. Values of pressures at the isomorphic (P_{itr}) and polymorphic (P_{ptr}) transitions according to [75, 76] are following:

	SmS	SmSe	SmTe	YbO	YbS	YbSe	YbTe
P_{itr}	1.5	9	7	8	10	15	15
P_{ptr}	42	30	13.5				

Table 10.1 Pressures of phase transitions (GPa) in the MX compounds

N_c	M	F	Cl	Br	I
$4 \to 6$	Cu		10.6[a]	8.4[a]	17.7[b]
$6 \to 8$	Na	27[c]	30[d]	35[c]	25[c]
	K	4[e]	1.95[d,f]	1.75[d,f]	1.8[d,f,g]
	Rb	3[e]	0.5[d,f]	0.45[d,f]	0.4[d,f,g]
	Ag	2.7[h]			
N_c	M	O	S	Se	Te
$4 \to 6$	Be		51[i]	56[i]	35[i]
	Mg				~2[j]
	Zn[Φ]	9.1[k]	15[l]	13.6[l]	9.5[l]
	Cd[Φ]		2.6[m]	2.0[n]	3.4[o]
	Hg	14[p]	13[q]		8.1[q]
$6 \to 8$	Ca	61[r]	40[r]	38[r]	33[r]
	Sr	36[s]	18[s]	14[s]	12[s]
	Ba	14[t]	6.5[t]	6[u]	4.8[v]
	Pb		21.5[w]	16[w]	13[w]
	Y		53[W]	36[W]	14[W]
	La		27[W]	19[W]	7[W]
	Th		20[x]	15[x]	
	U		10[x]	20[x]	9[x]
N_c	M	N	P	As	Sb
$4 \to 6$	B	>106[y]	40[z]		
	Al	17[y,z,α]	9.5[β]	7[β]	5.3[β]
	Ga	48[y,γ]	26[δ]	14[ε]	6.2[η]
	In	12[y,Θ]	11.6[z,λ]	8.5[μ]	2.1[ν]
$6 \to 8$	Th		30[x]	18[x]	9[x]
	U	29[x]	28[x]	20[x]	9[x]

[a][20], [b][21], [c][22], [d][23, 24], [e][25], [f][26], [g][27], [h][28], [i][29, 30], [j][31], [k][32], [l][33, 34], [m][35], [n][36], [o][37–39], [p][40], [q][41, 42], [r][43], [s][44], [t][45], [u][46], [v][47], [w][48], [W][49, 50], [x][51, 52], [y][53], [z][54], [α][55], [β][56], [γ][57–59], [δ][60, 61], [ε][62], [η][63], [Θ][64], [λ][65], [μ][66, 67], [ν][68], [Φ]ZnO and CdO transform from B1 to B2 at 261 and 83 GPa, respectively [69]

We shall consider now the geometric features of the polymorphic transformations under pressure in the mercury and cadmium chalcogenides. Thus, at normal conditions HgO and HgS have the cinnabar structure with different lengths of the Hg–X bonds along the *a*, *b* and *c* crystal axes: 2.03 (HgO) and 2.36 Å (HgS) along (*a*), 2.79 and 3.10 Å along (*b*), and 2.90 and 3.30 Å along (*c*), respectively [77]. HgSe, HgTe and CdTe adopt the same structure under the pressures of 1.5, 3.6 and 3.6 GPa, respectively. In CdTe, two bonds have the lengths of 2.724 Å (as in the ZnS structure), and two other 2.971 Å (as in the NaCl structure). Thus, the cinnabar structure is intermediate between the structural types of ZnS and NaCl. Further compression lengthens the short bonds in the cinnabar structure, converting it into the NaCl type [78]. This equalization of interatomic distances under compression is the general phenomenon, see below.

Compression of crystalline compounds MX_2 leads to a consequential change in the structure types with an increase of density and coordination numbers (in

Table 10.2 Pressures of phase transitions (GPa) in MX_2 compounds

TiO_2 ($N_c = 6$) → $CaCl_2$ ($N_c = 6$)[a]							
TiO_2	SiO_2	GeO_2	SnO_2	PbO_2	CrO_2	MnO_2	RuO_2
7	50	26.7	11.8	4	12.2	0.3	11.8
TiO_2 ($N_c = 6$) → CaF_2 ($N_c = 8$)							
MgF_2	ZnF_2	MnF_2	FeF_2	CoF_2	NiF_2	PdF_2	
14	8	4	4.5	6.5	8.5	6	
TiO_2	SnO_2	PbO_2	MnO_2	RuO_2			
20[b]	20[b]	7[b]	25[b]	18[b]			
CaF_2 ($N_c = 8$) → $PbCl_2$ ($N_c = 9$)							
CaF_2	SrF_2	$SrCl_2$	$SrBr_2$	BaF_2	EuF_2	MnF_2	
10	6	4	3	5	11	15	
CeO_2	ThO_2	ZrO_2	HfO_2	TeO_2	UP_2	UAs_2	
31[c]	40[d]	30[e]	30[e]	12[b]	22[f]	15[f]	

[a] [79], [b] [80], [c] [81], [d] [82], [e] [83], [f] [84]

brackets): SiO_2 ($N_c = 4$) → TiO_2 ($N_c = 6$) → $CaCl_2$ ($N_c = 6$) → PbO_2($N_c = 6$) → ZrO_2 ($N_c = 7$) → CaF_2($N_c = 8$) → $PbCl_2$($N_c = 9$) → Co_2Si ($N_c = 10$) → Ni_2In ($N_c = 11$). The phase transition pressures for halides and oxides of metals (Table 10.2) tend to increase with the cation charges. On the contrary, the volume change of the solid at phase transitions depends only on geometrical parameters: on increase of N_c from 4 up to 6 the volume decreases on average by 17 %, on a $N_c = 6 \rightarrow N_c = 8$ transformation it decreases by 10 %.

Generally speaking, an increase in density can occur upon a polymorph transformation even without change in coordination of the nearest atoms in the structure. An example of such densification is presented by the SiO_2 modifications, where a more compact packing of silicon-oxygen tetrahedra reduces the molar volumes in the succession: cristobalite (25.9 cm^3) → keatite (24.0 cm^3) → quartz (22.7 cm^3) → coesite (20.0 cm^3). However, the transition under pressure from coesite ($N_c = 4$) to stishovite ($N_c = 6$) reduces this volume much more drastically, to 13.8 cm^3.

As above mentioned, usually the points of phase transitions are indicated *in situ* by a change of the XRD pattern or of IR or Raman spectra. Usually, the high-pressure phases do not survive after the pressure is relaxed and revert to the initial phase, but occasionally are stabilized in an intermediate phase. Thus, SnO_2 at 25 GPa and 1,000 °C has the structure of fluorite, and in the unloaded state the α-PbO_2 structure [85] which is denser than the initial, rutile-type structure. Dioxides under high pressures undergo the following transformations of structural types:

$$TiO_2 \rightarrow CaCl_2 \text{ and/or } Fe_2N, \alpha\text{-}PbO_2 \rightarrow Pa\bar{3}$$

So, SiO_2 at 35–40 GPa transforms into the Fe_2N type, then at 53 GPa to a structure of the $CaCl_2$ type, and after 68 GPa exists in the α-PbO_2 structure. GeO_2 converts into the $CaCl_2$ type at 25 GPa, the α-PbO_2 type at 44 GPa and above 70–90 GPa adopts the FeS_2 type (the deformed structure CaF_2) which is the densest phase of

Table 10.3 Phase transitions in the alkali chalcogenides [92–97]

Li_2O	*Fm3m* (antifluorite, af)	*Pnma*		
Li_2S	*Fm3m*	*Pnma* + $Pn2_1a$	*Pnma*	Distorted Ni_2In (*Cmcm*)
Na_2S	*Fm3m*	*Pnma*	Ni_2In	
K_2S	*Fm3m*	Distorted Ni_2In (*Pmma*)		
Rb_2S	af	Ni_2In		
Cs_2S	*Pnma*	Ni_2In ($P6_3/mmc$ or *Cmcm*)		
P, GPa	15	30	45	60

GeO_2 [86–88]. SnO_2 at 11.8 GPa converts into the $CaCl_2$ type, at 14 GPa into the α-PbO_2 structure, at 18–21 GPa into $Pa\bar{3}$ one, then in the ZrO_2 structure and, finally, at 74 GPa pass into the cottunite phase ($PbCl_2$ type) [89, 90]. PbO_2 at 4 GPa transforms into the $CaCl_2$ type, and at 71 GPa undergoes a transition into the $Pa\bar{3}$ structure. The general dependence of pressures in the $TiO_2 \rightarrow CaCl_2$ phase transition is given in Table 10.2. Besides the examples presented in this table, there are compounds with the anti-fluorite type structures (α-CaF_2, $N_c = 8$), which under pressure transform to the structures of anti-cottunite type (α-$PbCl_2$, $N_c = 9$), and then into the Ni_2In structure ($N_c = 10$). Table 10.3 shows the pressures of such phase transitions in Li_2O, Li_2S, Na_2S, K_2S, Rb_2S, Cs_2S [91] which are the function of the cation sizes.

An outstanding achievement of structural chemistry of high pressures is the discovery of transformations of CO_2 under compression [98, 99]. At present, there are established six high pressure phases of CO_2, five of them molecular and the sixths non-molecular (covalent polymeric). At room temperature CO_2 solidifies at 0.5 GPa into a cubic phase I commonly known as dry ice; it is a van der Waals crystal with strong intra-molecular and weak intermolecular bonds. Between 12 and 22 GPa phase I undergoes a sluggish transition into the orthorhombic phase III which is a molecular crystal too. The transition III → V begins at $P > 40$ GPa and $T = 1{,}800$ K. On XRD data, phase V has a tridymite structure where each carbon atom is tetrahedrally bonded with four O, and the CO_4 tetrahedra share their corner oxygen atoms to form layers of distorted hexagonal rings.This latter phase was temperature-quenched and pressure-recovered down to a few GPa where it transforms back to the molecular CO_2. Known also are phase II which contains dimers $(CO_2)_2$, and phase IV which exists in the 11–50 GPa range, and having structure with bent molecules, is intermediate between the molecular and atomic types. While in phase II the intra-molecular C=O distance is 1.33 Å, and the intermolecular one 2.33 Å, in phase IV the former increases to 1.5 Å, and the latter is reduced to 2.1 Å. This is a trend towards equalization of interatomic distances. The transition I → III occurs without change of volume, but on the transition III → V the volume decreases by 15.3 % (at 40 GPa). CO_2-IV has been obtained at 11.7 GPa and 830 K, its structure is of $R\bar{3}c$ type with linear molecules, $d(C{=}O) = 1.155$ Å at 15 GPa. It is interesting that CO_2 in the SiO_2 structure is a very hard material, close in its bulk modulus to the diamond-like *c*-BN.

However, a further increase of pressure up to 80 GPa at 1,500–3,000 K causes CO_2 to dissociate into oxygen and diamond [98]. Above 40 GPa the dissociation is preceded by the formation of a new non-molecular phase, CO_2-VI, which is the precursor of the dissociation itself. An amorphous nonmolecular, silica-like carbon dioxide material 'a-carbonia' (a-CO_2), was discovered by compressing phase III above 40–48 GPa [99]. It was established that a-CO_2 is the disordered counterpart of crystalline phase V; 'a-carbonia' is also a very hard glass material as compared to a-SiO_2. The pressurization of CO_2 at room temperature causes the transformation from the molecular solid CO_2 to nonmolecular (polymeric) amorphous state at 65 GPa, and the polymeric CO_2 is not returned to the molecular down to 30 GPa on decompression [100]. The formation of the nonmolecular CO_2 without heating allows to suggest that the usually inert CO_2 molecule becomes rather reactive at high pressure and may react with other elements or compounds, yielding exotic and useful materials.

As noted in Sect. 5.1.2, under pressure of 21 and 80 GPa the molecular structures of I_2 and Br_2, respectively, undergo a dissociation to form monatomic (metal-like) structures. Transition of molecular structures into atomic lattices under high pressure were observed in HCl (at 51 GPa), HBr (at 42 GPa), H_2O (at 60 GPa) [101] and H_2S (molecular dissociation at 44 and metallization at 96 GPa) [102, 103]. HI turns from an insulator into a molecular conductor at 45 GPa, and then to a monatomic metal at 51 GPa [104].

We have calculated [105, 106] the phase transition pressures in A_2 molecules supposing that shortening of vdW contacts $A\cdots A$ causes a corresponding lengthening of covalent A–A bonds until the full equalization of inter- and intra-molecular distances, which corresponds to a transition of a molecular structure into a lattice of a monatomic metal. This model assumes monotonic change of interatomic distances, however in fact a phase transition of the first order means a jump of the parameters: as soon as the reduction of vdW contacts under compression reaches a critical value, these contacts and the covalent bonds sharply change their lengths. Hence, the problem is to determine the critical value of the intermolecular distance after which further reduction is impossible without general rearrangement of the atomic structure.

Substances under high pressures change not only their atomic, but also the electronic structure, e.g. changing from the insulators to metallic phases, due to delocalization of valence electrons. The most accurate optical measurements of the band gaps under high pressures were carried out by Ruoff et al. [107], who found the following metallization pressures P_{ME} (in GPa):

KI	RbI	NH_4I	CsI	CsBr	BaSe	BaTe
170	*ca.* 130	*ca.* 120	105	*ca.* 250	*ca.* 52	*ca.* 24

Later the same group has corrected some of the P_{ME} values (BaTe 20, BaSe 61, CsI 110 GPa) and added new ones: Xe 160 GPa (later corrected to 121–138 GPa [108]), I_2 16, O_2 95, S 95 GPa. Transitions into the metallic state in PbX occur at the pressures: PbS 18, PbSe 16, PbTe 12 GPa [109], i.e. as in CsX and BaX, the metallization pressure drops as the size of the anion increases. In isostructural MnO and FeO, metallization occurs at 110 and 70 GPa, respectively [110]; in ZnS, ZnSe, ZnTe at 14.7, 13.0, 9.5 GPa, respectively; in HgS, HgSe, HgTe at 27, 15.5, 8.4 GPa, respectively [111]. P_{ME} depends also on polymorphic modifications of the initial substance: trigonal, rhombohedral and monoclinic phases of Se transform into a metal state, respectively at 12, 16 and 23 GPa, in inverse proportion to the intermolecular distances in their structures [112]. A study of CH_4 under high pressures has revealed that a phase transformation of the insulator → semiconductor type might have occurred by 288 GPa [113]. SiH_4 remains thermodynamically unstable with respect to decomposition over a wide range of pressure. A small pressure window of stability exists between 6 and 16 GPa, and above it silane becomes unstable again with respect to decomposition. In the absence of decomposition SiH_4 remains partially transparent and nonmetallic to at least 150 GPa with $E_g = 0.6–1.8$ eV [114]. YH_3 undergoes an *hcp-fcc* structural transition near 12 GPa, but when the high-pressure *fcc* phase is compressed beyond 23 GPa, the band gap abruptly closes without a structural change in the *fcc* metal lattice. The energy gap closure (metallization) of ScH3 occurs about 50 GPa [115, 116]. Xe transforms into the metal state at 132 GPa [117].

MgH_2 has been studied up to 16 GPa using a high-pressure synchrotron X-ray diffraction technique. Several pressure-induced phase transitions have been identified in this pressure range. Owing to the close structural similarity between the α and γ modifications, the high-pressure γ form can be stabilized as a metastable phase after pressure release. The experimentally observed structural transition sequence and the volume changes at the transition points as well as bulk moduli are found to be in good agreement with theoretically calculated data [118]. CaH_2, SrH_2, and BaH_2 adopt the cotunnite structure at ambient conditions. Due to greater relative compressibility of anions, the cation–anion radius ratio is increased under pressure and these alkaline-earth hydrides transform into the Ni_2In structure at 15, 10, and 2.5 GPa, respectively, with the volume reduction of 6.6 %. SrH_2 retains the Ni_2In structure at pressures up to 72 GPa but BaH_2 undergoes the second transition (in the post-Ni_2In structure, named SH) at 50−65 GPa, with $\Delta V/V = 0.15$. Upon decompression, the SH phase returns to the cotunnite structure. The volume of BaH_2 is smaller than that of pure Ba at ambient pressure ($\Delta V = 6.5$ Å^3), which is common for metal hydrides with ionic bonds. Crystal lattices of the transition metals expand when incorporated hydrogen atoms form local bonding with surrounding metal atoms by hybridization of the H 1*s* and metal *d* states. The extraordinarily large volume reduction at the Ni_2In–SH transition should result from the change in the bond nature as well as an increase in the packing efficiency. Thus, in the SH phase, Ba atoms form a simple hexagonal lattice and the hydrogen atoms form a two-dimensional honeycomblayer sandwiched between hexagonal Ba layers [119].

LnH_3 (Ln = Sm, Gd, Ho, Er, Lu) and YH_3 undergo *hcp–fcc* phase transitions under high pressures from 2 up to 12 GPa; P_{tr} increases with the decrease of the cation radius [120]. In search of the cause of the observed *hcp–fcc* transition, XRD study of hydrogen separations in the structure before and after transition was performed. The H–H distance is considered to be important since it affects repulsive interactions between hydrogen species. As is known, H–H hydrogen repulsion influences the increase of the ideal *c/a* ratio from 1.63 to approximately 1.80 of *hcp* structure that takes place in lanthanide and yttrium trihydrides. Thus repulsive interactions increase dramatically if this distance is less than the critical one for negatively charged hydrogen particles. The fact that the minimal H–H distance before the transition is close to 2 Å can account for the instability of the initial hexagonal phase of the trihydrides under investigation. After the transition the minimum H–H distance sharply increases. YH_3 undergoes the *hcp–fcc* transition at 12 GPa, and attains $E_g = 0$ at $P \approx 23$–26 GPa without a structural change [115]; an extrapolation of data [121] gives $P_{ME} = 55$ GPa.

The structure of the α phase of Li_3N (layers of Li_2N connected by Li^+ ions) at $P = 0.5$ GPa transforms into the β phase with a hexagonal LiN network where each atom N is connected with ions Li^+ above and under the layer; at $P = 36$–45 GPa there occurs a transition from the β to the γ phase in which each N is surrounded by $14\,Li^+$ at distances of $8 \times 1.95 + 6 \times 2.25$ Å. This phase is stable up to 200 GPa. The values of P_{ME} obtained by extrapolation for NaCl, Li_3N, MgO and Ne, are 0.5, 8, 21 and 134 TPa, respectively [122]. The phase transition pressures of alkali azides MN_3 decrease with increasing cation radii: Li has no transitions up to 62 GPa, Na undergoes one at 19, K at 15.5, Rb at 0.5 and Cs at 0.4 GPa [123].

Very interesting phase transitions have been observed in RuO_2 and FeI_2. In the former the TiO_2 type structure at 11.8 GPa transforms into the $CaCl_2$ type without changing the coordination number, i.e. as a second-order phase transition [124]. In FeI_2 a sluggish structural phase transition takes place, which starts at about 20 GPa and is completed about 35 GPa; it results in the doubling of lattice parameters and the formation of a new Fe sublattice replacing the original CdI_2-type structure. The latter alterations in the Fe sublattice may indicate a trend of the Fe sites towards disorder in the new high-pressure phase. This first-order phase transition is characterized by a significant change of the unit cell parameters, a reduction in volume, and a change of the Fe-I distances. The substantial reduction of the Fe-I distances with minimal changes in the Fe–Fe bond lengths at the transition, suggests a charge-transfer-type gap closure mechanism involving the I*p*-Fe*d* bands. At P > 40 GPa, an overturn of the structural transition is observed resulting in the return of the original, CdI_2-type structure [125].

Structures of organic [126, 127] and especially biomolecular [128] compounds are usually studied at moderately high pressures, so the observed phase transitions, whether of the first or second order, result in a change of the molecular packing while the intramolecular structure is only insignificantly affected. However, this limitation is essentially a matter of choice. There is every reason to expect that at higher pressures organic molecules will also show interesting structural behavior [129, 130]. Thus, for benzene theoretical calculations [129] predict a rearrangement into a saturated, tetracoordinate polymer via an intermediate metallic state (stable in the 180–200 GPa range).

The thermodynamic aspect of the metallization of substances under pressure is considered in [131, 132]. It is well-known (see Fig. 2.2 in Chap. 2), that the potential curve crosses the zero line of the dissociation energy (E_D) twice: at thermal expansion when the induced heat = E_D (thermal dissociation) and at the compression when the repulsion energy of approaching atoms = E_D (pressure dissociation). From the quantum-chemical calculations of Slater [133] it follows that when atoms in a molecule are brought closer to one another, the valence electrons are shifted from the region of chemical bonds into anti-bonding orbitals. For crystals this is equivalent to a transition of bond electrons from a valence zone into a conductivity band. It follows that

$$D = P_{ME} \Delta V \tag{10.1}$$

where ΔV is the volume change under transition to $N_c = 12$. Crystal-chemical estimations of ΔV allow to calculate P_{ME} with the accuracy of *ca.* 10 %.

10.2 Pressure-induced Amorphization and Polyamorphism

As mentioned above, phase transitions can occur by coherent displacement of atoms or by transformation of the order-disorder type, as was observed in ice [134] where as a result of breaking of H-bonds under high pressure a full disorganization of the long-rage order occurs. The phenomenon of pressure-induced amorphization is now known to occur in nearly a hundred crystalline compounds, including structures with tetrahedral building blocks, e.g. quartz and silicates, GeO_2, $LiKSO_4$, $AlPO_4$, [135], SnI_4 and GeI_4 [136–138], GaP, GaAs, Ge [139, 140], ZnTe [141], BeTe [28, 29], in wolframates of Sc, Lu and Zr [142]; in layered structures of graphite [143] and $Ca(OH)_2$ [144, 145]; or in a chained structure of sulfur [146]. Note that in silicates, SiO_2 and GeO_2 the amorphization is accompanied by an increase of $N_c{}^*$, i.e. simultaneously with desintegration of the long-range order there occurs a transformation of the short-range order, and on unloading the N_c returns to 4. $BaSi_2$ undergoes an amorphization under $P = 13$ GPa with a decrease of d(Ba–Si) but with almost no change of the Ba–Ba distance [147]. $Sc_2(WO_4)_3$ has a negative coefficient of thermal expansion β and it is noteworthy that other crystals with tetrahedral structures which undergo an amorphization under pressure, also have $\beta < 0$ [148]. A number of factors may be responsible for this, viz. metastable melting, a kinetic hindrance of phase transition, dynamical instability, polytetrahedral packing and orientational disorder of polyatomic ions; some of these factors may be interrelated [149]. Generally speaking, amorphization of crystals under pressure can procede both as a phase transition to a more dense structure, or as a breakdown of the initial structure [150, 151]. The reason of amorphization under pressure is that the N_c is forcibly increased to a value uncharacteristic for the given structure, which upon a fast unloading therefore loses the crystalline order and forms an amorphous material. Low speed of annihilation of defects in substances with high covalency of bonds can stabilize the amorphous state at ambient conditions. It is likely that by faster unloading of the high pressure phases with simultaneous rapid cooling, more materials can be made amorphous.

The topic of polyamorphism, i.e. phase transitions between amorphous phases of different structure, has been briefly outlined above (Sect. 9.1.1) but is worth mentioning here because practically all proven cases of it involve high pressure experiments. While it has been accepted for a long time that amorphous solids can resemble the structure of different crystalline polymorphs, until recently it was usually taken for granted that the liquid state is unique for a given substance. High-pressure studies helped to refute this assumption completely. Polyamorphic transformations were found in liquid alkaline metals, Ga, Si, Ge, Sn, P, Sb, Bi, S, Se, Te and iodine; for a detailed discussion see [152, 153]. The measurements have revealed that compressions of liquid alkali metals is uniform, whereas that of covalent liquids is mostly anisotropic and for these substances the pressure dependence of interatomic distances deviate from the $(V/V_o)^{-1/3}$ behavior. Some elements show different types of the $d = f(P)$ dependence in different pressure ranges. Although most of the observed structural changes are continuous, the abrupt structural change in liquid phosphorous (completed at $P \approx 1$ GPa and $T = 1{,}050\,°\mathrm{C}$) suggests a first-order phase transition [154, 155]. As in crystalline solids, the coordination number increases with pressure, as discussed for a-Si in Sect. 9.1.1. Amorphous Ge films, studied under high pressure by Raman scattering and X-ray absorption spectroscopies, convert at 7.9–8.3 GPa from a low-density amorphous to a high-density amorphous phase with an increase of N_c(Ge) from 5 ± 1 at 8.3 GPa to 6 ± 0.5 at 9.8 GPa [156]. The As_2O_3 glass has $N_c(\mathrm{As}) = 3.1$ at ambient conditions but $N_c(\mathrm{As}) = 4.6$ at 32 GPa [157]. In the GeO_2 glass the N_c(Ge) increases monotonically from 4.2 at 1 GPa to 5.5 at 15.7 GPa [158]. Melted AsS undergoes a phase transition under high pressure with a structural change [159]. In the structure of liquid CdTe the N_c grows from 3.6 to 5.1 as the pressure increases from 0.5 to 6.2 GPa [160], in liquid GaSb the proportion of the *bcc*-type structure increases from 8 % up to 43 % with the increase of P from 1.7 to 20 GPa (the rest is a structure of β-Sn with $N_c = 6$) [161]. Unusual change has been observed in the structure of molten Cs: under $P = 3.9$ GPa there is a discontinuity in density of the liquid caused by a drop of N_c from 12 to 8, which marks the change to a non-simple liquid. The specific volume measurements and XRD analysis of liquid Cs suggest the existence of dsp^3 electronic hybridization above 3.9 GPa similar to that reported on compression in the crystalline phase [162].

In such substances as S, Se, Te, P, I_2, As_2S_3, and As_2Se_3, the pressure of semiconductor-to-metal transitions for liquid is much lower than that for the corresponding crystals (see discussion in Sect. 9.1.1) [163, 164], viz. for Se at 4 GPa *vs* 14 GPa, for S at 19 GPa *vs* 90 GPa, for I_2 at 4 GPa *vs* 16 GPa, for H_2 at 140 GPa *vs* > 300 GPa, respectively. In liquid nitrogen a liquid-liquid polymer transition occurs at 50 GPa and 1,920 K, while the phase transformation in solid nitrogen takes place at $P > 110$ GPa and $T = 300$ K [165].

10.3 Effect of the Crystal Size on the Pressure of Phase Transition

Experiments have shown that the stability ranges of polymorphs in nano-sized crystals can be very different from those in large crystals. Remarkably, for some solids the pressure of phase transition (P_{tr}) decreases with the decrease of particle size, but for others it increases. Thus, ZnO transforms into the structure of NaCl at 9.9 GPa in the bulk and at 15.1 GPa in 12 nm crystals [166]. The same transition in occurs in ZnS at 13 GPa (bulk), 15 GPa (25.3 nm size) and 18 GPa (2.8 nm size) [167, 168], in CdS at 2.7 GPa (bulk) and 8.0 GPa (4 nm size) [169], in CdSe at 2.0 GPa (bulk) and 3.6–4.9 GPa (1–2 nm size) [35]. In PbS, P_{tr} increases from 2.4–5.8 GPa to 3.3–9.0 GPa as the particle size decreases from 8.8 to 2.6 nm [170, 171].

The anatase form of TiO_2 rearranges into the structure of PbO_2 at 2.6–4.5 GPa in the bulk material, at 16.4 GPa in 30 nm particles, and at 24 GPa in 9 nm grains [172], in [173] it has been shown that the pressure-induced amorphization happens only if the starting crystallite size is < 10 nm. A study [174] of two different samples of anatase, the mechanically prepared polycrystals and the chemically prepared amorphous particles with the same particle sizes of 6 nm, has shown that the reversible transformation to high-density amorphous state happens in the range of 13–16 GPa in the former and around 21 GPa in the latter. Further compression leads to a transformation between two amorphous states at 30 GPa. The onset of the transformation of ZrO_2 from the orthorhombic to the monoclinic phase, increases from 3.4 to 6.1 GPa upon reduction of the particle size from 10 to 3 nm [175]. Amorphization of a quartz-like GeO_2 occurs at $P = 11$–15 GPa in 40 nm particles, at 9.5–12.4 GPa in 260 nm crystals, and about 6 GPa in the bulk material [176]. In SnO_2 the phase transition pressure increases from 23 GPa (for the bulk material) to 29, 30, and over 39 GPa at the progressive reduction of the particle sizes to 14, 8 and 3 nm, respectively [177]. When two different morphologies of nanostructured SnO_2, nanobelts and nanowires, were compressed to 38 GPa, they behaved drastically different from the bulk material: a phase transition from the rutile to the $CaCl_2$ type structure occurs at 11.8 GPa in the bulk, at 15.0 GPa in nano-belts, and at 17.0 GPa in nano-wires. These examples illustrate the peculiar pressure behavior of nanomaterials and are relevant for production of desirable structures for new applications by combined pressure-morphology tuning [178].

Similar phenomena were observed in AlN: the onset of the B3 $\rightarrow$ B1 type transformation changes from 20–23 GPa in the bulk material, to 14.5 GPa in nano-crystals and 24.9 GPa in nano-wire samples [179]. Bulk and nanocrystalline GaN also have been studied by high-pressure X-ray diffraction, revealing the phase transitions from the wurtzite to the NaCl-type phase at 40 GPa for the bulk and 60 GPa for the nanocrystals [180].

However, in Fe_2O_3 the opposite result has been observed: diminishing particle size led to a decrease of the γ (magnetite) $\rightarrow$ α(hematite) phase transition pressure from 35 GPa for the bulk material to 27 GPa for 9 nm particles [181]. In [182] this result was confirmed: the γ-Fe_2O_3 $\rightarrow$ α-Fe_2O_3 transition in nano-particles was observed at $P = 13.5$–26.6 GPa. Similar effects were observed also for CeO_2 where nano-crystals transform from the CaF_2 to the α-$PbCl_2$ structure at 22.3 GPa, but in the bulk material $P_{tr} = 31$ GPa [183]. The bulk rutile TiO_2 transforms to the structure of ZrO_2 at 13 GPa, but already at 8.7 GPa in 30 nm particles [175]. A decrease of the phase transition pressure has been observed in nano-particles of Se [184]. Pressure-induced phase transitions in a 12 nm sample of ReO_3 have been investigated by X-ray diffraction using synchrotron radiation [185], revealing that the ambient-pressure cubic-I phase transforms to the monoclinic phase (with a *ca.* 5 % change of volume) then to a rhombohedral-I phase, and finally to another rhombohedral phase at 20.3 GPa. The transition pressures are generally lower than those known for bulk ReO_3. Thus, nanocrystals are more compressible than bulk ReO_3.

Chemical explosives have detonation velocities of up to 9–10 km/s, a projectile can be accelerated by a light gas gun can to 7–8 km/s and by an electromagnetic accelerator to 10–12 km/s (cf. the first cosmic velocity of 7.9 km/s). Impulse irradiation can produce the shock-front velocity of 15 km/s and more. Assuming that $U_p \approx ½ U_s$, the shock velocities of 9, 12, and 15 km/s will generate the pressures of 100, 180, and 280 GPa, respectively, in a material with $\rho_0 = 2.5$ g/cm^3 (roughly the density of Al, graphite or most common rocks). Note that the dynamic pressure is proportional to the density (Eq. 10.2), thus for steel ($\rho_0 \approx 7.8$ g/cm^3) the pressures will be more than three times higher. Therefore in practice the pressure is often enhanced by adding heavy inert substances, such as tungsten powder ($\rho_0 = 21$ g/cm^3), to the investigated sample. The very first dynamic experiments (up to 1960) easily produced pressures of several hundred GPa. Today, with better understanding of compression physics and more sophisticated devices, thousands or even tens of thousands of GPa (i.e. tens of TPa) are achievable. Another important advantage of the dynamic method is that it imposes no strict limitation on the sample size, unlike the static method. On the other hand, the dynamic method has important disadvantages. The first problem was to preserve the sample during the explosion and to recover it afterwards for investigation and applications. It was first solved in 1956 by Riabinin [199] who invented for this purpose a thick-walled steel vessel (known as the recovery ampoule), subsequently much improved and modified [200–204]. This signified the proper beginning of the shock chemistry.

The second problem is the extremely short duration of the dynamic pressure (see Fig. 10.2). Initially, there was a widespread disbelief that a rearrangement of a crystal structure can be accomplished within such a short time. The skepticism was apparently confirmed by Riabinin's initial failure to achieve a graphite → diamond transition (although he shock-compressed numerous carbon-containing substances specifically for this purpose) but was refuted by by Bancroft et al. [205] who observed the α-Fe → ε-Fe polymorphic transformation (at 13 GPa) while studying the shock compressibility of iron. The controversy about the possibility of shock-induced phase transitions was resolved by a freak of nature: in 1961 a meteorite hit a coal-bed in Canyon Diablo (USA), and traces of diamond were later found at the point of impact. By an incredible coincidence, the coal there had the structure of the so-called Ceylon graphite, a thermodynamically unstable rhombohedral form with an ABCABC stacking sequence of the layers, which can be considered as an extended stacking fault of the ordinary hexagonal graphite (see Sect. 6.3.1). Today this form is known to be best suited for diamond synthesis—but it was not tested by Riabinin because this mineral was not available in the USSR. In 1962 this natural graphite → diamond transformation was reproduced in laboratory experiments by De Carli and Jamieson [206]. Since then, shock-wave induced phase transitions in elemental solids, alloys and compounds attract intense scientific and industrial interest.

Now let us consider specific features of these transformations. The pressure of phase transition, P_{tr}, depends on the direction in which the shock wave front propagates through a crystal. Thus, the transitions from a NaCl to a CsCl type structure occur at lower pressures if the crystallographic axis <111> is normal to the shock wave front, i.e. the unit cell is compressed along its spacial diagonal, because this is the optimal route of the structural reorganization. Thus, in KCl samples the compression along <100>, <110> and <111> directions, gave $P_{tr} = 2.5, 2.2$ and 2.1 GPa, respectively [207]. However, no coherent atomic displacements can transform the graphite lattice into the diamond one. This phase transition requires amorphization of the original crystalline graphite. Accordingly, the highest yield of diamond was obtained in the detonation-induced transformation of the most defect-rich material, namely, carbon black. According to Ahrens [208], the SiO_2 glass is generally the most useful source of the rutile phase SiO_2, as it is not necessary to destroy the initial crystal structure. Contrary, in the graphite–hexagonal diamond and h-BN–w-BN systems both the initial and final modifications have the same symmetry and the phase transition occurs without hindrance. Analysis of these experimental findings, gave an impetus to studies on geometric modelling of phase transitions that became a separate avenue of scientific research [209].

The kinetics and mechanism of such transformations are considered in [210] from the standpoint of the evolution of structures. Imagine the following model. The potential surface of a crystal has a minimum corresponding to a particular structure. As thermodynamic conditions change during shock compression, the depth of this potential well is reduced; however, at the same time, a new potential well is initiated on the former surface or the potential surface itself is transformed. Therefore, the atom vibrating with increasing amplitude can jump from one potential minimum to the other. It is known that at room temperature the vibration amplitudes of atoms in solids are of the order of 10 % of the bond length, increasing to 10–20 % on heating to the melting point in accordance with Lindemann's law (Sect. 6.2). Allowing for the fact that interatomic distances vary by 20 % when N_c varies from 1 to 12, i.e. from a molecular to a metallic structure, it is easy to see that possible changes in the vibration amplitude of atoms in solids cover the whole range of variation of bond lengths at polymorphic transformations. Let us estimate the time of the atomic jump from one position (the initial phase) to the other (the high pressure phase). The frequency of thermal vibrations of atoms is

$$w = \frac{1}{2\pi \, \Delta r}\sqrt{\frac{3kT}{m}} \tag{10.4}$$

where k is the Boltzmann's constant, m is the atom mass Δr is the mean-square displacement of the atom ($\Delta r \approx 0.1d$). For a hydrogen atom at $T = 300$ K, $w \approx 6 \times 10^{13}$ s^{-1}. For a substance with $m = 100$, shocked by $P = 10$ GPa, $T \approx 1{,}000$ K and $w \approx 1 \times 10^{13}$ s^{-1}, i.e. the time of the atomic jump $\tau \approx 10^{-13}$ s, which is close to experimental estimations of the minimum duration of phase transitions in shock waves.

The studies of the shock transitions by the X-ray pulsed analysis have shown [211–214] that at low pressure the compression of a lattice is mono-axial, at medium pressure the compression evolves toward isotropic, and at high pressure the compression is strictly isotropic, being of a purely hydrostatic character. It results in similar values of the pressure for phase transitions under static and dynamic regimes, which in fact is the case. Generally speaking, the dynamic values usually must exceed the static characteristics because the sample under shock loading is heated, whereas

Table 10.4 Minimum sizes (nm) of domains (D) in shocked crystals

Solid	Diamond	TiO_2	BN	AlN	Mo	MgO	LiF	CdF_2	Al_2O_3
hkl	–	–	110	100	110	100	100	110	113
D	10	10	15	15	15	16	16	18	19
Solid	CaF_2	BaF_2	UO_2	Y_2O_3	LaB_6	ZrC	ZrO_2	Cu	NaF
hkl	220	111	200	–	–	–	–	111	110
D	19	22	23	25	30	30	35	35	44

the pressures at static phase transitions are often determined at ambient temperature [215]. However, this difference is compensated by the different methods of measuring P_{tr}: on static compression it is usually defined as the average of the pressures of the direct (P_{dtr}) and reverse (P_{rtr}) transitions, and $P_{dtr} > P_{rtr}$ (hysteresis) [216]. On the other hand, the dynamic phase transition pressure is measured only at the loading stage. Note also that under dynamic loading, crystals' defects and vacancies in their structures play very important role: they can define the degree and even the possibility of a polymorphic transformation of the martensite type, blocking the growth of grains of the new phase [199, 217]. This is especially important for such materials as graphite and *h*-BN, which are aways structurally imperfect.

In fact, when shock waves are propagated through a solid, crystal domains suffer severe fragmentation and misalignment at the shock front due to the steep pressure gradient. The domain size in shocked crystals can be reduced to as little as 100–250 Å (Table 10.4) and the density of defects and dislocations may increase by several orders of magnitude. These characteristics are studied by analyzing the XRD profiles of both single crystals and polycrystals. It is convenient to divide the objects for microstructure study into two groups, metals and dielectrics. The common characteristics of the substructure in various metals subjected to shock loading is a high density of defects, in particular, the density of disclocations reaches 10^{10}–10^{11} cm^{-2} on loading to 10 GPa [199].

Let us consider now the results of investigations of the substructure in nonmetallic compounds. The density of dislocations of shocked single crystals in this case also increases by several orders of magnitude, disorientation of blocks increases by 2–5° and microhardness by several times. It is interesting to note that an increase of the dislocation density in many shocked solids is accompained by retaining the dislocation configurations which existed prior to shock loading. Only in KBr a new dislocation picture was observed, and that because the reversible phase transition had completely rebuilt the dislocation structure.

Domains of real crystals have different sizes in different crystallographic directions; in the general case they are ellipsoidal. It was observed that in shocked solids the longer axis of the ellipsoid is reduced stronger than the shortest one. Table 10.5 presents for comparison the results of one (D_1) and two successive (D_2) shock compressions of several crystals. As follows from this table, multiple shock compression alters an anisotropic structure in the direction of the isotropic one. These results are due, to a certain limit, to the shock wave fragmentation effect, which is naturally related to the characteristic depth of the shock front (of the order of 100 Å for crystalline media).

Table 10.5 Domain sizes after a single and double shock compression, in nm

MgO		Al_2O_3		CdF_2		BaF_2	
D_1	D_2	D_1	D_2	D_1	D_2	D_1	D_2
25	19	20	19	18	18	22	24
160	29	80	18	60	38	38	36

Extreme grinding of solids, by creating the maximum concentration of defects, tends to transform an anisotropic crystal into an amorphous isotropic solid. Thus, quartz and silicate crystals undergo a complete or partial conversion into the amorphous state (*diaplectic* glass) under shock compression. Quartz crystals become amorphous at $P = 36$ GPa [218], i.e. after shock loading so many defects were formed that a return of such disordered material into a crystalline state was not feasible. This effect has been studied in greatest detail in connection with the lunar rock samples analysis. These samples contained glassy intrusions which could not have formed thermally; these must have formed as a result of meteorite shocks. Later, similar glass formations were discovered also in volcanic craters on Earth.

If defects in a shocked crystal appear in a regular pattern, then the lattice of defects can be created. Although a high concentration of defects usually decreases the density, shocked solids with an ordered system of vacancies may have a denser structure. Thus, shock compression of the Ln_2S_3 compounds, where Ln = Tb, Dy, Ho, Er and Y, led to a transition from the monoclinic form of the Ho_2S_3 type to a denser cubic form of the Th_3P_4 type where every ninth atomic site in the anionic sublattice was vacant [219]. Phase transitions with an increase of density were observed also in the shocked compounds MF_4, where M = Ce, Th or U. Whereas the starting tetrafluorides had the structure of the ZrF_4 type, the shocked compounds have structures of the LaF_3 type where 25 % of sites in the cationic sublattice are vacant (therefore, the formula of the new modification should be properly written as $M_{0.75}F_3$) [220, 221]. The structures with subtraction of atoms from the sublattices can be formed on the quenching of a shock-compressed solid; the absence of such phase transitions in MF_4 under static compression agrees with this hypothesis. Similar phenomena were subsequently observed upon shock compression of Nb_2O_5 and Ta_2O_5 which yielded the TiO_2 type structures ($M_{0.8}O_2$) [222, 223].

The above-mentioned phase transitions conform to the Le Chatelier principle, the sample volume decreasing under high pressure. They are not basically different from those observed in the static method, under conditions of thermodynamic equilibrium. There is, however, a class of anomalous phase transitions, which occur *only* in dynamic experiments and in which the shock compression gives rise to *lower* densities. The first of such phases was obtained in 1965 by shock treatment of the turbostratic BN [224]; the new phase differed from both the graphite-like (*h*-BN) like (*c*-BN) polymorphs of boron nitride and was named E-BN (E standing for the 'explosion phase'). Later, it appeared that the lattice parameters of E-BN are nearly identical to one of the phases of fullerene C_{60} [225, 226], viz. $a = 11.14$, $b = 8.06$, $c = 7.40$ Å for E-BN, cf. $a = 11.16$, $b = 8.17$, $c = 7.58$ Å for C_{60}, with similar densities of 2.50 g/cm^3. Thus, the BN-fullerene was obtained by explosion (though not recognized as such) some 25 years before the carbon fullerene was identified. Later on,

Table 10.6 Anomalous phase transitions effected by shock waves

Substance	Before shock compression		After shock compression	
	Structure, phase	ρ, g/cm^3	Structure, phase	ρ, g/cm^3
PbO	Orthorhombic	9.71	Tetragonal	9.43
SmF_3	LaF_3	6.93	YF_3	6.64
HoF_3	LaF_3	7.83	YF_3	7.64
Tm_2S_3	Ho_2S_3	6.27	Novel type	6.06
Nd_2O_3	A	7.42	C	6.29
ZrO_2	Tetragonal	5.86	Monoclinic	5.74
GeSSe	α	4.0	β	3.6
GeSTe	α	5.1	β	4.7
GeSeTe	α	5.6	β	5.2

several researchers have obtained E-BN by different methods, always under strongly non-equilibrium conditions [227]. Other substances, which transform to low-density modifications under shock treatment, are listed in Table 10.6 [204].

Numerous syntheses of E-BN by different techniques gave materials with ρ between 2.5 and 2.6 g/cm^3 and with similar, but not the same, structural and spectroscopic properties. As have been established, E-BN is an intermediate step in the *h*-BN ↔ *w*, *c*-BN transformation under irradiation, high pressures or temperatures. Since both diamond and *c*-BN were known to form fullerenes under electronic irradiation, could E-BN also be a fullerene? However, the cages with 6- *and* 5-membered rings, typical for carbon fullerenes, are topologically impossible for BN without allowing some homoatomic bonds. Since both B–B and N–N bonds are weaker than B–N, such topologies are thermodynamically forbidden: any model to be considered should contain only B–N bonds. The principal change came in 1993 when several authors [228–230] predicted the possibility of a fullerene cage with hexagonal and tetragonal rings containing only B–N bonds, thus circumventing the thermodynamic ban. Indeed, it was soon found [231] that under intense electron beams (in transmission electron microscopes) atomic layers of BN, like those of carbon, tend to curl and form concentric shells in onion-like structures. In BN, spherical basal planes develop which, in contrast to carbon analogues, are not completely closed. It has been suggested that the formation of spherical clusters under electron irradiation is a general phenomenon in substances which crystallize in graphite-type structures. Later Golberg et al. [232–234] observed isolated fullerene-type BN particles with reduced numbers of layers; the high-resolution transmission electron microscopy images indicated an octahedral geometry for these BN clusters.

Olszyna et al. [235] were the first to propose a molecular BN fullerene model which appeared to have a multi-atom unit cell. The nitrogen and boron atoms are arranged in parallel layers of strongly deformed hexagons. Topological analysis of the structure suggests that the E-BN molecule contains equal numbers of sp^3 and sp^2 bonds. The molecular and crystal structures of E-BN have been determined by Pokropivny et al. [236], who for the first time interpreted all the main peaks of the observed XRD patterns and demonstrated that E-BN has the diamond-type crystal lattice (the space group *Fd3m* or O_h^7) and is built of $B_{12}N_{12}$ cages polymerized through their hexagonal faces. This phase, which was termed 'extra-diamond', has the framework of faujasite type and can be regarded as a $[B_{12}N_{12}]$-zeolite. The same group [237] suggested a possible mechanism for polymerization of BN molecules in the solid state. The mode of polymerization depending on the applied pressure and other experimental conditions, the differences between the properties of the E-phase reported by different authors become understandable.

Before attempting to explain these transitions, it is necessary to prove that they are in fact caused by shock loading rather than the residual heat. After all, the decrease of density and coordination number is normal for a *thermal* phase transition, e.g. diamond → graphite above 1,000 °C or B2 → B1 in CsCl at 432 °C. A crucial test is provided by certain compounds, like PbO, MnS, Ln_2S_3, Ln_2O_3, ZrO_2, and LnF_3, whose high-density polymorphs (1) are thermodynamically more stable than the low-density ones (2) at all temperatures up to the melting point. Thus, mere heating of these solids will not decrease N_c and density. On the contrary, it will precipitate the transition (which is kinetically hindered at low temperature) into the deeper minimum of the denser phase (1). Such transitions are indeed known, and higher pressure can only facilitate them. Apparently, no thermodynamic conditions can induce the reverse transition, (1) → (2). However, this is precisely what happens under shock compression—hence we have encountered a novel effect indeed.

It can be explained by the peculiar feature of the dynamic compression, that the high-pressure zone is followed by a regime of sharp decompression, which generates extensive plastic deformations and tensile (stretching) stresses, making possible even a break-up of a monolithic body. It saturates the crystal with defects and dislocations, ultimately converting it into an amorphous phase (3) with the density lower than that of any crystalline polymorph. Therefore the conditions are created for rapid nucleation of the new phase. According to Ostwald's rule of stages, the most kinetically accessible phase (i.e. the first one to crystallize) is never the most stable one, but rather the least thermodynamically stable one, closest to the unstable starting phase in energy. Therefore phase (2) with the lowest (after the amorphous phase) density will crystallize first. If the residual heat persists long enough, phase (2) can be annealed into the denser phase (1), otherwise the process can be kinetically 'frozen' at stage (2).

$$\text{HDP}\,(1) \xrightarrow{P_{\text{dyn}}} \text{Amoph}\,(3) \xrightarrow{T_{\text{res}}} \text{LDP}\,(2)$$

For example, shock compression of A-Nd_2O_3 ($\rho = 7.42\,\text{g/cm}^3$) converts it completely into the C-phase ($\rho = 6.29\,\text{g/cm}^3$), but on heating to 400–600 °C the A-form reappears in increasing concentration, and at 800 °C the reverse C → A transition is complete. To imitate the proposed mechanism stage-by-stage, we subjected the crystalline A-Nd_2O_3 to grinding for 20 h in a ball mill and obtained the X-ray amorphous phase, which on heating to 400–600 °C crystallized into C-Nd_2O_3. Evidently, if the residual temperature in the recovery ampoule is within this range, the net result of the shock compression will be the A → C transition. Obviously, the probability of incidentally satisfying the necessary thermodynamic conditions is very small, therefore this type of transition was such a rarity. However, once the mechanism of anomalous phase transformations is clarified, the quest for them can be more successful [204].

Phase transitions with decreasing density were achieved also in experiments involving an irradiation of substances [238], ball milling of crystals [239, 240], and static compression with shear [241–243]. Each of these processes involves a partial or total amorphization of the solid, by generating numerous defects, creating ultra-fine particles with highly active surfaces, or causing plastic flows, respectively.

Ball-milling can also yield a high pressure phase, since it involves high dynamic pressures (on direct collision of balls), plastic deformations of crystalline grains (between sliding balls) and high temperatures (caused by friction). Thus, a phase transition of the C $\rightarrow$ B type for Y_2O_3, Dy_2O_3, Er_2O_3,Yb_2O_3, previously achieved under high static pressures and temperatures (25–40 kbar, 900–1,020 °C) [244] and on shock compression [245], was also observed under ball milling [246]. Later it was shown that this treatment can generate traces of the E-BN and *c*-BN [247]. We studied [248] the mechanism of the structural transformation in *h*-BN under ball-milling and found that the yield of the high pressure phase ($\sim$ 20 %) reaches a maximum after 12 h and then remains constant for any grinding time, i.e. the phase transformation is of reversible nature. This was confirmed by carrying out the reverse *w*-BN $\leftrightarrows$ *h*-BN transition by ball-milling. Besides the known forms of BN, several new phases were obtained.

On shock compression, some crystal compounds have been observed to dissociate into components. Thus, Schneider and Hornemann observed a shock decomposition of $Al_2Si_2O_5$ into Al_2O_3 and stishovite (the high-pressure phase of SiO_2) [249–251]. Staudhammer et al. discovered the formation of *c*-BN (the high-pressure phase) by shock decomposition of $BH_3 \cdot NH_3$ [252], and most remarkably, detonation of some explosives yields substantial amounts of diamond nano-particles [253, 254] (see below). The stability of methane, ethane, octane, decane, octadecane, and nonadecane were studied in a CO_2-laser heated diamond anvil cell at pressures up to 25 GPa and temperatures up to about 7,300 K. Methane and ethane were found to decompose to form hydrogen and diamond. Substantially greater yields of diamond were obtained from longer-chain alkanes [255]. A survey of publication on the pressure dissociations of chemical compounds into components can be found in [256]. Results and perspectives of studies of the physical and chemical transformations in shock waves are considered in [204].

No high-pressure phase transition attracted as much attention as that of graphite into diamond and of hexagonal BN into cubic BN (borazon), due to their great technological importance and (one may suggest) the aesthetic fascination with diamond. And no other example can better illustrate the difficulties created for the dynamic-pressure studies by the problem of the residual heat. This problem became the subject of a veritable scientific race between Soviet and American researchers, the major result of which was to show that the graphite $\rightarrow$ diamond transition proceeded quantitatively in the shock wave but in the decompressed state only a few per cent of diamond remained, due to the effect of the residual heat. Diamond and borazon were annealed and reverted to the graphite modification in the aftermath of unloading, when the pressure already had returned to ambient but the temperature remained above 1,000 °C in recovery ampoules. Therefore, most researchers sought the solution in rapid removal of heat during or after unloading. Various techniques

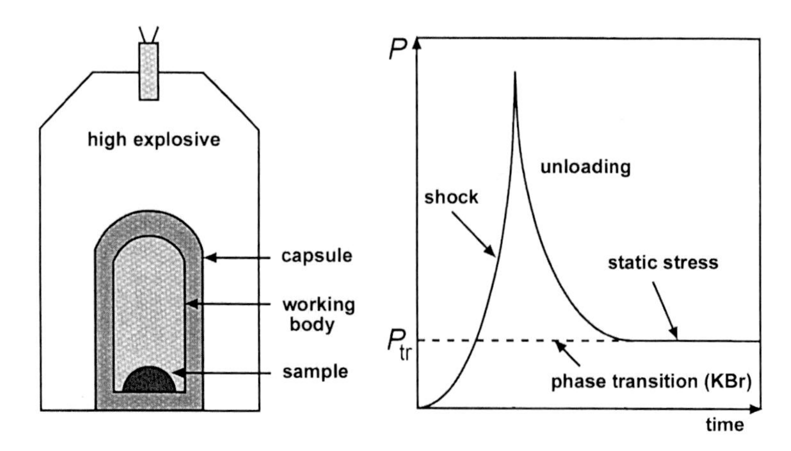

Fig. 10.5 Schemes of the dynamic-static compression. *Left*: experimental set-up. *Right*: evolution of the pressure inside the capsule with time. (Adapted with permission from [258]. Copyright 1997 Springer Science and Business Media)

were tried (and patented), viz. the experiments were carried out at low temperature, in water, in copper matrices; the starting material was mixed with some inert component to act as a heat sink, etc. None of these helped to preserve more than 10 % of diamond. However, in any case the unreacted graphite (or the annealed diamond) had to be removed chemically, leaving the diamond residue in the form of fine powder which had to be compacted in a static apparatus in order to obtain bulk super-hard material. Thereby the intrinsic advantages of the dynamic method were largely lost.

In the author's laboratory this problem was solved by a principally new approach: rather than trying to reduce the temperature, we extended the duration of high pressure by creating a high *hydrostatic* pressure inside the recovery ampoule [257]. This was achieved by placing inside a high-strength ampoule a working body of KBr (Fig. 10.5). During shock loading the volume of the ampoule is reduced and KBr undergoes the B1 $\rightarrow$ B2 type phase transition (at $P = 1.8$ GPa) with a 15 % contraction in volume, the transition being reversible. On the unloading stage at this pressure KBr must return to the initial state but, being prevented by the rigid (after a shock strengthening) ampoule from expanding, creates as a result a residual static pressure inside it. This pressure can be maintained indefinitely as long as the ampoule is closed and the sample is allowed to cool to the ambient temperature *while still under high pressure*. This method we named the dynamic-static compression (DSC) technique, permitted to obtain a 100 % yield of the diamond phase of BN in the form of a monolithic specimen with good mechanical properties [258]. Figure 10.6 shows the hardness distribution on the surface of a w-BN sample. The maximum hardness in the centre of the disk corresponds to the maximum pressure along the cylindrical capsule axis (in the Mach field), hardness increases as the size of crystallites falls and the largest value one can expect for the compacted nanocrystalline w-BN and diamond.

It is interesting that although *w*-BN on heating to 700–1,100 °C partially transforms into *h*-BN and the yield increases with temperature, it never reaches 100 %,

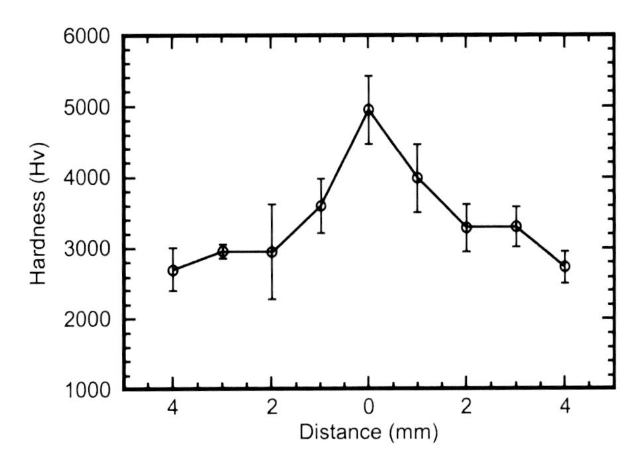

Fig. 10.6 Distribution of Vickers Hardness on the top surface of w-BN along the cross-section of the cylinder axis. (Reprinted with permission from [258]. Copyright 1997 Springer Science and Business Media)

because the nascent *h*-BN, having the volume 50 % greater than *w*-BN, creates strong external pressure upon the remaining *w*-BN [259]. Furthermore, at ambient pressure and 600–970 °C, a partial reverse transition *h*-BN → *w*-BN occurs [260]. Similar processes have been observed (and similarly explained) upon annealing of diamond powders at temperatures 300–1,870 °C where even at 1,530 °C the degree of transformation does not exceed 0.85 [261].

10.5 Detonation Transformation and Synthesis of Diamond and *c*-BN

As follows from geometrical considerations, the graphite to diamond phase transformation is impossible without complete destruction of the original crystal lattice. This transformation is known to proceed only by the mechanism of diffusion which can be realized in static conditions and not under shock loading in recovery ampoules: in recovery shock experiments the yield of diamond is below a few per cent. The diamond yield up to 50–80 % was achieved only upon detonation of graphite-containing charges of high explosives in large chambers (no recovery ampoule) where the starting material became strongly atomized upon expansion of the detonation products. Detonation of high explosives which yielded elemental carbon among the products, or mixtures of graphite powder with an explosive, generated $P = 20$–30 GPa and $T = 2{,}000$–$3{,}000$ °C, whereupon the scattering of detonation products rapidly cooled them and the yield of diamond reached 20 %. Using explosives in the ice shell (Fig. 10.7) for better cooling of the products, increases the yield of diamond to 80 %.

After 1988, when the first syntheses of the detonation diamond were reported[1] in the USA and the USSR [253, 254], the structure and properties of this material

[1] To the authors' knowledge, studies of detonation diamond in the USSR were going on since 1960's but were kept secret. However, we believe that in the absence of any open publications prior to 1988, any attempts to re-establish the scientific priority now are pointless

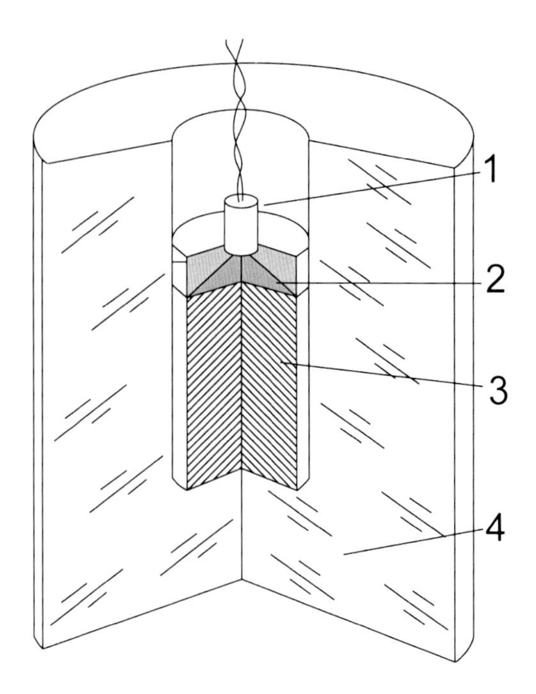

Fig. 10.7 Set-up for the detonation synthesis of diamond: *1* = detonator, *2* = generator of the plane wave, *3* = explosive (e.g. TNT + RDX), *4* = ice. (Reprinted with permission from [266]. Copyright 2009 John Wiley and Sons)

have been studied in numerous works, however the *mechanism* of the formation of diamond clusters remains enigmatic. It was found that the diamond cluster size does not depend on the size of the explosive charge and therefore on the duration of dynamic pressure [262, 263], i.e. there is some mechanism limiting the growth of the cluster. This mechanism was identified [264] by features of the vdW interaction of molecules in the detonation cloud; see also [265].

The typical products of detonation of the high explosives are H_2O, N_2, C, CO, CO_2 (and minor amounts of H_2, O_2, NO, NO_2, NH_3). If we restrict ourselves even to the five major products, the probability of a C_n cluster forming by successive collisions with carbon atom (or other C_n clusters) in a homogenous detonation cloud is 5^{-n}, which makes the formation of a diamond nanoparticle of *ca.* 10^4 atoms practically impossible. This means that the detonation products are segregated and diamond clusters grow within a carbon-rich region. In Sect. 4.2 we show that the vdW heterogenous intermolecular distances are always larger than additive values

$$D(\mathrm{A}\cdots\mathrm{B}) > \frac{1}{2}[D(\mathrm{A}\cdots\mathrm{A}) + D(\mathrm{B}\cdots\mathrm{B})]$$

Hence the volume of a mixture of heterogenous molecules interacting by the vdW mechanism exceeds the additive value, and according to the Le Chatelier principle such mixture under high pressure will be separated. It follows that the diamond cluster size depends on the separation extent of components of the detonation cloud. In this context, it seemed reasonable to perform a detonation synthesis of *c*-BN under similar conditions (in a large chamber). No attempt in this direction has been reported so far, despite the fact that the first publications on the detonation synthesis of diamond appeared 24 years ago. In our opinion, the main reason for the absence of any progress in this direction is the strongly different attitude of carbon and boron nitride to water vapor that is present in detonation products: carbon is chemically inert with respect to water while BN undergoes decomposition through the reaction

$$2BN + 3H_2O = B_2O_3 + 2NH_3.$$

Therefore, successful realization of detonation-induced *h*-BN → *c*-BN phase transition requires that the hydrolysis of BN is suppressed, but detonation gases usually contain water vapour (see above) at high temperature. However, using the hydrogen-free explosive benzotrifuroxan, $C_6N_6O_6$ (and thus excluding water in the detonation cloud) we have achieved the detonation synthesis of *c*-BN with a 80 % yield [266].

10.6 Equations of State of Solids

Results of experimental studies of solids under pressure can be described by the equations of state (EOS). So many equations of state involving the temperature, volume and pressure of a condensed phase have been proposed and used that a full review of them is out of question. Besides books dealing with the subject in general, many papers are more concerned with some specific element such as carbon, or aluminium, or the rare-gas solids, or the behavior of solids under extreme pressures, etc. Useful compilations of EOSs are available [267–269]. The most popular among physicists are the EOSs of Murnaghan [270], Birch [271], Vinet et al. [272, 273] and Holzapfel et al. [274–277]. The latter two EOSs have similar forms:

$$P(x) = 3B_o \left[\frac{(1-x)}{x^2}\right] \exp\left[1.5(B_o{}' - 1)(1-x)\right] \tag{10.5}$$

$$P(x) = 3B_o \left[\frac{(1-x)}{x^5}\right] \exp\left[1.5\left(B_o{}' - 3\right)(1-x)\right] \tag{10.6}$$

where B_o is the bulk modulus (the inverse of compressibility), $B_o{}' = \partial B_o/\partial P$, $x = (V/V_0)^{1/3}$, V_0 and V are the initial and final volumes of the body. Eq. 10.6 has been formulated for metals, but Eq. 10.5 is applied to any matter and therefore it named 'the universal equation of state'. EOSs permit to calculate interatomic distances in solids for any pressures, *if* the bulk modulus is known for a given phase. However, this implies the absence of polymorphic transformations, because on such a transition the compressibility (bulk modulus) usually changes. The bulk modulus is formally the second derivative of the bond energy with respect to volume

$$B = V\left(\frac{\partial^2 E}{\partial V^2}\right) \tag{10.7}$$

If the potential (atomization) energy is presented as

$$E = \frac{A}{d^n} = \frac{A}{V^{n/3}} \tag{10.8}$$

then twice differentiating E by V in Eq. 10.8 gives

$$BV = \frac{n\,(n+3)}{9}E \tag{10.9}$$

Knowing E, B_0 and V, one can find the values of n in different types of compounds [278, 279]. The thus-derived n values are much greater than the n magnitudes derived from thermal expansion (see [280]). The discrepancy between the thermal expansion and pressure data does not arise from an experimental error but reflects the properties of the potential curve (see Fig. 2.4) on either side of the minimum: as the bond length shortens, the repulsive force increases sharply, as the bond lengthens, the potential curve slopes gently out and approaches $E = 0$ asymptotically. This point is very important for understanding the chemical bonding at various thermodynamic conditions.

Bulk moduli for ambient conditions (with index 'o') are listed in Tables: 10.7 for elements, S10.1 for compounds MX, S10.2 for MX_2, S10.3 for MX_3, S10.4 for binary oxides, S10.5 for binary nitrides, S10.6 for binary borides, S10.7 for binary carbides and silicides, S10.8 for binary phosphides and arsenides, S10.9 for ternary oxides and coordination compounds, S10.10 for molecular substances and polymers, S10.11 for characteristics of polymorphous modifications of elements and the MX compounds, S10.12 for various phases of MX_2 crystals.

Fluorene ($C_{13}H_{10}$) undergoes a transformation at 36 GPa with a very big change in volume (922.5 → 862.5 Å^3) and in mechanical properties: the low pressure phase has $B_o = 5.9$ GPa and $B_o' = 7.5$, while the high pressure phase has $B_o = 11.3$ GPa and $B_o' = 5.4$ [335]. Cyclohexane undergoes a phase transition from the monoclinic into the triclinic phase at 7–8 GPa, $B_o = 22.9$ GPa, $B_o' = 4.76$ [298].

Chalcogenides of the M_2X_3 type are studied much less; we have found in the literature only a few examples, viz. As_2S_3 has $B_o = 12.9$ GPa, $B_o' = 7.5$; in As_2Se_3 $B_o = 14.0$ GPa, $B_o' = 7.9$ [299, 300]; in Sb_2S_3 $B_o = 26.9$ GPa, $B_o' = 7.9$, and in Bi_2S_3 $B_o = 36.6$ GPa, $B_o' = 6.4$ [301].

Taking into account that the tabulated data of bulk moduli cover only a part of compounds that are significant for solving theoretical and applied problems of high-pressure physics and material science, attempts were made to derive B_o from the composition and structure of substances. First such dependence for metals and alkali halides was established by Bridgman [302–305]:

$$B_o V_o^{4/3} = const \tag{10.10}$$

Later it was concluded [50, 51] that similar equations work well for isostructural sulfides, selenides and covalent solids, whereas Eq. 10.11 was proposed for alkali halides and Eq. 10.12 for oxides and sulfides [306]:

$$B_o V_o = a Z_c \tag{10.11}$$

$$B_o V_o = b Z_c^{3/4} \tag{10.12}$$

Table 10.7 Bulk moduli (GPa) and pressure derivatives of elements

A	B_0	B_0'	A	B_0	B_0'	A	B_0	B_0'
Li	11.8	3.6	Ti[c]	126	2.6	Mn	131	5.8
Na	6.5	3.9	Zr	89	3.8	Tc	281	5.7
K	3.2	4.0	Hf	109	4.0	Re	368	5.4
Rb	2.6	3.9	C[d]	456	3.8	H	0.166	7.3
Cs	1.8	3.8	C[e]	34.8	9.0	F[j]	8.1	4.4
Cu[a]	133	5.4	C[f]	9.5	11.5	Cl	13.1	5.2
Ag[a]	101	6.15	Si	97.8	4.1	Br	14.3	5.2
Au[a]	167	6.2	Ge	75.8	4.5	I	14.5	5.2
Be[b]	106	3.5	Sn	54.2	5.2	Fe	170	6.1
Mg[b]	35.0	4.2	Pb	43.9	5.4	Co[k]	199	3.6
Ca	17.6	3.2	V	160	4.3	Ni	183	5.2
Sr	11.7	2.9	Nb[F]	168	3.4	Ru	316	6.6
Ba	9.3	2.6	Ta[g]	194	3.25	Rh	269	4.5
Zn[b]	56	6.1	N	2.8	3.9	Pd	184	6.4
Cd[b]	43	6.6	P[h]	36	4.5	Os[l]	395	4.5
Hg[b]	36	6.4	As	59.0	4.3	Ir	364	4.8
Sc	60	2.8	Sb	39.8	5.1	Pt	280	5.3
Y	40	2.4	Bi	31.6	5.7	He[m]	0.0193	9.215
La	22.6	3.9	Cr	180	5.2	Ne[n]	1.08	8.4
B	187	3.3	Mo	267	4.2	Ar[n]	2.83	7.8
Al	75.2	4.8	W[i]	322	3.8	Kr[n]	3.31	7.8
Ga	58.8	3.0	O	2.2	4.4	Xe[n,o]	3.61	7.9
In	42.0	5.6	S	9.5	6.0	Th	60	4.6
Tl	35.4	5.3	Se	16.9	2.7	U[p]	104	6.2
			Te	21.0	5.3	Pu[q]	54.4	

[a][281], [b][282], [c][283], [d]diamond, [e]graphite, [f]*fcc*-phase of C_{60}; *fcc*-phase of C_{70}: $B_o = 7.9$, $B_o' = 16$ [284, 285], [F][286], [g][287], [h][288], [i][289], [j][290], [k][291], [l]calculated of [292], [m][293], [n][294], [o]$B_0 = 3.6$ and $B_o' = 5.5$ for *fcc*-phase, $B_0 = 4.3$ and $B_o' = 4.9$ for *hcp*-modification [295], [p][296], [q][297]

where a and b are the constants, and Z_c is the cation valence. For ionic crystals it follows from Born's theory [307] that

$$B_o V_o = k_M Z_c Z_a \frac{(n-1)}{9d} \tag{10.13}$$

where k_M is the Madelung constant, Z_c and Z_a are the valences of cation or anion, n is the mean value of Born's coefficient of repulsion for a given pair of ions, and d is the interatomic distance. From Eq. 10.13 it also follows

$$B_o' = \frac{\partial B}{\partial P} = \frac{(n+7)}{2} \tag{10.14}$$

Cohen [308, 309], using the Phillips's dielectric theory for calculations of B_o, for crystals of the tetrahedral structure obtained

$$B_o = (A - B\lambda)\, d^{-3.5} \tag{10.15}$$

where A and B are the constants, $\lambda = 0$ corresponds to the purely covalent elements of Group 14. The ionic compounds of Group 1 and Group 17 elements are described by the equation

$$B_o = Cd^{-3} \tag{10.16}$$

where C is a constant. Al-Douri et al. [310] studied the bulk moduli of semiconductors and proposed an empirical relation for B_o in terms of the transition pressure P_{tr}.

$$B_o = [a - (b + \lambda)]\,(cP_{tr})^{1/3} \tag{10.17}$$

where a, b, c are constants, P_{tr} is the transition pressure (in GPa) from ZnS to β-Sn type structures and $\lambda = 1$ for Group 14, $\lambda = 5$ for Group 13/Group 15, and $\lambda = 8$ for Group 12/Group 16 semiconductors. The expressions relating bulk moduli and cohesive energy of these semiconductors, as well as alkali halides, alkaline-earth chalcogenides, transition metal nitrides, rare-earth mono-chalcogenides have been suggested [311]. The bulk moduli and cohesive energy of B1 and B3 type structure compounds exhibit a linear relationship when plotted on a log–log scale versus d, but conform to different linear dependences according to the ionic charge product of the compounds. The relation between the lattice energies and the bulk moduli of binary inorganic crystals was studied and the concept of lattice energy density (LED) was introduced [312]. The LED values correlate linearly with the bulk moduli of crystals, the slopes of the lines depending on the valence and coordination number of cations in the crystal structures.

Makino [313] related B_o to d in the general form

$$B = kd^{-m} \tag{10.18}$$

where k and m depend on the type of bonds (Table S10.8). *Ab initio* calculations of B_o for some covalent crystals have been carried out [314–316]; this approach can be used to study the details of the structural, bonding, and electronic properties of diamond and B3 type solids. The B_o of inorganic compounds have been calculated [317] as

$$B_o = \frac{1}{2}\,(I_+ I_-) \tag{10.19}$$

where I_+ and I_- are the elastic moduli of ions with different charges and coordination numbers.

Thus, it is not possible to establish the general dependence of B on V, since the energy of the real chemical bond is a superposition of ionic, covalent and polarizing components which depend differently on the distance. In our work [278, 279] such characteristics of polar compounds as the bond ionicity, the atomic valence, the effective nuclear charge, the Born's factor of repulsion and the interatomic distance are used to derive an equation suitable for approximate calculations of B_o in crystal substances with N_c from 4 to 8 with a 10 % accuracy.

The discussed expressions $B_0 = f(d)$ have been deduced from properties of classes of related substances supposing for them the same type of function, but one can determine the $B = f(P,V)$ function for a particular substance. Since the bulk modulus for pressure P is equal to $B = B_o + PB_o'$, the relations $k_{BP} = B/B_o$ and $k_{VP} = V/V_o$ show the change of the bulk modulus and volume under compression. Then for $P = 1$ GPa we have

$$\frac{\partial B}{\partial V} = \frac{k_{BP} - 1}{1 - k_{VP}} = \frac{B_o' V_o}{B_o \Delta V} \tag{10.20}$$

and from the equality $\Delta V/V_o = 1/B$, it follows that

$$B_{VP}' = B_o' + \frac{\left(B_o'\right)^2}{B_o} \tag{10.21}$$

In Tables S10.9 and S10.10 are given these derivatives for metals and ionic crystals MX, which show that elastic moduli change directly proportional to the densities of solids; the average coefficients of proportionality for metals and ionic crystals are equal to 5.0 and 5.4, respectively, for compresing, and 4.2 for heating. Knowing these derivatives we can estimate sound speeds, Debye's temperatures, and other properties of substances under high pressures. In Table S10.13 are presented the experimental bulk moduli of elements at different temperatures.

The above mentioned EOSs are essentially empirical. Recently Setton [318] proposed a EOS directly derived from the thermodynamic definitions of compressibility and thermal expansion and therefore applicable to all homogeneous condensed matter. Indeed, the definitions of the compressibility k $(1/B)$

$$k = \frac{-(\partial V/V)_T}{\partial P} = \frac{-\partial(\ln V)_T}{\partial P} \tag{10.22}$$

and of the volume thermal expansion β

$$\beta = \frac{(\partial V/V)_P}{\partial T} = \frac{\partial(\ln V)_P}{\partial T} \tag{10.23}$$

are valid for both the liquid and the solid states. Because k and β are two physically independent parameters, which can both affect the volume V of the phase, simple mathematical operations give the expression

$$\partial(\ln V) = k\partial P + \beta\partial T. \tag{10.24}$$

Hence,

$$V = V_o e^{-kP} e^{\beta T} \tag{10.25}$$

where V_o is the volume at reference values of P and T. This EOS was applied to 100 solid elemental species, both typical elements whose compressibility and expansivity depend on the packing coefficient of the crystal structures and isovalent elements whose values of k and β depend on the valence of the element in the solid state.

Bulk moduli and pressure derivaties allow not only to definite volumes of solids at high static pressures, but also to calculate the parameters of shock adiabats (Hugoniot

Table 10.8 Relations between bulk moduli of bulk solids and nano-materials

B(nano) > B(bulk)		B(nano) ≈ B(bulk)		B(nano) < B(bulk)	
Ag, Au	[320]	CuO	[327]	MgO	[329]
Ag, Pb	[321]	MgO	[327]	PbS	[328]
Mo, Ni	[322]	ε-Fe	[327]	CdSe	[328]
c-In_2O_3	[323]	W	[328]	W_2N	[328]
Ga_2O_3	[324]			Al_2O_3	[330]
Fe_2O_3	[178]			SnO_2	[173]
CeO_2	[179]			ReO_3	[185]
TiO_2	[325]			SiC	[331]
AlN	[175]				
GaN	[176]				
Si_3N_4	[326]				

equations) [319], viz.

$$c_o = \left(\frac{B_o}{\rho}\right)^{1/2} \quad and \quad s = \frac{1}{4}\left(1 + B_o{}'\right) \tag{10.26}$$

Experimental values of Hugoniot parameters for elements are listed in Table S10.14 and S10.15 for molecular compounds and polymers.

B of bulk solids can be both higher and lower than those of the corresponding nano-materials (Table 10.8). As the atomic density in the surface layer is smaller, its bulk modulus should be lower than that in a perfect crystal lattice. Indeed, the surface layer in Fe is shown to be more compressible than the interior [332]; similar effects were observed in nanocrystals of Ni [333] and MgO [334]. Since the contribution of the surface layer in the total compressibility of a material increases with the decrease of the grain sizes, this effect may explain the B(nano) < B(bulk) cases; the origins of the opposite effect are unclear. Note also that the pressure of phase transition in some nano-materials decreases for lower particle size (CeO_2, γ-Fe_2O_3, TiO_2-rutile, AlN) and in others increases (Si, CdS, CdSe, PbS, ZnO, ZnS). Such different behavior of nanophases is caused by two factors acting in contrary directions: the growth of the surface energy (which is equivalent to an additional pressure) decreases the pressure necessary for a given phase transition, but the increased proportion of surface atoms in the sample decreases the average coordination number and hence requires higher pressure to achieve a phase transition with an *increase* of N_c.

Appendix

Supplementary Tables

Compilations of EOSs are available in [10.1–10.5]; these values of B_o and $B_o{}'$ are given without references, data from other sources are given with references.

Table S10.1 Bulk moduli (GPa) and pressure derivatives of the MX compounds

M	F		Cl		Br		I	
	B_o	B_o'	B_o	B_o'	B_o	B_o'	B_o	B_o'
Li	73.0	5.2	32.4	5.0	24.7	5.6	19.2	6.0
Na	48.0	5.1	25.3	5.2	20.9	5.2	16.5	5.3
K	31.7	5.2	18.2	5.4	15.8	5.4	12.2	5.4
Rb	27.7	5.5	16.7	5.4	14.1	5.4	11.3	5.3
Cs	25.0	5.5	18..2	5.4	15.7	5.4	12.2	5.6
NH_4			19.7	5.3	16.6	4.7	14.9	4.2
Cu			39.3	4.1	38.7	4.0	35.5	4.0
Ag	61[a]	(5.2)	43.9	6.4	40.6	6.8	24.0	6.6
Tl	19.1[b]	4.2	23.5	5.3	21.6	5.4	17.1	5.5
M	O		S		Se		Te	
Be[c]	228	4.5	105	3.5	92.2	4.0	66.8	4.0
Mg	163	4.1	78.9	3.7	55.2	4.5	60.6	4.1
Ca	114	4.5	64	4.2	51	4.2	42	4.3
Sr	91	5.1	58	5.5	45	5.5	40	5.6
Ba	68.5	5.7	55.6	5.5	41.5	6	35.8	7
Zn	141.4	4.5	76.3	4.4	64.0	4.4	50.0	4.9
Cd	140	5.7	64.3	4.3	55	5.3	43.7	6.0
Hg	44[d]	7[d]	68.6	(5.8)	53.8	3.0	45.5	3.0
Y[e]			93	4.0	82	4.0	67	4.0
La[e]			89	6.5	74	4.7	60.6	
Ga[f]			37	5.2	92	5		
In[f]					24	8.6		
Ge			40.7[g]	5.0[g]			49.9[g]	3.7[g]
Sn	35	6.1	36.6[g]	5.5[g]	50.3	6.3	35.9	5.7
Pb	23.1[h]	7.0	52.9	6.3	50.0	4.9	40.1	5.2
Cr							45.2[i]	(5.3)
Mn	155.3	4.9	72	4.2			54	4.3
Fe	168[j]	4[j]	73[k]	4[k]	31[l]			
Co	180[m]	3.8[m]						
Ni	191[n]	3.9[n]	156[o]				103[o]	4.9[o]
Th			145	5	125	4	102	3.8
U			105	5	74	5	48	4.9
M	N		P		As		Sb	
B	375[p]	4.9	173[q]		148	3.9		
Al	200	4.8	86	4.6	77	4.7	56	4.6
Ga	188	3.8	88.3	4.7	75.0	4.6	56	4.5
In	144	4.1	72.9	4.5	58.4	5.0	46.2	4.7
La[r]			67	4	92	2.5	71.5	4
Th	175	4.0	137	5	118	3.4	84	5
U	203	6.3	102	4	90	4.4	73	4

[a]For the B1 type [10.6], [b][10.7], [c][10.8], [d][10.9], [e][10.10], [f][10.11], [g]for α-SnS, for γ-SnS $B_o = 86.0$ GPa, $B_o' = 4$ [10.12], [h][10.13], [i][10.14], [j][10.15], [k][10.16], [l][10.17], [m][10.18], [n][10.19], [o][10.20], [p]for *w*-BN; for *c*-BN $B_o = 377$ GPa, $B_o' = 4.1$; for *h*-BN $B_o = 36.7$ GPa, $B_o' = 5.6$; for t-BN $B_o = 17.5$, $B_o' = 11.4$ [10.21], [q][10.22], [r][10.23]; ScSb: $B_o = 58$ GPa, $B_o' = 9.5$; YSb $B_o = 58$ GPa, $B_o' = 6.2$

Table S10.2 Bulk moduli (GPa) and pressure derivatives of the MX_2 hydrides, halides, oxides and chalcogenides

MX_2	B_o	B_o'	MX_2	B_o	B_o'	MX_2	B_o	B_o'
BeH_2[a]	14.2	5.3	$MnBr_2$	18.7		ThO_2[b]	195.3	5.4
MgH_2[c]	45	3.35	$FeBr_2$	21.7		UO_2[b]	207	4.5
SrH_2[d]	57	3.13	$CoBr_2$	24.9		NpO_2[b]	200	
ScH_2	114	2.9	$NiBr_2$	26.8	4.8	PuO_2[b]	178	
YH_2[e]	86.4		CdI_2	17.4		AmO_2[b]	205	
TiH_2[e]	142	3.3	HgI_2[f]	33.9	8.6	$ThOS$[b]	201	3.0
ZrH_2[e]	125		PbI_2	15.3		$UOSe$[b]	154	1.8
MgF_2[g]	101	4.2	VI_2	27.4		CS_2	1.7	5.6
CaF_2[h]	74.5	4.7	NiI_2[i]	27.7	4.8	GeS_2	11.8	6.8
SrF_2[h]	74	4.7	$Mg(OH)_2$[i]	54	4.7	SnS_2[k]	27.9	10.7
$SrFCl$	61	5	$Ca(OH)_2$[j]	37.8	5.2	TiS_2[m]	45.9	9.5
BaF_2[l]	57	4	$Ba(OH)_2$[n]	40	5.0	HfS_2	31.9	
$BaFCl$	62	4	$Co(OH)_2$[j]	73.3	4	NbS_2[p]	57	8.6
$BaFBr$[o]	38	7.6	$Ni(OH)_2$[j]	88	4.7	MoS_2[q]	53.4	9.2
$BaFI$[o]	47	5	$AlOOH$[j]	134	4.7	WS_2[q]	61	9.0
ZnF_2	105	4.7	$FeOOH$[n]	111	4	MnS_2[r]	76	5.4
CdF_2	106	6.1	Li_2O[u]	90	3.5	MnS_2[s]	214	5.0
PbF_2[t]	66	7	Cu_2O	121	3.5	ReS_2[v]	23	29
MnF_2	86		CO_2	0.6		FeS_2	133.5	5.73
FeF_2	100	4.6	SiO_2[w]	37.4	5.1	ThS_2[b]	195	2
CoF_2	102		GeO_2[w]	30.5	6.1	US_2[b]	155	
NiF_2	119	5.0	SnO_2[y]	205	7.4	Li_2S[x]	52	2.1
$SrCl_2$	36.2		PbO_2	170	3.8	Cs_2S	6.7	16
$BaCl_2$[α]	47.2	7.4	BaO_2[β]	105	3	$GeSe_2$[z]	46.6	4.0
$ZnCl_2$[γ]	15.1	8	TiO_2	214	6.2	$NbSe_2$	41	
$SnCl_2$[δ]	31	4.9	ZrO_2[ε]	212	6.9	$TaSe_2$	44	
$PbCl_2$[δ]	34	7.4	HfO_2[ε]	284	8	$MoSe_2$[θ]	45.7	11.6
VCl_2	22.5		NbO_2	235	5	WSe_2[κ]	72	4.1
$MnCl_2$	17.8		CrO_2[λ]	138	5.8	$MnTe_2$[r]	32.4	12
$FeCl_2$[μ]	35.3	4	TeO_2[ρ]	44.4	5.8	$MnTe_2$[s]	36.0	4.0
$CoCl_2$	22.9		TeO_2[σ]	51.3	4.3	$RuTe_2$[τ]	255	0.35
$NiCl_2$	25.3		MnO_2	328	4	$IrTe_2$[φ]	132	2.0
$BaBr_2$[α]	38.6		RuO_2	270	4	$IrTe_2$[φ]	126	5.6
VBr_2	23.9		CeO_2[π]	220				

[a][10.24], [b][10.25], [c]for the TiO_2 type, other phases see in [10.26], [d][10.27], [e][10.28], [f]tetragonal phase, for orthorhombic one $B_o = 74.2$, $B_o' = 10$, for hexagonal one $B_o = 70.4$, $B_o' = 7.7$ [10.29], [g]the TiO_2 type; for the α-PbO_2 phase $B_o = 69$, $B_o' = 4$; for PdF_2 one $B_o = 123$, $B_o' = 4$; for $PbCl_2$ one $B_o = 163$, $B_o' = 7$ [10.30], [h][10.31], [i][10.32], [j][10.33], [k][10.34], [l]the CaF_2 type, for the $PbCl_2$ structure $B_o = 79$ GPa, for Ni_2In one $B_o = 133$ GPa, $B_o' = 4$ for all phases [10.35], [m][10.36], [n][10.37], [o][10.38], [p][10.39], [q][10.40], [r]pyrite, [s]markasite, [t]the cottunite type; for the $SnCl_2$ type: $B_o = 91$, $B_o' = 4$ [10.41], [u]the anti-fluorite structure, for the anti-cottunite form $B_o = 188$ GPa, $B_o' = 4$ [10.42], [v][10.43], [w]quartz, [x]the anti-CaF_2 type, for anti-$PbCl_2$ form $B_o = 137$ и $B_o' = 4.0$ [10.44], [y]the rutile type [10.45], [z][10.46], [α][10.47], [β][10.48], [γ]the low-pressure phase, for the high-pressure form (the distorted type CdI_2) $B_o = 45$ GPa, $B_o' = 4$ [10.49], [δ][10.50], [ε][10.51], [θ][10.52], [κ][10.53], [λ]for the CdI_2 structure $B_o = 132$, for the FeS_2 type $B_o = 126$ [10.54], [μ][10.55], [ρ]tetragonal phase, [σ]orthorhombic one, [τ][10.56], [φ][10.57], [π][10.58]

Table S10.3 Bulk moduli (GPa) and pressure derivatives of the MX_3 hydrides, halides, and hydroxides

MH_3	B_o	B_o'	MX_3	B_o	B_o'	MX_3	B_o	B_o'
AlH_3[a]	40	3.1	LaF_3[e]	104.9	4.5	CrF_3[h]	29.2	10.1
YH_3[b]	77.5	4	$LaCl_3$[f]	30.1	6.0	$Al(OH)_3$[i]	49	4
LaH_3[c]	90	4	AsI_3[F]	62.5	9.55	$In(OH)_3$[i]	99	
UH_3[d]	33		rhom-SbI_3[F]	113.0		NbO_2F[j]	24.8	4
			mono-SbI_3[F]	226.3		TaO_2F[j]	36	4

[a][10.59], [b][10.60], [c][10.61], [d][10.62], [e]the tisonite type, for the high pressure phase $B_o = 160.5$ GPa, $B_o' = 4$ [10.63], [f][10.64], [F][10.206], [g][10.65], [h][10.66], [i][10.67], [j][10.68]

Table S10.4 Bulk moduli (GPa) and pressure derivatives for binary oxides of the M_nO_m type

M_nO_m	B_o	B_o'	M_nO_m	B_o	B_o'	M_nO_m	B_o	B_o'
H_2O[a]	24.5	4	*c*-In_2O_3[C]	178.9	5.15	Fe_2O_3	227	3.5
H_2O_2[a]	13.6	4	*h*-In_2O_3[C]	215.1	4.74	Fe_3O_4	183.4	5.0
Cu_2O	105	4.2	Sc_2O_3[d]	154	7	V_2O_5[f]	50	12
B_6O[b]	181	6.0	Y_2O_3	168	4.1	As_2O_5[g]	78.1	5.3
B_2O_3[b]	170	2.5	Ti_2O_3	208	4.1	α-Sb_2O_4[h]	143	4
Al_2O_3	252.5	4.1	V_2O_3	173.5	7.5	β-Sb_2O_4[h]	105	4
β-Ga_2O_3[c]	202	4.3	Cr_2O_3	232.5	2.0	WO_3[i]	27	9.4
α-Ga_2O_3[c]	271	5.9	Mn_2O_3[e]	169.1	7.35	ReO_3[j]	96	4
			Mn_3O_4[e]	137.4		ReO_3[k]	129	4

[a][10.69], [b][10.70], [c][10.71], [C][10.207], [d][10.72], [e][10.73], [f][10.74], [g][10.75], [h][10.76], [i][10.77], [j]for monoclinic phase, [k]for rhombohedral phase [10.78]

Table S10.5 Bulk moduli (GPa) and pressure derivatives for binary nitrides

$M(N_3)_2$[a]	B_o	MN_3[a]	B_o	MN_2/M_2N	B_o	B_o'	MN	B_o	B_o'	M_3N	B_o	B_o'	M_3N_4	B_o	B_o'
Ca	31	Li	33	OsN_2[b]	358	4.67	Ti[j]	277		Li[g]	71	3.9	α-Si[j]	228.5	4.0
Sr	26	Na	28	IrN_2[b]	428	4	Zr[i]	248	4	Re[f]	397		α-Ge[j]	178	2.1
Ba	42	K	18.6[α]	Ta_2N[c]	360	4	Hf[i]	260	4				Sn[k]	186	
Cd	29	Rb	24	Cr_2N[d]	275	2.0	Ta[d]	288	4.7				Zr[i]	250	4
Pb	27	Cs	18	Mo_2N[e]	301	4	Mo[d]	345	3.5				Hf[i]	227	5.3
		Cu	23	W_2N	408		PtN[b]	372	4						
		or-Ag	39[β]	Re_2N[f]	401										
		te-Ag	57[β]												
		Tl	24												
		NH_4	23												

[a][10.79], [α][10.208], [β][10.209], [b][10.80], [c][10.81], [d][10.82], [e][10.83], [f][10.84], [g][10.85], [h][10.86], [i][10.87], [j]for α-Si_3N_4, for β-phase $B_o = 270$, $B_o' = 4$, for γ-phase $B_o = 308$, $B_o' = 4$; for α-Ge_3N_4, for β-phase $B_o = 218$, $B_o' = 4$, for γ-phase $B_o = 296$, $B_o' = 4$, [10.88], [k][10.89]

Table S10.6 Bulk moduli (GPa) and pressure derivatives for binary borides

MB_2	B_o	B_o'	MB_2	B_o	B_o'	Compound	B_o	B_o'
Mg[a]	151	4	V[b]	322	4.0	B_{28}[f]	285	1.8
Zn[b]	317	4.0	Re[d]	334	4.0	B_4C[g]	199	1
Al[a]	170	4.8	Os[e]	365		Fe_2B[h]	164	4.4
Ti[c]	237	2.2	U	216	3.8	LiB[i]	48	4

[a][10.90], [b][10.91], [c][10.92], [d][10.93], [e][10.94], [f][10.95], [g][10.96], [h][10.97], [i][10.98]

Table S10.7 Bulk moduli (GPa) and pressure derivatives for binary carbides and silicides

MC	B_o	B_o'	M_nC_m	B_o	B_o'	M_nSi_m	B_o	B_o'
Si[a]	260	2.9	ThC[e]	109	4.0	Mg_2Si[h]	57.0	4.0
Ti[b]	242		UC[e]	160	3.6	$MoSi_2$	210	
Zr	187		UC_2	216	3.8	WSi_2	222	
V	195		Mo_2C[f]	307	6.2	Fe Si_2[i]	243.5	3.2
W[c]	384	4.65	Al_4C_3[g]	233	3.4			
Pt[d]	301	5.2						

[a][10.99], [b][10.86], [c][10.100], [d][10.101], [e][10.102], [f][10.103], [g][10.104], [h][10.105], [i][10.106]

Table S10.8 Bulk moduli (GPa) and pressure derivatives for binary phosphides and arsenides

Comp.	B_o	B_o'	Comp.	B_o	B_o'	Comp.	B_o	B_o'
UP_2[a]	124	9	Th_3P_4	126	4.0	α-$AsNa_3$[b]	24.5	1
UAs_2[a]	101	4.7	U_3P_4	160	4	β-$AsNa_3$[b]	35.4	4
UAsS[a]	105	3.7	U_3As_4	121	4	$ZnAs_2$	56	
UAsSe[a]	99	3.8						

[a][10.25], [b][10.107]

Table S10.9 Bulk moduli (GPa) and pressure derivatives for ternary oxides

Compound	B_o	B_o'	Compound	B_o	B_o'	Compound	B_o	B_o'
$MgCO_3$ [a]	117	2.3	Mg_2SiO_4 [m]	127	4	$KReO_4$ [o]	18	
$CaCO_3$	71	3.1	Mg_2SiO_4 [n]	212	4.1	$AgReO_4$ [o]	31	4
$SrCO_3$ [b]	101	4	Cr_2SiO_4 [o]	95	8.3	$TlReO_4$ [o]	26	
$ZnCO_3$ [c]	124		Mn_2SiO_4	141		$KClO_4$	19.3	
$CdCO_3$ [c]	97		Fe_2SiO_4 [p]	139		$RbClO_4$	17.1	
$MnCO_3$ [c]	107		Fe_2SiO_4 [n]	201	5.6	$CsClO_4$	15.0	
$FeCO_3$ [c]	117		Co_2SiO_4	161		NH_4ClO_4	16.1	
$CoCO_3$ [c]	125		Ni_2SiO_4 [n]	225	7.6	$TlClO_4$	18.0	
$NiCO_3$ [c]	131		$ZrSiO_4$	228	8.2	$BeAl_2O_4$	242	4.0
Li_2SiO_3	207		Mg_2GeO_4	179	4.2	$MgAl_2O_4$	196.2	4.4
Na_2SiO_3	163		$ZrGeO_4$ [o]	238	4.5	$ZnAl_2O_4$	201.7	7.6
K_2SiO_3	80		$HfGeO_4$ [o]	242	4.8	$FeAl_2O_4$	211	
$MgSiO_3$ [d]	212	5.6	Fe_2GeO_4	196	4.9	$CuMn_2O_4$ [u]	198	
$MgSiO_3$ [e]	260	3.7	Co_2GeO_4	192	5.6	$ZnMn_2O_4$ [u]	197	
$CaSiO_3$	279	4.1	Ni_2GeO_4	203	4.8	$TiMn_2O_4$	167.3	
$CuGeO_3$ [f]	40.0	6.5	YVO_4 [o]	138		$NiMn_2O_4$ [u]	206	
$MgGeO_3$ [e]	229	3.7	$GaPO_4$ [p]	12.1	11.5	$CoFe_2O_4$	185.7	
$CaGeO_3$ [g]	194	6.1	$BiVO_4$ [o]	150		$NiFe_2O_4$	198.2	
$FeGeO_3$ [h]	90	9.1	$LaNbO_4$ [o]	111		Al_2SiO_5	156	5.6
$CaSnO_3$ [g]	163	5.6	Cs_2SO_4 [q]	28.4	5.1	$Li_2Si_2O_5$	156	
$CdSnO_3$ [i]	185	5.1	$SrSO_4$ [o]	82		$Na_2Si_2O_5$	151	
$CaTiO_3$ [g]	176	5.6	$BaSO_4$ [r]	63		$K_2Si_2O_5$	111	
$CdTiO_3$ [l]	214	6.4	K_2CrO_4 [r]	26	6.0	$Rb_2Si_2O_5$	88	
$CaZrO_3$ [g]	154	5.9	$CaCrO_4$ [s]	103.7	4.0	$Cs_2Si_2O_5$	76	
$LiNO_3$	40.0		$BaCrO_4$	53	6.8	TiP_2O_7 [v]	42	6.0
$NaNO_3$	26.7	5.9	$CaMoO_4$ [o]	83	4.2	ZrP_2O_7 [v]	39	2.8
KNO_3	26.8	16.6	$SrMoO_4$ [o]	73		β-ZrV_2O_7 [v]	20.8	4
$CsNO_3$	19.2		$CdMoO_4$ [o]	104	4	α-$ZrMo_2O_8$ [v]	17.0	4
$AgNO_3$	28.8	6.2	$PbMoO_4$	64	4.0	δ- $ZrMo_2O_8$ [v]	19.1	4
$TlNO_3$	23.5		$CuWO_4$	139	4	ε-$ZrMo_2O_8$ [v]	74	4
$YAlO_3$ [j]	192	7.3	$CaWO_4$ [o]	78	5.7			
$GdAlO_3$ [i]	203	5.1	$SrWO_4$ [o]	64	5.4			
$EuAlO_3$ [i]	213	4.9	$BaWO_4$ [t]	57	3.5			
$YCrO_3$	208.4	3.7	$PbWO_4$ [t]	67	8			
Bulk moduli (GPa) and pressure derivatives for coordination compounds								
Compound	B_o	B_o'	Compound	B_o	B_o'	Compound	B_o	B_o'
$NaBH_4$ [w]	15	5	$(NH_4)_2SiF_6$	13.1	9	K_2SnBr_6 [z]	12.5	
NH_4BF_4	15.7		K_2SnCl_6 [z]	14.3		K_2SeBr_6 [z]	16.7	
$RbBH_4$ [x]	14.5	4	K_2ReCl_6 [z]	16.1	8.1	$(NH_4)_2TeBr_6$	12.5	
$TlBF_4$	16.8		Cs_2CuCl_4 [α]	15.0	12.2	K_2PtBr_6 [z]	15.2	
KPF_6 [y]	15		Cs_2CoCl_4 [α]	17.0	4.0			

[a][10.108], [b][10.109], [c][10.110], [d]Ilmenite [10.111], for the quartz-like phase B_o = 308 GPa, B_o' = 4; [e]perovskite [10.112], [f][10.113], [g][10.114], [h][10.115], [i][10.116], [j][10.117], [k]forsterite [10.118], [l]spinel, [m][10.119], [n]olivine, [o][10.120], [p][10.121, [q][10.122], [r][10.123], [s][10.124], [t][10.125], [u][10.126], [v][10.127], [w]for the cubic form, for the tetragonal B_o = 22.2 GPa, B_o' = 3.5 [10.128], [x][10.129], [y][10.130], [z][10.131], [α][10.132]

Table S10.10 Bulk moduli (GPa) and pressure derivatives for molecular substances and polymers. (Experimental data from review [5.354], except where specified)

Molecule	B_o	B_o'	Molecule[k]	B_o	B_o'	Molecule[k]	B_o
HNO_3[a]	7.1	9.5	LiC_5H_5[l]	7.7	7.1	$(C_6H_5CH)_2$	6.13
HCl[b]	0.79	8.6	KC_5H_5[l]	4.9	11.1	$(C_6H_5C)_2$	5.18
CCl_4[c]	3.30		CsC_5H_5[l]	18.0		$CO(NH_2)_2$	14.4
CBr_4[d]	3.66		C_6H_{12}[m]	15.0	6.9	$C_3H_6N_6$	11.05
CH_4[e]	6.4	5.7	C_6H_6	4.14		$C_6H_{12}N_4$	8.77
SiH_4[f]	7.8	4	$C_{10}H_8$	6.73	7.1	$C_4H_5NO_2$	7.43
$CH_4.H_2O$[g]	15.4	4	$C_{12}H_{10}$	6.11		$C(CH_2OH)_4$	17.1
C_6Cl_6[h]	8.39	8.2	$C_{14}H_{10}$[n]	6.08	9.8	$CH_2(COOH)_2$	11.7
C_6Br_6[h]	9.08	8.7	$C_{20}H_{12}$	10.14	7.7	$(CH_3)_2C(COOH)_2$	7.75
C_6I_6[i]	9.1	9.0	$(C_6H_5CH_2)_2$	4.95		$(CHCOOH)_2$	13.3
SF_6[j]	2.62		$C_3H_6N_6O_6$[o]	9.8	11.4	$C_6H_5CONHCH_2COOH$	9.44
						$He@C_{60}$[p]	35.1

Polymer	B_o	Polymer	B_o	Polymer	B_o
Polypropylene	4.10	Teflon $(CF_2)_n$	2.77	Polyvinyl chloride	5.14
Plexiglas	5.85	Polyurethane	5.43	Polyvinylidene fluoride	6.09

[a][10.133], [b][10.134], [c][10.135], [d][10.136], [e][10.137], [f][10.138], [g][10.139], [h][10.140], [i][10.141], [j][10.142], [k][10.143], [l][10.144], [m][10.145], [n][10.146], [o][10.147], [p][10.148]

Table S10.11 Characteristics of polymorphous modifications of elements and MX compounds

Substance	N_c	B_o	B_o'	Substance	N_c	B_o	B_o'
NaF[a]	6	46.4	4.9	InTe[u]	6	69.7	2.3
	8	103	4.0		8	90.2	2.3
NaCl[a]	6	23.8	4.0	PbSe[v]	6	28.8	4.1
	8	121	4.0		8	197	5.0
KH[b]	6	15.6	4.0	PbTe[v]	6	38.9	5.4
	8	28.5	4.0		8	38.1	5.4
KF[c]	6	29.3	5.4	FeO[w]	6	174	4.9
	8	37.0	5.4		8	205	6
KCl[d]	6	18.2	5.4	ThSe[x]	6	125	4.0
	8	28.7	4.0		8	149	4.0
KI[e]	6	12.2	5.4	UTe[x]	6	48	4.9
	8	24.2	4.3		8	62	5.1
RbH[b]	6	10.0	3.9	BN[ψ]	3	36.7	5.6
	8	18.4	3.9		4	369	4.5
RbCl[f]	6	16.7	5.4	AlN[yz]	4	194	4.4
	8	17.9	5.2		6	253	4.7
RbI[g]	6	11.3	5.3	AlSb[α]	4	57.3	5.3
	8	15.7	4.8		6	105	3.3
NH_4I[h]	6	14.9	4.2	GaN[β]	4	189	3.0
	8	18.8	4.2		6	235	4.6
CsH[i]	6	8.0	4.0	GaAs[γ]	4	75.4	4.5
	8	14.2	4.0		6	57.0	4.8
AgF[j]	6	61		GaSb[δ]	4	56.1	4.8
	8	110			6	58.8	7.7
CaO[k l]	6	116	4.7	InP[ε]	4	76	4
	8	160	3.8		6	130	1.6
SrSe[m]	6	45.2	4.5	InAs[ϕ]	4	59.2	6.8
	8	46.5	4.5		6	40.6	7.3
BaO[n]	6	66.2	5.7	UAs[φ]	6	100.7	2.7
	8	33.2	6.0		8	121.6	2
BaS[n]	6	55.1	5.5	USb[x]	6	62	4.0
	8	21.4	7.8		8	84	4.0
BaTe[o]	6	29.4	7.4	ThSb[x]	6	84	5.2
	8	27.5	4.6		8	99	5.1
ZnO[p]	4	142.6	3.6	Si[θ]	4	98.4	4.2
	6	202.5	3.5		6	180	4.2
ZnS[q]	4	75.0	4	Ge[θ]	4	75	4.4
	6	103.6	4		6	149	4.4
CdSe[r]	4	37	11	Sn[σ]	6	50.2	4.9
	6	74			8	82	5.5
CdTe[s]	4	42.0	6.4	P[τ]	3	36	4.5
	6	68.7	5.1		6	70.7	4.7
HgTe[t]	4	42.3	2.1	Xe[π]	*fcc*,12	3.6	5.5
	2+2	16.0	7.3		*hcp*, 12	4.3	4.9

[a][10.149], [b][10.150], [c][10.151], [d][10.152], [e][10.153], [f][10.154], [g][10.155], [h][10.156], [i][10.157], [j][10.158], [k][10.159], [l][10.160], [m][10.161], [n][10.162], [o][10.163], [p][10.164], [q][10.165], [r][10.166], [s][10.167], [t][10.168], [u][10.169], [v][10.170], [w][10.171], [x][10.172], [y][10.173], [z][10.174], [α][10.175], [β][10.176], [γ][10.177], [δ][10.178], [ε][10.179], [ϕ][10.180], [φ][10.181], [θ][10.182], [σ][10.183], [τ][10.184], [Ψ][10.185], [π][10.186]

Table S10.12 Characteristics of polymorphous modifications of MX_2 compounds

MX_2	Phase or structural type, coordination number/B_oGPa, B_o'									
CO_2[a]	I	$N_c=2$	II	$N_c=2$	III	$N_c=2$	IV	$N_c=2$	V	$N_c=4$
	6.1	6	131	2.1	87	3.3			365	0.8
SiO_2	Quartz	$N_c=4$	Coesit[b]	$N_c=4$	Rutile[c]	$N_c=6$				
	37.4	6.1	101	1.8	298	5.0				
GeO_2[c]	Quartz	$N_c=4$	$CaCl_2$	$N_c=6$	Rutile	$N_c=6$				
	30.5	6.8	241	4	250	5.6				
SnO_2[d]	Rutile	$N_c=6$	$CaCl_2$	$N_c=6$	FeS_2	$N_c=6$	ZrO_2	$N_c=7$	$PbCl_2$	$N_c=9$
	205	7.4	204	8	246	4	259	4	417	4
PbO_2[e]	Rutile	$N_c=6$	II	$N_c=6$	III	$N_c=6+2$	CaF_2	$N_c=8$	$PbCl_2$	$N_c=9$
	175	3.7	141	3.9	223	3.7	181	3.8	225	3.8
TiO_2[f]	Anatas	$N_c=6$	Rutile	$N_c=6$	PbO_2	$N_c=6$	ZrO_2	$N_c=7$		
	179	4.5	211	6.5	258	4.1	290	4.0		
ZrO_2[g]	ZrO_2	$N_c=7$	OI	$N_c=7$	OII	$N_c=9$				
	210	4	290	4	316	4				
HfO_2[g]	ZrO_2	$N_c=7$	CaF_2	$N_c=8$	$PbCl_2$	$N_c=9$				
	284	5	281	4.2	340	2.6				
MgF_2[h]	Rutile	$N_c=6$	PbO_2	$N_c=6$	PdF_2	$N_c=6+2$	$PbCl_2$	$N_c=9$		
	101	4.2	69	4	123	4	163	7		
MgH_2[i]	Rutile	$N_c=6$	CaF_2[j]	$N_c=6$	PbO_2	$N_c=6$	$AuSn_2$			
	45	3.35	47.4	3.5	44.0	3.2	49.8	3.5		
CaF_2[k]	CaF_2	$N_c=8$	$PbCl_2$	$N_c=9$	Ni_2In	$N_c=11$				
	82	4.8	74.5	4.7	118	4.7				
BaF_2[k]	CaF_2	$N_c=8$	$PbCl_2$	$N_c=9$	Ni_2In	$N_c=11$				
	57	4	51	4.7	67	4.7				

[a][10.187], [α]non-linear phase, [b][10.188], [c][10.189], [d][10.190], [e][10.191], [f][10.192], [g][10.193], [h][10.194], [i][10.26], [j]distorted cub, [k][10.31]

Table S10.13 Bulk moduli (GPa) of elements at different temperatures (K)

M	Li[a]	Na[a]	K[a]	Rb[a]	Cs[a]	Cu[b]	Ag[b]	Au[c]
T = 0	12.5	7.2	3.7	2.9	2.1	142.0	108.9	180.0
T = 300	11.6	5.9	2.9	2.3	1.7	137.1	103.8	167.5
M	Ni[b]	Rh[b]	Os[d]	Pt[c]	M	α-Pu[g]	M	δ-Pu[g]
T = 0	187.4	268.6	415.4	288.4	T = 18	72.3	T = 14.6	37.8
T = 300	183.6	266.5	405.2	276.4	T = 407	48.2	T = 496	24.1
M	Cu[e]	Ag[e]	Au[e]	Fe[f]				
T = 0	142.3	110.8	180.9	170.3				
T = 500	125.2	93.1	155.7	159.8				

[a][10.195], [b][10.196], [c][10.197], [d][10.198], [e][10.199], [f][10.200], B(Be) = 120.4 GPa at 300 K and 117.9 GPa at 1000 K[10.201], B(Ti) = 106.4 GPa at 300 K and 87.7 GPa at 1273 K[10.202], [g][10.203]

Table S10.14 Coefficients in the Hugoniot equation of elements; c in km/s [10.204]

A	c	s	A	c_o	s	A	c	s	A	c	s
Li	4.760	1.065	Ge	3.151	1.79	Sb	1.98	1.63	Tc	*4.97*	
Na	2.624	1.188	Sn	2.437	1.688	Bi	1.861	1.520	Re	4.068	1.375
K	1.991	1.17	Pb	1.981	1.603	V	5.050	1.227	Fe	4.63	1.33
Rb	1.232	1.184	Ti	4.937	1.04	Nb	4.472	1.114	Co	4.743	1.227
Cs	0.363	1.583	Zr	3.812	0.977	Ta[a]	3.293	1.307	Ni	4.501	1.627
Cu	3.899	1.520	Hf	2.954	1.121	O	2.327	1.215	Ru	*5.055*	*1.90*
Ag	3.178	1.773	B	*9.06*	*1.075*	S	2.334	1.588	Rh	4.775	1.331
Au	3.063	1.563	Al	5.333	1.356	Se	*1.87*	*0.925*	Pd	3.955	1.701
Be	7.993	1.132	Ga	2.501	1.560	Te	3.242	0.888	Os	4.170	1.375
Mg	4.540	1.238	In	2.560	1.477	Cr	5.153	1.557	Ir	3.930	1.536
Ca	3.438	0.968	Tl	1.809	1.597	Mo	5.100	1.266	Pt	3.605	1.560
Sr	2.10	0.94	Sc	4.496	0.955	W	4.015	1.252	He	0.712	1.36
Ba	1.108	1.369	Y	3.381	0.725	H	*1.37*	*2.075*	Ne	*0.895*	*2.55*
Zn	3.031	1.608	La	2.064	1.012	F	*2.18*	*1.35*	Ar	1.249	1.588
Cd	2.434	1.759	N	1.572	1.365	Cl	*2.53*	*1.55*	Kr	0.70	1.72
Hg	1.75	1.72	P	*3.584*	*1.575*	Br	1.51	1.24	Xe	1.16	1.40
C	12.16	1.00	As	*3.195*	*1.325*	I	1.50	1.46	Th	2.18	1.24
Si	7.99	1.42				Mn	*4.185*	*1.70*	U	2.51	1.51

The values of c_o and s are italicized if calculated from B_o and B_o'

Table S10.15 Coefficients in Hugoniot equations for molecular compounds; c in km/s [10.205]

Compound	c	s	Compound	c	s
N_2	1.59	1.36	$(C_2H_5)_2O$	1.67	1.455
NH_3	2.00	1.51	$CHBr_3$	1.265	1.533
H_2O	1.50	2.00	C_2H_5Br	1.58	1.36
CO	1.54	1.40	$C_2H_2Cl_2$	1.11	1.75
CO_2	2.16	1.465	CH_2Br_2	1.0	1.6
CS2	1.64	1.46	CH_2I_2	0.96	1.54
CCl_4	1.47	1.57	Formic acid	1.982	1.406
CH_4 (115 K)	2.841	1.168	Acetic acid	2.299	1.267
C_6H_{14}	1.738	1.446	Butyric acid	2.128	1.384
C_7H_{16}	1.808	1.450	Aniline	2.506	1.275
$C_{10}H_{22}$	1.970	1.458	Anthracene	3.21	1.445
$C_{13}H_{28}$	2.013	1.460	Phenanthracene	3.097	1.417
$C_{14}C_{30}$	2.095	1.463	Pyrene	3.031	1.457
$C_{16}H_{34}$	2.127	1.464	Polystyrene	2.48	1.63
C_6H_6	1.88	1.58	Polyethylene ($\rho = 0.92$)	2.83	1.408
$C_6H_5CH_3$	1.72	1.66	Polypropylene	3.00	1.42
$C_6H_5NO_2$	2.01	1.59	Polyurethane	2.24	1.71
$C_6H_3(NO_2)_3$	2.318	2.025	Paraffin	2.960	1.531
$C_6H_2CH_3(NO_2)_3$	2.390	2.050	Plexiglass ($\rho = 1.18$)	3.08	1.308
CH_3NO_3	2.07	1.34	Teflon ($\rho = 2.18$)	2.18	1.580
CH_3OH	1.73	1.50	Nylons	2.500	1.747

Table S10.15 (continued)

Compound	*c*	*s*	Compound	*c*	*s*
C_2H_5OH	1.785	1.575	Rubber	1.84	1.44
C_4H_9OH	1.868	1.497	Foam plastic		
$C_5H_{11}OH$	1.988	1.487	$\rho = 0.30$	0.15	1.290
$C_6H_{13}OH$	2.324	1.378	$\rho = 0.65$	1.07	1.340
$C_3H_5(OH)_3$	3.07	1.34	$\rho = 0.70$	1.19	1.350
$CO(CH_3)_2$	1.91	1.38	Acetonitrile	2.754	1.088

Supplementary References

10.1 Schlosser H, Ferrante J (1988) Phys Rev B 37:4727
10.2 Freund J, Ingalls R (1989) J Phys Chem Solids 50:263
10.3 Pucci R, Piccitto G (eds) (1991) Molecular systems under high pressures. North-Holland, Amsterdam
10.4 Holzapfel WB (1998) High Press Res 16:81; Holzapfel WB (2001) Z Krist 216:473; Ponkratz U, Holzapfel WB (2004) J Phys Cond Matter 16:S963
10.5 Batsanov SS (1999) Inorg Mater 35:973; Batsanov SS (2002) Russ J Inorg Chem 47:660
10.6 Hull S, Berastegui P (1998) J Phys Cond Matter 10:7945
10.7 Berastegui P, Hull S (2000) J Solid State Chem 150:266
10.8 Narayana C, Nesamony VJ, Ruoff AL (1997) Phys Rev B 56:14338
10.9 Zhou T, Schwarz U, Hanfland M et al (1998) Phys Rev B 57:153
10.10 Vaitheeswaran G, Kanchana V, Svane A et al (2011) Phys Rev B83:184108; Vaitheeswaran G, Kanchana V, Heathman S et al (2007) Phys Rev B 75:184108
10.11 Pellicer-Porres J, Machado-Charry E, Segura A et al (2007) Phys Stat Solidi 244 b:169
10.12 Ehm I, Knorr K, Dera P et al (2004) J Phys Cond Matter 16:3545
10.13 Eremets MI, Gavriliuk AG, Trojan IA (2007) Appl Phys Lett 90:171904
10.14 Takagaki M, Kawakami T, Tanaka N et al (1998) J Phys Soc Japan 67:1014
10.15 Ono S, Ohishi Y, Kikegawa T (2007) J Phys Cond Matter19:036205
10.16 Kusaba K, Syono Y, Kikegawa T, Shimomura O (1997) J Phys Chem Solids 58:241
10.17 Millican JN, Phelan D, Thomas E L et al (2009) Solid State Commun 149:707
10.18 Sakamoto D, Yoshiasa A, Yamanaka T et al (2002) J Phys Cond Matter 14:11369
10.19 Noguchi Y, Uchino M, Hikosaka H et al (1999) J Phys Chem Solids 60:509
10.20 Onodera A, Minasaka M, Sakamoto I et al (1999) J Phys Chem Solids 60:167
10.21 Solozhenko VL, Häusermann D, Mezouar M, Kunz M (1998) Appl Phys Lett 72:1691; Will G, Nover GJ, von der Gönna (2000) J Solid State Chem 154:280; Solozhenko VL, Solozhenko EG (2001) High Pres Res 21:115
10.22 Lundström T (1997) J Solid State Chem 133:88
10.23 Hayashi J, Shirotani I, Hirano K et al (2003) Solid State Commun 125:543; Shirotani I, Yamanashi K, Hayashi J et al (2003) Solid State Commun 127:573
10.24 Ahart M, Yarger JL, Lantzky KM et al (2006) J Chem Phys 124:014502
10.25 Idiri M, Le Bihan T, Heathman S, Rebizant J (2004) Phys Rev B70:014113; Olsen JS (2004) J Alloys Comp 381:37; Gerward L, Olsen JS, Benedict U et al (1994) High-pressure X-ray diffraction studies of ThS_2, US_2 and other AnX_2 and AnXY compounds. In: Schmidt SC, Shaner JW, Samara GA, Ross M (eds) High-pressure science and technology. AIP Press, New York
10.26 Vajeeston P, Ravindran P, Hauback BC et al (2006) Phys Rev B 73:224102
10.27 Smith JS, Desgreniers S, Klug DD, Tse JS (2009) J Alloys Comp 468:830
10.28 Ito M, Setoyama D, Matsugawa J et al (2006) J Alloys Comp 426:67
10.29 Karmakar S, Sharma SM (2004) Solid State Commun 131:473
10.30 Haines J, Legar JM, Gorelli F et al (2001) Phys Rev B 64:134110

10.31 Dorfman SM, Jiang F, Mao Z et al (2010) Phys Rev B 81:174121
10.32 Pasternak MP, Taylor RD, Chen A et al (1990) Phys Rev Lett 65:790
10.33 Grevel KD, Burchard M, Fasshauer DW, Peun T (2000) J Geophys Res B 105:27877; Garg N, Karmakar S, Surinder SM et al (2004) Physica B 349:245
10.34 Knorr K, Ehm L, Hytha M et al (2001) Phys Stat Solidi 223b:435
10.35 Leger JM, Haines J, Atouf A et al (1995) Phys Rev B 52:13247
10.36 Aksoy R, Selvi E, Knudson R, Ma Y (2009) J Phys Cond Matter 21:025403
10.37 Nagai T, Kagi H, Yamanaka T (2003) Amer Miner 88:1423
10.38 Subramanian N, Shekar NVC, Sahu PC et al (2004) Physica B 351:5
10.39 Ehm L, Knorr K, Depmeir W (2002) Z Krist 217:522
10.40 Aksoy R, Ma Y, Selvi E et al (2006) J Phys Chem Solids 67:1914; Selvi E, Ma Y, Aksoy R et al (2006) J Phys Chem Solids 67:2183
10.41 Haines J, Leger J M, Schulte O (1998) Phys Rev B 57:7551
10.42 Lazicki A, Yoo CS, Evans WJ, Pickett WE (2006) Phys Rev B 73:184120
10.43 Hou D, Ma Y, Du J et al (2010) J Phys Chem Solids 71:1571
10.44 Grzechnik A, Vegas A, Syassen K et al (2000) J Solid State Chem 154:603; Santamaria-Perez D, Vegas A, Muehle C, Jansen M (2011) J Chem Phys 135:054511
10.45 Shieh SR, Kubo A, Duffy TS et al (2006) Phys Rev B 73:014105
10.46 Grzechnik A, Stølen S, Nakken E et al (2000) J Solid State Chem 150:121
10.47 Leger JM, Haines J, Atouf A (1995) J Appl Cryst 28:416
10.48 Efthimiopoulos I, Kunc K, Karmakar S et al (2010) Phys Rev B 82:134125
10.49 Brazhkin VV (2005) In: 20th AIRAPT—43th EHPRG June 27 Karlsruhe T10-KN
10.50 Leger JM, Haines J, Atouf A (1996) J Phys Chem Solids 57:7
10.51 Desgreniers S, Lagarec K (1999) Phys Rev B 59:8467
10.52 Aksoy R, Selvi E, Ma Y (2008) J Phys Chem Solids 69:2137
10.53 Selvi E, Aksoy R, Knudson R (2008) J Phys Chem Solids 69:2311
10.54 Maddoz BR, Yoo CS, Kasinathan D et al (2006) Phys Rev B 73:144111
10.55 Rozenberg GKh, Pasternak MP, Gorodetsky P et al (2009) Phys Rev B 79:214105
10.56 Fjellvag H, Grosshans WA, Honle W, Kjekhus A (1995) J Magn Magn Mater 145:118
10.57 Leger JM, Pereira AS, Hines J et al (2000) J Phys Chem Solids 61:27
10.58 Gerward L, Staun Olsen J, Petit L et al (2005) J Alloys Comp 400:56
10.59 Graetz J, Chaudhuri S, Lee Y et al (2006) Phys Rev B74:214114
10.60 Ohmura A, Machida A, Watanuki T et al (2006) Phys Rev B 73:104105
10.61 Palasyuk T, Tkacz M (2009) J Alloys Comp 468:191
10.62 Halevy I, Salhov S, Zalkind S et al (2004) J Alloys Comp 370:59
10.63 Chrichton WA, Bouvier P, Winkeln B, Grzechnik A (2010) Dalton Trans 39:4302
10.64 Benedict U, Holzapfel WB (1993) High-pressure studies – structural aspects. In: Gschneidner KA Jr, Choppin GR (eds) Handbook on the physics and chemistry of rare earths, vol 17. North-Holland, Amsterdam
10.65 JØrgensen J-E, Marshall WG, Smith RI (2004) Acta Cryst B 60:669
10.66 Liu H, Hu J, Xu J et al (2004) Phys Chem Miner 31:240 4
10.67 Gurlo A, Dzivenko D, Andrade M et al (2010) J Am Chem Soc 132:12674
10.68 Cetinkol M, Wilkinson AP, Lind C et al (2007) J Phys Chem Solids 68:611
10.69 Cynn H, Yoo C-S, Sheffield SA (1999) J Chem Phys 110:6836
10.70 Nieto-Sanz D, Loubeyre P, Crichton W, Mezouar M (2004) Phys Rev B 70:214108
10.71 Yusa H, Tsuchiya T, Sata N, Ohishi Y (2008) Phys Rev B 77:064107
10.72 Liu D, Lei W, Li Y et al (2009) Inorg Chem 48:8251
10.73 Yamanaka T, Nagai T, Okada T, Fukuda T (2005) Z Krist 220:938
10.74 Loa I, Grzechnik A, Schwarz U et al (2001) J Alloys Comp 317–318:103
10.75 Locherer T, Halasz I, Dinnebier R, Jansen M (2010) Solid State Comm 150:201
10.76 Orosel D, Balog P, Liu H et al (2005) J Solid State Chem 178:2602
10.77 Bouvier P, Crichton WA, Boulova M, Lucazeau G (2002) J Phys Cond Matter 14:6605
10.78 Biswas K, Muthu DVS, Sood AK et al (2007) J Phys Cond Matter 19:436214

10.79 Belomestnykh VN (1993) Inorg Mater 29:168
10.80 Young AF, Sanloup C, Gregoryanz E et al (2006) Phys Rev Lett 96:155501
10.81 Lei WW, Liu D, Li XF et al (2007) J Phys Cond Matter 19:425233
10.82 Soignard E, Shebanova O, McMillan PF (2007) Phys Rev B 75:014104
10.83 Soignard E, McMillan PF, Chaplin TD et al (2003) Phys Rev B 68:132101
10.84 Zhang RF, Lin ZJ, Mao H-K, Zhao Y (2011) Phys Rev B 83:060101
10.85 Lazicki A, Maddox B, Evans WJ et al (2005) Phys Rev Lett 95:165503
10.86 Yang Q, Lengauer W, Koch T et al (2000) J Alloys Comp 309:L5
10.87 Dzivenko DA, Zerr A, Boehler R, Riedel R (2006) Solid State Commun 139:255
10.88 Soignard E, Somayazulu M, Dong J et al (2001) J Phys Cond Matter 13:557
10.89 Shemkunas MP, Petuskry WT, Chizmeshya AVG et al (2004) J Mater Res 19:1392
10.90 Loa I, Kunc K, Syassen K, Bouvier P (2002) Phys Rev B 66:134101
10.91 Pereira AS, Perottoni CA, da Jordana JAH et al (2002) J Phys Cond Matter 14:10615
10.92 Dandekar DP, Benfanti DC (1993) J Appl Phys 73:673
10.93 Wang Y, Zhang J, Daemen LL et al (2008) Phys Rev B7 8:224106
10.94 Robert WC, Michelle BW, John JG et al (2005) J Am Chem Soc 127:7264
10.95 Zarechnaya E, Dubrovinskaya N, Caracas R et al (2010) Phys Rev B 82:184111
10.96 Zhang Y, Mashimo T, Uemura Y et al (2006) J Appl Phys 100:113536
10.97 Chen B, Penwell D, Nguyen JH et al (2004) Solid State Comm 129:573
10.98 Lazicki A, Hemley RJ, Pickett WE, Yoo C-S (2010) Phys Rev B 82:180102
10.99 Yoshida M, Onodera A, Ueno M et al (1993) Phys Rev B 48:10587
10.100 Litasov KD, Shatskiy A, Fei Y et al (2010) J Appl Phys 108:053513
10.101 Ono S, Kikegawa T, Ohishi Y (2005) Solid State Commun 133:55
10.102 Benedict U (1995) J Alloys Comp 223:216
10.103 Haines J, Leger JM, Chateau C, Lowther JE (2001) J Phys Cond Matter 13:2447
10.104 Ji C, Chyu M-C, Knudson R, Zhu H (2009) J Appl Phys 106:083511
10.105 Hao J, Zou B, Zhu P et al (2009) Solid State Comun 149:689
10.106 Takarabe K, Ikai T, Mori Y et al (2004) J Appl Phys 96:4903
10.107 Beister H, Syassen K (1990) Z Naturforsch b 45:1388
10.108 Ross NL (1997) Amer Miner 82:682
10.109 Ono Sh, Shirasaka M, Kikegawa T, Ohishi Y (2005) Phys Chem Miner 32:8
10.110 Zhang J, Reeder RJ (1999) Am Miner 84:861
10.111 Reynard B, Figuet G, Itie JP, Rubie DC (1996) Amer Miner 81:45
10.112 Runge CE, Kubo A, Kiefer B et al (2006) Phys Chem Miner 33:699
10.113 Ming LC, Kim YH, Chen J-H et al (2001) J Phys Chem Solids 62:1185
10.114 Ross NL, Chaplin TD (2003) J Solid State Chem 172:123; Ross NL, Downs RT (2003) High-pressure crystal chemistry: "stuffed" framework structures at high pressure. In: Katrusiak A, McMillan P (eds) High pressure crystallography. Kluwer, Dordrecht
10.115 Hattori T, Tsuchiya T, Naga T, Yamanaka T (2001) Phys Chem Miner 28:377
10.116 Kung J, Rigden S (1999) Phys Chem Miner 26:234
10.117 Ross NL, Zhao J, Angel RJ (2004) J Solid State Chem 177:1276
10.118 Downs RT, Zha C-S, Duffy TS, Finger LW (1996) Amer Miner 81 51; Zha C-S, Duffy TS, Downs RT et al (1996) J Geophys Res B 101:17535
10.119 Miletich R, Nowak M, Seifert F et al (1999) Phys Chem Miner 26 446
10.120 Errandonea D, Pellicer-Porres J, Manjón FJ et al (2005) Phys Rev B72 174106; Panchal V, Garg N, Achary SN et al (2006) J Phys Cond Matter 18:8241
10.121 Ming LC, Nakamoto Y, Endo S et al (2007) J Phys Cond Matter 19:425202
10.122 Winkler B, Kahle A, Griewatsch C, Milman V (2000) Z Krist 215:17
10.123 Edwards CM, Haines J, Butler IS, Leger J-M (1999) J Phys Chem Solids 60:529
10.124 Long YW, Yang LX, You SJ et al (2006) J Phys Cond Matter 18:2421
10.125 Grzechnik A, Crichton WA, Marshall WG, Friese K (2006) J Phys Cond Matter 18:7
10.126 Gerward L, Jiang JZ, Olsen JS et al (2005) J Alloys Comp 401:11
10.127 Carlson S, Anderson AMK (2001) J Appl Cryst 34:7

10.128 Sundqvist B, Andersson O (2006) Phys Rev B73:092102
10.129 Kumar RS, Cornelius AL (2009) J Alloys Comp 476:5
10.130 Sowa H, Ahsbahs H (1999) Z Krist 214:751
10.131 Lundin A, Soldatov A, Sundquist B (1995) Europhys Lett 30:469
10.132 Xu Y, Carlson S, Söderberg K, Norrestaam R (2000) J Solid State Chem 153:212
10.133 Allan DR, Marshall WG, Francis DJ et al (2010) Dalton Trans 39:3736
10.134 Shimizu H, Kamabuchi K, Kume T, Sasaki S (1999) Phys Rev B 59:11727
10.135 Zuk J, Kiefte H, Clouter MJ (1990) J Chem Phys 92:917
10.136 Zuk J, Brake DM, Kiefte H, Clouter MJ (1989) J Chem Phys 91:5285
10.137 Sun L, Zhao Z, Ruoff AL et al (2007) J Phys Cond Matter 19:425206
10.138 Degtyareva O, Canales MM, Bergara A et al (2007) Phys Rev B 76:064123
10.139 Hirai H, Tanaka T, Kawamura T et al (2003) Phys Rev B 68:172102
10.140 Vaidya SN, Kennedy GC (1971) J Chem Phys 55:987
10.141 Nakayama A, Fujihisa H, Aoki K, Charlon RP (2000) Phys Rev B 62:8759
10.142 Kiefte H,, Penney R Clouter MJ (1988) J Chem Phys 88:5846
10.143 Haussühl S (2001) Z Krist 216:339
10.144 Dinnebier RE, van Smaalen S, Olbrich F, Carlson S (2005) Inorg Chem 44:964
10.145 Pravica M, Shen Y, Quina Z et al (2007) J Phys Chem B 111:4103
10.146 Oehzelt M, Resel R, Nakayama A (2002) Phys Rev B 66:174104
10.147 Davidson AJ, Oswald IDH, Francis DJ et al (2008) Cryst Eng Comm 10:162
10.148 Kawasaki S, Hara T, Iwata A (2007) Chem Phys Lett 447:3169
10.149 Sato-Sorensen Y (1983) J Geophys Res B 88:3543
10.150 Duclos SJ, Vohra YK, Ruoff AL et al (1987) Phys Rev B 36:7664
10.151 Yagi T (1978) J Phys Chem Solids 39:563
10.152 Campbell A, Heinz D (1991) J Phys Chem Solids 52:495
10.153 Köhler U, Johannsen PG, Holzapfel WB (1997) J Phys Cond Matter 9:5581
10.154 Campbell A, Heinz D (1994) J Geophys Res B 99:11765
10.155 Vohra YK, Brister KE, Weir ST et al (1986) Science 231:1136
10.156 Quadri SB, Yang J, Ratna BR, Skelton EF (1996) Appl Phys Lett 69:2205; Jiang JZ, Gerward L, Secco R et al (2000) J Appl Phys 87:2658
10.157 Ghandehari K, Luo H, Ruoff AL et al (1995) Phys Rev Lett 74:2264
10.158 Hull S, Berastegui P (1998) J Phys Cond Matter 10:7945
10.159 Boslough M, Ahrens TJ (1984) J Geophys Res B 89:7845
10.160 Richet P, Mao H-K, Bell P (1988) J Geophys Res B 93:15279
10.161 Luo H, Greene RG, Ruoff AL (1994) Phys Rev B 49:15341
10.162 Weir ST, Vohra YK, Ruoff AL (1986) Phys Rev B 33:4221
10.163 Grzybowski TA Ruoff AL (1984) Phys Rev Lett 53:489
10.164 Desgreniers S (1998) Phys Rev B 58:14102
10.165 Ves S, Schwarz U, Christensen N E et al (1990) Phys Rev B 42:9113
10.166 Strössner K, Ves S, Dietrich W et al (1985) Solid State Commun 56:563
10.167 Werner A, Hochmeir HD, Strössner K, Jayaraman A (1983) Phys Rev B 28:3330
10.168 Chattopadhyay T, Santandrea R, von Schnering H-G (1986) Physica B 139–140:353
10.169 Chattopadhyay T, Werner A, von Schnering H-G (1984) Rev Phys Appl 19:807
10.170 Jackson I, Khanna S, Revcolevschi A, Berthon J (1990) J Geophys Res B 95:21671
10.171 Gerward L, Staun Olsen J, Steenstrup S et al (1990) J Appl Cryst 23:515
10.172 Staun Olsen J, Gerward L, Benedict U, Dabos-Seignon S (1990) High Press Res 2:35
10.173 Dankekar D, Abbate A, Frankel J (1994) J Appl Phys 76:4077
10.174 Van Camp PE, Van Doren VE, Devreese J (1991) Phys Rev B 44:9056; Xia Q, Xia H, Ruoff AL (1993) J Appl Phys 73:8198
10.175 Greene RG, Luo H, Ghandehari K, Ruoff AL (1995) J Phys Chem Solids 56:517
10.176 Konczewicz L, Bigenwald P, Cloitre T et al (1996) J Cryst Growth 159:117; Perlin P, Jauberthie-Caillon C, Itie JP et al (1992) Phys Rev B 45:83; Xia H, Xia Q, Ruoff AL (1993) Phys Rev B 47:12925

10.177 Weir ST, Vohra YK, Vanderborgh CA, Ruoff AL (1989) Phys Rev B 39:1280
10.178 Weir ST, Vohra YK, Ruoff AL (1987) Phys Rev B 36:4543
10.179 Menoni CS, Spain IL (1987) Phys Rev B 35:7520
10.180 Soma T, Kagaya H-M (1984) Phys Stat Solidi b 121:K1
10.181 Leger JM, Vedel I, Redon A et al (1988) Solid State Commun 66:1173
10.182 Menoni C, Hu J, Spain I (1984) In: Homan C, MacCrone RK, Whalley E (eds) High pressure in science and technology, vol 3. North-Holland, New York
10.183 Liu M, Liu L-G (1986) High Temp-High Press 18: 79
10.184 Holzapfel WB, Hartwig M, Sievers W (2001) J Phys Chem Ref Data 30:515
10.185 Solozhenko VI, Will G, Elf F (1995) Solid State Comm 96:1
10.186 Cynn H Yoo CS, Baer B et al (2001) Phys Rev Lett 86:4552
10.187 Yoo CS, Kohlmann H, Cynn H et al (1999) Solid State Comm 83:5527; Iota V, Yoo CS (2001) Solid State Comm 86:3068; Yoo CS, Kohlmann H, Cynn H et al (2002) Phys Rev B 65:104103; Park J-H, Yoo CS, Iota V et al (2003) Phys Rev B 68:014107; Datchi F, Giordano VM, Munsch P, Saito AM (2009) Phys Rev Lett 103:185701
10.188 Angel RJ (2000) Phys Earth Planet Inter 124:71
10.189 Ono S, Ito E, Katsura T et al (2000) Phys Chem Miner 27:618
10.190 Haines J, Leger JM (1997) Phys Rev B 55:11144
10.191 Haines J, Leger JM, Schulte O (1990) J Phys Cond Matt 8:1631
10.192 Arlt T, Bermejo M, Blanco MA et al (2000) Phys Rev B 61:14414
10.193 Al-Khatatbeh Y, Lee KKM, Kiefer B (2010) Phys Rev B 81:214102
10.194 Haines J, Legar JM, Gorelli F et al (2001) Phys Rev B 64:134110
10.195 Vinet P, Rose JH, Ferrante J, Smith JR (1989) J Phys Cond Matt 1:1941
10.196 Çağın T, Dereli G, Uludoğan M, Tomak M (1999) Phys Rev B 59:3468
10.197 Yokoo M, Kawai N, Nakamura KG, Kondo K-I (2009) Phys Rev B 80:104114
10.198 Pantea C, Stroe I, Ledbetter H et al (2009) Phys Rev B 80:024112
10.199 Holzapfel WB, Hartwig M, Sievers W (2001) J Phys Chem Ref Data 30:5151
10.200 Adams JJ, Agosta DS, Leisure RG et al (2006) J Appl Phys 100:113530
10.201 Nadal M-H, Bourgeois L (2010) J Appl Phys 108:033512
10.202 Ledbetter H, Ogi H, Kai S, Kim S (2004) J Appl Phys 95:4642
10.203 SuzukiY, Fanelli VR, Betts JB et al (2011) Phys Rev B 84:064105
10.204 Dai C, Hu J, Tan H (2009) J Appl Phys 106:043519
10.205 Batsanov SS (2006) Rus Chem Rev 75:601
10.206 Hsueh HC, Chen RK, Vass H et al (1998) Phys Rev B 58:14812
10.207 Qi J, Liu JF, He Y et al (2011) J Appl Phys 109:063520
10.208 Ji C, Zhang F, Hou D et al (2011) J Phys Chem Solids 72:609
10.209 Hou D, Zhang F, Ji C et al (2011) J Appl Phys 110:023524

References

1. Loubeyre P, LeToullec R (1995) Stability of O_2/H_2 mixtures at high pressure. Nature 378:44–46
2. Loveday JS, Nelmes RJ (1999) Ammonia monohydrate VI: A hydrogen-bonded molecular alloy. Rev Phys Lett 83:4329–4332
3. Sihachakr D, Loubeyre P (2006) High-pressure transformation of N_2/O_2 mixtures into ionic compounds. Phys Rev B 74:064113
4. Katrusiak A (2008) High-pressure crystallography. Acta Cryst A 64:135–148
5. Frevel LK (1935) A technique for X-ray studies of substances under high pressures. Rev Sci Instrum 6:214–215
6. Jacobs RB (1938) X-ray diffraction of substances under high pressures. Phys Rev 54:325–331
7. Lawson AW, Tang TY (1950) A diamond bomb for obtaining powder pictures at high pressures. Rev Sci Instrum 21:815

8. Eremets MI (1996) High pressure experimental methods. Oxford University Press, Oxford
9. McMillan PF (2003) Chemistry of materials under extreme high pressure-high temperature conditions. J Chem Soc Chem Commun 919–923
10. McMillan PF (2006) Chemistry at high pressure. Chem Soc Rev 35:855–857
11. San-Miguel A (2006) Nanomaterials under high pressure. Chem Soc Rev 35:876–889
12. Goncharov AF, Hemley RJ (2006) Probing hydrogen-rich molecular systems at high pressures and temperatures. Chem Soc Rev 35:899–907
13. McMahon MI, Nelmes RJ (2006) High-pressure structures and phase transformations in elemental metals. Chem Soc Rev 35:943–963
14. Wilding MC, Wilson M, McMillan PF (2006) Structural studies and polymorphism in amorphous solids and liquids at high pressure. Chem Soc Rev 35: 964–986
15. Horvath-Bordon E, Riedel R, Zerr A et al (2006) High-pressure chemistry of nitride-based materials. Chem Soc Rev 35:987–1014
16. Angel RJ (2004) Absorption crorrections for diamond-anvil pressure cells. J Appl Cryst 37:486–492
17. Mujica A, Rubio A, Munōz A, Needs RJ (2003) High-pressure phases of group IV, III-V, and II-VI compounds. Rev Modern Phys 75:863–912
18. Hochheimer HD, Strössner K, Honle V et al (1985) High Pressure X-Ray Investigation of the Alkali Hydrides NaH, KH, RbH, and CsH. Z phys Chem 143:139–144
19. Duclos SJ, Vohra YK, Ruoff AL et al (1987) High-pressure studies of NaH to 54 GPa. Phys Rev B 36:7664–7667
20. Hull S, Keen DA (1994) High pressure polymorphism of the copper(I) halides. Phys Rev B 50:5868–5885
21. Hofmann M, Hull S, Keen DA (1995) High-pressure phase of copper(I) iodide. Phys Rev B 51:12022
22. Leger JM, Haines J, Danneels C, de Oliveira LS (1998) The TlI-type structure of the high-pressure phase of NaBr and NaI; pressure-volume behavior to 40 GPa. J Phys Cond Mater 10:4201–4210
23. Sata N, Shen G, Rivers ML, Sutton S R (2001) Pressure-volume equation of state of the high-pressure B2 phase of NaCl. Phys Rev B 65:104114
24. Ono S, Kikegawa T, Ohishi Y (2006) Structural property of CsCl-type sodium chloride under pressure. Solid State Commun 137:517–521
25. Demarest HH, Cassell CR, Jamieson JC (1978) High-pressure phase-transitions in KF and RbF. J Phys Chem Solids 39:1211–1215
26. Vaidya SN, Kennedy GC (1971) Compressibility of 27 halides to 45 kbar. J Phys Chem Solids 32:951–964
27. Köhler U, Johannsen PG, Holzapfel WB (1997) Equation of state data for CsCl-type alkali halides. J Phys Cond Matt 9:5581–5592
28. Hull S, Berastegui P (1998) High-pressure structural behavior of silver(I) fluoride. J Phys Cond Matter 10:7945–7955
29. Luo H, Ghandehari K, Greene RG et al (1995) Phase transformation of BeSe and BeTe to the NiAs structure at high pressure. Phys Rev B 52:7058–7064
30. Narayana C, Nesamony VJ, Ruoff AL (1997) Phase transformation of BeS and equation-of-state studies to 96 GPa. Phys Rev B 56:14338–14343
31. Li T, Luo H, Greene RG, Ruoff AL (1995) High-pressure phase of MgTe. Phys Rev Lett 74:5232–5235
32. Desgreniers S (1998) High-density phases of ZnO:structural and compressive parameters. Phys Rev B 58:14102–14105
33. Ves S, Schwarz U, Christensen NE et al (1990) High-pressure Raman study of the chain chalcogenide $TlInTe_2$. Phys Rev B 42:9113–9118
34. Ves S (1991) Band-gaps and phase transitions in cubic ZnS, ZnSe and ZnTe. In: Hochheimer HD, Etters RD (eds) Frontiers of high-pressure research. NATO ASI Ser B 286:369–376
35. Tang Z, Gupta Y (1988) Shock-induced phase transformation in cadmium sulfide dispersed in an elastomer. J Appl Phys 64:1827–1837

36. Tolbert SH, Alivisatos AP (1995) The wurtzite to rock-salt structural transformation in CdSe nanocrystals under high-pressure. J Chem Phys 102:4642–4656
37. Hu J (1987) A new high-pressure phase of CdTe. Solid State Commun 63:471–474
38. Nelmes RJ, McMahon MI, Wright G, Allan DR (1995) Structural studies of II-VI semiconductors at high-pressure. J Phys Chem Solids 56:545–549
39. Nelmes RJ, McMahon MI, Wright G, Allan DR (1995) Phase-transitions in CdTe to 28 GPa. Phys Rev B 51:15723–15731
40. Zhou T, Schwarz U, Hanfland M et al (1998) Effect of pressure on the crystal structure, vibrational modes, and electronic excitations of HgO. Phys Rev B 57:153–160
41. San-Miguel A, Wright NG, McMahon MI, Nelmes RJ (1995) Pressure evolution of the cinnabar phase of HgTe. Phys Rev B 51:8731–8736
42. Huang T-L, Ruoff AL (1983) Pressure-induced phase-transition of HgS. J Appl Phys 54:5459–5461
43. Luo H, Greene RG, Ghandehari K et al (1994) Structural phase transformations and the equations of state of calcium chalcogenides at high pressure. Phys Rev B 50; 16232–16237
44. Luo H, Greene RG, Ruoff AL (1994) High-pressure phase transformation and the equation of state of SrSe. Phys Rev B 49:15341–15343
45. Weir ST, Vohra YK, Ruoff AL (1986) High-pressure phase transitions and the equations of state of BaS and BaO. Phys Rev B 33:4221–4226
46. Grzybowski TA, Ruoff AL (1983) High-pressure phase transition in BaSe. Phys Rev B 27:6502–6503
47. Grzybowski TA, Ruoff AL (1984) Band-overlap metallization of BaTe. Phys Rev Lett 53:489–492
48. Chattopadhyay T, Werner A, von Schnering HG, Pannetier J (1984) Temperature and pressure-induced phase transition in IV-VI compounds. Rev Phys Appl 19:807–813
49. Vaitheeswaran G, Kanchana V, Svane A et al (2011) High-pressure structural study of yttrium monochalcogenides from experiment and theory. Phys Rev B 83:184108
50. Vaitheeswaran G, Kanchana V, Heathman S et al (2007) Elastic constants and high-pressure structural transitions in lanthanum monochalcogenides from experiment and theory. Phys Rev B 75:184108
51. Gerward L, Staun Olsen J, Steenstrup S et al (1990) The pressure-induced transformation B1 to B2 in actinide compounds. J Appl Cryst 23:515–519
52. Staun Olsen J, Gerward L, Benedict U, Dabos-Seignon S (1990) High-pressure studies of thorium and uranium compounds with the rocksalt structure. High Pressure Res 2:335–338
53. Ueno M, Yoshida M, Onodera A et al (1994) Stability of the wurtzite-type structure under high-pressure: GaN and InN. Phys Rev B 49:14–21
54. Endo S, Ito K (1982) Triple-stage high-pressure apparatus with sintered diamond anvils. Adv Earth Planet Sci 12:3–12
55. Dankekar D, Abbate A, Frankel J (1994) Equation of state of aluminium nitride and its shock response. J Appl Phys 76:4077–4085
56. Greene RG, Luo HA, Ghandehari K, Ruoff AL (1995) High-pressure structural study of AlSb to 50 GPa. J Phys Chem Solids 56:517–520
57. Konczewicz L, Bigenwald P, Cloitre T et al (1996) MOVPE growth of zincblende magnesium sulphide. J Cryst Growth 159:117–120
58. Perlin P, Jauberthie-Carillon C, Itie JP et al (1992) Raman scattering and X-ray absorption spectroscopy in gallium nitride under high pressure. Phys Rev B 45:83–89
59. Xia H, Xia Q, Ruoff AL (1993) High-pressure structure of gallium nitride: wurtzite-to-rock-salt phase transition. Phys Rev B 47:12925–12928
60. Soma T, Kagaya H-M (1984) High-pressure NaCl-phase of tetrahedral compounds. Solid State Commun 50:261–263
61. Itie JP, Polian A, Jauberthie-Carillon C et al (1989) High-pressure phase transition in gallium phosphide. Phys Rev B 40:9709–9714
62. Weir ST, Vohra YK, Vanderborgh CA, Ruoff AL (1989) Structural phase transitions in GaAs to 108 GPa. Phys Rev B 39:1280–1285

63. Weir ST, Vohra YK, Ruoff AL (1987) Phase transitions in GaSb to 110 GPa. Phys Rev B 36:4543–4546
64. Xia Q, Xia H, Ruoff AL (1994) New crystal structure of indium nitride: a pressure-induced phase. Modern Phys Lett B 8:345–350
65. Menoni CS, Spain IL (1987) Equation of state of InP to 19 GPa. Phys Rev B 35:7520–7525
66. Menoni CS, Spain IL (1989) Structural phase transitions in GaAs to 108 GPa. Phys Rev B 39:1280–1285
67. Soma T, Kagaya H-M (1984) NaCl-type lattice of GaAs and InSb under pressure. Phys Stat Solidi b121:K1–K5
68. Nelmes RJ, McMahon MI, Hatton PD et al (1993) Phase-transitions in InSb at pressures up to 5 GPa. Phys Rev B 47:35–54
69. Liu H, Tse J S, Mao H-k (2006) Stability of rocksalt phase of zinc oxide under strong compression. J Appl Phys 100:093509
70. Ghandehari K, Luo H, Ruoff AL et al (1995) New high-pressure crystal structure and equation of state of cesium hydride to 253 GPa. Phys Rev Lett 74:2264–2267
71. Knittle E, Rudy A, Jeanloz R (1985) High-pressure phase transition in CsBr. Phys Rev B 31:588–590
72. Brister KE, Vohra YK, Ruoff AL (1985) High-pressure phase transition in CsCl at $V/V0 = 0.53$. Phys Rev B31:4657–4658
73. Mao H-K, Hemley RL, Chen LC et al (1989) X-ray diffraction to 302 GPa: high-pressure crystal structure of cesium iodide. Science 246:649–651
74. Mao HK, Wu Y, Hemley RJ et al (1990) High-pressure phase transition and equation of state of CsI. Phys Rev Lett 64:1749–1752
75. Le Bihan T, Darracq S, Heathman S et al (1995) Phase transformation of the monochalcogenides SmX (X = S, Se, Te) under pressure. J Alloys Comp 226:143–145
76. Svane A, Strange P, Temmerman WM et al (2001) Pressure-induced valence transitions in rare earth chalcogenides and pnictides. Phys Stat Solidi b 223:105–116
77. Wright NG, McMahon MI, Nelmes RJ, San-Miguel A (1993) Crystal structure of the cinnabar phase of HgTe. Phys Rev B 48:13111–13114
78. McMahon MI, Nelmes RJ, Wright NG, Allan DR (1993) Phase transitions in CdTe to 5 GPa. Phys Rev B 48:16246–16251
79. Maddox BR, Yoo CS, Kasinathan D et al (2006) High-pressure structure of half-metallic CrO_2. Phys Rev B 73:144111
80. Ming L, Manghnani M (1982) High pressure phase transformations in rutile-structured dioxides. In: Akimoto S, Manghnani M (eds) High-pressure research in geophysics. Center for Acad Publ, Tokyo
81. Gerward L, Staun Olsen J (1993) Powder diffraction analysis of cerium dioxide at high pressure. Powder Diffr 8:127–129
82. Dancausse J-P, Gering E, Heathman S, Benedict U (1990) Pressure-induced phase transition in ThO_2 and PuO_2. High Press Res 2:381–389
83. Desgreniers S, Lagarec K (1999) High-density ZrO_2 and HfO_2: crystalline structures and equations of state. Phys Rev B 59:8467–8472
84. Gerward L, Staun Olsen J, Benedict U et al (1990) Crystal structures of UP·U_2, UAs·U_2, UAsS and UAsSe in pressure range up to 60 GPa. High Temp-High Press 22:523–532
85. Liu L (1978) Fluorite isotype of SnO_2 and a new modification of TiO_2 – implications for Earth's lower mantle. Science 199:422–425
86. Haines J, Leger JM, Chateau C, Pereira AS (2000) Structural evolution of rutile-type and $CaCl_2$-type germanium dioxide at high pressure. Phys Chem Miner 27:575–582
87. Ono S, Tsuchiya T, Hirose K, Ohishi Y (2003) High-pressure form of pyrite-type germanium dioxide. Phys Rev B 68:014103
88. Micoulaut M, Cormier L, Henderson GS (2006) The structure of amorphous, crystalline and liquid GeO_2. J Phys Cond Matter 18:R753–R784
89. Ono S, Ito E, Katsura T et al (2000) Thermoelastic properties of the high-pressure phase of SnO_2. Phys Chem Miner 27:618–622

90. Shieh SR, Kubo A, Duffy TS et al (2006) High-pressure phases in SnO_2 to 117 GPa. Phys Rev B 73:014105
91. Lazicki A, Yoo C-S, Evans WJ, Pickett WE (2006) Pressure-induced antifluorite-to-anticotunnite phase transition in lithium oxide. Phys Rev B 73:184120
92. Iota V, Yoo CS, Cynn H (1999) Quartzlike carbon dioxide: an optically nonlinear extended solid at high pressures and temperatures. Science 283:1510–1513
93. Yoo CS, Cynn H., Gygi F et al (1999) Crystal structure of carbon dioxide at high pressure. Phys Rev Lett 83:5527–5530
94. Iota V, Yoo C-S (2001) Phase diagram of carbon dioxide: evidence for a new associated phase. Phys Rev Lett 86:5922–5925
95. Yoo CS, Kohlmann H, Cynn H et al (2002) Crystal structure of pseudo-six-fold carbon dioxide phase II at high pressures and temperatures. Phys Rev B 65:104103
96. Park JH, Yoo CS, Iota V et al (2003) Crystal striucture of bent carbon dioxide phase IV. Phys Rev B 68:014107
97. Datchi F, Giordano VM, Munsch P, Saito AM (2009) Structure of carbon dioxide phase IV: breakdown of the intermediate bonding state scenario. Phys Rev Lett 103:185701
98. Tschauner O, Mao H-K, Hemley RJ (2001) New transformations of CO_2 at high pressures and temperatures. Phys Rev Lett 87:075701
99. Santoro M, Gorelli FA (2006) High pressure solid state chemistry of carbon dioxide. Chem Soc Rev 36:918–931
100. Kume T, Ohuya Y, Nagata M et al (2007) Transformation of carbon dioxide to nonmolecular solid at room temperature and high pressure. J Appl Phys 102:053501
101. Aoki K, Katoh E, Yamawaki H et al (1999) Hydrogen-bond symmetrization and molecular dissociation in hydrogen halids. Physica B 265:83–86
102. Sakashita M, Yamawaki H, Fujihisa H, Aoki K (1997) Pressure-induced molecular dissociation and metallization in hydrogen-bonded H_2S solid. Phys Rev Lett 79:1082–1085
103. Fujihisa H, Yamawaki H, Sakashita M et al (2004) Molecular dissociation and two low-temperature high-pressure phases of H_2S. Phys Rev B 69:214102
104. van Straaten J, Silvera IF (1986) Observation of metal-insulator and metal-metal transitions in hydrogen iodide under pressure. Phys Rev Lett 57:766–769
105. Batsanov SS (1994) Equalization of interatomic distances in polymorphous transformations under pressure. J Struct Chem 35:391–393
106. Batsanov SS (1997) Effect of high pressure on crystal electronegativities of elements. J Phys Chem Solids 58:527–532
107. Jeon S-J, Porter RF, Vohra YK, Ruoff AL (1987) High-pressure X-ray diffraction and optical absorption studies of NH_4I to 75 GPa. Phys Rev B 35:4954–4958
108. Eremets MI, Gregoryanz EA, Struzhkin VV et al (2000) Electrical conductivity of xenon at megabar pressure. Phys Rev Lett 85:2797–2800
109. Ovsyannikov SV, Shchennikov VV, Popova SV, Derevskov AYu (2003) Semiconductor-metal transitions in lead chalcogenides at high pressure. Phys Stat Solidi b 235:521–525
110. Mita Y, Izaki D, Kobayashi M, Endo S (2005) Pressure-induced metallization of MnO. Phys Rev B 71:100101
111. Hao A, Gao C, Li M et al (2007) A study of the electrical properties of HgS under high pressure. J Phys Cond Matter 19:425222
112. Kawamura H, Matsui N, Nakahata I et al (1998) Pressure-induced metallization and structural transition of rhombohedral Se. Solid State Commun 108:677–680
113. Sun L, Ruoff AL, Zha C-S, Stupian G (2006) Optical properties of methane to 288 GPa at 300 K. J Phys Chem Solids 67:2603–2608
114. Strobel TA, Goncharov AF, Seagle CT et al (2011) High-pressure study of silane to 150 GPa. Phys Rev B 83:144102
115. Ohmura A, Machida A, Watanuki T et al (2006) Infrared spectroscopic study of the band-gap closure in YH_3 at high pressure. Phys Rev B 73:104105
116. Kume T, Ohura H, Takeichi T et al (2011) High-pressure study of ScH_3: Raman, infrared, and visible absorption spectroscopy. Phys Rev B 84:064132

117. Goettel KA, Eggert JH, Silvera IF, Moss WC (1989) Optical evidence for the metallization of xenon at 132(5) GPa. Phys Rev Lett 62:665–668
118. Vajeeston P, Ravindran P, Hauback BC et al (2006) Structural stability and pressure-induced phase transition in MgH_2. Phys Rev B 73:224102
119. Kinoshita K, Nishimara M, Akahama Y, Kawamura H (2007) Pressure-induced phase transition of BaH_2: post Ni_2In phase. Solid State Commun 141:69–72
120. Palasyuk T, Tkacz M (2007) Pressure-induced structural phase transition in rare-earth trihydrides. Solid State Commun 141:354–358
121. Wijngaarden RJ, Huiberts JN, Nagengast D et al (2000) Towards a metallic YH_3 phase at high pressure. J Alloys Comp 308:44–48
122. Lazicki A, Maddox B, Evans WJ et al (2005) New cubic phase of Li_3N: stability of the N^{3-} ion to 200 GPa. Phys Rev Lett 95:165503
123. Hou D, Zhang F, Ji C et al (2011) Series of phase transitions in cesium azide under high pressure studied by in situ X-ray diffraction. Phys Rev B 84:064127
124. Rosenblum SS, Weber WH, Chamberland BL (1997) Raman-scattering observation of the rutile-to-$CaCl_2$ phase transition in RuO_2. Phys Rev B 56:529–533
125. Rozenberg GK, Pasternak MP, Xu WM et al (2003) Pressure-induced structural transformation in the Mott insulator FeI_2. Phys Rev B 68:064105
126. Boldyreva EV (2008) High-pressure diffraction studies of molecular organic solids: a personal view. Acta Cryst A 64:218–231
127. Fabbiani FPA, Pulham CR (2006) High-pressure studies of pharmaceutical compounds and energetic materials. Chem Soc Rev 35:932–942
128. Meersman F, Dobson CM, Heremans K (2006) Protein unfolding, amyloid fibril formation and configurational energy landscapes under high pressure conditions. Chem Soc Rev 35:908–917
129. Wen X-D, Hoffmann R, Ashcroft NW (2011) Benzene under high pressure. J Am Chem Soc 133:9023–9035
130. Fanetti S, Citroni M, Bini R (2011) Pressure-induced fluorescence of pyridine. J Phys Chem B 115:12051–12058
131. Batsanov SS (1982) Crystallochemical calculation of pressure of metallization of inorganic substances. Russ J Phys Chem 56:196–197
132. Batsanov SS (1991) Crystal-chenical calculations of the metallization pressure of inorganic substances. Russ J Inorg Chem 36:1265
133. Slater JC (1963) Quantum theory of molecules and solids, vol 1, Electronic structure of molecules. McGraw-Hill, New York
134. Mishima O, Calvert L, Whalley E (1984) Melting ice-I at 77 K and 10 kbar—a new method of making amorphous solids. Nature 310:393–395
135. Yamanaka T, Nagai T, Tsuchiya T (1997) Mechanism of pressure-induced amorphization. Z Krist 212:401–410
136. Onodera A, Fujii Y, Sugai S (1986) Polymorphism and amorphism at high pressure. Physica B 139:240–245
137. Chen A, Yu PY, Pasternak MP (1991) Metallization and amorphization of the molecular crystals SnI_4 and GeI_4 under pressure. Phys Rev B 44:2883–2886
138. Grocholski B, Speziale S, Jeanloz R (2010) Equation of state, phase stability, and amorphization of SnI_4 at high pressure and temperature. Phys Rev B 81:094101
139. Polian A, Itie JP, Jauberthie-Carillon C et al (1990) X-ray absorption spectroscopy investigation of phase transition in Ge, GaAs and GaP. High Pressure Res 4:309–311
140. Vohra YK, Xia H, Ruoff AL (1990) Optical reflectivity and amrphization of GaAs during decompression from megabar pressure. Appl Phys Lett 57:2666–2668
141. Nelmes RJ, McMahon MI, Wright NG, Allan DR (1994) Crystal structure of ZnTe III at 16 GPa. Phys Rev Lett 73:1805–1808
142. Liu H, Secco RA, Imanaka N, Adachi G (2002) X-ray diffraction study of pressure-induced amorphization in $Lu_2(WO_4)_3$. Solid State Commun 121:177–180
143. Goncharov AF (1990) Observation of amorphous phase of carbon at pressures above 23 GPa. Sov Phys JETP Lett 51:418–421

144. Madon M, Gillet Ph, Julien Ch, Price GD (1991) A vibrational study of phase transitions among the GeO_2 polymorphs. Phys Chem Miner 18:7–18
145. Meade C, Jeanloz R (1990) Static compression of $Ca(OH)_2$ at room-temperature – observations of amorphization and equation of state measurements to 7 GPa. Geophys Res Lett 17:1157–1160
146. Luo H, Ruoff AL (1993) X-ray-diffraction study of sulfur to 32 GPa—amorphization at 25 GPa. Phys Rev B 48:569–572
147. Nishii T, Mizuno T, Mori Y et al (2007) X-ray diffraction study of amorphous phase of $BaSi_2$ under high pressure. Physica Status Solidi b244:270–273
148. Secco RA, Liu H, Imanaka N, Adachi G (2001) Pressure-induced amorphization in negative thermal expansion $Sc_2(WO_4)_3$. J Mater Sci Lett 20:1339–1340
149. Arora AK, Yagi T, Miyajima N, Mary T A (2005) Amorphization and decomposition of scandium molybdate at high pressure. J Appl Phys 97:013508
150. Arora AK (2000) Pressure-induced amorphization versus decomposition. Solid State Commun 115:665–668
151. Arora AK, Sastry VS, Sahu PCh, Mary TA (2004) The pressure-amorphized state in zirconium tungstate: a precursor to decomposition. J Phys Cond Matter 16:1025–1031
152. Brazhkin VV, Lyapin AG (2003) High-pressure phase transformations in liquids and amorphous solids. J Phys Cond Matter 15:6059–6084
153. Katayama Y, Tsuji K (2003) X-ray structural studies on elemental liquids under high pressures. J Phys Cond Matter 15:6085–6103
154. Katayama Y, Mizutani T, Utsumi W et al (2000) A first-order liquid-liquid phase transition in phosphorus. Nature 403:170–173
155. Katayama Y, Inamura Y, Mizutani T et al (2004) Macroscopic separation of dense fluid phase and liquid phase of phosphorus. Science 306:848–851
156. Di Cicco A, Congeduti A, Coppari F et al (2008) Interplay between morphology and metallization in amorphous-amorphous transitions. Phys Rev B 78:033309
157. Soignard E, Amin SA, Mei Q et al (2008) High-pressure behavior of As_2O_3: Amorphous-amorphous and crystalline-amorphous transitions. Phys Rev B 77:144113
158. Mei Q, Sinogeikin S, Shen G et al (2010) High-pressure x-ray diffraction measurements on vitreous GeO_2 under hydrostatic conditions. Phys Rev B 81:174113
159. Brazhkin VV, Gavrilyuk AG, Lyapin AG et al (2007) AsS: bulk inorganic molecular-based chalcogenide glass. Appl Phys Lett 91:031912
160. Kinoshita T, Hattori T, Narushima T, Tsuji K (2005) Pressure-induced drastic structural change in liquid CdTe. Phys Rev B 72:060102
161. Hattori T, Tsuji K, Taga N et al (2003) Structure of liquid GaSb at pressures up to 20 GPa. Phys Rev B 68:224106
162. Falconi S, Lundegaard LF, Hejny C, McMahon MI (2005) X-ray diffraction study of liquid Cs up to 9.8 GPa. Phys Rev Lett 94:125507
163. Giefers H, Porsch F, Wortmann G (2006) Thermal disproportionation of SnO under high pressure. Solid State Ionics 176:1327–1332
164. Giefers H, Porsch F, Wortmann G (2006) Structural study of SnO at high pressure. Physica B 373:76–81
165. Eremets MI, Gavriliuk AG, Trojan IA (2007) Single-crystalline polymeric nitrogen. Appl Phys Lett 90:171904
166. Jiang JZ, Staun Olsen J, Gerward L, Mørup S (2000) Structural stability in nanocrystalline ZnO. Europhys Lett 50:48–53
167. Quadri SB, Skelton EF, Dinsmore AD, Hu JZ (2001) The effect of particle size on the structural transitions in zinc sulfide. J Appl Phys 89:115–119
168. Wang Z, Guo Q (2009) Size-dependent structural stability and tuning mechanism: a case of zinc sulfide. J Phys Chem C 113:4286–4295
169. Haase M, Alivisatos AP (1992) Arrested solid-solid phase transition in 4-nm-diameter cadmium sulfide nanocrystals. J Phys Chem 96:6756–6762

170. Quadri SB, Yang J, Ratna BR et al (1996) Pressure induced structural transition in nanometer size particles of PbS. Appl Phys Lett 69:2205–2207
171. Jiang JZ, Gerward L, Secco R et al (2000) Phase transformation and conductivity in nanocrystal PbS under pressure. J Appl Phys 87:2658–2660
172. Wang Z, Saxena SK, Pischedda V et al (2001) X-ray diffraction study on pressure-induced phase transformation in nanocrystalline anatase/rutile (TiO_2). J Phys Cond Matter 13:8317–8323
173. Swamy V, Kuznetsov A, Dubrovinsky LS et al (2006) Size-dependent pressure-induced amorphization in nanoscale TiO_2. Phys Rev Lett 96:135702
174. Machon D, Daniel M, Pischedda V et al (2010) Pressure-induced polyamorphism in TiO_2 nanoparticles. Phys Rev B 82:140102
175. Kawasaki S, Yamanaka T, Kume S, Ashida T (1990) Crystallite size effect on the pressure-induced phase transformation. Solid State Commun 76:527–530
176. Wang H, Liu JF, Wu HP et al (2006) Phase transformation in nanocrystalline α-quartz GeO_2 up to 51.5 GPa. J Phys Cond Matter 18:10817–10824
177. He Y, Liu JF, Chen W et al (2005) High-pressure behavior of SnO_2 nanocrystals. Phys Rev B 72:212102
178. Dong Z, Song Y (2009) Pressure-induced morphology-dependent phase transformations of nanostructured tin dioxide. Chem Phys Lett 480:90–95
179. Shen LH, Li XF, Ma YM et al (2006) Pressure-induced structural transition in AlN nanowires. Appl Phys Lett 89:141903
180. Jørgensen J-E, Jakobsen JM, Jiang JZ et al (2003) High-pressure X-ray diffraction study of bulk- and nanocrystalline GaN. J Appl Cryst 36:920–925
181. Jiang JZ, Staun Olsen J, Gerward L, Mørup S (1998) Enhanced bulk modulus and reduced transition pressure in γ-Fe_2O_3 nanocrystals. Europhys Lett 44:620–626
182. Wang Z, Saxena SK (2002) Pressure induced phase transformations in nanocrystalline maghemite (γ-Fe_2O_3). Solid State Commun 123:195–200
183. Wang Z, Saxena SK, Pischedda V et al (2001) *In situ* X-ray diffraction study of the pressure-induced phase transformation in nanocrystalline CeO_2. Phys Rev B 64:012102
184. Liu H, Jin C, Zhao Y (2002) Pressure induced structural transitions in nanocrystalline grained selenium. Physica B 315:210–214
185. Biswas K, Muthu DVS, Sood AK et al (2007) Pressure-induced phase transitions in nanocrystalline ReO_3. J Phys Cond Matter 19:436214
186. Liu D, Yao M, Wang L et al (2011) Pressure-induced phase transitions in C_{70} nanotubes. J Phys Chem C 115:8118–8122
187. Guo Q, Zhao Y, Mao WL et al (2008) Cubic to tetragonal phase transformation in cold-compressed Pd nanocubes. Nano Letters 8:972–975
188. Wang Y, Zhang J, Wu J et al (2008) Phase transition and compressibility in silicon nanowires. Nano Letters 8:2891–2895
189. Jiang JZ (2004) Phase transformations in nanocrystals. J Mater Sci 39:5103–5110
190. Wang Z, Guo Q (2009) Size-dependent structural stability and tuning mechanism: a case of zinc sulfide. J Phys Chem C 113:4286–4295
191. Takemura K (2007) Pressure scales and hydrostaticity. High Pressure Res 27:465–472
192. Jenei Zs, Liermann HP, Cynn H et al (2011) Structural phase transition in vanadium at high pressure and high temperature: influence of nonhydrostatic conditions. Phys Rev B 83:054101
193. Batsanov SS (2004) Determination of ionic radii from metal compressibilities. J Struct Chem 45:896–899
194. Batsanov SS (2005) Chemical bonding evolution on compression of crystals. J Struct Chem 46:306–314
195. Batsanov SS (2006) Mechanism of metallization of ionic crystals by pressure. Russ J Phys Chem 80:135–138
196. Zel'dovich YB, Raizer YP (1967) Physics of shock waves and high temperature hydrodynamics phenomena. Academic, New York

197. Kormer SB (1968) Optical study of the characteristica of shock-compressed condensed dielectrics. Sov Phys-Uspekhi 11:229–254
198. Kinslow R (ed) (1970) High-velocity impact phenomena. Academic, New York
199. Riabinin YuN (1956) Sov Phys Dokl 1:424
200. Adadurov GA (1986) Experimental study of chemical processes under dynamic compression conditions. Russ Chem Rev 55:282–296
201. Batsanov SS (1986) Inorganic chemistry of high dynamic pressures. Russ Chem Rev 55:297–315
202. Prümmer R (1987) Explosivverdichtung pulvriger Substanzen. Springer, Berlin
203. Batsanov SS (1994) Effects of explosions on materials. Springer, New York
204. Batsanov SS (2006) Features of solid-phase transformations induced by shock compression. Russ Chem Rev 75:601–616
205. Bancroft D, Peterson EL, Minshall S (1956) Polymorphism of iron at high pressure. J Appl Phys 27:291–298
206. DeCarli PS, Jamieson JC (1961) Formation of diamond by explosive shock. Science 133:1821–1822
207. Mashimo T, Nakamura K, Tsumoto K et al (2002) Phase transition of KCl under shock compression. J Phys Cond Matter 14:10783–10786
208. Kleeman J, Ahrens TJ (1973) Shock-induced transition of quartz to stishovite. J Geophys Res 78:5954–5960
209. Zahn D, Leoni S (2004) Mechanism of the pressure induced reconstructive transformation of KCl from the NaCl type to the CsCl type structure. Z Krist 219:345–347
210. Batsanov SS (1983) Phase transformation of inorganic substances in shock compression. Russ J Inorg Chem 28:1545–1550
211. Egorov LA, Nitochkina EV, Orekin YuK (1972) Registration of Debyegram of aluminum compressed by a shock wave. Sov Phys JETP Lett 16:4–6
212. Johnson O, Mitchell A (1972) First X-ray diffraction evidence for a phase transition during shock-wave compression. Phys Rev Lett 29:1369–1371
213. Müller F, Schulte E (1978) Shock-wave compression of NaCl single crystals observed by flash X-ray diffraction. Z Naturforsch 33a:918–923
214. Zaretskii EB, Kanel GI, Mogilevskii PA, Fortov VE (1991) X-ray diffraction study pf phase transition mechanism in shock-compressed KCl single crystal. Sov Phys Dokl 36:76–78
215. Batsanov SS (1985) Some features of phase transition under shock comptrssion. Sov J Chem Phys 2:1104–1112
216. Semin VP, Dolgushin GG, Korobov VK, Batsanov SS (1980) Phase transitions in halides of potasium and rubidium. Inorg Mater 16:1128–1129
217. Batsanov SS (1986) Maximum possible defect concentration in a solid. Inorg Mater 22:913–913
218. De Carli PS, Jamieson JC (1959) Formation of an amorphous form of quartz under shock conditions. J Chem Phys 31:1675–1676
219. Batsanov SS, Ruchkin ED, Travkina IN et al (1975) Polymorphic conversions of rare earth metal sulfides by impact compression. J Struct Chem 16:651–653
220. Batsanov SS, Kiselev YM, Kopaneva LI (1979) Polymorphic transformation of ThF_4 in shock compression. Russ J Inorg Chem 24:1573–1573
221. Batsanov SS, Kiselev YM, Kopaneva LI (1980) Polymorphic transformation of UF_4 and CeF_4 in shock compression. Russ J Inorg Chem 25:1102–1103
222. Adadurov GA, Breusov ON, Dremin AN et al (1971) Phase-transitions of shock-compressed t-Nb_2O_5 and h-Nb_2O_5. Comb Exp Shock Waves 7:503–506
223. Adadurov GA, Breusov ON, Dremin AN et al (1972) Formation of a Nb_xO_2 ($0.8 \le x \le 1.0$) phase under shock compression of niobium pentoxide. Dokl Akad Nauk SSSR 202:864–867
224. Batsanov SS, Blokhina GE, Deribas AA (1965) The effects of explosions on materials. J Struct Chem 6:209–213
225. Blank VD, Buga SG, Serebryanaya NR et al (1996) Phase transformations in solid C_{60} at high-pressure-high-temperature treatment and the structure of 3D polymerized fullerites. Phys Lett A 220:149–157

226. Batsanov SS (1998) The E-phase of boron nitride as a fullerene. Comb Expl Shock Waves 34:106–108
227. Batsanov SS (2011) Features of phase transformations in boron nitride. Diamond Relat Mater 20:660–664
228. Stankevich IV, Chistyakov AL, Galperin EG (1993) Polyhedral boron-nitride molecules. Russ Chem Bull 42:1634–1636
229. Jensen F, Toftlund H (1993) Structure and stability of C_{24} and $B_{12}N_{12}$ isomers. Chem Phys Lett 201:89–96
230. Silaghi-Dumitrescu I, Haiduc I, Sowerby DB (1993) Fully inorganic (carbon-free) fullerenes—the boron-nitrogen case. Inorg Chem 32:3755–3758
231. Banhart F, Zwanger M, Muhr H-J (1994) The formation of curled concentric-shell clusters in boron nitride under electron irradiation. Chem Phys Lett 231:98–104
232. Golberg D, Bando Y, Eremets M et al (1996) Nanotubes in boron nitride laser heated at high pressure. Appl Phys Lett 69:2045–2047
233. Golberg D, Bando Y, Stéphan O, Kurashima K (1998) Octahedral boron nitride fullerenes formed by electron beam irradiation. Appl Phys Lett 73:2441–2443
234. Golberg D, Rode A, Bando Y et al (2003) Boron nitride nanostructures formed by ultra-high-repetition rate laser ablation. Diamond Relat Mater 12:1269–1274
235. Olszyna A, Konwerska-Hrabowska J, Lisicki M (1997) Molecular structure of E-BN. Diamond Relat Mater 6:617–620
236. Pokropivny AV (2006) Structure of the boron nitride E-phase: diamond lattice of $B_{12}N_{12}$ fullerenes. Diamond Relat Mater 15:1492–1495
237. Pokropivny VV, Smolyar AS, Pokropivny AV (2007) Fluid synthesis and structure of a new boron nitride polymorph—hyperdiamond fulborenite $B_{12}N_{12}$ (E phase). Phys Solid State 49:591–598
238. Meyer K (1968) Physikalische-chemische Kristallographie. Deutsche Verlag Grundstoffindustrie, Leipzig
239. Lin I, Nadiv S (1979) Review of the phase transformation and synthesis of inorganic solids obtained by mechanical treatment. Mater Sci Eng 39:193–209
240. Musalimov LA (2007) Effect of mechanical activation on the structure of inorganic materials with different bonding types. Inorg Mater 43:1371–1378
241. Herak R (1970) A stress induced α-U_3O_8-β-U_3O_8 transformation. J Inorg Nucl Chem 32:3793–3797
242. Bokarev VP, Bokareva OM, Temnitskii IN, Batsanov SS (1986) Influence of shear deformation on phase transitions progess in BaF_2 and SrF_2. Phys Solid State 28:452–454
243. Batsanov SS, Serebryanaya NR, Blank VD, Ivdenko VA (1995) Crystal structure of CuI under plastic deformation and pressures up to 38 GPa. Crystallogr Reports 40:598–603
244. Hoekstra HR, Gingerich KA (1964) High-pressure B-type polymorphs of some rare-earth sesquioxides. Science 146:1163–1164
245. Ruchkin ED, Sokolova MN, Batsanov SS (1967) Optical properties of oxides of the rare earth elements. J Struct Chem 8:410–414
246. Michel D, Mazerolles L, Berthet P et al (1995) Nanocrystalline and amorphous oxide powders prepared by high-energy ball-milling. Eur J Solid State Inorg Chem 32:673–682
247. Gasgnier M, Szwarc H, Ronez A (2000) Low-energy ball-milling: transformation of boron nitride powders. J Mater Sci 35:3003–3009
248. Batsanov SS, Gavrilkin SM, Bezduganov SV, Romanov PN (2008) Reversible phase transformation in boron nitride under pulsed mechanical action. Inorg Mater 44:1199–1201
249. Schneider H, Hornemann U (1981) Shock-induced transformation of sillimanite powders. J Mater Sci 16:45–49
250. Kawai N, Nakamura KG, Kondo K-i (2004) High-pressure phase transition of mullite under shock compression. J Appl Phys 96:4126–4130
251. Kawai N, Atou T, Nakamura KG et al (2009) Shock-induced disproportionation of mullite ($3Al_2O_3{\cdot}2SiO_2$). J Appl Phys 106:023525

252. Liepins R, Staudhammer KP, Johnson KA, Thomson M (1988) Shock-induced synthesis: cubic boron-nitride from ammonia borane. Matter Lett 7:44–46
253. Lyamkin AI, Petrov EA, Ershov AP et al (1988) Production of diamonds from explosives. Soviet Physics Doklady 33:705–707
254. Greiner NR, Fillips D, Johnson J (1988) Diamonds in detonation soot. Nature 333:440–442
255. Zerr A, Serghiou G, Boehler R, Ross M (2006) Decomposition of alkanes at high pressures and temperatures. High Pressure Res 26:23–32
256. Shen AH, Ahrens TJ, O'Keefe JD (2003) Shock wave induced vaporization of porous solids. J Appl Phys 93:5167–5174
257. Batsanov SS (1994) Dynamic static compression – controlling the conditions in containment systems after explosion. Comb Expl Shock Waves 30:126–130
258. Batsanov SS, Gavrilkin SM, Kopaneva LI et al (1997) h-BN → w-BN phase transition under dynamic-static compression. J Mater Sci Lett 16:1625–1627
259. Gavrilkin SM, Kopaneva LI, Batsanov SS (2004) Kinetics of the thermal transformation of w-BN into g-BN. Inorg Mater 40:23–25
260. Gavrilkin SM, Bolkhovitinov LG, Batsanov SS (2007) Specifics of the thermal transformations of the wurtzite phase of boron nitride. Russ J Phys Chem 81:648–650
261. Butenko YV, Kuznetsov VL, Chuvilin AL et al (2000) Kinetics of the graphitization of dispersed diamond at "low" temperatures. J Appl Phys 88:4380–4388
262. Titov VM, Anisichkin VF, Mal'kov IY (1989) Synthesis of ultradispersed diamond in detonation-waves. Comb Expl Shock Waves 25:372–379
263. Vyskubenko BA, Danilenko VV, Lin EE et al (1992) The effect of the scale factors on the size and yield of diamonds during detonation synthesis. Fizika Gorenia i Vzryva 28(2):108–109 (in Russian)
264. Batsanov SS (2009) Thermodynamic reason for delamination of molecular mixtures under pressure and detonation synthesis of diamond. Russ J Phys Chem A 83:1419–1421
265. Ree FH (1986) Supercritical fluid phase separations—implications for detonation properties of condensed explosives. J Chem Phys 84:5845–5856
266. Gavrilkin SM, Batsanov SS, Gordopolov YA, Smirnov AS (2009) Effective detonation synthesis of cubic boron nitride. Propellants Explos Pyrotech 34:469–471
267. Schlosser H, Ferrante J (1993) High-pressure equation of state for partially ionic solids. Phys Rev B 48:6646–6649
268. Freund J, Ingalls R (1989) Inverted isothermal equations of state and determination of B_0, B'_0 and B''_0. J Phys Chem Solids 50:263–268
269. Holzapfel WB (1991) Equations of state and scaling rules for molecular-solids under strong compression. In: Pucci R, Piccitto G (eds) Molecular systems under high pressures. North-Holland, Amsterdam
270. Murnaghan FD (1944) The compressibility of media under extreme pressures. Proc Nat Acad Sci US 30:244–247
271. Birch F (1978) Finite strain isotherm and velocities for single crystal and polycrystalline NaCl at high-pressures and 300 K. J Geophys Res B 83:1257–1268
272. Vinet P, Ferrante J, Rose JH (1987) Compressibility of solids. J Geophys Res B 92:9319–9325
273. Schlosser H, Ferrante J, Smith JR (1991) Global expression for representing cohesive-energy curves. Phys Rev B 44:9696–9699
274. Winzenick M, Vijayakumar V, Holzapfel WB (1994) High-pressure X-ray diffraction on potassium and rubidium up to 50 GPa. Phys Rev B 50:12381–12385
275. Holzapfel WB (1998) Equations of state for solids under strong compression. High Pressure Res 16:81–126
276. Holzapfel WB (2001) Equations of state for solids under strong compression. Z Krist 216:473–488
277. Ponkratz U, Holzapfel WB (2004) Equations of state for wide ranges in pressure and temperature. J Phys Cond Matter 16:S963–S972
278. Batsanov SS (1999) Bulk moduli of crystalline A(N)B(8−N) inorganic materials. Inorg Mater 35:973–977

279. Batsanov SS (2002) Internal energy of metals as a function of interatomic distance. Russ J Inorg Chem 47:660–662
280. Batsanov SS (2007) Ionization, atomization, and bond energies as functions of distances in inorganic molecules and crystals. Russ J Inorg Chem 52:1223–1229
281. Holzapfel WB, Hartwig M, Sievers W (2001) Equations of state for Cu, Ag, and Au for wide ranges in temperature and pressure up to 500 GPa and above. J Phys Chem Ref Data 30:515–529
282. Velisavljevic N, Chesnut GN, Vohra YK et al (2002) Structural and electrical properties of beryllium metal to 66 GPa studied using designer diamond anvils. Phys Rev B 65:172107
283. Vohra YK, Spencer PT (2001) Novel γ-phase of titanium metal at megabar pressures. Phys Rev Lett 86:3068–3071
284. Sundqvist B (1999) The structures and properties of C_{60} under pressure. Physica B 265:208–213
285. Sundquist B (2000) Fullerenes under high pressures. In: Kadish KM, Ruoff RS (eds) Fullerenes: chemistry, physics, technology. Wiley, New York
286. Kenichi T, Singh AK (2006) High-pressure equation of state for Nb with a helium-pressure medium. Phys Rev B 73:224119
287. Dewaele A, Loubeyre P, Mezouar M (2004) Refinement of the equation of state of tantalum. Phys Rev B 69:092106
288. Akahama Y, Koboyashi M, Kawamura H (1999) Simple-cubic–simple-hexagonal transition in phosphorus under pressure. Phys Rev B 59:8520–8525
289. Ruoff AL, Rodriguez CO, Christensen NE (1998) Elastic moduli of tungsten to 15 Mbar, phase transition at 6.5 Mbar, and rheology to 6 Mbar. Phys Rev B 58:2998–3002
290. Etters RD, Kirin D (1986) High-pressure behavior of solid molecular fluorine at low temperatures. J Phys Chem 90:4670–4673
291. Fujihisa H, Takemura K (1996) Equation of state of cobalt up to 79 GPa. Phys Rev B 54:5–7
292. Takemura K (2004) Bulk modulus of osmium: High-pressure powder x-ray diffraction experiments under quasihydrostatic conditions. Phys Rev B 70:012101
293. Khairallah SA, Militzer B (2008) First-principles studies of the metallization and the equation of state of solid helium. Phys Rev Lett 101:106407
294. Vinet P, Rose JH, Ferrante J, Smith JR (1989) Universal feature of the equation of state of solids. J Phys Cond Matter 1:1941–1964
295. Cynn H, Yoo CS, Iota-Herbei V et al (2001) Martensitic fcc-to-hcp transformation observed in xenon at high pressure. Phys Rev Lett 86:4552–4555
296. Le Bihan T, Heathman S, Idiri M et al (2003) Structural behavior of α-uranium with pressures to 100 GPa. Phys Rev B 67:134102
297. Ledbetter H, Migliori A, Betts J et al (2005) Zero-temperature bulk modulus of α-plutonium. Phys Rev B 71:172101
298. Pravica M, Shen Y, Quine Z et al (2007) High-pressure studies of cyclohexane to 40 GPa. J Phys Chem B 111:4103–4108
299. Gerlich D, Litov E, Anderson OL (1979) Effect of pressure on the elastic properties of vitreous As_2S_3. Phys Rev B 20:2529–2536
300. Ota R, Anderson OL (1977) Variations in mechanical properties of glass induced by high-pressure phase change. J Non-Cryst Solids 24:235–252
301. Lundegaard LF, Miletich R, Ballic-Zunic T, Makovicky E (2003) Equation of state and crystal structure of Sb_2S_3 between 0 and 10 GPa. Phys Chem Miner 30:463–468
302. Bridgman PW (1912) Water, in the liquid and five solid forms, under pressure. Proc Am Acad Arts Sci 47:441–558
303. Bridgman PW (1931) The physics of high pressures. G Bell & Sons, London
304. Bridgman PW (1950) Physics above 20,000 kg/cm^2. Proc Roy Soc London A 203:1–17
305. Bridgman PW (1952) The resistance of 72 elements, alloys and compounds to 100,000 kg/cm^2. Proc Am Acad Arts Sci 81:167–251
306. Anderson OL, Nafe JE (1965) The bulk modulus-volume relationship for oxide compounds and related geophysical problems. J Geophys Res 70:3951–3963

307. Anderson DL, Anderson OL (1970) The bulk modulus-volume relationship for oxides. J Geophys Res 75:3494–3500
308. Cohen ML (1985) Calculation of bulk moduli of diamond and zinc-blende solids. Phys Rev B 32:7988–7991
309. Zhang B, Cohen ML (1987) High-pressure phases of III-V zinc-blende semiconductors. Phys Rev B 35:7604–7610
310. Al-Douri Y, Abid H, Aourag H (2005) Correlation between the bulk modulus and the transition pressure in semiconductors. Mater Lett 59:2032–2034
311. Verma AS, Bhardwaj SR (2006) Correlation between ionic charge and ground-state properties in rocksalt and zinc blende structured solids. J Phys Cond Matter 18:8603–8612
312. Zhang C, Li H, Li H et al (2007) Calculation of bulk modulus of simple and complex crystals with the chemical bond method. J Phys Chem B 111:1304–1309
313. Makino Y (1996) Empirical determination of bulk moduli of elemental substances by pseudopotential radius. J Alloys Comp 242:122–128
314. Wentzcovitch RM, Cohen ML, Lam PK (1987) Theoretical study of BN, BP and BAs at high pressures. Phys Rev B 36:6058
315. Cohen ML (1988) Theory of bulk moduli of hard solids. Mater Sci Engin A 105/106:11–18
316. Van Camp PE, Van Doren VE, Devreese JT (1990) Pressure dependence of the electronic properties of cubic III-V In compounds. Phys Rev B 41:1598–1602
317. Morosin B, Schriber JE (1979) Remarks on the compressibilities of cubic materials and measurements on Pr chalcogenides. Phys Lett A 73:50–52
318. Setton R (2010) An equation of state for condensed matter, application to solid chemical elements. J Phys Chem Solids 71:776–783
319. Ruoff AL (1967) Linear shock-velocity-particle-velocity relationship. J Appl Phys 38:4976–4980
320. Gu QF, Krauss G, Steurer W et al (2008) Unexpected high stiffness of Ag and Au nanoparticles. Phys Rev Lett 100:045502
321. Cuenot S, Frétigny C, Demoustier-Champagne S, Nysten B (2004) Surface tension effect on the mechanical properties of nanomaterials measured by atomic force microscopy. Phys Rev B 69:165410
322. Vennila S, Kulkarni SR, Saxena SK et al (2006) Compression behavior of nanosized nickel and molybdenum. Appl Phys Lett 89:261901
323. Qi J, Liu JF, He Y et al (2011) Compression behavior and phase transition of cubic In_2O_3 nanocrystals. J Appl Phys 109:063520
324. Wang H, He Y, Chen W et al (2010) High-pressure behavior of β-Ga_2O_3 nanocrystals. J Appl Phys 107:033520
325. Swamy V, Dubrovinsky LS, Dubrovinskaia NA et al (2003) Compression behavior of nanocrystalline anatase TiO_2. Solid State Commun 125:111–115
326. Kiefer B, Shieh SR, Duffy ThS, Sekine T (2005) Strength, elasticity, and equation of state of the nanocrystalline cubic silicon nitride γ-Si_3N_4 to 68 GPa. Phys Rev B 72:014102
327. Wang Z, Pischedda V, Saxena SK, Lazor P (2002) X-ray diffraction and Raman spectroscopic study of nanocrystalline CuO under pressures. Solid State Commun 121:275–279
328. Ma Y, Cui Q, Shen L, He Z (2007) X-ray diffraction study of nanocrystalline tungsten nitride and tungsten to 31 GPa. J Appl Phys 102:013525
329. Marquardt H, Speziale S, Marquardt K et al (2011) The effect of crystallite size and stress condition on the equation of state of nanocrystalline MgO. J Appl Phys 110:113512
330. Chen B, Penwell D, Benedetti LR et al (2002) Particle-size effect on the compressibility of nanocrystalline alumina. Phys Rev B 66:144101
331. Zhu H, Ma Y, Zhang H et al (2008) Synthesis and compression of nanocrystalline silicon carbide. J Appl Phys 104:123516
332. Trapp S, Limbach CT, Gonser U et al (1995) Enhanced compressibility and pressure-induced structural changes of nanocrystalline iron: in situ Mössbauer spectroscopy. Phys Rev Lett 75:3760–3763

333. Zhang J, Zhao Y, Palosz B (2007) Comparative studies of compressibility between nanocrystalline and bulk nickel. Appl Phys Lett 90:043112
334. Marquardt H, Gleason CD, Marquardt K et al (2011) Elastic properties of MgO nanocrystals and grain boundaries at high pressures by Brillouin scattering. Phys Rev B 84:064131
335. Heimel G, Hummer K, Ambrosch-Draxl C et al (2006) Phase transition and electronic properties of fluorene: a joint experimental and theoretical high-pressure study. Phys Rev B 73:024109
336. Batsanov SS (2004) Solid phase transformations under high dynamic pressures. In: Katrusiak A, McMillan P (eds) High-pressure crystallography. Kluwer, Dordrecht

Chapter 11
Structure and Optical Properties

A century ago optical methods of analysis were limited to a narrow range of wavelengths at, and immediately adjacent to, the visible range. Today, this range extends from X-ray to radio frequencies. A systematic description of various techniques is beyond the scope of the present book. Instead, we shall concentrate on applying optical methods to such cases which are difficult for diffraction techniques, such as amorphous materials or transient species with life-spans on the order of ms or even μs. Optical methods permit to study substances in different aggregate and thermodynamic states and give important information on the nature of chemical bonding. Particular attention here is paid to refractometry, because the achievements and potential of this method are not properly appreciated today. In fact, refractometry was the earliest (since mid-nineteenth century) physical method of investigating chemical structure and bonding. The initial high hopes of making it into a method of structure determination were generally not fulfilled, because the relation between refraction and structure proved more complicated than expected, and spectacular development of diffraction methods and NMR spectroscopy in the second half of the twentieth century seemed to make refractometry irrelevant. However, in this chapter it will be shown that this method does provide unique information about chemical bonding, not accessible by other means. Furthermore, modern techniques (e.g. atomic and molecular beams) can measure polarization directly where classical refractometry could only guess it, complementing optical data. Finally, a few words about refractometry of single crystals. From being an indispensable skill for a general chemist (let alone a crystallographer) a century ago, now it became almost a lost art, except as a means of fast identification of minerals under field conditions by geologists. Nevertheless we believe that this technique can flourish again, especially if reorganized on a modern technological level.

S. S. Batsanov, A. S. Batsanov, *Introduction to Structural Chemistry*,
DOI 10.1007/978-94-007-4771-5_11, © Springer Science+Business Media Dordrecht 2012

11.1 Refractive Index

11.1.1 Definitions, Anisotropy, Theory

When electromagnetic waves (including light) travel through a substance, the latter is subject to an alternating electric field and undergoes a (dynamic) polarization, which affects the speed of the waves. The speed of light (v) being different in different media, a beam crossing (non-perpendicularly) an interface between them is refracted. The law of refraction, discovered by Snelius and Descartes in the seventeenth century, states than the refractive index n (RI) which relates the angles of the incident and refracted beams, is constant for any pair of media. It the nineteenth century it was realized that n equals the ratio of light speeds in these media (see below).

$$n_{12} = \frac{\sin \theta_1}{\sin \theta_2} = \frac{v_1}{v_2}$$

If one of the media is vacuum, where the speed of light is constant (c), we measure the *absolute* RI[1]:

$$n = c/v$$

RI is routinely used in laboratory practice as a 'fingerprint' of a substance. Usually it is measured in relation to atmospheric air, but since the latter has $n = 1.00027 \approx 1$, the measurements are close enough to absolute RI for all chemical purposes.

Gases, crystals of cubic symmetry and (with a few important exceptions) liquids and non-crystalline solids are optically isotropic: v (and hence n) is equal in all directions. Light spreading from a point source in such a medium, will have a spherical wave-surface. In all other crystals optical properties depend on the crystallographic direction. A light beam entering a crystal of medium symmetry (hexagonal, trigonal or tetragonal) splits into two, one of which ('ordinary' beam) has identical speed, v_o, in all directions while the other ('extraordinary' beam) has direction-dependent speed, v_e. The former beam produces a spherical wavesurface, the latter that of a rotation ellipsoid. The two surfaces touch each other in two points only, where the rotation axis crosses the ellipsoid (Fig. 11.1). This axis, coinciding with the main crystallographic axis, is called the optical axis. Along this direction the ordinary and extraordinary beams travel with identical speed. Such crystals are called optically mono-axial.

The optical properties of an anisotropic crystal can be conveniently described by the optical indicatrix (Fig. 11.2), which is derived from the optical surfaces by

[1] It is often stated that $n > 1$ always, as nothing can travel faster than light. This is a misconception. The speed featuring in Eq. 11.1 is the phase velocity with which the crests of the waves move. This velocity carries neither energy nor information and therefore it *can* exceed c. Thus, X-rays traveling through matter are weakly scattered forward, with a $-\pi/2$ shift. This forward-scattered wave interferes with the incident beam to create a wave with v slightly exceeding c. Thus X-rays have $n = 1 - \delta$, where δ ranges from 10^{-4} to 10^{-6} for various materials [1].

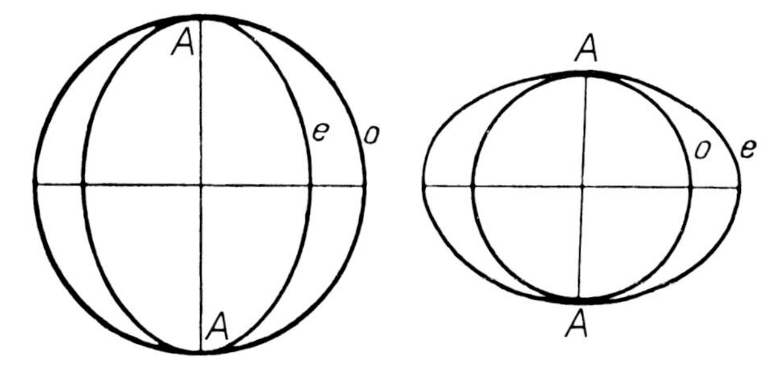

Fig. 11.1 Optical surfaces of the ordinary (*o*) and extraordinary (*e*) beams in mono-axial crystal: the optically positive (*left*, $v_o > v_e$) and the optically negative (*right*, $v_e > v_o$). AA is the optical axis

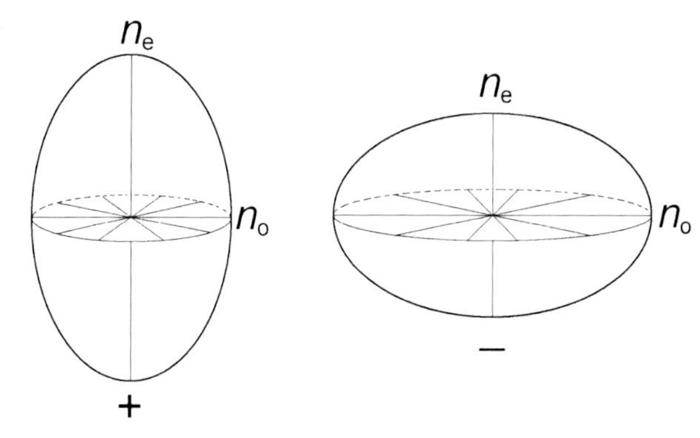

Fig. 11.2 Optical indicatrices of mono-axial crystals: + for $n_e > n_o$, − for $n_o > n_e$

replacing the radius-vector v with its inverse (as RI is proportional to $1/v$). Thus, all the properties of the optical surfaces are also reversed (compare Figs. 11.1 and 11.2). The indicatrix is also a rotation ellipsoid around the main axis of symmetry, prolate in the case of optically positive ($n_e > n_o$) and flattened in the case of optically negative ($n_o > n_e$) crystals. The difference $\Delta n = n_o - n_e$ is a measure of the optical anisotropy. In crystals of orthorhombic, monoclinic and triclinic symmetry a beam of light also splits into two components, but here *both* beams are 'extraordinary', they are polarized in mutually perpendicular planes. The RI surface is an ellipsoid of a general type with all three principal half-axes different, designated n_g, n_m, n_p (from the French *grand, moyen* and *petite*). In such ellipsoids there are two directions (optical axes) the perpendicular cross-sections to which have the form of a circle; therefore such crystals are named optically biaxial. Crystals with $n_g - n_m > n_m - n_p$ are regarded as optically positive, those with $n_g - n_m < n_m - n_p$ as negative. Note that optically mono-axial crystals can be described as biaxial with $n_o = n_m$ and $n_e = n_g/n_p$, depending on the optical sign of the crystal.

An important corollary is that measurements of RI can distinguish between three major types of crystal symmetry: high—cubic (optically isotropic crystals), medium—tetragonal, trigonal or hexagonal (anisotropic crystals with n_o and n_e), and lower—orthorhombic, monoclinic or triclinic (crystals with three indices of refraction).

The average RI of an anisotropic crystal can be calculated by converting the ellipsoid into a sphere of equal volume,

$$\bar{n} = (n_e n_o^2)^{1/3} \quad \text{or} \quad \bar{n} = (n_g n_m n_p)^{1/3} \tag{11.1}$$

Refraction depends on the wavelength, this property is known as dispersion. For uniformity, RI is commonly determined using the D line of sodium (589 nm or 2.10 eV) and referred to as n_D. In general, according to the theory of Drude,

$$n^2 = 1 + \frac{N_1 e^2}{\pi^2 m} \Sigma \frac{c_i}{\nu_i^2 - \nu^2} \tag{11.2}$$

where N_1 is the particle density, e and m are the charge and the mass of the electron, c_i is the oscillator force, ν_i is the absorption frequency of the sample, ν is the frequency of measurement. It follows that RI is the lowest at $\nu = 0$, i.e. at infinitely long wavelength (n_∞), and increases with the incident frequency towards an absorption band, where $n = \infty$ at $\nu = \nu_i$. In the $\nu < \nu_i$ range, ν decreases with increasing n (normal dispersion) but at $\nu > \nu_i$ the opposite is true (anomalous dispersion). It will be shown below, how the anomalous dispersion in crystals can be measured by spectroscopic method. If there is a single, main oscillator, Eq. 11.2 takes the form

$$n^2 = 1 + \frac{C_o}{E_o^2 - (\hbar\nu)^2} \tag{11.3}$$

where C_o is the oscillator force, E_o is the single oscillator energy, $\hbar\nu$ is the photon energy. For $\nu = 0$, Eq. 11.3 converts to a simpler form:

$$n^2 - 1 = \frac{C_o}{E_o{}^2}. \tag{11.4}$$

Wemple and DiDomenico [2–4], using the well-known Kramers–Kronig relation, replaced C_o in Eq. 11.3 by the product $E_o E_d$ where E_d is the dispersion energy, obtaining

$$n^2 - 1 = \frac{E_d E_o}{E_o^2 - (\hbar\nu)^2} \tag{11.5}$$

They have analyzed RI-dispersion data in more than a hundred different solids and liquids and established that the parameter E_d, which is a measure of the strength of inter-band optical transitions, obeys a simple empirical rule

$$E_d = \beta N_c N_e Z_a \tag{11.6}$$

where N_c is the coordination number of the cation, Z_a is the formal charge of the anion, N_e is the effective number of valence electrons per anion (usually $N_e = 8$) and β is the constant equal to 0.26 ± 0.04 eV for ionic substances and 0.37 ± 0.05 eV for covalent ones. E_d is about 1.5 times the width of the band gap (E_g), hence

$$(n^2 - 1)E_g = \frac{N_c}{a} N_e Z_a \tag{11.7}$$

where $a = 4$ for covalent substances and $= 6$ for ionic ones, i.e. the ratio N_c/a is the normalized (by the typical covalent or ionic values) coordination number. Substituting the n and E_g of vitreous As_2S_3, Se, and Te into Eq. 11.7, gives for these amorphous materials the effective coordination numbers (see Sect. 5.4) $N_c^* = 3.4$, 2.8, and 3.0, respectively [2–4] in good agreement with X-ray diffraction results.

Shannon et al. [5] analyzed RIs of 509 oxides and 55 fluorides using Eq. 11.5 and an alternative form of Sellmeier's equation,

$$\frac{1}{n^2 - 1} = A - \frac{B}{\lambda^2} \tag{11.8}$$

where λ is the wavelength, and calculated n for $\lambda = \infty$ (n_∞). They established that dispersion is controlled by the combined effects of E_o and E_d. The effects of formal valence, Z_a, were seen in low E_d and the relatively high dispersion of CuCl, Na_2SbF_5, TlCl and NiF_2. The effects of cation coordination were observed in Cu_2O, ZnO, arsenates, vanadates, iodates, molybdates. All hydrates have a relatively low E_d. However, using the Wemple and DiDomenico scheme, it is not possible to calculate E_d for most compounds because of the uncertainty in estimating N_e for s^2 and certain d^{10} compounds, nor a cation coordination when there are two or more cations with different N_c, for example YVO_4, $CaTiO_3$, $CaMoO_4$, $Y_3Fe_5O_{12}$, and for more complex compounds such as crystalline hydrates. Authors of [5] compared E_d values calculated from the properties of such anions as BO_3^{3-}, BO_4^{5-}, PO_4^{3-}, VO_4^{3-}, SO_4^{2-}, SeO_4^{2-}, MoO_4^{2-} and WO_4^{2-}, and found that the counter-ion almost always affects E_d. These uncertainties, added to those concerning N_e and N_c, make it difficult to calculate E_d values in the majority of multi-ion compounds. Several empirical dependences of n on E_g are known, viz.

$n^4 E_g = a$ [6], $n^2 = b - c \ln Eg$ [7], $n = d - eE_g$ [8], $n^2 E_g = f$ [9], and

$$n^2 = 1 + \frac{g}{E_g + h} \tag{11.9}$$

[10] where a, b, c, d, e, f, g, h are constants. All these formulae give qualitatively the same result: RI increases when E_g decreases. Because decreasing E_g amounts to increasing metallic character of bonding and in metals $n \to \infty$, this conclusion seems obvious. However, it is not always true, viz. n(ZnO) = 1.94 and n(ZnS) = 2.30, although E_g(ZnO) = 3.4 and E_g(ZnS) = 3.9 eV.

Electron energy loss spectroscopy experiments made possible to determine the dielectric function $\epsilon(\omega) = \epsilon_1(\omega) + i\epsilon_2(\omega)$ of a material (see Sect. 11.3), from which the refractive index n and the extinction coefficient k, can be derived as

$$n_o = 1 + \frac{2}{\pi} \int_0^\infty \frac{k(\omega)}{\omega} d\omega \tag{11.10}$$

The zero-frequency limit of the real component $\epsilon_1(\omega)$ is equal to n_o^2 and is related to the integrated $\epsilon_2(\omega)$ weighted by $1/\omega$

$$n_o^2 = \varepsilon_1(0) = 1 + \frac{2}{\pi} \int_0^\infty \frac{\varepsilon_2(\omega)}{\omega} d\omega \tag{11.11}$$

Table 11.1 Refractive indices n_D (at 589 nm) of polymorphous modifications with different coordination numbers N_c (for anisotropic crystals mean n_D are given)

Crystal	N_c	n_D	Crystal	N_c	n_D	Crystal	N_c	n_D
HgS	2	2.29	GeO_2	4	1.708	CsBr	6	1.582
	4	3.37		6	2.016		8	1.698
C	3	2.03	Al_2O_3	4, 6	1.696	CsI	6	1.661
	4	2.42		6	1.766		8	1.788
BN	3	1.952	Nd_2O_3	6	1.93	SrF_2	8	1.435
	4	2.117		7	2.10		9	1.482
As_2O_3	3	1.755	Er_2O_3	6	1.953	BaF_2	8	1.475
	6	1.93		7	2.022		9	1.518
Sb_2O_3	3	2.087	Y_2O_3	6	1.915	PbF_2	8	1.766
	6	2.29		7	1.97		9	1.847
MnS	4	2.43	RbCl	6	1.51	ThF_4	8	1.530
	6	2.70		8	1.80		11	1.612
SiO_2	4	1.547	CsCl	6	1.534	UF_4	8	1.576
	6	1.812		8	1.642		11	1.685

It has been shown [11] that the observed refractive indices of $TiOF_2$, TiF_4 and the seven allotropic phases of TiO_2 (columbite, rutile, brookite, anatase, ramsdellite, bronze, hollandite) can be explained not by their E_g, but rather by the total absorption power per unit of volume

$$I(\varepsilon_2) = \frac{2}{\pi} \int_0^\infty \varepsilon_2(\omega) d\omega \tag{11.12}$$

The refractive indices of MX (M = Zn, Cd) decrease steadily in the succession MO < MS < MSe < MTe, as $I(\epsilon_2)$ increases in the same order. This is due to the fact that the degree of covalent bonding in the Zn–X and Cd–X bonds increases in the succession O < S < Se < Te [12].

11.1.2 Influence of Composition, Structures and Thermodynamic Conditions on Refractive Indices

The RI, of course, depends on the atomic and electronic structures of the substance. These dependences can be highlighted by comparing different polymorphs. RI is known to increase together with N_c (see Table 11.1) [13, 14]; this provides an estimate of N_c in glasses and films where diffraction methods are of limited use. Thus, crystalline GeO_2 converts to glass under shock compression; the heterogenous product contains grains with $n = 1.608–1.610$ alongside those with $n = 1.8–2.0$, suggesting an increase of N_c of Ge to 6 [15], which was later confirmed in static-compression experiments on vitreous GeO_2 [16].

It is well known that the first layers of a crystalline substance deposited on a substrate with a different crystal structure, emulate the latter (epitaxy), i.e. undergo a phase transition. Thus were obtained thin epitaxial films of Al_2O_3 with $n = 1.632$ (corresponding to $N_c = 4$) and of CaF_2 with $n = 1.217$ (corresponding to $N_c = 6$) [17]. Such transformations it is impossible to trace by other methods.

Measurements of optical anisotropy require good-quality crystals. However, when the concentration of defects increases and/or the crystal size decreases, symmetry is increasingly blurred and the specimen can become pseudo-isotropic. This is most evident in crystals after shock-wave treatment. Thus, shocked silicates displayed both the absolutely lower *and* less anisotropic values of RI [18]. The effect was the stronger, the higher the pressure. A microscopic investigation traced it down to the shocked crystals containing strongly disordered grains and even amorphous areas; after annealing the defects the optical anisotropy was restored. Shock-treated quartz samples showed a simultaneous decrease of density, RI and anisotropy due to emergence of vitreous areas, while the surviving crystalline areas showed an expanded unit cell [19]. Later it was shown [20] that a shock compression of quartz glass up to $P = 26$ GPa increases both its density (by up to 11 %) and RI, whereas more intense shock wave reduces both. XRD of the condensed glass revealed that N_c and d(Si–O) did not change, and the density increased due to more compact packing of SiO_4 tetrahedra. A decrease of density and RI was also observed in the shocked polycrystalline LnF_3 due to formation of defects (including those of electronic type) in crystals [21].

Measurements of RI of crystals during heating allow to study the nature of chemical bonds in real time. Thus, the factor $\delta = \frac{1}{n}\frac{\partial n}{\partial t}$ is different for ionic and semiconductor crystals (Table S11.1). In the former, RI decreases only due to decreasing density (thermal expansion), whereas in semiconductors there is a simultaneous increase of polarizability due to emergence of free electrons. Considering the expansion and electronic factors together, it was possible to classify materials on the basis of their δ [22]. If a material changes its composition on heating, then RI can indicate the structural role of its various components. Thus, removing water from ordinary crystalline hydrates increases their RI as the structure is rearranged into a denser one, while dehydration of zeolites decreases RI as the density is reduced but the framework remains essentially the same.

Compression of the condensed rare gases, hydrogen and water by pressures up to 35 GPa produces a monotonic increase of RI, and fusion of these solids is not associated with a drastic change of polarizability [23]. RI of ionic and semiconductor crystals also increase with density, except for MgO and Al_2O_3 where a reduction of RI along the c axis was found after a shock compression [24], and differ in this respect only because of their different compressibility [25–27]. The greatest increase of RI in ionic crystals, from 1.28 at normal pressure to 3.20 at $P = 253$ GPa, was observed in CsH [28]. The RI of H_2 under $P \leq 130$ GPa [29] changes as $n = -0.687343 + 0.00407826P + 1.86605(0.29605 + P)^{0.0646222}$, that of BeH_2 [30] as $n = 1.474 + 0.0868P - 0.00245P^2$. The RI of CH_4 shows a sharp increase between 208 and 288 GPa, which indicates a phase transformation of the insulator → semiconductor type [31]. RI of SiH_4 changes as $n = 1.5089 + 0.00349 \times 10^{-4}P$ between 7 and 109 GPa, but as $n = 0.33955 + 0.02332P$ from 109 to 210 GPa, which also indicates an insulator → semiconductor transition at $P = 109$ GPa,

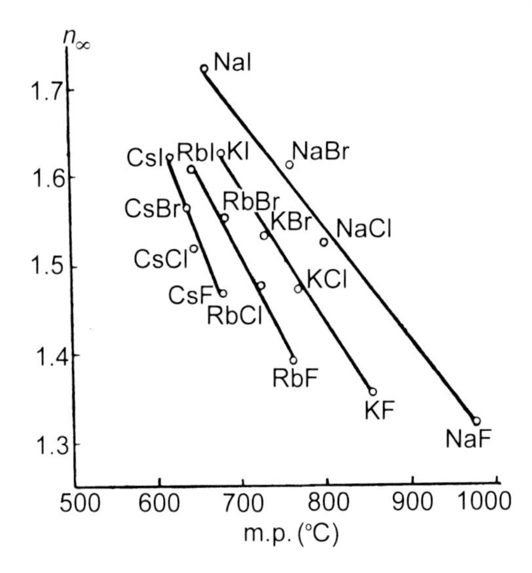

Fig. 11.3 Dependence of refractive indices (n_∞) of crystalline alkali halides on the melting temperature

accompanied by a sharp increase of the RI [32]. RI of CO_2 varies above 0.6 GPa, as $n = 1.41P^{0.041}$ [33].

RIs of elemental solids are listed in Table S11.2 and those of binary compounds in Tables S11.3–S11.7. For substances with the same structure, RI increases with the covalence and metallicity of bonding, as follows from Eq. 11.3: as on transition from ionic to covalent and then to metallic substances the absorption frequency (ν_i) decreases and approaches ν, the denominator of Eq. 11.3 approaches 0, hence $n \to \infty$. Later we shall make use of this circumstance to define atomic polarizabilities and estimate the pressures of metallization for elementary solids and compounds.

RI is related to molecular-physical properties of the substance. Thus, RIs of crystalline alkali halides decrease linearly as their melting temperatures increase (Fig. 11.3) [34], while RIs of organic liquids correlate similarly with their boiling temperatures [35].

11.2 Polarization and Dipole Moments

The refractive index is measured in the visible range of the spectrum. If such measurements are carried out in the infrared region or with still longer waves, the obtained characteristic is the dielectric constant (ε). For covalent substances, according to Maxwell's law, $\varepsilon = n^2$, for polar compounds $\varepsilon > n^2$. The difference is due to dissimilar mechanisms of polarization: in the former case only the electronic clouds of atoms/molecules are deformed (electronic polarization, P_e), in the latter whole atoms, possessing net charges, are shifted (atomic polarization, P_a). In general, according to the theory of Debye, the molecular polarization (P_m) can be represented as

$$P_m = V\frac{\varepsilon - 1}{\varepsilon + 2} = \frac{4}{3}\pi N\left(\alpha + \frac{\mu^2}{3kT}\right) \qquad (11.13)$$

where V is the molecular volume, N is Avogadro's number, α is the polarizability of the molecule and μ is the dipole moment. For non-polar substances where $\varepsilon = n^2$, Eq. 11.13 takes the form

$$P_e = V\frac{n^2 - 1}{n^2 + 2} = \frac{4}{3}\pi N\alpha \tag{11.14}$$

known as the Lorentz–Lorenz equation. From Eq. 11.13 it follows that by measuring P_m at various T, the polarizability and the dipole moment can be found separately. Dipole moments of gases can be measured also by spectroscopic methods. Information about dipole moments is of independent interest and there is a considerable literature on this topic, see reviews [36–43]. Table 11.2 lists the dipole moments of diatomic molecules and bonds (from these reviews, unless specified). Recently the dipole moments of metallo-fullerenes, $M \cdot C_{60}$, were measured, viz. $\mu = 12.4$ D for M = Li, 16.3 for Na, 21.5 for K, 20.6 for Rb, 21.5 for Cs [77].

In the first approximation, in the definition of the dipole moment

$$\mu = e^*d \tag{11.15}$$

we can identify e^* with the effective charges of the atoms, and the dipole length d with the bond distance. Hence the μ/d ratio should characterize the polarity of the chemical bond (p). This ratio was used by Fajans to estimate the bond polarities in HCl and HBr [78], and later Pauling [79] used this parameter to determine the ionicity of many chemical bonds. The polarities calculated as $p = \mu/4.8d$, with μ in Debye units (D) and d in Å, are listed together with μ in Table 11.2. As one can see, the expected dependence of bond polarities on atomic electronegativities is observed for saturated molecules, but not for those where atoms have lone electron pairs. Thus, μ of NO is small (0.16 D) and that of CO is practically nil, due to mutual compensation of charges [80]. Thus Eq. 11.15 is a rather crude simplification. Coulson [81], having analyzed the charge distribution in the C–H bond by the MO method, concluded that the dipole moment of a bond consists of several terms, resulting from the electronegativity difference (μ_i), the electron density difference (μ_ρ), hybridization of the atomic orbitals (μ_h) and the effect of non-bonding electrons (μ_e):

$$\mu = \mu_i + \mu_\rho + \mu_h + \mu_e.$$

Hence, the bond ionicity can be calculated from the measured dipole moments only if other components of μ were known theoretically or empirically. To circumvent this problem, we can compare a series of similar compounds, where terms other than μ_i are absent or compensate each other. This approach was applied by Pauling, who used the molecules of HX and alkali halides to elucidate the dependence of μ/d on the difference of electronegativities.

In a polyatomic molecule the dipole moment is the sum of all bond moments and therefore $\mu = 0$ in a symmetrical molecule (e.g. MX_4) but $\mu \neq 0$ in an asymmetric one. Thus, $\mu \leq 0.2$ D in $M(C_6H_6)_2$ complexes with M = Sc, Ti, V, Nb, Ta, while $Co(C_6H_6)_2$ and $Ni(C_6H_6)_2$ have $\mu = 0.7$ and 1.3 D, respectively, reflecting their less

Table 11.2 Dipole moments (D) and polarities of bonds in the MX molecules

M	F		Cl		Br		I	
	μ	p	μ	p	μ	p	μ	p
H	1.91	0.41	1.08	0.18	0.79	0.12	0.38	0.05
Li	6.33	0.84	7.13	0.73	7.27	0.70	7.43	0.65
Na	8.155	0.88	9.00	0.79	9.12	0.79	9.24	0.71
K	8.59	0.82	10.27	0.80	10.63	0.78	10.82	0.74
Rb	8.55	0.78	10.51	0.78	10.86	0.77	11.48	0.75
Cs	7.88	0.70	10.39	0.74	10.82	0.73	11.69	0.73
Cu	5.77[a]	0.69	5.2[a]	0.53				
Ag	6.22[a]	0.65	6.08[a]	0.55	5.62	0.49	5.10	0.42
Mg	3.2	0.38						
Ca	3.07	0.33	4.26	0.36	4.36	0.35	4.60	0.34
Sr	3.47	0.38						
Ba	3.17	0.34					5.97[d]	0.40
Al	1.53	0.19						
Ga	2.45	0.29						
In	3.40	0.36	3.79	0.33				
Tl	4.23	0.42	4.54	0.38	4.49	0.36	4.61	0.34
Sc	1.72[d]	0.20						
Y	1.82[d]	0.20	2.59[d]	0.23				
La	1.81[d]	0.18						
C	1.41	0.22	1.46	0.17	1.38	0.13	1.19	0.12
Ge	1.70[e]	0.20						
P	0.77	0.10	0.81	0.08	0.36	0.03		
F	0	0	0.884	0.11	1.36	0.16	1.95	0.21
Cl	0.884	0.11	0	0	0.61[f]	0.06	1.24[g]	0.10
Br	1.36	0.16	0.61[f]	0.06	0	0	0.73	0.05
I			1.24[g]	0.10	0.73	0.05	0	0
Co	4.51[A]	0.54						
Ru	5.34[B]	0.58						
Ir	2.82[C]	0.32						
O	Cu		Mg		Ca		Sr	
	4.45[h]	0.54	6.2[i]	0.74	4.57[l]		8.90	0.96
	Ba		Sc		Y		La	
	7.95	0.84	4.55[h]	0.57	4.52[j]	0.53	3.21[j]	0.37
	Sm		Ce		Dy		Ho	
	3.52[j]		3.12[J]		4.51[j]		4.80[j]	
	Yb		Nd		Th		U	
	5.89[j]		3.31[j]		2.78[k]	0.31	3.36	
	Ti		Zr		Hf		C	
	3.34[h]	0.38	2.55[j]	0.31	3.43[j]	0.41	0.12[K]	0.02
	Si		Ge		Sn		Pb	
	3.10[l]	0.43	3.28	0.42	4.32	0.49	4.64	0.50
			V		Nb		N	
			3.35[h]	0.44	3.50[m]	0.43	0.16	0.03
	Cr		W		Fe		Pt	
	3.88[h]	0.50	1.72	0.21	4.50[h]	0.60	2.77[n]	0.33

Table 11.2 (continued)

S	Cu	4.31[h]	0.44	Ba	10.86[s]	0.90	Sc	5.64[h]	0.44	Y	6.10[o]	0.56
	Ti	5.75[h]	0.57	Zr	3.86[r]	0.37	C	1.97[p]	0.27	Si	1.73[p]	0.19
	Ge	2.00	0.21	Sn	3.18	0.30	Pb	3.59	0.33	Pt	1.78[n]	0.18
N	Ti	3.56[d]	0.36	V	3.07[h]	0.41	Nb	3.26[q]	0.41	Cr	2.31[d]	0.31
	Mo	2.44[d]	0.31	W	3.77[c]	0.48	Re	2.45[c]	0.31			
	Ru	1.89[b]	0.25	Ir	1.66[d]	0.21	Pt	1.98[n]	0.24			
C	Mo	6.07[w]	0.75	W	3.90[x]	0.46	Fe	2.36[s]	0.31	Ni	2.98[t]	0.38
	Ir	1.60[v]	0.20	Pt	0.99[n]	0.12	Ru	1.95[u]	0.53			

[a][44], [b][45], [c][46], [d][47], [e][48], [f][49], [g][50], [A][51], [B][52], [C][53], [h][54], [i][55], [I][56], [j][57], [J][58], [k][59], [K][60, 61], [l][62], [m][63], [n][64], [o][65], [p][66, 67], [q][68], [r][69], [s][70, 71], [t][72], [u][73], [v][74], [w][75], [x][76]

symmetric structures [82]. Remarkably, gas-phase complexes formed from neutral atoms and molecules, can have very substantial dipole moments, e.g. X(C_6H_6) with X = Cl, Br, I have μ = 1.712, 1.715, 1.625 D, respectively [83]. The dipole moments of mixed rare gas dimers have been measured [84], viz. μ (Ne · Ar) = 0.003, μ (Ne · Kr) = 0.011, and μ (Ne · Xe) = 0.012 D; with the interatomic distances of 3.49, 3.63 and 3.86 Å, respectively, these correspond to negligible bond polarity.

Despite the qualitative validity of such comparisons, the precise relation between polarization and the electronic structure of a substance has been found only within the spectroscopic approach, which has permitted to determine separately the atomic (P_a) and electronic (P_e) polarization and also the effective charges of atoms by measuring the intensities of infrared (IR) absorption bands in molecules. Evidently, the ratio P_a/P_e must qualitatively characterize the polarity of a bond. However, one should keep in mind, that the spectroscopic values of the atomic charges are also approximate owing to the neglect of anharmonism and changes in the orbital overlaps during vibrations of atoms.

Very important is the application of the polarization concept to solids including the ionic compounds. Thus, using experimental values of ε and V, the molar polarizations of 129 oxides and 25 fluorides of various metals were calculated according to Eq. 11.13 [85], from which the polarizations of 61 ions (Table S11.8) were determined by the additive method. Using tabulated values of ionic polarizations, one can calculate the molar polarizations of numerous ionic crystals, and by the deviation from the normal value one can conclude about the appearance of specific features such as piezo- or ferro-electrics, deformations of cations or changes of the bond character.

Mechanism of the crystal polarization was considered in the theory of Born [86] who had suggested that in an inorganic crystal the atomic interactions are the sums of ionic attractions and electronic repulsions. Assuming that the electric field acting on each ion is equal to the applied external field, Born related ε and n with the frequency of the transverse lattice vibrations (ω_t):

$$\varepsilon = n^2 + \left(\frac{4\pi e^2}{\omega_t{}^2 \bar{m} V}\right) \quad (11.16)$$

where $\bar{m}$ is the reduced mass of the vibrating particle, other symbols are as above mentioned. Comparison of the experimentally measured and theoretical (calculated from the Born formula) values of ε showed that usually the calculated ε were less than the experimental ones. This discrepancy is caused by this fact that ions are not rigid spheres, and P_a and P_e are not additive. Actually, the electron clouds of the ions in the crystals are deformed and overlapped at the lattice vibrations, and as a result the atomic and electronic polarizations are no longer independent additive quantities.

Szigeti [87] showed that the deformations of the electron clouds of ions due to the transverse vibrations of the lattice increase the atomic polarization. Taking into account that the field strength in the dielectric is less than the external field due to the polarization of the specimen, Szigeti deduced the following equation:

$$\varepsilon = n^2 + \frac{4\pi e^2}{\omega_o^2 \bar{m} V} \frac{(n^2 + 2)^2}{9} \quad (11.17)$$

The Szigeti equation differs from the Born formula by the factor $(n^2 + 2)^2/9$, which takes into account the contribution of the electronic polarization; this term plays the greater part the higher is the RI, that is, the greater the covalency of the bond. The values of ε calculated by Eq. 11.17 are always higher than those measured experimentally. This led Szigeti to the idea of replacing the formal charge of the ion by a parameter s which characterizes the deviation of the real charge e^* from the theoretical value. It is easy to see, that in the physical sense the s parameter is the effective charge of the atom: $s = e^*$. Taking into account that $\omega_t = 2\pi c \nu_t$, where c is the velocity of light and ν_t is the transverse vibration frequency of the lattice, Szigeti found that

$$e^* = \frac{3\nu_t}{Ze(n^2 + 2)}[\pi(\varepsilon - n^2)\bar{m}V]^{1/2} \quad (11.18)$$

where Z is the valence of the atom.

The Szigeti equation was used initially to calculate the effective charges of atoms for simple cubic crystals, but later more complex crystalline compounds were studied, both isotropic and anisotropic. The development of the Szigeti method is linked with improvements of the measuring technique and with the refinement of the method of calculating the optical characteristic of anisotropic crystals. For the latter, Eq. 11.18 should be used with the RIs averaged according to Eq. 11.1 and dielectric constants according to

$$\bar{\varepsilon} = (\varepsilon_\perp{}^2 \varepsilon_\parallel)^{1/3} \quad \text{or} \quad \bar{\varepsilon} = (\varepsilon_\alpha \varepsilon_\beta \varepsilon_\gamma)^{1/3} \quad (11.19)$$

where ⊥ and || mean that measurements are carried out in the directions perpendicular or parallel, respectively, to the major axes of a crystal of the medium system of

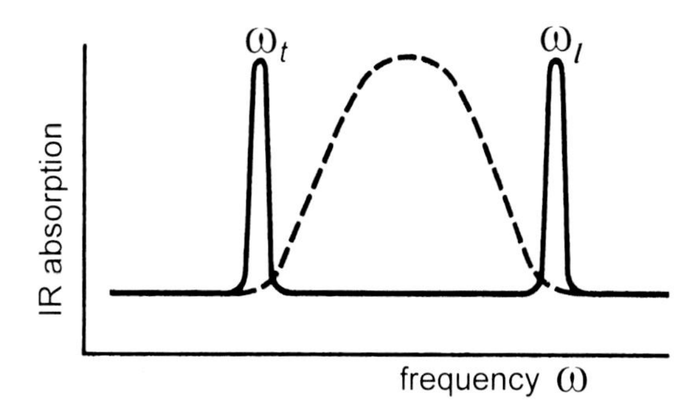

Fig. 11.4 Typical IR absorption of a single crystal (*solid line*) and powder (*dotted line*)

symmetry; α, β, γ mean the maximum, medium and minimum values of ε for the crystals of the lower system of symmetry. Frequencies of transverse vibrations, strictly speaking, can be determined only on single crystals, their determination on powder materials is very difficult since the IR-absorption band changes monotonically in all spectral ranges from ν_l to ν_t vibrations (Fig. 11.4).

There are several empirical relations which link the maximum of IR absorption with the ω_l and ω_t frequencies of the lattice vibrations; however, of practical significance is the equation of Lyddane, Sachs and Teller (LST) [88],

$$\left(\frac{\omega_l}{\omega_t}\right)^2 = \frac{\varepsilon}{n^2} \tag{11.20}$$

From this formula it follows that if $\varepsilon = n^2$, the transverse and longitudinal frequencies are the same, i.e. the width of the IR absorption band is zero. In fact, spectroscopic theory states that the intensity of IR-absorption is zero if the atomic vibrations concerned do not affect the dipole moment, but also if $\mu = 0$, i.e. in the case of purely covalent substances for which $\varepsilon = n^2$. Another approach is based on the assumption that all the valence (outer) electrons of atoms in a crystal are delocalized just as in metals and the laws of an electronic gas are applicable to them. In accordance with the Penn's theory [89]:

$$n^2 = 1 + \left(\frac{h\omega_p}{E_g}\right)^2 \tag{11.21}$$

where E_g is the band gap width and ω_p is the plasma frequency of vibrations of the valence electrons,

$$\omega_p = \left(\frac{4\pi e^2 N}{mV}\right)^{1/2} \tag{11.22}$$

where N is the number of valence electrons, m is their mass, V is the volume of the formula unit of the crystal. Phillips and van Vechten [90–94] developed on the basis of these relations the 'dielectric theory of a chemical bond'. Because for such elements as carbon (in diamond) and silicon, $\varepsilon = n^2$ and $E_g = E_h$ (the subscript '*h*' stands for a covalent, i.e. *homeopolar* crystal), Eq. 11.21 takes the form:

$$\varepsilon = 1 + \left(\frac{h\omega_p}{E_h}\right)^2 \tag{11.23}$$

Assuming Eq. 11.23 to be valid for all crystals and E_g to comprise the Coulomb, or ionic (C) and covalent (E_h) terms, Phillips derived the equation

$$E_g{}^2 = E_h{}^2 + C^2 \tag{11.24}$$

Equations 11.21 and 11.23 giving good agreement with the experiment, C can be determined from Eq. 11.24. C can be expressed as a function of the charges and radii of atoms,

$$C = a\left(\frac{Z_A}{r_A} - \frac{Z_B}{r_B}\right) f \tag{11.25}$$

where a is a constant, Z is the valence, r is the radius, and f is the Thomas-Fermi screening factor. The latter signifies that valence electrons in a semiconductor are considered to be free (as in metals) and therefore Thomas-Fermi statistics are completely applicable to them. Eq. 11.25 was used by Phillips to calculate the atomic electronegativity as

$$\chi = 4.0\left(\frac{Z}{r}\right) f + 0.5 \tag{11.26}$$

where the coefficients 4.0 and 0.5 were chosen to bring χ to the Pauling's scale (and get Pauling's values for C and N). The homeopolar component of E_g depends only on the bond length and for Group 14 elements fits the empirical equation

$$E_h = r^{-2.5} \tag{11.27}$$

Knowing E_h and C, we can determine the bond ionicity as

$$f_i = \left(\frac{C}{E_g}\right)^2 \tag{11.28}$$

The theory of Phillips and van Vechten has become an indispensable tool of solid-state chemistry and physics, and it is now difficult to list all the studies where the dielectric theory of chemical bonding is used or developed. Here we have considered only the foundation of this approach; some other aspects and applications will be described in the next section.

11.3 Molecular Refraction: Experiment and Calculation

The electronic polarization (P_e) of a molecule is not only useful to calculate the atomic component of the molar polarization and the dipole moment, but is of independent interest for studying the bonding type and structure of a substance. Among chemists the P_e determined according to Eq. 11.14 is commonly known as the 'refraction', R, and is expressed in non-SI units, cm^3/mol, whereas physicists prefer to use polarizability α, expressed in $Å^3$, so that formally $\alpha = 0.3964\ R$ (α is also expressed in atomic units, 1 a. u. = 0.148185 $Å^3$). For uniformity, P_e is measured with the 589 nm Na line (R_D) or is extrapolated to infinitely long waves (R_∞), unless the entire dispersion spectrum of RI is determined.

11.3.1 Formulae of Refraction

Besides Eq. 11.14, there are other formulae for refraction. Newton [95] introduced the concept of refraction into science at the beginning of eighteenth century and gave in his 'Optics' the first formula expressing the refraction in terms of refractive index and density (volume),

$$R_1 = V(n^2 - 1) \quad (11.29)$$

He subjected this formula to a great number of tests, some of which involved a thousand fold variation in density, and found the refraction remaining constant to within one order of magnitude, with a mean deviation of 33 %. In 1853, Beer [96] expressed the opinion that for gases the quantity

$$R_2 = V(n - 1) \quad (11.30)$$

is more independent of thermodynamic conditions than R_1. Later, Gladstone and Dale [97] concluded that the same is true for liquids. These authors showed convincingly in this and later papers that R_2 is affected little by changes in temperature and the aggregate state, by dissolution, by mixing with other liquids, or even, within certain limits, by a chemical reaction. In 1875–1880, Eq. 11.14 (see above) was derived by L. Lorenz [98] from his theory of the propagation of light, a precursor to Maxwell's theory, and from the classical electromagnetic theory by H. Lorentz [99–101]. According to Maxwell's theory, the electric field within the particle, E_1, and the external electric field (E) in the dielectric are related as

$$E_1 = E + \frac{4\pi P}{3} \quad (11.31)$$

where the polarization $P = N\alpha E_1$ (N is a number of particles and α is their polarizability). The electric field within the dielectric can be characterized by the electrostatic induction, D, which, as Maxwell has shown, is also linked with the polarization,

$$D = E + 4\pi P \quad (11.32)$$

combining Eqs 11.31 and 11.32 gives the Mossotti–Clausius equation,

$$\frac{\varepsilon^2 - 1}{\varepsilon^2 + 2} = \frac{4}{3}\pi N\alpha \qquad (11.33)$$

and, after substituting n^2 for ε (for $\lambda = \infty$), the Lorentz-Lorenz formula, Eq. 11.14. Equations 11.29, 11.30 and 11.14 mark the principal steps in the formal development of refractometry, although other formulae have been proposed (for the historical outline of this field see [13, 14]). Note that the refractions according to Newton (R_1),Gladstone and Dale (R_2) and Lorentz-Lorenz (R_3) differ only by their denominators, viz. 1, $n+1$ and n^2+2, respectively; therefore for gases (where $n \approx 1$) they relate as $R_1 \approx 2R_2 \approx 3R_3$. However, for liquids and solids the three expressions give substantially different trends; Eq. 11.14 is used here as the most accurate, both theoretically and empirically.

11.3.2 Dependence of Refractions on the Structure and Thermodynamic Parameters

The refraction of a substance decreases very slowly on heating (by a factor of $10^{-4} - 10^{-5}$ K^{-1}, practically in line with density) as long as the chemical composition and structure is unaltered, in which case the effect can be substantial. Thus, the rupture of hydrogen bonds reduces R from 13.41 cm^3/mol for the acetic acid dimer (at 120 to 190 °C) to 13.21 for the monomer (at 192 to 300 °C). Melting affects the refraction only if the coordination number is altered, as in alkali halides (Table S11.9). Refraction decreases with an increase of N_c, as can be seen in polymorphs with different coordination (Table 11.3), which makes it a useful structural-chemical tool (see below). Compression can alter R either way, depending on how the tighter intermolecular contacts affect the intramolecular electron density distribution. Thus, R_D of CO_2 decreases from 6.46 cm^3/mol at 1 GPa to 6.15 at 6 GPa [102], those of liquid H_2S and NH_3 decrease from 9.63 and 5.57 cm^3/mol at ambient pressure to 9.53 and 5.30 cm^3/mol at 3 GPa. However, the R_D of liquid HCl first increases from 6.59 cm^3/mol at ambient pressure to 7.04 at 1.5 GPa and then decreases to 6.96 at 3 GPa and 6.78 at 4.5 GPa [103]. To highlight the structural effects on refraction, Müller [104] introduced the factor

$$\Lambda_o = \frac{\Delta R}{R} : \frac{\Delta V}{V} \qquad (11.34)$$

$\Lambda_o = 1$ if the refraction varies only due to volume change but $\neq 1$ if the bonding also changes. As shown in Table 11.3, Λ_o is a sensitive indicator of structural change: an increase of N_c in a phase transition decreases the volume more than the refraction. The overestimated Λ_o of the diamond-graphite pair is an exception, because of aromatic nature of bonding in graphite layers. Polymorphs with the same N_c have refractions identical within *ca.* 2 % (Table S11.10).

Table 11.3 Change of molar refractions R_D and Müller's factors (Λ_o) with coordination numbers

Composition	Polymorph type	N_c	R_D (cm^3/mol)	Λ_o
C	Diamond	4	2.11	0.79
	Graphite	3	2.70	
BN	Diamond	4	3.83	1.08
	Graphite	3	5.27	
MnS	B1	6	14.5	0.68
	B3	4	16.3	
SiO_2	Rutile	6	6.01	0.51
	Quartz	4	9.53	
GeO_2	Tetragonal	6	8.47	0.40
	Hexagonal	4	9.53	
Al_2O_3	α-phase	6	10.6	0.55
	γ-phase	4, 6	11.3	
Nd_2O_3	Hexagonal	7	24.2	0.34
	Cubic	6	25.4	
Y_2O_3	Hexagonal	7	20.2	0.60
	Cubic	6	20.1	
CsCl	B2	8	15.20	0.33
	B1	6	16.15	
CsBr	B2	8	18.44	0.14
	B1	6	18.84	
CsI	B2	8	24.19	0.22
	B1	6	25.01	
SrF_2	Cottunite	9	7.63	0.20
	Fluorite	8	7.78	
BaF_2	Cottunite	9	9.78	0.32
	Fluorite	8	10.09	
PbF_2	Cottunite	9	12.94	0.13
	Fluorite	8	13.08	

Table 11.4 Müller's factors for solids. (From [13, 14, 22, 105, 106] and the authors' unpublished measurements)

Substance	Λ_o	Substance	Λ_o	Substance	Λ_o	Substance	Λ_o
CeO_2	6.92	TiO_2 (n_o)	0.92	$CaCO_3$	0.50	CO_2	0.22
SiC	1.64	MgF_2	0.73	$CaMoO_4$	0.46	HCl	0.18
C (diamond)	1.60	RbCl	0.67	$NH_4H_2PO_4$	0.41	SiO_2 glass	0.17
Ge	1.37	LiF	0.62	Al_2O_3	0.40	CsH	0.16
GaAs	1.33	AgCl	0.57	BaF_2	0.39	H_2 [a]	0.16
ZnO	1.31	CaF_2	0.56	NaCl	0.38	He [a]	0.07
CdS	1.25	PbF_2	0.54	KH_2PO_4	0.33	Ne [a]	0.08
TiO_2 (n_e)	1.21	$CaWO_4$	0.52	KBr	0.24	Ar [b]	0.12
Si	1.17	Nd_2O_3	0.51	KCl	0.23	Kr [b]	0.10
ZnS	1.10					Xe [b]	0.07

[a] [23], [b] [107]

It is noteworthy that the Müller's factor of a given phase usually remains fairly constant under different thermodynamic conditions, whereas Λ_0 of different substances vary widely, as shown in Table 11.4. For some solids (listed in the left column) the

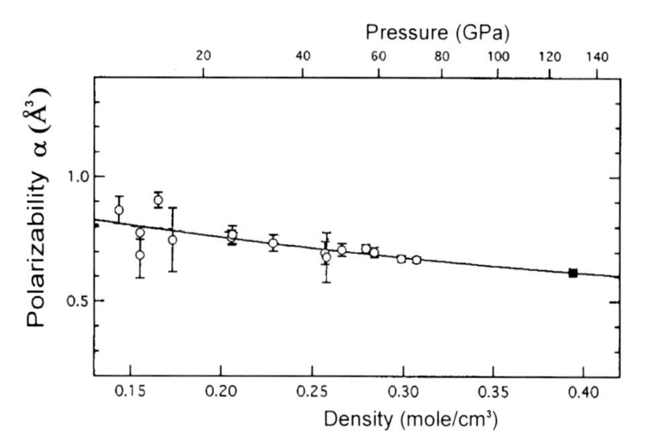

Fig. 11.5 Experimental density/polarizability plot for solid hydrogen ($T = 80$ to 90 K), from polarizability data (○) and H_2/diamond RI matching (■). Near-linearity of the curve indicates the constancy of the Λ_o factor. (Adapted from [29] with permission of the American Physical Society (Copyright 1998) and Lawrence Livermore national Laboratory)

refractions decrease under pressure faster than volumes, more often it is the other way round. Rutile (TiO_2) belongs to both groups: it has $\Lambda_o > 1$ for the extraordinary and $\Lambda_o < 1$ for the ordinary beam, reflecting different changes of the bond character along and perpendicular to the major crystallographic axis. The lowest Λ_o were observed for the solid rare gases and hydrogen under compression (Fig. 11.5), where spherical atoms are bound together by weak van der Waals interactions with very small electron density accumulation between the atoms, hence the polarization is little affected by compression.

11.3.3 Atomic and Covalent Refractions

The application of refractions to the study of structures is based on comparing the experimental values with those calculated on various structural assumptions, of which the most important is *additivity* (Landolt, 1862): in the first approximation (within *ca* 10 %), the refraction of a compound is the sum of constant increments of different atoms, ions and bonds. Refractions of some isolated atoms can be measured by the deviation of an atomic beam in an inhomogeneous electric field or by spectroscopic methods. In other cases electronic polarizabilities of free atoms were calculated by *ab initio* methods. All available experimental and the best of the computed refractions of free atoms are presented in Table 11.5. These values can be used to calculate the energy of van der Waals interactions, magnetic susceptibility, or to establish correlations with atomic and molecular-physical properties. The formation of covalent bonds changes the refractions of isolated atoms and their values transform into the covalent refractions, which are different for isolated molecules and for crystals. Direct measurements of RI of A_2 molecules or elemental solids give the most accurate information on the covalent refractions, in other cases the latter have to be calculated from molecular refractions by the additive method.

Experimental refractions of elements of Groups 14–17 are close to the additive values of atomic increments derived by Eisenlohr [131, 132], Vogel et al. [133, 134]

Table 11.5 Refractions (cm^3) of atoms in free state (*upper lines*), in diatomic molecules (*middle lines*), in elemental solids (*lower lines*); R(H) = 1.68 in the free state, 1.02 in H_2, R(He) = 0.52 [108]

Li	Be	B	C	N	O	F	Ne		
59.1[a]	14.0	7.6	4.22[m]	3.05[b]	1.97[b]	1.40	1.00		
41.4[c]	10.8	4.3	2.07	2.20	1.99	1.45			
13.0	4.9	3.5	2.07						
Na	Mg	Al	Si	P	S	Cl	Ar		
60.2[d]	27.5[e]	21.9[F]	13.9[m]	9.2	7.3	5.5	4.14		
49.9[c]	19.3	11.5	9.05	8.57[o]	7.7	5.69			
23.6	14.0	10.0	9.05	8.75	7.7				
K	Ca	Sc	Ti	V	Cr	Mn	Fe	Co	Ni
110[a]	61.6[e]	44.9	36.8	31.3	29.3	23.7	21.2	18.9	17.1
93.1[g]	46.2	32.5	27.3	21.1	20.9	19.5	14.0	13.3	12.5
45.6	26.3	15.0	10.6	8.4	7.2	7.3	7.1	6.7	6.6
Cu	Zn	Ga	Ge	As	Se	Br	Kr		
28.5[F]	14.5[f]	20.5	14.7[m]	10.9	9.5	7.7	6.27		
8.1	12.9	17.2	11.3	10.9[p]	10.8	8.17			
7.1	9.2	11.7	11.3	10.3	10.6	8.75			
Rb	Sr	Y	Zr	Nb	Mo	Tc	Ru	Rh	Pd
121[a]	72.2[e]	57.2	45.1	39.6	32.3	28.8	24.2	21.7	20.1
99.6[j]	55.0	38.0	27.6	23.8	21.4	19.0	13.3	14.0	14.7
55.9	33.9	20.0	14.0	10.9	9.4	8.6	8.2	8.2	8.8
Ag	Cd	In	Sn	Sb	Te	I	Xe		
28.7[q]	8.6[h]	25.7	15.8[m]	16.6	13.9	12.5[i]	10.20		
11.9	14.5	1.6	16.3	17.7	14.4	13.0[k]			
10.3	12.9	15.8	16.3	17.7	15.4				
Cs	Ba	La	Hf	Ta	W	Re	Os	Ir	Pt
150[β]	95.6[e]	78.4	40.9	33.0	28.0	24.5	21.4	19.2	16.4
131[j]	71.8	55.3	26.7	16.7	15.6	15.5	11.2	11.0	10.0
69.7	37.9	22.5	13.5	10.8	9.6	8.8	8.4	8.5	9.1
Au	Hg	Tl	Pb	Bi	Th	U	Rn		
20.9[F]	12.7[l]	19.2	17.6[m]	18.7	81.0	51.2[n]	13.4		
11.0	12.8	16.7	18.4	21.3	57.3	38.2			
10.2	13.9	17.1	18.3	21.3	19.8	12.5			

[a] [109, 110, 111–113], [β] [114], [b] [115], [c] [109, 116, 117], [d] [113, 118], [e] [111, 112, 118, 119], [F] [120], [f] [121], [g] [109, 110, 113], [h] [122], [i] [123], [j] [110], [k] [124], [l] [125], [m] [126], [n] [127], [o] [128], [p] [129], [q] [130]

from the molecular refractions of organic compounds. These works resulted in the system of 'organic' covalent refractions (Table S11.11), where increments per double and triple bonds of carbon correspond to an increase of R(C) with the decrease in N_c (Table 11.3) but different refractions of the same atom in different functional groups reflect varying effects of chemical bonding. Covalent refractions of several metals were determined by the additive method from the measurements on organometallic substances [13, 14, 135–139]. The deviations of molecular refractions from the sums of atomic increments show good correlation (up to 0.988 in some cases) [140] with the dissociation enthalpy of the AB bond,

$$\Delta H_{AB} = a + b[R(AB) - R(A) - R(B)] \tag{11.35}$$

where a and b are constants. Alternatively, since the dimensionality of pressure is the

Table 11.6 Atomic refractions R (cm^3) in alkali metal clusters [109, 110, 146]

M	M_1	M_2	M_8	M_∞ (bulk)
Li	59.1	41.4	26.2	13.0
Na	60.2	49.9	41.4	23.6
K	110	93.1	68.1	45.6

energy (Z^*/r) per unit volume (r^3) and refraction is broadly proportional to volume, then for a homo-atomic A_2 molecule

$$\frac{Z^*}{r^4} = k\frac{E}{2R(A) - R(A_2)} \tag{11.36}$$

where E is the dissociation energy of the A–A bond, and k is an empirical constant [141]. The covalent refractions of some metals were calculated by Eq. 11.36 (Table 11.5, middle lines) and agree within 25 % with the additive [13, 14] values. Equation 11.36 was later applied to describe the 'atomic compressibility' [142, 143]. Refraction also depends on the hybridization state of atoms [144, 145],

$$\alpha_A = \frac{4}{N}\left(\sum_A h_A\right)^2 \tag{11.37}$$

where N is the number of electrons in a molecule, h is the hybridization-dependent atomic increment. Equation 11.37 gives $\alpha = 1.574$, 1.673 and 1.283 Å^3 for sp^3, sp^2 and sp hybridized carbon atoms, respectively, and describes well the refractions of organic molecules containing C, H, N, O, S, P and halogen atoms.

In inorganic crystals atoms have higher N_c and, as follows from Table 11.3 and the polarizabilities of clusters (Table 11.6), their refractions must be lower than in the molecular state. Because the Lorentz–Lorenz function approaches 1 when $n \to \infty$, and metals have very high RIs (at $\lambda = 10$ μm, Cu has $n = 29.7$, Ag 9.9, Au 8.2, Hg 14.0, V 12.8, Nb 16.0, Cr 21.2, etc [147]), we assumed that $R = V$ for solid metals [148]. These refractions of metals, R_M (Table 11.5, lower lines) in some cases are close to the additive values [13, 14]. R_M cannot be applied directly to calculate molar refractions of crystalline inorganic compounds because of the differences in N_c, but can be used [149] to calculate refractions of metals for such N_c as they have in the structures of their compounds, using the formula

$$R = \frac{V_o}{\rho} \tag{11.38}$$

where V_o is the inner volume of an atom ($4\pi r^3/3$) and ρ is the packing density. Combining all the constant terms, we obtain $R = ar^3$, where $a = 7.419, 4.472, 3.710$, respectively, for $N_c = 4, 6, 8$. Using the crystalline covalent radii for polyvalent metals in low oxidation states, the crystalline covalent refractions of metals can be calculated for a given N_c by Eq. 11.38 [150], the results are listed in Table 11.7. R(Sn) for $N_c = 4$ and 6 was equaled to volumes for both polymorphs; the same relation was used for Si and Ge. The obtained atomic refractions were corrected according to effective nuclear charges of elements; the corrections did not exceed several percent.

Table 11.7 Covalent atomic refractions (cm^3) in crystalline solids (N_c are specified in brackets, unless constant for a given element under normal conditions, e.g. 4 for diamond, 8 for Cs and 6 for other alkali metals, etc.)

Li 15	Be 7	B 4	C 2.4	N 2.1	O 2.0	F 1.5			
Na 27	Mg 15	Al [6] 11.5 [4] 18	Si [6] 7 [4] 9	P 8.3	S 7.6	Cl 5.4			
K 46	Ca 27	Sc 17	Ti 13	V 13	Cr 13.5	Mn 13	Fe 13	Co 12.5	Ni 12
Cu 16	Zn 15	Ga [6] 10.5 [4] 16	Ge [6] 8.5 [4] 11	As 10.3	Se 8.5	Br 7.8			
Rb 55	Sr 35	Y 23	Zr 19	Nb 14	Mo [6] 14 [4] 16.5	Tc 12.5	Ru 23	Rh 23	Pd 25
Ag [6] 17 [4] 22	Cd [6] 16 [4] 21	In [6] 16 [4] 24	Sn [6] 16 [4] 20	Sb [6] 16 [4] 21	Te 14	I 13			
Cs 68	Ba 44	La 28	Hf 16	Ta 15	W [6] 15 [4] 17	Re 11.5	Os 24	Ir 24	Pt 24
Au 21	Hg 21	Tl 25	Pb [6] 19 [4] 23	Bi [6] 18 [4] 23	Th 25	U 25			

Table 11.8 Spectroscopic atomic and ionic refractions (cm^3/mol)

Atom	R	Atom	R	Atom	R	Atom	R
Mg^0	27.8	Zn^0	14.0	Cd^0	14.84	Hg^0	8.82
Al^+	9.05	Ga^+	6.78	In^+	7.03	Tl^+	4.75
Si^{2+}	4.36	Ge^{2+}	3.99	Sn^{2+}	4.45	Pb^{2+}	2.92
P^{3+}	2.36	As^{3+}	2.31	Sb^{3+}	2.69	Bi^{3+}	2.13
S^{4+}	1.68	Se^{4+}	1.72	Te^{4+}	2.05	Po^{4+}	1.61
Cl^{5+}	1.03	Br^{5+}	1.26	I^{4+}	1.51	At^{5+}	1.24
Ar^{6+}	0.75	Kr^{6+}	1.01	Xe^{6+}	1.23	Rn^{6+}	0.99

11.3.4 Ionic Refractions

Like the covalent radii, the covalent refractions are strictly applicable only to covalent compounds. For polar inorganic substances, the ionic refractions are more suited. However, refractions of free ions are not available from direct experimental measurements. Several methods exist for deriving these from spectroscopic data. Thus, it has been shown [151] that electrons bound to positive atomic ions in non-penetrating high-L Rydberg states, are uniquely sensitive to the long-range interactions of those ions. If only the fully-screened Coulomb interaction were present, all high-L Rydberg levels would be degenerate except for small relativistic effects. However, the presence of permanent and induced electric moments in the core ion results in systematic fine structure patterns which lift the Coulomb degeneracy between states of different L. Although these patterns are not large in size, when they are measured carefully they can give reliable and precise determinations of ionic refractions. So far, the following values of refractions have been reported (cm^3/mol): Si^{2+} 4.36, Si^{3+} 2.77, and Ne^+ 0.487 [151], Ba^+ 46.3 [152], Pb^{2+} 5.09, Pb^{4+} 1.35 [153], Th^{4+} 2.885 [154]. Another spectroscopic approach uses the equation

$$R = \sum_{n'} \frac{C}{E^2} \tag{11.39}$$

where C and E are the oscillator strength and the energy of excitations from the ground state to accessible bound and continuum excited states, respectively (compare with Eq. 11.4); the results for the Mg-, Zn-, Cd- and Hg-series [155] are listed in Table 11.8. Comparing Tables 11.5 and 11.8, one can see that refractions of atoms very strongly depend on their charges.

The problem in establishing an empirical system of ionic refractions is the same as with ionic radii: the observations give only the sum (the refraction of the compound, or the internuclear distance, respectively) but no definite rule how to apportion this between different elements, unless the increment of at least one ion is known from some external source. This difficulty is compounded in both cases by the additivity itself being imperfect. In 1922, Wasastjerna [156] based the first system of ionic refractions on the assumption that $R(H^+) = 0$, as this ion has no electrons and hence zero electronic polarizability. Then the refraction of, e.g. Na^+, can be found simply as the difference between the refractions of a strong acid and its sodium salt. The weakness of this approach is that proton does not exist in solution by itself but is usually solvated, e.g. in the form of a hydronium ion H_3O^+ in aqueous solutions. More correct method of determining ionic refractions was suggested by Fajans [157–162], based in the fact that the larger the positive charge of an iso-electronic ion, the more resistant its electron shells are to deformation. Thus the ionic refractions must satisfy the conditions: $R^{2-}/R^- > R^-/R^0 > R^0/R^+ > R^+/R^{2+}$ where R^+ and R^- are the refractions of the cation and anion and R^0 is the *experimental* refraction of an iso-electronic rare-gas atom. In this way Fajans found $R(Na^+) = 0.47$ cm^3 and used it as the key to calculate the refractions of other ions by additivity. Later, Markus et al. [163] suggested a new system of ionic refractions for diluted solutions, based on $R(Na^+) = 0.65$ cm^3.

An independent semi-empirical method of determining ionic refractions was suggested by Pauling [164], using the relation between the polarizability and the second-order Stark effect,

$$R = const \frac{(n^*)^6}{(Z^*)^4} \tag{11.40}$$

where n^* is the effective total quantum number and Z^* is the effective nuclear charge. Using the atomic refractions of rare gases and molar refractions of salt solutions,

Pauling determined the constant of Eq. 11.40 and calculated ionic refractions. The systems of Fajans and Pauling are similar, being based on the same experimental data, namely, the refractions of rare gases.

Other authors have calculated refractions of a number of ions by starting from the principle of additivity and adopting the Pauling's value of $R(Li^+)$ since this value is so small that even large relative errors would not affect the results too much. However, because of the strong polarizing effect of Li^+ such calculations can lead to unreal results, as were obtained in some cases by Born and Heisenberg [165], or by Tessman et al. [166]. Generally speaking, the ionic refractions are diminished in the field of a positive charge and are enhanced in that of a negative charge. This effect was taken into account by Jörgensen [167, 168], who pointed out that the approach of Fajans, Markus and Pauling overestimated the values of refractions for cations and understated those of anions. Instead, he proposed to depart from constant magnitudes of ionic refractions and to compute specific values for different groups of compounds and aggregate states. Later similar results were obtained by Iwadate et al. [169–171] in refractometric studies of the melted halides and nitrates of mono-, di- and trivalent cations.

These empirical results stimulated extensive theoretical studies of the effect of the structure on the refraction values. Thus, Magan [172] calculated the refractions of ions in NaCl-type crystals, showing that electronic polarizabilities of cations change with varying composition less than those of anions. Therefore the system of ionic refractions for the crystalline state can be based on assuming the refractions of cations as constant. Other researchers [173–178] theoretically computed refractions of some cations, while the anionic refractions were obtained from the experimental molar refraction by the additive method. They confirmed that the Madelung field increases the cation refraction and decreases that of the anion. Ionic refractions were also calculated for the solutions and crystalline states (for different structural types), revealing the dependence of R on N_c. Both theoretical and empirical results are summarized in recommend system of the crystalline ionic refractions (Table S11.12) which allow to calculate the molar refraction of solid inorganic compounds with good accuracy.

A set of the ionic electronic polarizabilities based on the data for 387 oxides, hydroxides, oxyfluorides, and oxychlorides [179] also showed that free anions have considerably higher polarizabilities than in crystals and free cations have somewhat lower polarizabilities than in crystals. Most of the cation polarizabilities found in this study are certainly much greater than *ab initio* free-ion polarizabilities. These larger cation polarizabilities allow an excellent fit between calculated and observed total polarizabilities for most oxides, hydrates, hydroxides, oxyfluorides, and oxychlorides. Systematic comparisons of the deviations of Born's effective charges from the formal charges (ΔZ^*) with deviations of certain ions in α–r^3 plots, as well as comparisons of ΔZ^* with differences between empirical and free-ion α indicate good correlations of refractions with hybridization and covalence of bonds. Assuming that these correlations represent the effects of covalence and charge transfer, the differences between the empirical polarizabilities and the free-ion values can be attributed to charge transfer, which effectively increases the cation polarizabilities and decreases

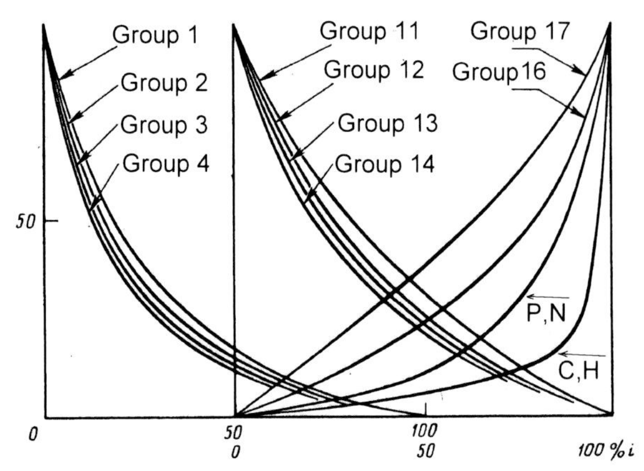

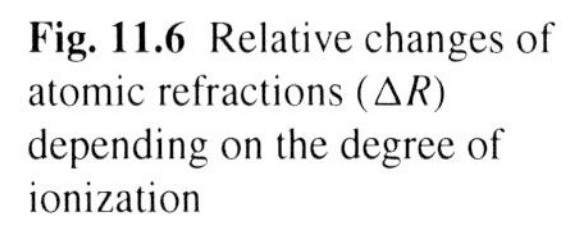
Fig. 11.6 Relative changes of atomic refractions (ΔR) depending on the degree of ionization

those of anions. Jemmer et al. [180] obtained a dependence of the anion refractions on the cell parameters that was compared with refractive indices of substances under pressure, by representing the Müller's factor in the form

$$\Lambda_o = 1 - \left(\frac{6n}{(n^2+2)(n^2-1)}\right)\left(\rho\frac{\partial n}{\partial \rho}\right) \tag{11.41}$$

where Λ_o can be determined from the $n = f(P)$ dependence. Besides crystals and solutions, refractometric studies were carried out for vitreous solids, important for glass industry, since RI and their dispersion are controlled by ionic refraction of the components. It has been shown [181] that in vitreous state the average refraction of the oxygen ion increases with RI, and the dependence of this parameter on the composition of the glass and the additive nature of the Müller's factor have been established [182].

The comparison of the covalent and ionic refractions of the same elements shows that the addition or removal of the valence electrons influences the atomic refraction most. The changes of atomic refractions due to bond ionicity in molecules and crystals have been calculated as

$$\Delta\bar{R}_+ = \frac{R_{\text{cov}} - R_+}{R_{\text{cov}} - R_{\text{c}}}, \quad \Delta\bar{R}_- = \frac{R_- - R_{\text{cov}}}{R_a - R_{\text{cov}}}, \tag{11.42}$$

where R_+ and R_- are the refractions of atoms with the positive or negative effective charges, R_{cov} is the covalent refraction, R_{c} and R_{a} are the cation and anion refractions, respectively [11]. The results are presented in Fig. 11.6 and Table S11.13.

The accuracy of calculated molar refractions of inorganic compounds has been much improved by taking into account the polarizing effect of atoms *(g)* and bond metallicity *(m)* [150]. The former was described proceeding from the van der Waals interactions, $g = [(R_A - R_B)/R_A]^2$ where $R_A < R_B$ (see Eq. 4.24). If the smaller ion (A) is the cation, this will lower the refraction, otherwise will increase it. The bond metallicity *m*, i.e. the delocalized fraction of covalent electrons of A–A bonds in AB

Table 11.9 Ionic refractions (cm^3/mol); anions: the upper lines R_∞, the low lines R_D

+1		+2				+3		+4		+5		−1	
Li	0.07	Be	0.02	Cr	1.8	B	0.01	Si	0.10	P	0.06	F	2.5
Na	0.45	Mg	0.25	Mn	1.6	Al	0.15	Ge	0.5	As	0.3		2.5
K	2.2	Ca	1.5	Fe	1.5	Ga	0.6	Sn	1.4	Sb	1.2	Cl	8.0
Rb	3.5	Sr	2.5	Co	1.4	In	1.8	Pb	1.8	Bi	1.5		8.5
Cs	6.2	Ba	4.6	Ni	1.3	Tl	2.0	Ti	1.0	V	0.8	Br	11
Cu	1.5	Zn	0.9	Ru	2.6	Sc	1.2	Zr	1.7	Nb	1.2		11.8
Ag	3.5	Cd	2.5	Rh	2.4	Y	2.0	Hf	1.6	Ta	1.1	I	17
Au	4.5	Hg	3.0	Pd	2.9	La	3.5	Cr	0.9	+6			18
Tl	10.0	Cu	1.1	Os	2.6	V	1.7	Mo	1.4	S	0.04	−2	
		Sn	8.0	Ir	2.5	As	4.5	W	1.4	Se	0.25	O	7.5
		Pb	9.0	Pt	2.4	Sb	6.5	Te	5.0	Te	0.8		8.0
						Bi	7.5	Mn	0.9	Cr	0.6	S	17
						Cr	1.6	Pt	1.4	Mo	1.2		18
						Mn	1.5	Th	4.5	W	1.2	Se	21
						Fe	1.4	U	4.0				22.5
						Co	1.3					Te	29
						Ni	1.2						31
						U	4.5						

compounds, was calculated as

$$m = c\frac{\chi_A}{\chi_A + \chi_B} \tag{11.43}$$

where c is the bond covalency. Taking into account these factors, the accuracy of calculations can be improved. Given the refractions of atoms in polar bonds (Tables 11.5, 11.7 and S11.12), the refractions of molecules and crystals can be calculated with the accuracy of a few per cent, i.e. much better than by consistent application of atomic or ionic refractions [13, 14]. The recommended values of refractions for free ions are given in Table 11.9.

11.3.5 Bond Refractions

An alternative to improving atomic/ionic refractions was to express molecular refraction through bond increments. A system of bond refractions is definitely superior to the system of atomic refractions, as it allows to account for chemical interactions explicitly. The concept of 'bond refraction' was introduced by Bachinskii [183] who suggested that the molar refraction (as well as volumes, heats of combustion, *etc*) of organic compounds can be calculated of bond increments. According to Bachinskii, $R_{C-C} = 1/4R_C + 1/4R_C$, $R_{C-H} = 1/4R_C + R_H$, etc. This method is not quite consistent, as it uses atomic increments and so does not eliminate the intrinsic defects of the atomic approach. More efficient methods of calculating bond refractions directly from molar refractions, e.g. as $R(A{-}X) = R(AX_n)/n$, were proposed by Steiger [184], Smyth [185], and especially Denbigh [186, 187]. The Denbigh method has been used to evaluate a great number of bond refractions in organic and organometallic compounds; the most exact values obtained by Vogel et al. [188, 189], with later additions [13, 14, 190], are listed in Table S11.14. These bond refractions can be applied only to molecules having similar structures and bond character. This is immediately evident from the fact that the difference in the refractions of isomers is considerably greater than the experimental error, so that the evaluation of these through constant bond refractions is always inaccurate. The solution of this problem (in the case of hydrocarbons) was found by Huggins [191, 192] who took into account that the refraction of a given bond must vary depending on the adjacent atoms. Palit and Somayulu proposed to calculate molar refractions using bond refractions and bond polarities [193]:

$$R_{\mathrm{AB}} = \frac{1}{2}(R_{\mathrm{AA}} + R_{\mathrm{BB}}) - c\Delta\chi_{\mathrm{AB}} \qquad (11.44)$$

where $c = 0.37$ for single bonds.

The anisotropy of bond refractions, which can be determined by measuring the Kerr constant, is very useful for studying the nature of the chemical bond [194, 195]. It shows that the valence electron cloud is ellipsoidal (with the longest axis in the internuclear direction) and its parameters depend on the polarity and the bond order. The experimental anisotropic bond refractions in molecules A_2, AX, AX_n, A_2X_n, and $M(C_5H_5)_2$ (Table S11.15) which conform to the formula:

$$\bar{R}_{bond} = \frac{1}{3}(R_{\parallel} + 2R_{\perp}) \qquad (11.45)$$

where $R_{\parallel}$ and $R_{\perp}$ are the components of refractions along and perpendicular to the bond direction, $\overline{R}_{bond}$ is the averaged bond refraction; $\gamma = (R_{\parallel} \colon R_{\perp})^{1/3}$ is the linear factor of the optical anisotropy. In the majority of heteronuclear molecules γ is small (average 1.2), i.e. the ellipsoid is nearly spherical and only for essentially covalent bonds it becomes really anisotropic. In A–H and A–F bonds the anisotropy is very small and the valence cloud is quasi-spherical. Recently the anisotropy of the polarizability ($\Delta\alpha = \alpha_{\parallel} - \alpha_{\perp}$) was determined in molecules of rare gases Ar_2, Kr_2, Xe_2, as 0.5, 0.7 and 1.3 Å^3, respectively [196].

Yakshin [197] developed the concept of bond refractions and for the first time made it applicable to coordination compounds. For the latter, bond refractions are not additive because each bond is strongly affected by its *trans* counterpart. Instead, Yakshin proposed to use 'coordinate refractions' referring to various combinations of *trans*-ligands, i.e. ligands lying on the same Werner coordinate. Thus, R(X–M–X) can be calculated as $R(MX_4)/2$ or $R(MX_6)/3$. The refractions along (mutually) perpendicular coordinates proved additive with high accuracy, indicating that *cis*-influence between ligands is negligible. This approach allowed to give quantitative characteristic of the mutual influence of atoms in coordination compounds.

The formation heats of the organic compounds correlate with differences between experimental molar refractions and sums of the corresponding atomic refractions (ΔR) [198]. Hohm [199] obtained a similar relationship between the atomization energy of inorganic compounds and $\Delta R^{1/3}$. However, this correlation can be used only for compounds of the same structures, since polymorphs with different coordination numbers have similar atomization energy but very different molar refractions (see Table 11.3).

11.4 Structural Application of Refractometry

The applications of refractometry for determining the structural formulae in organic chemistry, geometrical isomers in coordination compounds, chemical structures of silicates *etc.* are now only of historical interest, but in studying hydrogen bonding and the mutual influence of atoms it is still relevant. As was mentioned in Sect. 4.5, as a result of the formation of a hydrogen bond (H-bond) in the X–H$\cdots$Y system the H$\cdots$Y separation decreases, but the X–H one increases. However, as have been shown above, the other crystal chemical parameters also affect these distances and separating different effects is not easy. Therefore, finding new methods of exploring H-bonds is an important task. Useful information can be obtained from the comparison of molecular refractions of substances with and without H-bonds. Thus, the difference $\Delta R = R(NH_4X) - R(KX)$ is 1.5 cm^3/mol in the absence of H-bonds, but increases by 0.2–1.0 cm^3 if the anion X^- is a good hydrogen bond acceptor [10]; a similar comparison of $MX_n \cdot mH_2O$ and MX_n gives $R(H_2O) = 3.5$ to 3.9 cm^3/mol for a good and 3.40 cm^3/mol for a poor H-bond acceptor X. Surprisingly, comparison of $K_2M(CN)_4 \cdot 3H_2O$ and its dehydration product gave still lower $R(H_2O)$, viz. 2.0 cm^3/mol for M = Pt and 2.4 cm^3/mol for M = Pd. This turned out to be a spurious effect, due to formation of M–NC bonds on dehydration; taking these into account gave sensible $R(H_2O)$ values of 3.5 and 3.9 cm^3/mol, respectively. The rearrangement, guessed from refractometric data, was later confirmed by NMR- and IR-spectroscopy.

The refraction of the liquid water is 3.66 cm^3/mol, indicative of H-bonding. Under shock compression the RI of water increases to 1.60 at $P = 22$ GPa, cf. 1.33 at ambient conditions. This increase, combined as it is with a steep increase of electric conductivity, was once explained by metallization [200]. In fact, it is due simply to higher density of the compressed water: the $R(H_2O) = 3.2$ cm^3/mol is actually *lower* than the ambient and indicates the rupture of H-bonds [201], this results in higher mobility of protons and hence higher conductivity. After the refractometric refutation of the 'metallic water', this mechanism of conductivity was confirmed directly [202].

Table 11.10 presents the major results of refractometric studies of H-bonds in inorganic compounds, which reveal that the refractions of these bonds change in the succession: acid > acid salts > ammonium salts > crystallo-hydrates. This sequence is caused by variation of the effective charges of atoms: an accumulation of hydrogen atoms in the outer sphere of a complex ion in the acid salts and acids increases the polarity of the oxygen or nitrogen atoms and thereby raises the strength of the H-bonds with the latter. Effective charges of the hydrogen atoms in H_2O and NH_4^+ are similar and therefore the bonds of the O–H$\cdots$X and N–H$\cdots$X types with the same anions have similar refractions.

Table 11.10 Refractions (cm^3/mol) of hydrogen bonds

Acids	$R_{XH\cdots X}$	*Acid salts*	$R_{H\cdots X}$
HF	0.38	KHF_2	0.43
HNO_3	0.56	$KHCO_3$	0.35
H_2SO_4	0.70	$KHSO_4$	0.41
H_3PO_4	0.75	K_2HPO_4	0.53
$H_3Fe(CN)_6$	0.64	KH_2PO_4	0.66
$H_4Fe(CN)_6$	1.08		
Ammonium salts	$R_{NH\cdots X}$	*Crystallo-hydrates*	$R_{OH\cdots X}$
NH_4F_2	0.09	$KF{\cdot}2H_2O$	0.09
NH_4NO_3	0.07		
NH_4HCO_3	0.09	$Na_2CO_3{\cdot}10H_2O$	0.18
$(NH_4)_2SO_4$	0.09	$Na_2SO_4{\cdot}10H_2O$	0.18
$(NH_4)_2HPO_4$	0.13		
$NH_4H_2PO_4$	0.13	$Na_2HPO_4{\cdot}12H_2O$	0.19
$(NH_4)_3Fe(CN)_6$	0.12		
$(NH_4)_4Fe(CN)_6$	0.29	$K_4Fe(CN)_6{\cdot}3H_2O$	0.27

The refractions of H-bonds are *ca.* 0.2 cm^3/mol in aliphatic and from 0.2 to 0.4 cm^3/mol in aromatic alcohols; this is also due to different charges on the oxygen atoms in these compounds. The H-bond refraction in alcohols is enhanced by addition of such electrophylic substituents as OH or Cl, which is consistent with the fact that the boiling points in alcohols are increased by the introduction of electrophylic groups into the chain. In aliphatic acids the H-bond refractions diminish with increasing length of the carbon chain [13, 14]. It is necessary to emphasize that the refractions of H-bonds comprise only a few per cent of the intrinsic refractions of substances, which corresponds to the correlation between the H-bond energy and the atomization energy.

Refractometric methods are useful for investigating vitreous or polycrystalline silicates to determine the structural role of anions or to define N_c of cations, since the refraction of the oxide component of the silicate depends on N_c as

$$R_N = R_1 N_c^{-1/3} \tag{11.46}$$

where R_1 is the refraction of the same oxide in the molecular form ($N_c = 1$) [203].

Refractometric method is especially useful for the study of the *trans*-effect in coordination compounds (see Chap. 3). In Pt(II) complexes this increases in the succession

$$\mathrm{CO, CN, C_2H_4 > NO_2, I, SCN > Br, Cl > OH, F > NH_3, H_2O,}$$

which evidently must correlate with the negative effective charges on the ligands influenced by the *trans*-partners, since the increase in the bond polarity will labilize these ligands in aqueous solutions. The first quantitative characteristic of the *trans*-effect was given by the optical method: the refraction of the bond electrons, i.e. the difference between the atomic and ionic refractions of ligands and the platinum, changes in the sequence corresponding to the *trans*-effect succession: H > S > SCN > I > Br > CN > Cl > NCS > NC > NO_2 > OH, CO_3 > NO_3, SO_4 >

$F > NH_3 > H_2O$. These radicals can be neatly grouped according to the atom which is bonded to Pt:

$$F < O\ (ONO, SO_4, NO_3, CO_3, OH) < N\ (NO_2, NCS, NC) < S\ (SCN, S).$$

The essentially new feature in the above sequence is the difference in *trans*-activity shown by those radicals which are capable to isomerize: the activity of the SCN, CN, and NO_2 groups markedly depends on which end of the ligand is bonded to the central atom. These radicals can be linked differently in the complex compounds of various metals, or even in various compounds of the same metal, and in these cases the order of *trans*-activity would be considerably different.

A quantitative evaluation of the *trans*-effect can also be made by comparing the coordinate refractions. Thus, $\Delta R = R(X{-}M{-}Y) - ½\ [R(X{-}M{-}X) + R(Y{-}M{-}Y)]$ corresponds to the *trans*-activity of the X or Y atoms. By this method it has been established that in the Pt^{II} compounds a NO_2 ligand has stronger *trans*-effect than Cl, while in the coordination compounds of Pt^{IV} it is the other way round [204]. Similar conclusions were made also for the complex compounds of Co^{III} and Pd^{II}, which contradicted the paradigm of that time but was subsequently confirmed by other physical methods [205].

The ratio of the refraction of a ligand experiencing the *trans*-effect, to the intrinsic atomic refraction in the complex compound in question, is known as the 'mobility coefficient' (MC). The calculations of MC are described in [13, 14]. They give the following succession of ligands according to their susceptibility to the *trans*-effect,

$$H \gg NH_3 > H_2O > Cl > NO_2 > SCN > Br,$$

which has been confirmed by chemical experiments, although no other physical method could give a comparable characterization.

An important part of structural refractometry is the determination of atomic sizes. The expression $\alpha = r^3$ (as required by the Clausius-Mossotti theory) was first used to calculate ionic radii by Wasastjerna [206]. Later, Goldschmidt [207] used the oxygen and fluorine ionic radii calculated by Wasastjerna to derive other radii by the additive method. In 1939 Kordes began his works [208–212] leading to the empirical relation between the univalent ionic radius (according to Pauling) and the refraction,

$$R_i = k r_i^{\ 4.5} \tag{11.47}$$

where $k = 0.603$ for ions of the rare-gas type and 1.357 for other ions. Assuming r_1 to be the characteristic of a free ion and using these ionic radii, Kordes found the refractions of free ions, and later (using the crystal radii) calculated crystalline ionic refractions. A correlation between refractions and radii of the free and crystalline ions was established by Wilson and Curtis [213], and Vieillard [214]. Similar dependence has been obtained for melted alkali fluorides [215]. Interdependence of polarizabilities and van der Waals radii (r_W) of isolated atoms was described by Miller [145]: since according to the Slater-Kirkwood theory

$$\alpha = \frac{4}{a_o}\left(\frac{r_w^2}{3}\right)^2 \tag{11.48}$$

then, knowing the atomic refractions, one can calculate the vdW radii which differ from the conventional values by 10–30 %.

Interesting results were obtained for fullerenes and their derivatives. Thus, $R_\infty(C_{60}) = 193$ and $R_\infty(C_{70}) = 257\,cm^3/mol$ [216], in agreement with the additive model. C_{60} formally has 30 double and 60 single bonds, for the standard bond refractions (see Table S11.14) of 1.254 (C–C) and $3.94\,cm^3/mol$ (C=C), this gives $R(C_{60}) = 193.4\,cm^3/mol$. Sodium adducts of fullerene have the refraction much greater than either the pristine fullerene or atomic sodium ($60.2\,cm^3$), which decreases in the succession $Na \cdot C_{60}$ 4410, $Na_2 \cdot C_{60}$ 2520, $Na_3 \cdot C_{60}$ $1{,}890\,cm^3/mol$, and then gradually increases with the number of sodium atoms, to reach $5{,}040\,cm^3/mol$ for $Na_{34} \cdot C_{60}$ [217]. This effect was explained by a charge transfer from Na to fullerene and the formation of a permanent dipole.

11.5 Structural Applications of Spectroscopy

Various spectroscopic methods are important sources of information about geometry and energetic parameters (bond energies, ionization potentials, etc.) which have been discussed in the previous chapters; there are also many reference books available. In this section we shall discuss some specific questions of structural spectroscopy.

Spectroscopic methods are invaluable in studying molecules, radicals and ions which either can exist only in the gas phase (e.g. alkali metal molecules) or are unstable and have very short time-span. Of the latter, van der Waals molecules were discussed earlier. Another example is the IR-spectroscopic studies of the short-lived $CH_2{=}MHF$ molecules which gave the bond distances M=C of 1.812, 1.996, 1.979 and 2.129 Å for M=Ti, Zr, Hf and Th, respectively [218], that of Co_2O_2 molecule [219] estimated the Co–O bond force constant at 2.435 N/cm and the OCoO bond angle at $93 \pm 5°$. In comparison with the CoO molecule, this suggests a near square-planar structure with the Co–O distance of 1.765 ± 0.01 Å. Wang et al. [220] have reported the first detection of the HgF_4 molecule in the photochemical reaction of Hg and F_2 in solid neon and argon at 4 K. Experiments in both neon and argon matrixes show IR absorptions at the positions predicted for HgF_4 by quantum-chemical calculations. HgF_4 is a square-planar low-spin d^8 transition-metal complex (Fig. 11.7), in which the 5*d* orbitals of mercury are strongly involved in bonding. Mercury should therefore be viewed as a genuine transition-metal element rather than as a post-transition metal, and thus the observation of HgF_4 affects the way we should view the Periodic Table.

In the last decades, photoelectron spectroscopy also began to be applied to solving structural problems, using a dependence of the form and frequency of the absorption band on the symmetry of molecules and groups. For example, it proved that the short-lived Al_4C species (in the gas phase) has tetrahedral geometry with carbon as the central atom, while the anion Al_4C^- is square planar [221], as well as Al_5C and Al_5C^- derived from the latter by an addition of Al^+ or Al^0 to one of vertices of the square [222]. For the first time, planar complexes with carbon as the central atom have

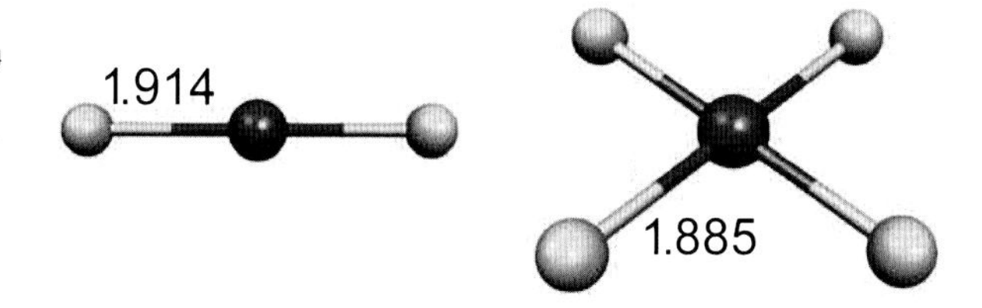

Fig. 11.7 Computed structures of HgF_2 and HgF_4 (distances in Å). (Reprinted with permission from [216]; Copyright 2007 John Wiley and Sons)

been identified in the gas phase, viz. CAl_3Si, CAl_3Si^-, CAl_3Ge, CAl_3Ge^- [223]. The square structure was found for the anions $Al_4{}^{2-}$, $Ga_4{}^{2-}$, $In_4{}^{2-}$ and $XAl_3{}^-$ (X = Si, Ge, Sn, Pb, but not C) [224] which are aromatic. Planar structures were established for MO_4 radicals where M = Li, Na, K, Cs [225]; while the $Pn_5{}^-$ anions with Pn = P, As, Sb and Bi were shown to be isostructural with $C_5H_5^-$ [226]. By experimental, as well as theoretical evidence, $SiAu_n$ clusters ($n = 2-4$) are structurally and electronically similar to SiH_n [227]. A combination of photoelectron spectroscopy with *ab initio* calculations [228] elucidated the structures of B_4O_2 and $B_4O_2{}^-$ clusters. Both possess highly stable linear O–B–B–B–B–O structures, the central B–B bond shortening from 1.514 Å in the neutral to 1.482 Å in the anionic cluster, as the bond order increases from 2 to 2.5.

Another important task of the vibration spectroscopy is the determination of force constants (f). For diatomic molecules, these can be obtained directly from the vibration frequencies (ω) and the reduced mass ($\bar{m}$) of oscillating atoms by the harmonic oscillator equation

$$f = 4\pi^2 c^2 \omega^2 \bar{m} \tag{11.49}$$

where c is the velocity of light. The force constant being the second derivative of the bond energy with respect to the distance at the equilibrium state,

$$f = \left(\frac{\partial^2 E}{\partial r^2}\right)_{r=r_o} \tag{11.50}$$

it gives important information about the strength (multiplicity) of a chemical bond. Thus, the force constants of M_2 molecules were used to define the bond multiplicity (q) as the ratio of the actual f to that of a standard single bond [229]. The frequencies and force constants in molecules A_2 are listed in Table 11.11. For polyatomic molecules the precise solution of the vibration problem requires the knowledge of the structure, character and strengths of the bonds, as well as sterical interactions between different fragments of the molecule. Therefore the theoretically rigorous results are not many, e.g. the review [236] contains the data for only 300 chemical bonds of covalent or ionic types. The most successful determination of the molecular structure can be achieved by the combination of spectroscopic methods with electron diffraction (see [230, 236–239]).

For the first time, the empirical relationship between stretching vibration frequencies of atoms and their distances in molecules were established by

Table 11.11 Vibration frequencies (ω, cm^{-1}) and force constants (*f*, *m*dyn/Å) of A_2 molecules. (From [229, 230], except where specified)

A	ω	*f*	A	ω	*f*	A	ω	*f*
Ag	192.4	1.18	Hf	176	1.63	Rb	57	0.08
Al	286[a]	0.65	Hg	18.5	0.02	Re	338	6.26
As	430	4.09	I	214	1.72	Rh	284	2.44
Au	191	2.12	In	142	0.68	Ru	347	3.59
B	1051	3.52	K	92.0	0.10	S	726	4.99
Be	498[b]	0.27	Kr	24.2	0.01	Sb	270	2.62
Bi	173	1.85	Li	351	0.25	Sc	240	0.76
Br	325	2.49	Mg	51.1	0.02	Se	385	3.45
C	1855	12.2	Mn	76	0.09	Si	509[c]	2.15
Ca	64.9	0.05	Mo	473	6.33	Sn	186[d]	1.21
Cd	23	0.02	N	2359	23.0	Ta	300	4.80
Cl	560	3.28	Na	159	0.17	Te	247	2.30
Co	297	1.53	Nb	420	4.84	Ti	408	2.35
Cr	481	3.54	Ni	259	1.16	Tl	39[e]	0.09
Cs	42.0	0.07	O	1580	11.8	V	537	4.33
Cu	266	1.33	P	781	5.58	W	337	6.14
F	917	4.71	Pb	110	0.74	Zn	26	0.01
Fe	300	1.48	Pd	210	1.38	Zr	306	2.51
Ga	165	0.56	Pt	222	2.84	Y	184	0.90
H	4401	5.72						

[a][231], [b][232], [c][233], [d][234], [e][235]

Birge and Mecke [240, 241]:

$$\omega d^2 = const \tag{11.51}$$

Later, Morse [242] suggested the formula

$$f = \frac{c}{d^6} \tag{11.52}$$

where c is the constant, that was used and refined in [243, 244–246]. More recently, a simple relationship between vibration frequencies, bond lengths, and reduced masses was reported [247] for many families of stable diatomic molecules of alkali metals, elements of Groups 15 and 16, and LiX,

$$\omega = \frac{a}{d\bar{m}^{1/2}} + b \tag{11.53}$$

where a and b are the constants. After Morse the next important step was made by Badger [248, 249], who suggested the equation

$$f = A(d - B)^{-3} \tag{11.54}$$

where A and B are constants. The parameters of Eq. 11.54 were repeatedly modified [250, 251], e.g. it was proposed to use the power of 4.33 instead of 3, which greatly improved the correlation [251]. Nevertheless, Badger's equation has been used in its original form in most works on structural and physical chemistry. Murrel [252] analyzed the nature of polar M–X bonds in terms of the ionic model and derived the formula

$$f_b = 2\frac{Z_M Z_X}{d_e^3}(1 - \sigma) \tag{11.55}$$

where σ is the screening factor. Pearson has simplified this formula, replacing numerator with the product of Z^*_M and Z^*_X [253]

$$f_b = 2\frac{Z_M^* Z_X^*}{d_e^3} \tag{11.56}$$

Z^* of main-group elements calculated by this formula, agree qualitatively with those obtained using modified Slater's rules (Table S11.16), but for atoms with incomplete *d* shells the results are underestimated, by a factor of up to 1.5. Gordy [254] was the first to establish the dependence of the force constants in molecules on atomic electronegativities, χ,

$$f = aq\left(\frac{\chi_M \chi_X}{d^2}\right)^{3/4} + b \tag{11.57}$$

where a and b are substance-specific parameters and $q = v/N_c$ is the bond multiplicity, v being the valence and N_c the coordination number. From Eq. 11.57 it follows that the force constant and, hence, the vibration frequency (ω) will increase with valence, which was proven experimentally. Compilations of d and ω for various valence states of atoms are given in [255–257] and, in an abridged form, in Table 11.12. As can be seen from this table, ω increases on average by 30 % as v of M increases by one unit. For molecules of the alkali-earth halides:

$$f = k\Delta\chi^n \tag{11.58}$$

where $k = 0.5883$ and $n = 3.8404$ [258]. Attempts to improve the dependence of f on χ, which were made from time to time, are reviewed in [259]. Later Pearson [260] suggested the equation

$$fd = aq(\chi_M \chi_X) + b \tag{11.59}$$

for MX-type molecules. This expression describes polar compounds well; e.g. for alkali halides the correlation coefficient is 0.995, however, for covalent molecules it is only 0.88.

On phase transitions with an increase of N_c, ω decreases. Table 11.13 lists the resulting ω of atoms in gases and crystals [261–263], which characterize the effect of the coordination numbers. Change of N_c also occurs when polar molecules associate via hydrogen bonds (e.g. in HX) or when MX_n molecules become $[MX_{n+1}]^-$ ions. Thus, on condensation of HF, HCl, HBr and HI molecules, ω(H–X) decreases thus: 3,962 → 2,420 cm^{-1}, 2,886 → 2,746 cm^{-1}, 2,558 → 2,438 cm^{-1}, and

Table 11.12 Influence of valence on stretching vibration frequencies (cm^{-1})

Crystals	ω	Crystals	ω	Crystals	ω	Crystals	ω
CuCl	172	TiF_3	452	MnF_2	407	$[FeBr_4]^{2-}$	219
$CuCl_2$	312	TiF_4	560	MnF_3	560	$[FeBr_4]^-$	285
CuBr	137	VF_3	511	MnF_4	622	$FeCl_2$	350
$CuBr_2$	224	VF_4	583	MnO	361	$FeCl_3$	370
HgCl	252	VF_5	715	Mn_2O_3	480	RuF_3	497
$HgCl_2$	370	CrF_2	481	MnO_2	606	RuF_4	581
HgSCN	207	CrF_3	528	$[FeCl_4]^{2-}$	286	RuF_5	640
$Hg(SCN)_2$	311	CrF_4	556	$[FeCl_4]^-$	378	RuF_6	735

Table 11.13 Influence of coordination numbers on bond vibration frequencies (cm^{-1})

Substance	ω (gas)	ω (solid)	N_c change	Substance	ω (polymorph 1)	ω (polymorph 2)	N_c change
LiF	910	305	1→6	$ZnCl_2$	516	280	2 → 4
LiCl	643	203		$ZnBr_2$	413	172	
LiBr	563	173		ZnI_2	340	122	
LiI	498	142		HgI_2	200	132	
NaF	536	246		$CdCl_2$	425	250	2 → 6
NaCl	366	164		$CdBr_2$	315	163	
NaBr	302	134		CdI_2	265	117	
NaI	258	116		$GaCl_3$	462	346	3 → 4
CuCl	415	172	1 → 4	$GaBr_3$	343	210	
CuBr	315	139		GaI_3	275	145	
CuI	264	124		BN	1385	1140	
BeO	1487	750		$AlCl_3$	616	257	3 → 6
BN	1515	1056		$InCl_3$	394	255	
				$InBr_3$	280	180	
				$FeCl_3$	370	280	
				MnS	295	220	4→ 6
				SiO_2	1090	888	
				GeO_2	890	720	
				Al_2O_3	800	590	

2,230 → 2,120 cm^{-1}, respectively. Similarly, an increase of N_c reduces ω(N–H) from 3,378 cm^{-1} in NH_3 to 3,145 cm^{-1} in NH_4^+.

If the atomic valence does not change, or its change is known, IR-spectra can give important information about the N_c of atoms. Thus, IR-spectra of the shock-compressed GeO_2 glass revealed an increase of N_c(Ge) from 4 to 6 [11]. Additional

Table 11.14 Influence of coordination number and valence on the absorption band width (cm^{-1})

Substance	v	N_c	$\Delta\omega_{1/2}$	Substance	v	N_c	$\Delta\omega_{1/2}$
BN	3	3	110	Cu_2O	1	2	35
		4	180	CuO	2	4	145
MnS	2	4	50	GeS_2	4	4	60
		6	150	GeS	2	6	280
GeO_2	4	4	65				
		6	140				

information can be obtained from the form of the IR absorption bands. Generally speaking, the intensity of the absorption spectra of a powder material depends on a number of factors, viz. the sizes of particles, the difference (Δn) between the RI of the sample and that of the immersion medium, the concentration of defects. Nevertheless, Δn always plays the key role. On this fact the method of measuring RIs in the region of the anomalous dispersion of crystals is based. In the absorption band of a substance $n \rightarrow \infty$, therefore if the IR absorption spectrum is recorded using a succession of immersion media with increasing n, e.g. the matrices of KBr, CsI, CuCl, AgCl, TlCl and TlBr, at certain frequencies there comes the moment when $\Delta n = 0$ and the background becomes zero (filter-effect of Christiansen). Hence, varying ω, one can determine the RI corresponding to the filter-effect, i.e. the anomalous dispersion curve; this method allows also to distinguish isotropic ions (e.g. NH_4^+) from anisotropic structural units (e.g. NO_3^-) in crystals [264–266].

One of the most important factors influencing the shape of absorption bands of powders in IR-spectra, is the coordination number, N_c (Table 11.14). Since the bond polarity increases with N_c, the difference $\varepsilon - n^2$ increases (see Sect. 11.2) and according to Eq. 11.20 the absorption band, extending as it is from ω_l to ω_t, should widen. For example, γ-MnS has $N_c = 4$, so from $\varepsilon = 8.2$ and $n^2 = 6.0$ we obtain $(\omega_l/\omega_t)^2 = 1.2$. In the α-polymorph, $N_c = 6$, so $\varepsilon = 20$ and $n^2 = 6.2$ give $(\omega_l/\omega_t)^2 = 3.3$. Therefore, the $\gamma \rightarrow \alpha$ phase transition should almost triple the width of the ω(Mn–S) band, which is confirmed experimentally.

Interesting changes of the IR-spectra of $NaNO_3$ and KNO_3 are observed on heating [267]. Figure 11.8 shows the positions and shapes of the ω(N–O) bands of absorption in samples heated up to 400 °C, which show breaks at 270–275 °C for $NaNO_3$ and at 128–130 °C for KNO_3, corresponding to phase transitions in these substances. In $NaNO_3$ this signifies the onset of NO_3-ions' rotation, in KNO_3 the transition from the aragonite into the calcite-type structure, with a decrease of N_c from 9 to 6.

In organic compounds N_c, as a rule, does not change, but the character of chemical bonding varies substantially with composition. Thus, in a E_nAB molecule where $\chi_A > \chi_B$, replacing the atom (or group) E with a more electronegative one, increases the effective χ_A and the difference $\Delta\chi = \chi_A - \chi_B$ and hence reduces ω_{AB}. If $\chi_A < \chi_B$, the same replacement will reduce $\Delta\chi$ and increase ω_{AB}. Empirical dependences confirm these conclusions. In the former case we have

$$\omega_{EOH} = \omega^o{}_{OH} - a\chi_E;\ \omega_{EE'CH_2} = \omega^o{}_{CH} - b\sum\chi_E,$$

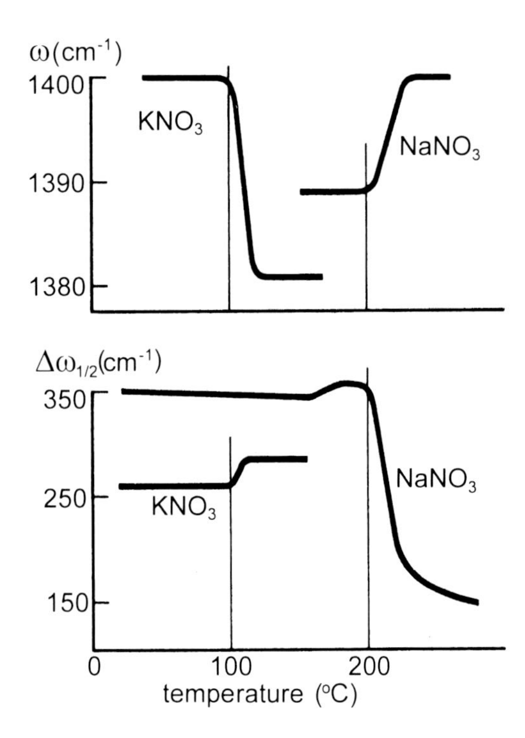

Fig. 11.8 Variations of ω(N–O) and the half-width $\Delta\omega_{1/2}$ of the IR peak upon phase transitions

and in the latter,

$$\omega_{\mathrm{ESiF}} = \omega^{\circ}{}_{\mathrm{SiF}} + c\Sigma\chi_{\mathrm{E}}, \quad \omega_{\mathrm{EAB}} = \omega^{\circ}{}_{\mathrm{AB}} + d\chi_{\mathrm{E}},$$

where A = P, As, Sb, B = O, S, Se, Te;

$$\omega_{\mathrm{EE'CO}} = \omega^{\circ}{}_{\mathrm{CO}} + e\Sigma\chi_{\mathrm{E}}; \quad \omega_{\mathrm{EE'E''PO}} = \omega^{\circ}{}_{\mathrm{PO}} + f\Sigma\chi_{\mathrm{E}};$$

$$\omega_{\mathrm{EE'E''PF}} = \omega^{\circ}{}_{\mathrm{PF}} + g\Sigma\chi_{\mathrm{E}}$$

where *a, b, c, d, e, f, g* are the constants. Similar dependences have been established also for inorganic compounds with mixed ligands. So, a decrease of ω_{MO} with decreasing χ_{X} is observed in MO_nX_m where M = Th, Sn, P, V, Nb, Cr, Mo, W, S, Se, and X = halogens. Similar results were obtained for chalcogenides-halides, hydroxides-halides and mixed halides of Au, Hg, Ti, Sn, Pb and Pt. Force constants of identical bonds obviously decrease when a less-electronegative ligand is introduced. For example, *f* (Sn–C) decreases on transition from complexes $[SnX_6]^{2-}$ to $[SnX_4(CH_3)_2]^{2-}$: for X = F from 2.8 to 1.2 *m*dyn/Å, for Cl from 1.5 to 0.8 *m*dyn/Å, for Br from 1.2 to 1.1 *m*dyn/Å, accordingly. Similarly, *f*(Au–X) in complexes $[AuX_4]^-$ and $[AuX_2(CH_3)_2]^-$, respectively, are equal to 2.2 and 1.4 *m*dyn/Å for X = Cl, 1.8 and 1.2 *m*dyn/Å for X = Br.

In the succession of molecules $CH_3I \rightarrow CH_3Br \rightarrow CH_3Cl \rightarrow CH_3F$, the intensity of the ω(C–H) absorption increases monotonically, as the ionicity of the C–H bond increases. The ω(Hg–S) absorption band in the crystalline $Hg_3S_2Cl_2$, $Hg_3S_2Br_2$ and $Hg_3S_2I_2$ also widens with increasing polarity of the Hg–S bond. An increase of the integrated intensity of the ω(N–O) absorption is observed in the succession $M_2RuNOCl_5 \rightarrow M_2RuNOBr_5 \rightarrow M_2RuNOI_5$, sympathetically with the increase of polarity of the Ru–NO bond. The ω(Sn–X) intensity in $CsSnCl_{3-x}Br_x$ varies with x, as replacing Cl with Br increases the polarity of the remaining Sn–Cl bonds in comparison with $CsSnCl_3$ but reduces the polarity of Sn–Br bonds in comparison with $CsSnBr_3$, due to competition between different halogens for attracting the electron density from Sn [268].

The spectroscopic study of MNO_3 revealed a regular increase in splitting of the ω_3(N–O) absorption band with increasing covalency in the M–O bonds. The absorption band splitting helps to solve some crystallographic problems also, in particular to clarify the local symmetry of ions. Here spectroscopy is complementary to XRD which gives the averaged symmetry of a crystal. In fact, vibration spectroscopy can determine 206 out of 230 crystallographic space groups [269, 270].

11.6 Optical Electronegativities

A few examples of applying electronegativities (EN) of atoms for interpreting their optical properties has already been given above. In fact, a system of EN has been developed specifically for the optical field. Gordy used Eq. 11.57 to calculate the EN of atoms [254], which was later applied to atoms with multiple bonds [271]. Other dependences of force constants on ENs were subsequently suggested [258, 272–276]. However, an efficient calculation of f in polyatomic molecules requires knowledge of the structure and the character of bonding. The result therefore depends on the correctness of the model used, and the discrepancies between f values reported by different authors for the same bonds usually amounts to a few tenths of the total. The equation [277]

$$f = q\left(\frac{\chi_M \chi_X}{d}\right)^{(1.2-0.1v)} \tag{11.60}$$

where v is the valence and q is the multiplicity of bonds, is applicable to any molecule of the MX_n type. It allows to calculate χ_M from known χ_X and d, in agreement with the thermochemical EN. ENs of univalent metals in M_2 molecules can be calculated as

$$f = q\left(\frac{\chi_M^2}{d}\right)^{1.1} \tag{11.61}$$

The experimental values of force constants are presented in Tables S11.17 and S11.18, together with the χ_M calculated by Eqs 11.60 and 11.61. The force constants of MX crystalline compounds cannot be calculated by the harmonic oscillator formula from the lattice vibrations of crystals, because their calculations require taking into account crystal field effects and the polarizabilities and/or the effective charges of oscillating atoms. For this reason, the data on the optical force constants of crystalline compounds (especially the polyatomic ones) are incomplete and insufficiently accurate for calculating ENs. A principally different solution was suggested by Waser and Pauling [278] who showed that f of cubic crystals can be found from their mechanical characteristics as

$$f_{\mathrm{c}} = \frac{9B_{\mathrm{o}}V_{\mathrm{o}}}{N_{\mathrm{c}}d^{2}} \tag{11.62}$$

where V_{o} is the molar volume and B_{o} is the bulk modulus. The force constants of cubic crystals of the MX type, calculated by Eq. 11.62, are given in Table S11.19, together with the ENs of atoms in crystals (χ^{*}), obtained via the equation

$$f_{c} = k_{M}q\left(\frac{\chi_{M}^{*}\chi_{X}^{*}}{d}\right)^{(1.2-0.1v)} \tag{11.63}$$

which can be derived from Eq. 11.60 by introducing the Madelung constant, k_{M} (see [253]). ENs of metals, calculated from optical characteristics of their compounds with different ligands, are similar or identical, which allows to create a system of averaged spectroscopic χ and χ^{*} [279], Table 11.15.

There are also numerous works where ENs were calculated by the $\omega = \varphi(\chi)$ equations considered in Sect. 11.5. Significantly, these formulae are applied to the radiowave spectra [280]. The ENs of radicals averaged over all published data are listed in Table 11.16. Optical ENs agree well with thermochemical ENs. Jörgensen for the first time used electron-spectroscopy data on coordination compounds to define EN [281, 282]. It was established that the frequency of an electronic transition within a complex (ν) depends on $\Delta\chi$ of the central atom (M) and the ligand (L):

$$hv = a(\chi_{\mathrm{M}} - \chi_{\mathrm{L}}) \tag{11.64}$$

where the constant a is defined by scaling to Pauling's system.

Later the approach of Jörgensen was applied for defining ENs of such molecules as water, pyridine, bipyridine, bipyrazine, *etc.* [283], yielding the values of ENs from 1.0 to 1.3, in agreement with the ionization and geometrical systems of EN of van der Waals atoms. Duffy extended Jörgensen's method to binary crystalline compounds [284–286] by replacing $h\omega$ in Eq. 11.58 with E_{g},

$$E_{\mathrm{g}} = b(\chi_{an} - \chi_{cat}) \tag{11.65}$$

To define $\chi(\mathrm{O})$ in different compounds, Duffy suggested the empirical formula

$$\chi_{\mathrm{O}} = 4.30 - 1.51(\chi_{\mathrm{M}} - 0.26) \tag{11.66}$$

where χ_{M} is the EN of the metal on Pauling's scale. Eq. 11.66 allows to establish the dependence of $\chi(\mathrm{O})$ from $f(N_{\mathrm{c}})$. For the oxides of Mg, Ca, Sr and Ba, crystallizing in the structural type Bl, it gives $\chi_{\mathrm{O}} \approx 2.3$; for the oxides of Cu, Be, Zn and Hg, which have B3 structures, $\chi_{\mathrm{O}} \approx 3.3$. ENs of metals, calculated by Duffy's method, correspond to a solid state and therefore, are usually lower than the canonical values of Pauling: Zn 1.1, Cd 1.45, Hg 1.55, Al 0.95, Ga 1.15, In 1.45, P 1.75, As 1.55, Sb

Table 11.15 Spectroscopic electronegativities of atoms in molecules (χ, *upper lines*) and crystals (χ^*, *lower lines*)

M^{I}	χ	χ^*	M^{II}	χ	χ^*	M^{II}	χ	χ^*	M^{III}	χ	χ^*	M^{III}	χ	χ^*	M^{IV}	χ	χ^*
Li	0.9	0.5	Cu	1.7		Hf	1.9		Sc	1.3	1.0	N	3.2		Ti	1.9	1.0
Na	0.8	0.5	Be	1.5	1.05	Sn	1.5	0.65	Y		0.9	P	1.9		Zr	1.8	1.0
K	0.75	0.4	Mg	1.2	0.7	Pb	1.3	0.65	La	1.5	0.9	As	1.6		Hf	1.9	1.1
Rb	0.7	0.4	Ca	1.0	0.65	Cr	1.5	0.5	B	1.9	1.3	Sb	1.5		Si	2.0	1.05
Cs	0.7	0.4	Sr	0.9	0.6	Mn	1.4	0.75	Al	1.7	1.0	Bi	1.4		Ge	1.9	1.0
Cu	1.6	1.0	Ba	0.9	0.6	Fe	1.5	0.7	Ga	1.7	0.95	Th		1.1	Sn	1.7	0.9
Ag	1.5	0.8	Zn	1.7	0.9	Co	1.5	0.7	In	1.9	1.18	U		0.9	Pb	1.6	0.9
Au	1.9	1.4	Cd	1.6	0.85	Ni	1.4	0.75	V	1.6		Rh		0.8	Mo	1.85	1.2
Tl	1.15	0.6	Hg	1.8	0.9	Pd		0.9	Nb	1.9	1.3	Ir		1.0	W	1.9	1.25
NH_4		0.5	Ti	1.6		Pt	1.6	1.1	Ta		0.95				Th	1.7	1.3
			Zr	1.8		Th	1.9	0.70							U	1.6	1.0

Table 11.16 Spectroscopic electronegativities of radicals

E	χ	E	χ	E	χ
CF_3	3.2	NO_3	3.9	SO_4	3.6
CCl_3	3.0	NO_2	3.6	SO_3	3.6
CBr_3	2.7	N_3	3.4	SCN	2.8
CH_3	2.5	NH_2	3.1	SH	2.5
C_6H_5	2.8	CHO	2.8	SeH	2.3
$CH{=}CH_2$	3.0	COOH	2.9	PH_2	2.3
$C{\equiv}CH$	3.2	OCN	3.6	AsH_2	2.1
CN	3.3	OH	3.6	SbH_2	1.8

1.35 whereas for halogens these ENs are identical to the 'molecular' values. Thus, spectroscopic data confirm the conclusion made on the basis of the energetic and geometrical characteristics of substances: a transition from a molecule to a crystal with an atomic lattice lowers the EN of the metal without affecting that of the nonmetal, which results in an increase of the bond polarity. Duffy has established also the relationships between the optical basicity and thermo-chemistry of silicates, between the cationic charge, coordination number and polarizability in oxides [287, 288], and also proposed a general formula for determining the optical basicity and ionicity of bonds in oxides of many metals and nonmetals. Finally, Reddy et al. [289] derived from optical EN the molar refraction, ionicity and density values for binary oxides, SiO_2 polymorphs, silicates and minerals.

Appendix

Supplementary Tables

Table S11.1 Temperature coefficients of refractive indices

Substance	$-\delta \times 10^5$	Substance	$-\delta \times 10^5$	Substance	$+\delta \times 10^5$
LiF	1.2	CaF_2	0.8	Diamond	0.5
NaF	0.9	BaF_2	0.9	Si	4.5
NaCl	2.0	PbI_2	8.0	Ge	6.9
KCl	2.1	Ar	27	GaP	3.7
KBr	2.6	Kr	32	GaAs	4.5
KI	3.0	Xe	29	GaSb	8.2
CsBr	4.7	SiO_2	0.4	InP	2.7
CsI	5.2	$CaSO_4 \cdot 2H_2O$	1.0[a]	InSb	12
AgCl	3.0		2.8[b]	ZnSe	2.0
			1.7[c]	CdS	5.0

[a] For n_p, [b] for n_m, [c] for n_g

Table S11.2 Refractive indices of elementary solids. (From [11.25] and the authors' own unpublished measurements, except where specified)

M	n_∞	M	n_∞	A	n_g	n_m	n_p	A	n_g	n_m	n_p
Cu	29.7	Fe	6.41	B		3.08		P[i]	3.21	3.20	3.11
Ag	9.91	Co	6.71	C[a]		2.42		As		3.6	
Au	8.17	Ni	9.54	C[b]		2.15	1.81	Sb		10.4	
Hg	14.0	Ru	11.7	Si[c]		3.42		S[j]		2.02	
Al	98.6	Rh	18.5	Ge[c]		3.99		S[k]	2.24	2.04	1.96
Nb	16.0	Pd	4.13	Sn[d]		4.8		S[k]		2.06	
Cr	21.2	Os	4.08	Pb[e]		13.6		Se[l]	2.91	2.84	
Mo	18.5	Ir	28.5	P[f]		2.12		Se[m]	4.04	3.00	
W	14.1	Pt	13.2	P[g]	3.15	2.72		Te[c]	4.82	2.61	
Re	4.25[c]			P[h]	3.20	2.72		I[n]		3.34	

[a] diamond, [b] graphite, [c] n_m, [d] [11.1], [e] at $\lambda = 10\ \mu$, [f] cubic, [g] tetragonal, [h] hexagonal, [i] triclinic, [j] glass, [k] orthorhombic, [l] monoclinic, [m] trigonal, [n] at $\lambda = 1.8\ \mu$, for liquid iodine (114°C) $n = 1.934$; for liquid bromine (19°C) $n = 1.604$

Table S11.3 Refractive indices of crystalline substances of the MX type (n_∞ from [11.2])

M	F		Cl		Br		I	
	n_D	n_∞	n_D	n_∞	n_D	n_∞	n_D	n_∞
Li	1.392	1.386	1.662	1.646	1.784	1.752	1.955	1.906
Na	1.326	1.320	1.544	1.528	1.641	1.613	1.774	1.730
K	1.362	1.355	1.490	1.475	1.559	1.537	1.667	1.628
Rb	1.396	1.389	1.494	1.472	1.553	1.523	1.647	1.605
Cs[a]	1.478	1.469	1.534	1.517	1.582	1.558	1.661	1.622
Cs[b]	1.578	1.566	1.642	1.619	1.698	1.669	1.788	1.743
NH_4	1.315	1.312	1.643	1.614	1.712	1.672	1.701	1.633
Cu		1.973	1.891	2.116	2.014	2.345	2.217	
Ag	1.80	1.73	2.071	2.004	2.252	2.179	2.20[c]	2.13
Tl	2.055		2.247	2.162	2.418	2.302	2.78	2.60
M	O		S		Se		Te	
Be	1.724	1.679	2.275				2.65[d]	
Mg	1.737	1.718	2.271	2.084	2.42			
Ca	1.837	1.804	2.137	2.020	2.274	2.148	2.51	
Sr	1.870	1.802	2.107	1.927	2.220	2.092	2.41	
Ba	1.980	1.883	2.155	2.075	2.268	2.146	2.44	
Zn	2.018	1.922	2.368	2.267[e]	2.611	2.429[e]	3.060	2.698[e]
Cd	2.55	2.15	2.514	2.31[g]	2.650	2.42	2.91	2.685[g]
Hg	2.504[v]		2.927[w]	2.512	3.46		3.90	
Cu	2.84	2.54[h]						
Ga				2.26[i]	2.39[i]			
Eu	2.35[j]		2.43[j]	2.20	2.51[j]	2.29	2.70[j]	2.42
Ge			3.267[k]					
Sn	2.78[f]		3.61		3.80	3.595[l]	6.70[x]	
Pb	2.621[m]			4.17		4.69[x]		5.73[x]
Mn	2.18		2.67[n]		3.12[f]	2.83		
Fe	2.32							
Co	2.30							
Ni	2.27[o]		2.325[p]					
M	N		P		As		Sb	
B	2.117[q]		3.25					
Al	2.19		2.75		2.86		3.19[x]	
Ga	2.43[r]	2.30[s]		3.02[g]		3.30[g]		3.74[t]
In			3.54[u]	3.10[g]	3.89[u]	3.50[g]	4.10[u]	3.96[g]

[a]for $N_c = 6$, [b]for $N_c = 8$, [c]for cubic phase; for hexagonal one $n_o = 2.218$, $n_e = 2.229$, [d][11.3], [e][11.4], [f][11.5], [g][11.6], [h][11.7], [i][11.8], [j]for $\lambda = \infty$, [k][11.9], [l][11.10], [m]average value of $n_o = 2.665$ and $n_e = 2.535$ for hexagonal phase; for orthorhombic form $n_g = 2.71$, $n_m = 2.61$, $n_p = 2.51$, [n]for the cubic phase (average value for hexagonal würtzite phase $n = 2.45$), [o]for $\lambda = 671$ nm, [p]average value of $n_g = 3.22$, $n_m = 2.046$, $n_p = 1.908$, [q]for the cubic phase (for the hexagonal graphite-type phase, $n_o = 2.08$, $n_e = 1.72$), [r]average value of $n_o = 2.44$, $n_e = 2.40$ [11.11], [s][11.12], [t][11.13], [u][11.14], [v]average value of $n_g = 2.65$, $n_m = 2.50$, $n_p = 2.37$, [w]average value of $n_o = 2.822$, $n_e = 3.149$, [x][11.15]

Table S11.4 Refractive indices ($\lambda = D$) of crystalline substances of the MX_2 type

Substance	n_g	n_m	n_p	Substance	n_g	n_m	n_p
BeH_2		1.648		$Pt(SCN)_2$		1.93	
BeF_2[a]		1.275		TlClS		2.18	
BeF_2		1.328		TlBrS		2.46	
BeF_2[c]		1.345		TlIS		2.7	
BeI_2[d]		1.99		TlClSe		2.30	
BeI_2[e]	1.988	1.954	1.952	TlBrSe[z]		2.51	
MgH_2	1.96	1.95		TlSeBr[α]		2.34	
MgF_2	1.389	1.377		TlISe		2.7	
$MgCl_2$		1.675	1.590	NdOF		1.82	
$Mg(OH)_2$	1.595	1.584		SmOF		1.82	
CaF_2		1.434		SmOOH	1.924	1.855	
$CaCl_2$	1.613	1.605	1.600	DyOF		1.83	
CaI_2		1.743	1.652	HoOF		1.785	
$Ca(OH)_2$	1.577	1.550		ErOF		1.79	
SrF_2[f]		1.438		YbOF		1.80	
SrF_2[g]		1.482		LaSF	>2.14	2.06	
SrFCl		1.651	1.627	CeSF	>2.14	2.03	
$SrCl_2$		1.691		PrSF	>2.14	2.10	
$Sr(OH)_2$	1.610	1.599	1.588	NdSF	>2.14	2.04	
BaF_2[f]		1.474		CO_2		1.41	
BaF_2[g]		1.518		SiO_2[a]		1.459	
BaFCl		1.640	1.633	SiO_2[n]	1.473	1.469	1.468
$BaCl_2$	1.742	1.736	1.730	SiO_2[o]		1.484	
$BaBr_2$		1.793		SiO_2[p]		1.487	1.484
BaO_2	1.85	1.775		SiO_2[q]	1.522	1.513	
ZnF_2	1.525	1.495		SiO_2[r]	1.540	1.533	
ZnFCl		1.70		SiO_2[s]	1.553	1.544	
$ZnCl_2$	1.713	1.687		SiO_2[c]	1.599	1.595	
$ZnBr_2$		1.842	1.825	SiO_2[j]	1.835	1.800	
CdF_2		1.562		GeO_2[a]		1.607	
$CdCl_2$		1.850	1.714	GeO_2[b]	1.735	1.695	
$CdBr_2$		2.027	1.866	GeO_2[j]	2.07	1.99	
CdI_2		2.36	2.17	GeS_2[v]	2.2	1.95	
$HgCl_2$	1.965	1.859	1.725	$GeSe_2$	3.32	2.83	2.65
$HgBr_2$	2.095	1.922	1.879	SnO_2	2.097	2.001	
HgI_2		2.748	2.455	SnS_2		2.85	2.16
Hg_2Cl_2	2.656			$SnSe_2$		3.26	2.88
CuF_2		1.527	1.515	PbO_2		2.30	

Table S11.4 (continued)

SmF_2		1.636		TiO_2[t]		2.561	1.488
EuF_2		1.555		TiO_2[u]	2.700	2.584	2.583
YbF_2		1.618		TiO_2[j]	2.908	2.621	
SnF_2[x]	1.878	1.831	1.800	TiS_2		3.79	
PbF_2[i]		1.767		ZrO_2	2.20	2.19	2.13
PbF_2[g]	1.853	1.844	1.837	ZrS_2		3.14	1.74
PbFCl		2.145	2.006	$ZrTe_2$		3.13	
$PbCl_2$	2.260	2.217	2.199	HfS_2		2.49	
$PbBr_2$[h]	2.560	2.476	2.439	$HfSe_2$		2.84	
PbI_2		2.80	2.13	CeO_2		2.31	
CrF_2	1.525	1.511		ThO_2[v]		2.170	
MnF_2[j]	1.501	1.472		UO_2		2.39	
MnF_2[e]	1.492	1.490		PuO_2		2.402	
$MnCl_2$		1.80		MoS_2		4.336	2.035
FeF_2	1.524	1.514		$MoSe_2$		4.22	
$FeCl_2$		1.567		WSe_2		4.5	
CoF_2	1.547	1.524		TeO_2	2.430	2.247	
NiF_2	1.561	1.526		Li_2O		1.644	
$PdCl_2$[k]		2.17	2.145	Cu_2O[w]		2.787	
$PdCl_2$[l]	2.50	2.04	1.75	Cu_2S		3.52	3.49
$PtCl_2$	2.14	2.14	2.055	Ag_2S		3.55	

[a]glass, [b]quartz-like, [c]coesite type, [d]tetragonal, [e]orthorhombic, [f]CaF_2 structure, [g]$PbCl_2$ type, [h][11.16], [i]cubic, [j]rutile, [k]own structural type, [l]monoclinic (?), [m]for $\lambda = 514$ nm [11.17], [n]tridymite, [o]β-cristobalite, [p]α-cristobalite, [q]keatite, [r]β-quartz, [s]α-quartz, [t]anatase, [u]brookite, [v]for λ = D, $n_\infty = 2.119$ [11.18], [w]for $\lambda = \infty$ $n = 2.557$ [11.18], [x]for $\lambda = \infty$ $n_g = 1.8105$, $n_m = 1.7749$, $n_p = 1.7505$ [11.19], [y][11.20], [z]for: Se=Tl[III]–Br, [α]for Tl[I]–Se–Br

Table S11.5 Refractive indices ($\lambda = D$) of crystalline substances of the MX_3 type

Substance	n_g	n_m	n_p	Substance	n_g	n_m	n_p
ScF_3		1.401		SbI_3		2.78	2.36
YF_3[a]	1.570	1.553	1.539	BiF_3		1.86	
LaF_3[a]		1.602	1.597	CrF_3	1.582	1.568	
CeF_3		1.605	1.598	$CrCl_3$		2.0	
PrF_3		1.608	1.602	FeF_3	1.552	1.541	
NdF_3[a]		1.618	1.612	CoF_3	1.726	1.703	
SmF_3		1.595		UCl_3	2.08	1.965	
EuF_3		1.590		$B(OH)_3$	1.462	1.461	1.337
GdF_3[a]	1.598	1.581	1.565	$Al(OH)_3$	1.587	1.566	1.566
DyF_3[a]	1.602	1.590	1.564	$Y(OH)_3$	1.714	1.676	
HoF_3[a]	1.599	1.580	1.562	$La(OH)_3$		1.760	
ErF_3[a]	1.600	1.579	1.566	$Nd(OH)_3$		1.800	1.755
YbF_3[a]	1.593	1.580	1.569	$Sm(OH)_3$		1.800	1.758
LuF_3		1.525		$Eu(OH)_3$		1.735	
UF_3	1.738	1.732		MoO_3	2.37	2.27	2.25
AlF_3	1.377	1.376		WO_3	2.703	2.376	2.283
GaF_3		1.457		NLi_3[b]		2.12	
InF_3		1.453		PLi_3[b]		2.19	
VF_3	1.544	1.536		$AsLi_3$[b]		2.28	
SbF_3	1.667	1.620	1.574	AsI_3	2.59	2.23	

[a][11.21], [b][11.22]

Table S11.6 Refractive indices ($\lambda = D$) of crystalline substances of the M_2X_3 type

M_2O_3	n_m	n_p	M_2X_3	n_g	n_m	n_p
B_2O_3[a]	1.447		B-Tm_2O_3[e]		2.015	
B_2O_3[b]	1.468		C-Tm_2O_3[e]		1.951	
B_2O_3[c]	1.458		B-Yb_2O_3[e]		2.00	
B_2O_3[d]	1.648		C-Yb_2O_3[e]		1.947	
Al_2O_3[d]	1.769	1.760	B-Lu_2O_3[e]		1.99	
Al_2O_3[d]	1.67	1.64	C-Lu_2O_3[e]		1.927	
Al_2O_3[c]	1.696		Sb_2O_3[c]		2.087	
Ga_2O_3[c]	1.927		Sb_2O_3[h]	2.35	2.35	2.19
In_2O_3[c]	2.08		Bi_2O_3[g]		2.45	
Sc_2O_3[k]	1.994		Bi_2O_3[c]		2.42	
B-Y_2O_3	1.97		Cr_2O_3[d]		2.49	2.47
C-Y_2O_3	1.915		Mn_2O_3		2.33	
La_2O_3	2.03		Fe_2O_3[d]		3.22	2.94
Pr_2O_3	1.94		Y_2S_3[f]		2.61	
Nd_2O_3	2.00		La_2S_3[f]		2.85	
B-Sm_2O_3[e]	2.08		Ho_2S_3[f]		2.63	
B-Eu_2O_3[e]	2.07		Yb_2S_3[f]		2.61	
C-Eu_2O_3[e]	1.983		As_2S_3[i]	3.02	2.81	2.40
B-Gd_2O_3[e]	2.04		As_2S_3[a]		2.59	
C-Gd_2O_3[e]	1.977		As_2Se_3[a]		3.3	
Tb_2O_3[k]	1.964		Sb_2S_3[h]	4.303	4.046	3.194
B-Dy_2O_3[e]	2.035		Sb_2Se_3[i]		3.20	
C-Dy_2O_3[e]	1.974		Sb_2Te_3[i]		9.0	
B-Ho_2O_3[e]	2.03		As_2O_3[g]	2.01	1.92	1.87
C-Ho_2O_3[e]	1.963		As_2O_3[c]		1.755	
B-Er_2O_3[e]	2.025		Bi_2Te_3[i]		9.2	

[a]Glass, [b]densed glass, [c]cubic, [d]hexagonal, [e][11.23], [f][11.24], [g]monoclinic, [h]orthorhombic, [i]trigonal

Table S11.7 Refractive indices ($\lambda = D$) of crystalline substances of the MX_n type

MX_4	Structure	n_g	n_m	n_p	MX_4	Structure	n_g	n_m	n_p
CeF_4	ZrF_4	1.652	1.613	1.607	PuF_4	ZrF_4	1.629	1.612	1.577
LaF_3		1.632	1.629		$GeBr_4$	Cubic		1.627	
ZrF_4	ZrF_4	1.60	1.57		SnI_4	Cubic		2.106	
HfF_4	ZrF_4	1.58	1.54		$SeCl_4$	Cubic		1.807	
ThF_4	ZrF_4		1.53		$PtCl_2Br_2$			2.07	
LaF_3		1.613	1.610		$PtBr_2I_2$		2.09	2.01	
UF_3Cl	Monocl	1.755	1.745	1.725	PCl_5	Tetra	1.708	1.674	
UCl_4	$ThCl_4$		2.03	1.92					

Table S11.8 Polarization of ions (Å^3) according to Shannon

M^+	M^{2+}		M^{3+}		M^{4+}	M^{5+}	A^{n-}
Li 1.20	Be 0.19	Zn 2.04	Sc 2.81	B 0.05	Ti 2.93	V 2.92	F^- 1.62
Na 1.80	Mg 1.32	Cd 3.40	Y 3.81	Al 0.79	Zr 3.25	Nb 3.97	OH^- 2.26
K 3.83	Ca 3.16	Mn 2.64	La 6.07	Ga 1.50	Si 0.87	Ta 4.73	O^{2-} 2.01
Rb 5.29	Sr 4.24	Fe 2.23	Cr 1.45	In 2.62	Ge 1.63	P 1.22	
Cs 7.43	Ba 6.40	Co 1.65	Fe 2.29	Sb 1.63	Sn 2.83	As 1.72	
Tl 7.28	Cu 2.11	Ni 1.23		Bi 6.12	Te 5.23		

Table S11.9 Effect of the aggregate state on molecular refractions (cm^3/mol)

Molecule	H_2O	CCl_4	$SiCl_4$	$SnCl_4$	$SnBr_4$	C_6H_6
$R_{gas}{}^a$	3.665	25.88	28.20	34.59	47.71	26.02
$R_{liq}{}^a$	3.656	25.78	27.98	33.92	46.18	25.10
Molecule	NH_4	N_2O	HCl	Cl_2	H_2S	SO_2
$R_{gas}{}^b$	5.62	7.71	6.68	11.55	9.46	10.25
$R_{liq}{}^b$	5.57	7.35	6.59	11.90	9.63	9.71
Molecule	H_2	O_2	N_2	CO_2	CH_4	CF_4
$R_{gas}{}^c$	2.09	4.05	4.47	6.68	6.59	7.33
$R_{liq}{}^c$	2.08	3.98	4.41	6.60	6.50	7.10
Molecule	Ar	Kr	Xe	H_2O	CO_2	
$R_{liq}{}^c$	4.160	6.466	10.512	3.72	6.60	
$R_{sol}{}^c$	4.180	6.379	10.387	3.79	6.77	
Salt	$LiNO_3$	$NaNO_3$	KNO_3	$RbNO_3$	$AgNO_3$	$TlNO_3$
$R_{liq}{}^b$	10.74	11.39	13.44	15.31	16.17	21.38
$R_{sol}{}^b$	10.28	11.04	12.73	14.37	15.96	21.05
Salt	KCl	KBr	KI	RbCl	RbBr	
$R_{liq}{}^a$	11.24	14.79	20.77	13.04	16.40	
$R_{sol}{}^a$	10.85	14.22	19.83	12.72	15.76	
Salt	CsCl	CsBr	CsI	AgI	$CdCl_2$	$CdBr_2$
$R_{liq}{}^a$	15.91	19.30	25.41	19.82	20.3^d	27.3^d
$R_{sol}{}^a$	15.51	18.53	23.76	13.31	19.4^d	25.3^d

$^a\lambda = \infty$, $^b\lambda = 589$ nm, $^c\lambda = 546$ nm, ^{d}for $\lambda = 592$ nm

Table S11.10 Refractometric constants of the polymorphic modifications of silica [11.25, 11.26]

Modification	ρ, g/cm^3	$\bar{n}_D$	R_D, cm^3/mol
Deca-dodecasil 3R[a]	1.760	1.376	7.83
Silicalite 1	1.805	1.382	7.75
Dodecasil 1H[a]	1.843	1.386	7.66
Dodecasil 3C[a]	1.858	1.393	7.72
Silica-ZSM-12[a]	1.907	1.403	7.69
Nonasil[a]	1.936	1.407	7.64
Silica-ZSM-22	1.969	1.415	7.64
Silica -ZSM-48	1.997	1.416	7.55
Glass	2.203	1.461	7.48
Low-temperature tridymite	2.26	1.477	7.51
Low-temperature cristobalite	2.318	1.485	7.43
Keatite	2.502	1.519	7.29
Moganite	2.52	1.52	7.25
High-temperature quartz	2.53	1.535	7.39
Low-temperature quartz	2.649	1.547	7.19
Coesite	2.920	1.596	7.00
Stishovite ($N_c = 6$)	4.291	1.806	6.02

[a]Guest-free porous materials

Table S11.11 Atomic refractions (cm^3/mol) according to Vogel [11.27] and Miller [11.28]

Atoms, groups	R_D^a	R_∞	R_D^b	Atoms, groups	R_D	R_∞
H	1.03	1.01	0.98	CN	5.46	5.33
C	2.59	2.54	2.68	NO_3	9.03	8.73
O	1.76	1.72	1.61	CO_3	7.70	7.51
OH	2.55	2.49	2.58	SO_3	11.34	11.04
F	0.81	0.76	0.75	SO_4	11.09	10.92
Cl	5.84	5.70	5.84	PO_4	10.77	10.63
Br	8.74	8.44	7.60	CH_2	4.65	4.54
I	13.95	13.27	13.66	CH_3	5.65	5.54
N (amino)	2.74	2.57	2.43	Formation of: 3-member cycle	0.60	0.53
N (aromatic)	4.24	3.55	2.75			
ONO (nitrito)	7.24	6.95		Formation of: 4-member cycle	0.32	0.28
NO_2 (nitro)	6.71	6.47		Formation of: 5-member cycle	− 0.19	− 0.19
S	7.92	7.60	7.57	Formation of: 6-member cycle	− 0.15	− 0.15
SCN (thiocyanate)	13.40	12.98		Double bond	1.58	1.42
NCS (isothiocyan.)	15.62	14.85		Triple bond	1.98	1.85

[a][11.27], [b][11.28]

Table S11.12 Empirical values of the crystalline ionic refractions (cm^3/mol)

+1		+2				+3		+4		+5		+6		−1		−2		−3	
Li	0.1	Be	0.05	V	2.8	B	0.03	C	0.01	P	0.25	S	0.04	F	2.0	O	3.8	N	5.5
Na	0.8	Mg	0.5	Cr	2.6	Al	0.4	Si	0.3	As	1.4	Se	0.25	Cl	7.6	S	11.3	P	16.7
K	3.1	Ca	2.3	Mn	2.3	Ga	1.6	Ge	1.5	Sb	3.5	Te	0.8	Br	11.3	Se	14.8	As	19.8
Rb	4.9	Sr	3.8	Fe	2.1	In	3.8	Sn	3.6	Bi	3.8	Cr	0.6	I	15.8	Te	23.2	Sb	27.6
Cs	8.0	Ba	6.3	Co	2.0	Tl	4.7	Pb	4.0	V	1.5	Mo	1.2						
Cu	2.6	Zn	1.7	Ni	1.9	Sc	1.9	Ti	1.7	Nb	2.7	W	1.2						
Ag	6.1	Cd	4.3	Ru	3.7	Y	3.2	Zr	2.9	Ta	2.7	U	6.0						
Au	8.3	Hg	6.1	Rh	3.5	La	5.4	Hf	3.0	U	6.5								
Tl	11.2	Cu	1.5	Pd	3.3	V	2.2	Mo	4.0										
		Ti	3.0	Os	4.2	As	6.0	W	4.0										
		Sn	8.2	Ir	4.0	Sb	7.0	Te	6.0										
		Pb	10.8	Pt	3.8	Bi	10.1	Ru[a]	2.4										
						Cr	2.0	Os	2.8										
						Mn	1.8	Ir	2.7										
						Fe	1.6	Pt	2.6										
						Co	1.5	Th	7.0										
						Ni	1.4	U	7.0										

[a] For Rh and Pd the same values

Table S11.13 Relative change of refractions of atoms on ionization

i, %	10	20	30	40	50	60	70	80	90
In molecules									
Anion	0.062	0.125	0.192	0.218	0.355	0.455	0.572	0.692	0.83
Cation[a]	0.52	0.33	0.225	0.155	0.11	0.074	0.045	0.027	0.015
Cation[b]	0.70	0.52	0.39	0.305	0.23	0.175	0.125	0.078	0.037
In crystals									
Anion	0.036	0.078	0.126	0.183	0.233	0.333	0.434	0.563	0.737
Cation[a]	0.58	0.385	0.26	0.18	0.125	0.078	0.048	0.024	0.012
Cation[b]	0.73	0.55	0.42	0.32	0.235	0.17	0.115	0.07	0.027

[a]For *a*-subgroups, [b]for *b*-subgroups

Table S11.14 Bond refractions (cm^3/mol)

Systems of Vogel[a] *and Miller*[b]

Bond	R_D[a]	R_∞[a]	R_D[b]	Bond	R_D	R_∞
C–H	1.676	1.644	1.645	C=O (ketone)	3.49	3.38
C–C	1.296	1.254	1.339	C–S	4.61	4.42
C–C (cyclopropane)	1.50	1.44		C=S	11.91	10.79
C–C (cyclobutane)	1.38	1.32		C–N	1.57[c]	1.49
C–C (cyclopentane)	1.28	1.24		C=N	3.75	3.51
C–C (cyclohexane)	1.27	1.23		C≡N	4.82	4.70
C_{ar}–C_{ar}	2.69	2.55	2.74	N–N	1.99	1.80
C=C	4.17	3.94	4.14	N=N	4.12	3.97
C≡C	5.87	5.67	5.13	N–H	1.76[d]	1.74
C–F	1.55	1.53	1.40	N–O	2.43	2.35
C–Cl	6.51	6.36	6.51	N=O	4.00	3.80
C–Br	9.39	9.06	8.27	O–H (alcohols)	1.66[e]	1.63
C–I	14.61	13.92	14.33	O–H (acids)	1.80	1.78
C–O (ethers)	1.54	1.49	1.47	S–H	4.80	4.65
C–O (acetals)	1.46	1.43		S–O	4.94	4.75
C=O	3.32	3.24	2.57	S–S	8.11	7.72

Additions to Vogel's system

Bond	R_D	Bond	R_D	Bond	R_D
O–O	2.27	Si–Br	10.24	Sn–Sn	10.7
Se–Se	11.6	Si–O	1.80	B–H	2.15[f]
P–H	4.24	Si–S	6.14	B–F	1.68
P–F	3.56	Si–N	2.16	B–Cl	6.95
P–Cl	8.80	Si–C_{alkyl}	2.47	B–Br	9.6[f]
P–Br	11.64	Si–C_{aryl}	2.93	B–O	1.61
P–O	3.08	Si–Si	5.87	B–S	5.38
P–S	7.56	Ge–H	3.64	B–N	1.96
P=S	6.87	Ge–F	2.3	B–C_{alkyl}	2.03
P–N	2.82	Ge–Cl	7.65	B–C_{aryl}	3.07
P–C	3.68	Ge–Br	11.1	Al–O	2.15
As–O	4.02	Ge–I	16.7	Al–N	2.90
As–C	4.52	Ge–O	2.50	Al–C	3.94
As–Cl	9.23	Ge–S	7.02	Hg–Cl	7.63[f]
As–Br	13.3	Ge–N	2.33	Hg–Br	9.77[f]
As–I	20.4	Ge–C	3.05	Hg–C	7.21

Table S11.14 (continued)

Systems of Vogel[a] *and Miller*[b]					
Sb–H	3.2	Ge–Ge	6.85	Zn–C	5.4
Sb–Cl	10.6	Sn–H	4.83	Cd–C	7.2
Sb–Br	13.6	Sn–Cl	8.66	In–C	5.9
Sb–I	20.8	Sn–Br	11.97	Pb–C	5.25
Sb–O	5.0	Sn–I	17.41	Sb–C	5.4
Sb–C	5.4	Sn–O	3.84	Bi–C	6.9
Si–H	3.0	Sn–S	7.63	Se–C	6.0
Si–F	2.1	$Sn–C_{alkyl}$	4.17	Te–C	7.9
Si–Cl	7.92	$Sn–C_{aryl}$	4.55		

[a][11.27], [b][11.28], [c]1.48, [d]1.79, [e]1.78 cm^3/mol, [f]R_∞

Table S11.15 Anisotropy of bond refractions (cm^3/mol): $\gamma = (R_{||}/R_{\perp})^{1/3}$

Molecule	$R_{\|\|}$	$R_{\perp}$	γ	Molecule	$R_{\|\|}$	$R_{\perp}$	γ
H_2[a,b]	2.54	1.709	1.13	CO[b,e]	5.832	4.495	1.09
Cl_2[c]	16.0	9.44	1.19	CO_2[e]	5.04	4.91	1.01
I_2[d]	37.3	20.45	1.22	CS_2[e]	19.03	13.85	1.11
O_2[a]	5.678	3.102	1.22	SO_2[f]	14.63	9.055	1.04
N_2[b]	5.53	3.816	1.13	C_2H_2[e]	11.88	7.265	1.18
NO[e]	5.885	3.564	1.17	C_2H_6[e]	11.27	1.69	
Bond[g,h]	$R_{\|\|}$	$R_{\perp}$	γ	Bond[g]	$R_{\|\|}$	$R_{\perp}$	γ
B–F	2.62	2.06	1.08	Ge–H	6.03	2.08	1.42
B–Cl	9.26	6.10	1.15	Ge–Cl	11.85	5.55	1.29
B–Br	13.5	8.38	1.17	Ge–Br	15.9	7.98	1.26
B–I	20.6	12.6	1.18	Ge–I	23.35	11.6	1.26
C–H	1.74	1.59	1.03	Ge–C	6.19	1.18	1.74
C–Cl	9.57	4.76	1.22	Sn–Cl	13.15	6.08	1.29
C–Br	14.0	6.60	1.28	Sn–Br	17.6	8.43	1.28
C–C	2.27	0.75	1.45	Sn–I	25.85	12.7	1.27
Si–H	4.56	2.18	1.28	Sn–C	8.68	1.52	1.79
Si–F	2.43	1.93	1.08	N–H[i]	2.24	1.99	1.04
Si–Cl	9.86	5.51	1.21	Fe–Cp[j]	55.59	42.92	1.09
Si–Br	14.3	7.55	1.24	Ru–Cp[j]	58.50	45.38	1.09
Si–I	21.3	11.5	1.23	Os–Cp[j]	60.64	47.33	1.09
Si–C	4.38	1.36	1.48				

[a][11.29], [b][11.30], [c][11.31], [d][11.32], [e][11.33], [f][11.34], [g][11.35], [h][11.36], [i][11.37], [j][11.38]

Table S11.16 Effective nuclear charges of atoms in AH molecules according to Pearson and Slater. (See Chap. 1, Table 1.7)

A	Pearson	Slater	A	Pearson	Slater	A	Pearson	Slater
Li	1.8	1.3	Mg	2.8	2.8	Si	3.6	4.0
Na	2.3	2.2	Ca	3.4	2.8	N	2.9	3.7
Rb	2.7	2.2	Sr	3.6	3.3	P	4.0	4.6
K	3.0	2.7	Ba	3.8	3.3	O	3.1	4.3
Cs	3.1	2.7	B	2.5	2.5	S	4.4	5.2
Be	2.3	1.9	Al	3.1	3.4	F	3.2	4.9
			C	2.8	3.1	Cl	4.6	5.8

Table S11.17 Force constants (*mdyn*/Å) and electronegativities of metals in molecular halides

MX_n	M	F		Cl		Br		I	
		f	χ	*f*	χ	*f*	χ	*f*	χ
MX	Li	2.48	0.94	1.415	0.92	1.19	0.91	0.962	0.92
	Na	1.76	0.85	1.10	0.86	0.960	0.86	0.763	0.87
	K	1.38	0.77	0.865	0.78	0.702	0.73	0.612	0.78
	Rb	1.29	0.75	0.767	0.73	0.699	0.76	0.577	0.77
	Cs	1.22	0.74	0.756	0.75	0.658	0.75	0.543	0.76
	Cu	3.34	1.38	2.31	1.46	2.07	1.50	1.74	1.55
	Ag	2.51	1.20	1.855	1.33	1.66	1.36	1.465	1.44
	Au	4.42	1.95	2.60	1.75	2.34	1.79	2.12	1.95
	Sc[a]	4.25	1.77	2.34	1.65	1.846	1.52	1.50	1.51
	Al[b]	4.232	1.62	2.094	1.39	1.700	1.33	1.311	1.30
	Ga[c]	3.403	1.42	1.852	1.18	1.560	1.26	1.244	1,26
	Tl	2.33	1.18	1.43	1.15	1.25	1.15	1.04	1.17
MX_2	Be	5.15	1.60	3.28	1.75	2.53	1.59	1.96	1.55
	Mg	3.23	1.35	2.05	1.39	1.67	1.33	1.45	1.41
	Ca	1.90	0.99	1.18	0.95	1.03	0.96	0.86	0.99
	Sr	1.62	0.90	1.04	0.90	0.89	0.88	0.74	0.91
	Ba	1.51	0.89	0.97	0.90	0.77	0.84	0.65	0.85
	Zn	4.20	1.69	2.63	1.66	2.33	1.70	1.77	1.61
	Cd	3.67	1.69	2.14	1.50	1.93	1.55	1.65	1.61
	Hg	3.87	1.85	2.63	1.81	2.28	1.82	1.85	1.79
	Sn			1.82	1.35	1.72	1.47	1.72	1.77
	Pb	3.36	1.61	1.68	1.30	1.41	1.27	1.41	1.53
	Cr	3.28	1.39						
	Mn	3.75	1.59	2.21	1.50	1.72	1.36		
	Fe	4.04	1.66	2.18	1.44	1.99	1.52		
	Co	4.05	1.65	2.49	1.60	2.09	1.55		
	Ni	4.76	1.79	2.50	1.48	2.49	1.66		
MX_3	B	7.41	2.13	3.74	1.93	3.20	1.95	2.20	1.74
	Al	4.94	1.84	2.84	1.76	2.38	1.74	1.83	1.70
	Ga	4.70	1.85	2.68	1.71	2.09	1.57	1.72	1.61
	In			2.36	1.65	2.03	1.68	1.54	1.56
	Sc	3.68	1.59	1.93	1.37	1.61	1.33	1.10	1.14
	As	4.50	1.76	2.42	1.61	1.94	1.52	1.51	1.49
	Sb	4.73	2.03	2.20	1.59	1.81	1.53	1.42	1.49
	Bi	3.60	1.67	1.93	1.47	1.61	1.45	1.25	1.37
MX_4	Si	6.57	2.26	3.37	2.03	2.67	1.91		
	Ge	5.57	2.09	2.92	1.86	2.41	1.81	1.75	1.67
	Sn			2.63	1.82	2.12	1.72	1.60	1.63
	Ti	4.91	1.96	2.78	1.83	2.47	1.90	1.90	1.82
	Zr	4.08	1.78	2.56	1.82	2.20	1.80	1.82	1.83
	Hf	4.33	1.89	2.61	1.85	2.26	1.84	1.86	1.87
	Th	3.24	1.60						
	U	3.30	1.57	2.05	1.61	1.75	1.59		

[a][11.39], [b][11.40], [c][11.41]

Table S11.18 Force constants (*mdyn*/Å) and electronegativities of metals in molecular oxides and chalcogenides

M^{II}	O		S		M^{II}	O		S	
	f	χ	f	χ		f	χ	f	χ
Cu	3.08	1.56	2.16	1.70	Sn	5.62	1.51	3.53	1.50
Be	7.51	1.47	4.13	1.38	Pb	4.55	1.29	3.02	1.33
Mg	3.50	0.90	2.28	0.94	Bi[e]			2.735	1.22
Ca	3.61	0.97	2.24	1.00	Cr	5.64[a]	1.34	4.51	1.80
Sr	3.40	0.96	2.09	0.98	Mn	5.055[a]	1.22	2.87	1.14
Ba	3.79	1.08	2.20	1.06	Fe	5.56[a]	1.32	3.07[b]	1.20
Zn	4.49	1.14			Co	5.375[a]	1.29		
Hf[d]	8.206	2.08			Ni	5.08[a]	1.22	3.14[c]	1.18
Th[d]	7.077	2.02			Pt	6.31	1.60	4.24	1.66

CuSe $f = 1.89$, $\chi = 1.59$; CuTe $f = 1.60$, $\chi = 1.71$; SnSe $f = 3.06$, $\chi = 1.42$; SnTe $f = 2.44$, $\chi = 1.40$; PbSe $f = 2.60$, $\chi = 1.25$; PbTe $f = 2.09$, $\chi = 1.23$ [11.42], [a][11.43], [b][11.44], [c][11.45], [d][11.46], [e][11.47], [11.48]: BiSe $f = 2.368$, $\chi = 1.17$

Table S11.19 Elastic force constants (*mdyn*/Å) and electronegativities of metals in MX crystals

M(I)	F		Cl		Br		I	
	f_c	χ^*	f_c	χ^*	f_c	χ^*	f_c	χ^*
Li	0.265	0.54	0.150	0.52	0.123	0.50	0.104	0.52
Na	0.201	0.48	0.129	0.49	0.113	0.50	0.096	0.52
K	0.153	0.43	0.103	0.45	0.094	0.46	0.078	0.47
Rb	0.141	0.42	0.099	0.45	0.087	0.45	0.075	0.47
Cs	0.136	0.44	0.068	0.43	0.061	0.43	0.050	0.43
NH_4			0.117	0.53	0.103	0.53	0.098	0.60
Cu			0.397	0.87	0.403	0.99	0.383	1.13
Ag	0.272	0.67	0.220	0.79	0.212	0.85	0.239	0.79
Tl			0.082	0.47	0.078	0.45	0.077	0.58
M(II)	O		S		Se		Te	
Be	1.700	1.39	0.919	1.25	0.855	1.28	0.676	1.26
Mg	0.620	0.88	0.370	0.84	0.272	0.68	0.335	1.00
Ca	0.495	0.80	0.328	0.82	0.273	0.74	0.241	0.81
Sr	0.422	0.73	0.315	0.83	0.253	0.72	0.241	0.83
Ba	0.343	0.64	0.320	0.90	0.247	0.74	0.226	0.82
Zn	1.159	1.13	0.745	1.13	0.654	1.08	0.550	1.11
Cd	0.593	0.93	0.678	1.11	0.601	1.06	0.511	1.10
Hg			0.725	1.19	0.590	1.05	0.528	1.13
Sn			0.395	0.92	0.271	0.74	0.202	0.66
Pb			0.283	0.74	0.276	0.77	0.234	0.78
Mn	0.624	0.93	0.730	1.15			0.287	0.87
M(III)	N		P		As		Sb	
B	2.482	1.88	1.234	1.66	1.276	1.91		
Al	1.578	1.37	0.846	1.31	0.778	1.31	0.621	1.17
Ga	1.527	1.36	0.870	1.35	0.766	1.28	0.616	1.15
In	1.297	1.26	0.773	1.28	0.639	1.12	0.541	1.06
Th	0.817	1.27	0.721	1.32	0.637	1.79	0.479	1.46
U	0.897	1.33	0.522	1.43	0.522	1.39	0.347	1.00

[a] χ refers to crystal (rather than molecule)

Supplementary References

11.1 Herve P, Vandamme LKJ (1994) Infrared Phys Technol 4:609
11.2 Balzaretti NM, Da Jordana JAH (1996) J Phys Chem Solids 57:179
11.3 Wagner V, Gundel S, Geurts J et al (1998) J Cryst Growth 184/185:1067
11.4 Yamanaka T, Tokonami M (1985) Acta Cryst B41: 298
11.5 Batsanov SS, Grankina ZA (1965) Optica i Spectr 19:814 (in Russian)
11.6 Sharma SB, Sharma SC, Sharma B, Bedi S (1992) J Phys Chem Solids 53:329
11.7 Ito T, Yamaguchi H, Masumi T, Adachi S (1998) J Phys Soc Japan 67:3304
11.8 Gauthier M, Polian A, Besson J, Chevy A (1989) Phys Rev B40: 3837
11.9 Ren Q, Ding L-Y, Chen F-S et al (1997) J Mater Sci Lett 16:1247
11.10 Elkorashy A (1989) Physica B 159:171
11.11 Elkorashy A (1990) J Phys Chem Solids 51:289
11.12 Yamaguchi M, Yagi T, Azuhata T et al (1997) J Phys Cond Matter 9:241
11.13 Azuhata T, Sota T, Suzuki K, Nakamara S (1995) J Phys Cond Matter 7:L129
11.14 Uribe MM, de Oliveira CEM, Clerice JHM et al (1996) Elect Lett 32:262
11.15 Sun L, Ruoff AL, Zha C-S, Stupian G (2006) J Phys Chem Solids 67:2603
11.16 Sysoeva NP, Ayupov BM, Titova EF (1985) Optika i Spectr 59:231
11.17 Shimizu H, Kitagawa T, Sasaki S (1993) Phys Rev B 47:11567
11.18 Medenbach O, Shannon RD (1997) J Opt Soc Amer B 14:3299
11.19 Acker E, Haussühl S, Recker K (1972) J Cryst Growth 13/14:467
11.20 Gavaleshko NP, Savchuk AI, Vatamanyuk PP, Lyakhovich AN (1981) Neorg Mater 17:538
11.21 Batsanova LR (1963) Izv Sib Otd AcadSciSU Ser Khimiya 3:83
11.22 Nazri GA, Julien C, Mavi HS (1994) Solid State Ionics 70/71:137
11.23 Ruchkin ED, Sokolova MN, Batsanov SS (1967) Zh Struct Khim 8:465
11.24 Kustova GN, Obzherina KF, Kamarzin AA et al (1969) Zh Struct Khim 10:609
11.25 Batsanov SS (1966) Refractometry and chemical structure. Van Nostrand, Princeton; Batsanov SS (1976) Structural refractometry, 2nd edn. Vyschaya Shkola, Moscow (in Russian)
11.26 Marler B (1988) Phys Chem Miner 16:286; Guo YY, Kuo CK, Nicholson PS (1999) Solid State Ionics 123:225
11.27 Vogel AI (1948) J Chem Soc 1833; Vogel AI, Cresswell WT, Jeffery G, Leicester J (1952) J Chem Soc 514
11.28 Miller KJ (1990) J Am Chem Soc 112:8533
11.29 Hohm U (1994) Chem Phys 179:533
11.30 McDowell SAC, Kumar A, MeathWJ (1996) Canad J Chem 74:1180
11.31 Bridge NJ, Buckingham AD (1966) Proc Roy Soc A295:334
11.32 Maroulis G, Makris C, Hohm U, Goebel D (1997) J Phys Chem A 101:953
11.33 Baas F, van den Hout KD (1979) Physica A95:597
11.34 Gentle IR, Laver DR, Ritchie GLD (1990) J Phys Chem 94:3434
11.35 Allen GW, Aroney MJ (1989) J Chem Soc Faraday Trans II 85:2479
11.36 Keir RI, Ritchie GLD (1998) Chem Phys Lett 290:409
11.37 Ritchie GL, Blanch EW (2003) J Phys Chem A107:2093
11.38 Goebel D, Hohm U (1997) J Chem Soc Faraday Trans 93:3467
11.39 Gurvich LV, Ezhov YuS, Osina EL, Shenyavskaya EA (1999) Russ J Phys Chem 73:331
11.40 Hargittai M, Varga Z (2007) J Phys Chem A 111:6
11.41 Singh VB (2005) J Phys Chem Ref Data 34:23
11.42 Batsanov SS (2005) Russ J Phys Chem 79:725
11.43 Zhao Y, Gong Y, Zhou M (2006) J Phys Chem A110:1077
11.44 Takano S, Yamamoto, Saito S (2004) J Mol Spectr 224:137
11.45 Yamamoto T, Tanimoto M, Okabayashi T (2007) PCCP 9:3774
11.46 Merritt JM, Bondybey VE, Heaven MC (2009) J Chem Phys 130:144503
11.47 Setzer KD, Meinecke F, Fink EH (2009) J Mol Spectr 258:56
11.48 Setzer KD, Breidohr R, Meinecke F, Fink EH (2009) J Mol Spectr 258:50

References

1. Toney MF, Brennan S (1989) Observation of the effect of refraction on x-rays diffracted in a grazing-incidence asymmetric Bragg geometry. Phys Rev B 39:7963–7966
2. Wemple SH, DiDomenico M Jr (1969) Optical dispersion and the structure of solids. Phys Rev Lett 23:1156–1160
3. Wemple SH, DiDomenico M Jr (1971) Behavior of the electronic dielectric constant in covalent and ionic materials. Phys Rev B 3:1338–1351
4. Wemple SH (1973) Refractive-index behavior of amorphous semiconductors and glasses. Phys Rev B 7:3767–3777
5. Shannon RD, Shannon RC, Medenbach O, Fischer RX (2002) Refractive index and dispersion of fluorides and oxides. J Phys Chem Ref Data 31:931–970
6. Moss T (1950) A relationship between the refractive index and the infra-red threshold of sensitivity for photoconductors. Proc Phys Soc B 63:167–176
7. Dionne G, Wooley JC (1972) Optical properties of some $Pb_{1-x}Sn_xTe$ alloys determined from infrared plasma reflectivity measurements. Phys Rev B 6:3898–3913
8. Ravindra N, Auluck S, Srivastava V (1979) Penn gap in semiconductors. Phys Status Solidi B 93:K155–K160
9. Grzybowski TA, Ruoff AL (1984) Band-overlap metallization of BaTe. Phys Rev Lett 53:489–492
10. Herve P, Vandamme LKJ (1994) General relation between refractive-index and energy-gap in semiconductors. Infrared Phys Technol 4:609–615
11. Rocquefelte X, Goubin F, Montardi Y et al (2005) Analysis of the refractive indices of TiO_2, $TiOF_2$, and TiF_4: Concept of optical channel as a guide to understand and design optical materials. Inorg Chem 44:3589–3593
12. Rocquefelte X, Whangbo M-H, Jobic S (2005) Structural and electronic factors controlling the refractive indices of the chalcogenides ZnQ and CdQ (Q = O, S, Se, Te). Inorg Chem 44:3594–3598
13. Batsanov SS (1966) Refractometry and chemical structure. Van Nostrand, Princeton NJ
14. Batsanov SS (1976) Structural refractometry, 2nd edn. Vyschaya Shkola, Moscow
15. Batsanov SS, Lazareva EV, Kopaneva LI (1978) Phase transformation in GeO_2 under shock compression. Russ J Inorg Chem 23:964–965
16. Itie JP, Polian A, Galas G et al (1989) Pressure-induced coordination changes in crystalline and vitreous GeO_2. Phys Rev Lett 63:398–401
17. Hacskaylo M (1964) Determination of refractive index of thin dielectric films. J Opt Soc Amer 54:198
18. Stöffler D (1974) Physical properties of shocked minerals. Fortschr Miner 51:256–289
19. Schneider H, Hornemann U (1976) X-ray investigations on deformation of experimentally shock-loaded quartzes. Contrib Mineral Petrol 55:205–215
20. Shimada Y, Okuno M, Syono Y et al (2002) An X-ray diffraction study of shock-wave-densified SiO_2 glasses. Phys Chem Miner 29:233–239
21. Batsanov SS, Dulepov EV, Moroz EM et al (1971) Effect of an explosion on a substance. Impact compression of rare-earth metal fluorides. Comb Expl Shock Waves 7:226–228
22. Tsay Y-F, Bendow B, Mitra SS (1973) Theory of temperature derivative of refractive-index in transparent crystals. Phys Rev B 8:2688–2696
23. Dewaele A, Eggert JH, Loubeyre P, Le Toullec R (2003) Measurement of refractive index and equation of state in dense He, H_2, H_2O, and Ne under high pressure in a diamond anvil cell. Phys Rev B 67:094112
24. Jones SC, Robinson MC, Gupta YM (2003) Ordinary refractive index of sapphire in uniaxial tension and compression along the *c* axis. J Appl Phys 93:1023-1031
25. Balzaretti NM, da Jornada JAH (1996) Pressure dependence of the refractive index of diamond, cubic silicon carbide and cubic boron nitride. Solid State Commun 99:943–948
26. Balzaretti NM, da Jornada JAH (1996) Pressure dependence of the refractive index and electronic polarizability of LiF, MgF_2 and CaF_2. J Phys Chem Solids 57:179–182

27. Johannsen PG, Reiss G, Bohle U et al (1997) Effect of pressure on the refractive index of 11 alkali halides. Phys Rev B 55:6865–6870
28. Ghandehari K, Luo H, Ruoff AL et al (1995) Band-gap and index of refraction of CsH to 251 GPa. Solid State Commun 95:385–388
29. Evans WJ, Silvera IJ (1998) Index of refraction, polarizability, and equation of state of solid molecular hydrogen. Phys Rev B 57:14105–14109
30. Ahart M, Yarger JL, Lantzky KM et al (2006) High-pressure Brillouin scattering of amorphous BeH_2. J Chem Phys 124:014502
31. Sun L, Ruoff AL, Zha C-S, Stupian G (2006) Optical properties of methane to 288 GPa at 300 K. J Phys Chem Solids 67:2603–2608
32. Sun L, Ruoff AL, Zha C-S, Stupian G (2006) High pressure studies on silane to 210 GPa at 300 K: optical evidence of an insulator-semiconductor transition. J Phys Cond Matter 18:8573–8580
33. Shimizu H, Kitagawa T, Sasaki S (1993) Acoustic velocities, refractive-index, and elastic-constants of liquid and solid CO_2 at high-pressures up to 6 GPa. Phys Rev B 47:11567–11570
34. Batsanov SS (1956) Relationship between melting points and refraction indices of ionic crystals. Kristallografiya 1:140–142 (in Russian)
35. Samygin MM (1938) On the relation between boiling temperatures and refraction indices. Zhurnal Fizicheskoi Khimii 11:325–330 (in Russian)
36. Sorriso S (1980) Dielectric behavior and molecular-structure of inorganic complexes. Chem Rev 80:313–327
37. Batsanov SS (1982) Dielectric method of studying the chemical bond and the concept of electronegativity. Russ Chem Rev 51:684–697
38. Törring T, Ernst WE, Kändler J (1989) Energies and electric-dipole moments of the low-lying electronic states of the alkaline-earth monohalides from an electrostatic polarization model. J Chem Phys 90:4927–4932
39. Ohwada K (1991) Application of potential constants—charge-transfer and electric-dipole moment change in the formation of heteronuclear diatomic-molecules. Spectrochim Acta A 47:1751–1765
40. Sadlej AJ (1992) Electric properties of diatomic interhalogens—a study of the electron correlation and relativistic contributions. J Chem Phys 96:2048–2053
41. Steimle TC, Robinson JS, Goodridge D (1999) The permanent electric dipole moments of chromium and vanadium mononitride: CrN and VN. J Chem Phys 110:881–889
42. Medenbach O, Dettmar D, Shannon RD et al (2001) Refractive index and optical dispersion of rare earth oxides using a small-prism technique. J Opt A 3:174–177
43. Vereschagin AN (1980) Molecular polarizability. Nauka, Moscow (in Russian)
44. Thomas JM, Walker NR, Cooke SA, Gerry MCL (2004) Microwave spectra and structures of KrAuF, KrAgF, and KrAgBr; ^{83}Kr nuclear quadrupole coupling and the nature of noble gas-noble metal halide bonding. J Am Chem Soc 126:1235–1246
45. Steimle TC, Virgo W (2003) The permanent electric dipole moments and magnetic hyperfine interactions of ruthenium mononitride. J Chem Phys 119:12965–12972
46. Steimle TC, Virgo WL (2004) The permanent electric dipole moments of WN and ReN and nuclear quadrupole interaction in ReN. J Chem Phys 121:12411–12420
47. Steimle TC (2000) Permanent electric dipole moments of metal containing molecules. Int Rev Phys Chem 19:455–477
48. Liao D-W, Balasubramanian K (1994) Spectroscopic constants and potential-energy curves for GeF. J Mol Spectr 163:284–290
49. Ogilvie JF (1995) Electric polarity $_{+}BrCl^{-}$ and rotational g factor from analysis of frequencies of pure rotational and vibration–rotational spectra. J Chem Soc Faraday Trans 91:3005–3006
50. Bazalgette G, White R, Loison J et al (1995) Photodissociation of ICl molecules oriented in an electric-field—direct determination of the sign of the dipole-moment. Chem Phys Lett 244:195–198
51. Wang H, Zhuang X, Steimle TC (2009) The permanent electric dipole moments of cobalt monofluoride, CoF, and monohydride, CoH. J Chem Phys 131:114315

52. Steimle TC, Virgo W L, Ma T (2006) The permanent electric dipole moment and hyperfine interaction in ruthenium monoflouride. J Chem Phys 124:024309
53. Zhuang X, Steimle TC, Linton C (2010) The electric dipole moment of iridium monofluoride. J Chem Phys 133:164310
54. Zhuang X, Steimle TC (2010) The permanent electric dipole moment of vanadium monosulfide. J Chem Phys 132:234304
55. Büsener H, Heinrich F, Hese A (1987) Electric dipole moments of the MgO $B^1\Sigma^+$ and $X^1\Sigma^+$ states. Chem Phys 112:139–146
56. Zhuang X, Frey SE, Steimle TC (2010) Permanent electric dipole moment of copper monoxide. J Chem Phys 132:234312
57. Heaven MC, Goncharov V, Steimle TC, Linton C (2006) The permanent electric dipole moments and magnetic g factors of uranium monoxide. J Chem Phys 125:204314
58. Linton C, Chen J, Steimle TC (2009) Permanent electric dipole moment of cerium monoxide. J Phys Chem A 113:13379–13382
59. Wang F, Le A, Steimle TC, Heaven MC (2011) The permanent electric dipole moment of thorium monoxide. J Chem Phys 134:031102
60. Cooper DL, Langhoff SR (1981) A theoretical-study of selected singlet and triplet-states of the CO molecule. J Chem Phys 74:1200–1210
61. Scuseria GE, Miller MD, Jensen F, Geertsen J (1991) The dipole moment of carbon monoxide. J Chem Phys 94:6660–6663
62. Langhoff SR, Arnold JO (1979) Theoretical-study of the $X^1\Sigma^+$, $A^1\Pi$, $C^1\Sigma^-$, and $E^1\Sigma^+$ states of the SiO molecule. J Chem Phys 70:852–863
63. Suenram RD, Fraser GT, Lovas FJ, Gilles CW (1991) Microwave spectra and electric dipole moments of VO and NbO. J Mol Spectr 148:114–122
64. Steimle TC, Jung KY, Li B-Z (2002) The permanent electric dipole moment of PtO, PtS, PtN and PtC. J Chem Phys 103:1767–1772
65. Steimle TC, Virgo W (2002) The permanent electric dipole moments for the $A^2\Pi$ and $B^2\Sigma^+$ states and the hyperfine interactions in the $A^2\Pi$ state of lanthanum monoxide. J Chem Phys 116:6012–6020
66. Pineiro AL, Tipping RH, Chackerian C (1987) Rotational and vibration rotational intensities of CS isotopes. J Mol Spectr 125:91–98
67. Pineiro AL, Tipping RH, Chackerian C (1987) Semiempirical estimate of vibration rotational intensities of SiS. J Mol Spectr 125:184–187
68. Steimle TC, Gengler J, Hodges Ph J (2004) The permanent electric dipole moments of iron monoxide. J Chem Phys 121:12303–12307
69. Bousquet R, Namiki K-IC, Steimle TC (2000) A comparison of the permanent electric dipole moments of ZrS and TiS. J Chem Phys 113:1566–1574
70. Steimle TC, Virgo WL, Hostutler DA (2002) The permanent electric dipole moments of iron monocarbide. J Chem Phys 117:1511–1516
71. Tzeli D, Mavridis A (2001) On the dipole moment of the ground state $X^3\Delta$ of iron carbide, FeC. J Chem Phys 118:4984–4986
72. Borin AC (2001) The $A^1\Pi$–$X^1\Sigma^+$ transition in NiC. Chem Phys 274:99–108
73. Virgo WL, Steimle TC, Aucoin LE, Brown JM (2004) The permanent electric dipole moments of ruthenium monocarbide in the $^3\pi$ and $^3\delta$ states. Chem Phys Lett 391:75–80.
74. Marr AJ, Flores ME, Steimle TC (1996) The optical and optical/Stark spectrum of iridium monocarbide and mononitride. J Chem Phys 104:8183–8196
75. Wang H, Vigro WL, Chen J, Steimle TC (2007) Permanent electric dipole moment of molybdenum carbide. J Chem Phys 127:124302
76. Wang F, Steimle TC (2011) Electric dipole moment and hyperfine interaction of tungsten monocarbide. J Chem Phys 134:201106
77. Antoine R, Rayane D, Benichou E et al (2000) Electric dipole moment and charge transfer in alkali-C_{60} molecules. Eur Phys J D 12:147–151
78. Fajans K (1928) Deformation of ions and molecules on the basis of refractometric data. Z Elektrochem 34:502–518

79. Pauling L (1960) The nature of the chemical bond, 3rd edn. Cornell Univ Press, Ithaca
80. Liu Y, Guo Y, Lin J et al (2001) Measurement of the electric dipole moment of NO by mid-infrared laser magnetic resonance spectroscopy. Mol Phys 99:1457–1461
81. Coulson CA (1942) The dipole moment of the C–H bond. Trans Faraday Soc 38:433–444
82. Rayane D, Allouche A-R, Antoine R et al (2003) Electric dipole of metal-benzene sandwiches. Chem Phys Lett 375:506–510
83. Dorosh O, Bialkowska-Jawarska E, Kisiel Z, Pszczólkowski L (2007) New measurements and global analysis of rotational spectra of Cl-, Br-, and I-benzene: spectroscopic constants and electric dipole moments. J Mol Spectr 246:228–232
84. Xu Y, Jäger W, Djauhari J, Gerry MCL (1995) Rotational spectra of the mixed rare-gas dimers Ne-Kr and Ar-Kr. J Chem Phys 103:2827–2833
85. Shannon RD (1993) Dielectric polarizabilities of ions in oxides and fluorides. J Appl Phys 73:348–366
86. Born M (1921) Electrostatic lattice potential. Z Physik 7:124–140
87. Szigeti B (1949) Polarizability and dielectric constant of ionic crystals. Trans Faraday Soc 45:155–166
88. Lyddane RH, Sachs RG, Teller E (1941) On the polar vibrations of alkali halides. Phys Rev 59:673–676
89. Penn D (1962) Wave-number-dependent dielectric function of semiconductors. Phys Rev 128:2093–2097
90. Phillips JC (1967) *A posteriori* theory of covalent bonding. Phys Rev Lett 19:415–417
91. Phillips JC (1968) Dielectric definition of electronegativity. Phys Rev Lett 20:550–553
92. Phillips JC, Van Vechten JA (1970) New set of tetrahedral covalent radii. Phys Rev B 2:2147–2160
93. Phillips JC (1970) Ionicity of chemical bond in crystals. Rev Modern Phys 42:317–356
94. Phillips JC (1985) Structure and properties—Mooser-Pearson plots. Helv Chim Acta 58:209–215
95. Newton I (1704) Opticks: or a treatise of the reflexions, refractions, inflexions and colours of light. Smith S & Walford B, London
96. Beer M (1853) Einleitung in hohere Optik. Vieweg und Sohn, Brunswick
97. Gladstone JH, Dale TP (1863) Researches on the refraction, dispersion, and sensitiveness of liquids. Philos Trans Roy Soc London 153:317–343
98. Lorenz L (1880) Ueber die Refractionsconstante. Wied Ann Phys 11:70–103
99. Lorentz HA (1880) Ueber die Beziehung zwischen der Fortpflanzungsgeschwindigkeit des Lichtes und der Körperdichte. Wied Ann Phys 9:641–665
100. Lorentz HA (1895) Versuch einer Theorie der electrischen und optischen Erscheinungen in bewegten Körpern. Brill EJ, Leiden
101. Lorentz HA (1916) The theory of electrons and its applications to the phenomena of light and radiant heat. G E Stechert, New York
102. Shimizu H, Kitagawa T, Sasaki S (1993) Acoustic velocities, refractive-index, and elastic-constants of liquid and solid CO_2 at high-pressures up to 6 GPa. Phys Rev B 47:11567–11570
103. Shimizu H, Kamabuchi K, Kume T, Sasaki S (1999) High-pressure elastic properties of the orientationally disordered and hydrogen-bonded phase of solid HCl. Phys Rev B 59:11727–11732
104. Müller H (1935) Theory of the photoelastic effect of cubic crystals. Phys Rev 47:947–957
105. Yamaguchi M, Yagi T, Azuhata T et al (1997) Brillouin scattering study of gallium nitride: elastic stiffness constants. J Phys Cond Matter 9:241–248
106. Setchell RE (2002) Refractive index of sapphire. J Appl Phys 91:2833–2841
107. Dewaele A, personal communication
108. Hohm U, Kerl K (1990) Interferometric measurements of the dipole polarizability α of molecules between 300 K and 1,100 K: monochromatic measurements at $\lambda = 632.99$ nm for the noble gases and H_2, N_2, O_2, and CH_4. Mol Phys 69:803–817
109. Müller W, Meyer W (1986) Static dipole polarizabilities of Li_2, Na_2, and K_2. J Chem Phys 85:953–957

110. Brechignac C, Cahuzac P, Carlier F et al (1991) Simple metal clusters. Z Phys D 19:1–6
111. Miller TM, Bederson B (1977) Atomic and molecular polarizabilities. Adv Atom Mol Phys 13:1–55
112. Miller TM (1995-1996) Atomic and molecular polarizabilities. In: Lide DR (ed) Handbook of chemistry and physics, 76th edn. CRC Press, New York
113. Rayane D, Allouche AR, Benichou E et al (1999) Static electric dipole polarizabilities of alkali clusters. Eur Phys J 9:243–248
114. Amini JM, Gould H (2003) High precision measurements of the static dipole polarizability of cesium. Phys Rev Lett 91:153001
115. Wettlaufer DE, Glass II (1972) Specific refractivities of atomic nitrogen and oxygen. Phys Fluids 15:2065–2066
116. Goebel D, Hohm U (1997) Comparative study of the dipole polarizability of the metallocenes $Fe(C_5H_5)_2$, $Ru(C_5H_5)_2$ and $Os(C_5H_5)_2$. J Chem Soc Faraday 93:3467–3472
117. Tarnovsky V, Bunimovicz M, Vuskovic I et al (1993) Measurements of the DC electric-dipole polarizabilities of the alkali dimer molecules, homonuclear and heteronuclear. J Chem Phys 98:3894–3904
118. Ekstrom CR, Schmiedmayer J, Chapman MS et al (1995) Measurement of the electric polarizability of sodium. Phys Rev A 51:3883–3888
119. Kowalski A, Funk DJ, Breckenridge WH (1986) Excitation-spectra of CaAr, SrAr and BaAr molecules in a supersonic jet. Chem Phys Lett 132:263–268
120. Sarkisov GS, Beigman IL, Shevelko VP, Struve K W (2006) Interferometric measurements of dynamic polarizabilities for metal atoms using electrically exploding wires in vacuum. Phys Rev A 73:042501
121. Goebel D, Hohm U, Maroulis G (1996) Theoretical and experimental determination of the polarizabilities of the zinc ${S_1}^0$ state. Phys Rev A 54:1973–1978
122. Goebel D, Hohm U (1995) Dispersion of the refractive-index of cadmium vapor and the dipole polarizability of the atomic cadmium S_1^0 state. Phys Rev A 52:3691–3694
123. Braun A, Holeman P (1936) The temperature dependence of the refraction of iodine and the refraction of atomic iodine. Z phys Chem B 34:357–380
124. Hohm U, Goebel D (1998) The complex refractive index and dipole-polarizability of iodine, I_2, between 11,500 and 17,800 cm^{-1}. AIP Conf Proc 430:698–701
125. Goebel D, Hohm U (1996) Dipole polarizability, Cauchy moments, and related properties of Hg. J Phys Chem 100:7710–7712
126. Thierfelder C, Assadollahzadeh B, Schwerdtfeger P et al (2008) Relativistic and electron correlation effects in static dipole polarizabilities for the Group 14 elements from carbon to element Z = 114: theory and experiment. Phys Rev A 78:052506
127. Kadar-Kallen MA, Bonin KD (1994) Uranium polarizability measured by light-force technique. Phys Rev Lett 72:828–831
128. Hohm U, Loose A, Maroulis G, Xenides D (2000) Combined experimental and theoretical treatment of the dipole polarizability of P_4 clusters. Phys Rev A 61:053202
129. Hohm U, Goebel D, Karamanis P, Maroulis G (1998) Electric dipole polarizability of As_4. J Phys Chem A 102:1237–1240
130. Hu M, Kusse BR (2002) Experimental measurement of Ag vapor polarizability. Phys Rev A66:062506
131. Eisenlohr FZ (1910) A new calculation for atom refractions. Z phys Chem 75:585–607
132. Eisenlohr FZ (1912) A new calculation for atom refraction II. The constants of nitrogen. Z phys Chem 79:129–146
133. Vogel AI (1948) Investigation of the so-called co-ordinate or dative link in esters of oxy-acids and in nitro-paraffins by molecular refractivity determinations. J Chem Soc 1833–1855
134. Vogel AI, Cresswell WT, Jeffery GH, Leicester J (1952) Physical properties and chemical constitution: aliphatic aldoximes, ketoximes, and ketoxime O-alkyl ethers, NN-dialkylhydrazines, aliphatic ketazines, mono-di-alkylaminopropionitriles and di-alkylaminopropionitriles, alkoxypropionitriles, dialkyl azodiformates, and dialkyl carbonates—bond parachors, bond refractions, and bond-refraction coefficients. J Chem Soc 514–549

135. Strohmeier W, Hümpfner K (1956) Das Dipolmoment zwischen gelosten metallorganischen Verbindungen und organischen Losungsmittelmolekulen mit Elektronendonatoreigenschaften. Z Elektrochem 60:1111–1114
136. Strohmeier W, Hümpfner K (1957) Dipolmoment und Elektronenakzeptorstarke der Metalle der III-Gruppe in metallorganischen Verbindungen. Z Elektrochem 61:1010–1014
137. Strohmeier W, Nützel K (1958) Der Einfluss der Substituenten R auf die Elektronenakzeptorstarke des Metalles Me in Verbindungen MeRX. Z Elektrochem 62:188–191
138. Strohmeier W, von Hobe D (1960) Dipolmomente und Elektronenakzeptoreigenschaften von Cyclopentadienylmetallverbindungen und Benzolchromtricarbonyl. Z Elektrochem 64:945–951
139. Phillips L, Dennis GR (1995) The electronic polarizability distribution of several substituted ferrocenes and di(η^6-benzene)chromium. J Chem Soc Dalton Trans 26:1469–1472
140. Hohm U (1994) Dipole polarizability and bond dissociation energy. J Chem Phys 101:6362–6364
141. Batsanov SS (2003) On the covalent refractions of metals. Russ J Phys Chem 77:1374–1376
142. Noorizadeh S, Parhizgar M (2005) The atomic and group compressibility. J Mol Struct Theochem 725:23–26
143. Donald KJ (2006) Electronic compressibility and polarizability: origins of a correlation. J Phys Chem A 110:2283–2289
144. Miller KJ, Savchik JA (1979) New empirical-method to calculate average molecular polarizabilities. J Am Chem Soc 101:7206–7213
145. Miller KJ (1990) Additivity methods in molecular polarizability. J Am Chem Soc 112:8533–8542
146. Antoine R, Rayane D, Allouche AR et al (1999) Static dipole polarizability of small mixed sodium-lithium clusters. J Chem Phys 110:5568–5577
147. Lide DR (ed) (1995-1996) Handbook of chemistry and physics, 76th edn. CRC Press, New York
148. Batsanov SS (1957) Atomic refractions of metals. Zhurnal Neorganicheskoi Khimii 2:1221–1222 (in Russian)
149. Batsanov SS (1961) Covalent refractions of metals. J Struct Chem 2:337–342
150. Batsanov SS (2004) Molecular refractions of crystalline inorganic compounds. Russ J Inorg Chem 49:560–568
151. Komara RA, Gearba MA, Fehrenbach CW, Lundeen SR (2005) Ion properties from high-L Rydberg fine structure: dipole polarizability of Si_2^+. J Phys B 38:S87–S95
152. Snow EL, Lundeen SR (2007) Fine-structure measurements in high-L n = 17 and 20 Rydberg states of barium. Phys Rev A 76:052505
153. Hanni ME, Keele JA, Lundeen SR et al (2010) Polarizabilities of Pb^{2+} and Pb^{4+} and ionization energies of Pb^+ and Pb^{3+} from spectroscopy of high-L Rydberg states of Pb^+ and Pb^{3+}. Phys Rev A 81:042512
154. Keele JA, Lundeen SR, Fehrenbach CW (2011) Polarizabilities of Rn-like Th^{4+} from rf spectroscopy of Th^{3+} Rydberg levels. Phys Rev A 83:062509
155. Reshetnikov N, Curtis LJ, Brown MS, Irwing RE (2008) Determination of polarizabilities and lifetimes for the Mg, Zn, Cd and Hg isoelectronic sequences. Physica Scripta 77:015301
156. Wasastjerna JA (1922) About the formation of atoms and molecules explained using the dispersion theory. Z phys Chem 101:193–217
157. Fajans K (1923) The structure and deformation of electron coating in its importance for the chemical and optical properties of inorganic compounds. Naturwiessenschaft 11:165–172
158. Fajans K, Ioos G (1924) Mole fraction of ions and molecules in light of the atom structure. Z Phys 23:1–46
159. Fajans K (1934) The refraction and dispersion of gases and vapours. Z phys Chem B 24:103–154
160. Fajans K (1941) Polarization of ions and lattice distances. J Chem Phys 9:281–282
161. Fajans K (1941) Molar volume, refraction and interionic forces. J Chem Phys 9:282
162. Fajans K (1941) One-sided polarization of ions in vapor molecules. J Chem Phys 9:378–379

163. Marcus Y, Jenkins HDB, Glasser L (2002) Ion volumes: a comparison. J Chem Soc Dalton Trans 3795–3798
164. Pauling L (1927) The theoretical prediction of the physical properties of many electron atoms and ions—mole refraction—diamagnetic susceptibility, and extension in space. Proc Roy Soc London A 114:181–211
165. Born M, Heisenberg W (1924) The influence of the deformability of ions on optical and chemical constants. Z Phys 23:388–410
166. Tessman JR, Kahn AH, Shockley W (1953) Electronic polarizabilities of ions in crystals. Phys Rev 92:890–895
167. Salzmann J-J, Jörgensen CK (1968) Molar refraction of aquo ions of metallic elements and interpretation of optical refraction measurements in inorganic chemistry. Helv Chim Acta 51:1276–1293
168. Jörgensen CK (1969) Origin of approximative additivity of electric polarisabilities in inorganic chemistry. Rev Chimie minerale 6:183–191
169. Iwadate Y, Mochinaga J, Kawamura K (1981) Refractive-indexes of ionic melts. J Phys Chem 85:3708–3712
170. Iwadate Y, Kawamura K, Murakami K et al (1982) Electronic polarizabilities of Tl^+, Ag^+, and Zn^{2+} ions estimated from refractive-index measurements of $TlNO_3$, $AgNO_3$, and $ZnCl_2$ melts. J Chem Phys 77:6177–6183
171. Shirao K, Fujii Y, Tominaga J et al (2002) Electronic polarizabilities of Sr^{2+} and Ba^{2+} estimated from refractive indexes and molar volumes of molten $SrCl_2$ and $BaCl_2$. J Alloys Comp 339:309–316
172. Mahan GD (1980) Polarizability of ions in crystals. Solid State Ionics 1:29–45
173. Fowler PW, Pyper NC (1985) In-crystal ionic polarizabilities derived by combining experimental and *ab initio* results. Proc Roy Soc London A 398:377–393
174. Fowler PW, Madden PA (1985) In-crystal polarizability of O^{2-}. J Phys Chem 89:2581–2585
175. Pyper NC, Pike CG, Edwards PP (1992) The polarizabilities of species present in ionic-solutions. Mol Phys 76:353–372
176. Pyper NC, Pike CG, Popelier P, Edwards PP (1995) On the polarizabilities of the doubly-charged ions of group IIB. Mol Phys 86:995–1020
177. Pyper NC, Popelier P (1997) The polarizabilities of halide ions in crystals. J Phys Cond Matter 9:471–488
178. Lim IS, Laerdahl JK, Schwerdtfeger P (2002) Fully relativistic coupled-cluster static dipole polarizabilities of the positively charged alkali ions from Li^+ to 119^+. J Chem Phys 116:172–178
179. Shannon RD, Fischer RX (2006) Empirical electronic polarizabilities in oxides, hydroxides, oxyfluorides, and oxychlorides. Phys Rev B 73:235111
180. Jemmer P, Fowler PW, Wilson M, Madden PA (1998) Environmental effects on anion polarizability: Variation with lattice parameter and coordination number. J Phys Chem A 102:8377–8385
181. Dimitrov V, Komatsu T (1999) Electronic polarizability, optical basicity and non-linear optical properties of oxide glasses. J Non-Cryst Solids 249:160–179
182. Duffy JA (2002) The electronic polarisability of oxygen in glass and the effect of composition. J Non-Cryst Solids 297:275–284
183. Bachinskii AI (1918) Molecular fields and their volumes. Bull Russ Acad Sci 1:11 (in Russian)
184. von Steiger AL (1921) An article on the summation methodology of the molecular refractions, especially among aromatic hydrocarbons. Berichte Deutsch Chem Ges 54:1381–1393
185. Smyth C (1925) Refraction and electron constraint in ions and molecules. Phil Mag 50:361–375
186. Denbigh KG (1940) The polarisabilities of bonds. Trans Faraday Soc 36:936–947
187. Vickery BC, Denbigh KG (1949) The polarisabilities of bonds: bond refractions in the alkanes. Trans Faraday Soc 45:61–81
188. Vogel AI, Cresswell WT, Jeffery G, Leicester J (1950) Bond refractions and bond parachors. Chem Ind 358

189. Vogel AI, Cresswell WT, Leicester J (1954) Bond refractions for tin, silicon, lead, germanium and mercury compounds. J Phys Chem 58:174–177
190. Yoffe BV (1974) Refractometric methods in chemistry. Khimia, Leningrad (in Russian)
191. Huggins ML (1941) Densities and refractive indices of liquid paraffin hydrocarbons. J Am Chem Soc 63:116–120
192. Huggins ML (1941) Densities and refractive indices of unsaturated hydrocarbons. J Am Chem Soc 63:916–920
193. Palit SR, Somayajulu GR (1960) Electronic correlation of molar refraction. J Chem Soc 459–460
194. Hohm U (1994) Dispersion of polarizability anisotropy of H_2, O_2, N_2O, CO_2, NH_3, C_2H_6, and cyclo-C_3H_6 and evaluation of isotropic and anisotropic dispersion-interaction energy coefficients. Chem Phys 179:533–541
195. McDowell SAC, Kumar A, Meath WJ (1996) Anisotropic and isotropic triple-dipole dispersion energy coefficients for all three-body interactions involving He, Ne, Ar, Kr, Xe, H_2, N_2, and CO. Canad J Chem 74:1180–1186
196. Minemoto S, Tanji H, Sakai H (2003) Polarizability anisotropies of rare gas van der Waals dimers studied by laser-induced molecular alignment. J Chem Phys 119:7737–7740
197. Yakshin MM (1948) On atomic polarization and bond refraction of complex compounds of platinum. Izvestia Sektora Platiny 21:146–156 (in Russian)
198. de Visser SP (1999) On the relationship between internal energy and both the polarizability volume and the diamagnetic susceptibility. Phys Chem Chem Phys 1:749–753
199. Hohm U (2000) Is there a minimum polarizability principle in chemical reactions? J Phys Chem A 104:8418–8423
200. Zeldovich YaB, Raizer YuP (1967) Physics of shock waves and high temperature hydrodynamics phenomena. Academic, New York
201. Batsanov SS (1967) The physics and chemistry of impulsive pressures. J Engin Phys 12:59–68
202. Goncharov AF, Goldman N, Fried LE et al (2005) Dynamic ionization of water under extreme conditions. Phys Rev Lett 94:125508
203. Poroshina IA, Berger AS, Batsanov SS (1973) Determination of coordination number of metals of groups I and II in silicates from refractometric data. J Struct Chem 14:789–793
204. Bokii GB, Batsanov SS (1954) About quantitative characteristics of trans-influence. Doklady Academii Nauk SSSR 95:1205–1206 (in Russian)
205. Kukushkin YuN, Bobokhodzhaev RI (1977) Chernyaev's law of trans-influence. Nauka, Moscow (in Russian)
206. Wasastjerna JA (1923) On the radii of ions. Comm Phys-Math Soc Sci Fenn 1(38):1–25
207. Goldschmidt VM (1929) Crystal structure and chemical constitution. Trans Faraday Soc 25:253–282
208. Kordes E (1939) The discovery of atom displacement from refraction. Z phys Chem B 44:249–260
209. Kordes E (1940) Calculation of the ion radii with help from physical atom sizes. Z phys Chem B 48:91–107
210. Kordes E (1955) Ionengrosse, Molekularrefraktion bzw Polarisierbarkeit und Lichtbrechrechung bei anorganischen Verbindungen.1. AB Verbindungen mit einwertigen edelgasahnlichen Ionen (Alkalihalogenide). Z Elektrochem 59:551–560
211. Kordes E (1955) Direkte Berechnung von Ionenradien aus der Molekularrefraktion bei AB Verbindungen mit einwertigen edelgasahnlichen Ionen. Z Elektrochem 59:927–932
212. Kordes E (1955) AB Verbindungen mit edelgasahnlichen einwertigen und zweiwertigen Ionen. Z Elektrochem 59:932–938
213. Wilson JN, Curtis RM (1970) Dipole polarizabilities of ions in alkali halide crystals. J Phys Chem 74:187–196
214. Vieillard P (1987) A new set of values for Pauling's ionic radii. Acta Cryst B43:513–517
215. Iwadate Y, Fukushima K (1995) Electronic polarizability of a fluoride ion estimated by refractive indexes and molar volumes of molten eutectic LiF-NaF-KF. J Chem Phys 103:6300–6302

216. Compagnon I, Antoine R, Broyer M et al (2001) Electric polarizability of isolated C_{70} molecules. Phys Rev A 64:025201
217. Dugourd P, Antoine R, Rayane D et al (2001) Enhanced electric polarizability in metal C_{60} compounds: Formation of a sodium droplet on C_{60}. J Chem Phys 114:1970–1973
218. Lyon JT, Andrews L (2005) Formation and characterization of thorium methylidene $CH_2=ThHX$ complexes. Inorg Chem 44:8610–8616
219. Danset D, Manaron L (2005) Reactivity of cobalt dimer and molecular oxygen in rare gas matrices: IR spectrum, photophysics and structure of Co_2O_2. Phys Chem Chem Phys 7:583 591
220. Wang XF, Andrew L, Riedel S, Kaupp M (2007) Mercury is a transition metal: The first experimental evidence for HgF_4. Angew Chem Int Ed 46:8371–8375
221. Li X, Wang L-S, Boldyrev AI, Simons J (1999) Tetracoordinated planar carbon in the Al_4C^- anion. A combined photoelectron spectroscopy and ab initio study. J Am Chem Soc 121:6033–6038
222. Boldyrev AI, Simons J, Li X, Wang L-S (1999) The electronic structure and chemical bonding of hypermetallic Al_5C by ab initio calculations and anion photoelectron spectroscopy. J Chem Phys 111:4993–4998
223. Wang L-S, Boldyrev AI, Li X, Simons J (2000) Experimental observation of pentaatomic tetracoordinate planar carbon-containing molecules. J Am Chem Soc 122:7681–7687
224. Kuznetsov AE, Boldyrev AI, Li X, Wang L-S (2001) On the aromaticity of square planar Ga_4^{2-} and In_4^{2-} in gaseous $NaGa_4^-$ and $NaIn_4^-$ clusters. J Am Chem Soc 123:8825–8831
225. Zhai H-J, Yang X, Wang X-B et al (2002) In search of covalently bound tetra- and penta-oxygen species: a photoelectron spectroscopic and *ab initio* investigation of MO_4^- and MO_5^- (M = Li, Na, K, Cs). J Am Chem Soc 124:6742–6750
226. Zhai H-J, Wang L-S, Kuznetsov AE, Boldyrev AI (2002) Probing the electronic structure and aromaticity of pentapnictogen cluster anions Pn_5^- (Pn = P, As, Sb, and Bi) using photoelectron spectroscopy and ab initio calculations. J Phys Chem A 106:5600–5606
227. Kiran B, Li X, Zhai H-J et al (2004) [$SiAu_4$]: aurosilane. Angew Chem Int Ed 43:2125–2129
228. Li S-D, Zhai H-J, Wang L-S (2008) $B_2(BO)_2^{2-}$ – diboronyl diborene: a linear molecule with a triple boron-boron bond. J Am Chem Soc 130:2573–2579
229. Jules JL, Lombardi JR (2003) Transition metal dimer internuclear distances from measured force constants. J Phys Chem A 107:1268–1273
230. Huber KP, Herzberg G (1979) Constants of diatomic molecules. Van Nostrand, New York
231. Fu Z, Lemire GW, Bishea GA, Morse MD (1990) Spectroscopy and electronic-structure of jet-cooled Al_2. J Chem Phys 93:8420–8441
232. Merritt JM, Kaledin AL, Bondybey VE, Heaven M C (2008) The ionization energy of Be_2. Phys Chem Chem Phys 10:4006–4013
233. Kitsopoulos TN (1991) Study of the low-lying electronic states of Si_2 and Si_2^- using negative-ion photodetachment techniques. J Chem Phys 95:1441–1448
234. Ho J, Polak ML, Lineberger WC (1992) Photoelectron-spectroscopy of group IV heavy metal dimers Sn_2^-, Pb_2^-, and $SnPb^-$. J Chem Phys 96:144–154
235. Stangassinger A, Bondybey VE (1995) Electronic spectrum of Tl_2. J Chem Phys 103:10804–10805
236. Van Hooydonk G (1999) A universal two-parameter Kratzer potential and its superiority over Morse's for calculating and scaling first-order spectroscopic constants of 300 diatomic bonds. Eur J Inorg Chem 1617–1642
237. Vilkov LV, Mastryukov VS, Sadova NI (1978) Determination of geometrical structure of molecules. Khimia, Moscow (in Russian)
238. Giricheva NI, Lapshin SB, Girichev GV (1996) Structural, vibrational, and energy characteristics of halide molecules of group II-V elements. J Struct Chem 37:733–746
239. Gurvich LV, Ezhov YuS, Osina EL, Shenyavskaya EA (1999) The structure of molecules and the thermodynamic properties of scandium halides. Russ J Phys Chem 73:331–344
240. Birge RT (1925) The law of force and the size of diatomic molecules, as determined by their band spectra. Nature 116:783–784

241. Mecke R (1925) Formation of band spectra. Z Physik 32:823–834
242. Morse PM (1929) Diatomic molecules according to the wave mechanics: vibrational levels. Phys Rev 34:57–64
243. Clark CHD (1934) The relation between vibration frequency and nuclear separation for some simple non-hydride diatomic molecules. Phil Mag 18:459–470
244. Ladd JA, Orville-Thomas WJ (1966) Molecular parameters and bond structure: nitrogen-oxygen bonds. Spectrochim Acta 22:919–925
245. Zallen R (1974) Pressure-Raman effects and vibrational scaling laws in molecular crystals—S_8 and As_2S_3. Phys Rev B 9:4485–4496
246. Hill FC, Gibbs GV, Boisen MB (1994) Bond stretching force-constants and compressibilities of nitride, oxide, and sulfide coordination polyhedra in molecules and crystals. Struct Chem 5:349–355
247. Zavitsas AA (2004) Regularities in molecular properties of ground state stable diatomics. J Chem Phys 120:10033–10036
248. Badger RM (1934) A relation between internuclear distance and bond force constant. J Chem Phys 2:128–131
249. Badger RM (1935) Between the internuclear distances and force constants of molecules and its application to polyatomic molecules. J Chem Phys 3:710–714
250. Cioslowski J, Liu G, Castro RAM (2000) Badger's rule revisited. Chem Phys Lett 331:497–501
251. Kurita E; Matsuura H; Ohno K (2004) Relationship between force constants and bond lengths for CX (X = C, Si, Ge, N, P, As, O, S, Se, F, Cl and Br) single and multiple bonds: formulation of Badger's rule for universal use. Spectrochim Acta A 60:3013–3023
252. Murell JN (1960) The application of perturbation theory to the calculation of force constants. J Mol Spectr 4:446–456
253. Pearson RG (1977) Simple-model for vibrational force constants. J Am Chem Soc 99:4869–4875
254. Gordy WR (1946) A relation between bond force constants, bond orders, bond lengths, and the electronegativities of the bonded atoms. J Chem Phys 14:305–320
255. Batsanov SS, Derbeneva SS (1969) Effect of valency and coordination of atoms on position and form of infrared absorption bands in inorganic compounds. J Struct Chem 10:510–515
256. Voyiatzis GA, Kalampounias AG, Papatheodorou GN (1999) The structure of molten mixtures of iron(III) chloride with caesium chloride. Phys Chem Chem Phys 1:4797–4803
257. Bowmaker GA, Harris RK, Apperley DC (1999) Solid-state ^{199}Hg MAS NMR and vibrational spectroscopic studies of dimercury(I) compounds. Inorg Chem 38:4956–4962
258. Spoliti M, de Maria G, D'Alessio L, Maltese E (1980) Bonding in and spectroscopic properties of gaseous triatomic-molecules: alkaline-earth metal dihalides. J Mol Struct 67:159–167
259. Kharitonov YuA, Kravtsova GV (1980) Empirical correlations between molecular constants and their use in coordination chemistry. Koordinatsionnaya Khimiya 6:1315 (in Russian)
260. Pearson RG (1993) Bond-energies, force-constants and electronegativities. J Mol Struct 300:519–525
261. Bhar G (1978) Trends of force constants in diamond and sphalerite-structure crystals. Physica B 95:107–112
262. Batsanov SS (1986) Experimental foundations of structural chemistry. Standarty, Moscow (in Russian)
263. Kanesaka I, Kawahara H, Yamazaki A, Kawai K (1986) The vibrational-spectrum of $AlCl_3$, $CrCl_3$ and $FeCl_3$. J Mol Struct 146:41–49
264. Batsanov SS, Derbeneva SS (1964) Infrared spectra of strontium and lead nitrates pressed into various media. Optika i Spectroskopiya 17:149–151 (in Russian)
265. Batsanov SS, Derbeneva SS (1965) Infrared spectra of anisotropic carbonates imbedded in different media. Opt Spect-USSR 18:342–343
266. Batsanov SS, Derbeneva SS (1967) Effect of anisotropy on diffuse light scattering in polycrystals. Opt Spect-USSR 22:80–81

267. Batsanov SS, Tleulieva KA (1978) Infrared spectroscopic study of structural transformations in sodium and potassium nitrates. J Struct Chem 19:329–330
268. Donaldson JD, Ross SD, Silver J (1975) Vibrational spectra of some cesium tin(II) halides. Spectrochim Acta A 31:239–243
269. Arkhipenko DK, Bokii GB (1977) On the possibility of the space group refinement by the vibration spectroscopy method. Sov Phys Cryst 22:667–671
270. Yurchenko EN, Kustova GN, Batsanov SS (1981) Vibration spectra of inorganic compounds. Nauka, Novosibirsk (in Russian)
271. Somayajulu GR (1958) Dependence of force constant on electronegativity, bond strength, and bond order. J Chem Phys 28:814–821
272. Hussain Z (1965) Dependence of vibrational constant of homonuclear diatomic molecules on electronegativity. Canad J Phys 43:1690–1692
273. Szöke S (1971) Approach of equalized electronegativity by molecular parameters. Acta Chim Acad Sci Hung 68:345
274. Spoliti M, De Matia G, D'Allessio L, Maltese E (1980) Bonding in and spectroscopic properties of gaseous triatomic molecules: alkaline-earth metal dihalides. J Mol Struct 67:159–167
275. Pearson RG (1993) Bond energies, force constants and electronegativities. J Mol Struct 300:519–525
276. van Hooydonk G (1999) A universal two-parameter Kratzer potential and its superiority over Morse's for calculating and scaling first-order spectroscopic constants of 300 diatomic bonds. Eur J Inorg Chem 1999:1617–1642
277. Batsanov SS (2005) Metal electronegativity calculations from spectroscopic data. Russ J Phys Chem 79:725–731
278. Waser J, Pauling L (1950) Compressibilities, force constants, and interatomic distances of the elements in the solid state. J Chem Phys 18:747–753
279. Batsanov SS (2011) System of metal electronegativities calculated from the force constants of the bonds. Russ J Inorg Chem 56:906–912
280. Reynolds W (1980) An approach for assessing the relative importance of field and σ-inductive contributions to polar substituent effects. J Chem Soc Perkin Trans II 985–992
281. Jörgensen CK (1963) Optical electronegativities of 3d group central ions. Mol Phys 6:43–47
282. Jörgensen CK (1975) Photo-electron spectra of non-metallic solids and consequences for quantum chemistry. Structure and Bonding 24:1–58
283. Dodsworth ES, Lever ABP (1990) The use of optical electronegativities to assign electronic-spectra of semiquinone complexes. Chem Phys Lett 172:151–157
284. Duffy JA (1977) Variable electronegativity of oxygen in binary oxides—possible relevance to molten fluorides. J Chem Phys 67:2930–2931
285. Duffy JA (1980) Trends in energy gaps of binary compounds—an approach based upon electron-transfer parameters from optical spectroscopy. J Phys C 13:2979–2989
286. Duffy JA (1986) Chemical bonding in the oxides of the elements—a new appraisal. J Solid State Chem 62:145–157
287. Duffy JA (2004) Relationship between cationic charge, coordination number, and polarizability in oxidic materials. J Phys Chem B 108:14137–14141
288. Duffy JA (2006) Ionic-covalent character of metal and nonmetal oxides. J Phys Chem A 110:13245–13248
289. Reddy RR, Gopal KR, Ahammed YN et al (2005) Correlation between optical electronegativity, molar refraction, ionicity and density of binary oxides, silicates and minerals. Solid Sate Ionics 176:401–407

Index

S. S. Batsanov, A. S. Batsanov, *Introduction to Structural Chemistry*,
DOI 10.1007/978-94-007-4771-5, © Springer Science+Business Media Dordrecht 2012